English-Chinese
DICTIONARY OF Construction
Surveying & Civil Engineering

牛津英汉双解
土木工程手册

（英）克里斯托弗·戈斯（CHRISTOPHER GORSE）
大卫·约翰斯顿（DAVID JOHNSTON）
马丁·普理查德（MARTIN PRITCHARD） 编

陈思诺 译

编委会 陈思诺 刘 珺 黄燕平
吕焕鑫 邓蒙蒙 马悦龄

·广州·

著作权合同登记号 图字：19－2018－028号
图书在版编目（CIP）数据

牛津英汉双解土木工程手册/（英）克里斯托弗·戈斯（Christopher Gorse），（英）大卫·约翰斯顿（David Johnston），（英）马丁·普理查德（Martin Pritchard）编；陈思诺译. —广州：华南理工大学出版社，2018.11
ISBN 978－7－5623－5820－6

Ⅰ.①牛… Ⅱ.①克… ②大… ③马… ④陈… Ⅲ.①土木工程－词汇－手册－英、汉 Ⅳ.①TU－62

中国版本图书馆CIP数据核字（2018）第244574号

A DICTIONARY OF CONSTRUCTION，SURVEYING，AND CIVIL ENGINEERING
by Christopher Gorse，David Johnston，Martin Pritchard
ISBN 978－0－1995－3446－3

牛津英汉双解土木工程手册
（英）克里斯托弗·戈斯（Christopher Gorse），（英）大卫·约翰斯顿（David Johnston），（英）马丁·普理查德（Martin Pritchard） 编
陈思诺 译

出 版 人：卢家明
出版发行：华南理工大学出版社
（广州五山华南理工大学17号楼，邮编510640）
http://www.scutpress.com.cn E-mail：scutc13@scut.edu.cn
营销部电话：020－87113487 87111048（传真）
责任编辑：林起提
特邀编辑：陈广元 温少芝
印 刷 者：广州市新怡印务有限公司
开 本：787mm×1092mm 1/16 **印张**：34 **字数**：1 347千
版 次：2018年11月第1版 2018年11月第1次印刷
定 价：100.00元

作　者　简　介

克里斯托弗·戈斯（Christopher Gorse）是利兹城市大学建成环境中心的负责人。他在建筑、法律和可持续性发展方面进行研究及咨询工作。

大卫·约翰斯顿（David Johnston）是利兹城市大学低碳建筑方面的讲师，主要讲授环境科学、服务和建筑技术等课程。他曾从事低碳建筑方面的研究和咨询工作，其研究项目对最近英国建筑条例的修改作出贡献。

马丁·普理查德（Martin Pritchard）是利兹城市大学土木工程的讲师。他也是一名土木工程师，在水净化系统方面有丰富的国际合作经验，为发展中国家基础设施工程提供可持续方案。他对岩土工程也有所涉猎。

原著前言

本书旨在为建筑、测量和土木工程等方面的术语提供最新最全面的参考。书中的词条能满足学习建成环境课程或从事专业实操人士的需求，并能为那些对建筑或施工有兴趣的人提供许多有用的信息。建筑商、专业供应商和承包商可将本词典作为一本随身手册，用于快速轻松地检索不熟悉的术语。

由于该行业的规模和性质，即使是知识渊博的专业人士也很难及时跟踪最新出现的议题，更何况是记录下来并编成建筑术语的百科全书。因此，本书主要起提醒的作用，用于确认词义，使交流更有保证。本书适用于建成环境专业人士，包括建筑师、建筑测量师、建筑服务工程师、施工管理人员、土木工程师、电气工程师、设备管理人员、机械工程师和工料测量师。本书为支撑我们所生活的环境，不论是人造的还是天然的，提供了一个快速简易的指南。为了涵盖这些术语，作者对建筑和工程方面所涉及的建筑物理等科学领域，以及它们对设计、结构、材料和实操方面的应用都作了深入的研究；与地质、地理、气候以及建筑物和构筑物所处的自然环境相关的学科同时亦涵盖在内；还包括为建成环境提供管理和流程的政策及法律框架，以及与在行业内有重要影响的专业组织有关的议题。

可持续性发展议程和建筑工程、测量等相关领域的重要性意味着越来越多的人对建筑产生兴趣，需要让他们快速熟悉所使用的语言和术语。本书对于在行业内实操的人员和对建成环境及工程感兴趣的人都是一本很好的参考书。作为世界上自然资源的主要消耗行业，建成环境及其对环境的影响在政府议程中占据重要位置，所有决策者都不能对此忽视。预计本书的术语和参考资料将继续增加，并且随着与建成环境相关的领域越来越重要，其需求量也会增长。

致 谢

得益于许多专家和专业人员的经验和知识上的帮助，我们多年来对建筑、测量和土木工程有了更深入的认识。其中一些人对本书的出版有重大的贡献，其他的专家也帮助我们对议题和领域有所了解。虽然不能向每一个对知识库有贡献的人逐一致谢，但我们非常感谢他们的帮忙。在此，作者特别感谢以下对本书出版有重大贡献的专家：

索尔福德大学的查尔斯·艾格布教授、斯蒂芬·埃米特教授；

诺丁汉特伦特大学的阿尼·雷登教授、大卫·海菲尔德、约翰·布拉德利、约翰·斯图尔奇、约瑟夫·康瓦博士、迈克·贝茨、保罗·怀特海德；

利兹城市大学的杰弗里·霍布迪博士。

另外，作者还要感谢以下做出额外贡献的人士：

利兹城市大学的阿什·艾哈迈德和圣戈班的珍妮佛·慕斯顿。

译 者 前 言

《牛津英汉双解土木工程手册》是英国牛津大学出版社最新出版的土木工程专业词典，词条来自牛津参考资源（Oxford Reference）数据库，是全球最受信赖的参考资源。全书收集词条超过 8000 条，内容涵盖房屋、道路、桥梁、隧道、运河、堤坝和给排水工程建设中所涉及的结构、岩土、力学、机电、材料、勘测、设计、施工、法律和商务等方面内容，覆盖面广泛，专业性较强。

本书译者毕业于理工院校，曾赴英国深造，回国后十几年来从事海外建筑工程工作。自从 2013 年我国提出“一带一路”倡议以来，海外工程加速了走出去的步伐，相关从业人员已达上百万。但在处理涉外工程项目时，很多人常常苦于没有一本称心的专业工具书，很多的工程专业词汇通过一般的字典普遍存在查找不到甚至翻译错误的现象。因此，本书译者通过华南理工大学出版社引进本书的版权，希望以自身的工作经验和知识积累，翻译这本书，为中国广大海外工程从业人员提供更好的参考。

为了能对读者提供更大的帮助，本书采用词条全部内容英汉双解、对照互译的形式。读者除了可以在需要时很方便地检索词条外，亦可以当作土木工程方面的科普书籍阅读，甚至能在词条释义当中发掘到新的专业词汇。本书在翻译撰写过程中，先后有多名工程方面的专业译员、华南理工大学教授、外聘专家参与编译、审校的工作，可以说是凝聚了众多涉外工程工作者的心血。特别致谢家人对我一直以来的支持和关怀，尤其是谭慧娜和陈盈熙，她们的爱和激励是我能一路前行的最大力量！

因译者自身水平有限，以及建筑工程领域发展日新月异，部分词汇中外表达方式不同等的原因，本书难免存在错误和不足之处，恳请读者不吝指正，为我国海外工程事业贡献一份力量。

译者

2018 - 10 - 1

abandonment of work To leave site and refuse to continue work. Must entail a complete stoppage of work with a clear intention not to continue or complete the work. The term is often used in arbitration and adjudication and marks a point where reference to legal proceedings can commence.
放弃工作 离开现场且拒绝继续工作。须导致工作的完全停止且具有使工作不再继续或完成的明确意图。该词语常用于仲裁和裁决中，且标明一个时点，自该时点起即可启动法律程序。

abatement of action An interruption of legal proceedings. A party, normally the defendant, serves an application of abatement giving reasons why proceedings should not continue.
中止诉讼 对法律程序的终止。某一当事方（一般为被告）提出关于中止的申请，说明程序为何不应继续的理由。

abeyance 1. A state of suspension, temporarily inactive, cessation or put to one side for a period of time. 2. When the ownership of a property has not yet been decided or the condition of a property without an owner.
1. 中止 一种暂停、暂时停滞、停止或搁置一段时间的状况。**2. 所有权待定** 当某项财产的拥有权并未确定时或某项财产并无拥有人的情况。

abioseston Non-living particulate matter found floating in watercourses.
非生物悬浮物 被发现漂浮于河道中的无生命颗粒状物质。

abiotic Non-living chemical and physical matter (not biological). Abiotic components are parts of the environment; water, sunlight, oxygen, temperature, soil, and climate.
非生物的/无生命的 无生命的化学或物理物质（非生物）。无生命成分指环境的组成部分；水、阳光、氧气、温度、土壤及气候。

ablution fitting A large communal sanitary fitting used to perform ritual washing.
洗礼装置 用于进行仪式性洗浴的大型公用卫生装置。

Abney level A hand-held instrument to measure vertical angles. *See also* CLINOMETER OF ACTION.
阿布尼水准仪 测量对顶角的手持式工具。另请参考“测斜仪”词条。

above grade A higher elevation than ground level. *See also* ABOVE GROUND LEVEL.
地面高程以上 比地平面高的高程。另请参考“地面高程以上”词条。

above ground level (US above grade) 1. The height above ground level. 2. Work undertaken on a building after the building has come out of the ground.
1. 地面高程以上（美国 above grade） 地面高程以上的高度。**2. 出地面施工** 当建筑物超出地面之后所实施的建筑物施工。

Abram's law A rule that states the strength of concrete (or mortar) is inversely related to the water/cement (w/c) ratio. Provided there is suffcient water to ensure that the *hydration reaction occurs, the lower the water/cement ratio, the higher the strength of the concrete/mortar.
阿伯拉姆定律 一条说明混凝土（或灰浆）的强度与其水/灰（w/c）比呈相反关系的规则。如果所加的水足以确保发生*水化反应，则水/灰比越低，混凝土/灰浆强度越高。

abrasion The wearing away of one material against another by *friction.
磨损 一种材料经与另一种材料摩擦所致的损耗。

abrasion resistance The ability of a finish to withstand wear.

耐磨性 某种饰面经受磨损的能力。

abrasive The use of friction/roughness to clean, grind, or polish a surface.
研磨 利用摩擦/粗糙以清洁、磨制或打磨某个表面。

abrupt wave An unexpected increase in flow caused by a sudden change in flow conditions.
陡浪/突变波 由于流动状况的突然改变所致的流量异常增大。

ABS (acrylonitrile butadiene styrene) A polymeric material composed of acrylonitrile, butadiene, and styrene to give a good balance of mechanical properties, used commonly as lavatory seats.
ABS（丙烯腈－丁二烯－苯乙烯共聚物） 由丙烯腈、丁二烯及苯乙烯组成的聚合材料，具有良好均衡的机械性能，常用作马桶座。

abscissa The horizontal or x-coordinate within the *Cartesian coordinate system.
横坐标 笛卡尔坐标系统中的水平坐标或 x 坐标。

abseil survey An inspection undertaken on a tall building/structure by a qualified person suspended via a rope.
绳降调查 通过以绳索吊挂有资质的专业技术人员对较高大楼的建筑/结构进行检查。

absolute Not depending on anything else.
绝对的 不取决于其他任何事物。

absolute assignment The irrevocable transfer of the entire debt or all legal rights.
绝对转让/无条件转让 不可撤销转让全部债务或所有合法权利。

absolute zero The lowest possible temperature. Defined as zero degrees *Kelvin, which is equivalent to 273. 15 degrees on the Celsius scale.
绝对零度 可能的最低温度。定义为开氏零度，等于零下 273. 15 摄氏度。

absorbent The ability to absorb, i. e. to soak up a liquid.
吸液性 吸收（即吸取液体）的能力。

absorption The process of absorbing a liquid.
吸液 吸收液体的过程。

absorption rate (absorption coefficient of water) The rate of water absorption, ingress by capillary action, into a material when exposed to a water medium over a period of time. The method for determining the water absorption property is carried out in accordance with BS EN 772-11. After drying to constant mass, a face of the masonry material is immersed in water for a specific period of time, both short-and long-term and the increase in mass is usually determined. The water absorption of the face of the unit exposed is measured, after drying to constant mass. The minimum immersion time is usually one hour, as specified in BS EN 771-4. The coefficient of water absorption is normally stated at 10, 30, and 90 minutes for masonry materials (bricks and blocks).
吸液率（对水的吸收系数） 当在一段时间内把某种材料暴露于水介质时，介质通过毛细管作用进入材料中的吸水比率。吸水性的判定方法根据 BS EN 772—11 标准进行。把砖石材料干燥至恒定质量后，在特定时限内把材料的一面浸入水中，通常对短时间和长时间以及质量增加都予以测定。在干燥至恒定质量后，即测得材料该面的吸水量。浸渍时间通常至少为一小时，如 BS EN 771—4 标准所规定。通常列出砖石材料（砖块和砌块）在第 10 分钟、第 30 分钟及第 90 分钟时的吸水系数。

absorption refrigerator A refrigerator that operates using the absorption cycle rather than the vapour compression cycle. In such systems, a heat source, usually waste heat, is absorbed and used to evaporate the refrigerant rather than using a compressor to compress the refrigerant. Quiet in operation and capable of being driven by solar energy.
吸收式制冷机 采用吸收循环而非蒸汽压缩循环的制冷机，在该系统中，热源（通常为余热）被吸收并用于蒸发制冷剂，而非使用压缩机压缩制冷剂。吸收式制冷机运行时较安静且可由太阳能驱动。

absorptive form *See* PERMANENT FORMWORK.

一次性消耗模板 参考“永久性模板”词条。

absorptivity (absorptive power) A property of a material that determines the amount of incident heat, light, or sound energy that is absorbed by the material.
吸收性（吸收能力） 材料的属性，用以判定材料所吸收的入射热能、光能或声能的量值。

ABT *See* ASSOCIATION OF BUILDING TECHNICIANS.
ABT 参考“建筑技术员协会”词条。

abut To adjoin; be next to; touch.
毗邻/紧接 毗邻、邻接、紧贴。

abutment 1. A solid structure, usually a pier or wall, which provides support to an arch, bridge, or vault. It enables the loads from the structure to be transmitted to the *foundations. 2. The point where a roof slope intersects a wall that extends above the roof slope. 3. The intersection between two building elements. 4. In dam construction, the sides of the valley against which the dam is constructed.
1. 支座/拱座/桥台 一种牢固的结构，通常为桥台或墙壁，为拱门、桥梁或穹隆提供支持，能使负荷由该结构传递至基础。**2. 屋顶与外墙交界处** 斜屋面与超出斜屋面的墙壁相交的点位。**3. 接界** 两个建筑构件的接合处。**4. 坝肩** 在水坝建设中，指峡谷的一面，水坝即依此而建。

abutment flashing A *flashing at an *abutment, usually made from lead, although copper and mortar have been used.
屋顶与外墙交界处防水 位于屋顶与外墙交界处的泛水板，虽然也有用铜和灰浆制成的，但通常是用铅制成。

abutment piece (US) A *sill or *sole plate.
底槛（美国） 槛或底板。

abutment wall The wall that extends beyond an *abutment.
桥台墙 突出至桥台外的墙壁。

ACA Form of Building Agreement The Association of Consultant Architects construction contract.
ACA 建筑协议格式 咨询过建筑师协会的建筑合同。

ACAS (Advisory, Conciliation, and Arbitration Service) An independent publicly funded organization founded in 1975 that acts to prevent and resolve employment disputes. It also provides professional advice and training.
ACAS（咨询、调解及仲裁局） 英国一个受公共资助的独立组织，设立于 1975 年，用于预防和解决劳动纠纷，另外也提供专业咨询和培训。

accelerated weathering (artificial ageing) Normally a cyclic process that simulates adverse climatic conditions (such as temperature, ultraviolet radiation, and moisture), which is used to assess the durability of materials.
加速风化（人工老化）试验 一般指模拟恶劣气候状况（例如温度、紫外辐射及湿度）的循环过程，用于评估材料的耐久性。

acceleration A measure of the rate at which *velocity is changing, i. e. a change in speed over time.
加速度 对速度变化率的计量，即速度随时间而发生的变化。

acceleration of work Agreement to complete the work in a shorter time. Within general law, the architect, contract administrator, or client has no power to instruct the contractor to complete the work in a shorter period of time than the expressed completion date, unless there is a provision (term) within the contract that allows for the acceleration of works.
加快施工 在较短时间内完成工程的协定。在普通法范围内，除非合同中存在允许加快施工的条文（条款），否则建筑师、合同管理人或业主无权指示承包商在比明示竣工日期更短的时间内完成工程。

accelerator An admixture that enhances early strength (hardening) but the long-term strength remains unaffected. Regularly used in cold weather when urgent repair work is needed. Calcium chloride is the most com-

mon accelerator, however, it reduces corrosion protection.
加速剂/促进剂/早强剂 提高早期强度（硬化）但并不影响长期强度的外加剂。常在需要进行紧急修复工作的寒冷天气中使用。氯化钙是最常见的加速剂，但会使防腐性降低。

accelerometer An instrument for measuring acceleration.
加速计 测量加速度大小的仪器。

acceptance An agreement, based on an offer, which forms a contract. An offer and acceptance are used in contract law to determine whether a contract exists. For a contract to exist, there needs to be an offer, acceptance of that offer, and consideration (the thing of value at the centre of the contract). Acceptance can be made orally, in writing, or by conduct. Acceptance must be unqualified.
承诺/接纳 基于一项要约达成的一项协议。要约和承诺在合同法中用于确定是否存在一份合同。要使合同存在，需要有要约、对该要约的承诺，以及对价（居于合同中心位置的价值事项）。承诺可由口头、书面或行动做出。承诺须为不受限制的承诺。

accepted programme The schedule of works that has been agreed, under the contract terms, to be used to schedule activities and resources. The * Engineering and Construction Contract, previously the New Engineering Contract, makes reference to the accepted programme.
已确认的进度 根据合同条款获得认可的施工计划，将用于活动和资源的规划。工程和施工合同（原为新工程合同）中会提到已确认的进度。

accepted risk Risks accepted by the client or employer which can influence the works but are beyond the contractor's control or scope of works. Term used in GC (General Contract) /Works Contract.
已确认的风险 已被业主或雇主接受的、可能使工程受到影响的，但超出承包商控制或工程范围的风险。该术语用于 GC（主合同）/施工合同中。

access 1. The method of gaining entry to a building, a room, a site, or services. 2. The right or permission to use something (access to documents).
通道/通行权/访问权 1. 进入某个建筑、房间、场所或设施的入口。2. 使用某物的权利或许可（对文件的存取）。

access chamber *See* INSPECTION CHAMBER.
检修间 参考“检查井”词条。

access cover (access eye, inspection cover) The removable cover on an * inspection chamber.
检修盖 检查井上可拆卸的盖子。

access floor (raised floor) A floor that is suspended above the structural floor. Usually consists of removable panels supported on a metal grid that is raised off the floor by adjustable pedestals or battens. The space between the two floors can be used to route various services such as data, telephone, power, lighting, heating and cooling pipes, and ventilation.
活动地板（高架地板） 悬在结构层之上的地板，通常由可拆卸的板材构成，这些板材由金属格栅支撑，而金属格栅则由可调节的支座或板条抬高。活动地板与结构层之间的空间可收纳数据、电话、电力、照明、供暖与制冷管道以及通风等各种设施。

access hole An opening, usually large enough for a person, that enables access to be gained to an area, an installation, or services.
检修孔 一个开口，通常足以容纳一个人。检修孔可使人出入于某个区域、装置或设施。

accessories 1. Components not included as part of the original product. Often used to enhance or modify the performance of the original product. 2. Components that are used to connect various building elements or services together.
1. 配件 并未被包含为原产品组成部分的零件。常用于提高或改变原产品的性能。**2. 连接件** 用于把各种建筑构件或设施连在一起的零件。

accessory box (accessory enclosure) A box that is used to store accessories.

配件箱 用于储存配件的箱子。

Access to Neighbouring Land Act 1992 Act designed to address problems that arise when entry to a neighbour's land is required to carry out work. The Act applies to basic preservation works only and would include maintenance and repair of buildings and property that cannot be reasonably undertaken within the boundary of the property. Application is made to the courts. The courts can only award access if it does not interfere with the enjoyment of an owner's land, nor should it cause hardship.
1992 年相邻土地通行法 为解决施工时须进入相邻土地所产生问题的法律。该法仅适用于基本保护工程，也包括不能在房产边界以内合理地实施对建筑物和房产的维护和维修。经向法院提出申请后，仅当业主对土地的享有权不受影响，且通行不致造成困难时法院才可裁定（给予）通行权。

access to works The contractual right to have access to the place where the work is to be undertaken. The employer has to give the contractor access to the site otherwise it would be impossible for him/her to carry out the work. It is not a breach of contract where access is prevented by third parties over whom the employer has no control.
对工程的通行权 合同权利，赋予对将要施工的地点的通行权。业主须给予承包商对现场的通行权（承包商无此通行权则不可能施工）。对于业主并无控制力的、第三方对通行权的阻碍，并不属于对合同的违约。

accident An identifiable and specific event that was not anticipated and was unexpected. Generally, the fact that something was an accident is not an adequate defence to an action brought in tort.
事故 一桩可辨识但不可预计的具体事件，是意料之外的。一般而言，某事为一桩事故的事实在侵权诉讼中并不足以作为抗辩理由。

accidental error A mistake happening by chance that would influence accuracy; for instance, taking an incorrect reading without realizing the mistake.
随机错误 随机发生并将影响精确性的错误；例如，读取了不正确的读数而未认识到该错误。

acclivity An upward slope or incline.
上斜坡 向上的斜坡或斜面。

accommodation works Activities that are undertaken or required to maintain structures, equipment, or land that is the property or under the control of a statutory undertaker; such work may include: bridges, fences, and gates to protect railways, substations, etc. Where there is a statutory obligation to protect such works it must be undertaken.
配套工程 为了对属于法定承担者或处于其控制下的财产的结构、设备或土地进行维护而承担的或必要的活动；此类工程可能包括：桥梁、栅栏及用于保护铁路的大门、变电站等。如存在需保护此类工程的法定义务则需承担。

accord and satisfaction The purchase of a release from an *obligation. In *contract law an obligation arising under the contract may be purchased by means of consideration, without having to undertake the obligation. The accord is the agreement and the satisfaction represents the consideration taken into account to discharge the obligation and makes the agreement valid.
和解与清偿 通过购买解除一项义务。根据合同法，可通过对价的方式购买合同项下所产生的义务，即无须履行义务。和解即达成协定，而清偿则指为解除义务和使协定有效而需支付的对价。

accordion door A door comprising a number of sections that fold together like the bellows of an accordion when opened.
风琴折叠门 由折叠在一起的多截组成的门，打开时犹如（拉动）手风琴的风箱。

accreditation A certification process by which an organization is awarded a standard of competency and credibility.
认证 向某个组织授予某种能力和认可标准的证明过程。

accrued rights The rights and duties of the parties following a contractor's determination of employment under the contract (remedies of either party). Such rights also give powers to contract administrators to issue instruc-

tions, in relation to work already completed, e. g. for defective work.
应计权利 在根据合同确定对承包商的雇佣后，签约各方享有的权利和义务（任何一方享有的救济）。该权利也是赋予合同管理人的权力，可发出与已完成工程有关（例如针对有缺陷的工程）的指示。

accumulation The increase of something over time.
累积 某种事物随时间而增长。

acequia An irrigation canal/ditch used in Spain and the American Southwest.
灌溉水渠 用于西班牙和美国西南部的灌溉渠/沟。

acetone A flammable liquid from the family of ketones.
丙酮 一种属于酮类的易燃液体。

acetylene A hydrocarbon substance from the alkynes family.
乙炔 一种属于炔烃类的碳氢化合物。

acid A liquid with a pH value of less than 7. 0.
酸 pH 值低于 7. 0 的液体。

acidic A chemical compound whose properties are typical of an *acid.
酸性物 具有酸的典型属性的化合物。

acknowledgement of service The formal step, by way of a letter, of responding to a claim. The letter would state that a defendant intends to dispute part or all of the claim, state the particulars of the claim, and record whether the court's jurisdiction is accepted. The procedure is governed by the Civil Procedure Rules.
送达确认 通过信函方式对一项诉讼申索做出正式回应的步骤。信函将说明被告方有意对申索的部分或全部做出抗辩，说明申索的详情，并记录是否接受该法院的管辖权。该程序属于民事诉讼规则。

acoustic board A board used on walls, floors, and ceilings that is designed to improve *sound absorption or *sound insulation.
隔音板 用于门、地板及天花的板材，旨在改善吸音或隔音效果。

acoustic clip A clip that is attached to the floor joists to provide support for an acoustic floor. Used to reduce *impact sound transmission on floating timber floors.
隔音夹 附着于地板托梁上并为隔音地板提供支撑的夹子，用于减弱活动木地板的碰撞声传输。

acoustic construction Any type of construction that improves the *acoustics of the construction. For instance, reduces *sound transmission, improves *sound insulation, or increases *sound absorption.
隔音构造 使隔音效果得以改善的任何类型的构造。例如，减弱声音传输、改善隔音或促进吸音。

acoustic finish (acoustic finishing) A finish that reduces sound energy by absorption.
隔音饰面 通过吸收来降低声能量的饰面。

acoustic floor A floor that is designed to reduce sound transmission.
隔音地板 被设计为减弱声音传输的地板。

acoustic lining A lining material that is designed to absorb noise and reduce *sound transmission. Used on pipes and ductwork.
隔音衬里 被设计为吸收噪音并减弱声音传输的衬里材料，用于管道和导管中。

acoustic plaster Plaster that has high sound absorption properties.
隔音灰泥 具有较高吸音性能的灰泥。

acoustics 1. A branch of physics that relates to the study of *sound generation, sound transmission, absorption, and reflection. 2. In construction, the effect that sound has on a room or space.
1. 声学 与对声音产生、声音传输、声音吸收及声音反射的研究有关的物理学分支。**2. 音效** 在建筑中，声音对房间或空间的影响。

acoustic tile A tile manufactured from *acoustic board that is usually used on ceilings to improve *sound absorption or *sound insulation.
隔音瓦 通常用于天花的由隔音板制成的瓦，用以改善吸音或隔音。

acre An area of land equal to 4840 square yards (0.405 hectares).
英亩 土地面积单位，（一英亩）等于4840平方码（0.405公顷）。

acrylated rubber paints Special type of paints that are suitable for internal and external applications exposed to chemical attack, or wet and humid conditions. May be applied to metal or masonry by either brushing or spraying.
丙烯酸化橡胶漆 特殊类型的漆，适用于暴露在化学侵蚀或潮湿条件下的内部和外部涂装，可使用刷涂或喷涂的方法涂到金属或石材上。

acrylic A compound derived from acrylic acid.
丙烯酸酯/亚克力 一种源自丙烯酸的化合物。

acrylic paints A type of paint in an acrylic polymer emulsion renowned for its quick drying properties.
丙烯酸漆 一种丙烯酸类高聚物乳状漆，以其快干性而知名。

acrylic resin *See* *POLYMETHYL METHACRYLATE.
丙烯酸树脂 参考“聚甲基丙烯酸甲酯”词条。

acrylic sealants A type of *sealant derived from acrylic acid.
丙烯酸密封剂 一种源自丙烯酸的密封剂。

acrylic sheet A polymer material derived from acrylic acid, usually transparent.
丙烯酸板/亚克力板 一种源自丙烯酸的高聚物材料，通常为透明的。

acrylonitrile butadiene styrene *See* *ABS.
丙烯腈-丁二烯-苯乙烯共聚物 参考“ABS”词条。

action 1. The physical operation of something. 2. A civil legal proceeding brought by one party against another.
1. 行动 某事物的实际实施。**2. 诉讼** 由一方针对另一方提起的民事法律程序。

active earth pressure (Pa) The horizontal pressure that is exerted on a retaining wall by the mass of soil it retains. Active pressure develops when the wall moves away from the retained mass and the soil expands sufficiently to mobilize its shear strength. The minimum horizontal stress (Pa) occurs when the failure strength of the soil is fully mobilized, represented by the minor principal stress, and is calculated by multiplying the *coefficient of active earth pressure (K_a) by the vertical stress at the point of consideration.
主动土压力（Pa） 被拦挡的土体对挡土墙施加的水平压力。当把挡土墙移开且土体充分扩张至激发其抗剪强度时，则产生主动压力。当土体的破坏强度被充分激发时，则产生最小水平应力（Pa），以最小主应力表示，并通过以主动土压力系数（K_a）乘以该点位的垂直应力计算得出。

active fire protection A fire protection system that uses electrical and/or mechanical equipment to detect or suppress fires, such as *fire detectors, *alarms, *sprinklers, or *spray systems. They require regular maintenance to ensure that they are fully operational.
主动消防 使用火警探测器、警报、喷淋头或喷淋系统等电气/或机械设备探测或扑救火灾的消防系统，须受到定期维护以确保其正常运行。

active leaf The most frequently used opening leaf on a double door. When closed, it is held in position by latching to the *inactive leaf.
活动门扇 双扇门中使用最频繁的开启门扇。在关门时，被门闩锁闭在固定门扇上。

active solar heating A heating system that utilizes electrical and/or mechanical equipment, such as solar collectors, pumps, and fans to collect solar energy from the sun to heat water.
主动式太阳能供暖 使用太阳能收集器、泵及风扇等电气/或机械设备收集来的太阳能加热水的供暖系统。

activity An item of work or task that is needed to complete a project. Used in *critical path Method scheduling. An **activity arrow** is used to represent an activity graphically. The time that is expected to be taken to complete an activity is known as the **activity duration**. An **activity schedule** is a list of all of the activities that are required to be carried out to complete a project.
活动 为完成某个项目而需要进行的一项工作或任务。该词被用于关键路径法的时序安排中。活动箭头以图形方式代表一项活动。预计完成某项活动的时间被称为活动持续时间。活动日程表就是为完成某个项目而须进行的所有活动的清单。

Act of God A legal term meaning an occurrence caused by natural forces where foresight and prudence could not reasonably anticipate it and guard against it. The term is not generally used in construction contracts, *force majeure is the term usually used to cover such occurrences.
天灾 法律术语，指由不可通过远见、审慎、合理的预计和防范的自然力所导致的事件。该术语一般并不用在建筑合同中，不可抗力才是常被用于涵盖此类事件的术语。

Act of Parliament A statute that sets out the law, the legal will of parliament in written form. The acts provide the broad principles, with the detail of the law covered by regulations in the form of a statutory instrument.
议会的法案 以书面形式载列法律、议会法律意志的法令。法案将规定广泛的原则，以及以法定文书的形式由法规涵盖的法律详情。

actuator A mechanical device that is used to convert electrical, hydraulic, or pneumatic energy into mechanical energy. Used in flow control valves, meters, motors, pumps, switches, and relays.
致动器 一种把电能、水能或风能转为机械能的机械装置。常用在流量控制阀、仪表、电机、泵、开关及继电器中。

adaptor A device that enables the characteristics of one component to be matched to another component. Used for joining *pipework, *ductwork, and electrical cables.
适配器 能使一种组件的特性与另一种组件相匹配的装置，用于接合管道、导管及电缆。

addendum bills Bills of quantities produced to alter or modify parts of the original bills. Problems often occur where additional documents are used to modify the original bills as the addendum will refer to the original document, but the original document does not make reference to the addendum. If not cross-referenced properly such documents can be problematic. Addendums may be used to reduce the lowest tender so that it is within the client's budget or to cater for minor alterations.
补遗清单 为对部分原始清单补遗加以修改或修订而制作的工程量清单。由于修订原始清单的补遗文件将参照原始文件，但原始文件并没有提到，因此常发生问题。如果有关文件不是交叉参考，可能会有问题。补遗文件可用于降低报价最低的投标，使其符合业主的预算，或被用于细微的变更。

additional work Work not identified or work over and above that specified in the contract. In standard building contracts additional work is dealt with under the various terms associated with *variations. *See also* *EXTRA.
额外工作 合同中并未指定的工作或超过合同所述的工作。在标准建筑合同中，额外工作是通过与变更有关的各种条款加以处理。另请参考“额外工程”词条。

additive A substance that is added to another to alter (normally improve) properties or performance, e. g. an *admixture to concrete.
添加剂 被添加到另一种物质中用以改变（一般是改善）其属性或性能的物质，例如混凝土的添加剂。

addressable system (intelligent fire alarm) A fire alarm system incorporating detectors that are capable of sending signals back to the main control unit. Such a system can be used to determine the location of a fire and activate the fire protection system.
可寻址系统（智能火警） 一种火警系统，该系统含有能向主控制装置发回信号的探测器。该系统用于确定火灾位置和启动消防系统。

adds Quantities of work, of a similar nature, that are added together under the same description. Adds are

brought together before deducts are taken off, e. g. quantities of brickwork added together before the wall openings, windows and doors, are taken off (deducts).
增项/加项 具有相似性质并在同一说明项下加总的工程量。把加项加总后才会扣除减项，例如，把砌砖量加总后才会扣除墙壁开口、窗户及门（减项）。

adhesion A process whereby the molecules of two separate substances cling together to form a bond.
黏合 两个不同物质的分子聚合在一起形成黏结的过程。

adhesive (glue) A substance used to bond two objects or materials together.
黏合剂（胶水） 一种用于把两个物体或材料黏结在一起的物质。

ad hoc Something that is undertaken only when necessary for the specific purpose in question and is not planned in advance.
临时的/特设的 某事项仅因有需要才进行的，不是事先规划好的。

ad idem A Latin expression that means at the same point, agreed, or of the same mind. Parties that have agreed to a contract are said to be ad idem when agreement has been reached on the contract terms.
一致 一个拉丁词语，指具有相同观点、意见一致，或想法相同。当合同条款达成协定后，即可称在议定合同的签约方之间已达成一致。

adjacent Adjoining, next to, near, or close to.
临近 毗邻、邻接或靠近。

adjoining property A building that has a common boundary or is attached to another property.
相邻物业 与另一物业具有共同边界或附属于另一物业的一座建筑物。

adjudication Method used to resolve disputes. An adjudicator is empowered by contract or statute to settle a dispute. A party to a construction contract has the right to refer a dispute to an adjudicator, where the parties to the contract have failed to resolve their contractual differences. The adjudicator must act independently, hearing the evidence and facts from both sides. Adjudicators are often lawyers or experts within the industry, but can be any person that the parties agree to. Adjudication was introduced as a quick and cost effective method of resolving disputes, when compared to *arbitration or litigation. The Housing Grant Construction Regeneration Act 1996 states that provisions for adjudication must be included within construction contracts. Where contracts do not comply with the Act, the Scheme for Construction Contracts will apply, thus adjudication is forced into contracts even if the construction contract does not have provisions to include it. The Construction Industry Council (CIC) procedures and the Technology and Construction Court (TeCSA) rules are often adopted as part of the contractual machinery associated with adjudication.
审判、裁判 用于解决争议的方法。审裁员经合同或法规授权以解决争议。如建筑合同的签约方之间未能解决其合同分歧，则任何一方有权把争议递交予审裁员。审裁员须独立行事，聆讯双方的证据和事实。审裁员常为律师或业内专家，但也可为双方同意的任何人士。与仲裁或诉讼相比，审裁是作为快速和具有成本效益的争议解决方法而被引入的。英国《1996 年房屋许可、建设及重建法》规定，建筑合同中必须包含审裁条款。如合同并未遵守该法，则将适用建筑合同方案（Scheme for Construction Contracts），因此即使建筑合同中并无把审裁包括在内的条款，审裁仍强制适用于合同。英国建筑业委员会（CIC）的程序和技术与建筑初级律师联合会（TeCSA）的规则常被采纳作为与审裁有关的合同机制部分。

adjustable steel prop (ASP) A steel support that is adjustable along its length. Used to provide vertical support to floors and temporary beams.
可调钢支柱（ASP） 可沿其长度方向调节的钢支架，用于为地板和临时梁提供竖向支撑。

administrative charges Costs associated with off-site secretarial, coordination, and management provisions. With off-site resources it may be difficult to identify an exact figure, for one contract, so a percentage for the *overheads, which includes administration, insurance, and banking charges are included in the *preliminaries.
管理费 场外文秘、协调及管理条款有关的成本。由于可能难以为单一合同确定场外资源的确切金额，因此把日常开支（包括行政、保险及银行收费）的一定百分比计入费用中。

administrative receiver A person or group appointed to act on behalf of the debenture holders to manage the company as a going concern, often before the company goes into *liquidation.
管理接收人 被任命作为债权人的代表行事，以便管理公司（通常在公司被清算前）使其持续经营的一个人或群体。

admissibility of evidence The degree of consideration that can be given to evidence. Information, hearsay, or opinion of an expert that a tribunal finds useful in coming to its decision. Where issues are irrelevant or considered to be biased, overstated, or glorified, the tribunal may be restricted in the use of such evidence. It is more common in criminal cases for evidence to be restricted for fear that the nature of the evidence will result in undue prejudice.
证据的可采性 可对证据给予的考量程度。法庭认为有助于其达成裁决的信息、传闻或专家意见。如为无关内容或被认为有偏见、被夸大或被美化，则法庭对有关证据的使用可能会受到限制。由于担心证据的性质将导致不当的偏见，证据受到限制的情况在刑事案件中较为常见。

admixture A chemical product which is added to *concrete (<5% by mass) during mixing or by additional mixing, prior to placing, such that the normal concrete properties are changed, e. g. the speed of *strength development, improvement of *permeability, protection from *fungicidal attack, improvement in *workability.
外加剂 一种在搅拌时加入或在浇筑前通过额外添加的方式加入混凝土中（以质量计低于5%）以便改变普通混凝土性能（例如强度发展的速度、渗透性的提高、对真菌侵袭的防护、和易性的提高）的化学产品。

adobe A building material made from water, sand, and soil containing clay that is dried in the sun. Straw or other organic fibrous materials may be added to reduce cracking, aid drying, and to help bind the material together. An **adobe brick**, or mud brick, is made by placing adobe into a mould.
土坯 一种由水、砂及含黏土的泥土制成并经日晒烘干的建筑材料，其中可能加入稻草或其他有机纤维材料，以减少开裂有利于烘干，使材料结合在一起。黏土砖则是通过把重黏土放入模具中制成。

adsorption The accumulation of a substance (gas or liquid) on a surface of another substance (liquid or solid), forming a molecular film.
吸附 物质（气体或液体）聚集于另一物质（液体或固体）的表面，形成一层分子膜。

advance (advance payment) Payment or part payment made before work is undertaken. Some standard forms of contract have provisions for advance payment. To be binding, the amount agreed and the date for payment must be inserted in the *appendix of the contract, together with the times and amounts of repayment. *Bonds are normally required for advanced payments.
预付款 在施工之前付款或部分付款。某些标准格式的合同含有预付款条文。为了具有约束力，关于预付款的约定金额与日期和偿还的时间与金额须列入合同的附录中。预付款一般需要出具保函。

advances on account A provision (term) used in construction contracts which refers to periodic payments during the progress of the work, normally at monthly intervals. The housing Grants Construction and Regeneration Act 1996 now requires that all construction contracts make provision for periodic payments. Periodic payments may be covered under the provisions for *interim certificates.
定期预付款 建筑合同中所用的一项条文（条款），指在工作期间的定期付款，一般按月做出。英国《1996 年住房许可、建设及重建法》目前规定一切建筑合同均须载有定期付款的条文。定期付款可能包含于中期证书的条文项中。

adverse possession Where the title of a person's land transfers to another without compensation, often called squatter's rights. Title of land may be acquired under the Limitations Act 1980. If a person occupies or remains in possession of land without rent or payment for a period of twelve years for private land or 30 years for Crown land, and makes uses of the land which are inconsistent with the owner's enjoyment or intended use, then the land may be acquired.
逆权侵占 指土地的业权在业主未获补偿的情况下被强制转让予侵占者，这常被称为侵占者权利。

根据《1980年时效法》可获得土地的业权。如某人在12年（适用于私人土地）或30年（适用于公有土地）的期间内在并未付租或付款的情况下即占用或持续占有土地，且其使用土地的方式与业主的权利或预定用途并不相符，即可获得该土地。

adverse weather conditions Weather conditions that restrict or preclude work from being undertaken on site, such as high winds and heavy rain.
恶劣天气状况 使现场施工受到限制或阻碍的天气状况，例如狂风和暴雨。

advertisement for bids *See* INVITED BIDDER.
招标公告 参考“特邀投标人”词条。

adze A carpenter's tool used for shaping wood. It has an axe-like head, long handle, and a thin blade perpendicular to the handle.
扁斧/锛子 一种用于木材成形加工的木工工具，具有斧状头部、长把手及与把手垂直的薄刃。

aerated concrete *See* AIRCRETE.
充/加气混凝土 参考“充气混凝土”词条。

aeration The introduction of air (or gas) into a substance (e. g. water, concrete, sewage).
充气/加气 把空气（或气体）导入某一物质（例如水、混凝土、污水）中。

aerial Relating to the air, e. g. an aerial photograph can be taken from an aeroplane to produce maps and scale drawings of the ground.
空气的/空中的/航空的 与空气有关的，例如航空摄影即可在飞机上进行，以便制作关于地面的地图和比例图。

aerobic Any treatment, activity, or process that requires the presence of air (oxygen).
有氧的/需氧的 任何需要有空气（氧气）存在的处理方法、活动或过程。

aerochlorination Treatment of wastewater (e. g. *sewage) by a mixture of compressed air and *chlorine gas.
压气氯化法 通过压缩空气和氯气的混合物进行的废水（即污水）处理。

aerodynamics The study of airflow around objects and the movement of objects through air, with particular reference to the forces experienced by an object as a result of the movement of air.
空气动力学 对物体周围的气流和物体在空气中运动的研究，尤其是针对由于空气的运动而使物体受到的力。

aerotriangulation The use of angles and distances to orient and align overlapping *aerial photographs.
空中三角测量 利用角度和距离对重叠的航空摄影进行定向和排列。

affidavit A factual or written statement or evidence made under oath. Documents sworn in under an affidavit are called exhibits.
宣誓书 根据宣誓做出的实际或书面的声明或证明。在宣誓书项下宣誓的文件称为证据。

affirmation of contract Where there is a breach of contract, the innocent party may choose to continue with the contract and affirm the contract, treating it as still being in force.
对合同的主张 如发生对合同的违约，无过失方可选择继续执行合同，确认合同仍然合法有效。

AFNOR (Association Franc, aise de Normalisation) The French National Standards Organization.
AFNOR (Association Franc, aise de Normalisation) 法国国家标准协会。

a fortiori argument Latin term used in law to show that the case being presented falls well within the scope of something that has gone before, thus there is even greater reason that the situation under examination should be treated with at least the same consideration. The situation can be treated as a maiore ad minus e. g. if the situation applies to a large group then it should also hold true for a smaller group or if 8 m^3 of concrete is contained in a dumper then 6 m^3 would also fit into the same dumper. Alternatively, but less common, a minor ad maius argument can be used, which means from smaller to bigger, e. g. if two packs of bricks overloaded the scaffolding, then six packs of the same type of brick would also overload it. All of these arguments rely on the first assumption holding true, thus if two packs of bricks would not overload the scaffolding then there would be no evidence to check the second part of the statement. Where illogical arguments are used to justify a claim, it falls under the term petitio principii.

顺理成章的论证 拉丁术语，用在法律中以表示所阐述的情况恰好属于之前已有的某事物的范围以内，因此有较大理由把正在经受验证的该情况，按照相同的考虑因素予以处理。该情况可按以大推小（a maiore ad minus）的论证处理，例如，如果该情况适用于某个较大的群体，则其对较小的群体而言也应属实，或如果某辆自卸卡车能装 8 立方米混凝土，则 6 立方米混凝土应该也可装入这辆车。从另一方面而言（但不太常见），也可使用以小推大（a minor ad maius）的论证，即从较小者推导出较大者。例如，如果两堆砖将使脚手架超载，则六堆同样类型的砖也会超载。上述全部论证均有赖于前一个假设属实。因此，如果两堆砖不会使脚手架超载，则并无凭据可核实该论述的第二部分。但如果以不合逻辑的论据来论证某项主张为合理，则此被称为预期理由（petitio principii）。

A-frame building A triangular-shaped building with a steeply sloping roof that extends to ground level.
A 形架构建筑物 三角形的建筑物，具有延伸至地平面的陡峭斜屋顶。

African mahogany Common name for genus Khaya of the meliaceae family and genus Afzelia of the fabaceae (legumes) family trees.
非洲桃花心木 楝（meliaceae）科卡雅（Khaya）属树木和蝶形花（fabaceae）科［即豆科（legumes）］缅茄（Afzelia）属树木的通用名。

after-flush The water remaining in a cistern after the toilet has been flushed. This water provides the water seal between flushing.
冲后水封 马桶冲水后水槽中剩余的水，这些水将在冲水的间隔期间起到水封作用。

aftershock A small earthquake, normally one of many, that occurs after a large earthquake.
余震 小型地震，一般是多次地震中的一次，常发生在大型地震之后。

aftertack (after-tack) A defect in a paint film where the film remains tacky and sticky over an extended period of time.
回粘 漆膜的一种缺陷，即漆膜经过较长时间仍然发黏和黏结。

ageing *See* PRECIPITATION HARDENING.
老化 参考“沉淀硬化”词条。

agency An organization that acts for and on the behalf of others. For instance, a construction recruitment agency would recruit staff for various construction companies.
代理 一个代理他人行事的组织。例如，建筑招聘代理会为各种建筑公司招聘员工。

agent 1. A person who has the authority to act on behalf of others. 2. The *site manager. 3. A substance that changes the characteristics of another substance when added to it.
1. 代理人 有权代理他人行事的人。**2. 代表** 现场经理。**3. 作用剂** 一种物质，在被加入另一种物质时会改变另一种物质的特性。

agglomerate To collect together in a round (ball) mass. **Agglomerates** are formed from volcanic fragments, consisting of various rock types, sizes, and shapes, fused together.
团块/团聚 聚合为圆（球）块。团聚物由火山岩碎屑形成，包括熔合在一起的各种类型、尺寸及形状的岩石。

aggregate Granular material used in concrete, road construction, mortars, and plaster. Aggregates can originate from natural rocks, gravel and sands (e.g. limestone and sandstone), formed from artificial sources (via a thermal process), or be recycled inorganic construction waste. From June 2004 aggregate size is detailed by a lower (d) and upper (D) sieve sizes (expressed as d/D) in accordance with the appropriate European Standards: e.g. BS EN 12620 ‘aggregate for concrete’; BS EN 13043 ‘aggregates for bituminous mixtures and surface treatments for roads, airfields, and other trafficked areas’; BS EN 13285 ‘unbound mixtures’; and BS EN 13139 ‘aggregates for mortar’.

骨料/集料 用于混凝土、道路建设、砂浆及灰泥中的颗粒材料。骨料可源自天然岩石、砾石及砂（例如石灰岩和砂岩），或来自人工来源（通过热加工），或通过回收的无机建筑废物。自2004年6月起，骨料尺寸按筛网尺寸（以 *d/D* 表示）的小（*d*）和大（*D*）细分，依据为适用的欧洲标准：例如 BS EN 12620《混凝土骨料》；BS EN 13043《用于道路、机场及其他交通区域的沥青拌和料和表面处理的骨料》；BS EN 13285《松散拌和料》；及 BS EN 13139《砂浆用骨料》。

aggregate cement ratio Weight of aggregate divided by the weight of cement.
骨料水泥比 骨料质量与水泥质量的比例。

aggressive Exhibiting attacking or hostile behaviour, which can lead to damage.
侵略性的 表现出可能导致损害的攻击或敌意行为。

aggressive environments Locations where acid rain, corrosive gas/liquids, chemical/industrial pollutants, etc occur or are present. These elements can cause damage to materials such as concrete, steel, and timber.
腐蚀环境 发生或存在酸雨、腐蚀性气体/液体、化学/工业污染物等的地点。上述因素可导致对混凝土、钢材及木材等材料的损害。

agreement 1. The sense of a meeting of minds between two people. 2. An essential part of a valid contract.
1. 合意 两人之间达成一致的意见。**2. 协议** 一份有效合同的基本部分。

Agreement for Minor Building Works Standard form of contract, appropriate for use on projects up to a value of £ 90,000.
小型建筑工程协议 标准格式的合同，适用于最高价值为90 000英镑的项目。

agreement to negotiate Agreement to make a contract. Although parties can agree to negotiate rates and make a *contract, due to the uncertainty that such practices bring, English law does not recognize a ‘contract to negotiate’ as a contract.
磋商意向书 关于达成协议的意向书。虽然当事方可据此磋商并达成协议，但由于此类做法带有不确定性，故英国法律并不承认“磋商协议书”为合同。

Agrément Certificate A certificate issued by the *British Board of Agrément that contains information on a product's durability, installation, and compliance with Building Regulations. Only issued once the product has successfully been assessed.
许可证书 一种由英国许可委员会颁发的证书，其中载有关于产品耐用性、安装及对建筑条例合规情况的信息，仅当产品顺利通过评估后才予颁发。

ahead of schedule (advance) An activity that is ahead of schedule is one that has been completed prior to the date recorded in the *programme or *activity schedule.
超前（提前） 超前活动即在计划表或活动日程表中所记录日期之前完成的活动。

air-admittance valve A device that allows air to be admitted into sanitary pipework to equalize the pressure and maintain the water seal in the trap of the appliance.
进气阀 一种使空气得以进入卫生管道以便平衡压力并使器具的存水弯处的水密性得以保持的装置。

air balancing The process of adjusting the air flows through the supply and extract grilles in ventilation or air-conditioning systems to provide the required air flow rates.
风量平衡 在通风系统或空调系统中对经过送风和排风格栅的气流进行调节以便提供所需气流速率的过程。

air barrier The part of the building envelope that has been designed and constructed to resist *air leakage from a conditioned to an unconditioned space. Air barriers can be constructed from a wide range of materials that are impervious to air, such as wet plaster and polythene sheeting. Materials that have been connected together to form an air barrier are termed an air barrier system.
气障 为防止使空调空间变为无空调空间的漏气而设计和建设的建筑物围护部分。气障可由湿灰浆及聚乙烯板等各种不透气材料建成。被连接在一起构成气障的材料称为空气屏障系统。

air-blast cleaning High pressure cleaning operation using air alone. Pneumatic hoses are often used to re-

move rubbish, dust, and debris that has become trapped between * steel reinforcement and * formwork. Formwork should be cleaned before any concrete is poured.
压缩空气清洁 仅使用空气进行的高压清洁操作。常以压气软管来清洁落入钢筋与模板之间的垃圾、灰尘及渣土。在浇筑混凝土之前，应对模板加以清洁。

airborne noise Noise or sound that is admitted into the atmosphere. Typical sources of airborne sound include human voices, radios, musical instruments, and audio-visual equipment. *See also* AIRBORNE SOUND, IMPACT SOUND.
空气噪音 进入大气中的噪音或声音。空气噪音的常见来源是人类语音、广播、乐器及视听设备。另请参考“空气传声，撞击声”词条。

airborne sound Noise that propagates through air. *See also* AIRBORNE NOISE.
空气传声 通过空气传播的噪音。另请参考“空气噪音”词条。

air brick (ventilation brick) A brick containing a series of holes that run through the brick to allow ventilation. Usually built into a wall to ventilate rooms or a space under a ground floor level. Available in a range of different sizes.
空心砖（通风砖） 含有一系列孔洞的砖，这些孔洞贯穿砖身以供通风。空心砖一般被筑入通风房间或地面以下空间的墙壁中。空心砖有各种不同规格可供使用。

air brush A small, air-operated spray gun that is used to produce a fine spray of paint.
气动刷/喷笔 一种以空气操作的小型喷枪，用于进行精细喷涂。

air change The replacement of air within a space with new air, usually from outside.
换气 （通常是从外部）以新鲜空气替换某一空间内的空气。

air change rate The rate at which replacement air enters a space divided by the volume of that space. Expressed in air changes per hour (ac/h). Used to measure ventilation.
换气率 进入某一空间的替换空气除以空间体积的比率，以换气量/小时（ac/h）表示，用于计量通风情况。

air compressor A machine that compresses (pressurizes) air, typically to power pneumatic tools such as drills, jackhammers, pumps, etc.
空气压缩机 一种压缩（增压）空气的机器，常用于为钻机、手提钻、泵等气动工具提供动力。

air conditioner (air conditioning) A mechanical device used to control temperature, relative humidity, and air quality within a space. Usually comprises an evaporator, a cooling coil, a compressor, and a condenser. Also commonly used to describe any system where refrigeration is included to provide cooling. A wide range of air-conditioning systems are available and they can be categorized according to their function:
空调机（空气调节） 一种用于调节某一空间内的温度、相对湿度及空气质量的机械装置，通常由蒸发器、冷却盘管、压缩机及冷凝器组成。空调机一词也常被用于描述包含制冷过程以提供冷却的所有系统。空调系统的种类繁多，可根据功能分类为：

· **Full air-conditioning systems**—provide all of the ventilation and fresh air requirements of the building. Temperature and humidity can be controlled within predetermined limits. The air supply is filtered, heated, or cooled to the required temperature, and is humidified or dehumidified to maintain acceptable levels of relative humidity.
全空调系统——为建筑提供所需的全部通风和新鲜空气。温度和湿度可控制在预设限度内。送风已经过过滤、加热（或冷却）至所需温度，并加湿（或除湿）以保持可接受程度的相对湿度。

· **Close control air-conditioning systems**—similar to full systems, but are capable of maintaining temperature and relative humidity within close control limits, typically within ±0.5 ℃ and ±5% RH.
精密控制空调系统——与全空调系统类似，但能够把温度和相对湿度保持在精密控制的限度内，一般在±0.5 ℃和±5%相对湿度以内。

· **Comfort cooling systems**—use refrigeration to reduce internal temperatures when required to produce more acceptable conditions for occupants. Ventilation and fresh air are not always supplied from a comfort cooling

system, and may need to be provided separately.
舒适冷却系统——使用制冷设备，降低室内温度，为使用者提供舒适的温度环境。通风和新鲜空气并不一定由舒适冷却系统提供，而可能需要单独提供。

aircrete A low-density porous material extensively utilized in the construction industry, usually in block form. The air content (*porosity) is typically between 60% and 85%. It is produced by mixing cement and/or pulverized fuel ash (PFA), lime, sand, water, and aluminium powder. The final process involves autoclaving for approximately 10 hours at high temperature and pressure. Aircrete (in 2008) accounted for a third of all concrete blocks in the UK. Such blocks are suitable as vertical load-bearing elements and provide the thermal insulation expected from typical UK wall construction. They may also be used as non-load-bearing outer leaves of masonry walls, external walls, and walls below ground level, where adequate care is essential to ensure their durability and protection from effects of the environment. The lightweight porous structure creates an effective moisture barrier with desirable thermal insulation properties. Their porous cellular structure and durability make them a recognized alternative in most below-ground situations. Aircrete originated from Scandinavia in the 1950s as an alternative to building with timber. Aircrete is also known commercially as **AAC (Autoclaved Aerated Concrete)**, **Celcon**, **Durox**, **Thermalite**, and **Topblock**. Aircrete is normally categorized as low, medium, and high density.
充气混凝土 在建筑业中广泛使用的低密度多孔材料，一般为方块形。含气量（孔隙率）一般介于60%至85%。充气混凝土是通过把水泥和/或粉煤灰（PFA）、石灰、砂、水及铝粉混合而制成。最后的过程是在高温高压下进行大约10个小时的蒸压。充气混凝土（在2008年）占英国全部混凝土砌块的三分之一。此类砌块适于作为竖向承重构件并为英国标准墙体结构提供隔热。还可用作砌筑墙、外墙及地面以下墙壁的非承重外叶部分，在此情况下足够的养护是至关重要的，以便确保其耐久性并抵御来自环境的影响。轻量级的多孔结构造成了具有适当隔热性的有效防潮层。多孔的蜂窝结构和耐久性使充气混凝土成为大多数地下场合的公认选项。充气混凝土起源于20世纪50年代的斯堪的纳维亚，作为木材的替代品用于建筑。充气混凝土的商品名有AAC（蒸压充气混凝土）、Celcon、Durox、Thermalite及Topblock等。充气混凝土一般按低、中、高密度分类。

air curtain A mechanical device that produces a narrow jet of high velocity air that is directed down and across an opening. Commonly used on exterior doors, loading platforms, and refrigeration display cases to prevent air from one side of the opening mixing with the air from the other.
空气幕 一种机械装置，形成对向下并穿过一个开口的高速空气的狭长喷射，一般用于外部门户、装料台及冷冻展示柜以防来自开口的一侧的空气与来自另一侧的空气混合。

air diffuser An outlet or *grille designed to distribute air within a space. Commonly located in a *suspended ceiling.
散流器 被设计用于向某个空间送风的出口或格栅，一般位于吊顶天花中。

air distribution The movement of air from one space to another.
风量分配 空气从一个空间向另一个空间的移动。

air drain An empty space left around the external *perimeter of a *foundation to prevent moisture from the surrounding ground causing *dampness.
通风道 围绕基础的外部周长留出的清空空间，用于防止来自四周地面的水汽导致的潮湿。

air-dried timber *See* **AIR-SEASONING**.
风干木材 参考“风干”词条。

air eliminator An automatic device used in plumbing to release air from the pipework.
除气器 一种在给排水设施中用于把空气从管道中排出的自动装置。

air-entrained concrete A type of *concrete with purposely incorporated minute air bubbles to improve its freeze-thaw resistance.
掺气混凝土 一种特意含有微小气泡以提高其抗冻融性的混凝土。

air-entraining agent An *admixture which causes minute air bubbles to be incorporated into concrete, used

to improve workability and frost resistance.
引气剂 一种使混凝土中含有微小气泡的外加剂，用于提高混凝土的和易性与抗冻性。

air exfiltration The uncontrolled exchange of air out of a building through a wide range of * air leakage paths in the * building envelope.
空气渗出 不受控制的空气交换，空气通过建筑围护中的大量漏气通道，渗出建筑物。

airfield soil classification A soil classification system for engineering purposes developed by A. Casagrande in the early 1940s, which became known as the **Unified Soil Classification System** in 1952. It described a standard system for classifying mineral and organo-mineral soils for engineering use based on the particle size, liquid limit, and plasticity values.
机场土壤分类 用于工程目的的土壤分类体系，由 A. Casagrande 于 1940 年代初制定，并于 1952 年被称为统一土壤分类体系。该体系规定了一个为工程用途而根据颗粒尺寸、液限及塑性值对矿质土壤和有机矿质土壤进行分类的标准体系。

air gap A gap between two components in a construction that is filled with air.
气隙 构造中的两个构件之间由空气填满的空隙。

air grille (air grating) A framework of bars used to cover an air outlet that enables air to pass through.
通风格栅（通风篦子） 用于覆盖在出风口使空气通过的格栅框架。

air-handling equipment The equipment used to move air from one conditioned space to another in an air-conditioning system. Usually comprises a series of supply and extract fans and terminal units.
空气处理设备 用于把空气从一个有空调的空间移至空调系统内另一个空间的设备，一般由一系列送风机和排风扇以及终端装置组成。

air-handling luminaire A * luminaire that is designed to allow air to be extracted through it. Used to reduce the heat gains from lighting and reduce the cooling load on the air-conditioning system.
风口灯盘 一种设计为可抽取空气的灯具，用于降低来自照明的热增益并减少空调系统的制冷负荷。

air-handling unit (AHU, air handler) A packaged item of air-conditioning equipment that treats the incoming air prior to it being distributed around the building. Comprises a fan, combinations of heating and cooling coils, filters, humidifiers, and control dampers. Usually located on the roof with * chillers and * boilers located nearby.
空气处理机（AHU） 空调设备的封装组件，对进气先加以处理，随后才被送风至建筑物各处。空气处理机包括风机、加热与冷却盘管、过滤器、加湿器及调节风门，通常位于屋顶的位置，与冷水机组和锅炉相邻。

air heater *See* UNIT HEATER.
空气加热器 参考“暖风机”词条。

air house *See* AIR-INFLATED STRUCTURE.
充气屋 参考“充气式结构”词条。

air infiltration The uncontrolled exchange of air into a building through a wide range of air leakage paths in the * building envelope.
空气渗透 不受控制的空气交换，空气通过建筑围护中的大量漏气通道渗入建筑物。

air-inflated structure An air-inflated structure has a double membrane supported by a series of tubes that are inflated using high pressure air. **Air-supported structures** have a single membrane that is supported by low pressure air. Both types of structures can be rapidly erected or dismantled and are mainly used as temporary structures.
充气式结构 具有双层膜的充气结构，而膜则由一系列以高压空气填充的导管支撑。气承式结构具有由低压空气支撑的单层膜。两种结构均可快速搭建或拆卸，主要用作临时结构。

air leakage The uncontrolled fortuitous exchange of air both into (* infiltration) and out of (* exfiltration) a * building envelope, space, or component through cracks, discontinuities, and other unintentional openings. It is driven by the wind and the stack effect. Typical air leakage points occur at: cracks, gaps, and joints in the structure; plasterboard dry-lining; windows, doors, and their surrounds; joist penetrations of external walls; boundary/wall junction; internal partition walls; loft service entries, ducts, and electrical components; permanent ventilators; loft hatches and skirting boards.

漏气 不受控制的偶发空气交换，空气通过裂缝、间断处及其他非有意的开口进入（渗透）和流出（渗出）建筑围护、空间或构件中。漏气是由风力和烟囱效应所导致。典型的漏气点会发生在：结构中的裂缝、空隙及接缝处；石膏板干衬里；窗、门及其周围；外墙的托梁插入处；边界/墙壁交会处；内部隔墙；阁楼设备入口、导管及电气构件；固定通风筒；阁楼上人孔和踢脚板。

air leakage audit An inspection of a building that is carried out to identify the main *air leakage paths and any areas where there may be discontinuities in the *air barrier. Can be undertaken during construction or after the building is completed.

漏气检查 为确定主要的漏气通道以及气障中可能不连续的任何区域而对建筑物进行的检查，可在施工时或竣工后进行。

air leakage index The volume flow of air either into or out of a space, per square metre of *building envelope area, at a given pressure difference between the inside and the outside of the building (in the UK 50 Pa is used). For air leakage index, the area of the lowest floor is included in the building envelope area if it is not ground supported. Expressed in $m^3/(h.m^2)$. *See also* AIR LEAKAGE RATE.

漏气指数 在建筑物内部和外部的给定压力差（在英国使用的是50帕）下，进入或流出某一空间的空气体积流量除以建筑围护面积的平方米数字。对于漏气指数而言，如最底层并非由地面所支撑，则其面积将被计入建筑围护面积中。漏气指数以 $m^3/(h\cdot m^2)$ 为单位表示。另请参考“漏气率”词条。

air leakage path The path that air takes when leaking into or out of a building envelope or component.

漏气通道 空气进入或流出建筑围护或构件时所经过的通道。

air leakage point The point at which air leaks into or out of a building envelope or component.

漏气点 发生空气进入或流出建筑围护或构件的点位。

air leakage rate The rate at which air leaks into or out of a *building envelope, space, or component, per unit volume, at a given pressure difference across the building envelope, space, or component (in the UK 50 Pa is used). Traditionally, expressed in ac/h, however, nowadays $m^3/(h\cdot m^2)$ is more commonly used as it takes into consideration the effects of shape and size. *See also* AIR LEAKAGE INDEX.

漏气率 在建筑围护、空间或构件内外部的给定压力差（在英国使用的是50帕）下，进入或流出建筑围护、空间或构件的空气量与单位体积的比率。传统上，漏气率以ac/h为单位表示，但现在 $m^3/(h\cdot m^2)$ 则更为常用，因为后者把形状和尺寸的作用也纳入考虑。另请参考“漏气指数”词条。

airless spraying A method of painting that involves forcing paint through a nozzle at high pressure. Results in less overspray than *compressed-air spraying.

无气喷涂 一种通过高压喷嘴把涂料压出的涂漆方法，与压缩空气喷涂相比，其所导致的过喷现象较少。

air lift pump A deep (up to 200 m) *well pump, which is located below the *groundwater table. *Compressed air, from the surface, is pumped down the well where it bubbles into a large diameter pipe, lifting a *slurry mix to the surface. A key element to this pump is that there are no mechanical moving parts in the well, which would otherwise be abraded by the slurry causing maintenance issues.

空气升液泵 一种较深（最深200米）的井泵，位于地下水位以下。来自地面的压缩空气被泵入井下，在大直径管道中成为气泡，把一种泥浆混合物提升至地面。这种泵的优点是在井中并无机械运动部件，因此不会因泥浆所致磨损而带来维护问题。

air monitoring Sampling and measuring the quality (normally the level of pollutants) of the atmosphere/ambient air at regular intervals.

大气监测 按固定间隔时间对大气/环境空气的质量（一般为污染物水平）进行取样和测量。

air permeability The volume flow of air either into or out of a space, per square metre of * building envelope area, at a given pressure difference between the inside and the outside of the building (in the UK 50 Pa is used). For air permeability, the area of the ground floor is included in the building envelope area. Expressed in $m^3/(h \cdot m^2)$.

透气性 在建筑围护的内、外部的给定压力差（在英国使用的是 50 帕）下，进入或流出某一空间的空气体积流量除以建筑围护面积的平方米的数值。对于透气性而言，底楼面积将被计入建筑围护面积中。透气性以 $m^3/(h \cdot m^2)$ 为单位表示。

air permeability test A test that is undertaken to determine the air permeability of a building envelope, space, or component.

透气性测试 为确定建筑围护、空间或构件的透气性而进行的测试。

airport A location where non-military aircraft (i. e. only civil aeroplanes, helicopters, and airships) take off and land, equipped with a * runway and facilities for handling passengers and cargo.

航空港 非军用飞机（即仅有民航飞机、直升机及飞船）起飞和着陆的场所，设有跑道和处理乘客和货物的设施。

air quality The composition of * ambient air with respect to the * pollution levels.

空气质量 关于污染水平方面的环境空气组成情况。

air receiver (air vessel) A pressure vessel used in compressed air applications to store the compressed air and to equalize the pressure in the system.

储气罐（空气罐） 一种储存压缩空气并平衡系统压力的压力容器。

air-release valve (pet cock) A device that is used to release trapped air from pipework or a fitting.

放气阀（小旋塞阀） 一种用于释放管道或管件中的滞留空气的装置。

air-seasoned timber *See* AIR-SEASONING.

风干木材 参考“风干”词条。

air-seasoning (air-dried timber, air-drying) The process of removing moisture from timber before it is used and put into service by drying in air. Two main methods of seasoning are air-seasoning (natural seasoning) and kiln-seasoning (artificial seasoning). In a living tree, the weight of water in the tree's vessel cells will frequently be greater than the dry weight of the tree itself. Seasoning is the removal of most of this moisture, and the stabilization of the moisture content before putting the timber into service. The primary aim of seasoning is to render the timber as dimensionally stable as possible. This ensures that once put into service as flooring, furniture, doors, etc, movement will be negligible. Seasoning involves removing most but not all of the moisture from the timber, and when this is undertaken other advantages accrue. Most wood-rotting fungi can grow in timber only if the moisture content is above 20% ～ 22%. Drying of timber occurs because of differences in vapour pressure from the centre of a piece of wood outwards. As the surface layers dry, the vapour pressure in these layers falls below the vapour pressure in the wetter wood further in, and a vapour pressure gradient is built up which results in the movement of moisture from centre to surface; further drying is dependent on maintaining this vapour pressure gradient. The steeper the gradient, the more rapidly the drying (seasoning) progresses, but in practice, too steep a gradient can cause the wood to split. For large-sized timbers, a combination of the two methods is often used. Air-seasoning is still practised in some countries where the cost of kiln-drying is high (particularly in the developing world), and it is still used to at least partially dry timbers of large cross-sectional area. Where such timber would take more than 4 – 5 weeks to kiln-dry, it is often air-dried to a moisture content of 25% ～ 30% first. In the UK, hardwood planks are usually air-dried for 18 – 24 months before being kiln-dried.

风干（风干木材、晾干） 木材在被使用和投入服务之前，通过在空气中干燥除去水分的过程。两种主要的风干方式为干燥（自然风干）和窑干（人工风干）。在活的树木中，树木导管细胞里的水分重量一般大于树木自身的干重量。干燥就是在木材投入使用前把大部分水分除去，并使水分含量保持稳定。干燥的主要目的是尽可能地使木材保持固定尺寸，这将确保木材一旦被用作地板、家具、门等时，出现的形变将可予以忽略。干燥会除去木材中大部分水分但非全部，且在干燥时将保留木材的其他优点。只要水分含量高于20%～22%，大部分木腐菌均可在木材上生长。木材的干燥是由于木材中心到外部的蒸汽压力存在差异产生的。当表面层干燥后，其蒸汽压力会低于较湿的内部木材，因而形成压力差使水分由内部往表层移动；保持该蒸汽压力差可使干燥过程持续进行。压力差越大干燥（风干）的速度越快，但实际操作中太大的压力差会导致木材开裂。规格较大的木材通常组合使用两种方法。某些国家仍在使用自然风干，因为窑干的成本较高（发达国家通用），自然风干还用于木材横截面大的区域的部分干燥。这种木材窑干会耗时4～5周，所以通常先自然风干到水分含量25%～30%。在英国，实木板材通常先自然风干18～24个月的时间再窑干。

air shaft A vertical shaft within a building or tunnel that is used for ventilation.
通风井 建筑物或隧道内用于通风的竖井。

air space (air cavity) A space between two elements that is filled with air. Commonly known as a *cavity in masonry cavity construction.
气室（气穴） 两个构件之间被空气充满的空间，在砖石空腔构造中常称为空腔。

air terminal unit A mechanical device used in *air-conditioning and *ventilation systems to supply and extract air from a space. *Fan coil units, *induction units, and *variable air volume units are all examples of air terminal units.
空气终端装置 用于空调和通风系统中向空间送风或从空间中排气的机械装置。风机盘管装置、诱导器及变风量装置均为空气终端装置的实例。

air termination network (air termination system) The part of a lightning protection system that is designed to intercept a lightning strike. Comprises a series of vertical rods that are placed on a roof or equipment, known as AIR TERMINALS.
接闪器网络（接闪器系统） 防雷系统的组成部分，旨在拦截雷击，由安放在屋顶或设备上的一组竖杆（称为避雷针）构成。

air test *See* FAN PRESSURIZATION TEST.
空气测试 参考“风扇加压法测试”词条。

airtight inspection cover An *inspection cover that incorporates an airtight seal.
密封井盖 一个含有空气密封的检查井盖。

airtightness (air tightness) Used to describe the air leakage characteristics of a building. Frequently expressed in terms of a whole building leakage rate at an artificially induced pressure (in the UK 50 Pa is used). The smaller the air leakage rate at a given pressure difference across the building envelope, the greater the airtightness. The airtightness of a building determines the uncontrolled background ventilation or *leakage rate of a building which, together with ventilation caused by occupancy (opening windows, etc), makes up the total ventilation rate for the building.
气密性 用于描述某个建筑物的漏气特征，通常为整栋建筑物在人工诱发的压力（在英国使用的是50帕）下的漏气率。在整个建筑围护的给定压力差下漏气率越小，气密性就越高。建筑物的气密性决定了不受控制的背景通风或建筑物的漏气率，而漏气率与由于有人入住而导致的通风情况（开窗等）构成建筑物的整体通风率。

airtightness layer A layer within the building envelope, space, or component that resists air leakage. This may or may not be the same layer as the *air barrier.
气密层 建筑物围护、空间或隔间内的一层，能防止漏气，有可能与气障是同一层也有可能不是。

airtightness test *See* FAN PRESSURIZATION TEST.
气密性测试 参考“风扇加压法测试”词条。

air-to-air heat-transmission coefficient *See* THERMAL TRANSMITTANCE.
空气对空气热传导系数 参考“传热系数”词条。

air treatment The process of treating the air within an air-conditioning system. This may involve filtering the

air to remove contaminants, and/or heating, cooling, humidifying, or dehumidifying the air. *See also* AIR CONDITIONER.
空气处理 空调系统内对空气进行处理的过程，这可能涉及把空气过滤除去污染物和/或对空气加热、制冷、加湿或除湿。另请参考“空调机”词条。

air valve A *valve that controls the flow of air into or out of a system.
空气阀 对进出某个系统的空气流进行控制的阀门。

air vent An opening that allows air to flow from one space to another.
通气孔 使空气得以从某个空间流至另一个空间的开口。

air void An enclosed space filled with air.
气穴 一个充满空气的封闭空间。

air washer A mechanical device used to cool, clean, and dehumidify air. No longer used in air-conditioning systems due to health concerns.
空气洗涤器 一种用于冷却、清洁及除湿空气的机器设备，由于卫生问题已不再用于空调系统中。

airy stress function A function used to solve elastic analysis of plane stress distribution.
艾里应力函数 一个函数，用于解决平面应力分布的弹性分析。

alarm Noise, created by a bell, gong, buzzer, horn, electric signal, or speaker to indicate a problem or danger where immediate action is required. A **fire alarm** is used to alert occupants of a building or a building site to the risk of a fire and requires the action of all persons evacuating the building (or building site) to go to an area that has been designated safe.
警报 由铃、锣、蜂鸣器、喇叭、电讯号或扬声器发出的响声，提示存在必须立即采取行动的问题或危险。火灾警报被用于向在建筑物或建筑工地的人员警示火灾风险并需要动员全部人员撤离建筑物（或建筑工地）至指定的安全区域。

albic A *soil horizon from which clay and free iron oxides have been removed or iron oxides have been separated.
白浆土 一种土壤层，其中的黏土和游离氧化铁已被除去或氧化铁已被分离。

alcove A recessed or partially enclosed portion of a room.
凹室 房间中一处凹陷或部分封闭的部分。

alidade A sighting device (telescope and attached parts) used on a *plane table for angular measurements.
照准仪 一种用于在平板仪上进行角度测量的照准装置（望远镜和附件）。

alien enemy A person whose state (nationality) is at war with the land in which they reside. Such people may be allowed to leave the country or may be interned within the country. The term is occasionally used in construction contracts.
敌侨 其所属的国家（国籍）与其所居住的国家正在进行战争的人，此人可能被允许离开该国或可能被拘禁于该国。该词偶尔会被用在建筑合同中。

alignment The correct positioning of different components relative to something else, e.g. *setting out a road or railway on the ground.
对齐 不同构件相对于其他构件的准确定位，例如对地面上道路或铁路的放线。

alkali-aggregate reaction A chemical reaction between the aggregates in concrete and cement resulting in damage to the material.
碱－骨料反应 一种在混凝土中骨料和水泥之间会导致对材料损害的化学反应。

alkali-resistant glass fibre A type of fibre utilized in concrete for protection against alkali reaction.
抗碱玻璃纤维 一种用于混凝土中以防碱性反应的纤维。

alkali-resistant paint Fast-drying paint for metals, concrete, and masonry which provides a tough, fiexible film with excellent resistance to moisture, salt water, and alkali.

抗碱漆 用于金属、混凝土及砌体的快干漆，将形成对水汽、盐水及碱具有优异抗性的牢固、柔韧漆膜。

alkali-resistant primer A protective sealer used in masonry against stains and blistering caused by salts in the plaster.
抗碱底漆 一种用于砌体的保护性底漆，以防由于灰泥中的盐所致的污渍和起泡。

alkyd paint A type of paint based on an oil-modified polyester resin (alkyd) which provides a surface that is highly resistant to wear.
醇酸漆 一种基于油改性聚酯树脂（醇酸树脂）的漆，能形成具有较高耐磨损性的表面。

alkyd plastics A type of polymer based on esters.
醇酸塑料 一种基于酯类的聚合物。

alkyd resin A synthetic resin used in paints.
醇酸树脂 一种用于油漆中的合成树脂。

all-in aggregate A mixture of coarse and fine aggregates.
未筛分骨料/统货骨料 粗骨料和细骨料的混合物。

all-in contract A *design and build or *turnkey contract, where the design and construction of the building is included in one contract.
全包合同 一种设计与建造合同（总承包）合同，建筑物的设计和建造均包括在此合同中。

allowable To permit or be acceptable. Allowable *bearing capacity is an acceptable (factored) pressure the ground can withstand from a foundation.
容许的/可容许的 将允许或可接受的。容许承载能力为地基可承受来自基础可接受（分解）的压力。

alloy Ametallic material comprising two or more elements that can be either a metal or non-metal. Any metal that is not 100% pure is classified as an alloy. Examples of alloys are *steel, *stainless steel, *galvanized steel, *cast iron (all based on iron), *bronze, *brass (both based on copper), and *solder (a mixture of lead and tin). The primary objective of alloys is to obtain a metal which comprises a mixture of the main attributes of each constituent.
合金 由两种或以上的金属或非金属元素组成的金属材料。并非100%纯度的任何金属均被归类为合金。合金的实例有钢、不锈钢、镀锌钢、铸铁（以上均以铁为基础）、青铜、黄铜（两者均以铜为基础）及焊锡（一种铅和锡的混合物）。（制成）合金的主要目的在于获得具有各成分的主要特性的金属混合物。

alloy steel A *ferrous metal containing carbon and several other alloying elements.
合金钢 一种含有碳和若干其他合金元素的黑色金属。

all risks The list of risks that the JCT 98 (*Joint Contracts Tribunal) form of contract identifies as needing works insurance.
一切险 JCT 98（合同审定联合会）合同格式确认为需要投保工程险的风险清单。

alluvium Sediments (silt, sand, gravel, etc.) deposited by flowing water (rivers and streams).
冲积层 被流水（河流和溪流）所堆积的沉积物（粉土、砂、砾石等）。

alms house A sheltered house or building for people without satisfactory income or support.
救济院/济贫院 针对没有满意收入或抚养的人的庇护所。

alteration or amendment of contract Alterations made to the contract that are written into the contract, or otherwise agreed, before the contract is executed (signed by the parties). Standard forms of contract are regularly updated to keep in line with court decisions and new statutes. The standard forms can also be amended to include specific requirements to suit either of the parties, so long as they are included prior to the contract being agreed and signed.
合同的变更或修订 在合同履行（由签约方签字）之前以书面写入合同或另行议定的方式对合同做

出的变更。合同标准格式会定期更新，以符合法院判决和新的法规的规定。标准格式也可予修订以便纳入适合任一订约方的具体规定，只要该规定是在合同被议定或签署之前被纳入即可。

alternate bay construction A traditional method of concreting that uses a ‘twostage’ process. In the first stage, the area to be concreted is divided into bays using formwork. Concrete is then poured into each alternative bay. As soon as is possible, the formwork between the bays is then removed. In the second stage, concrete is poured up against the existing concrete to fill the remaining bays. It is, however, becoming more common for concrete to be laid in a continuous slab.
隔仓施工法 一种使用“两阶段”程序的混凝土浇筑传统方法。在第一阶段，会用模板把将要浇筑混凝土的区域分为隔仓，再把混凝土倒入各个隔仓，随后在可能的情况下尽快把隔仓之间的模板抽除。在第二阶段，会把混凝土倾倒在现有的混凝土上，以填充隔仓的剩余空间。但目前在连续模板中浇筑混凝土已越来越普遍。

alternate lengths work The underpinning of foundations that is carried out in small sections to ensure that support for the building remains in place at all time. To avoid settlement, one section of foundation should be underpinned, with the adjacent section of foundation remaining in place providing support until the new concrete foundation has gained sufficient strength. Thus, new underpinning is carried out in alternate sections with adjacent existing sections of foundation providing temporary support.
分段托换施工 通过较小分段的基础托换，以确保保持对建筑物的支撑。为免沉降，基础的一个分段将进行托换，而邻近的基础分段则保持原位并提供支撑直至新浇筑的混凝土获得足够的强度。因此，轮流进行新的基础托换时，邻近的既有基础分段提供临时的支撑。

alternate proposal (US) An alteration or variation to the contract different from that described in the tender or bid documents.
变更建议（美国） 对合同的变更或更改，使合同不同于招标或标书文件所述。

alternating current (AC) An *electric current that reverses direction in a circuit at a constant *frequency. If the frequency is 60 Hertz, the direction in the flow of the current changes 60 times per second.
交流电（AC） 一种按恒定频率在电路中反转方向的电流。如频率为 60 Hz，则电流的流动方向每秒变化 60 次。

altitude The height of an object above a known *datum, normally sea level.
高度/海拔 物体高于已知基准点（通常为海平面）的高度。

aluminium A non-ferrous metal/alloy, chemical formula Al. Aluminium has the merit of being light (density 2,700 kg/m^3 as compared to 7,900 kg/m^3 for steel), ductile, and easily rolled into sheets and thin strips, and extruded into complex sections. It resists corrosion, especially if it is anodized. However, it has only one-third of the stiffness of steel, and it melts at the relatively low temperature of 550 – 600℃ (in comparison to approximately 1500℃ for steels). Its performance in fire is therefore not good, and so it cannot be used structurally in building. **Aluminium alloys** are easily formed due to its high ductility; this is evidenced by the thin aluminium foil sheet into which the relatively pure material may be rolled. In construction aluminium is used externally for window frames and structural glazing systems, roofing, and claddings, flashings, rainwater goods, etc. It is used internally for ceilings, panelling, ducting, light fittings, vapour barriers (as aluminium foil), architectural hardware, walkways, handrails, etc. Aluminium can be alloyed with lithium to further enhance its properties. These materials have lower densities (2500 – 2600 kg/m^3) in comparison with other metals, especially ferrous alloys, high specific moduli (elastic modulus-specific gravity ratios), and excellent fatigue and low temperature toughness properties. Aluminium-lithium alloys are mainly utilized in the aerospace industry.
铝 一种有色金属/合金，化学式为 Al。铝具有重量轻（密度为 2 700 千克/立方米，而钢为 7 900 千克/立方米）、易延展与轧制为板材和薄条及挤出为复杂型材的优点。铝耐腐蚀，尤其是在经过阳极氧化处理后。但铝的硬度仅为钢的三分之一，且其熔点温度为相对较低的 550 ～600℃（钢的熔点为1500℃），因此铝在火中的性能不佳，不能用于建筑物的结构方面。**铝合金** 易于成型是由于铝的延展性高；可以由相对较纯的材料轧制出的细铝箔片证明。在建筑中，铝在外部用于窗框和结构玻璃系统、屋顶与覆层、防雨板、雨水管件等；铝在内部用于天花、镶板、导管、灯具、隔汽层（以铝箔的形式）、建筑五金、走道、扶手等。铝可与锂形成合金以增强性能，此类材料具有比其他金属（尤其是黑色合金）更低的密度（2500～2600千克/立方米）、较高的比模量（弹性模量与密度比）及优异的疲劳性能和低温韧性。铝锂合金主要用于航空业。

aluminium cable An aluminium/aluminium alloy conductor material utilized for the transmission and distribution of electrical power. Since the late 1990s aluminium has replaced copper conductors for these applications, and is the standard material for electrical conductors. These conductor designs have consistently provided a superior combination of strength and conductivity for distribution and transmission applications. Aluminium and aluminium alloy conductor materials are excellent choices for wire and cable products in various electrical applications because of aluminium's excellent physical and electrical properties. Aluminium is light-weight—about a third as heavy as copper; it is an excellent conductor of heat and electricity, an excellent reflector of heat and light, it is highly resistant to corrosion, strong, and flexible, and can be made stronger or more flexible by alloying and/or heat treatments.
铝线/铝缆 用于电力传输和分配的铝/铝合金导体材料。自20世纪90年代后期起，铝在此类应用方面已取代铜导体，成为电导体的标准材料。此类导体的设计常常是对分配和传输用途方面的强度和导电性的卓越综合。由于铝具有优异的物理和电气性能，铝导体及铝合金导体材料在各种电气应用方面是良好选择。铝的重量较轻——约为铜重量的三分之一；铝是热和电的优良导体、热和光的优良反射体，具有较强耐腐蚀性，坚固且柔韧，可通过制成合金和/或经热处理而变得更坚固或更柔韧。

aluminium foil Aluminium sheets with a typical thickness of less than 0.2 mm. Typically laminated to other materials, e.g. plaster ceiling boards.
铝箔 标准厚度低于0.2毫米的铝片，一般与石膏吊顶板等其他材料叠轧。

aluminium paint A special paint for various applications particularly where a bright reflective finish is required; especially for hot pipes.
铝粉漆 一种多用途的专用漆，主要是用于当需要明亮反光的饰面（特别是热管道）。

aluminium primer An aluminium-based *pigment used particularly on timber surfaces.
铝底漆 一种铝基颜料，用于木料表面。

aluminium roofing Sheet roofing material used to provide the protective finish or cladding to an aluminium roof.
铝屋面 用于为铝屋顶提供保护性饰面或覆层的薄板屋面材料。

aluminium windows External windows manufactured from extruded sections of *aluminium that are either screwed or mechanically *cleated together. They are usually anodized, powder-coated, or organically coated. Advantages are: durability, narrow frame sections, require little maintenance, and have good resistance to corrosion. Disadvantages are: aluminium is a good conductor of heat so frames need to incorporate a *thermal break, more expensive than steel, and they have a high embodied energy content.
铝窗 以铝质挤出型材制作并通过螺丝或机械方式契合的外窗，一般经过阳极氧化、粉末喷涂或涂有机涂层。优点有：耐用、细框架、仅需较少维护且具有较好耐腐蚀性。缺点有：铝是热的良导体，故窗框中需含有热障，与钢相比价格昂贵，而且铝窗具有较高的内涵能源量。

aluminium-zinc coating Anti-corrosion treatment of metal-coated steel. An aluminium-zinc alloy coating applied to steel provides enhanced corrosion protection and can improve the lifetime of steel by up to four times that of galvanized steel under the same conditions.
铝锌涂层 对金属涂层钢材的抗腐蚀处理，把铝锌合金涂在钢材表面形成强化防腐保护层，可在同等条件下提高钢材寿命，达至镀锌钢的四倍。
ambient The immediate surrounding area. The **ambient conditions**, i.e. air temperature, wind speed, and humidity, can be defined for a given location.
周边/周围 直接邻近的区域。一旦给定某个位置，则周围情况即空气温度、风速及湿度可予界定。

ambiguity Something which is unclear, or has a degree of uncertainty or has more than one meaning. The

A

standard forms of contract *ACA, *JCT 98, and *IFC 98 have their own provisions to deal with discrepancies and ambiguities. Each contract has different interpretations.
模糊 某事物不清楚，或具有一定程度的不确定性或具有一个以上的含义。ACA、JCT 98 及 IFC 98 标准合同格式均有条款处理不一致和模糊的情况。各份合同具有不同的解释。

amendment Change or correction to something, usually to documentation or a contract. Updates and revisions to standard forms of contract are often published and attached to the contract so that the contract encompasses changes in *statutes and *case law.
修订 对某事物（通常为文件或合同）的改动或改正。对标准合同格式的更新和修订常被刊发并附录于合同，以便把法规和案例法的修订纳入合同中。

American bond *See* ENGLISH GARDEN WALL BOND.
美式砌（墙）法 参考“三顺一丁砌墙法”词条。

ammeter An instrument used to measure electric *current (amps) in a circuit.
电流表/安培计 一种仪器，用于测量电路中的电流（安培）。

ammonia A hazardous gas at ambient temperature with a very strong odour. Chemical formula NH_3.
氨气 一种有害气体，在室温下具有极强烈的气味，化学式为 NH_3。

ampere (amp A) The unit of *current. One ampere is the amount of electric current for one *coulomb of charge per second.
安培（amp A） 电流的单位。一安培的电流量即每秒通过一库仑的电荷。

amplifier A device that changes (normally increases) the *amplitude of a *signal. A sound amplifier is a device that makes *sound louder.
放大器 一种使信号的振幅改变（一般是增大）的装置。例如扩音器即为使声音更响亮的装置。

amplitude The distance from the mean (mid-point) value to the *crest or *trough of a wave or *oscillating signal.
振幅 从平均（中间点）值到波或振荡信号的波峰或波谷的距离。

anaerobic Without needing oxygen. An **anaerobic organism** does not need oxygen for *metabolism.
厌氧的 不需要氧。**厌氧生物** 进行新陈代谢时不需要氧气。

analogy A comparison made between two things, used for the purpose of explanation. One thing, which is less understood, is compared with something that is well understood.
类比 出于解释目的而在两个事物之间做出的比较。较少为人所知的一个事物被与人所共知的某个事物相比较。

analysis A thorough investigation of something in order to understand.
分析 为了了解某事物而对其进行的透彻调查。

anchor 1. To fix firmly and securely in position. 2. A mechanical device that is built into a structure to prevent a component from moving.
1. 锚固 牢固和稳固地固定到位。**2. 锚** 一种机械装置，被制成用于防止构件移动的结构。

anchorage 1. The process of anchoring. 2. In a suspension bridge, the part of the bridge where the suspension cables are connected.
1. 锚固 锚固的过程。**2. 锚碇** 在悬索桥中指桥梁与悬索相连的部分。

anchor block A block of wood set within a masonry wall, in place of a brick, that provides a surface for connecting other wooden items.
锚块 在砌筑墙中取代砖块的一个木块，用于提供一个平面以便连接其他木制件。

anchor plate (floor plate) A plate attached to a component that enables other components to be connected to it.
锚定板（底板） 附着于构件上的一块板，使其他构件可与该构件相连。

ancient lights Windows or openings in walls that have access to natural light for a period of twenty years or more have a right of access to unobstructed light. Under the Prescription Act 1932 the light cannot be obstructed without the owner's permission.
采光权 墙壁上的窗户或开口如可接触自然光达二十年或以上的时间，则享有对光线的不可阻碍的接触权。根据 1932 年惯例法，未经业主允许，不可阻碍其采光权。

ancient monument A site, feature, or building of historical or archaeological importance listed under the Ancient Monuments and Archaeological Areas Act 1979. It is an offence to undertake any work on listed monuments without consent.
古迹 根据 1979 年古迹和考古区域法列出的具有历史或考古重要性的场所、景观或建筑物。未经同意，对列名古迹实施的任何工程将构成一项犯罪。

anemometer An instrument that is used to measure wind speed and direction.
风速计 一种用于测量风速和风向的仪器。

aneroid barometer An instrument for measuring atmospheric pressure.
空盒气压表/无液气压计 一种用于测量大气压力的仪器。

angle bead (plaster bead) A perforated metal strip in the shape of an angle that is used to protect and reinforce the corner of a plaster or plasterboard wall. The perforations are used to nail the strip to the surrounding plaster or plasterboard and act as a *key when the bead is plastered over.
护角条（灰泥护角） 角形的有孔金属条，用于保护和增强灰泥或石膏板墙的边角。打孔是为了把金属条钉在外围灰泥或石膏板并在为护角抹灰时作为毛面。

angle block (glue block) A small wooden block, usually triangular in shape, that is fastened into the interior angle between two surfaces to strengthen and stiffen the joint. The interior angle is usually 90°.
斜垫块（胶角垫） 一个小木块，通常为三角形，被固定在两个平面之间的内角中以便加固和强化接合处。这种内角通常为 90°。

angle board A board cut at a predetermined angle. Used as a guide for cutting other boards at the same angle.
角板 一块按预定角度切割的板，用作模板以便按相同角度切割其他的板。

angle brace (angle tie) A *brace fixed across an interior angle in a frame to improve the frame's rigidity. The brace can be temporary or permanent.
角撑（隅撑） 装在框架中横跨内角的支撑，以便提高框架的刚性。支撑可为临时性或永久性的。

angle cleat (clip anchor) A short L-shaped angle used to connect components to structural members, e. g. attaching *pre-cast concrete cladding panels to the main structural frame, or *purlins to *roof trusses.
连接角钢（角码） 短 L 形角钢，用于把构件连接至结构元件，例如把预制混凝土覆板附加于主结构框架，或把檩条附加至屋顶架。

angle closer *See* CLOSER.
墙角砍砖 参考“过渡砖”词条。

angled (angling) Set at an angle.
有角度的 按照角度设置的。

angled tee A T-shaped pipe fitting where the main inlet of the *tee is at a shallow angle.
斜三通 一种 T 形管件，其中三通的主管与支管成小角度。

angle fillet A thin strip inserted into the internal angle between two surfaces to cover the joint. Triangular in cross-section.
阴角嵌条 插入两个平面之间的内角中以便覆盖接缝的细条，横截面为三角形。

angle float A plasterer's smoothing tool that is shaped to enable the plaster to shape, smooth, and form internal corners.
阴角抹子 泥水匠的平整工具，其形状可用于使灰泥成形、平整及形成内角。

angle gauge A template that has been cut or made to set out or check corners.
角规/角度计 经过切割或制作用于设定或核对边角的模板。

angle grinder (disc grinder) A hand-held power tool with a small *abrasive rotating disc used to cut or grind masonry, concrete, or steel.
角磨机（圆盘磨光机） 手持式电动工具，带有较小的研磨转盘，用于切割或研磨砖石、混凝土或钢材。

angle joint A joint between two surfaces that results in a change of direction.
角接 在两个表面之间形成的可使方向改变的接头。

angle of friction (φ, angle of internal friction, angle of shearing resistance) Ratio between the shear stress and normal stress at failure, originally proposed by Coulomb in 1773。
摩擦角（φ，内摩擦角，剪切阻力角） 破坏时剪切应力与正应力之间的比率，最初于 1773 年由哥伦布提出。

angle of repose The steepest angle a granular heap of material would make to the horizontal when poured.
休止角 当倾倒颗粒状材料时材料堆与水平方向形成的最大角度。

angle rafter *See* HIP RAFTER.
角椽 参考“四坡屋顶面坡椽”词条。

angle section (angle bar, angle iron) A steel L-shaped angle.
角型材（角钢、角铁） L 形角的钢材。

angle staff *See* ANGLE BEAD.
护角线 参考“护角条”词条。

angle strut A structural member in the shape of an angle that acts in *compression.
角铁支柱 角形的结构元件，在受压环境下工作。

angle tie *See* ANGLE BRACE.
隅撑 参考“角撑”词条。

angle tile (angular tile, arris tile) A tile in the shape of an angle that is used for tiling the corners of a wall, a *hip, or a *ridge.
角砖（角形砖、脊瓦） 角形的砖瓦，用于铺砌墙壁、斜脊或屋脊的边角。

angle trowel (twitcher) A plasterer's trowel that has been shaped to work with awkward edges and angles. Some trowels may have a ‘V’ cut out of them for external angles; others may have edges that are upturned to create internal angles and rounds.
角抹子 一种泥水匠抹子，其形状适用于棘手的边角，某些抹子可能具有“V”字形切口以便适用于阳角；其他的则可能具有上翘的边缘以便适用于阴角和倒圆角。

angle valve A valve whose inlet is at 90° to the outlet. Used to control flow.
角阀 入口与出口成 90°角的阀门，用于控制流动。

angstrom (Å) A unit of length equal to 10^{-10} metres, used to measure wavelengths of electrometric radiation, i. e. very short wavelengths.
埃（Å） 长度单位，等于 10^{-10}米，（常）用于计量电磁辐射的波长，即非常短的波长。

angular Relating to angles. **Angular distortion** is the relative rotation of a point, within a structure, about another point.
角的/角度的 与角度有关的。**角变形** 是某个结构内某点相对于另一点的相对转动。

anhydrite A mineral also known as calcium sulphate commonly used as floor *screeding. It is anhydrous sulphate of lime, and differs from gypsum in not containing water.
硬石膏/无水石膏 一种矿物质，即硫酸钙，通常用作地面刮平砂浆层。其为石灰的无水硫酸盐，因不含水而区别于石膏。

anicut A *dam built in a river or stream for *irrigation purposes.
灌溉堰 建在河流或溪流上用于灌溉目的的坝。

animal black A paint derived from animal products such as bones.
动物炭黑 一种来自动物产品（例如骨头）的涂料。

anisotropic Having different properties in different directions. Timber is anisotropic because it is stronger loaded in the direction of the *grain rather than *perpendicular to the grain.
各向异性的 在不同方向上具有不同属性。木材即为各向异性的，因为其在纹理方向上较垂直于纹理的方向上负荷能力更强。

anneal (annealing) A heat treatment applied to materials, especially metals, to alter the chemistry, hence mechanical properties.
退火 一种用于材料（尤其是金属）上以改变其化学性质从而改变其机械属性的热处理方式。

annealing point (glass) A specific temperature for a glass which corresponds to *zero residual stress.
退火点（玻璃） 对玻璃来说，符合零剩余应力的某个特定温度。

annual ring (growth ring) A layer of wood that represents the annual growth in a tree's diameter.
年轮（生长轮） 树木的一层，代表树木直径的年度增长。

annular In the shape of a ring.
环形的/有环路的 圆环形的。

annular nail A *nail that contains a series of circular rings on the shaft. The rings improve grip, holding the nail firmly in position. Used for fixing plywood and other materials.
环纹钉 钉杆有一串圆环的钉子。圆环可提高钉子的夹持力，使钉子牢固地处于位置上。环纹钉用于固定胶合板及其他材料。

annunciator A device that can be used to visually or audibly indicate whether an event has taken place. For instance, these are used in *burglar alarm systems to indicate when an intrusion has taken place.
信号器/报警器 一种装置，可被用于通过声音或视觉指示某事件是否发生。例如，被用于防盗报警系统以便在入侵发生时做出指示。

anode 1. The electrode in an electrochemical cell through which electric current flows. 2. The positive terminal in an electrolytic cell. 3. The negative terminal of a galvanic cell (battery). The electrode in a galvanic couple that experiences oxidation, or gives up electrons. *See also* SACRIFICIAL ANODE AND ANODIC PROTECTION.
1. 板极 电化学电池中的电极，电流经此流过。**2. 阳极** 电解槽中的正极端。**3. 正极** 原电池的负极端。电偶中氧化反应或给出电子的电极。另请参考“牺牲阳极”和“阳极保护”词条。

anodic protection (aP) A process utilizing *electrolysis to improve the corrosion resistance of metals.
阳极保护（aP） 利用电解作用来提高金属抗腐蚀性的过程。

anodize (anodizing) A process using *electrolysis to coat a metal with a protective film.
阳极化 利用电解作用为金属覆盖一层保护膜的过程。

anodized A metal which is corrosion-protected by *electrolysis.
阳极化的/阳极处理的 金属通过电解作用受到防腐保护。

anoxic A lack of oxygen.
缺氧的 缺乏氧气的。

ANSI American National Standards Institute.
ANSI 美国国家标准学会。

anti-bandit glass A type of security glazing that is designed to be resistant to manual attack using a blunt instrument, such as a hammer. A wide range of glazing materials can be used, the most common being 11.3 mm 5-ply laminate and 11.5 mm 3-ply laminate. The specification for anti-bandit glazing is covered in BS5544:

1978 (1994).
防暴玻璃 一种安全玻璃，旨在抵抗使用锤子等钝器的人为攻击。可用作防暴玻璃的材料范围较广，最常见的是11.3毫米的5层玻璃和11.5毫米的3层玻璃。防暴玻璃的规范载于BS5544：1978（1994）标准。

anticipatory breach of contract Where it is known that a party will not perform its obligations to a contract. An example of such breaches would include a party stating that it will not carry out part or the whole of its work. The other party may act on the statement and *sue for damages or *repudiate the contract.
对合同的先期违约 合同的一方已知另一方将不会继续履行合同义务的情况。此类违约的实例之一包括某一方申明其不会实施部分或全部工程的情况。另一方可就该申明提起诉讼，或就损害赔偿或拒不履行合同而起诉对方。

anti-climb paint A type of paint applied to pipes, fences, etc to make it virtually impossible to climb (for intruders).
防爬漆 一种涂在管道、栅栏等处使其几乎不可能被攀爬（针对入侵者）的漆。

anti-condensation paint A type of paint used to reduce condensation.
抗凝固漆 一种用于降低凝结的漆。

anti-corrosive paints A paint utilized to protect iron and steel surfaces that contains a corrosive-resistant pigment (commonly *zinc-based) and a chemical-and moisture-resistant binder.
防腐漆 一种用于保护铁和钢表面的漆，其中含有防腐颜料（一般为锌基）及化学黏合剂与防潮黏合剂。

antifreeze A liquid used to reduce the freezing point of liquids and as a heat transfer medium.
防冻剂 用于降低液体凝固点并作为传热介质的液体。

anti-frost agent An additive used as frost protection.
防霜剂 用于防霜的添加剂。

anti-graffiti coating A coating that can be applied to a surface that either resists the application of graffiti or enables the graffiti to be removed easily. A wide variety of coatings are available, such as paints and resins.
防涂鸦涂料 一种可涂在表面的涂料，既可防止涂鸦又可使涂鸦被轻易抹除。可用作防涂鸦用途的涂料范围较广，例如油漆和树脂。

anti-intruder chain link fencing A security fence constructed from galvanized or coated steel wire mesh, usually 1.8 m to 3 m in height.
反入侵钢丝网围栏 由镀锌钢或涂层钢的钢丝网建成的安全围栏，通常为1.8米至3米高。

anti-siphon pipe A pipe in a drainage system that emits air, preventing *induced siphonage.
防虹吸管 排水系统中的管道，可排出空气，以防诱导虹吸作用。

anti-siphon trap A trap that maintains a water seal by preventing *siphonage.
防虹吸存水弯 保持水封以防虹吸作用的存水弯。

anti-slip paint A tough, durable, high opacity floor coating providing anti-slip properties. Applied to concrete, masonry, tiles, wood, metal, and many other surfaces.
防滑漆 一种坚韧、耐用、高遮光度的地面涂料，具有防滑属性，可涂在混凝土、砖瓦、木料、金属及其他很多表面上。

anti-static flooring [electrostatic discharge (ESD) flooring, static control flooring] A floor covering that is designed to reduce *static electricity, such as carpet, epoxy coatings, rubber tiles, or vinyl tiles.
防静电地板［静电释放（ESD）地板、静电控制地板］ 一种地板，旨在减少地毯、环氧涂料、橡胶板或乙烯板的静电。

anti-sun glass *See* SOLAR-CONTROL GLAZING.
防晒玻璃 参考“遮阳玻璃”词条。

anti-vibration mounting (flexible mounting, resilient mounting) A mounting, usually in the form of a pad or mat, that is designed to prevent the noise from vibrating machinery being transmitted to the structure. Materials used include cork, felt, plastic, and rubber.
减震底座(柔性底座、弹性底座) 一种底座，通常为衬垫或垫子的形式，旨在防止来自机器振动的噪音被传递至结构。减震底座所用材料包括软木、毛毡、塑料及橡胶。

apartment A small dwelling housed in a *multi-storey development. May also be called a flat or maisonette.
公寓 位于多层开发项目中的小住宅。也可被称为 flat 或 maisonette。

apex The tip or highest point of a building or structure.
尖端/顶点 建筑物或结构的顶点或最高点。

apparent cohesion Fine-grained soil particles (e. g. silt and clay) being held together by capillary/surface water tension.
表观黏聚 细粒土壤颗粒（例如粉土和黏土）通过毛细管力/表面水张力而聚在一起。

appeal A person or body has a right to make an application to a court or tribunal higher than that where the case was first considered for reconsideration of a decision made, where Acts of Parliament allow.
上诉 在法律所允许的情况下，如案件已经过一审裁决或判决，某个人或实体有权向更高一级的法院或法庭申请对该裁决予以重新审议。

appendix An additional section of a document or book that normally contains additional or supporting information and is placed at the back of the document. In some standard contracts the appendix is an integral part of the contract and cannot operate unless the sections within the appendix are completed.
附录 文件或书籍的补充部分，一般包含额外或证明信息并置于文件的后面。在某些标准合同中，附录是合同的组成部分，但除非附录中的各条均已完备否则不发生效力。

appliance ventilation duct *See* DUCTED FLUE.
设施通风管 参考“锅炉进气管”词条。

application The appliance of a material to another material. For example, painting a coat of primer on the surface of a steel frame.
应用 把某种材料用于另一种材料。例如，在钢框架的表面涂上一层底漆。

applicator A tool used to place materials into their desired position. For example, a silicone gun is used to place silicone around windows to provide a sealant.
敷料器 一种工具，用于把材料放置于预定位置。例如，用硅酮枪在窗户四周涂硅酮胶以提供密封。

appraisal 1. An estimate of value. For a property it could be the current market value for insurance purposes or functional value for business purposes. 2. Staff appraisals identify how a person has performed, whether existing goals have been met, whether he or she has established new business targets and areas of personal and *professional development.
1. 估值 对价值的估计。对某项房产而言，估值可能是出于保险目的的现行市值或出于商业目的的功能价值。**2. 评价** 对员工的评价说明该员工的表现如何、是否已达成既定目标、是否已设定个人职业发展的目标和领域。

apprentice A trainee who works under the stewardship of a skilled or trained person to learn a trade or profession for a period of time, described as an *apprenticeship.
学徒 在熟练或经训练的人的管理下工作，以便在一定期限（称为学徒期）内学习某个行业或职业的受训者。

apprenticeship The period of time served by an *apprentice to gain the required skills needed to become a qualified *craftsperson or *professional. The written agreement that engages the apprentice and describes the training is called an indenture.
学徒期 一个学徒为成为合格工匠或专业人士所需的必要技能而花费的时间期限。聘请学徒并说明培训内容的书面协议被称为学徒契约。

appropriation of payments In the law of debtors and creditors, it is the application for a payment to pay a particular debt.

指定付款 在关于债务人和债权人的法律中，指定付款是指对偿付某项具体债务的付款申请。

approval and satisfaction The act to state that the standard of processes or products are such that these meet specified conditions or a profession's requirements, or are perceived as being satisfactory. Contracts have provisions for a contract administrator or named professional to approve materials, workmanship, and operations. The nature and process associated with approval varies depending on the *standard form of contract.

批准和满意 表明过程或产品的标准即为满足指定条件或行业要求的标准，或被认为是令人满意的标准的行动。合同中的某些条文载明由合同管理人或指定专业人士核准材料、工艺及操作。与批准有关的性质和过程根据合同标准形式的不同而变化。

approved document Qualified guidance, approved by a government body, giving detailed information on how the *Building Regulations can be accommodated within building design, construction, and maintenance.

经批准的文件 经政府机构批准的合格指南，其中详细列出关于任何能在建筑设计、施工及维护中符合建筑条例的信息。

Approved Installer Scheme Professional body and organization schemes to list those contractors that satisfy training and skill requirements to undertake work properly. Such schemes aim to remove or reduce the possibility of rogue traders and poor service.

经批准的安装商方案 专业机构和组织的方案，列出可满足培训和技能要求而妥善施工的有关承包商。该方案旨在消除或降低出现流氓交易商和劣质服务的可能性。

approximate quantities The preliminary list of items that is difficult to measure and know exact quantities. The items are identified and listed in the preliminary *bills of quantities, and sometimes provisional costs or rates may be put against the work. When work is underway and can be measured, the approximate quantities are properly quantified and costed.

近似数量 对难以计量和知晓确切数量的单项的初步清单。各单项在初步工程量清单中指出和列出，且暂定成本或费率有时有可能针对工程而被列出。当施工开始且能够计量时，将对近似数量妥为定量和计值。

approximation An inexact number, cost, rate, or measurement that is useful to work with. Where there is insufficient information to produce an exact figure, it is necessary to use the information available to give a prediction. The symbol used to show that something is approximately equal to something is ≈.

近似值 一个在工作时有用的不精确数字、成本、比率或计量值。如无足够信息生成精确数字，则有必要使用能够获得的信息做出预计。用于显示某物大致相等于某物的符号为≈。

appurtenant works (US) All additional works undertaken to complete a job even though these have not been identified or described in the bills or original quantities. Appurtenant means to belong to or be part of the whole.

附属工程（美国） 即使并未在清单或原始工程量中被列出或描述，但为了工作的完成而实施的其他一切工程。附属是指属于整体的或为整体的一部分。

apron 1. A horizontal piece of trim inserted underneath a window-sill on the inside. 2. The tarmac or concrete area at an airport where aircraft stand when not in use.

1. 窗台挡板 插在内侧窗台下的水平镶边。**2. 停机坪** 机场中停放未使用的飞机的柏油或混凝土区域。

apron flashing An L-shaped *flashing used to prevent water penetration at the junction between a vertical wall and a sloping roof. Commonly found on the lowest side of a chimney.

披水板 一条L形的泛水板，用于防止竖直墙与斜屋顶的接合处渗水。披水板一般位于烟囱的最低侧。

apron lining The horizontal facing used to cover the opening in the floor at a stair well.

楼梯井侧面板 用于对楼面的楼梯井开口进行覆盖的水平饰面。

arbitration A private legal process used to settle disputes. Arbitration is often adopted in favour of *judicial proceedings, because the proceedings are conducted in a private forum, unless the parties agree the case is not available to the general public. Parties may prefer to have the case decided by an arbitrator or arbitrators that have technical or business expertise in the field of the dispute, since such expertise is not always available in the courts. There are often less grounds for appeal in arbitration, which means that once a decision is made it is easier to seek enforcement of an award. Rules of arbitration may be agreed to speed up the process, and although not always the case, arbitration is generally viewed as a quicker, more cost-effective way of settling disputes, when compared with going to court. However, arbitration is considered to be more costly and slower than *adjudication. Adjudication has many similarities to arbitration and has now been adopted by the industry as the main way of settling construction disputes.
仲裁 一种用于解决争议的非公开法律程序。仲裁常常优先于诉讼程序被采纳，因为仲裁是在非公开的仲裁庭进行，除非各当事方同意，否则公众不会得知案件。由于法院并不总是能够提供争议领域的技术或业务专业知识，故此各当事方可能倾向于由一名或多名在争议领域具有技术或业务专业知识的仲裁员裁定案件。在仲裁中能用于上诉的理由一般不多，这意味着裁决一旦做出，则易于寻求对仲裁结果的强制执行。仲裁规则可以约定以便加快仲裁过程，且（虽然并非总是如此）与向法院起诉相比，仲裁一般更快、更具有成本效益的争议解决方式。但与审裁相比，则仲裁花费更多、速度更慢。审裁与仲裁具有很大相似之处，且现已被业界采纳作为解决建筑争议的主要方法。

arbitration agreement clause (agreement to refer) Provision or term contained within the contract that binds the parties to submit future disputes to arbitration.
仲裁协议条款（同意提交仲裁） 合同中所载关于约束各当事方把未来的争议提交仲裁的条文或条款。

arbitrator An impartial referee appointed by the parties concerned to make decisions on disputes. The jurisdiction (scope of work) and decision-making powers are agreed upon by the appointing parties. Arbitrators are often appointed because of their technical expertise and knowledge, and powers and procedures are often agreed within the arbitration rules *agreement and the Arbitration Act 1996.
仲裁员 由有关的各当事方任命的裁判者，以便对争议做出裁决，其管辖权（工作范围）和裁决权由做出任命的各当事方授权。对仲裁员的任命一般是由于其技术专长和知识，且权力和程序通常已在仲裁规则协议及1996年仲裁法中约定。

arbitrator's award The award made by the arbitrator, which must deal with all matters referred, comply with any directions given in the submission, be final, have clear, certain meaning, and be consistent.
仲裁结果 由仲裁员做出的判决结果，该结果须处理被提起仲裁的所有事项、符合任何在提交仲裁时发出的指示，且为具有清晰和确定含义及具有一致性的最终结果。

arc 1. A curve or a section of a circle. 2. A luminous discharge caused by an electrical current bridging the gap between two electrodes in a circuit.
1. 弧形 一段曲线，或一个圆形的一部分。**2. 电弧** 由跨接电路中两个电极之间的电流所引起的发光放电。

arcade An arched or sometimes covered passage or alley lined with shops.
拱廊 一段拱形或有时是带遮盖的通道或小径，与各家店铺相连。

arch A curved structure that is designed to span an opening and support weight. Many different types of arches exist, varying in shape and style.
拱形 被设计为跨越某个开口和支撑重量的弯曲结构。拱形的种类有很多，其形状和样式各有不同。

arch bar A curved steel or iron bar that supports the brickwork above a fireplace. Found in older properties in place of a *lintel.
拱形条 对壁炉上方的砖加以支撑的弯曲钢条或铁条。常在老旧房屋中见到，用于取代过梁。

arch brick 1. A wedge-shaped brick used to construct an arch. 2. A very hard over-fired brick.
1. 楔形砖/拱砖 楔形的砖块，用于构筑拱形。**2. 过烧砖** 非常坚硬的过烧结砖。

arch centre The temporary formwork used to support an arch under construction.
拱架/拱轴 用于支撑在建拱形的临时模板。

arch dam A curved water-retaining structure. The arch action is used in the horizontal plane to withstand water *pressure, which is transferred to the *abutments.
拱坝 弯曲的挡水结构，在地平面使用拱形效应设计以便抗衡传导至坝肩的水压力。

architect A professional who designs buildings in the UK, and is a member of the *Royal Institute of British Architects (RIBA). In the UK the title 'registered architect' can only be used by qualified members of RIBA. Other countries have similar requirements.
建筑师 在英国指设计建筑物并且身为英国皇家建筑师协会会员（RIBA）的专业人士。在英国，只有 RIBA 的合格会员才能使用“注册建筑师”的称号。其他国家也有类似要求。

architect's appointment A rather dated standard form of agreement published by the *Royal Institute of British Architects that contains provisions for the agreement of fees, services, and responsibilities. The form is occasionally used, although it has largely been replaced by other standard forms and procedure.
建筑师委任状 由英国皇家建筑师协会刊发的、历史极为悠久的标准格式协议，其中载有关于费用、服务及责任的协定条文。虽然该委任状已被其他标准格式和程序所取代，但偶尔也会被用到。

architect's instruction Written direction given to undertake work, normally issued where there is a change or variation to the *contract drawings. Depending on the *profession of the person appointed as the client's representative or contract administrator, an *architect may not have the power to give instruction. Traditionally, the architect acted as the client's representative and contract administrator; however, it is quite common for other professions to undertake this role.
建筑师的指示 为实施工程而发出的书面指令，一般是在出现对合同图纸的改动或变更的情况下发出。依据被任命为业主代表人或合同管理人的人员的专业性，建筑师可能无权发出指示。传统上，业主代表人和合同管理人由建筑师担任；但由其他专业人员担任该职务的情况也相当普遍。

Architects' Registration Council (ARCUK) Operating under the Architects Act 1997, the organization was established to maintain a record of all 'registered architects'.
建筑师注册委员会（ARCUK） 根据《1997 年建筑师法》的规定而设立的组织，以便保存对所有“注册建筑师”的记录。

architectural drawings Drawings produced by the architect that provide a detailed image of the proposed building.
建筑图纸 由建筑师制作的图纸，其中载列拟议建筑物的详细图形。

architectural ironmongery (architectural metalwork) Decorative or ornamental products made from iron or other metals, such as banisters, screens, and railings.
建筑五金（建筑金属制品） 用铁或其他金属制成的建筑用制品如栏杆、屏风及护栏等装饰性或装潢性的产品。

architectural sections Details that cut through the building to show how the components fit together within the building.
建筑截面图 从建筑物中间剖开的详图，显示建筑物中的构件如何组装在一起的。

architecture The science, art, or field of study associated with the construction, detail, style, and aesthetics of buildings.
建筑学 与建筑物的建设、细节、样式及审美有关的科学、技艺或研究领域。

architrave The moulded trim that is used internally around door *sill and *jambs. Used to conceal the gap between the sill and jambs and the wall finish.
门头线/门贴脸板 用在沿门槛和门边梃四周内部的线脚镶边，用于遮掩门槛和门边梃与墙壁饰面之间的空隙。

architrave bead A *stop bead used internally around door *sill and *jambs covered by the architrave.

门框压条 用在沿门槛和门边梃四周内部的压条，被门贴脸板遮盖。

architrave block A block-shaped trim used internally at the junction between the jamb and sill on a door.
门脚 用在门边梃与门槛接合处内部的块状镶边。

architrave trunking An architrave with a removable or accessible cover that is designed to carry wiring.
门框线槽 配有可拆卸或可启闭盖子的门头线，被设计用于进行布线。

arch stone A wedge-shaped stone used to construct an arch. Also known as a voussoir.
拱石 用于建设拱形的楔形石头，也称为楔形拱石。

arc welding An inexpensive and widely used welding process, it involves an electric arc to melt the metals at the melting point. Sometimes the welding region is protected by an inert gas.
弧焊 一种廉价和广泛使用的焊接过程，以电弧熔化熔点的金属。有时会以惰性气体保护焊接区域。

area The size of a surface confined within a boundary.
面积 限定于某一边界内的表面的尺寸。

area way An area outside a basement, providing access, light, and air from outside.
地下室前空地 地下室外的一个区域，提供来自外部的通路、光线及空气。

arenaceous Sandy or having the appearance of sand.
砂质的/多砂的 砂质或具有砂子的外观。

argillaceous Clayey or having fine silt or clay particles.
黏土的/似黏土的 黏土质的或含有细粉粒或黏土颗粒的。

argon An inert gas, chemical formula Ar. One of the Noble gases used in lightbulbs and to create an inert rich atmosphere for electric welding (argon weld).
氩气 一种惰性气体，化学式为 Ar。是用在灯泡中的惰性气体之一，也用于电焊，在电焊周围通上氩气形成富集层，防止氧化。

arid Little or no rainfall.
干旱的 较少或没有降雨。

arm The mechanical part of an excavator that is attached to the bucket.
臂 挖掘机上与铲斗相连的机械部件。

armoured cable (US armored cable) An electrical cable that is designed to resist accidental breakage. Consists of a core of two to four plastic-sheathed insulated cables, which are encased in a flexible steel cable, and then sheathed in plastic. Used to transmit mains electricity underground.
铠装电缆（美国 armored cable） 为抵御意外破损而设计的电缆，由二至四根外包塑料的绝缘线构成电缆芯，包裹在柔性钢缆中，然后再以塑料封装。铠装电缆一般用作地下输电的主线。

array A group of photovoltaic modules that are connected together.
阵列 一组连接在一起的光伏模块。

arris The sharp edge formed at an external corner where two surfaces meet. Commonly found in plasterwork and joinery.
棱 在两个表面交会的外角处形成的锐边。常见于抹灰或细木工制品中。

arris fillet A triangular-shaped piece of wood that is nailed across the rafters and used to raise the roofing slates where they abut a chimney or wall. Similar to a *tilting fillet or *eaves board.
三角垫木 跨越椽条而钉入的三角形木块，用于撑起连接烟囱或墙壁处的屋顶板。三角垫木作用类似于披水条或檐口板。

arris gutter A V-shaped *gutter fixed to the *eaves of a building.
V 形檐沟 安装至建筑物屋檐上的 V 形沟槽。

arris-wise (arris-ways) The diagonal laying of slates, tiles, bricks, or timber.

A

对角铺砌 按对角方向铺砌石板、瓦片、砖块或木料。

arrow A sign or marker to point the direction.
箭头 一种指示方向的符号或标志。

arrow diagram Project management technique for working out a logical arrangement of events, used to calculate the sequence of *activities, duration of the project, and *critical path. Circles or rectangles are used to represent events or activities and these are connected together by arrows. The duration, start, and finish times for each event are written within the activity boxes.
箭头图 标出各事件的逻辑安排的一种项目管理技术，用于计算活动的顺序、项目持续时间以及关键路径。圆形或矩形被用于表示事件或活动，且以箭头将其联系到一起。各事件的持续时间、开始时间及结束时间在活动框中标出。

Art Deco A 1930s style of art and architecture, characterized by precise geometric shapes, bold colour, and outlines. The style was particularly prominent between 1925 and 1935 and was considered elegant, functional, and modern with its long, bold, and tall geometrical shapes, raised, exaggerated, or strong lines, and sweeping curves (unlike the more natural curves of *Art Nouveau). The Chrysler Building in New York is considered to be an example of Art Deco.
装饰艺术 一种20世纪30年代的艺术和建筑风格，以精确的几何形状、大胆的色彩及轮廓为特点。该风格尤其盛行于1925年至1935年，以其较长、大胆及高大的几何形状，抬升、扩张或强劲的线条，以及大幅度的曲线（不同于新艺术风格的较自然的曲线）而被视为高雅、实用及现代化的风格。纽约的克莱斯勒大厦即被视为装饰艺术的示例之一。

artefact An object made by a human, often associated with objects that have *archaeological interest.
工艺品/人工品 由人类制作的物品，通常与具有考古趣味的物品有关。

artery Main route in a road, rail, river, and drainage system.
干道/干线 道路、铁路、河流及排水系统的主要线路。

artesian well A well where water flows out at the surface under *hydrostatic pressure, the water being drawn from a *confined aquifer.
自流井 水在液体静压力下流出地面的井，其中的水抽取自承压蓄水层。

articles of agreement Normally refers to the opening provisions within the contract; the core components of the contract.
协议条款 一般指合同中的起始条文，是合同的核心组成部分。

articulated A joint that allows angular movement.
铰接的 可容许角运动的连接。

artificial Synthetic, not occurring naturally. Examples of artificial construction materials include: bricks, concrete, and plastics.
人工的/人造的 人造的，并非自然形成的。人工建材的实例包括：砖、混凝土及塑料。

artificial ageing *See* ACCELERATED WEATHERING.
人工老化 参考“加速风化”词条。

artificial aggregate Most lightweight aggregates with the exception of pumice.
人工骨料 除浮石以外的最轻量级骨料。

artificial lighting (artificial illumination) Lighting provided by artificial means, such as electric lighting, gas lighting, or an open fire.
人工照明 通过人工方式提供的照明，例如电灯、气灯或篝火。

artificial person A local authority, corporate body, or other legal entity that is recognized as a legal person capable of taking on rights and duties.
法人 被承认为权利能力和可承担具有民事责任的地方机构、法人团体或其他合法实体。

artificial seasoning An improved method of preventing bowing, cupping, twisting, or springing of boards or planks of lumber. During the latter portion of the timber-drying schedule, the boards or planks are passed between spaced pairs of rollers arranged to hold the lumber in a fixed plane. The lumber is therefore held in this plane while it is setting. Preferably the drying schedule includes intermittent exposure of the lumber to microwave radiation in an electronic kiln dryer.
人工风干 一种防止木板或木片顺弯、翘弯、扭弯或横弯的加工方法。在木材干燥过程中的后期阶段，让木板或木片通过具有间隔并成对排列的滚轴，滚轴使木材处于固定平面中。木材就此被夹在该平面中静置。干燥过程最好还包括间歇性地把木材暴露于电子烘干器的微波辐射中。

artificial stone A synthetic product imitating stone. Various kinds have been used from the 19th century in building and for industrial purposes such as grindstones.
人造石 一种模仿石头的人造产品。自19世纪以来，各种人造石被用于建筑和磨石等工业用途。

artists and tradespeople A contractual phrase used in early *JCT contracts to refer to parties employed by the client under a separate contract.
匠人和从业者 早期的JCT合同所用的合同用语，指客户根据某份单独合同而雇用的工作人员。

Art Nouveau A form of art and architecture that was most popular between 1890 and 1905. It is characterized by flowing forms and lines that are said to resemble natural forms such as flowers, leaves, and vines, and other organic forms. Some of the work of Antoni Gaudi, although very individual, can be said to be floral and natural Art-Nouveau style.
新艺术 一种主要流行于1890年至1905年的艺术和建筑形式，据称以模仿花朵、叶子、藤蔓及其他有机物等天然形式的流动形状和线条为特点。安东尼·高迪的某些作品（虽然非常个性化）被称为仿花和自然的新艺术风格。

asbestos A naturally occurring heat-and fire-resistant fibre previously used extensively in construction, however, asbestos was outlawed in the UK in 1978 due to danger to human lungs.
石棉 一种天然的耐热防火纤维，过去曾专用于建筑，但由于对人类肺部的危害，1978年英国已在建筑禁用。

asbestos encapsulation Process whereby asbestos is sprayed with a coating to minimize danger to human beings.
石棉封闭 为降低对人类的危害在石棉上喷涂一层涂层的工艺。

asbestos-free materials A material with 0% asbestos content, e. g. asbestos-free board.
无石棉材料 石棉含量为0的材料，例如无石棉板。

asbestos removal Work undertaken to remove the banned construction material asbestos from buildings and equipment. The specialized work must be carried out by competent workers wearing *protective clothing and breathing apparatus.
石棉清除 为了把建筑物和设备中的被禁止建材——石棉清除而实施的工作。专业工作须由穿戴合格的防护服和呼吸装置的工人进行。

as-built drawing Drawings that show how a building has been constructed; the true position of service. Such information is essential for the safe maintenance and operation of the building.
竣工图 显示建筑物竣工时状况的图纸；设施的实际位置。有关信息是建筑物的安全维护和运行所需的基本信息。

ascertain Within standard forms of contract, it is used to direct a person to find out for certain. Would preclude making a general assessment or making an estimate of quantities.
确认 在标准格式合同中，“确认”被用于指示某人以便发现情况以确定是否会妨碍对工程量做出大致评估或做出估计。

ASCII (American Standard Code for Information Interchange) A standard code that identifies letters, numbers, and certain symbols by code numbers for representing information on computer systems.
ASCII码（美国信息交换标准代码） 以代码编号标示字母、数字及某些符号的标准代码，用于在电脑系统中表示信息。

as drawn wire Steel wire drawn through a die to harden it.
拉拔钢丝 经模具拉拔而硬化的钢丝。

as-dug aggregate Quarried material delivered directly to site without any modification or treatment.
原状骨料 开采后直接送至现场而未经任何改变或处理的石料。

aseismic Not having or able to withstand earthquakes. Also denotes a fault where no earthquakes have occurred.
无震的/耐震的 并无地震或能够经受地震的。也指一种并无地震发生的断层。

ashlar The stonework that is evenly dressed with 3 mm joints and provides the facings for buildings.
琢石 均匀分布着 3 毫米接缝并可用作建筑物表面的石制品。

ashlaring 1. The short vertical members (ashlar pieces) that span between the *ceiling joist and the *rafter. 2. A low *attic wall, usually 1 m in height and constructed in blockwork, that extends from attic floor to the ceiling. 3. The process of bedding *ashlar in mortar.
1. 琢石镶面 跨越天花龙骨与椽条之间的较短竖直构件（琢石件）。**2. 琢石墙** 一种较低的*阁楼墙，一般高度为 1 米，由砌块筑成，从阁楼楼面延伸至天花。**3. 砌琢石** 把琢石铺设于砂浆中的过程。

ASHRAE American Society of Heating, Refrigeration, and Air-Conditioning Engineers.
ASHRAE 美国采暖、制冷及空调工程师协会。

ASHVE American Society of Heating and Ventilation Engineers.
ASHVE 美国采暖通风工程师协会。

Asiatic closet (Eastern closet) A *WC designed to be used in the squatting position, and comprises an elongated bowl that is positioned either at or just above the floor level. Rarely found in the UK.
亚洲式大便器（蹲式大便器） 一种设计有蹲位的厕所，包括一个与地面齐平或稍高于地面的细长便池。在英国很少见。

ASP *See* ADJUSTABLE STEEL PROP.
ASP 参考“可调钢支柱”词条。

aspect The orientation of a building on site.
方位 现场建筑物的朝向。

aspect ratio The ratio between two dimensions, e. g. width to height ratio in terms of a building width to its height; a computer screen; or ratio of wing length to breadth in terms of an aircraft's wing.
纵横比 两个维度之间的比例，例如就某一建筑物的宽度与高度而言的宽度与高度比；电脑屏幕；或就飞机机翼而言的机翼长度与宽度之比。

asphalt A material comprised mainly of bitumen, sand, clay, and limestone. Used in road construction.
柏油/沥青 一种用于道路建设的材料主要由沥青、砂子、黏土及石灰石构成。

asphalt mixer Equipment used to heat and mix asphalt.
沥青搅拌器 用于加热和搅拌沥青的设备。

asphalt paver (asphalt plant) A self-propelled or towed wheeled or tracked machine used for laying asphalt in *flexible pavement construction.
沥青摊铺机 一种自行式或牵引式的轮式或履带式机器，用在柔性路面建设中铺设沥青。

asphalt roofing Normally, flat roofing made up of two or three coats of asphalt and binding materials.
沥青屋顶 一般指由两到三层沥青与黏合材料制成的平屋顶。

asphalt soil stabilization The addition of asphalt to soil to improve its properties, i. e. reduce *permeability and increase *bearing capacity.

沥青稳定土壤 向土壤中添加沥青以改善其性能，即降低其渗透性和提高其承载力。

assembly The joining of the components that make up the final structure.
组装 把部件连接起来组成最终结构。

assembly gluing The gluing together of pre-assembled construction elements on site.
组装黏合 在现场把预组装的建筑构件黏合到一起。

assent To agree or comply.
同意 认可或遵守。

assessment of tender responses (vetting) Detailed assessment and evaluation of tenders received in order to advise the client.
对招标答复的评估（审核） 为了向业主提出建议对所收到的投标进行的详细评估和评价。

assignment and subletting The transfer of a contract to another party. A party can transfer benefits to a contract, but cannot transfer a burden without express agreement from the other contracting parties.
出让与转包 合同一方当事人把合同转让给第三方的行为。如果合同一方当事人转让合同辖下的权利与义务，但未经其他签订约方的明确同意，则不得转让合同义务。

Association of Building Technicians (ABT) A powerful *union of trade workers and technicians that was particularly prominent in the Victorian era. The association was formerly known as the Architects' and Surveyors' Assistants' Professional Union and as the Association of Architects, Surveyors, and Technical Assistants. The society amalgamated with the Amalgamated Union of Building Trade Workers (AUBTW) to form the Amalgamated Society of Woodworkers, Painters, and Builders (ASWPB).
建筑技术员协会（ABT） 行业工人和技术员的强大工会，在维多利亚时代尤其举足轻重。该协会原名为建筑师和测量师助理专业联盟以及建筑师、测量师及技术助理协会。该协会与建筑业工人联合工会（AUBTW）合并组成木工、漆工及建筑工联合会（ASWPB）。

as soon as possible Contractually, a direction given to use all means possible, giving due consideration of other work a party is committed to, to get a job done or materials delivered. When used in contracts or instructions, it has stronger connotations than in a reasonable time or forthwith.
尽快 就合同而言，尽快是指发出指示，要求以一切可能的方法，并经适当考虑有关方所做的其他工作后，使工作完成或使材料送达。当用于合同或指示时，比"合理的时间"或"即时"的含义要强烈。

ASTM (American Society for the Testing of Materials) A standards development organization that provides an independent source of technical standards for materials, products, systems, and services. The society was originally formed to address problems in standards in the railroad industry, which had previously been plagued with rail breaks and different standards.
ASTM（美国材料试验协会） 一个标准制定组织，提供关于材料、产品、系统及服务的技术标准独立来源。该协会最初是为处理铁路行业标准方面的问题（曾深受铁路中断和不同标准的困扰）而成立的。

astragal 1. A small semicircular convex moulding. 2. A vertical member attached to the *inactive or fixed leaf of a double door, which projects over the *active leaf when it is closed. Prevents the active leaf of the door being opened past the inactive leaf and seals the doors when they are in the closed position. 3. (Scotland and US) A *glazing bar.
1. 半圆小线脚 较小的半圆形凸线脚。**2. 门扇盖缝条** 附着在双开门的不活动或固定门扇上的竖直构件，当活动门扇关闭时凸出至活动门扇上。门扇盖缝条可防止活动门扇被打开至超过不活动门扇的程度，并在门处于闭合位置时把门封闭。**3. 格条（苏格兰和美国）** 玻璃窗格条。

asymmetric Not symmetric, unequal.
非对称的 不对称的、不均匀的。

atmometer An instrument used to measure the rate of evaporation.

蒸发计 用于测量蒸发率的仪器。

atmosphere A gaseous region (air) that surrounds celestial bodies, such as the Earth.
大气层 环绕于天体（例如地球）的气体区域（空气）。

atmospheric pressure The downward pressure caused by the weight of air above.
大气压 由于上方空气的重量而引起的向下压力。

at rest earth pressure The in-situ stress state of a soil where no horizontal or vertical strains occur. The ratio of horizontal stress (σ'_H) to vertical stress (σ'_V) is denoted by K_0, the coefficient of earth pressure at rest.
静止土压力 在并无水平或垂直形变发生时土壤的原位应力状况。水平应力（σ'_H）与垂直应力（σ'_V）的比率以静止土压力系数K_0表示。

atrium A tall internal courtyard spanning a number of floors that has a glazed roof.
中庭 跨越数个楼层的高大内部庭院，拥有玻璃屋顶。

attachment of debts A procedure employed in the High Court where a judgement has been made against a debtor, and in order for that debtor to deliver sums of monies owed, third parties that owe money to the debtor can become part of the order to pay off the debts.
债务扣押 当针对某个债务人做出裁决后，由高等法院采取的程序，以便使该债务人交出所欠款项的金额，结欠该债务人款项的第三方也可能成为清偿债务的命令的对象。

attendance Generally, the act of persons attending; however, different standard forms of contract can refer to general attendance as being plant, equipment, and temporary roads necessary to undertake works. Items described as special attendance are listed in the contract or *Standard Method of Measurement.
参与/出场/出席 一般是指由某些人员参加的行为；但不同的标准格式合同也可指称施工所需设施、设备及临时道路的一般性参与。被描述为特定参与的单项会被列在合同或标准计量方法中。

attenuation The reduction in sound level as a sound travels from one place to another. Measured in decibels.
衰减 当声音从一处传至另一处时声级的降低，以分贝计量。

attenuator A mechanical device installed within ductwork that is primarily designed to control fan noise. Comprises a circular or rectangular casing of acoustically absorbent material. It is normally located as close to the fan as possible; however, secondary units may be used where close control of noise is required, such as in auditoriums.
衰减器 安装在管道中主要旨在控制风机噪音的机械装置，包括由吸音材料制成的圆形或矩形外壳。衰减器一般尽可能设置于紧靠风机的位置；但如需对噪音进行密切控制时，也可能采用次级装置，如设置在观众席中。

Atterberg limits (consistency limits) The physical properties of a fine-grained cohesive soil is normally directly related to its water content. There are four states in which the soil can exist, the boundaries between these states are known as the consistency limits.
阿太堡界限（稠度界限） 细粒黏性土壤的物理性质一般与其含水量直接相关，土壤可存在四种状态，有关状态之间的界限即称为稠度界限。

attic (garret) A space in the roof that can be converted into a habitable room.
阁楼 屋顶能变为可居住房间的空间。

ATTMA The Air-Tightness Testing and Measurement Association.
ATTMA 气密性测试与计量协会。

auger A screw-shaped tool used to cut into wood or soil to produce a hole. The tools used to extract soil can be hand-held or mounted on tracked vehicles.
螺旋钻 螺旋形的工具，用于切入木头或土壤中以便形成孔洞的工具，该工具可为手持式或安装于履带式车辆上。

austenite A phase found in ferritic metals and alloys. Austenite is a *face-centred cubic (FCC) iron phase in iron and steel alloys with an FCC crystal structure.

奥氏体 在铁素金属和合金中发现的一种相。奥氏体是具有面心立方（FCC）晶体结构的铁和钢合金中的一种面心立方铁相。

austenizing A technique used to form the *austenite phase in a *ferrous metal alloy.
奥氏体化 一种用于在黑色金属合金中形成奥氏体相的技术。

autobahn *See* MOTORWAY.
高速公路 参考“高速公路”词条。

autoclave A pressurized device designed to heat aqueous solutions above their boiling point to achieve sterilization. The term ‘autoclave’ is also used to describe an industrial machine in which elevated temperature and pressure are used in processing materials, especially *autoclaved aerated concrete. The process can be used to sterilize equipment or materials.
高压釜 一种压力装置，旨在把水溶液加热至其沸点以上从而实现消毒。“autoclave”一词也用于描述以高温高压处理材料（尤其是蒸压充气混凝土）的工业机械。该过程可用于为设备或材料消毒。

autoclaved aerated concrete *See* AIRCRETE.
蒸压充气混凝土 参考“充气混凝土”词条。

automatic controls Electrical or mechanical devices that are capable of operating items of equipment automatically according to predefined conditions.
自动控制 能够根据预设条件自动操作各项设备的电气或机械装置。

automatic flushing cistern A cistern that flushes automatically, usually just after it has been used. Passive infrared sensors determine when the urinal or toilet has been used. Reduces water usage and ensures that the urinal or toilet is cleaned just after use.
自动冲水箱 一种在被使用后就会自动冲水的水箱。被动式红外传感器会判定小便器或大便器是否已被使用。可减少用水量并确保小便器或大便器使用后即可被清洁。

automatic level An optical instrument used for *levelling. Prisms within the instrument ensure that a horizontal line of *collimation will pass through the cross-hairs, provided that the instrument is approximately level.
自动水准仪 一种用于水平测量的光学仪器。如仪器近似于水平，则仪器内的棱镜将确保水平瞄准直线穿过十字丝。

autoroute *See* MOTORWAY.
高速公路 参考“高速公路”词条。

auto suppression system Part of an **active fire protection** system where detectors automatically activate the fire extinguishers to suppress the fire.
自动灭火系统 **主动消防**系统的组成部分，其中的探测器将自动激活灭火器以便扑救火灾。

auxiliary equipment Back-up equipment, for example, a power supply that runs from its own generator rather than the National Grid.
辅助设备 备用设备。例如，使用其自身发电机而非依靠国家电网运行的电源。

available (availability) Standard forms of contracts use the term to mean contract drawings, bills of quantities, descriptions of works, and other documents that should be accessible where they are to be used. Documents would not be considered available if they were stored at a head office, if it was clear that their use was required on site. (The use of *web portals for accessing documents could change how such documents are made available); terms within contracts will need to be adjusted to accommodate such access.
可用的（可用性） 标准格式合同使用该词指合同图纸、工程量清单、工程说明及其他文件在被使用时可被读取。如文件被存放在总部，而其使用明显需在现场进行，则不可被视为是可用的。网络门户为读取文件的使用可能改变有关文件的可用性；合同中的词语将需要予以调整以包括这种读取方式。

avalanche A sudden mass movement of snow down a mountainside.
雪崩 大量的雪突然滑下山坡的运动。

award *See* ARBITRATOR'S AWARD.
裁决 参考“仲裁结果”词条。

award of tender Notification by the *client or nominated body of acceptance of tender. Acceptance of the tender normally leads to the signing of the contract and binding agreement. Acceptance of the tender (offer) is normally a contract in itself; however, the finalities of the agreement are often negotiated after the tender has been accepted.
中标通知 由业主或指定机构发出接受投标的通知。对投标的接受一般会导致合同或有约束力协议的签署。对投标（要约）的接受通常也是一份合同；但协议的最终确定则常常是在投标被接受之后议定的。

awl A small, sharp-pointed tool used for marking hard surfaces or for punching small holes.
锥子 尖头的小工具，用于在坚硬表面做记号或扎出较小的孔洞。

awning (terrace blind) An external covering that projects out from a building to provide protection from the sun and rain and is usually retractable.
雨篷/遮阳篷 凸出于建筑物以便遮阳挡雨的外部遮盖物，一般是可伸缩的。

axe 1. A tool used for chopping timber. 2. A bricklayer's hammer used to cut and break bricks.
1. 斧头 一种用于劈木材的工具。**2. 斧锤** 瓦工的锤子，用于切开和切断砖块。

axed work Masonry (bricks and stonework) that is cut, formed, and shaped with a mason's hammer and chisel, and often used to form a brick or axed arch. When forming an axed arch, bricks are carefully shaped using a wooden template so that they are all the same size and fit together to form a neat brick arch. Joints in load-bearing brick arches are very fine, 3 - 4 mm, thus the bricks are cut precisely to form the arch.
斧琢石 用泥工锤和凿切割、加工制造的砖石制品（砖头和石制品），常用于制造砖拱形或斧琢拱形。在制造斧琢拱形时，会细心地使用木质模板，对砖块进行加工使其尺寸相同并组合在一起，从而形成齐整的砖拱。承重砖拱中的接缝非常细，为3至4毫米，因此组成拱形的砖块应予精确的切割。

axial Relating to or forming an axis.
轴的/轴向的 与轴线有关或构成轴线的。

axial flow machine A pump or turbine with rotating blades parallel to the *axis about which they rotate. An **axial pump** (also known as a **propeller pump**) discharges fluid parallel to the pump shaft; an **axial turbine** has air/water flow parallel to its rotor axis.
轴流机 与轴线平行、旋转叶片绕轴线旋转的泵或涡轮机。**轴流泵**（也称为**螺旋泵**）排出的液体平行于泵轴；**轴流涡轮机**的空气流/水流则平行于其转子轴。

axis 1. The vertical line and horizontal line (plural **axes**) on a graph, forming coordinates. 2. A line on which an object would rotate. 3. A line of symmetry around an object, e. g. the equator forms an axis around the centre of the Earth.
1. 轴 图表中的垂直线和水平线（复数为axes），构成坐标。**2. 轴线** 某物绕以旋转的直线。**3. 轴心线** 某物体的对称线，例如，经过地球中心，垂直于赤道平面的线就是赤道的轴心线。

axisymmetric Symmetrical about an axis.
轴对称的 沿轴心线对称的。

axonometric An image that is drawn slightly skewed so that more than one face of an object can be seen. With axonometric three-dimensional drawings, one of the axes is normally drawn vertically. Axonometric projections include *isometric, *dimetric, and trimetric.
三面投影的 一幅略为倾斜绘制的图形，以便能够看到物体的一个以上的面。在三面投影的三维图中，其中的一条轴线一般是垂直绘制的。三面投影图包括正等测、正二测及斜二测。

backacter *See* BACKHON.
反铲 参考“反铲”词条。

back boiler A *boiler located at the back of a fireplace or stove. The hot water produced from the boiler can be used to provide a hot-water supply or for space heating.
壁炉后热水炉 位于壁炉或炉子之后的锅炉。该锅炉产生的热水可被用于提供热水或空间加热。

back cutting Additional excavation that is required to form the correct formation level.
必要的超挖 为了形成正确的平整面标高而需进行的超挖。

backdraught The rush of air through a door or window that is opened into an oxygenstarved room where a fire is burning. The fire uses all of the oxygen in the room to fuel the fire, and if a window or door is opened, air rushes into the oxygen-starved room providing fuel for the fire. Prior to the door being opened, the fire may have become so starved of oxygen that it is reduced to a smouldering fire; however, glowing embers can burst into flame once oxygen is present. When it is suspected that there is a fire in a room, the temperature of the door should be checked. If the door is hot, indicating that there is a fire behind it, the door should be left closed.
逆通风 空气通过打开的门或窗冲入正在起火的缺氧房间内。火会利用房间内所有的氧气助燃，而如果窗或门是打开的，则冲入缺氧房间的空气将为火焰助燃。在门被打开之前，火焰可能由于缺乏氧气而减弱为阴燃火；但灼热的余火一旦遇到氧气就会爆发为火焰。如怀疑房间内起火，应检查门的温度。如门变热，则说明门后有火，应使门保持闭合。

backdrop (backdrop connection, drop connection, US sewer chimney) A vertical pipe, located in an *inspection chamber, which allows a foul water drain to be connected to a sewer pipe at a lower level.
垂直落水管（垂直落水管连接、垂落连接，美国污水竖管） 位于检查井中的垂直管道，使污水下水管能与较低高程的排污管道相连。

backer *See* BACKUP STRIP.
背垫 参考“背垫条”词条。

backfall An adverse slope in a drain or gutter causing water to flow in the opposite direction than intended, which can lead to *ponding of water.
反向斜坡 排水道或排水沟中的反方向斜坡，使水流向与预定相反的方向，导致水洼的蓄积。

backfill (backfilling) Soil returned to an excavation, after a foundation has been cast or a pipe placed; typically compacted in layers.
回填料 在基础打好或管道铺好后填回基坑的土壤；一般会分层压实。

back-flap hinge A *hinge with wide flaps—the flaps being the part of the hinge where the screw holes are located.
轻型铰链 具有宽铰链页（铰链页即为铰链上带有螺丝孔的部分）的铰链。

backflow Movement of liquid or gas in the opposite direction than intended. A **backflow valve** prevents such movement.
回流/逆流 液体或气体按照与预定相反方向进行的运动。 **回流阀**能阻止此类运动。

background (backing, base) The surface on which the first coat of plaster is applied.
背景/基底 将被涂上第一层灰泥的表面。

background drawing A single drawing used to help coordinate and distribute building services. All subcontractors and designers should use the same background drawings and information to ensure that services can be properly coordinated without service runs clashing, e. g. two or more different service runs trying to occupy the same space. Where transparent service drawings are all laid over the top of the same background drawing, distribution problems and clashes can be detected and avoided. With digital information and extensive use of CAD, the use of overlays and background information is becoming more common. With any information management and coordination system, it is essential that the latest information and revisions are used.
背景图 用于帮助协调和分配建筑设施的单一图纸。所有分包商和设计师均应使用相同的背景图和信息，以确保设施使用能被协调好，避免发生冲突（例如，将运行的两项或多项设施会占用同一空间）。当把透明的设施图全部铺在同一背景图上，分配冲突和问题即可被发现和避免。借助于数字信息以及 CAD 的广泛使用，对重合及背景信息的使用变得越来越普遍。对于任何信息管理和协调系统而言，使用最新的信息和修订都是至关重要的。

background heating The process of providing constant low-temperature space heating to a space. Used to reduce the risk of condensation and to maintain a constant internal temperature.
背景供暖 为某一低温空间提供持续的供暖过程，用于降低冷凝的风险并维持恒定的内部温度。

back gutter A *gutter that is installed on the 'back' or highest side of a chimney where it intersects the roof. Usually formed in lead, although other materials can be used.
烟囱排水背沟 设置于烟囱穿越屋顶处的“背面”或最高侧的排水沟，一般用铅制成，也可能使用其他材料。

backhoe (backacter, drag shovel, trench hoe) A wheeled tracker excavator (trademarked as a **JCB**), with buckets to its front and rear, each being independently connected to the tracker via a two-part hydraulic arm.
反铲（反铲挖掘机、反铲挖沟机） 一种轮式牵引挖掘机（商标名为 **JCB**），在其前部和尾部配有铲斗，两个铲斗均可通过一个两段式液压臂独立连接至牵引车。

backing (base) 1. Any material installed behind a facing material to provide a base or support to the facing material, such as brickwork or concrete behind stonework. 2. Any coat of paint or plaster that is not the final finishing coat.
1. 底衬 安装在面材下方为面材提供基础或支撑的任何材料，例如石制品下方的砖砌或混凝土。
2. 基底 并非最终饰面层的任何一层涂料或灰泥。

backing bevel The V-shaped bevel found at the top of a hip rafter where the two roof slopes join.
底衬斜角 可见于两个屋顶斜面交会处的角椽顶部的 V 形斜角。

back lintel A *lintel that is used to support the inner leaf of a wall.
隐蔽过梁 用于支撑墙壁内叶的过梁。

back observation *See* BACKSIGHT.
后视 参考“后视”词条。

back propping *See* SHORING.
背撑 参考“支撑”词条。

back sawing A hand tool used for cutting, with a stiffened back section.
背锯 用于切割的手工工具，具有加固的锯背。

backset (set back) The distance from the leading edge of the door to the centre of the hole that is drilled for the spindle of the door handle or latch to be housed.
门边距 从门的前缘到为了安装门把手或门闩的主轴而钻出的孔洞中心的距离。

back shore (jack shore) An outer bottom support of a racking shore.
顶撑 架式支撑的外部底部支撑。

backsight A reading taken in surveying (e. g. levelling) to a reference location after the instrument has been repositioned. For example, a *foresight is taken before the level is moved to a *staff that is visible from the new and old levelling station. The level is then set up at its new location and a backsight is taken.
后视 在仪器被重新定位后取自对参考位置的测量（例如水平测量）的读数。例如，前视为新的和旧的水平测量站看到的水平移动到标尺前的读数。水准仪在新地点设立后取后视读数。

backsplash *See* SPLASHBACK.
后挡板 参考“防溅挡板”词条。

backup strip (joint backing, backer) Compressible material, often foam strips, inserted to fill the majority of a movement joint so that the depth of sealant that is finally applied to seal the joint is limited. The backup strip usually has a finish that prevents a bond between the sealant and backup strip, allowing the sealant to stretch and compress within the movement joint without being inhibited by the filler strip.
背垫条 插入伸缩缝填充大部分空间，并对封闭接缝的黏合剂最终涂到的深度加以限制的压缩材料（一般为泡沫条）。背垫条一般具有可防止黏合剂与背垫条之间形成黏合的表面，使黏合剂在伸缩接缝中得以伸展和压缩而不受填充条的限制。

backwater curve The longitudinal profile of a liquid surface in an open channel when the depth of water has been increased by an obstruction such as a dam or a weir.
回水曲线 当由于受到坝或堰的阻碍而使水的深度增加时，明渠中的液体表面的纵向轮廓。

bacteria bed A layer of sand which *effluent is passed through in sewage treatment plants. The action of air and *microorganisms in the bed help to break down the effluent.
细菌床 污水处理厂中废水所流经的一个砂层。空气和微生物在细菌床中的作用有助于分解废水。

baffle (baffle plate, baffler) A device that controls or restricts the flow of air, water, or light.
挡板 控制或限制空气、水或光的流动或穿过的装置。

bag filter Air filter used in a balanced-flue gas fire.
袋式过滤器 用于平衡烟道燃气取暖炉的空气过滤器。

bagged joint A *flush joint that has been finished using a coarse cloth, rather than a piece of wood or pointing tool. Usually used with handmade bricks.
抹平缝 使用粗布（而非木片或尖头工具）勾缝的平灰缝，一般用于手工砖。

bagging (bag rendering) The process of applying cement mortar to external brickwork using a brush or a hessian bag to fill in any gaps and joints. This results in a rough cement-like finish that can be painted over.
打底 用刷子或麻袋把水泥砂浆涂到砖砌物外部以便填充空隙和接缝的过程。这会形成可供涂刷的粗糙水泥样饰面。

bag set Plaster or cement that has set in its bag, caused by storing the bag in damp conditions.
袋内结块 灰泥或水泥在其包装袋内凝固，是包装袋被储存于潮湿状况下所致的。

bailer A device that is used to remove sludge from the bottom of a mine.
抽泥筒 一种用于移除矿井底部淤泥的装置。

bailey The outer fortified walls surrounding a courtyard or keep; term associated with castles.
城堡外墙 围绕于庭院或要塞的外部壁垒墙；该词与城堡有关。

Bailey bridge A prefabricated lattice steel truss bridge designed for use by the military in the early 1940s. It can span up to 60 m and be assembled rapidly on site with basic hand tools.
活动便桥/贝雷桥 一种预制格栅钢桁架桥，于20世纪40年代设计用于军事目的。这种桥最多可跨越60米的距离，并能以基本的手工工具在现场快速组装。

bailment The legal process of placing goods with another person or organization on the condition that the goods will be returned. Goods are lent or deposited in the temporary custody of another. The bailment may be a simple gratuitous loan or for reward, as would be the case when something is hired. The holder is known as the **bailee**, the person who delivers or transfers the property is known as the **bailor**. For a bailment to be valid

the holder (bailee) of the property must have physical control of the property and intent to possess it.
寄托 以货物将被退还为条件，把货物存放于其他人员或组织之处的法定过程。货物被出借予他人或被置于他人的临时保管下。根据某物被占用的情况，寄托可为完全无偿或为取得回报而做出的借贷。物品的持有人被称为**受寄托人**，交付或转让该财物的人被称为**寄托人**。如要使一项寄托有效，则财物的持有人（受寄托人）必须实际控制该财物且有意领有此物。

bainite A phase found in ferrous metals, for example, steel and cast iron. Bainite is comprised of ferrite and cementite.
贝氏体 钢和铸铁等黑色金属中发现的一种相。贝氏体由铁素体和渗碳体构成。

Baker bell dolphin *See* DOLPHIN.
贝克式钟形系船柱 参考“码头系缆桩”词条。

balance beam A timber beam that is pushed to open a lock gate when the water level is equal at either side.
平衡梁 一种木梁，用于在两边水位相等时推开闸门。

balance box A counterbalance load that is located on a crane at the opposite side of the *jib.
配重箱 一个平衡负载，位于吊车上与吊臂相反的一侧。

balance bridge *See* BASCULE BRIGE.
竖旋桥 参考“竖旋桥”词条。

balanced construction Method used to ensure that there is the same amount of grain running in opposite directions in plywood. In three-ply timber, the thickness of the core timber would be twice that of either of the surface veneers. The grain of the two surface veneers would run in the same direction and the grain of core timber runs at 90 degrees to adjacent veneers, thus the amount of timber running in different directions is balanced.
对称结构 用于确保胶合板的不同方向具有相同数量的纹理的方法。在三层胶合板中，芯材的厚度应为表层木皮的两倍，两层表层木皮的纹理的走向应相同，而芯材的纹理走向应与相邻木皮成90°，由此使沿着不同方向的木纹数量均衡。

balanced flue A flue where the air is drawn in and exhausted through separate compartments of the same flue. This is achieved using a concentric arrangement. Used in room-sealed appliances.
平衡通风烟道 一种通过同一烟道的不同隔间吸入和排出空气的烟道，采用同心排布而实现。平衡通风烟道用于密闭式用具。

balanced sash A double-hung sash window that is balanced by weights hung on chains or ropes, or pre-tensioned springs. The balancing enables the sash to be opened smoothly and easily.
平衡式上下推拉窗 通过悬吊于链条或绳索上的重物或通过预张力弹簧保持平衡的双悬上下推拉窗。平衡状态下可流畅和轻便地打开窗扇。

balanced step (dancing step, French flier) A step on a stairway that is slightly tapered at one end to produce a curve in a flight of stairs. Most tapered steps are very narrow on this inside edge allowing them to be accommodated by a central newel post; however, the wider tapered steps are easier to negotiate but require more room, often accommodated by an open stairwell.
扇形踏步（法式梯级） 台阶上在一端略变细的踏步，从而使台阶的阶梯形成弧线。大多数变细的踏步均在内侧边缘变窄，使其可与楼梯中柱相配；而较宽一侧的踏步较易跨越，但需要较多空间，常与开敞式楼梯间相配。

balancing The process used when setting up *central heating systems and *commissioning them, where the inlet valve on each of the radiators is adjusted to ensure that the flow of hot water to each radiator is evenly distributed.
配平 在搭建集中供热系统和调试该系统时所用的过程，其中对每个暖气片的进水阀进行调整以确保通往每个暖气片的热水水流都是均匀分配的。

balancing valve A valve used for controlling the flow of a fluid through different legs or branches of a system

to ensure that all outlets are balanced and receive equal fluid pressure.
平衡阀　一种阀门，用于控制流经系统中分叉或分支的液体流，并确保所有出口流量均衡和液体压力相等。

balcony　An open or covered platform, usually enclosed with a railing, that projects out from the external wall of a building.
阳台　一个露天或被遮蔽的平台，通常由护栏包围，凸出于建筑物的外墙。

baling　A method of compressing waste to form bales to be either burnt or disposed of in landfill.
打包　一种把废弃物压缩为大包以便焚烧或填埋处置的方法。

balk (US)　*See* BAULK.
木方（美国）　参考“粗枋”词条。

ballast　1. Stone or gravel used as a foundation material on roads and rail tracks. 2. Unscreened gravel, containing small stones, sand, and grit, used to make concrete. 3. Heavy weights that are used to stabilize or hold something down, e. g. weights that are carried in the hold of a cargo ship to increase its stability when it is empty. 4. **ballast (choke)**　An electronic device that controls the current through discharge lighting.
1. 道砟　用作道路或轨道的基础材料的石头或砾石。**2. 石渣**　未过筛的碎石，含有小石头、砂及沙砾，用于制作混凝土。**3. 压载物/压舱物**　用于使某物保持平衡或使某物向下的重物，例如当货船空载时运入以提高货船稳定性的重物。**4. 镇流器（扼流器）**　一种对通过放电灯具的电流加以控制的电子装置。

ball catch (bullet catch)　A door fastener comprising a spring-loaded ball that is set within the mortise of the door. When the door is closed, the ball engages a striking plate on the jamb of the door that contains either a slight indentation or a hole that the ball sits within, thus keeping the door closed. The door is opened when applying enough force to the door so that the ball springs back out of the indentation or hole.
门碰珠　门上的一种紧固件，包含由弹簧承载的碰珠，固定于门的榫眼中。当关门时，碰珠碰到门框上的锁舌板，而锁舌板含有较浅的凹槽或孔洞可容碰珠驻留，从而使门保持闭合。当对门施加足够的力使碰珠弹簧缩回退出凹槽或孔洞时，门就会被打开。

ball cock　*See* BALL VALVE.
球形旋塞　参考“球阀”词条。

ball joint　A connection, consisting of a rounded end that fits into a cup-shaped socket, which allows rotational movement.
球形接头　一种连接装置，包括一个装在杯状凹槽中的圆形端，该圆形端使接头可做旋转运动。

balloon frame (balloon framing)　A timber-frame construction consisting of long continuous vertical *studs that run from ground to eaves-level of a house. The timberframe supporting walls extend the full height of the building, from the base plate to the roof plate. Each wall of the timber frame is normally prefabricated, typically in two-storey units, then simply erected and bolted to the other walls on site. Due to the units extending more than one storey, balloon framing can be a quicker method of construction than *platform frames which only extend one storey high. Platform frame units are easier to handle and transport, and because of this, they are often used in preference to balloon frames.
轻型木构架/轻便型骨架　一种木构架，包括较长的连续立柱，每根柱都从房屋的地面到屋檐层面。木构架支撑墙延伸至建筑物的整个高度，从底板到屋顶板。木构架的每面墙通常都是预制的，一般为双层单元，随后在现场直接搭建和拴接至其他的墙。由于各单元延伸超过一个楼层，因此与仅延伸至一个楼层的平台型底架相比，轻型木构架可作为一种更快的建筑方法。平台型底架更易于处理和运输，故此在使用上其常常比木构架更受青睐。

balloon grating (balloon wire)　A large strainer placed over a rainwater outlet.
雨水斗导流罩　一种置于雨水出口上的较大过滤器。

ball penetration test (Kelly ball)　A field test used on freshly poured concrete to check the consistency for control purposes. The test consists of determining the depth to which a hemispherical ball, having a diameter of

B

150 mm and a weight 13.6 kg, will sink to under its own weight.
球体贯入试验（凯氏球） 一种用在新浇筑混凝土上以便出于控制目的而检查稠度的现场试验。试验内容为测量一个直径为150毫米和重量为13.6千克的半球体在其自身重量作用下将陷入的深度。

ball support A support that permits rotational movement but prevents displacements.
滚珠支座 一种允许旋转运动但防止位移的支座。

ball test A ball slightly smaller in diameter (i.e. 12 mm) than the pipe is passed down the pipe to determine if any restrictions in the pipe exist.
滚球试验 用一个直径比管道稍微小（如12毫米）的球穿过管道以确定管道中是否不通。

ball valve (ball cock) A spherically shaped gate valve that provides a very tight shut-off for fluids in a high-pressure piping system.
球阀（球形旋塞） 一种球形闸阀，可在高压管道系统中非常严密地切断流体。

Baltimore truss A very strong *truss bridge; a subdivision of a *Pratt truss with additional bracing in the lower part of the bridge to help prevent *buckling and control *deflection. Typically it has been used for rail bridges.
巴尔的摩桁架 一种非常坚固的桁架桥；巴尔的摩桁架是普拉特桁架的一个子类，在桥的下部具有额外的支撑以防止屈曲并控制挠曲，一般被用于铁路桥。

baluster (banister) Edge guard to stairways, open edges of landings, roofs, and floors. Normally consists of a top-rail (coping or handrail), infill *spindles, or *balustrade that sit on the floor or bottom rail.
栏杆 楼梯、楼梯平台的开放边缘、屋顶及楼面的边缘防护物。一般由顶栏（栏顶或扶手）、插接栏板或竖立于楼面或底栏上的栏杆柱组成。

balustrade The vertical in-filling components of a structure that acts as a guard against an open edge of a roof, floor, landing, or stairway. A *baluster (edge guard) consists of vertical balusters, occasionally these may be supporting, but mostly they act as in-fill under the coping (top-rail or handrail).
栏杆柱 设于屋顶、楼面、楼梯平台或楼梯的开放边缘作为防护的结构物中的竖直插接构件。栏杆（边缘防护物）包括竖直的栏杆柱，有时栏杆柱可作为支撑，但大多数情况下是作为栏顶（顶栏或扶手）下方的插接物。

band course *See* STRING COURSE.
挑出层 参考“挑出层”词条。

banding *See* LIPPING.
封边 参考“门扇封边”词条。

band saw A power cutting device with a continuous vertically mounted blade.
带锯 一种电动切割装置，配有连续的竖直安装的刀片。

band screen A mesh wire sieve, in the form of an endless moving belt, used at water and sewage treatment plants to remove solids so as to protect downstream equipment.
回转格栅 一种钢丝网筛，形状像没有边的输送带，用在水处理厂和污水处理厂以除去水中的固体杂质，从而保护下游设备。

banister *See* BALUSTER.
栏杆 参考“栏杆”词条。

bank 1. A raised area of land, e.g. the side of a waterway. 2. The grouping of similar objects, e.g. a bank of lights.
1. 岸 土地的隆起区域，例如水道的侧边。**2. （一）排/（一）组** 成组的类似物品，例如一排灯具。

banker (gauging board) Profile, often made of timber, to show the position of each brick course and mortar joint. Can be used to assist bricklayers to lay bricks, blocks, and stonework to a consistent height with consistent mortar joints.

造型台（计量用板） 常以木材制成的模具，用以展示各砖层和灰浆接缝的位置，可帮助砖瓦工铺设砖块、砌块及石制品并使之达到与均匀灰浆接缝一致的高度。

bank guarantee A legally binding promise from a lending institution (bank) that the liabilities of a debtor will be met. The bank guarantee promises a sum of money to the beneficiary should something go wrong, e. g. if a buyer of goods experiences cash-flow problems, then the lending institution would still honour the transaction if a bank guarantee had been agreed.
银行保函 由放贷机构（银行）发出的承诺如果债务人不能履行债务，银行将代其履行的具有法律约束力的文书。银行保函中承诺如某事出错将向受益人支付一笔款项，例如，如果货物的买方遭遇现金流问题，则如已议定一份银行保函，放贷机构将代其履行。

banking Slope of a road surface.
埂 道路表面的斜坡。

bank material Soil, gravel, and rock before it is excavated or blasted from the ground.
筑堤材料 在其被从地面挖出或爆破之前的土壤、砾石及岩石。

bank of lifts A number of *lifts at the same location.
电梯组群 在同一地方的数台电梯。

bankruptcy Legal declaration that a company or individual is unable to pay their creditors. The state procedure protects the debtor from demands of creditors and ensures the assets of the individual or company are equally distributed among the creditors. The procedure is laid down in Part IX of the Insolvency Act 1986.
破产 关于一家公司或一名个人无能力向其债权人付款的法定宣告。该声明程序保护债务人免受债权人的索求并确保债务人的资产在债权人之间公平分配。有关程序载于1986年破产法第九部分。

bank seating The end support of a bridge, i. e. a bridge abutment.
桥台 桥梁的端点支撑，即桥台。

banksman (banksperson) A competent person who provides directions to a crane driver from where loads are attached or detached, because crane drivers, especially on tower cranes, do not always have a clear view of the loading and unloading areas.
起重机信号工 为吊车司机指示加载或卸除货物地点的符合资格人员，因为吊车司机，尤其是塔吊司机并不总是能够清晰地看到装载和卸载的区域。

bank storage Water that is held in the ground above the *phreatic surface, from *run-off during heavy rains and flood periods.
河岸调蓄量 来自暴雨和汛期的溢流，并被留滞于潜水面以上地层中的水。

banquette 1. A horizontal ledge. 2. A footbridge.
1. 长软座 水平平台。**2. 步行桥**。

bar 1. Unit of *pressure, equivalent to 100 kN/m^2. 2. A solid length of material, usually steel, round, rectangular, or hexagonal in cross-section. 3. Silt and sand deposited at a shallow or low *velocity part of a riverbed or the mouth of a river.
1. 巴 压强单位，等于100千牛/平方米。**2. 杆** 固体的长条材料，一般为钢，横截面为圆形、矩形或六边形。**3. 沙洲** 沉积于河流的河床或河口的较浅或低速部分的粉土和砂。

bar bender A machine that bends reinforcement bars to the desired shape before they are fixed in position and concrete is cast around them.
弯钢筋机 在周围浇筑混凝土之前，把钢筋布设到位和把钢筋折弯至所需形状的机器。

bar chain method An analysis method for *trusses. The deflected form of the truss is represented by a series of straight lines about the joints around which they rotate.
结点法 一种针对桁架的分析方法，桁架的偏斜形式以绕接缝旋转的一系列直线表示。

bar chair A special shaped reinforcement bar, which separates the top and bottom mats of reinforcement to provide the correct concrete cover.

马凳筋 一种特殊形状的钢筋，把上下两层隔开，以保证正确的混凝土覆层。

B

bar chart (Gantt chart) A schedule or programme of activities plotted on a graph with time plotted on the *x* axis. Each activity is represented by a solid or hatched bar, with the length of the bar extending for the proposed and actual duration of the activity, showing when the activity starts and finishes. With both estimated and actual durations shown on the same graph, any slippage of activities or changes to timing of events can be seen and tracked.
横道图（甘特图） *x* 轴标绘制于在有时间图的活动日程或计划。每项活动均以实心或阴影的条形表示，条形的长度对应活动的拟议和实际时限，显示活动在何时开始和结束。由于估计和实际的时限显示于同一图形中，对活动的任何滑移或对时间设置的任何变动均可被关注和追踪。

bar code (barcode) 1. A number given to steel reinforcement to identify shapes and characteristics of the reinforcement bar, also called **shape code** . 2. Series of lines or shapes that can be optically read by a scanner and correlated with data stored on a central processor. Most commonly used in supermarkets, but now is increasingly used to track parcels and materials in construction projects.
1. 钢筋代码 为钢筋赋予的一个数字，用以标明钢筋的形状和特性，也称为**形状码**。**2. 条形码** 可由扫描仪进行光学读取，且与储存在中央处理器的数据有关的一系列线条或图形。最常用于超市，但现在也日益在建筑项目中用于对包裹和材料的跟踪。

barcol hardness A hardness value obtained by measuring the resistance to penetration of a sharp, spring-loaded steel point. It is used to measure the degree of cure of a plastic.
巴氏硬度 一种通过计量对由弹簧承载的尖锐钢针刺入的抗性获得的硬度值，被用于计量塑料的固化程度。

bareface tenon The male part of a mortise-and-tenon timber joint with only one side of the tenon having a shoulder and the other side being a flush face.
裸面单肩榫 榫卯结合的公头部分，榫仅有一侧有榫肩，另一侧则为齐平的表面。

bareface tongue Part of a timber joint made by rebating the end of timber on one side only so that the other side is flush. The rebate is often half the thickness of the original plank or sheet of wood. The thin section of tongue fits into a recess to provide a neat strong joint.
裸面榫舌 木接头的组成部分，通过在木头末端的一边开企口，使另一边保持齐平而制成，开企口处的厚度一般为原有木板或木块厚度的一半，榫舌的细部可插入凹槽中，构成一个整齐的牢固连接。

barge board (verge board, gable board) Diagonal piece of timber used to provide a decorative cover to the edge of the roof on a gable end, and serves the same function as the *facia board. The barge board is either fixed under the edge of the roof tiles or slates, hiding the junction between the roof and the gable wall or is fixed to cover the side of the edge of roof tiles and wall. The barge board often meets the facia board and provides a neat continuation along the diagonal slope of the pitched roof.
博风板（封檐板、山墙封檐板） 用于山墙末端处的屋顶边缘的装饰性遮盖并发挥与挑檐立板相同作用的斜向木构件。博风板既可装在屋顶瓦或屋顶板的边缘以遮蔽屋顶与山墙的连接处，也可按照遮蔽屋顶瓦和墙壁边缘的方式安装。博风板常与挑檐立板相交，并沿着坡屋顶的斜坡形成整齐的连续线条。

barge course (verge course) Projecting brick coping or tiles next to the gable that protrude slightly.
山墙压顶（山墙檐瓦） 山墙旁边略微凸出的挑出砖顶盖或瓦。

barge flashing Flashing that extends over the gable tiles and barge board.
封檐泛水板 凸出于山墙瓦和博风板的防水板。

bar joist A flat truss structural member.
钢桁条 水平的桁架构件。

barometer An instrument for measuring atmospheric pressure.
气压计 一种用于测量大气压力的仪器。

barrage A lower height dam placed across a watercourse (river or estuary) to control the water level.
拦河坝 跨越水道（河流或河口）而设置的高度较低的坝，用于对水位进行控制。

barrel bolt A common type of door fastener comprising a metal rod set within a cylindrical case. When the door is closed, a knob attached to the metal rod is used to push it forwards by hand until it engages a separate smaller cylindrical case located on the jamb of the door.
圆插销 一种常见的门紧固件，包括套在一个圆柱形外壳中的一条金属棒。在关门时，把与金属棒相连的把手向前推，使其插入位于门边梃上的另一个较小的圆柱形外壳为止。

barrel light Roof light with either curved glazing or reflector unit that resembles a barrel vault.
圆拱形天窗 配有拱形玻璃或反光板（与筒形穹顶相似）的屋顶天窗。

barrel nipple (US shoulder fitting) A pipe fitting comprising a short hollow cylinder with a slightly tapered thread at each end.
螺纹接头 一种管道配件，由每端具有一个呈锥形螺纹较短的空心圆柱体构成。

barrel roof A semicircular arch capable of spanning long distances.
拱形屋顶 一个能够跨越较长距离的半圆拱顶。

barrette A large excavated pile, where the excavation is normally rectangular and undertaken by a grab.
方形桩 较大的挖孔桩，其基坑一般为矩形并由抓斗挖出。

barrier An obstacle obstructing access, e. g. a fence.
屏障 阻碍出入的障碍物，例如栅栏。

barrister A lawyer, being a member of one of the Four Inns of Court and who has been called to the Bar. The barrister is normally appointed by the client's solicitor once advocacy before a court is needed by the client. Barristers are engaged to provide specialist advice on points of law. There are an increasing number of barristers who specialize in construction law.
出庭律师 身为英国四大律师公会之一的成员并已取得出庭律师资格的律师。出庭律师一般是在客户需要出庭辩护前由客户的事务律师聘请。出庭律师受聘就法律要点提供专业建议。从事于建筑法律的出庭律师人数在日益增长。

barrow run A temporary path, found on construction sites, that is used by loaded wheelbarrows to gain access across unstable and soft ground. Usually constructed from scaffold boards or wooden planks.
手推车便道 建在建筑工地中的一种临时道路，供载重的手推车使用以便出入于不稳定和较软的地面，一般由脚手架板或木板建成。

bar scale A line on a map, drawn to scale, to represent a set distance.
条形比例尺 在地图上按比例绘制并代表一定距离的线条。

bar schedule (bending schedule) List of reinforcement needed, the size and shape (*See* BAR CODE) for each type of bar is specified and the quantities required are stated.
钢筋一览表（弯钢筋表） 所需钢筋的列表，其中说明各类钢筋的尺寸和形状（参考“钢筋代码”词条）并列出所需数量。

bar spacer A plastic object to position reinforcement in concrete, particularly in relation to maintaining the correct degree of cover for *formwork.
钢筋定位卡 在混凝土中使钢筋定位（尤其是关于保持保护层相对于模板的准确度数）的塑料物体。

basalt An extrusive *igneous rock; black to greyish-black in colour, aphanitic (fine-grained) and dense. Two principal types exist: olivine basalts (undersaturated in quartz), and tholeiitic basalts (quartz-saturated where mafic minerals are mainly pyroxenes). Typically used as an aggregate and in road construction.
玄武岩 一种喷出岩，颜色为灰黑色，是微晶（细密纹理）和致密的岩石。玄武岩主要有两类：橄榄玄武岩（含有未饱和石英）和拉斑玄武岩（饱和石英，其中的铁镁矿物质主要为辉石），玄武岩一般用作骨料和用于道路建设。

bascule bridge A bridge in which the span is moveable, e. g. a drawbridge. It can have single or double

spans, with each span counterbalanced to control the lowering and rising motions.
竖旋桥 跨度可变的桥，即开合桥。竖旋桥可为单跨或双跨，各跨可以平衡控制放下和抬起的运动。

B

base 1. The lowest part of something that normally provides support, e. g. a foundation. 2. A solvent in which the other compounds are held. 3. A component of paint that is the material producing the required opacity (ability to mask colour of surface). The body of the paint may be increased by the addition of inert extenders such as silica, calcium carbonate, or barytes. 4. A reference number used for counting digits, e. g. base 10 contains ten digits from zero to nine. 5. A logarithm reference number; a number raised to a power (indicated by a superscript number), e. g. $10^2 = 100$, where 10 is the base. 6. A substance that reacts with acids to form salts.
1. 基底 某物的下部，一般提供支撑，例如基础。**2. 打底剂** 一种其中含有其他化合物的溶剂。**3. 基料** 油漆的成分之一，即用于形成所需的不透明性（使表面颜色被遮盖的能力）的材料。油漆基料可通过添加硅土、碳酸钙或重晶石等惰性填充物而得以增强。**4. 基数** 用于计算数码的参考数字，例如基数 10 即包含从零到九的十个数码。**5. 底** 对数的参考数；自乘至一个幂（以上标表示）的某个数，例如 $10^2 = 100$，其中的 10 即为底。**6. 碱** 一种能与酸反应生成盐的物质。

base bid (US) The initial bid against which an alternative bid is offered.
基础投标（美国） 相对于所提出的备选方案投标而言的最初投标。

basecourse *See* BINDER COURSE.
基层 参考“黏结层”词条。

base date Used in standard form contracts, such as *JCT, and IFC (*Joint Contracts Tribunal, and Intermediate Form of Building Contract) to describe the date agreed and specified in the appendix by the parties. The base date is normally the same as the date of tender; other contracts may simply refer to the date of tender.
基准日期 常用在 JCT 和 IFC（合同审定联合会和中型工程建设合同）等标准的格式合同中，以说明双方已议定并在附录中指定的日期。基准日期一般与招标日期相同；在其他合同中也可能直接指招标日期。

base exchange A water-softening process. *Hard water is passed through a tank, at mains pressure, containing *zeolite. The zeolite, a mineral reagent, aborbs the salts that harden the water. At regular intervals a salt solution is flushed over the zeolite to regenerate it.
碱交换 水的软化过程。硬水在管道中受到压力通过含有沸石的水箱，而沸石是一种矿物质反应物，会吸收使水变硬的盐类。硬水按一定间隔流过沸石而被软化。

base flow The amount of flow in a river or stream emanating from groundwater, as opposed to run-off.
基流 河流或溪流中源自地下水的流量，与径流相对。

base gusset A stiff *rib that is located between a *base plate and steel column.
底部角撑板 位于底板和钢柱之间的刚性肋板。

baseline A datum mark, period, event, or standard from which all other things can be measured and compared.
基线 基准标志、期间、事件或标准，用于据此测量和比较所有其他事物。

base load The capacity to produce something continuously at a certain level, e. g. the minimum amount of electricity from a power station.
基本负荷 按某一水平持续产生某事物的能力，例如某个发电站的最低发电量。

base map An unsophisticated map on which data can be plotted. A base map tends to include boundaries, buildings, and roads with reference to geographic coordinates. An example of a base map is a topographic map.
底图 未经精细制作的地图，可在其上标绘数据。底图一般包括边界、建筑物及道路，并配有地理坐标作为参照。底图的示例之一是地形图。

basement An underground storey of a building.
地下室 建筑物的地下楼层。

baseplate 1. A steel plate located at the bottom of a column with holes predrilled for holding-down bolts, also known as the **stantion base**. 2. The lower part of a theodolite used to level the instrument and fix it to the tripod.
1. 柱脚底板 位于柱子底部的钢板，带有为地脚螺栓而预钻的孔洞，也称为**柱基板**。**2. 底座** 经纬仪的下部，用于把仪器放平并安装至三脚架上。

base shoe (shoe mould) A moulding, usually in the shape of a quarter circle, that is used to cover the junction between the skirting board and the floor.
踢脚线压条 一种线条，形状一般为四分之一个圆，用于遮盖踢脚板与地板交会处。

base slab A concrete slab beneath a structure.
承台 结构物之下的混凝土板。

base station A *GPS receiver positioned at a precise location.
基站 设置于精确位置的 GPS 接收器。

basic prices *See* SCHEDULE OF PRICES.
基础价 参考"价格清单"词条。

basin 1. A depression; in land that contains water, or at sea that contains sediment. 2. An area drained by a river. 3. (**washbasin**) A bathroom *sink used for face and hand washing. Usually made from *vitreous china.
1. 盆地 在陆地上蓄积水或在海洋里蓄积沉积物的洼地。**2. 流域** 河流流过的区域。**3. 盥洗盆** 用于洗脸和洗手的浴室水槽，一般由玻璃瓷制成。

basin mixer A *mixer tap on a washbasin.
洗脸盆混水龙头 洗脸盆上的混合龙头。

basket weave pattern A type of pattern used when laying brick patios. The bricks are laid in pairs with each pair laid at a right angle to the surrounding pair.
席纹图案 在铺设砖砌天井时所使用的一类样式。把砖块成对铺设，每对都和相邻的对成直角。

bat bolt *See* RAGBOLT.
棘螺栓 参考"棘螺栓"词条。

batch A quantity of something that is produced in one operation. **Batching plant** is the equipment used to make the batch, e. g. a **concrete batching plant**.
批次 在一次操作中生产的一定数量的某物。**分批配料装置**即用于分批的设备，例如**混凝土分批配料机**。

batch box (measuring frame, gauge box) A box without a top and bottom (i. e. only sides), used for measuring the volume of the constituent of a mix. Once the box is full it is lifted away leaving the contents on the mixing platform. For example, aggregate, sand, and cement can be measured out in a **batching box** to form the correct proportions for a concrete mix.
量斗（量料框，量料箱） 无顶无底（即仅有各边）的箱子，用于计量混合物成分的体积。一旦箱子装满就会被撤去，使内容物被留在搅拌台上。例如，骨料、砂及水泥，即在量斗中被计量，从而产生适用于搅拌混凝土的正确配比。

bath A sanitary fitting where the body or part of the body is immersed in water for washing.
浴缸 一种洁具，可在其中把身体或身体的一部分浸入水中以便清洗。

bathing waters Open waters (seawater or freshwater) having water quality, designated under an EC directive, suitable for public bathing (swimming).
洗浴水体 具有一定水质并根据欧盟相关法规被认定为适于公共洗浴（游泳）的开放水体（海水或淡水）。

bath mixer A *mixer tap on a bath.
浴缸混水龙头 浴缸上的混合龙头。

bath panel A board attached to the side or sides of a bath to conceal the pipework and the underneath of the bath.
浴缸侧板 附着在浴缸一侧或多侧用于遮蔽管道和浴缸底部的板子。

bathroom A room containing a bath, a toilet, and a sink. A **3/4 bathroom** contains a shower, toilet, and a sink, while a **1/2 bathroom** contains a toilet and a sink.
浴室 含有浴缸、马桶及水槽的房间。**3/4 浴室**指含有淋浴、马桶及水槽的浴室，而**1/2 浴室**则指含有马桶和水槽的浴室。

bathroom pod A bathroom where all of the plumbing, tiling, electrics, fixtures, and fittings have been *prefabricated in a factory ready for installation on site.
整体浴室 其中的全部给排水设施、贴砖、电气设备、固定装置及附件均已在工厂中预制可供现场安装的浴室。

bathymetry The measurement of the depths of water, e. g. the depth of lakes, oceans, and seas.
水深测量 对水体深度（例如湖泊、大洋及海洋的深度）的测量。

batt 1. A rectangular blanket of insulation material. Used in cavity walls, stud walls, floors, and loft spaces. 2. Short for battening.
1. 棉块 矩形毯状的保温材料，用于夹心墙、立柱墙、地板及阁楼空间。**2. 板条** battening 的简称。

batten A thin long strip of material (typically wood) used to strengthen or secure something.
板条 用于加固或固定某物的细长条状材料（一般是木材）。

battenboard A strong stable panel consisting of a *batten, *plywood, or *laminated core with a wood *veneer face.
条板芯细木工板 一种牢固稳定的板材，由板条、胶合板或层压芯构成并贴有薄木贴面。

batten door *See* LEDGED AND BRACED DOOR.
板条门 参考“直拼 Z 型撑门”词条。

battening Narrow strips of timber fixed to a wall or roof to provide a level surface. Used for fixing dry-lining or tiles.
钉板条 安装于墙壁或屋顶，以形成平整的表面的细木条，用于固定干内衬或砖块。

batten roll A traditional method of jointing used over battens in metal roofing.
板条卷缝 一种传统的拼接方法，用于金属屋顶上的板条。

batter The inclination from vertical of slope sides or of a ditch.
内倾角 斜边或水沟在竖直方向上的倾斜。

batterboard (US) A *profile.
龙门板（美国） 一种轮廓板。

batter peg A peg that is knocked into the ground to indicate the edge of a slope.
坡脚桩 打入地面的桩脚，用于指示斜坡的边缘。

battery 1. A device that uses a chemical reaction to produce electricity. 2. A heating or cooling coil in an HVAC system.
1. 电池 一种利用化学反应产生电力的装置。**2. 排管组** 暖通空调系统中的加热或制冷盘管。

battledeck Steel plates stiffened by a grillage of welded sections beneath.
加劲钢板 下方以焊制型材制成的格架加固的钢板。

baulk A large square piece of wood functioning as a beam.
粗枋 大块的方形木料，作用相当于梁。

Bauschinger effect A type of deformation in metals which results in an increase of the tensile yield strength

and a decrease of the compressive yield strength. *See also* STRAIN HARDENING.
包辛格效应 金属的一种变形，会导致拉伸屈服强度增大和压缩屈服强度减少。另请参考“应变硬化”词条。

bay 1. A section or division of a building, e. g. section between columns or beams. 2. Part of the building used for loading and unloading goods, e. g. loading bay. 3. Strip of concrete, where concrete is poured in sections. 4. Section of material plaster, concrete, or brickwork laid during a set period of time.
1. 架间 建筑物的一段或一部分，例如柱子之间或梁之间的部分。**2. 隔间** 建筑物中用于装卸货物的部分，例如装卸区。**3. 混凝土段** 混凝土的分段浇筑。**4. 工作段** 在一定时间段内铺设的灰泥、混凝土或砖结构材料部分。

bayonet fitting A tubular fixing unit that secures two parts together. A tube with studs on opposing faces that slides inside a slightly wider cylinder. The tube is located and locked in the cylindrical socket by twisting the studs into rebates.
卡口式组装件 管状的安装装置，可把两个部件固定到一起。在相对的两面上有栓头的管子滑入略宽的柱状套筒，通过把栓头扭入槽中即可使管子在柱状套筒中定位和锁紧。

bayonet holder Part of the fitting that receives the male part of the locking device, *See* BAYONT FITTING. Light bulbs not secured by a screw fitting are held in place by a bayonet fitting housed into a bayonet holder.
卡口底座 接受锁紧装置的公头部分的固定部件，参考“卡口式组装件”。非螺口安装的灯泡即通过卡口式组装件安装在卡口灯座中。

bay window A window that projects outwards from the external wall of a building and usually comprises three or more individual windows. Various types exist depending upon the shape of the projection, e. g. **bow window** (semicircular) and **square bay** (square). If the bay only projects from an upper floor, it is known as an **oriel window** .
飘窗 凸出于建筑物外墙的窗，一般包括三个或以上的独立窗户。视凸出形状的不同，凸窗有各种类型，例如**凸肚窗**（半圆形）和**方凸窗**（方形）。若飘窗仅是从上楼板凸出，则称为**上飘窗**。

bead 1. Semicircular timber moulding, used as a feature or to mask joints. 2. Defect in paint or varnish caused by excessive accumulation of liquid leading to uncontrolled flow during application.
1. 凸圆线脚 半圆形的木线条，用作造型装饰或遮盖接缝。**2. 流挂** 油漆或涂料中的缺陷，因液体的过度累积而导致涂刷时不受控制的流动。

beading *See* BEAD.
修边 参考“凸圆线脚”词条。

beading router (beader) Machine for making beading and cutting grooves into timber.
木铣机 用于制作线脚及在木料上开槽的机器。

beam A horizontal *structural member that resists *bending loadings. Typically formed from wood, steel, or concrete and used to support the storey or roof above.
梁 用于抵御弯曲载荷的水平结构件，一般由木、钢或混凝土制成并用于支撑其上的楼层或屋顶。

bearer A timber *beam that carries the floor joist.
托梁 用于承载地板龙骨的木梁。

bearing 1. Supporting load, at a particular location, over a given area. 2. A device to allow constrained motions between two parts.
1. 承载 在特定位置支撑某一给定区域所受的负载。**2. 轴承** 一种使两个部件之间运动得以约束的装置。

bearing capacity The ability of the ground to resist load. **Bearing pressure (q)** is the total stress created at the underside of a foundation in the soil. **Ultimate bearing capacity (q_f)** is the average contact pressure between the *footing and the soil which will produce shear failure in the soil. **Safe bearing capacity (q_s)** takes into account a factor of safety (FOS) against shear failure. **Allowable bearing pressure (q_a)** also includes a

FOS but takes into account the amount of settlement the structure can accommodate. Gross and net values can be obtained for q_f and q_a, to account for overburden pressure (i. e. the excavated soil).

地基承载力 地基承载负荷的能力。**承载压力（q）**即为浅土壤中基础底部所产生的总应力。**极限承载能力（q_f）**即为浅基础与土壤之间使土壤产生剪切破坏的平均接触的压力。**安全承载能力（q_s）**即把相对于剪切破坏的安全系数（FOS）纳入考虑后的承载能力。**容许承载压力（q_a）**也包括 FOS，但还考虑了结构物能够承受的沉降量。在说明上覆岩层压力（例如基坑土）时，总值和净值可取用q_f和q_a。

bearing capacity factors (N_c, N_q, N_g) Factors, obtained from charts or tables after **Terzaghi** (1943), used in the calculation to obtain the ultimate *bearing capacity.

地基承载力系数（N_c、N_q、N_g）取自**太沙基**（1943 年）的图形或表格的系数，用于计算出极限承载能力。

bearing pile A pile that transmits the load from the structure to the ground. An **end-bearing pile** transmits the load to the base of the pile, whereas a **friction-bearing pile** develops skin resistance around the pile to transmit the load. A combination of both end-and frictional-bearing resistance will occur in the majority of piles.

承重桩 把来自结构物的负荷传至地基的桩。**端承桩**把负荷传至桩的端部，而**摩擦桩**则在桩的周围产生表面阻力以传送负荷。大多数桩会同时发生端阻力和摩阻力。

bearing wall A wall that supports a load.

承重墙 支撑负荷的墙壁。

Beaufort scale An international scale for measuring wind speed. It ranges from force 0 (<1 km/h) for calm to force 12 (>120 km/h) for hurricane.

蒲福风级 一种用于计量风速的国际等级表，幅度从适用于静止的 0 级风力（ <1 千米/时）到适用于飓风的 12 级风力（ >120 千米/时）。

bedhead panel Electrical unit or socket, often used in hospitals and positioned above beds, providing an outlet and connection for communication services, nurse call, electric, and other medical or patient equipment.

医院床头面板 电子装置或插座，常用于医院中位于病床上方提供用于通讯服务、呼叫护士、供电及其他医疗或救治设备的插座和连接。

beetle maul A mallet with a wooden head.

大槌 木槌头的槌子。

Beggs deformeter A model to determine the reactions of a statically indeterminate structure, developed by **G. E. Beggs** in 1992. From its deflected shape, under the application of a load, the influence of the load along the structure (e. g. shear forces and bending moments) is deduced by using reciprocal theorem.

贝格斯变形测定仪 由 **G. E. 贝格斯**于 1992 年开发的一种模式，用于测定超静定结构的反应。通过其在承受负载时的偏向形状，沿着结构性使用互易定理而推理出负载沿产生的影响（例如剪切力和弯矩）。

beginning event The first task, fixture, or item of a project, also called a start event in a *network or *Gantt chart.

起始事件 某个项目的首个任务、装置或单项，在网络或甘特图中也称为开始事件。

behind schedule *See* SCHEDULE.

落后于进度表 参考“进度表”词条。

Belanger's critical velocity A condition of fluid flow in open channels in which the *velocity head equals half the mean depth.

伯朗格临界速度 在明渠中流动液体的速度等于平均速度的一半时的情况。

Belfast sink A large traditional rectangular-shaped ceramic sink with deep sides. Usually installed below the top of the work surface.

贝尔法斯特水槽 传统的带深边的矩形陶瓷水槽，通常安装在工作台面下方。

Belgian truss *See* FINK TRUSS.
比利时桁架 参考“芬克式桁架”词条。

bell and spigot A connection joint between two sections of pipe. The *spigot end of one pipe is inserted into a flared-out section (bell) of the adjoining pipe, with a rubber compressible ring or a caulking compound that seals the joint.
承插连接 两个管件间的连接接头。管件一端的插口插入相邻管件的承口内，用可压缩的橡胶环或嵌缝胶对接头进行密封。

bell-and-spigot joint (US) *See* SPIGOT-AND-SOCKET JOINT.
承插接头（美国） 参考“喇叭口接头”词条。

bellcast eaves A roof that is in the shape of a bell, i. e. curving with a more gradual outwards slope at the bottom.
钟形屋檐 形状像钟的屋顶，即底部逐渐向外倾斜弯曲。

bell curve A curve in the shape (outline) of a bell, e. g. a normal distribution curve.
钟形曲线 形状（轮廓）像钟的曲线，如正态分布曲线。

belled-out bored in-situ concrete pile *See* UNDER-REAMING.
扩底钻孔桩 参考“扩底”词条。

bellmouth overflow A water overflow structure situated in a reservoir, with the opening in the shape of an inverted bell. Used as an alternative to a side *spillway.
竖井溢洪道 一种安装于水库的溢水结构，溢水口为倒置钟形，用于代替侧溢洪道。

below ground level (US below grade) An elevation lower than ground level.
地平面以下（美国 below grade） 海拔低于地平面。

belt conveyor An endless belt driven by rollers used as a *conveyor.
带式输送机 一种用滚轴驱动的环形带进行输送的机器。

belt sander A *sander, with the *sandpaper in the form of a endless loop/belt.
砂带机 一种用环形圈/带作为砂纸的磨砂机。

benched foundation A foundation that is stepped.
阶形基础 一种阶梯式的基础。

benchmark 1. A *datum point used in surveying such as *ordnance datum. 2. A standard that is aimed at or used for comparison.
1. 水准点 用于水准测量的基准点，如水准零点。**2. 基准** 用于或旨在比对的标准。

bend 1. Something that is curved, e. g. a short length of pipe used for turning corners. 2. To *yield.
1. 弯 弯曲的物体，如用于拐弯处的短弯管。**2. 屈服**。

bender A tool used for bending bars and pipes.
弯管器 用于拗弯钢筋和管件的工具。

bender element A test to determine soil stiffness. Normally undertaken in a *triaxial cell using shear wave velocity.
三轴压缩试验 测定土壤抗剪强度的试验。通常在三轴压力室，采用剪切波速进行试验。

bending Caused by rotating either end of the length of a material in the opposite direction about its axis. **Bending moment** occurs in a structural element (beam) when a moment (force multiplied by a distance) is applied to the element. A **Bending moment diagram** shows the bending moment along the element.
弯曲度 通过将物体的任一末端按其轴线的相反方向进行旋转所形成。当力矩（力乘以距离）作用于结构件上，结构件（梁）就会产生**弯矩**。**弯矩图**用于表示结构件上的弯矩。

bending schedule (bar schedule) A list of the shapes and sizes of reinforcement bars required for a section of reinforced concrete.

B

钢筋弯折表（钢筋表） 列出钢筋混凝土截面所需的钢筋类型和尺寸的列表。

bending spring Coiled tool used by plumbers to exert an internal force on tubes to maintain the circular shape of copper and plastic tubes whilst the tube is bent. The spring is inserted into the tube to the point where the bend is desired. The tube is then bent and the spring is extracted by means of a draw chord, which is fitted prior to the spring being inserted. The resistance from the spring prevents the pipe folding; sand can be inserted into a pipe and used instead of a spring.

弯管弹簧 管道工人在弯曲管道时，用于向铜管和塑料管施加内部作用力以维持管道形状的螺旋工具。弹簧需插入到管道内部需要弯曲的位置，然后将管道拗弯，并用事先固定在弹簧上的拉绳将弹簧拉出。弹簧的阻力可以防止管道折叠，也可用砂代替弹簧灌入管道内。

bending stress (flexure stress) The stress along an element that results from bending it. As a result of the load, the beam will experience compressive stress along the top and tensile stress along the bottom. A neutral axis will exist in the middle of the beam where the stress level is zero. Thus, the bending stress will vary linearly with distance from the neutral axis (either in tension or compression). The equation to calculate bending stress is:

$$\sigma_b = M_y/I$$

where M = bending moment, y = vertical distance from neutral axis, and I = momentof inertia.

弯曲应力（挠曲应力） 由于构件弯曲而产生的应力。梁承受载荷时会在顶部产生压应力，在底部产生拉应力。梁的中部会出现应力为零的中性轴。因此，弯曲应力会随着离中性轴的距离（拉力或压力）而呈线性变化。计算弯曲应力的公式如下：

$$\sigma_b = M_y/I$$

其中，M = 弯矩；y = 到中性轴的垂直距离；I = 惯性矩。

bending tool A hand-operated tool used to bend pipes.

弯管工具 用于拗弯管道的手动工具。

bend test A test to determine the ductility of a metal without facture. The sample is bent through a specified arc, and if cracking is considered acceptable the metal is considered ductile.

弯曲试验 用于测定无裂纹金属的延展性的试验。将样品拗弯成指定的弧度，如果破裂程度是在可接受范围内，则金属被视为具有良好的延展性。

beneficial occupation An argument sometimes made by contractors that if an employer takes possession of part of a project before practical completion they are prevented from recovering liquidated damages; however, this seems to have no legal merit. Contrary to this argument case, law exists that suggests liquidated damages may be recovered even if the employer has taken possession (BFI Group of Companies vs DCB Integration Systems Ltd 1987 CILL 348).

实益占用 由承包商提出的说法，即如果在工程实际竣工之前业主已使用部分工程，则业主无权索要违约赔偿金；然而，这似乎不具有法律依据。与此说法相反的是，有案例表明即使业主提前使用项目，也可索取违约赔偿金（BFI 公司集团 vs DCB 集成系统有限公司 1987 年 CILL 348）。

bentonite A natural clay (aluminium phyllosilicate), which when mixed with water expands to form a *thixotropic gel. It can be used to support trenches, for drilling mud, in cements, and adhesives.

膨润土 一种天然黏土（铝硅酸盐），其与水混合时会膨胀形成触变性凝胶。可用于支撑沟槽，用作钻井泥浆，或用于制作水泥和黏合剂。

berm A long level section such as a shelf (narrow path) or raised barrier (low embankment) separating two areas, which normally contain water.

护堤 隔开两个区域（通常含有水）的长条形水平截面，如搁板（窄径）或凸起的屏障（低路堤）。

Bernoulli's theorem An expression for a steady flow of incompressible inviscid fluid, that, in its simplest form, states when the speed of a fluid increases the pressure decreases:

$$\frac{p}{\rho} + \frac{V^2}{2} + gz = \text{constant}$$

where p = pressure, ρ = fluid density, V = velocity of flow, g = acceleration due to gravity, and z = elevation.

伯努利定理 无黏性不可压缩流体的定常流的最简化表达式，指出当流体的速度增加时则压力会下降：

$$\frac{p}{\rho} + \frac{V^2}{2} + gz = \text{常量}$$

其中，p = 压强，ρ = 流体密度，V = 流动速度，g = 重力加速度，z = 标高。

berth A *jetty; a docking area for ships.

泊位 突堤式码头；泊船区。

bespoke Individual, purpose made. A range of individual purpose-made components, elements, or buildings would be termed a **bespoke system**.

定制 按照个人目的定做而成。按个人意愿定制而成的组件、构件或建筑物等，被称为定制系统。

Bessel functions (cylindrical functions) Canonical solutions $y(x)$ of Bessel's differential equation:

$$x^2 d^2y/dx^2 + xdy/dx + (x^2 - a^2)y = 0$$

Used in wave propagation and boundary value problems.

贝塞尔函数（柱函数） 贝塞尔方程的典型解 $y(x)$：

$$x^2 d^2y/dx^2 + xdy/dx + (x^2 - a^2)y = 0$$

适用于波的传播和边值问题。

best endeavours Duty of the contractor to use resources at its disposal to prevent delay. The carrying out of this duty is a precondition to awards of an extension of time under the *JCT and IFC forms of contract.

最大努力/勉尽全力 承包商利用其可支配的资源来按时完工，防止工期延误的职责。依照 JCT（合同审定联合会）和 IFC（中型工程建设合同），尽责的执行是延长工期的一个先决条件。

betz coefficient Theoretically the maximum efficiency a wind turbine can operate without slowing the wind down too much.

贝兹系数 理论上，风力涡轮机在不减慢风速的情况下可运行的最大效率。

bevel 1. A rounded or angled edge that is not at a right angle. 2. The point where two surfaces meet at an angle other than 90°. If the angle is at 45°, it is called a **chamfer**.

1. 斜角/斜面 圆形或有角的非直角边缘。**2. 割角** 两个表面形成非 90°角的点，如果形成的角度为 45°，则该角被称为**倒角**。

bevelled closer A brick that is cut at an angle such that it is a half brick's width at one end and a full brick's width at the other end.

斜削砖 一端被切为半个砖宽，另一端为整砖宽的砖。

bias External influences affect decisions or actions where consideration should be based on logic and merit alone. Under most standard forms of contract, the project administrator is under a duty to act fairly, reasonably, and independently when making decisions that affect matters between the client and contractor.

偏倚 本该根据逻辑和价值做出的决策或行为因受到外部因素影响而产生的偏差。大多数标准格式合同中，项目管理人有责任在有关业主和承包商的问题上做出公平、公正且独立的决策。

biaxial Having or about two axes.

双轴 有两个轴或与两个轴有关的。

bib (bibcock) A tap where the nozzle is bent down towards the ground.

弯嘴（星盆龙头） 喷嘴向下弯曲朝向地面的水龙头。

bib tap A tap where the water inlet is horizontal rather than vertical. Examples include a garden tap and traditional Belfast sink taps.

弯嘴水龙头 进水口呈水平而非垂直状的水龙头。例如花园水龙头和传统的贝尔法斯特水槽水龙头。

bid (US tender) Price submitted to carry out work described.
投标 执行所述工作所提交的价格。

bidet A sanitary fitting that is used to wash the genital area.
坐浴盆 用于清洗生殖器部位的卫生设备。

bi-directional Arranged in two directions, normally perpendicular to each other.
双向的 排列在两个方向上的，通常这两个方向相互垂直。

bi-fold door A door where the opening leaf comprises two panels that fold in two.
双向折叠门 包含两个可分别折叠的门扇的门。

bifurcation The division or forking of something into two parts, e. g. a stream splitting into two smaller streams.
分岔 某事物分开或分叉成两个部分，如一条溪流分离成两条较小的溪流。

billing Writing or recording the description and quantities in bills of quantities.
记账 在工程量清单上书写或记录的说明和数量。

bill of sale A document used to transfer property to another party without transferring possession. A bill of sale is a title of possession.
卖据/卖契 用于将财产转移给另一方但不转移所有权的文件。卖据是占有权的凭证。

bill of variations Final account and computation of adjustments to the contract sum.
工程变更单 对合同金额的决算和调整计算。

bills of quantities (BQ, BoQ) List of all the work, labour, and materials required to build a structure or complete the works. The purpose of the bills is to allow rates to be fixed to each item of work in a transparent and comparable method. If variations to the works occur, comparisons can be made between that described and the actual work.
工程量清单 建造一个结构或完成工程所需的所有工作、劳力和材料的清单。清单的目的是以一个透明和可比较的方法来确定各项工作的费率。如果工作发生变动，可以将所描述的和实际的工作进行比较。

bi-metal strip A metallic material, which is usually brass on one side and steel on the other, with the brass having the higher thermal expansion coefficient. When heat is applied, the strip curves toward the steel side. Conversely, when there is a decrease in temperature the material curves the other way. Bi-metallic strips are utilized in devices for detecting and measuring temperature changes.
双金属条 一种金属材料，通常一侧为黄铜另一侧为钢，黄铜具有较高的热膨胀系数。当施加热量时，金属带向钢的一侧弯曲；相反地，当温度降低时，材料向黄铜侧弯曲。双金属带常用于检测温度变化的装置。

binder A component of paint which produces the paint form on solidification. Binders are usually linseed oil or alkyd resins.
黏合剂/成膜物质 一种促使油漆凝固的油漆成分。黏合剂通常由亚麻籽油或醇酸树脂制成。

binder bar (binding) A secondary bar that holds the main bars together in a reinforced cage.
拉结筋（接筋） 用于将钢筋笼中的主要钢筋固定在一起的辅助钢筋。

binder course Previously termed **basecourse**; one of the layers in a *flexible pavement construction which is used to provide an even surface for the *surface (wearing) course and contributes strength.
黏结层 以前被称为"**基层**"；在柔性路面结构中用于为表面层（磨耗层）提供一个平滑并有助于增加强度的面层。

binding wire (tying wire, tie wire) Wire used for fixing (tying) reinforcement bars together to the correct shape.

绑扎钢丝（系结钢丝，扎线） 用于将钢筋固定（绑定）成特定形状的金属丝。

biofiltration A technique using microorganisms to remove and oxidize organic gases. Contaminated air is passed through a **biofilter** (a bed of compost containing indigenous microorganisms) to convert organic compounds to carbon dioxide and water.
生物过滤 一种利用微生物来氧化和去除有机物的技术。使受污染的空气通过一个生物滤池（含有土著微生物的堆肥床），将有机物转化为二氧化碳和水。

biological shield A large mass of material (such as concrete) placed around a nuclear reactor to reduce the release of radiation.
生物屏障 为减少辐射的释放，在核反应堆周围放置质量的材料（如混凝土）。

biological treatment A process that utilizes bacteria to break down waste into less polluting compounds, e. g. *trickling filters in sewage treatment.
生物处理 利用细菌将废物分解成污染较小的化合物的方法，如用于污水处理的滴滤池。

biomass 1. The number of organisms in a given space. 2. Plant and animal waste that is used as fuel.
1. 生物量 一定空间里的有机体的数量。**2. 生物质** 用作燃料的植物和动物粪便。

bioscrubber A *scrubber that uses microorganisms to convert organic compounds from contaminated air or gas to carbon dioxide and water.
生物洗涤器 利用微生物将受污染的空气或气体中的有机化合物转化为二氧化碳和水的一种涤气器。

biosolids Treated sludge from wastewater treatment plants.
生物固体 经污水处理装备处理后的淤泥。

bi-parting door A door where the opening leafs comprise two or more panels that move away from the centre of the door when opened.
双开推拉门 一种由两个或以上面板组成门扇的门，打开时由中心向两侧移动。

bird screen A wire mesh installed over an opening to prevent the entry of birds.
防鸟网 在开口处安装以防止鸟类进入的线网。

birdsmouth A V-shaped notch at the bottom of a rafter that allows the rafter to rest on the *wall plate.
承接角口 椽条底部的一个 V 形凹槽，可使椽条搭接在梁垫上。

bi-steel construction A composite construction material, consisting of two steel face plates connected by an array of transverse bar connectors, which can be in-filled with concrete, installing material, or left as a void. Applications include crane beams, lift shafts, and building cores.
双层钢板建筑 一种复合建筑材料，包括由一系列横向连接杆连接的两个钢面板，中间可填充混凝土，安装材料或留作空隙。可应用在吊车梁，电梯井和建筑筒体。

bit 1. A binary digit, 0 or 1, used to represent one of two outcomes, e. g. on or off. 2. A cutting end of a rotary tool.
1. 位 一种二进制位，0 或 1，用于表示两种结果中的一种，例如开或关。**2. 钻头** 一种旋转工具的切削端。

bittiness Paint finish where dust or other similar small-grained objects cause defects in the surface.
起（粗）粒 涂层饰面由于灰尘或其他类似小颗粒而造成的表面缺陷。

bitumen (bituminous) A highly viscous liquid used primarily for roads and pavements. Bitumen originates from crude oil.
沥青 主要应用于道路和人行道上的高黏稠度液体。沥青来源于原油。

Blackpool tables A series of tables that relate the amount of decompression time needed, to the time and pressure a person has spent breathing compressed air.
布莱克浦表格 与所需减压时间相关的一系列参数表格，包括人在呼吸压缩空气时所需的时间和压力。

blade A long thin flat part of a tool or machine, e. g. a blade of a propeller or wind turbine.
刀片/叶刃 一种工具或机器上的长薄片部分，如：螺旋桨或风力涡轮机上的叶片。

blank 1. End of a service run left ready for future connection. 2. Cover for electrical duct or conduit ready to receive a socket, if required.
1. 预接头 某一项设施末尾的预留，用于以后的连接。**2. 插座白板** 覆盖电线管槽以在需要时安装插座的盖板。

blank door (blank window) 1. False or painted door or window to give the impression of the real thing. 2. An opening previously used for a window or door that has been filled with masonry; walled up.
1. 假门（假窗） 假的或涂刷出来给人以真实的视觉感的门或窗。**2. 封堵门（窗）** 之前是门或窗的开口现在用砖石堵上；砌墙。

blank wall A wall without any door or window openings.
空白墙面 没有门或窗开口的墙。

blasting A process of extracting or mining rock from a quarry.
爆破 从采石场提取和挖掘岩石的过程。

bleach A corrosive liquid based on hydrogen peroxide utilized mainly for domestic cleaning applications.
漂白剂 一种基于过氧化氢为主要成分的腐蚀性液体，主要用于室内清洁。

bleed valve A valve that enables air or fluids to be released from a pressurized system. Commonly found on radiators in a central heating system.
放泄阀 一种可使空气或液体从增压系统中释放出的阀门。通常位于一个集中供暖系统的暖气片处。

blended water A mixture of hot and cold water, e. g. from a mixer tap.
混合水 一种冷热混合水，如：来自冷热混合龙头中的水。

blind fixing A concealed fixing system.
隐蔽式修复 一种隐蔽式修复系统。

blind mortise A mortise that only passes part-way through a piece of timber.
不贯通榫 一种只穿过部分木块的凹榫。

blind nailing A method of nailing where the head of the nail is not visible. Also known as **secret nailing**.
暗钉法 一种使钉子头部不可见的装钉方法。也称为“暗钉”。

blinds Window coverings that are used to control daylight, solar gain, and privacy. Usually comprise horizontal or vertical slats that can be raised, lowered, or rotated to achieve the desired level of control.
百叶窗 用于遮挡日光、太阳辐射和保持私密的窗户遮挡物。通常由横向或纵向板条构成，可升起、降低或旋转至预期的控制范围。

blind wall A wall that contains no openings.
闷墙 没有开口的墙。

bloated clay A light porous material, formed from clay that has expanded during firing.
膨胀黏土 一种轻质的渗透材料，由烧制过程中膨胀的黏土而来。

blob A soft mass of material.
坨 一团软质的材料。

block copolymer A combination or mixture of different polymers.
嵌段共聚物 不同聚合物的结合或混合。

blocklayer 1. Layer of road pavings, sets, or stones. 2. A semi-skilled tradesperson who specializes in the construction of concrete blocks walls, but may not be capable of constructing a wall of facing bricks. Where a person refers to themselves as a *bricklayer they are skilled and competent at laying *bricks and blocks.
1. 砌块层 人行道、花岗岩路或石路。**2. 砌块工** 一位专业于混凝土砌块墙施工但不精通墙表面贴砖施工的半技术工人。当某人自称为瓦工时，一般他能熟练精通铺设砖块和砌块。

blockmaker (blockmaking machine) Machine used for the manufacture of concrete or clay blocks.
压砖机 用于混凝土或黏土砖加工的机器。

blockout A pocket gap, space, or hole left in a wall or service run ready to receive equipment, structural insertion, or service.
预留孔堵块 墙面或设备上预留的凹槽空隙、间隔或穿孔，用于承接器械、结构插入或设备。

block plan A small-scale drawing that shows the position of buildings and proposed building, general layout of roads and services. Used for planning and as a working drawing, which shows the general outline of each building.
楼宇平面图 一种显示建筑和拟建筑位置，道路和设备一般布局的小比例图纸。用于规划用途和工程图纸，显示每一建筑物的整体略图。

block saw A hand saw for cutting lightweight concrete blocks.
轻质混凝土手板锯 用于切割轻质混凝土砌块的手锯。

blower door Equipment used to undertake a *fan pressurization test. Comprises a portable variable speed fan, an adjustable doorframe and panel, a fan speed controller, and a pressure and flow gauge.
鼓风门 用于进行风扇增压法测试的设备。组成部分包括一个便携式变速风扇、一个可调节的门框和门板、一个风扇速度控制器和一个压力与流量测量仪。

blower door testing *See* FAN PRESSURIZATION TEST.
鼓风门测试 参考“风扇增压法测试”词条。

blow gun (lance) A device in the form a long thin pipe that is connected to a compressed-air supply, to undertake *blowing out of formwork.
喷枪 一个长管状的设备，连接至压缩空气供应设备，用于进行模板吹尘。

blowing out A process using a *blow gun to clear rubbish out of formwork by blowing it out by compressed air.
吹尘 利用喷枪吹出压缩空气来清理出模板中垃圾的过程。

blowlamp (blowtorch) A portable gas burner used for heating materials, e. g. soldering copper pipes.
喷灯 一种便携式燃气燃具用于加热物质，如：焊接铜管。

BM *See* BENCHMARK.
BM 参考“水准点”词条。

board A flat piece of wood or other material.
板 木材或其他材料的一块薄片。

board foot The quantity of timber measuring one square foot with a thickness of one inch.
板英尺 根据一英寸厚度测量一平方英尺木材的体积量。

boarding joists *Joists for flooring.
板条地楞 地板龙骨。

board metre The quantity of timber measuring one square meter with a thickness of 25mm.
板米尺 根据 25 毫米厚度测量 1 平方米木材的体积量。

boaster A chisel used for dressing stone.
宽凿 用于修整石材的凿子。

boat scaffolding *See* FLYING SCAFFOLD.
船用脚手架 参考“悬挂式脚手架”词条。

bob *See* PLUMB BOB.
垂块 参考“铅垂”词条。

B

body Main (central) part of something.
主体 某物的主要（中心）部分。

body force The force that is distributed throughout the body, e. g. self-weight.
体积力 分布于实体各部分的力，例如自重。

body of deed The operative part of a contractual document which provides and sets out the terms of agreement between the parties to the contract.
契据正文 合同文件的执行部分，其中规定和列明合同订约方之间的协议条款。

body wave A *seismic wave that travels through the interior of the Earth.
体波 在地球内部传播的地震波。

bogie A small wheeled truck that runs on rails, used for carrying heavy loads from tunnel and mining works.
轨道小车 一种在轨道上行驶的小型轮式车，一般用于运送隧道和采矿工程的重物。

boil 1. To reach *boiling point. 2. When soil flows into the bottom of an excavation, normally as a result of an upward seepage force. *See also* PIPING.
1. 沸腾 达到沸点。**2. 管涌** 土壤流入基坑底部，一般是由向上的渗透力产生。另请参考“管涌”词条。

boiler (US furnace) An appliance used to heat water. The water is then used for space or water heating. Despite the name, the water does not necessarily boil.
锅炉 一种用于对水加热的装置，随后把水用于空间供暖或水暖。虽然锅炉的英语单词中带有 boil，但并不允许使水沸腾。

boiler failure A form of degradation associated with boilers, where the metal (usually a ferrous alloy) fails mainly due to the *creep mechanism.
锅炉失效 与锅炉有关的一种退化形式，其中的金属（一般为黑色合金）失效主要是由于蠕变机制。

bollard A strong post that is used to restrict access. Typically used in pedestrian zones to keep cars out; usually constructed from concrete or steel.
阻车柱 用于限制出入的牢固柱子，一般用于步行区内防止汽车的进入；通常由混凝土或钢制成。

bolster 1. A steel chisel with an enlarged head used for cutting masonry. Used to cut and trim bricks and blocks (**brick** or **masonry bolster**), or to chase walls (**electricians' bolster**). 2. A **crown plate** . 3. A short horizontal piece of timber on top of a column to support and decrease the span of a beam.
1. 砖石凿 具有较大凿头的钢凿，用于切削砖石结构。用于切削和修整砖块和砌块（**砖凿或石凿**），或用于雕镂墙壁（**电工凿**）。**2. 侧向支撑**。**3. 承板** 位于柱子顶部的水平小木块，用于对梁的支撑以减少梁的跨度。

bolt A bar with a head at one end and a thread section at the other to receive a nut.
螺栓 一端有栓头而另一端有螺纹部分的钢栓，用于承接螺母。

bolt box (bolt sleeve) Formwork in the form of an open-ended box that is used to provide protection for the alignment of holding down bolts in a pad foundation during a concrete pour. The gaps around the holding down bolts are then grouted up.
螺栓盒（螺栓套） 在浇筑混凝土时，用于为阶形基础上对齐的地脚螺栓提供保护的框架，其样式为开口的盒子，然后地脚螺栓周围的空隙被浇满。

bolt croppers A device, similar to a pair of shears, used for cutting bolts.
断线钳 一种装置，与剪刀相似，用于切割螺栓。

bombproof construction A structure that withstands the effects of a bomb exploding.
防炸结构 可经受炸弹爆炸影响的结构。

bond (bonds) 1. The way that courses of brick or blockwork are laid. *Stretcher bond, *English bond, *Flemish bond, *English garden wall, and *Flemish garden wall are all examples of brick bonds. 2. A writ-

ten agreement between two parties in which one party agrees to pay money to the other party. 3. Contract under a seal where a financial house such as a bank or insurance company takes on an obligation to pay a party (normally the client) should a specific event arise. Bonds generally act as a guarantee of payment should a party fail to perform a specific contractual duty. 4. A mechanism holding atoms together in molecules and crystals, involving the sharing and transferring of electrons. Common types include covalent, ionic, metallic, and van der Waals. 5. The adhesion that holds two items together, e. g. the forces that hold concrete and reinforcement or bricks and mortar together.
1. 砌合法 砖层或砌块层的铺设方法。顺砖式砌法、英式砌法、梅花丁砌法、三顺一丁砌墙法及变种荷兰花园墙式砌法均为砖块砌合法的例子。**2. 契约** 两个当事方之间的书面协议，其中一方同意向另一方付款。**3. 保函** 盖有印章的合同，一般由银行或保险公司等金融机构在发生约定事件的情况下，向某一当事方（一般为业主）付款的义务。保函一般作为某一当事方未能履行具体合同责任的情况下的付款保证。**4. 化学键** 一种使分子和晶体中的原子结合在一起的机制，涉及对电子的共享和转移。化学键的常见类型包括共价键、离子键、金属键。**5. 黏结力/结合力** 使两件物品结合在一起的附着力，例如使混凝土与钢筋或砖块与砂浆结合的力。

bond course The course in a brickwork or blockwork wall that bonds the construction together.
结合层 砖体或砌体墙中使构造接合在一起的层。

bonder (bonding brick) A *header.
丁砖 丁砖。

bonding capacity (US) A document provided by a bank or finance house that gives information on the financial stability of a contractor. A rating is provided to give an indication of the amount of work that the contractor is financially capable of handling.
担保能力（美国） 由银行或金融公司提供的文件，其中说明关于某家承包商的财务稳定性的信息。该文件中将给出评级，对该承包商在财务方面有能力处理的工程量予以标明。

boning 1. Checking alignment with the naked eye. 2. Using T-shaped rods for *setting out.
1. 目测 以裸眼检查对齐情况的方法。**2. 测平** 使用T形杆进行放线。

bonnet roof The roof over a bay window.
飘窗顶 飘窗上方的屋顶。

bonnet tile (bonnet hip) A roof tile that is used at the *ridge of a *hipped roof. *See also* HIP TILES.
屋脊瓦 用在四坡屋顶的屋脊上的屋瓦。另请参考“斜脊盖瓦”词条。

bonus Extra money awarded to workforce for good work, extra duties, or to help ensure retention of the workforce.
奖金 为了优秀的工作、额外的责任或为了确保能留住雇员而给予雇员的额外款项。

bonus clause A provision within a contract that acts as an incentive encouraging the contractor to complete the works before the contractual completion date; additional money is offered for early completion. The amount of money offered is usually provided as a set amount per day or week that the contractor completes before the end date. Being quite different from a *penalty clause, the amount of money specified does not need to bear any relationship to that which would be saved by the contractor for early completion. If the employer defaults, delays the contract, and prevents the contractor finishing earlier, the employer may have to pay the contractor the amount that they would have been entitled to had they not been delayed.
奖励条款 合同中，用作鼓励承包商在约定竣工日期之前完成工程的奖励条文，针对提前竣工而给出的额外款项。对于承包商在截止日期之前竣工而给出的款项金额一般是规定为按天或按周计的固定金额。与罚款条款大为不同的是，所规定的款项金额并不需要说明，如承包商提前竣工而导致节省成本的任何关系。如因业主违反、延迟合同并阻止承包商提前完工，则业主可能须向承包商支付在未受延误的情况下承包商原本有权获得的款项。

bonus system Incentive scheme supported by an explicit payment structure that establishes targets and goals, which if met, result in payment of extra money. Money could be awarded by achieving a set level of perform-

ance, producing the required quantity of work within a timeframe or before a deadline, for working a period of time without taking sick leave or unauthorized vacation. The extra money awarded is given on top of the normal salary or daily payment.
奖金制度 有明确的支付体系并确立了指标和目标的奖励方案。如达到有关指标和目标，则会支付额外的款项。一旦达到预设水平的表现、在时间表以内或最后期限之前产出规定的工程量、在一定时间内工作而没有休病假或未经批准的休假，则可受到奖赏的款项。奖赏的额外款项将在正常的工资或日薪之外给予。

bookmatching The slicing or splitting of timber so that the two faces of plywood are cut from adjacent timber resulting in the grain on each face of the plywood matching the other face.
正反配板 对木料的切片或剖削，以便使胶合板的两面都是从相邻木料上切下，从而使胶合板各面的纹理均与另一面相匹配。

boom 1. A *jib of a crane. 2. The *chord of a truss. 3. A floating barrier on a water surface, e. g. to prevent an oil slick for spreading.
1. 吊臂 起重机的臂。**2. 桁架的弦杆**。**3. 浮栅** 水面上的浮动屏障。例如，用于防止水面浮油扩散。

boom cutter (boom loader) A rock-tunnelling machine with a cutter device that projects out from the machine.
悬臂式掘进机 一种岩巷掘进机器，装有从机器上伸出的切割装置。

booster A device to increase the power or performance of something.
助力器/助推器 一种用于提高某物的功率或性能的装置。

boot 1. A computer start-up procedure. 2. The bottom of a bucket excavator. 3. A step around the perimeter of a concrete floor slab to carry the outer skin of a cavity wall.
1. 引导程序 一种电脑启动程序。**2.** 斗式挖掘机的底部。**3. 外叶墙拉结** 围绕于混凝土楼板四周的凸边，用于承载夹心墙的外层。

boot lintel A *lintel with a boot-shaped projection.
挑口过梁 具有靴状凸出的过梁。

border A strip around the edge of something.
边界线 围绕于某物边缘的线条。

bore 1. Internal diameter of a pipe. 2. To make or drill a hole. 3. A tidal wave in a river.
1. 内径 管道的内直径。**2. 开孔** 打孔或钻孔。**3. 涌潮** 河流中的浪潮。

bored pile (cast-in-situ pile, cast-in-place pile) Formed by excavating or boring a hole in the ground. The hole is then filled with concrete; it can be reinforced or *precast.
钻孔桩（现场浇筑桩，就地浇筑桩） 通过在地上挖孔或钻孔而做成，随后向孔中填充混凝土；钻孔桩可为加钢筋的或预制的。

borehole A hole drilled in the ground, used to investigate the underlying strata conditions of a site during a site/ground investigation. Also, a borehole can be used to extract water and other liquids such as oil and methane from the ground.
钻探孔 在地上钻的孔，用于调查现场/或地面的下方地质层的状况。钻探孔也可用于从土地中抽取水和其他液体，例如油和甲烷。

bore lock (key-in-knob set, tubular mortise) A door lock that has a T-shaped tubular case that slides into a mortise hole cut with a hole saw or special drill bit.
球形锁 一种具有T形管状外壳的门锁，管状外壳将装入用孔锯或专用钻头切出的榫眼中。

boring 1. Uninteresting. 2. The process of making holes in wood or the ground by animals or tools.
1. 乏味的。**2. 开孔** 由人工或工具在木头或地面打孔的过程。

borosilicate glass A type of glass based on boron oxide and silica. This type of glass is used mainly because

of its coefficient of thermal expansion which is about 60% less than ordinary glass; as a result borosilicate glass is more resistant to breaking and imparts superior thermal shock properties. In construction, borosilicate glass is popular for glazing applications.
硼硅酸盐玻璃 一种基于氧化硼和二氧化硅的玻璃，使用这种玻璃主要是由于其热膨胀系数比普通玻璃低60%；因此硼硅酸盐玻璃更能耐破裂和冲击，具有优异的热冲击性能。在建筑中，硼硅酸盐玻璃在嵌玻璃应用中较普遍。

borrow Material excavated from the ground and used as fill in another place.
借土 从地面挖出并作为填料用于它处的物料。

borrowed light A window in an internal wall that enables daylight to pass from a room that contains an external window to one that does not.
借光窗 内墙上的窗户，能使日光穿过开有外窗的房间进入没有外窗的房间。

bossed connection (bossed end) The end of a plastic pipe to which an *O ring is fitted.
承插连接（承口端） 放入O型密封圈的塑料管端。

bottled (bottled edge) A rounded edge. Can also be an internal angle that has been rounded and looks like it has been created by smoothing a bottle over the surface.
瓶形的（瓶形边缘） 圆边的（阴角边倒圆边）圆形的边缘，也可以是圆的阴角，看来像是用瓶子在其表面磨平的一样。

bottled gas *Propane, *butane, or a combination of both, compressed and stored under pressure in a steel container. Used as a portable gas supply.
瓶装气 在压力下压缩和储存于钢瓶中的丙烷、丁烷或两者的混合物，用作便携式的燃气供应。

bottle trap A section on a waste pipe from a sink or basin, having a bottle-shaped cap, designed to be removed to enable rubbish to be cleaned out of the system.
瓶式存水弯 来自水槽或水盆的排水管上的一段，具有瓶形的上盖，被设计为可以取下，以便从系统中把垃圾清理出去。

bottom chord The bottom-most truss member.
底弦杆 最底部的桁架构件。

bottom-hung window A window with a bottom-hinged *sash.
下悬窗 配有底部铰接式窗框的窗户。

bottom rail The horizontal lower rail of a door or a *sash.
下冒头 门或窗框的水平底栏。

bottom-up construction Standard method of construction, starting with the *substructure (foundations, basements, and services), and finishing with the *superstructure (walls, floors, and roof). This is the opposite of *top-down construction, where excavation for basements takes place at the same time as the superstructure.
顺筑法 标准建筑方法，从下部结构（基础、地下室及设施等）开始施工，并以上层结构（墙壁、楼面及屋顶）施工收尾。顺筑法与逆筑法相反，后者对地下室的挖掘与上层结构的建设同时进行。

boulder A large rounded rock.
巨砾 又大又圆的岩石。

boulder clay (till) Material from a glacier deposit, containing clay, boulders, gravels, sands, and silts.
冰砾泥（冰碛） 来自冰川沉积物的物料，含有黏土、巨砾、砾石、砂及粉土。

boundary (boundaries) The official line that divides areas of land. The **boundary conditions** relate to the condition at the edge of something.
边界 划分土地区域的正式界线。**边界状况**与某物边缘的情况有关。

Boundary Property Federation (BPF) system (BPF system) An organization that represents the interest

B

of property owners, which has drawn up a manual describing how contract management should be organized. The concept divides the process into five stages: concept; preparation of brief; development of design; tender documentation; and construction. The duties of the parties are also specified in the manual. Care should be taken when referring to any such documentation as new legislation may well make some clauses invalid.
边界地产联合会（BPF）系统（BPF系统） 一个已拟定了组织合同管理手册且代表业主利益的组织机构。该概念可划分为五个阶段：概念、扼要准备、深化设计、招标文件和建设；双方的责任明确在手册上。由于新出台法律可能致使一些合同条款失效，因此应注意涉及任何该文件的新法律。

bound construction Construction methods that use frame construction where the bricks, stonework, tiles, or blocks are bonded to the walling units.
组合施工 采用框架式的施工方法，如将砖块、石墙、瓷砖或砌块组合一起的墙结构单元。

bound water Water that has become adsorbed in the surface of a material.
结合水 指材料表面吸附的水分。

Bourdon gauge A mechanical pressure gauge.
波尔登管式压力计 一种机械压力表。

Boussinesq equation Used to calculate the stress beneath a point load (but can be extended to accommodate different shapes of foundation) on a semi-infinite, homogeneous, isotropic, elastic medium. The base equation is:

$$\sigma_z = 3P^3/2\pi R^5 = 3P/2\pi z^2\ [1 + (r/z)^2]^{5/2}$$

布西内斯克方程 用来计算在半无限的、均匀的、等向性的以及弹性介质上的点荷载作用下的应力的方程（点荷载可以延伸适用不同的基础形状）。基本方程为：

$$\sigma_z = 3P^3/2\pi R^5 = 3P/2\pi z^2\ [1 + (r/z)^2]^{5/2}$$

bow Bend or sag under the application of load.
弯曲 在荷载的作用下弯曲或下垂。

bowl The water-containing part of a sanitary appliance.
便池 洁具装置的存水部件。

bowlders Large rounded stones or pebbles used for paving and decoration to walls and floors.
巨砾 大圆形石头或鹅卵石，用于铺路、装饰墙壁和地板。

bowl urinal (pod urinal) A wall-hung, bowl-shaped urinal that can be installed at different heights to suit the users.
小便槽（小便池） 可安装在适应用户的不同高度位置的墙挂碗状尿盆。

bow saw A saw with a thin blade held in a U-shaped or H-shaped frame with a narrow handle at each end. Used for cutting curves .
弓锯 安装在U形或H形框架的薄锯片锯刀，框架两端各带一个小手柄。用于切割曲线结构。

bowser A tank, either fixed on a lorry or towed behind, used for refuelling or carrying water.
水槽车 在货车或拖车后面安装的水槽，用于加油或运水的车。

Bow's notation A graphical method of representing forces and members in a truss. Letters of the alphabet are used to label internal and external spaces between loads, reactions, and member forces. Two letters define the force that separates those two spaces.
鲍氏符号标注法 可表示桁架上的力和构件的图解法。以字母标注载荷、反应力和构建应力之间的内外部空间。两个字母定义两个空间分隔的力。

bowstring truss A truss where the top chord arches (like a bow) and meets the bottom straight chord at either side (like the string to the bow).
弓弦式桁架 上弦杆拱起（像一张弓）并与下直弦在任意一边交接的桁架（如弓的弦）。

bow window (bowed window) A *bay window that has a semicircular-shaped projection from the wall.
弓形窗 半圆形状从墙凸出的飘窗。

box beam (box girder) A beam in which the cross-section resembles a rectangular box made from either *steel, *pre-stressed concrete, or *reinforced concrete. Used in a box-girder bridge typically for *highway bridges.
箱型梁 横截面类似于矩形箱，由钢、预应力混凝土或钢筋混凝土制成的梁。通常用于箱型梁桥，如公路桥梁。

box frame The frame of a traditional sash window that contains the weights.
盒型窗框 自带配重的传统推拉窗框架。

box gutter (trough gutter) A U-shaped gutter.
方形排水槽（槽型沟） U 型排水沟。

brace A member within a structure to stiffen it.
支柱 结构内可加劲/加固的构件。

bracket A support to hold something.
支承 支撑某物的结构。

bracketed stairs A *string, diagonal-edge stair support that is cut to produce a more open string with ornamental brackets to support each *tread and *nosing.
护板托架 斜梁上对角线边的一种楼梯支撑，其被切割成更加露明的斜梁，以装饰性的支承结构来支撑每个踏步和前缘。

bracketing 1. Brackets used to support a plaster cornice or feature that is hung from the wall of ceiling. 2. Brackets or angles fixed to columns or walls to carry beams, floors, or stairs.
1. 线条托架 用于支撑石膏角线或线条等，从天花到墙悬挂的结构。**2. 牛腿** 固定在立柱或墙体的支架或角撑，以支撑横梁、地板或楼梯。

branch Any cable or pipework in an electrical or drainage system that is connected to the mains. A division off the main part of something, e. g. a secondary pipe in the water supply, or a branch sewer that feeds into the main sewer. The point where the branch joins the main in a drainage system is the **branch connection**.
支管 电力或排水系统中连接到主管的任何电缆或管道。离开某物主要部分的分支，如供水系统中的二级管或进入主下水道的分支下水道。排水系统中支管连接主管的点称为**支管连接**。

branched polymers (graft polymers) A type of polymer where the internal polymer chains are arranged like branches in a tree. Branched polymers usually exhibit higher stiffness and tensile strength.
支化聚合物（接枝聚合物） 一种内部的聚合物链像树枝排列的聚合物。支化聚合物通常具有较高的刚度和抗拉强度。

branch manhole An inspection chamber over a branch connection.
分支人孔 在支管连接上方的检修室。

branch pipe *See* DISCHARGE PIPE.
分支管道 参考“排水管”词条。

branch vent A vent on a *branch pipe that connects to a *vent stack in a drainage system.
通风支管 排水系统中分支管道连接到通风立管上的排气管。

brass A non-ferrous copper (Cu) rich copper—zinc alloy. The metal is copper based with zinc (Zn) as the alloying element. Brass is popular in bathroom applications.
黄铜 一种有色铜（Cu）、富铜的铜锌合金。金属以铜为主且含锌（Zn）作为合金元素。黄铜常应用于浴室中。

brazing A technique used whereby a molten metal is used to join other metal components. Brazing metals usually have a melting temperature greater than about 425°C.
钎焊 用熔化金属与其他金属连接的技术。钎焊金属的熔化温度通常大于 425°C。

breach of contract Failure to carry out contractual obligations; refusal to perform contractual obligations is

known as repudiation. Failure to perform a contractual duty can have varying consequences depending on whether the term broken is a condition or a warranty. Typically, the party wronged can sue for damages, however, if the condition affected goes to the root of the contract and is fundamental, the contract can be treated as a repudiatory breach and the contract discharged.
违约　没有履行合同的义务；拒绝履行合同义务被认为是违约。未履行合同义务可能会有不同的后果，这取决于违反的条款是一个条件还是担保。通常情况下，被侵权一方可以请求赔偿，但如果条件影响到合同根基，合同可以视为毁约，即合同解除。

breast　1. The protruding part of wall above a fireplace that contains the flue. 2. The portion of wall under a window that extends from the floor to the sill.
1. 壁炉腰　壁炉上方墙的突出部分包含烟道在内。**2. 窗下墙**　窗户的下方从地板延伸到窗台的墙体部分。

breather membrane (building paper)　A vapour-permeable *building paper.
呼吸膜（建筑防潮纸）　透气的建筑防潮纸。

BRE Digests　Technical journals on specific topics produced by the *Building Research Establishment.
《*BRE* 摘要》　由英国建筑研究院出版的特定主题的技术期刊。

breech fitting　A Y-shaped fitting connecting the flow from two pipes into one.
叉形装置　连接从两管流进一个管道的 Y 形接头。

bressummer (breastsummer)　A large timber *lintel; however, has largely been superseded by reinforced concrete or steel.
托墙梁　一根大木头过梁；然而已经很大程度上被钢筋混凝土或钢取代。

Brewster's fringes　An oily-looking rainbow effect (not a defect) in double-glazed insulating units, when both panes of glass are of equal thickness. When the incident light meets the reflected light from the other surface in such a way they are out of phase and thus cancel each other producing this oily-rainbow effect.
布儒斯特条纹　两片同等厚度玻璃制成的双层隔热玻璃内的油膜彩虹效果（不是缺陷）。当入射光遇到来自其他表面的反射光，它们干涉并互相抵消而产生这种油膜彩虹效果。

bribery and corruption　To offer money, reward, gifts, or secret commission to influence a party to give preferential treatment where contractual or legal dealings preclude such influence. Secret dealings of this nature can lead to discharge of contract, damages, and criminal prosecution.
贿赂和腐败　提供金钱、报酬、礼物或秘密佣金给对方工作人员施加影响，致使对方给予优惠待遇，该影响在合同或法律交易中是不被允许的。这种性质的秘密交易会导致合同终止、损害赔偿和刑事诉讼。

brick　A ceramic material commonly used in construction of dwellings and municipal buildings, and produced in a regular block form for building walls. Bricks are usually made from clay and calcium silicate, although other types are also used, e. g. concrete. Bricks are porous materials with up to 25% free volume with compressive strengths up to 100 MPa (N/mm^2) with common UK dimensions of 215 mm × 102. 5 mm × 65 mm. A common house brick is normally within a range of 20 ～40 MPa.
砖　常用于建筑住宅和市政建筑，在建筑墙体中以规格砖形式出现的陶质材料。砖通常是由黏土和硅酸钙制成，虽然也有其他类型的砖，如混凝土。砖通常为多孔材料，有最多 25% 的孔隙率与高达 100 MPa（N/mm^2）的抗压强度，英国砖的常见尺寸为 215 mm × 102. 5mm × 65 mm。一个普通砌房用砖的强度通常是在 20 ～40MPa 的范围内。

brick-and-half wall　*See* ONE-AND-A-HALF-BRICK WALL.
一砖半墙　参考“一砖半墙”词条。

brick axe (bricklayer's hammer)　A small hammer with a double axe-type blade used for cutting and shaping bricks.
砖刀（瓦工锤）　一种用于切割和塑造砌砖形状的双边斧式小锤。

brick bat 1. A brick that has been shortened along its length to complete a bond. For example, a brick that had been cut in half along its length would be a **half-brick bat**. 2. Specially designed bricks that are built into a wall to attract bats. Also known as a **bat brick**.
1. 砍砖/半砖 沿着长边切短的砖，用于完成砖块的砌合。如一块砖沿长边被切成两半，即是半砖。**2. 蝙蝠砖** 特制的砌砖，贴在墙上吸引蝙蝠。也被称为 bat brick。

brick building 1. The process of constructing load-bearing brick walls. Also known as **brick construction**. 2. A space that has been enclosed using load-bearing brick walls.
1. 砌砖 建造承重砖墙的施工工艺，也被称为 brick construction。**2. 砖屋** 用承重砖墙封闭的空间结构。

brick elevator A device used for transporting construction materials, traditionally bricks, vertically up scaffolding.
砖块升降机 一种用于垂直运送建材（一般是砖块）到脚手架上的装置。

brickie Slang term used to describe a *bricklayer.
砖瓦汉 瓦工的俚称。

bricklayer Skilled craftsperson capable of laying bricks and blocks to line and level. A craftsperson skilled at laying bricks will be able to work to various brick bonds, e. g. *stretcher, *English, and *Flemish bond, and would be able to do both *common and *facing brickwork.
瓦工 能够整齐水平铺设砖块和砌块的熟练工匠。熟手的瓦工能够进行顺砖式砌法、英式砌法、梅花丁砌法等各种砌砖法施工，并同时能做裸砖墙和贴面砖。

bricklayer's labourer Site operative who works alongside a gang of bricklayers supplying them with bricks and mixed mortar. Mortar needs to be mixed consistently and bricks delivered to the place where the individual bricklayers are working. The labourer's job is demanding and requires a high level of strength and fitness to ensure that the team of bricklayers are constantly supplied with the materials they need. Occasionally, the labourer works alongside the bricklayer observing practice as part of their training for bricklaying - although this is not that common since the labourer's job is demanding, leaving little time to watch over the bricklayer.
瓦工小工 与一群瓦工共事并为其提供砖块和混合砂浆的现场技工。砂浆需搅拌均匀，而砖块则应递到各瓦工正在工作的地方。小工的工作需要较强的力量与体力，以便确保瓦工团队获得所需物料的稳定供应。有时，小工会与瓦工共事，观察瓦工的做法以便作为自身在砌筑方面的培训——虽然这并不常见（因为小工的工作要求使其几乎没有时间观看瓦工的工作）。

bricklayer's line (builder's line) A line of material placed between two points that is used as a guide during bricklaying or building work.
瓦工线 一条有形的线，置于两点之间，用作砌砖或建筑工作时的导向。

bricklayer's scaffold Tubular supporting structure used by bricklayers to access walls at height. Traditionally *putlog scaffolding was used by bricklayers. Putlogs have a flat end to a steel or aluminium tube, which is built into a wall as the wall is built, the other end of the tube is fixed to and carried by a *ledger, which is supported on a standard.
瓦工脚手架 瓦工用于接近高处墙壁的管式支撑结构。传统上瓦工所用的是单排脚手架。连墙杆有与钢管或铝管相连的一个平端，在筑墙时即已筑入，管子的另一端固定至一条横杆并由横杆承载，而横杆则由立杆支撑。

bricklaying The process of laying bricks in courses to construct a wall.
砌砖 按层铺设砖块建造墙壁的过程。

brick-on-edge A course of *headers laid on their edges.
斗砖/侧置砖/竖砌砖 一层沿着自身一条边向下铺设的丁砖。

brick skin (brick veneer) The non-load-bearing brick cladding on an external cavity wall.
外层墙 外部夹心墙上非承重的砖覆面。

brick slip A thin (e. g. 15 mm) non-load-bearing slice of brick used as a cladding material. Designed to give the appearance of traditional brickwork.
砖片 用作覆面材料的非承重薄砖片（如15毫米厚），用于模仿传统砖墙的外表。

brick tie *See* WALL TIE.
拉结件 参考“拉结件”词条。

brick trowel (bricklayer's trowel, laying trowel) A trowel with a flat triangular-shaped steel blade, one side of which is slightly rounded. Used in bricklaying to pick up and spread mortar.
三角砌砖泥刀 一种具有三角形钢叶片的抹子，一侧略呈圆形，在砌筑时用于刮起和涂抹砂浆。

brick truck A small trolley used for transporting bricks horizontally around a site.
运砖小推车 一种用于在现场周围水平运送砖块的小推车。

brick veneer *See* BRICK SKIN.
外层墙 参考“外层墙”词条。

brickwork chaser (keyway miller) *See* WALL CHASER.
砖墙开槽机 参考“砖墙开槽机”词条。

bridge 1. A structure that spans a gap. It is used to carry highways, railways, and people on foot over depressions in the ground. 2. The point where an element crosses a gap in the construction. For example, a bridge in a cavity wall can occur if mortar droppings accumulate on the wall ties and cross the gap between the inner and outer leaf of the cavity. *See also* THERMAL BRIDGE.
1. 桥 一种跨越间隔的结构，用于承载越过地面上凹陷处的公路、铁路和人行道。**2. 桥接** 建筑中某一构件跨越间隙之处。例如，在夹心墙中当落下的砂浆在墙拉结件上累积并跨过夹心墙的内叶和外叶墙之间的间隙时即形成桥接。另请参考“热桥”词条。

bridging Timber or metal members that are placed perpendicular or diagonally to the *bridging joists to brace the joists and spread the loads from above.
搁栅撑 与横搁栅垂直或成对角安置的木质或金属构件，用于支撑横搁栅并分散来自上方的负荷。

bridging joist A *joist that spans across an opening from one support to another. Also known as a **common joist**.
横搁栅 从一个支撑物跨越一个开口至另一个支撑物的龙骨，也称为**通用搁栅**。

bridle (Scotland bridling) *See* TRIMMING JOIST.
横筋龙骨（苏格兰 bridling） 参考“横筋龙骨”词条。

bridle joint A mortise-and-tenon joint where the mortise and the tenon are the full width of the pieces of timber that are being joined. Also known as a **slot mortise-and-tenon**.
开口贯通榫接 一种榫卯连接，其中榫眼和榫头的宽度与连接后的木构件等长。也称为**滑槽榫**。

brine A water-based liquid saturated with salt (NaCl) used for presevation and transport of heat.
饱和盐水 一种食盐（氯化钠）含量饱和的水溶液，用于保留和输送热量。

Brinell hardness test A mechanical test utilized to ascertain the hardness of a metal. The hardness is determined by measuring the indentation by a hard steel or carbide ball on the surface of the material.
布氏硬度测试 一种用于确定金属硬度的机械测试。通过计量硬钢或碳化钨球在材料表面上形成的压痕而确定硬度。

brise soleil Shade used above windows with vertical and horizontal slats that reduce glare and solar gain.
百叶窗 用在窗户上的遮光窗，具有可减少眩光和太阳辐射的垂直和水平板条。

British Board of Agrément A body that issues quality certificates for products and installers, which state that the item or organization has achieved compliance with set standards and has been independently assessed.
英国认证委员会 一个为产品和安装商颁发质量证书的机构，其证书中说明有关物品或组织已符合既定标准并受到独立评估。

British Standard (BS) These are documents defining a standardized method or procedure such that specifications are met during testing or manufacture. The documents are produced by the **British Standards Institution (BSI)**, the UK's national standards organization.
英国标准（BS） 英国标准是界定在测试或制造时是否已符合标准化方法或程序（例如规范）的文件，有关文件由英国国家标准组织——**英国标准协会（BSI）**制定。

British Thermal Unit (bTU, bthU) A unit of energy, defined as the amount of heat required to raise one pound of water through one degree of *Fahrenheit. Replaced by the *joule in the *metric system.
英热单位（bTU、bthU） 一种能量单位，定义为使一磅水升温一华氏度所需的热量。在公制单位中已被焦耳所取代。

brittle A material that can break or shatter easily.
易碎的 指某种物料容易被打破或打碎。

brittle crack propagation Failure mechanism in brittle materials where the crack expands at an accelerated rate characterized by low energy.
脆裂纹扩展 易碎材料的破坏机理，其中的裂纹沿最低能量路线加速扩大。

brittle fracture A form of failure exhibited by brittle materials. Brittle material fracture is mainly characterized by rapid crack propagation, low energies, and little or no plastic deformation.
脆性破坏 脆性材料所表现出的一种破坏形式。脆性材料破坏的主要特征为裂纹扩展迅速、沿最低能量路线以及塑性变形较少或没有。

brittleness Measure of how brittle a material is, characterized by little or no plastic deformation.
脆性 对材料是否易碎的计量，特征为塑性变形较少或没有。

brittle to ductile transition temperature (for metals) The temperature at which a ductile metal transforms to a brittle material. In metals, this transition occurs at 0.1 to 0.2 of the absolute melting temperature T_m, whilst in ceramics the transition occurs at about 0.5 to 0.7 T_m.
韧脆转变温度（适用于金属） 使韧性金属变为脆性材料的温度。对于金属而言，这种转变发生于处在绝对熔解温度 T_m的 0.1 至 0.2 时，而对陶瓷而言，转变则发生在 T_m的 0.5 至 0.7。

broad-crested weir (wide-crested weir) A weir in an open channel, with a crest length in the direction of flow to ensure that a critical depth of flow occurs somewhere along the crest.
宽顶堰 明渠中的一段堰，其在水流方向的顶宽将确保在堰顶某处出现临界水深。

broken bond (irregular bond) Brickwork where the bond has been interrupted; for example, by the insertion of a *brick bat.
断口砌合（不规则砌合） 砌合中断的砖砌；例如，因砍砖的插入而中断。

bronze A non-ferrous alloy based on copper (Cu) with tin (Sn) as the main alloying element.
青铜 以铜（Cu）为基础加入锡（Sn）作为主要合金元素的有色金属合金。

brook A stream, small river, or creek.
溪 细流、小河或小溪。

brown clause Contractual provision that sets out conditions for liquidated damages associated with prolongation costs.
棕色条款 合同条款，其中载列与误期成本有关的违约赔偿责任。

brownfield site A previously developed site, which has been abandoned or has stood idle, or an underused industrial site, where redevelopment is complicated by environmental contamination issues.
棕色地块 原来已开发但现被废弃或被闲置或未被充分利用的工业地块，如果对其的重新开发会因环境污染问题变困难。

Brundtland Report A report published in 1987 by the World Commission on Environment and Development, entitled 'Our common future'. The report addresses sustainable development issues and is known for its definition of sustainable development, i. e. 'development that meets the needs of the present without compromising the ability of future generations to meet their own needs.'

布伦特兰报告 由世界环境与发展委员会于1987年发表的一份报告，题为《我们共同的未来》。该报告阐述可持续发展问题，并以其对可持续发展的定义而闻名，即“既满足目前需要且不会影响后代满足其自身需求的能力的发展”。

B

brush hand A painter who has not completed his or her apprenticeship or is not properly trained.

油漆刷手 尚未结束其学徒身份或未经适当培训的油漆工。

brushing out Initial application of paint distributed evenly over the surface in order to achieve a smooth finish.

初刷 首次涂漆，使油漆均匀地分布于表面以形成光滑饰面。

brush seal A narrow strip of brush-like bristles that are fixed to the opening portion of a door or a letterbox as a weatherstrip.

密封条刷 刷子式的细鬃条，固定于门或邮筒的开口作为防水条。

BS *See* BRITISH STANDARDS.

BS 参考“英国标准”词条。

BSI (British Standards Institution) *See* BRITISH STANDARDS.

BSI（英国标准协会） 参考“英国标准”词条。

bucket-handle joint (concave joint, bar joint) A concave * pointing finish in brickwork formed using a buckle handle, a length of hosepipe, or a trowel handle. The pointing tool concaves and compacts the mortar, improving weather resistance.

半圆凹缝（凹缝） 用桶柄、一段软管或抹子手柄在砖砌上做成的一种凹陷勾缝。勾缝工具把砂浆压低和压紧，提高了耐候性。

buckling A sudden mode of failure, when a structural member experiences high compressive stresses and the structure moves out of the line of the action of the load.

屈曲 一种突发破坏模式，即结构件经受高压应力且结构偏出荷载作用线。

budget price Estimate of the cost of work, and although not precise, is considered to cover the cost of all the contractors' work.

预算价格 对工程成本的估算，虽然不精确，但被视为已涵盖一切承包工程的成本。

buffer 1. A shock absorber, e. g. a railway buffer. 2. A solution that resists changes in pH when a small quantity of acid or alkali is added to the solution. 3. A temporary data storage area on a computer, where data is held before being transmitted to an external device such as a printer.

1. **缓冲器** 冲击吸收器，例如铁路缓冲器。2. **缓冲液** 一种溶液，当少量的酸或碱被加入其中时，该溶液的pH值不受改变。3. **缓冲** 电脑中的临时数据储存区，在被传输至打印机等外部设备之前，数据被保留在此。

buggy (US) *See* MOTORIZED BARROW.

巴吉车（美国） 参考“电动手推车”词条。

buildability The ease with which a design can be built. Some ideas and designs are not considered practical to build and would rate very low on a 'buildability' scale.

可建造性 设计可被建成的难易程度。某些想法和设计被视为不可被实际建成，则在“可建造性”等级上将被排至极低的位置。

builder 1. Contractor who undertakes building work. *See* BUILDING CONTRACTOR. 2. Craftsperson who engages in any building activity. Generally, the term is interchangeable referring to any company or operative that engages in the site-based activities.

1. **建筑商** 承担建筑工程的承包商，参考“建筑承包商”。2. **建筑工** 从事任何建造活动的工匠。一般而言，这个词在指从事现场活动的任何公司或技工时是可以互换使用的。

builder's rubbish (building rubbish) Waste material on-site originating from the process of building.

建筑垃圾 源自建筑过程的现场废料。

builder's work Work that needs to be carried out by a trade or skilled operative, organized by the contractor (builder), in order to allow a subcontractor to perform their part of the contract. An example of such work would include the construction of peers for beams or fittings to rest on or the breaking out of holes in walls and floors for the service contractor to feed its services through. The term is used in contractual documentation to describe the activities and work that the contractor must organize and complete before subcontractors or other parties can complete their work.
建筑商工程 需由专业或熟练技工承担，并由承包商（建筑商）组织分包商履行各自合同部分的工程。此类工程的实例包括为安放的梁或固件的建造在墙壁和楼面上开孔供承包商将设施送入。该词语被用于合同文件中以便说明在分包商或其他方完成其工程之前，须由承包商组织和完成的活动和工程。

builder's work drawing Illustration or detailed plan that clearly shows or highlights the *builders work. The drawing is used as a coordination tool between builder and subcontractor to ensure all necessary work is considered.
建筑商施工图 清楚显示或凸显建筑商工程的图示或详细平面图，该图纸被用作建筑商和分包商之间的协调工具，以便确保一切必要工程均被纳入考虑。

builder's work in connection Work that needs to be done to allow other trades or subcontractors to do their work. The builder's work must be complete to allow other connecting or associated works to commence.
前序工程 为使其他专业或分包商得以开展其工作而需要先完成的工程。该工程完成后，其他连接或相关工程才能开始的建筑商工程。

building A structure used to provide shelter to humans. The **building envelope** is the physical separator between the conditioned and unconditioned environment of a building including the resistance to air, water, heat, light, and noise transfer.
建筑物 用于为人类提供庇护的结构物。**建筑物围护**是分隔建筑物室内与室外环境的物理屏障，包括空气、水、热、光和噪音的传播。

Building Act 1984 Legislation that governs and provides the vehicle for the implementation of the Building Regulations.
1984年建筑法 制订建筑条例的依据，为建筑条例提供实施手段的法律。

Building Centre An organization set up to provide information to the construction industry.
建筑中心 为了向建筑业提供信息而成立的组织。

building code Set of guides or legislation that act as a minimum standard for construction or safety - in the UK these would be the *Building Regulations enforced by *building control.
建筑规范 作为适用于建筑或安全的最低标准的成套指引或法律——在英国即为通过建筑管控强制实施的建筑条例。

building contractor Organization that undertakes the management of the sitebased works and building construction. Traditionally, the organization would enter into a contract to construct the building only; it is now quite common for the *main contractor to enter into a *contract for both the design and construction of the building. Although the main contractor can enter into a contract to design the project, often much of the works are *subcontracted out to a design specialist, e. g. architects, structural engineers, and mechanical and electrical designers. Nowadays, main contractors have a limited pool of skilled and unskilled labour, and tend to subcontract a large portion of the actual building works. It is common for the main contractor to package ground works, steel frame fabrication and erection, wall cladding, roof covering, masonry, floors, and mechanical and electrical works into subcontracts.
建筑承包商 承担对现场工程和建筑施工的管理的组织。传统上，该组织订立的仅为施工合同；但现在则一般是由总承包商订立一份关于建筑物的设计和施工的合同。虽然总承包商可订立一份与项目的设计有关的合同，但较常见的情况是把该工作分包给设计专业人员，例如建筑师、结构工程师及机电设计师。现在，总承包商拥有的熟练和非熟练工人的数量有限，故趋向于把大部分的实际建筑工程分包出去。普遍的情况是总承包商把土方工程、钢架生产和搭建、外墙覆面、屋顶覆盖层、砖石砌筑、楼面及机电工程外包给分包商。

building control Enforced legislative procedures and processes that govern the minimum standards of building. Applications are made and designs submitted to the building control office to demonstrate and check compliance with legislation. Works are checked by the * building control officer at significant stages in the building process to ensure that the building is being constructed and is consistent with the approved drawings and plans.
建筑管控 对建筑最低标准进行监管的强制执法程序和过程。向建筑管控机构提出申请并递交设计，以证明及由其核查对法律的遵守情况。建筑管控官员在建设过程中的主要阶段对工程进行核查，以便确保建筑物是按照经过批准的图纸和规划建造。

building control officer Professionals employed by the local authority to inspect and check that buildings are compliant with the relevant legislation and the drawings that were submitted and approved by * building control. The officer ensures compliance with fire, safety, structural, thermal, acoustic, and environmental legislation. The term originates from the London Local Authority where principal building control officers were employed to inspect all buildings constructed in the area.
建筑管控官员 由地方当局聘用对建筑物是否遵守有关法律及图纸（已递交建筑管控机构并获得批准）的情况进行检查及核查的专业人员。该官员将确保对消防、安全、结构、热学、声学及环境等方面法律的遵守。该词语源自英国伦敦地方当局，该地聘用主要建筑管控官员对建于区域内的所有建筑物进行核查。

building craft A broad description of any one of or all of the skilled work or trade professionals associated with a construction project or industry.
建筑工艺 对于建筑项目或行业相关的任意一项或全部的熟练工作或行业专才的概括性的描述。

building industry All of the * contractors, * subcontractors, fabricators, materials suppliers, * plant and equipment companies, * labourers, and * professionals associated with * construction and * civil engineering works. The business and work directly associated with construction represents around 8% – 9% of * GDP (Gross Domestic Product) in the UK.
建筑业 所有承包商、分包商、制造商、材料供应商、机器和设备公司、劳工及与建筑和土木工程有关的专业人员。与建筑直接相关的业务和工作约占英国 GDP（国内生产总值）的 8%～9%。

building inspector Professionals who inspect the construction to ensure that works are properly constructed; can be an employee of the * local authority, bank, building society, or insurance company who checks that the building is compliant with legislation and regulations. Inspectors for finance organizations may inspect to assess the value of the property. Local authority inspectors may be called * building control officers although the origins of both terms are slightly different.
建筑检查员 对建筑进行检查以确保对工程的妥善施工的专业人员；可能是地方当局、银行、建房协会或保险公司的雇员，其对建筑是否符合法律和条例进行核查。金融组织的检查员可能进行检查以便评估房产的价值。虽然检查员和建筑管控官员这两个词的起源略有差异，但地方当局的检查员可被称为建筑管控官员。

building management system (bms) (building automation system, intelligent building) Automated central control unit that checks, monitors, adjusts the building's performance. Used to coordinate and operate the lighting, air conditioning, heating, lifts, security, and other building appliances and facilities.
建筑物管理系统（bms）（建筑物自动化系统、智能建筑物） 对建筑物的功能进行检查、监控、调整的自动化中央控制装置。用于协调和操作照明、空调、供暖、电梯、安保及其他建筑物的设备和设施。

building notice Notification to the local authority that building work of a particular nature is about to commence, e. g. excavation, or drainage, or foundations. This allows the * building control department to inspect and check the works for compliance with approved drawings and the * building regulations.
建筑通知 向地方当局发出的通知，其内容是关于某种具体性质的建筑工程即将开始，例如挖掘或排水或基础。该通知使建筑管控部门得以检查及核查有关工程是否符合经批准的图纸和建筑条例。

building operative Site-based person who undertakes the skilled or unskilled work involved in the actual construction of the building; does not include construction professionals.
建筑技工 承担与建筑物的实际建筑工作有关的熟练或非熟练的现场人员；不包括建筑专业人员。

building owner A *client who takes ownership of the building once it is completed.
建筑物业主 一旦建筑物竣工，则将获得该建筑物拥有权的业主。

building paper 1. Reinforced fabric, which is laid under concrete and used to prevent chemical attack from the ground and assist curing. 2. A breathable fabric membrane, which is fixed over timber walling panels and below roof tiles to allow water vapour to escape from the structure but prevent rain entering from the external environment. The paper helps to prevent condensation. 3. A waterproof sheathing paper, the degree of waterproofing varies depending on purpose.
1. 建筑防潮纸 被铺设在混凝土之下的加固织物纸，用于防止来自地面的化学侵蚀，有助于建筑物养护。**2. 防潮膜** 一种透气织物薄膜，被固定于木壁板上和屋顶瓦下面，可使水蒸气从建筑物中逸出但能阻止来自外部环境的雨水。这种膜有助于防止冷凝。**3. 柏油纸** 一种防水油纸，其防水级别取决于用途而有所差别。

building permit Authorization given by the local authority to allow works to commence.
建筑许可 由地方当局给予的允许工程开工的授权。

building regulations The approved documents and guides implemented under the *Building Act 1984 that enforce minimum standards of construction, safety, and performance of buildings. The legislation is enforced by *building control.
建筑条例 根据 1984 年英国建筑法而实施和经批准的文件与指引，强制规定了建筑、安全及建筑物性能的最低标准。该法规由建筑管控机构强制执行。

Building Research Establishment (BRE) An independent consultancy that specializes in undertaking work in the built environment.
英国建筑研究院（BRE） 一个独立咨询机构，专长于承担建成环境方面的工作。

building society Financial house, with very similar business activities to high-street banks.
建房协会 金融商行，拥有与主流银行非常类似的业务活动。

building surveyor Construction professional who provides advice and reports on the condition of houses or commercial buildings; the surveyor may also comment on the value of the house. The surveying professional will tend to specialize but can give advice on a range of issues including building condition, defects, restoration, renovation, fire protection, structural integrity, thermal performance, etc. It is normal for a survey to be requested by clients and financial institutions prior to the exchange of contracts for a building's sale. Surveys are particularly important in old or previously owned properties that are not covered by a warranty.
建筑测量员 就房屋或商业建筑物的状况提供建议和报告的建筑专业人员；测量员也可对房屋的价值提供报告。测量专业人员一般是专长于某一专业，但也可对包括建筑的状况、缺陷、修复、翻新、消防、结构完整性、热性能等在内的一系列专业问题提出建议。业主和金融机构在为买卖建筑物签订合同之前通常会聘请建筑测量员。对于古旧或很久之前拥有且无质保保障的房产而言，建筑测量员尤其重要。

building system The method of producing buildings. Buildings that are produced using modern methods of construction (MMC), with significant portions of the buildings being *prefabricated, may be classed as industrialized buildings.
建筑系统 建造建筑物的方法。使用现代建筑方法（MMC）制成的建筑物中有较多的部分是预制的，可被归类为工业化建筑。

building team Group of professionals associated with the design and manufacture of the whole process. As a minimum, the building team normally includes the client's representative, * architect, * contractor, * specialist contractors, and designers.

建筑团队 与整个过程的设计和制造有关的专业人员的团队。建筑团队一般至少包括业主的代表、建筑师、承包商、专业承包商及设计师。

building trades All skilled operatives and practices associated with building, including * joiners, * bricklayers, * ground workers, * steel fixers, etc.

建筑专业 与建筑有关的所有熟练技工和实务，包括细木工、瓦工、地基工、钢筋工等。

building-up of prices The gathering together of projected costs to produce an estimate for the total project.

价格归总 把预计成本归集到一起；计算出对整个项目的成本估算。

building work Work carried out during building.

建筑工程 在建筑时开展的工程。

Buildmark A 10-year warranty and insurance cover provided by the NHBC for newly built houses that have been inspected by their approved officers.

建筑标志 由英国全国住房建筑理事会（NHBC）为经其认可的职员检查合格的新建房屋提供的 10 年期质保和保险保障的标志。

built environment Man-made structures and facilities used to accommodate societies' activities. Any enclosures, spaces, structures, and infrastructure formed to convert the natural environment into a habitable and useable area for the purpose of living, working, and playing.

建成环境 用于容纳人类社会活动的人造结构物和设施。为使天然环境变为可供生活、工作及玩乐的宜居和可用的区域而建造所有围蔽、空间、结构及基础设施。

built-in (building-in) The process of inserting components into the construction as it progresses.

内置 在建设进程中把构件植入的过程。

built-up To make something that consists of different components.

组装 制成由不同构件组成的某物。

built-up roofing A flat-roof covering comprising multiple layers of sheet materials.

多层组装屋顶 由多层板材组成的平屋顶。

bulb A rounded spherical shape, such as a light bulb.

球泡 圆球形物体，例如灯泡。

bulk A large mass of something, e. g. a **bulk excavation** consists of removing a large mass of soil from the ground.

大宗/大量 较大数量的某物，例如**大规模开挖**即为从地面移去大量土壤。

bulk density The weight of a material per unit volume, which includes water, air voids, and solids.

体积密度 每单位容量的材料重量，包括水、气穴及固体。

bulkhead A dividing partition or barrier, e. g. a wall within a ship or aircraft; a seawall; a sheet pile.

隔板 分隔式的间壁或屏障，例如船舶或飞机中的壁，海堤，板桩。

bulking An increase in volume of a material (e. g. soil) when it is excavated from the ground.

膨胀 某种物料（例如土壤）被从地里挖出时其体积增大的情况。

bulk modulus (K) An elastic constant, defining the relationship between an increase in pressure with an associated decrease in volume.

体积模量（K） 一个弹性常数，定义了压力的增长和体积的相关减少之间的关系。

bulldog grip A rope clamp consisting of a U-shaped bar threaded at both ends.

钢丝绳夹 一种绳夹，其中包括一个两端有螺纹的 U 形杆。

bulldog plate connector *See* TOOTHPLATE CONNECTOR.
圆盘齿板连接件 参考“齿板连接件”词条。

bulldozer (dozer) A tracked earth-moving piece of equipment with an adjustable *mouldboard attached to the front.
推土机 一种履带式推土设备，该设备前部装有可调整的铲刀。

bullet catch *See* BALL CATCH.
门碰珠 参考“门碰珠”词条。

bullet-head nail (US finish nail) A nail with a small rounded head.
无头钉（美国 finish nail） 一种带有小圆头的钉子。

bullhead connector A T-shaped pipe fitting, where the branch is larger than the other two run openings.
异径三通接头 一种 T 形管件，其支管直径大于其他两个流向的开口。

bunched wires (bundled wires) Wires that have been manufactured for *trunking.
绞合线（束线） 为用于线槽而制造的线缆。

bund *See* EMBANKMENT.
堤岸 参考“路堤”词条。

bungalow A single-storey structure in which people live.
平房 供人居住的单层结构物。

bunker 1. A container used for storing materials, such as cement, sand, and aggregate. 2. An underground shelter.
1. 大柜子/大箱子 用于储存水泥、砂及骨料等材料的容器。**2. 地堡** 地下的掩体。

buoyancy The ability of an object to float. *See also* ARCHIMEDES' PRINCIPLE.
浮力 物体漂浮的能力。另请参考“阿基米德定律”词条。

bur A drill bit used for widening drilled holes rather than forming them.
扩孔钻头 用于使钻孔扩大而不是用于钻孔的钻头。

burden of a contract A party's obligation to perform, execute, and complete the contract. A party cannot assign (give its contractual obligation to another party) without the consent of the other contracted party.
合同义务 签约方履行、完成合同的义务。一方未经另一签约方的同意不得转让合同义务（将其合同义务交予其他方）。

burglar alarm A mechanical device comprising sensors and a siren designed to warn off intruders who enter a protected area.
防盗报警器 一种机械装置，由传感器和报警器组成，旨在警告进入受保护区域的入侵者退出。

burglar-resistant lock *See* SECURITY LOCK.
防盗锁 参考“安全锁”词条。

buried services (underground services) Gas, water, sewers, electric cables, etc., which are located (buried) *below ground level.
地下设施 位于（埋设于）地平面以下的燃气、水、下水道、电缆等。

burner A place where fuel is ignited and burns, e. g. a gas ring on a hob.
燃烧器 燃料被点燃并燃烧的设备，例如炉盘上的圆形煤气炉头。

burning off To remove a surface covering with heat, e. g. an old coat of paint can be burnt off with a blowlamp.
烧除 以热量清除某个表面的覆盖物。例如，旧的漆层即可用喷灯烧除。

B

burnt clay A *ceramic material.
烧结黏土 一种陶质材料。

burnt lime *See* QUICKLIME.
煅石灰 参考“生石灰”词条。

burring reamer A tool to remove jagged edges (caused by cutting) in pipes.
去毛刺铰刀/修边器 一种用于清除管道内（因切割所致）锯齿边缘的工具。

burst strength 1. The hydraulic pressure needed to burst a container of given thickness. 2. The pressure needed to break a fabric, by either inflating a diaphragm or forcing a smooth, spherical object against a clamped piece of fabric.
1. 爆裂强度 使给定厚度的容器爆裂所需的液体压力。**2. 胀破强度** 通过把薄膜吹胀或把一个光滑球体压入一块被夹紧的某种织物使其破裂所需的压力。

bus bar (busbar) A conductive bar, usually copper or aluminium, that enables a connection to be made between two or more electrical circuits. Used in consumer units, distribution boards, switchboards, and substations. Bus bars are connected together using a **bus coupler** and are usually enclosed within a **bus duct** or **busway**.
母线 一种导电排，通常为铜质或铝质，用于使两个或多个电路形成连接。母线通常用在配电箱、配电柜、开关柜及变电站中。母线用**母线连接器**连在一起且一般被包裹在**母线管**或**母线槽**中。

bush hammer A hand-held device that is either connected to compressed air or an electrical supply. Used to remove the surface layer of concrete to give an exposed aggregate finish.
凿毛机 一种与压缩空气或电源相连的手持式工具，用于除去表层的混凝土以便使骨料面外露。

bushing 1. A pipe fitting that contains both an internal and an external threaded connection. Used to join different diameters of pipe together. 2. An insulated sleeve used at openings to pass conductors through.
1. 内外螺纹接头 一种兼有内部和外部螺纹连接的管件。用于连接不同直径的管道。**2. 套管** 一种用于开口处的绝缘套管，以便使导体穿过。

business process re-engineering The redesign of work behaviour, processes, systems, and structure so that greater efficiency and value can be achieved with the resources available.
业务流程再造 对工作行为、过程、系统及结构的重新设计，以便用现有资源达到更高的效率和价值。

butt The process of placing components together without overlapping. The joint that is formed is known as a **butt joint**.
对接 把构件无重叠地拼合到一起的过程，如此形成的接缝即称为**对接接缝**。

butter coat A soft coat of render used to achieve a **pebble dash** finish. The render is applied to the wall over the *scratch coat and decorative aggregate is thrown and then pressed into the render.
黏结砂浆层 用于形成**干粘石**饰面的软灰浆层。在墙壁的底涂层涂刷灰浆，将装饰石碴撒上，并压入灰浆中。

butterfly gate Similar to a *sluice gate, however, the gate opens and closes on a central shaft located in the pipe, to save on the need for headroom to open the gate. A **butterfly valve** turns on the central shaft within the pipe to regulate flow.
蝶形闸 蝶形闸与水闸类似，但蝶形闸是通过位于管道内的中心轴进行启闭，以便留出开启闸门所需的净空。**蝶形阀**通过管道内中心轴的转动来调节流量。

butterfly wall tie A galvanized steel wire *wall tie in the shape of a figure eight.
蝶形墙拉结 8字形的镀锌钢丝拉结件。

buttering (buttering up) The process of spreading mortar on a material as one would spread butter on

bread.
抹灰 把灰浆涂在某种材料上的过程，类似于人在面包上涂黄油。

butt gauge (butt gage) A tool used for marking the position of hinges on doors and door jambs.
铰链画线规 一种用在门和门边梃上标记铰链位置的工具。

butt hinge A hinge comprising two leafs that are mortised into the door and the door jamb, and fold together when the door is closed. It is the most common type of hinge used for doors.
普通铰链 一种由嵌入门和门边梃的两个铰链页构成的铰链，在门闭合时会折叠在一起，是用于门的最常见铰链。

button *See* PUSH BUTTON.
按钮 参考“按钮”词条。

button-headed screw (half-round screw) A screw with a semi-circular head.
圆头螺钉（半圆头螺钉） 带有半圆形钉头的螺钉。

buttress A stone or brick structure built against a wall to support it, i. e. to resist active thrust.
扶壁 与墙壁相对而建的石质或砖砌结构，用于支撑墙壁，即抵御主动推力。

butyl rubber A polymeric material also known as polyisobutylene and PIB $(C_4H_8)_n$. Butyl rubber is a synthetic polymer used as a sealant in roofs, particularly for roof repair work.
丁基橡胶 一种聚合材料，也称为聚异丁烯和 PIB $(C_4H_8)_n$。丁基橡胶是一种合成聚合物，可用作屋顶密封剂，尤其是用于屋顶维修工程。

buzzer An electrical device that emits a buzzing sound when activated.
蜂鸣器 一种电子装置，在启动后会发出嗡嗡的声音。

buzz saw A circular saw.
圆锯 圆形的锯。

byatt A timber beam that carries a walkway in excavation.
基坑步道梁 承载基坑中的人行道的木梁。

bye channel A *spillway.
溢水道 溢洪道。

bypass To go around something, e. g. a bypass road with a route around a town rather than through it; a **bypass channel** diverts water.
迂回 绕过某物，例如围绕某个城镇的迂回线路，但不会直接穿过该城镇；**分水槽**使水流改向。

by-product A secondary product that is formed as a result of making something else.
副产品 因制造其他某物而生成的次要产品。

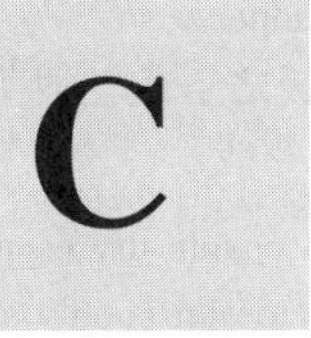

cabin Temporary site accommodation for offices, storage, and welfare facilities. The standard of accommodation can vary from something that is secure and provides cover from the rain, to offices that are fully fitted out with furniture, heated, air-conditioned, and insulated. Due to variation in levels that can be experienced on sites, most units are fitted with adjustable legs. Smaller units may be fitted with wheels enabling transportation without the need of a crane.

活动板房 可作建筑工地临时办公室、仓库和生活设施。活动板房安全可靠，屋顶可以避雨，室内还配备了各种家具、暖通空调系统和保温设施。因为工地不同位置的标高可能有差异，因此大部分活动板房都装有伸缩型支腿。小型活动板房则装有轮子以方便运输，而无须再借助吊车。

cable 1. A conductor of electricity usually made from copper and sheathed with insulation. 2. Steel rope used in winches and pulleys to lift and carry equipment and materials.

1. 电缆 由通常为铜导线和外包绝缘保护层制成的电导体。**2. 钢丝绳** 卷扬机和皮带轮上的钢丝绳，可用于提升设备和材料。

cable clip A plastic or metal fixing bracket that holds and secures electrical cables. The fixing is held in the wall or ceiling by a nail or screw.

电缆夹 用于固定电缆的一种塑料或金属固定架。可以用钉子或螺丝把电缆夹固定到墙上或天花板上。

cable duct (duct pipe, pipe duct, ducting) Tube or conduit which is mounted on surfaces or sunk into the ground, floors, ceilings, or walls and is used to carry service cables. The conduit is usually made from plastic or steel providing a robust containment to carry the services. The duct is laid as construction takes place and a draw cord is placed in the pipes as they are positioned. When needed, electrical cables can be pulled into the ducting with the draw cord. The duct serves a similar purpose to a *conduit.

电缆槽盒 明装或暗装在地下、地板、天花板或墙上，用于敷设电缆的线管或线槽。线槽通常为塑料或钢材制成的，外壳坚固、耐用。槽盒随建筑过程中安装，固定好后便可敷设电缆牵引绳。在需要时便用电缆牵引绳牵引电缆安装放线。槽盒的功能跟线槽类似。

cable gland A bulb of material used to seal the end of a cable or seal where the cable enters a casing, and prevents water, dust, and dirt penetrating into the cable or duct.

电缆接头/压盖 用于锁紧和固定电缆终端或进出线的球形材料，可起到防水防尘的作用。

cable sleeve A protective coating around the wire. The coating may be armoured to provide mechanical protection, or have a polymer-based coating to protect from the environmental conditions.

电缆编织层 电缆周围的防护层。可以铠装上去以增加机械保护，或加一层聚合物保护层，提高电缆防侵蚀能力。

cable tail 1. Additional cable left at the end of a run to ensure that there is sufficient material to make the connection to the fitting or equipment. 2. A length of electrical cable attached to a device to enable it to be wired in.

1. 电缆尾线 线路末端预留的额外一定长度的电缆，以确保电缆可以与装置或设备连接。**2. 电缆引线** 从设备引出连接设备到线路中的一截电线。

cable tray A long open container within which multiple services can be neatly positioned. Cable trays are often used below raised floors and above suspended ceilings to keep electrical and telecommunication cables together. The trays help to distribute the wires throughout the building in an organized manner.

电缆桥架 将多根电缆规范整理的长形无盖容器。电缆桥架通常敷设于架空地板之下或轻钢龙骨吊顶之上，用于汇集电力和通信电缆。电缆桥架的使用将有助于整栋建筑的标准化布线。

caisson A watertight structure (chamber) that is sunk through the ground or water to enable dry excavation and placing of the foundation. An **open-caisson** is open at the top and bottom. A **box-caisson** is only open at the top. A **compressed-air or pneumatic caisson contains** a chamber where air pressure is maintained above atmospheric pressure to prevent entry of water or mud.

沉井/沉箱 一种下沉到地底下或水下的一个挡水构筑物（箱形），可在井内挖土并浇筑基础。**开口沉井**是无底无盖的。**无压沉箱**是有底无盖的。**压气或气压沉箱**工作室的气压应维持在大气压之上，避免水或泥浆进入工作室。

calcination Heating a substance below its fusing and melting point, which leads to a loss of water or oxygen, resulting in a simpler form of the substance. Calcination is used in the extraction of metals from mineral ores.
煅烧 把某种物质在低于其熔点温度下加热，使其脱去水或氧，形成简单组分的物质。煅烧被用于从矿石中提取金属。

calcium A silvery, moderately hard metallic element, symbol Ca. Calcium is useful as an alloying element; the addition of calcium to steels result in improved mechanical properties. Calcium compounds are used in the production of Portland cement, plaster, and quicklime.
钙 一种银色、中等硬度的金属元素，符号为 Ca。钙作为合金元素用途较大；在钢里加入钙能使其机械性能提高。钙化合物可用于生产波特兰水泥、灰泥及生石灰。

calibre 1. The ability or quality of someone. 2. The internal diameter of a pipe or cylinder, particularly with reference to the barrel of a firearm.
1. 才干/素质 人的才干或素质。**2. 口径/内径** 管道或圆筒的内径，尤其是指手枪枪管的口径。

California bearing ratio test (CBR test) An empirical test, devised by the California Division of Highways in 1929, and used in pavement design to determine the strength of the lower layers in roads (*subgrade and *basecourse) and airfields. The ratio of force and penetration is monitored when a cylindrical plunger (49.6 mm diameter) is pushed into soil at a constant rate (1 mm per minute). At given values of penetration the ratio of applied force to standard force is expressed as a percentage. The higher of these two percentages is taken as the CBR value.
加州承载比测试（CBR 测试） 一种由加利福尼亚公路局于 1929 年发明，用于路面设计的实证测试，以便确定道路底层（路基和基层）及机场的强度。测试指标当以恒定速率（1mm/min）把一个圆柱（直径 49.6 毫米）压入土壤时所观测到的力与贯入程度的比例。所用力及标准力与给定贯入值的比例以百分比表示，这两个百分比中较高的一个即被取为 CBR 值。

calliper An instrument consisting of two hinged legs, which either turn out or in, to measure the internal or external dimensions of an object, e.g. the bore or diameter of a pipe.
卡尺 一种两脚可张合的量具，用于测量工件的内外径，如管道的孔径。

calorific value (CV) The amount of energy produced by a fuel that has undergone complete combustion, usually measured in J/kg for a solids or MJ/m^3 for gas.
热值（CV） 一种燃料在完全燃烧情况下产生的能量值，通常对固体以焦/千克为单位计量，对气体以兆焦/立方米为单位计量。

calorifier A vessel used for heating low-pressure hot water. The water within the vessel is heated indirectly by a coil immersed within the water. A domestic *indirect hot water cylinder is an example of a small storage calorifier.
水加热器 一种用于加热低压热水的容器。水箱里边的水是通过浸在水里的加热盘管直接加热的。家用间接热水箱就是标准的小型水加热器。

calyx boring (short drilling) A *boring method that uses steel shot as the cutting means.
钻粒钻进 使用钢粒破碎岩石的钻进方法。

camber A slight upward curvature of a surface, such as a cross-section of a highway, or in structural members (e.g. **camber beams**) to resist dead and imposed loads, such that the beam sits flat and prevents *sagging in service.
起拱 平面轻微向上弯曲，如高速公路或结构构件（如：拱形梁）的横截面，以承受恒载和外加荷载并保持平直，避免在使用中下垂。

canal An artificial waterway constructed for navigation, water power, or irrigation. Nowadays, canals are often used for recreational purposes.
运河 作航运、水力发电或灌溉用的人工水道。现今，运河常用作休闲用途。

candela (cd) The international standard (SI) unit of luminous intensity.
坎德拉（坎） 发光强度的国际单位制（SI）单位。

canopy 1. A roof over an open structure (one that is not enclosed by walls). 2. A hood above a cooker or fire used to direct steam or smoke to a chimney or flue. Sometimes hoods can be decorative and have no function other than aesthetics.
1. 雨棚 设在无外墙围护的构筑物上方的一种屋顶。**2. 排烟罩** 炊具或炉灶上方的烟罩，可把烟气导流到烟囱或烟道。有时候排烟罩只是起到装饰的作用，并无排烟功能。

C

cant (canted) 1. Any flat surface that has a slight incline or tilt that is not at right angles to the main plane. 2. External surface of brickwork that has a change of direction that is not a right angle. Bay windows often have brickwork that angles out to support the window frame. *See* CANT BAY.
1. 斜面 有一定倾斜度的平面，与水平方向不成直角。**2. 斜角砖** 外墙非直角的转角砖。通常，飘窗就有向外凸出的斜角砖支撑窗框。参考"斜角飘窗"词条。

cant bay The angled profile of brickwork used to form a bay window, *See also* CANT.
斜角飘窗 用于砌筑飘窗的异型砖，参考"斜角砖"词条。

cant brick A non-standard (special) brick with a slope or angle cut across the head of the brick. Such bricks are sometimes used to change the direction of the brickwork.
斜面砖 斜切或角切而成的非标准砖（异型砖）。这种砖有时用于转角处。

cantilever A projection, such as a beam, which is only supported or fixed at one end, the other end is free.
悬臂 一突出物，例如梁，一端被支撑或固定，另一端向外自由悬出。

cantledge *See* KENTLEDGE.
平衡块 参考"压载物"词条。

cant stop brick A specially trimmed brick used to neatly finish off the end of a wall where a *cant brick has been used. The brick is cut so that the end profile matches that of the cant brick to which it abuts.
斜面收口砖 一种用于墙头收口的异型砖。砖的轮廓与相邻斜面砖相匹配。

cap 1. The uppermost point, head or top component of a member. 2. Fixing that is pushed or screwed onto a container or vessel to close or seal the container. 3. A cover placed over the top of an object.
1.（建筑物）顶部构件 某一构筑物的最高点、头或顶部构件。**2. 容器盖** 可通过按压或旋转来密封容器的盖子。**3. 顶盖** 覆盖某物顶部的盖子。

capillarity (capillary action) The action of fluid (usually water in buildings) being drawn up or down pores or closely positioned surfaces due to the surface tension of a fluid. A **capillary break** is a gap between two surfaces that is large enough to prevent capillary action. **Capillary groove** is a rebate cut into a material to produce a surface that prevents capillary action. A **capillary joint** is a pipe joint formed by placing steel or copper pipe into a slightly larger diameter fitting that encourages solder or fixing solvent to be drawn between the tubes by capillary action. The contact surfaces of the pipe and fitting are cleaned with a flux ready to be soldered. The gap between the pipe and fitting is sufficiently close for capillary action to take place when solder is applied. When heated and molten solder is applied, the surface tension of the solder draws the fluid into the space, evenly distributing the solder and filling the joint. Once the solder solidifies it provides a watertight joint. Other adhesives are also available that work on capillary action. Plastic pipes can be joined using solvent-based adhesives, which also flow under capillary action.
毛细现象（毛细管作用） 是指液体（通常为建筑物中的水）在表面张力作用下，在毛细孔中被吸上或吸下或紧贴于表面的现象。**毛细缺口**是两个平面之间大得足以防止毛细作用的间隙。**毛细槽**是材料上的槽切口，用于形成可防止毛细作用的表面。**毛细接头**是通过把钢管或铜管插入直径略大的管件而形成的管接头，通过毛细作用促使焊料或固定溶剂渗入管间。管子和管件的接触面用焊剂清洁，以备焊接。管子和管件之间的空隙足够紧密，以便在涂焊料时发生毛细作用。当所涂的焊料被加热和熔化时，焊料的表面张力将把液体吸入缝隙中，使焊料均匀分布并填满接缝。一旦焊料固化，即形成防水接缝。其他黏接剂也适用于此类毛细作用施工。可用于把塑料管接合的溶剂型黏结剂的流动也遵循毛细作用。

capping layer Low-cost granular material used to provide a working platform when the *CBR of the *subgrade is weak (<5%). Capping is usually required to achieve a CBR of 15%.
路基改善层 当路基的 CBR 值（加州承载比）较弱（<5%）时，应铺设低成本的颗粒材料作为路基改善层。通常路基改善层的 CBR 值应达到 15%。

capstan A mechanical vertical rotating device, used to move heavy objects by employing ropes, cables, or chains, for example, a rotating drum located on a ship or shipyard to haul in ropes.
起锚绞盘 通过用绳索、缆或锚链来移动重物的立式机械转轮装置，如安装在船舶上或造船厂用于拖缆的绞缆筒。

carbonation A chemical process occuring when carbon dioxide is dissolved in water or an aqueous solution. This can be harmful for many materials used in constrution as carbonation usually causes a decrease in strength.
碳酸化作用 当二氧化碳溶解于水或水性溶液时发生的化学过程，这对很多用于建筑材料有害，因为碳酸饱和通常造成强度降低。

carbon steel A type of steel whose mechanical properties are dependent on the carbon content. The majority of steels contain between 0.1～1.0 weight % carbon and the steels are termed low (0.1～0.25 weight % carbon), medium (0.25～0.55), and high carbon (greater than 0.55) steels in accordance to the carbon content. Carbon has a very marked effect on the properties of steel; an addition of only a fraction of 1.0%, has a very great effect. Increasing the carbon content increases the *tensile strength, *yield strength and *hardness of the steel; however, increasing carbon content also results in decreased ductility. As ductility is an essential property for structural steel (for protection against earthquakes, explosions, and strong winds), most constructional steels contain between 0.15% and 0.4% of carbon.
碳钢 机械性能取决于碳含量的一种钢。大多数钢材碳含量介于重量的 0.1%～1.0%，且根据碳含量高低被称为低碳钢（碳含量占重量的 0.1%～0.25%）、中碳钢（0.25%～0.55%）及高碳钢（大于 0.55%）。碳对钢材性能具有极为明显的影响，仅添加不到 1.0% 的碳，就会有极大的作用。增大碳含量会提高钢材的抗拉强度、屈服强度及硬度；但碳含量的增大也将导致钢材延展性的降低。由于钢材延展性是结构钢的一项基本性能（以防地震、爆炸及强风），大多数建筑钢材的碳含量介于 0.15%～0.4%。

Carborundum A trademark used for an abrasive material of silicon carbide crystals.
金刚砂 用于碳化硅晶体研磨材料的商品名。

carcass The load-bearing structure, skeleton, or framework of a framed building. Used to describe the external load-bearing elements of timber frame construction.
承重结构/骨架/框架 框架建筑的承重结构、骨架或框架。通常用来描述木框架建筑的外部承重构件。

carcassing Rough framed timbers and structural work usually used to describe wall studs, panels, roof rafters, and floor joists.
木构件 木构架和木结构构件，通常指墙龙骨、墙板、椽条和地板龙骨。

carpenter Skilled craftsperson who works with wood, usually assembling the larger rough structural wooden components of a building, such as walls, floors, and roofs. The craftsperson may also be trained as a *joiner.
粗木工 从事木材加工的工匠，通常指墙体、地板和屋顶等大型木质建筑构件的施工工人。粗木工经过培训也可成为细木工。

carpenter's hammer (claw hammer) A hammer with a double head, one end for hitting and embedding nails, the other with two splayed tongues for extracting nails.
羊角锤 两头锤，一头是圆的，用于钉钉子；另一头呈张开的八字形，用于起钉子。

carpenter's helper A labourer who helps the carpenter, may be an apprentice carpenter.
粗木工小工 协助粗木工的力工，可能为学徒。

carpentry The craft of cutting and shaping timber to make and assemble structural frames.
木艺 制造木制品和组装木结构框架的工艺。

carriage (carriage piece, rough string) A beam that runs centrally underneath wide *staircases to carry the *treads and loads placed on them. Where staircases are narrow the treads are carried by the *strings; however, where the staircase is wide, excessive deflection would occur. The timber, which runs diagonally from the lower floor to the upper floor, may be toothed or stepped out to carry the treads or may have *brackets fixed on either side of a straight string which extend to the underside of the *tread.
楼梯踏步梁 楼梯下方中间设置的梁，用于承载踏步及其所受的负荷。如楼梯较窄，则踏步由斜梁承载；但如楼梯较宽，则会出现过大的挠度。在从下一楼层斜向延伸至上一楼层的木材上可以，做出锯齿或梯级以承载踏步，或可在一段笔直斜梁的两侧装上延伸至踏步下方的托架。

carriage bolt A fixing bolt with a rounded head and square section shank that is tightly fitted into the timber. The square shank embeds into the timber and prevents the bolt rotating when the nut is tightened.
马车螺栓 圆头方颈螺栓，可以牢固打入木质构件。螺母拧紧后，方颈螺杆可沉入木材中防止螺栓转动。

carriageway The part of the road that carries vehicles. *See also* DUAL CARRIAGEWAY.
车道 道路供车辆行驶的部分。另请参考“双程分隔车道”词条。

carrier wave (carrier) An electromagnetic waveform that is modulated in terms of frequency, amplitude, or phase, to carry a radio or television transmission signal.
载波 在频率、振幅或相位方面受到调制以承载无线电或电视传输信号的电磁波形。

carrying capacity The maximum current that an electrical cable can safely handle without overheating.
承载能力 指电缆能够承载而不会产生过热的最大电流。

carrying out The undertaking of specified work.
实施 执行某一工作。

carry up The laying of brickwork up to a given height.
砌砖 从下往上一层一层地把砖铺到规定的高度。

Cartesian coordinates A reference system used to define a point on a plane, in relation to an origin and two perpendicular axes (the x-axis and y-axis). An additional axis (the z-axis) can be introduced to define a point in space.
笛卡尔坐标 一种参照系，用于根据原点和两条正交轴线（x 轴和 y 轴）来界定平面上的点。还可再引入一条轴线（z 轴）用于界定空间中的点。

cartographer A person who makes maps and charts.
制图员 绘制地图和图表的人员。

cartridge filter Easily locatable and disposable filter with its own frame and carrying unit.
筒式滤芯 简易定位的一次性过滤器，自带框架和过滤单元。

cartridge fuse Metal *fuse in a non-combustible tube or container that is easily fitted and removed from an electrical circuit. The fuse is made from a metal strip or wire that melts or breaks when excessive current passes through it.
熔丝管 套在耐火管或容器内的金属保险丝，既容易安装又容易从电路中拆除。保险丝是由金属片或丝制成的，当通过电流过大时自身会熔断切断电流。

cartridge tool (fixing gun, nail gun, stud gun) A handheld pistol-shaped tool for firing nails into wood and steel.
射钉枪 手持式手枪形打钉工具，适用于木材和金属的打钉。

case hardening A process used to harden the outer surface of steel by exposing metal to a specific atmosphere. The hardening process improves the material's resistance to wear and fatigue.
表面硬化 通过把金属暴露于特殊的环境中，使钢的表面硬化的过程。硬化过程可提高材料耐磨损和抗疲劳的性能。

casement A vertically hinged window *sash that can be opened.
平开窗 可开启的竖铰链窗框窗。

casement door A hinged pair of glazed double doors. Also known as a **French door** or **French window**.
落地玻璃门 用铰链固定的双扇玻璃门。也称为 French door 或 French window。

casement fastener *See* SASH FASTENER.
窗户锁 参考“窗锁”词条。

casement stay A bar containing a series of holes that is used to hold a *casement window open in various positions.
平开窗风撑 使平开窗根据需要开启到不同位置的多孔金属杆。

casement window A window containing one or more hinged *sashes that can be opened.
平开窗 可开启的单扇或多扇铰链窗。

cash flow The movement of money coming into the project from clients and the money going out of the project by payment to subcontractors and suppliers. To ensure liquidity throughout the project a certain amount of funds are required to keep the project properly resourced. Failure to manage cash flow will mean additional money will need to be borrowed, incurring extra interest. If income is generated unexpectedly it may accrue without being properly invested. Good cash flow ensures income can be properly managed and invested.
现金流 款项从业主处流入项目及款项通过对分包商和供应商的付款从项目流出的运动。为确保项目始终具有流动资金，须保留一定金额的款项使项目具有适当资源。对现金流的管理失误将意味着需要借入额外款项和产生额外利息。如有意外的收入，即使没有进行适应的投资也会积累起来的。良好的现金流将可确保对收入进行适当的管理和投资。

cash flow chart Graph that shows the projected and actual expenditure and income over time.
现金流量图 显示一定期间内的预计和实际支出与收入的图形。

cash flow monitoring Checking of income and expenditure against that projected. As part of *earned value analysis, the amount of money earned, including profit, may also be monitored against each activity.
现金流监控 把收入和支出与预计数字核对。作为挣得值分析法的组成部分，也可按每项活动对挣得的款项金额（包括利润）进行监控。

casing 1. The timber or plastic enclosure used to carry a unit or provide an aesthetic finish. For example, the enclosed unit of a *staircase or *casement windows. 2. The framing or cover that hides the rough work such as structural beams, service pipes, and cables.
1. 套条 如楼梯或平开窗的木质或塑料密封套，可起到支承构件或装饰的作用。**2. 遮板** 可隐蔽结构梁、给水管或电缆的外壳。

castellated beam Steel beam, where the web is made from two sections of steel cut in a castellated pattern, a shape resembling the up-and-down profile of a castle turret wall. The two sections are joined with the uppermost parts welded together, leaving holes in the centre of the beam. As the neutral axis of the beam runs down the centre of the beam, the holes do not result in large strength losses. Where the greatest stresses occur the steel remains.
蜂窝梁 钢梁，其中的梁腹板由两段切为堞形的钢组成，形状类似于城堡角楼墙的高低造型。两段钢结合于高点并焊接在一起，在梁的中心留出孔洞。由于梁的中性轴向下延伸至梁的中心，孔洞并不会导致强度的大幅下降。当承受较大压力时钢材仍能维持不变。

casting A process used to shape metals that are difficult to form by other means. Can be used to cast complicated structural components for utilization in bridges and other structures.
铸造 一种金属加工工艺，其他方法难以加工。可用于铸造桥梁和其他构筑物的复杂结构构件。

cast iron A ferrous alloy with the carbon content typically between 3.0 and 4.5 weight % C. The chemical composition of cast iron is the same as steel, however, cast iron contains up to ten times more carbon. Carbon

gives it excellent casting properties, making it possible to produce many complex items and shapes.
铸铁 一种黑色合金，其碳含量一般介于重量的3.0%～4.5%。铸铁的化学成分与钢相同，但铸铁的碳含量最高可为钢的10倍。碳为铸铁带来优异的铸造性能，使其有可能生成多种复杂的制品和形状。

C

catalyst Something or someone that aids change without being altered itself. For example, a chemical that accelerates a chemical reaction without itself undergoing any change.
催化剂 有助于变化但自身并不发生改变的某物或某人。例如，可使化学反应加速但其自身却并无任何变化的化学物。

catch A fastener used on doors and windows to keep them closed.
门窗锁扣 安装在门窗上用于锁闭门窗的扣式固件。

catch basin *See* CATCH PIT.
雨水口 参考“集水井”词条。

catch bolt *See* RETURN LATCH.
回弹插销 参考“回弹插销”词条。

catch drain A channel to reroute streams or rivers to avoid flooding.
集水沟 用来分流溪水或河水以避免发生洪水的沟渠。

catch feeder A *ditch used for *irrigation.
灌溉沟 用于灌溉的沟渠。

catchment area An area of land that drains rainfall into a river or provides water to a *reservoir. The water from this area is referred to as **catchment water**.
集水区 把雨水排入某条河流或为某个蓄水池供水的陆地区域。来自该区域的水被称为**汇集水**。

catch pit (catch basin, gravel trap, sump) An area at the entrance of a sewer where grit and other obstructive material that would hinder the flow in the sewer, is collected. Periodically, the grit in the catch basin is pumped or dug out.
集水井（雨水口，沉沙池，集水坑） 下水道入口处的一个区域，将下水道中的沙砾和其他阻塞性材料聚集于此。集水沟中的沙砾会被定期抽出或挖出。

catenary The U-shaped curve formed by a heavy inextensible cable, chain, or rope of uniform density suspended from its endpoints. The curve is a hyperbolic cosine.
悬链线 由具有均匀密度的不可伸展的沉重线缆、链条或绳索从其端点处形成的U形曲线，该曲线是一个双曲余弦。

Caterpillar A brand name for a producer of large earth-moving equipment. When used on-site, the more generic term refers to a large vehicle that is propelled using *crawler tracks. The tracked vehicle is often used on construction sites to move earth.
卡特彼勒 一家大型推土设备生产商的品牌名。当在工地使用时，较常被用作以履带传动推进大型车辆的通用名，此类履带车辆常用在建筑工地上推土。

caterpillar gate A large steel gate used for regulating the flow over a *spillway.
履带式闸门 用于调节漫出溢洪道的水流的大型钢闸门。

cathode 1. The terminal from which a current leaves an electronic cell, voltaic cell, or battery. 2. The positive terminal of an electronic device, cell, battery, or voltaic cell; opposite to anode.
1. 负极 电子电池、原电池或蓄电池的端子，电流从此流出。**2. 阴极** 电子装置、电池、蓄电池或原电池的正极端子，与阳极相对。

cathode protection The protection of a metal structure from corrosion by making the metal act as an electrochemical cell. The metal being protected is placed in connection with another metal that is more easily corroded. The more easily corroded metal acts as the anode in the electrochemical cell. The anodic metal corrodes in preference to the other metal; eventually, the more easily corroded metal must be replaced.

阴极保护 通过电化学电池反应，使金属表面受到保护免受腐蚀。被保护的金属与另一种更易被腐蚀的金属连接放置在一起。更易被腐蚀的金属作为电化学电池中的阳极，阳极金属先被腐蚀；最后需对阳极的金属予以替换。

cation A positively charged *ion, formed when an atom loses one or more negatively charged electrons in a reaction as in *anions. In *electrolysis, cations move towards the negative electrode. **Cation exchange** is used to soften water. Hard water is passed over *zeolite and cations of like charge are exchanged.

阳离子 当原子在反应中失去一个或多个作为阴离子带负电荷的电子后，而形成带正电荷的离子。在电解作用中，阳离子向阴极移动。**阳离子交换**也可用于对水进行软化。硬水流过沸石，与具有相似电荷的阳离子发生交换。

cat ladder Steps used to climb on and over roofs. The ladder has cleats or angled legs fixed to it so that they can be hooked over the roof's eaves safely, holding the steps on the slope of the roof.

爬梯 用于攀爬和越过屋顶的阶梯，梯子配有角码或成角度固定于梯身的撑脚，以使梯子能安全地挂在屋檐上，固定踏步于屋顶斜面上。

catwalk A narrow walkway around the top of tall buildings, which provides access to the roof and eaves so that maintenance work can be undertaken.

天桥 围绕于高楼顶部的狭窄走道，提供前往屋顶和屋檐的通道，从而使维护工作得以进行。

caulk To make watertight or airtight by filling or sealing, e. g. caulk a pipe joint; caulking the cracks.

捻缝/填缝 通过填充或密闭形成水密性或气密性，例如对管道接缝进行捻缝和对裂缝的填缝。

cause and effect analysis Methods used to identify the possible causes of problems. The diagrams, maps, and charts encourage thought about all of the possible causes of a problem rather than focusing on just one aspect of the problem. Diagrams commonly used to pull together causes include Ishikawa and fishbone diagrams.

因果分析 用于确定问题的可能成因的方法。相较于仅聚焦在问题的一个方面，图解、地图及图表更有助于对问题的所有可能成因的思考。图解通常用于把成因归集到一起，包括石川图和鱼骨图。

causeway Raised road, track, or path through a marshland or low-lying area that is often waterlogged.

堤道 穿过常被水浸的沼泽地或低洼区域的高筑道路、轨道或路径。

cavity The space between two leaves of a wall.

空腔 两叶墙之间的空间。

cavity barrier A *fire barrier that is inserted horizontally and/or vertically in cavity walls. Usually consists of polythene sleeved mineral wool.

空腔防火屏障 水平/或竖直插入夹心墙中的防火屏障，通常由聚乙烯的矿物棉组成。

cavity batten A timber batten that is temporarily inserted into a cavity wall to collect mortar droppings during wall construction.

空腔板条 临时插入夹心墙内的木板条，用于收集墙壁建造过程中滴落的灰浆。

cavity closer A material used at openings in cavity walls to close the cavity.

空腔封口料 用在夹心墙的开口处封闭空腔的材料。

cavity fill A thermally insulating material used to fully fill the cavity within a cavity wall to reduce heat loss. Usually consists of blown fibre or glass fibre.

空腔填料 用于对夹心墙内的空腔进行完全填充降低热损耗的隔热材料，一般由吹制纤维或玻璃纤维组成。

cavity flashing *See* CAVITY TRAY.

空腔泛水 参考“空腔泛水”词条。

cavity inspection The process of inspecting a cavity wall for defects, such as *thermal bridging or dampness, either during or after construction.

空腔检查 在建造夹心墙之时或之后检查其缺陷（例如热桥或潮气）的过程。

C

cavity insulation Thermal insulation material that is placed within the cavity of a cavity wall. The insulation can be inserted to fully or partially fill the cavity.
空腔保温 在夹心墙的空腔中置入隔热材料，置入的材料可把空腔完全或部分填充。

cavity tie Plastic or galvanized strip used to add stability to the two skins of masonry either side of the * cavity wall. The tie provides a link that restrains the external leaf of masonry and holds it into the load-bearing internal partition. The ties are preformed of varying lengths to suite the width of the cavity. *See also* WALL TIE.
夹心墙连杆 塑料条或镀锌钢条，用于增加夹心墙两侧砖砌的两层外表的稳定性。连杆可产生使砖砌外层受到限制的连接力，使其拉结承重的内部隔墙。连杆有不同的长度以便与空腔宽度匹配。另请参考“墙拉结”词条。

cavity tray A tray-shaped damp-proof course built into a cavity wall above openings to prevent damp penetration. Can be preformed or made on-site using a flat roll material.
空腔泛水 建在夹心墙内，位于开口以上的托盘形成防水层，以防水汽渗入。空腔泛水可以是预制的，也可以使用平坦的卷材在现场制作。

cavity wall A wall constructed from two separate leaves separated by a * cavity and tied together using * wall ties. * Thermal insulation can be placed within the cavity to reduce heat loss.
夹心墙 由两叶被空腔隔开并以墙拉结绑缚在一起的墙扇组成的墙壁。空腔中可置入隔热材料以便降低热损耗。

ceiling The uppermost horizontal lining in a room.
天花 房间内最上部的水平衬里。

ceiling binder A tie that spans between the ceiling joists or trussed rafters.
天花格栅横撑 跨越天花龙骨或桁架椽条的连接件。

ceiling fan A fan installed on the ceiling to circulate air within a room.
吊扇 安装于天花的风扇，用于使房间内的空气流通。

ceiling hanger A hanger attached to the underside of the floor above to support a suspended ceiling.
天花吊杆 附着于上层楼板下方的吊件，用于支撑吊顶天花。

ceiling height (headroom) The distance between the finished floor level and the ceiling.
天花高度（净空） 竣工后地板的完成面与天花之间的距离。

ceiling insulation The insulation placed between and above the ceiling joists. *See also* LOFT INSULATION.
天花保温 在天花龙骨之间或之上置入的保温层。另请参考“阁楼保温层”词条。

ceiling joist A joist that supports the weight of the ceiling below.
天花龙骨 支撑下方天花重量的托梁。

ceiling outlet A surface-mounted electrical outlet in a ceiling used to wire in a light.
天花插座 天花上的面装式供电插座，用于为灯具布线。

ceiling strap *See* CEILING HANGER.
天花系板 参考“天花吊杆”词条。

ceiling switch A switch operated by a pull-cord that hangs down from the ceiling.
天花开关 由从天花悬吊下来的拉绳控制的开关。

cell ceiling A lightweight open-cell square grid suspended ceiling.
格栅天花 一种轻型开孔方形格栅吊顶天花。

cellular Materials or products that are cellular are produced with small internal voids or hollow internal structures. Cellular materials contain tiny voids (air pockets) that make them lightweight. The still air within the structure improves their thermal insulation performance, but reduces strength and resistance to impact damage. Cellular concrete (* aerated concrete) contains tiny voids, which makes it lightweight and improves its thermal insulation. Cellular core doors are hollow in structure, with the hollow structure being formed from honey-

combed cardboard or strips of timber that offer some stiffness and rigidity to the door. Cellular glass (often called foam glass) is an insulating material made from glass powder and an air entraining agent that results in a material full of equal-sized closed-cell voids (tiny pockets of air).
多孔的/蜂窝状的 多孔材料或产品在制成后具有较小的内部空隙或中空的内部结构。多孔材料含有细小的空隙（气穴），使其重量较轻。结构内的静止空气提高了多孔材料的隔热性能，但降低了强度和抗冲击破坏性。多孔混凝土（充气混凝土）含有细小的空隙，使其重量较轻并提高其隔热性。多孔芯门为中空结构，该中空结构由蜂窝纸板或木条（为门提供一定程度的硬度和刚性）构成。多孔玻璃（常被称为泡沫玻璃）是一种由玻璃粉和引气剂制成的保温材料，可使材料中充满等尺寸的闭孔气穴（微小的空气囊）。

CE mark A self-certification mark placed on products by manufacturers to state that the product meets relevant European Directives. The CE mark is not an independent certification that the product has met the safety standards, but is a statement by the manufacturer that the product does meet all the safety requirements imposed by the European Directives. CE is a mandatory mark, under many Directives, that needs to be placed on many products sold on the single market of European Economic Area (EEA). CE is not an official abbreviation, but it is linked to the European Community or *Conformité Européene*.
CE 标志 一种自我认证标志，由制造商置于产品上，表明产品符合有关的欧洲指引。CE 标志并不是关于产品已符合安全标准的独立认证，而是制造商关于产品确实符合由欧洲指引所规定的一切安全要求的标志。多项指引均规定 CE 为强制性标志，在欧洲经济区（EEA）的单一市场上出售的多种产品均需具有 CE 标志。CE 并不是一个正式缩写符号，但其与欧洲共同体或 *Conformité Européene* 有关。

cement One of the most widely used materials or substances in the construction industry. Cement is essentially a binder used in concrete and mortar. Most construction cements are based on *Portland cement (or ordinary Portland cement). The manufacture of cement is a highly energy intensive process which accounts for 7%-10% of global carbon emissions (2008).
水泥 建筑业中使用最广泛的材料或物质之一。水泥主要是用作混凝土和灰浆中的黏合剂。大部分建筑水泥都是基于波特兰水泥（或普通波特兰水泥）。制造水泥是高度能源密集型的过程，约占全球碳排放的 7%～10%（2008 年数据）。

CEN (Comité Européen de Normalisation) The European committee for the development of standardization across the European Union. By specifying common standards, the committee helps to ensure the flow of materials and working practices across Europe.
欧洲标准化委员会 一个推动欧盟境内标准化的欧洲委员会。该委员会通过规定通用标准，为确保欧洲范围内材料的流动和工作实践提供助力。

centesimal angular measure A system for angular measurement, where a right angle is divided into 100 centesimal degrees (known grades or GRADS when used on calculators), each grade divides into 100 minutes and each minute divides into 100 seconds. Used mainly in France.
百进制角度计量 一种角度计量体系，其中把直角划分为 100 个百分度（在计算器中使用时也称为度或 GRADS），每度分为 100 分，每分又分为 100 秒。主要在法国使用。

central reservation (US median strip) A strip of land that divides traffic in opposite directions on *dual carriageways or motorways.
中央隔离带（美国 分车带） 在双程分隔车道或公路上分隔相反方向的交通的一条土地。

centres (US centers, centre-to-centre) The spacing between the centre point (**centreline**) of one object to the centre point of the next object.
中心距（美国 centers、中心至中心） 某一物体的中心点（中心线）与相邻物体的中心点之间的间隔。

centrifugal compressor A *compressor in which compression is achieved using a centrifugal force. Suitable for applications that require high volume flow rates, such as large chilled water plants.
离心式压缩机 一种通过离心力来达成压缩目的的压缩机，适用于高容量流速的设备，例如大型冷冻水设备。

centrifugal fan (centrifugal blower) A fan that utilizes a *centrifugal force to increase the pressure of the gas moving through it. Used in air-conditioning systems, warm air-heating systems, and industrial processes.
离心式风扇（离心式鼓风机） 使用离心力使从通过其中的气体压力增加的风扇，用于空调系统、热风供暖系统及工业加工中。

C

centrifugal force An apparent force that appears to drag an object away from a centre point around which it is rotating. Roads and railways are designed with a super elevation so that cars and trains resist this force around bends.
离心力 一种似乎要把某物拖离其绕以转动的中心点的惯性力。道路和铁路被垫高设计，以便使汽车和火车在转弯时能经受住这种力。

centrifugal pump A pump that uses a centrifugal force to move a fluid.
离心泵 使用离心力抽取液体的泵。

centring (US centering) Temporary works used to provide support during the formation of brick arches or structures. Centring has now become an interchangeable term with *shuttering, *formwork, and *stagework.
拱架（美国 centering） 用于在构建砖拱或结构时提供支撑的临时工程。拱架现已成为一个可与模板、模板及模板工程互换的词语。

ceramic A material comprised of metallic and non-metallic constituents. Typically ceramics have high strength and *hardness but low *ductility.
陶 一种由金属和非金属成分组成的材料。陶瓷一般具有较高的强度和硬度，但延性较小。

ceramic disc valve A valve containing two polished ceramic plates, which move across each other allowing or preventing the flow of water. They are considered highly durable, drip-and maintenance-free.
陶瓷圆盘阀 一种含有两块抛光陶瓷板的阀门，这两块陶瓷板通过相互移动来容许或阻止水的流动。陶瓷板具有较高耐用性、防滴漏及免维护性的特点。

ceramic matrix composite (CMC) A composite material based on a ceramic matrix.
陶瓷基复合材料（CMC） 一种以陶瓷为基体的复合材料。

certificate Official document that records an achievement, test results, or position.
证书 记录某项成就、测试结果或状况的正式文件。

certificate of practical completion The formal statement that works have been finished to a satisfactory condition. The certificate is issued by a resident engineer, a client's representative, or the project administrator. The issuing of the certificate normally results in payments and triggers the start of the periods for withholding retention monies.
实际竣工证书 宣布工程已按照令人满意的状况竣工的正式说明。该证书由常驻工程师、业主代表或项目管理人签发。发出该证书一般将导致付款并作为扣留保留金时期的开始。

cesspit (cesspool) A pit or underground tank for collecting and temporarily storing wastewater, particularly sewage.
污水坑（粪坑） 用于收集和临时储存废水（尤其是污水）的坑或地下水箱。

chain-link fence A lightweight metal barrier formed from strands of wire threaded together so that they interlock to provide a fence. The fencing material is delivered in rolls.
铁丝网围栏 轻型的金属屏障，由铰接在一起而且互锁的钢丝构成，用于提供围栏。围栏材料是成卷交付的。

chain pump A water pump powered by a series of discs attached to a chain. The chain is drawn through a tube, part of which is open and exposed to water. As the chain is drawn along the tube, the discs are pulled through the water; the water becomes trapped and is drawn into the tube and then discharged.
链斗式水车 由一系列附着于一条链条上的板片驱动的水泵。链条穿过一根管子，管子的一部分是敞开并浸于水中的，当链条沿着管子被拖动时，板片即穿过水面；水被装入板片和拖进管子，然后被倒出。

chain saw A motorized saw where a chain with individual blades acts as the cutting edge. Each segment of the chain has small blades. The chain, similar to a bike chain, runs around a guide and is powered by a motor. As the chain rotates it is used to make rough cuts in timber and trees.
链锯 一种电动锯，其中由带有独立刀片的链条作为切削刃。链条的每一段都有小刀片，而链条则与自行车链条相似，围绕导向轮运行并由电机驱动，当链条旋转时可在木材和树木上形成粗糙切口。

chain survey A linear survey using a controlled framework of chain lines. The distance between survey points is measured by a chain. The **chain** consists of a connection of 100 metal links of equal length. A **Gunter's chain** or **surveyor's chain** is 20m (66 ft) and an **engineer's chain** is 30.5m (100 ft) The *offset distance of objects (tree, buildings, etc.) are measured perpendicular to the chain. The distance along the chain from one *station to the offset is known as the **chainage**.
测链测量 一种使用链式线条的受控框架的链条丈量。测量点之间的距离由一条链条来计量。该**链条**包含 100 个等长的金属链接。一条**冈特测链**或**测量员测链**为 20 米（66 英尺），而一条**工程师测链**为 30.5 米（100 英尺）。物体（树木、建筑物等）的支距按垂直于链条来计量。沿着链条从一个站点到支距处的距离被称为**链程**。

chair bolt (fang) A bolt fixing with a fishtail end that protrudes from the wall. The fishtail or bracket that is left exposed once the bolt is inserted and secured to the structure allows brickwork and other structural materials to be tied to the fixing.
鱼尾螺栓（棘螺栓） 一种螺栓，配有鱼尾状末端且该末端凸出于墙壁。一旦把螺栓插入并固定在结构物上，鱼尾端或支架即暴露在外，使砖砌和其他结构材料能够连结直至固定。

chair rail *See* DADO RAIL.
护墙板线条 参考"护墙板线条"词条。

chalk line A piece of string covered in chalk and used for marking straight lines on solid flat surfaces. A thin piece of string is pulled from a chalk dust container, then the chalk-covered string can be stretched between two points over a flat surface. Once pulled taut the string is lifted slightly and allowed to snap back onto the surface, resulting in a straight chalk line being transferred to the surface. A chalk line is often used to mark out concrete surfaces for brickwork, blockwork, or for cuts and rebates.
粉线 一条被白垩覆盖的细线，用于在实心平坦表面标记直线。细线被从白垩粉盒中抽出，被白垩覆盖的细线随后可在平坦表面上的两点之间拉直。把细线略微提起并使其弹回到表面上，表面上就会留下一条笔直的白垩线条。粉线常被用于为砖砌、砌块标记混凝土表面，或用于标记切口和槽口。

chamber (lock bay) 1. A bedroom. 2. An enclosed space.
1. 室 卧室。**2. 间** 一个封闭的空间。

chambered-level tube A *level-tube that has a chamber at one end to adjust the amount of air in the cylinder (i. e. bubble length) for temperature correction.
气室水准管 在一端有一个气室的水准管，该气室用于调整圆柱体中的空气量（即气泡长度），以便进行温度调整。

chamfer *See* BEVEL.
倒角 参考"割角"词条。

change An alteration to that which was planned, contracted, priced, or proposed. Standard terms of contract such as the *JCT and *ICE suite of contracts have specific definitions for activities and events that constitute a change or *variation. Changes to work are normally authorized by the client representative or project administrator using a change order.
变动 对于预定、约定、定价或拟议的一项修改。JCT 和 ICE 成套合同等合同标准术语对构成一项变动或变更的活动和事件有具体的定义。工程的变动一般由业主代表或项目管理人通过变动令而进行授权。

change agent 1. Something introduced into a substance to induce change. 2. Personnel brought into a pro-

ject or organization to manage change.
1. 促变剂 被引入某物质中以引发变化的某物。**2. 促变者** 被引入某个项目或组织中以便对变动加以管理的人员。

change face A procedure in surveying to minimize errors by rotating the telescope on a * theodolite through 180 °vertically and horizontally so that readings are taken from the opposite side of the circle.
正倒镜 测量中用于减少误差的一个步骤，通过 180 度垂直和水平旋转经纬仪上的望远镜，以便从圆圈的反向获得读数。

change management The planned process used to make alterations to business or activities. Procedures and guidance are often used to specify the course of action to be taken. New personnel with experience of managing change may be brought in to oversee and lead activities during this period.
变化管理 用于对业务或活动做出修改的预定过程。常使用步骤和指引说明将采取的行动。具有管理变动经验的新人可能被引入以便监督和领导在此期间的活动。

change point The position of the levelling * staff when a surveying instrument (level or theodolite) is repositioned. *See also* FORESIGHT and BACKSIGHT.
变动点 当测量仪器（水准仪或经纬仪）被重新安放时水平标尺的位置。另请参考“前视”和“后视”词条。

channel An object having a U-shaped cross-section. An **open channel** can be used to convey water for irrigation. A **channel bend** is an open channel which is curved on plan to guide the flow in an * inspection chamber from a * branch sewer. A **channel section** is a rolled * steel section in the shape of a channel. **Channel iron** is a bar with a U-shaped cross-section.
沟/槽/渠 横截面为 U 型的器具。**明渠**可用于输送灌溉用水。**明沟弯头**是明沟的拐弯部分，可引导下水道支管里边的水流流入检查井。**槽钢截面**是形状为槽形的轧制型钢截面。**槽钢**是截面形状为 U 形的钢条。

characteristic strength (characteristic stress) A value defined by a certain percentage of material specimens (typically steel or concrete) that exhibit values within a certain probability. Not more than 5% of test specimens should fall below the characteristic strength/stress and as such, is used by engineers as the material design strength.
（材料）强度标准值 材料样品（通常指钢筋或混凝土）通过试验取得统计数据后，根据其概率分布，取其中的某一分位值来作为标准值。一批试件中，屈服强度低于强度标准值的试件不得超过 5% 。因此，工程师把屈服强度当作材料强度标准值。

charge hand (leading hand) On-site gang leader or supervisor.
工长 驻在现场的班组领导或主管。

Charpy test (izod test, notched-bar impact test) A test utilized to measure the impact energy or notch toughness of a standard notched metal specimen. Many metals are ductile at room temperature, however, they become brittle at lower temperatures (e. g. -200℃). The charpy /izod/notched bar impact tests determine whether metals are brittle or ductile (or both) between -100 to -200℃. A small metal specimen (100mm in length) is supported as a beam in a horizontal position and loaded behind the notch by the impact of a heavy swinging pendulum. Impact pressure is applied to the specimen, which is forced to bend and fracture at a very high strain rate. The principal measurement from the impact test is the energy absorbed in fracturing the specimen, the higher the energy, the more ductile the material. Once the test bar is broken, the pendulum rebounds to a height that decreases as the energy absorbed in fracture increases. The principal advantage of the Charpy V-notch impact test is that it is a relatively simple test that utilizes a relatively cheap, small test specimen. *See also* IZOD TEST.
夏比测试（埃左测试、缺口杆冲击测试） 用于测量标准缺口金属样品的冲击能量或缺口韧性的测试。很多金属在室温时具有延展性，但在较低温度（例如 -200℃）下则变得脆弱。夏比/埃左/缺口杆冲击测试可确定金属在 -100 至 -200℃之间是否脆弱或具有延展性（或两者兼有）。一个小型金属样品（长度为 100 毫米）在水平位置作为一条梁受到支撑且被装在缺口后，受到重摆锤的冲击。冲击压力作用于样品，使其受力按极高的应变率弯曲和断裂。冲击测试的主要计量值为样品断裂时所吸收的能量，能量越高，则材料的延展性越大。一旦测试杆断裂，摆锤将回弹至某个高度，断裂时吸收的能量增加则该高度将降低。夏比 V 缺口冲击测试的主要优点在于其操作相对简单，使用相对廉价和小型的测试样品。另请参考“埃左测试”词条。

Chartered Engineer A professional engineer, registered with the Engineering Council UK and who has gained the post-nominal letters ‘CEng’. According to the Engineering Council UK ‘Chartered Engineers are characterized by their ability to develop appropriate solutions to engineering problems, using new or existing technologies, through innovations, creativity, and change. They might develop and apply new technologies, promote advanced designs and design methods, introduce new and more efficient production techniques, marketing and construction concepts, pioneer new engineering services and management methods.’ Typically, a Chartered Engineer would have gained an accredited academic qualification up to Masters level and have demonstrated professional competence through training and experience.
特许工程师 通过英国工程委员会认证的并带有‘CEng’头衔的专业工程师。根据英国工程委员会的说明，“特许工程师是能运用现有或新型技术制定适当的工程解决方案的科技人才，且具备良好的创新和应变能力。特许工程师应能开发和应用新技术，推广先进的设计和设计方法，引进新型和高效的生产技术、市场营销理念和施工理念，开拓新的工程服务和管理方法。”一般来说，特许工程师必须取得硕士学位证书，并已通过培训经历和工作经历证明自己的专业能力。

Chartered Institute of Arbitrators (CIArb) Body that accredits professional arbitrators. Provides governance and guidance in arbitration and allows access to its members who can advise, assist, or conduct arbitration.
特许仲裁员学会（CIArb） 对职业仲裁员进行认可的机构，提供对仲裁的管治和指导，并允许其会员参与对仲裁的建议、协助或实施。

Chartered Institute of Building (CIOB) Professional body that represents construction managers and other related professions. The organization was founded in 1834 as the Builder's Society. A professional builder who has membership (MCIOB) or fellowship (FCIOB) of the CIOB is known as a Chartered Builder.
特许建造学会（CIOB） 代表建筑经理人和其他相关行业的专业机构。该组织于 1834 年作为建造师协会而创立。具有 CIOB 会员（MCIOB）或准会员（FCIOB）资格的职业建造师被称为特许建造师。

Chartered Institute of Building Services Engineers (CIBSE) A UK professional body, which gained its Royal Charter in 1976, to support building services engineers. The CIBSE publishes guidance and codes on controls, equipment, and internal environments within buildings.
特许屋宇装备工程师学会（CIBSE） 英国的一个专业机构，为屋宇装备工程师提供支持，于 1976 年获得皇家特许。CIBSE 出版关于建筑物内控制、设备及内部环境的指南和规程。

chase (chased) A rebate or channel cut into walls to accommodate services. Specialist plant is available to rebate, brickwork, blockwork, and concrete. Ducting, wires, and pipework are clipped into the rebate. When the wall is plastered there is no evidence of the wire or ducts beneath the finish.
槽口 开在墙上以供安装设施用的槽口。可用专业的设备在砖墙砌块墙和混凝土上开槽口。电线和管道都可以安装在槽口里面。墙面抹上灰浆后，就看不到电线或管道了。

check 1. Examining work to ensure that it has been done correctly. 2. A *rebate (Scotland). 3. A crack or break in the surface of a material.
1. 检查 检查某一工程，看其是否施工得当。**2. 槽口**（苏格兰）。**3. 裂缝** 材料表面的裂纹或裂口。

checked back Recessed.
嵌入式的 把某物放在墙壁的凹处。

checking *See* CRAZING.
细裂 参考“小裂”词条。

check out The process of forming a rebate.

开槽口 加工槽口的过程。

check throat A recessed *throat on a *sill.
滴水槽 窗台或门槛下面的滴水槽。

check valve (clack valve, non-return valve, reflux valve) A *valve that only allows flow in one direction.
止回阀 介质只向一方向流动的阀门。

chemical degradation of timber A degradation mechanism for timber material. The higher the density and the greater the impermeability of the wood, the greater its resistance to chemical degradation. The acid resistance of impermeable timbers is greater than that of most common metals; iron begins to corrode below pH 5, whereas attack on wood commences below pH 2, and even at lower pH values, proceeds at a slow rate. In alkaline conditions wood has good resistance up to pH 11.
木材的化学降解 木材的降解机制。木头的密度越大和抗渗性越高，则其对化学降解的抗性则越强。抗渗木材的耐酸性高于大多数普通金属；铁在pH值为5以下时即开始腐蚀，而对木头的侵蚀是在pH值为2以下才开始，且即使在更低的pH值下，侵蚀速度也较慢。在碱性条件下，木头具有pH值高达11的良好抗性。

chequer-board construction (alternate-bay, chequer-board, hit-and-miss, chequer-board slabbing) Concrete floor slab cast in alternative bays giving an appearance of a chess board.
格子地板 由相间的格子组成的混凝土地板，形状像国际象棋盘。

chequer plate (chequerplate, checker plate, checkerplate) An anti-slip flooring material.
花纹板 一种防滑的地板材料。

cherry picker Mobile elevated working platform (MEWP) with an articulated pneumatic arm that allows personnel to work safely at heights.
车载式升降台 移动式高空作业平台（MEWP），带气动铰接式吊臂，可确保作业人员在上面安全工作。

chevron drain A V-shaped drain laid in a *herringbone fashion. Typically filled with free-draining granular material and used to collect surface water.
人字形排水沟 连续人字形排水沟。通常会在排水沟内填充透水良好的颗粒材料，可用于收集地表水。

chiller *See* REFRIGERATION UNIT.
冷水机组 参考“制冷机组”词条。

chimney A structure containing a vertical *flue.
烟囱 带垂直烟道的构筑物。

chimney block *See* FLUE BLOCK.
烟囱用耐火砖 参考“烟道用耐火砖”词条。

chimney breast The portion of a chimney that projects from a wall.
壁炉墙 烟囱凸出墙面的部分。

chimney shaft A free-standing industrial chimney that usually only contains one flue.
烟囱身 单烟道独立式工业烟囱。

chimney stack The portion of a chimney that projects from a roof. One or more flues can be contained within the stack.
烟囱筒身 烟囱伸出屋顶的部分。烟囱筒身里边可以容纳一个或多个烟道。

chipping To break or cut small pieces from the surface of a material.
刨/凿 从材料表面刮削出小碎片。

chipping chisel 1. A chisel bit for use in a pneumatic chipping hammer. 2. The chisel end of a chipping

hammer.
1. 錾子 气动凿锤的錾头。**2. 凿刃** 凿锤的凿刃部分。

chipping hammer 1. A portable pneumatic hammer. 2. A small portable rock hammer used by geologists to extract minerals, fossils, and fragments of rock.
1. 凿毛锤 一种便携式气动锤。**2. 地质锤** 小型便携式地质锤，可用于凿取矿物、化石或岩石碎片。

chippings Small pieces of a material, usually stone.
碎片 材料小碎片，通常指碎石片。

chlorination A process of adding chlorine to water as a method of water purification. Chlorination is also used to sterilize the water in swimming pools and also in sewage works. The term can also apply to the addition of chlorine to other elements, such as the addition of chloride to metals.
氯化 向水中加氯的过程，作为一种水净化方法。氯化也用于对泳池和污水工程中的水消毒。该词语也适用于向其他元素中加氯，例如向金属中加氯。

choke *See* BALLAST.
扼流器 参考“镇流器”词条。

chord 1. The horizontal part of a roof truss designed to resist tensile forces. 2. A straight line which connects two points on an arc or circle.
1. 弦杆 屋架的水平构件，可用来抵抗拉力。**2. 弦** 连接圆/弧上任意两点的直线。

chuck An adjustable device that is used to hold tools in the centre of a lathe (**lathe chuck**) or a drill (**drill chuck**). In a drill, the chuck is adjusted using a **chuck key**.
夹头/卡盘 一种可调装置，用于把工具夹持在车床（**车床夹头**）或钻机（**钻机夹头**）的中心。在钻机中，可用**夹头扳手**调节夹头。

chute An angled pipe or semi-circular channel that uses gravity to transport materials downwards.
流料槽/溜管 利用重力向下输送材料的弯管或半圆形斜槽。

cill *See* SILL.
窗台/门槛 参考“窗台/门槛”词条。

Cipolletti weir A trapezoidal-shaped weir with 1：4 side slopes. Used for measuring the flow in open channels such as streams and rivers.
西波勒梯堰/梯形堰 坡度比为1：4的梯形堰。梯形堰可用于计算明渠（如：溪流和河流）的流量。

circuit 1. A circular path of conductors that enables an electric current to flow from the power source to an electrical appliance or component (an **electrical circuit**). 2. The electrical wiring attached to the consumer unit that supplies electricity to electrical sockets, fixed appliances, and lights. 3. A series of pipework through which a fluid flows, i. e. a **hot-water circuit**.
1. 电路 由导线组成的回路，（电路）可以确保电流从电源处输送到电器或电子部件上。**2. 电气布线** 用户配电箱里的电气布线，可把电流输送给电气插座、电器和灯具。**3. 管路** 液体可流动的一系列管道，如：热水管路。

circuit breaker A safety device that automatically stops the flow of electricity through an electrical circuit if it becomes overloaded or if a fault develops. Provides greater control than a *fuse and can be manually reset once the overload has been removed. A series of **miniature circuit breakers (MCBs)** are used in a *consumer unit to protect various electrical circuits, such as the central heating system, lighting, cooker, electric shower, electric sockets, and smoke alarms. The number of MCBs contained within the consumer unit will depend upon the size of the building and the equipment installed. In dwellings, modern MCBs are rated at 6, 10, 20, 32, 40, 45 and 50 amps.
断路器 一种安全装置，当电路过载或电路出现故障时，断路器可以自动切断某一电路的电流。断路器的控制性能比熔断器好，一旦过载问题被解决就可以手动复位断路器。用户配电箱里会用到很多

微型断路器，这些微型断路器可以对各种电路（如：集中供热系统、照明系统、电炊具、电热水器、电气插座和烟雾警报器）起到过载保护的作用。用户配电箱里的微型断路器的数量取决于建筑或设备的大小。一般住宅现行使用的微型断路器，额定电流有 6A、10A、20A、32A、40A、45A 和 50A 的。

circulation The free movement of a fluid through a system.
循环 系统中流体的自由流动。

circulation pipe A pipe used to circulate gas or a liquid, for instance, a flow or return pipe in a central heating system.
循环管 燃气或液体的循环管道，如：集中供热系统的供水（气）管或回水（气）管。

circulation space (circulation area, circulating area) The space within a building that enables people to move freely from one part to another.
（建筑）交通空间 建筑物内部供人自由行走的空间。

CIRIA The Construction Industries Research and Information Association established to deliver services, information, and research across the sector.
CIRIA 建筑业研究与信息协会，其设立宗旨在于跨行业提供服务、信息及研究。

CI/SfB Classification system, used in libraries and offices to divide building information and literature into related sections. The system was developed by * RIBA and is used across the industry.
CI/SfB 用于图书馆和办公室的归类体系，以便把建筑信息和文献分别归入有关的部分。该体系由 RIBA 制订并在整个行业内使用。

cistern A tank used to store cold water at atmospheric pressure. Usually fitted with a * float valve and an * overflow pipe. Used to store rainwater, store the water used for flushing a WC, or as a feed and expansion tank in an * open-vented system.
水箱 在标准大气压条件下用来贮存冷水的水箱。通常装有一个浮球阀和一根溢流管。可用于贮存雨水以作冲厕水；也可用作开放式供热系统的膨胀水箱。

City & Guilds of London Institute (C&G) Awarding body of many vocational qualifications in the UK.
伦敦城市行业协会（C&G） 英国的多项职业资格的授予机构。

civil engineer A person who practises the profession of * civil engineering. He or she may deal with the design, construction, and maintenance of bridges, tunnels, roads, canals, railways, airports, tall structures, dams, docks, and clean/wastewater systems.
土木工程师 指从事土木工程的专业技术人员。土木工程师应能在桥梁、隧道、道路、运河、铁路、机场、高层建筑、堤坝、码头和给水排水领域从事设计、施工和维护等活动。

civil engineering This is a professional engineering discipline that deals with creating, improving, and protecting the environment. It provides the facilities for the built environment and includes environmental, geotechnical, materials, municipal, structural, surveying, transportation, and water engineering.
土木工程 这是一门创设、改善和保护环境的专业工程学科。土木工程应为建成环境提供各种设施，包括：环境工程、地质工程、材料工程、市政工程、结构工程、测量工程、交通运输工程和水利工程。

Civil Engineering Standard Method of Measurement (CESMM) A document containing set procedures to prepare and price the * bill of quantities, together with how the quantities of work should be expressed and measured for * civil engineering projects.
土木工程标准计量方法（CESMM） 给出工程量清单的编制和计价步骤的文件，文件还附有土木工程工程量的表述和计量方法。

cladding The non-load-bearing external envelope or skin of a building that provides shelter from the elements. It is designed to carry its own weight plus the loads imposed on it by snow, wind, and during maintenance. It is most commonly used in conjunction with a structural framework.

外墙板 为建筑构件提供围挡的非承重外墙。其设计为只需承受自身重量以及风雪荷载或外墙维护时施加在外墙上的荷载。外墙板通常与结构框架连接使用。

cladding rail A secondary beam that spans horizontally across the structural frame of the building to support the cladding.
干挂用横龙骨 横跨建筑物结构框架的用于支撑外墙板的次梁。

claim Allegation made against another person stating that something has or has not been performed or delivered. The party making the allegation normally takes the action in order to recover money or order a specific performance. The action is normally taken because a party has failed to perform as agreed under the contract.
索赔 针对其他人而提出的主张，说明某事已经或并未被履行或达成。提出主张的一方通常提起诉讼以便收回款项或取得有关具体履约的命令。诉讼的提起一般是由于一方未能按照合同的约定而履约。

claims conscious contracting Unofficial method of contracting where a *contractor enters into a contract with the intention of pursuing *claims for any extra work. Instead of identifying items not mentioned in the tender documents, but deemed necessary to compete the project, the contractor undertakes the contract and makes a claim for the *variations once work starts. Clients may enter into the contract believing everything is covered only to discover the cost of the contract increases as claims are served and granted in order to complete the works. Such procedures only work where the contract would allow for variations. Good practice would encourage contractors to identify the items missing from the documents at the tender stage, allowing the client to assess the extra costs that would be incurred.
恶意低价中标 非正规的承包方式，承包商订立合同时旨在寻求对其他额外工程的索赔。对于招标文件中并未提到，但视为完成项目所需的各项，承包商不会注明，而是在工程一旦开始时即实施合同并对变更提出索赔。业主可能是在认为一切工程均已被涵盖的想法下订立合同，但发现合同成本将随着为完成工程，而收到承包商的索赔而增长。该手段仅在合同允许工程变更的情况下才有用。良好诚信的承包商会在投标阶段把文件中所漏的各项注明，使业主得以评估将要产生的额外成本。

clamping screw A screw used to attach a vernier to levels and theodolites.
止动螺丝 用于把游标尺固定到水准仪和经纬仪上的螺丝。

clamshell (clamshell grab) A toothless grab bucket that is suspended from a crane jib for handling loose material.
蛤壳式抓斗 悬在起重机吊臂上的无齿抓斗，可用于散货的搬运。

clapotis A standing wave that does not break before it impacts on a near-vertical surface, e. g. a seawall.
驻波 一个竖立的波，在其与近似垂直的表面（例如防波堤）相撞前不会破碎。

classification test *See* ATTERBERG LIMITS.
分类测试 参考“阿太堡界限”词条。

clause A description setting down the agreements that form the component parts of a contract. Contracts are divided into individual items covering the various aspects of the agreement; these are the terms that make up the contracts.
条文 把构成合同组成部分的协定列出的一段说明。合同分为各项，涵盖了协定的不同方面；这些就是组成合同的条款。

clay puddle *See* PUDDLE.
黏土膏 参考“黏土膏”词条。

clearance The space between two objects, e. g. the space between a door and a doorframe to ensure that the door can open freely.
间隙 两个物体之间的空隙，如：门与门框之间的空隙，可确保门能自由开启。

clear span The unobstructed distance between two supports, e. g. the distance under a beam support between two columns.

净跨 指承重结构之间无障碍的距离，如梁下方两根支撑柱之间的距离。

cleat A bracket fixed to a structural component used to secure another component.
夹板 安装于一个结构构件的支承件，用于固定另一构件。

cleaving Splitting a material along its cleavage.
裂开 沿着材料的裂隙把材料分开。

clench nailing (clenching, clinching) A type of nailing where the point of the nail is driven through and out of the material, and then the projecting point is bent over and flattened against the material.
弯脚钉合 一种钉合方法，把钉尖敲入并穿透材料另一面，再把凸出的钉尖敲弯到与材料表面齐平。

clerestory (clerestorey, clearstorey) The upper part of a wall that contains windows to produce extra light and sometimes ventilation. Used in churches and tall building where the centre of the structure rises above the rooflines of perimeter sections of the structure.
高侧窗 在侧墙的上部预留的用来采光和有时候起通风作用的窗户。高侧窗常用于教堂和高层建筑，其中央结构比周边部分结构的屋顶线要高。

clerk of works (COW) An inspector used to check the project as it progresses. As a client's representative the COW may have to agree *day works items, check setting out, inspect excavations, and check levels. The COW takes no active part in running the project.
监理员（COW） 检查工程质量和安全问题的监督人员。作为业主代表，监理员应负责监督计日工项目、检查放线、基坑工程和各类标高。监理员不直接参与工程施工。

client (employer) The person or organization that requires and pays for the building or works.
业主（雇主） 需要及支付建筑物或工程项目的投资人或组织。

clinker 1. Large round lumps of cement-like substance, formed during the cement production and often ground down to become cement. It is produced during the cement kiln stage from sintering limestone and alumino silicate. 2. Lumps of vitrified brick or burnt clay fused together.
1. 水泥熟料 又大又圆的水泥状物质，在水泥生产中形成且通常再粉磨成水泥。水泥熟料是石灰石和硅铝酸盐在高炉中烧结而成的。**2. 缸砖** 玻化砖或用陶土烧结而成的砖块。

clinograph An instrument that records the derivation from vertical, used in boreholes, shafts, and wells.
钻孔测斜仪 用于记录钻孔、竖井和井道等的垂直度偏差的仪器。

clinometer A handled instrument for measuring the dip of a bedding plane or the angle of a slope, used in geology and surveying.
测斜仪 测量层理面倾斜度或坡体坡度的便携式仪器，用于地质学和测量学。

clip A type of fixing used to hold materials or components in position, such as cladding panels, electrical cables, pipes, or roof tiles.
夹具 用来固定建筑材料或建筑构件（如：外墙挂板、电线、管道或屋面瓦）的一种紧固件。

clip anchor *See* ANGLE CLEAT.
角码 参考“连接角钢”词条。

close boarded (close boarding) Timber boards laid abutting one another such that there is no gap in-between. Usually used for fencing.
无缝拼板 被紧挨着拼接在一起，没有缝隙的木板，通常用作栅栏。

close-coupled cistern A *WC where the *cistern is connected directly to the *pan.
连体式马桶 水箱与坐便器直接连在一起的冲水马桶。

closer (closing piece) A brick that is reduced in width by cutting it along its length. Used on a *course to complete the bond. Examples include a *king closer and a *queen closer.
过渡砖 被沿着砖的长度从中间切成两半的砖。在一层砖中能起到过渡的作用。包括砍角砖和竖半砖。

closing device A *fastener used to keep doors and windows closed.
闭门器 能将门窗关上的闭合装置。

closing error (error of closure, misclosure) The discrepancy between two readings in a survey, which should actually be the same. For example, in a closed loop *traverse the initial readings should be the same as the final reading. Any difference between these two readings (the closing error) is proportionally distributed over the traverse (*See* BOWDITCH'S METHOD).
闭合误差（闭合差） 在调查中，实际上本应相同的两个读数之间的差异。例如，在闭环导线测量中，最初的读数应与最终的读数相同。这两种读数的任何差异（闭合差）将按比例分布于导线（参考鲍迪奇法）。

closing face The part of a door leaf that is closed against the doorframe.
闭门侧 门扇关闭时与门框接触的侧边。

clough 1. The sloping sides of a deep narrow valley in a hill. 2. A *sluice for controlling the flow of water.
1. 深涧 指山间低凹而狭窄处。**2. 水闸** 控制水流的闸门。

clout nail A short galvanized nail with a large flat round head. Used for fixing plasterboard and sheet metal roofing.
大帽钉 大帽、平圆头、短杆的镀锌钉。用于石膏板和金属屋面板的固定。

coagulant An agent that causes a liquid or solution to semi-solidify.
凝结剂 一种能使液体或溶液半凝结的助剂。

coal A solid fossil fuel that is burnt providing heat and energy. Coal is usually a black combustible mineral formed about 300 million years ago.
煤炭 一种燃烧时提供热量和能量的固体化石燃料。煤炭通常是黑色并且可燃性的矿物，形成于约3亿年前。

Coanda effect Discovered in 1930 by aircraft engineer Henri Coanda, this effect is the tendency of a moving fluid (air or liquid) to adhere to a curved object as it passes over it. As a fluid passes over a surface it will be subjected to skin friction. This friction causes a decrease in flow rate and pulls the fluid towards the surface. If the surface is curved as the fluid leaves the surface it will have a tendency to continue the curved shape. Amongst other aerodynamic applications, the Coanda effect is employed in certain ceiling diffusers as it is known to be a very effective method for circulating air; these diffusers are also known to leave a dirty stain from the air on the surrounding ceiling.
康达效应 飞机工程师亨利·康达于1930年发现，该效应是指运动中的流体（空气或液体）在流过弯曲的物体时具有黏附于该物体上的倾向。由于流体流过某个表面，故将受到表面摩擦力的影响，该摩擦力导致流速降低并把流体拖向该表面。如该表面是弯曲的，当流体离开该表面时将具有保持该弯曲形状的倾向。除其他空气动力学方面的应用以外，康达效应也用于某些天花散流器，因为对于空气流通而言，该效应是已知极为有效的方法；（但）据知这些散流器也会在邻近的天花上留下污渍。

cob A mixture of clay and straw used as a building material. Cob is a very popular construction material in the developing world as it is cheap and easy to manufacture.
草泥垛 用黏土和稻草掺和而成的建筑材料。草泥垛是被发展中国家广泛使用的一种建筑材料，因为草泥垛易于生产且成本低。

cock *See* STOPCOCK.
旋塞 参考“旋塞阀”词条。

Code of Practice (CP) A description of work or procedure that establishes a minimum standard to be achieved or followed. *British Standards and *Eurocodes have now largely replaced CPs.
作业标准 施工或施工步骤说明，以确立最低质量标准供作业人员参考。现在作业标准基本上已经被英国标准和欧洲标准取代了。

coefficient 1. A number or constant multiplied by a variable or unknown in an algebraic equation, e.g. 4 is the coefficient in the equation 4*xy*. 2. A numerical measure of a physical or chemical property that is constant under specific conditions, various examples are presented below. **Coefficient of active earth pressure (K_a)** the ratio of minimum horizontal effective stress to vertical effective stress that would occur in a soil mass at some depth, e. g. behind a retaining wall where the surface of the wall is considered to move away from the soil it is retaining. It is defined by the equation

C

$$K_a = (1 - \sin\varphi) / (1 + \sin\varphi)$$

where φ is equal to the angle of internal friction of the soil. *See also* ACTIVE EARTH PRESSURE. **Coefficient of compressibility (m_v)**, the volume change per increase in effective stress, i. e. the inverse of pressure. **Coefficient of consolidation (c_v)**, the ratio of permeability to volume compressibility of a soil. It is used to estimate the rate at which settlement takes place. There are two methods to determine cv, i. e. root time or log time. **Coefficient of discharge (C_d)** ratio of actual discharge obtained through an orifice, weir, or pipe compared to the theoretical value. **Coefficient of dynamic viscosity (μ)**, a constant linking fluid flow to resistance. **Coefficient of earth pressure at rest (K_o)**, the ratio between horizontal effective stress and vertical effective stress in a soil mass at some depth where the horizontal strain is zero, e. g. behind a retaining wall where the surface is not considered to move. **Coefficient of expansion or coefficient of thermal expansion (α)**, the unit increase in length per degree increase in temperature. **Coefficient of friction (μ)**, the ratio between the force causing a body to slide along a plane and the force normal to the plane. **Coefficient of passive earth pressure (K_p)**, the ratio of maximum horizontal effective stress to vertical effective stress that would occur in a soil mass at some depth, e. g. behind a retaining wall where the surface of the wall is considered to move towards the soil it is retaining. It is defined by the equation

$$K_p = (1 + \sin\varphi) / (1 - \sin\varphi)$$

where f is equal to the angle of internal friction of the soil. *See also* PASSIVE EARTH PRESSURE. **Coefficient of permeability (k)**, the mean discharge velocity of water flow under the action of a unit *hydraulic gradient. **Coefficient of uniformity (C_u)**, the slope of a soil's grading curve, defined as

$$C_u = d_{60}/d_{10}$$

1. 系数 代数方程式中被一变数或未知数所乘的一个数字或常数，例如4在方程式4xy中即为系数。2. 对在特定条件下恒定的某个物理或化学性质的算术计量，各种示例说明如下。**主动土压力系数（K_a）**指在一定深度下（例如挡土墙后）当墙壁的表面被视为已从其所阻挡的土体处移开时，土体中将发生的最小的水平有效应力与垂直有效应力的比例，被下列方程式界定：

$$K_a = (1 - \sin\varphi) / (1 + \sin\varphi)$$

其中φ等于土壤内部摩擦力的角度。另请参考“主动土压力”词条。**压缩率系数（m_v）**指随着有效应力的增大而来的体积变化，即压力的倒数。**固结系数（C_v）**指一种土壤的渗透率与体积压缩率的比例，用于估计沉降发生的速度。确定c_v的方法有两种，即时间平方根法或时间对数法。**流量系数（C_d）**指通过一个孔、堰或管取得的实际流量与理论值的比例。**动态黏度系数（μ）**指流体流动及其阻力相关的常数。**静止土压力系数（K_o）**指在水平应变为零的一定深度下（例如挡土墙后）当表面被视为并未移动时，土体中将发生的水平有效应力与垂直有效应力的比例。**膨胀系数或热膨胀系数（α）**单位的长度随着温度升高而增加。**摩擦系数（μ）**指导致物体沿着平面滑动的力与垂直于平面的力的比例。**被动土压力系数（K_p）**指在一定深度下（例如挡土墙后）当墙壁的表面被视为向其所阻挡的土体移动时，土体中将发生的最大的水平有效应力与垂直有效应力的比例，被下列方程式界定：

$$K_p = (1 + \sin\varphi) / (1 - \sin\varphi)$$

其中φ等于土壤内部摩擦力的角度。另请参考“被动土压力”词条。**渗透系数（k）**指在某一单位水力梯度作用下水流的平均排放速度。**均匀系数（C_u）**指土壤级配曲线的斜率，定义为：

$$C_u = d_{60}/d_{10}。$$

coefficient of performance (COP) The ratio of total heat output to total electrical energy input (both in watts). Used to measure the efficiency of heat pumps. The higher the COP, the greater the efficiency of the heat pump.

性能系数（COP） 输出总热量与输入总电量（单位均为瓦特）之比，用于计量热泵的效率。性能系数越高，热泵的效率就越高。

cofferdam A temporary structure constructed from sheet piles. Used to support the surrounding ground and minimize pumping while construction within the ground takes place.
围堰 用钢板桩修建的临时结构。围堰既可以支撑基坑的坑壁又可以减少基坑施工时的排水量。

cohesion The molecular attractive forces that bind soil minerals (such as clays) together, sometimes referred to as the 'stickiness' of the material particles. It is a component of shear strength in *Coulomb's equation.
黏附力 使土壤材料（例如黏土）结合在一起的分子引力，有时被称为材料颗粒的"黏性"。在库仑公式中，黏附力是剪切强度的组成部分。

cohesionless soil (non-cohesive soil) *See* GRANULAR.
非黏性土 参考"颗粒"词条。

cohesive soil A soil that exhibits *cohesion, i. e. granular soil.
黏性土 有黏性的土，如：颗粒土。

coil 1. A length of material wound into a spiral. 2. A curved pipe used within air-conditioning systems or ductwork to heat (**heating coil**) or cool (**cooling coil**) the air passing across it. 3. An electrical device comprising a spiral of insulated electrical wire.
1. 卷/圈/匝 绕成螺旋形的一段材料。**2. 盘管** 空调系统中或通风管道中用来加热（加热盘管）或冷却（冷却盘管）流经的空气的弯曲管道。**3. 线圈** 由绝缘导线组成的环形绕组。

coincidence effect A reduction of the sound insulation of a sheet of material (such as partition wall) as a result of the critical frequency of a sound wave being the same as the flexural bending wavelength of the sheet of material. The resultant is an increase in the transmission of sound through the material.
吻合效应 因声波入射角度造成的声波作用与板材（如：隔墙）中弯曲波传播速度相吻合而使隔声量降低的现象，叫作吻合效应。吻合效应将导致入射声能大量透射到板材的另一侧去。

coir A fibrous material obtained from the outer husk of coconut. Used to manufacture rope, doormats, and floor coverings.
椰壳棕丝 从椰子外壳取得的纤维材料。可用来生产绳索、门垫和地毯。

coke A material derived from coal comprised mainly of carbon.
焦炭 一种源自于煤炭的材料，主要由碳构成。

cold bridge *See* THERMAL BRDGE.
冷桥 参考"热桥"词条。

cold joint A visible joint in concrete that occurs when concrete has been laid at different times.
冷缝 在先、后浇筑的混凝土之间所形成的明显接缝。

cold rolling (cold forming, cold working) A process that deforms steel or metal at room temperature (i. e. below its re-crystallization temperature) by squeezing it between a set of rollers. It causes defects to be formed in the crystalline structure, which in turn increases the yield strength and hardness of the metal.
冷轧（冷成型） 在室温条件下（即：低于金属再结晶温度条件下）辊轧挤压钢或金属的工艺。冷轧会导致金属的晶体结构出现缺陷，但会增加金属的屈服强度和硬度。

collapse The excessive and irregular shrinkage of hardwoods when dried in a kiln too quickly or for too long.
干缩 硬木窑干速度过快或时间过长而导致的过度和不规则收缩。

collapse grading The stability of a fire door during a rated period of resistance under specified fire conditions.
耐火等级 防火门在规定火灾条件下能阻止火势蔓延的能力（时长）。

collapsible form *Formwork that is designed to fold up or telescope inwards to take up less space or allow partial *stripping of the formwork to take place.
拼装式模板 可折叠或向内伸缩的模板，可以减少模板的占用空间，或允许部分模板先拆除。

collar A ring-shaped part of an object that guides, seats, or restricts another object.

轴环 物件上能起到定位、座接或拘束另一物件的作用的环形部件。

collar beam A horizontal tie, used in a **collar-beam roof**, to join similar *rafters on opposite sides of the roof to prevent the roof from spreading.
连系梁 在屋顶桁架中连接相对两边椽条，以避免屋架散开的水平拉杆。

C

collar boss A pipe fitting where sections can be drilled out to allow for future connections.
支管台 一种管接头，管座可以开孔作支管连接用。

collector well A vertical well with horizontal collection pipes radiating outwards at the base of the well.
集水井 井底有一圈外向的卧式雨水收集管的竖井。

Collimation line The line of sight through the cross-hairs of a surveying instrument (e. g. on a theodolite or level). **Collimation error** is the inaccuracy caused by the line of sight not being horizontal or out-of-alignment, e. g. error in corresponding *face left and face right readings. **Collimation method** is a technique used in levelling where the *staff readings are subtracted from the collimation line to determine the reduced levels. A **Collimation test** can be undertaken to determine the accuracy of the instrument. This test involves setting the instrument in the middle of two stations, with the stations being approximately 60m apart on level ground. A staff reading is taken from each station and a true level is obtained, and any Collimation errors will cancel each other out. The instrument is then placed about 3m beyond one of the stations, still in line with the stations. A staff reading is taken from the near station and using the true level the reading that should be obtained from the far station is calculated. The reading is then taken at the far station and if the two correspond, the instrument is considered accurate, if not adjustment to the instrument is required.
瞄准线 穿过测量仪器（例如经纬仪或水准仪）的十字准线的视线。**准直误差**即由于视线并未水平或未对齐而导致的误差，例如与盘左和盘右读数相对应的误差。**视准线法**即在瞄准线缺乏标尺读数以确定归化高程的情况下用于高程测量的技术。为了确定仪器的精确度，可进行**准直测试**，该测试中要把仪器放在两个站点中间，两个站点在水平面上相隔约 60 米。从每个站点各取一个标尺读数并获得真实的高程，且任何准直误差将相互抵消。随后把仪器放在距其中一个站点前约 3 米处，且仍与两个工作站成一直线。从较近的站点取一个标尺读数并使用真实高程对取自较远站点的读数进行计算。随后再从较远的站点取一个读数，如两个读数相互对应，则仪器被视为准确，如果不是则仪器需要调整。

collision load The load cause to a structure when a vehicle crashes into it.
碰撞荷载 车辆撞到构筑物上时对构筑物产生的荷载。

colloid A mixture where one element is mixed evenly with or throughout another. Examples of colloids include paint and pigmented ink.
胶体 一种成分与另一种成分均匀混合而成的混合物，胶质的实例包括油漆和有色墨水。

colonnade A row of evenly spaced columns. In reference to classical architecture these columns are joined by their *entablature.
柱廊 一排均匀布置的柱子。在古典建筑里，这些柱子上面通常伴有柱顶盘。

column A vertical structural component that acts as a strut or support.
柱子 起支撑作用的竖向结构构件。

column clamp Braces and yokes used to secure and tighten column formwork to ensure that no concrete leaks from the formwork.
混凝土柱模板支撑架 用来固定和拉紧柱子模板的支架和柱箍，以确保混凝土不会从模板中漏出。

column form (column formwork) *Formwork used to provide the mould or case for *in-situ concrete columns.
混凝土柱模板 用于为现浇混凝土柱子提供模子或模壳的模板。

column head Top of a vertical strut or support. The top of a column that supports a floor may be wider at the top to help distribute the loads and prevent the column puncturing through the floor.

柱头/柱帽 垂直支柱的顶部。柱头应比柱身宽，以更好地传递来自楼板的荷载，并避免柱子刺穿楼板。

column splice The joining of two columns end-on-end. Gusset supports and bolts or rivets are used to brace and hold the two columns together.
钢柱拼接 两节柱子端部对端部拼接。用角撑、螺栓或铆钉将两节钢柱拼接和固定在一起。

column starter (kicker) A concrete upstand cast on a concrete base used to secure column formwork. The raised piece of concrete is cast to the same size and profile as the main column. The formwork can then be fixed to the kicker and held firmly in position while the concrete column is being poured.
混凝土柱浇底 浇筑在混凝土基础上的混凝土柱的起始部分，可作固定模板用。混凝土柱的浇底部分应与柱身的尺寸和形状是一样的。模板可固定到混凝土柱的浇底部分；这样浇筑混凝土，模板不会晃动。

column yoke Brackets or braces that are clamped together to hold column formwork in position.
柱箍 夹在一起用来固定柱子模板的支架或支柱。

combed joint (tooth joint) A prefabricated joint between two comb- or saw-tooth pieces of material.
指接 两件梳齿或锯齿状接口材料之间的预制接缝。

comb hammer (scutch) A single- or doubled-headed hammer where small trimming combs can be inserted into the head. Used to cut and trim blocks to the required shape and size.
澳式石工锤 单头或双头锤，在锤头有塞入的细锯齿片。可用来修削砌块，使砌块达到要求的形状和尺寸。

combined system A below ground drainage system where the surface water and the foul water are discharged through a combined surface and foul water drain and sewer. Results in relatively low installation costs (less pipework) but increased sewerage costs. *See also* SEPARATE SYSTEM and PARTIALLY SEPARATE SYSTEM.
合流制排水系统 将地表水和污水混合通过地表和污水下水道和排水管道系统排放的地下排水系统。优点是安装成本低（较少的管道），缺点是污水处理费用高。另请参考“分流制排水系统”和“半分流制排水系统”词条。

combplate The tooth-shaped plate at the entrance and exit to an escalator.
梳齿板 设在自动扶梯入口和出口处的锯齿状板。

comfort index A single index that attempts to reflect human perceptions of thermal comfort. A number of different comfort indices are available including Predicted Mean Vote (PMV), globe temperature, *environmental temperature, *effective temperature, corrected effective temperature, *equivalent temperature, and *operative temperature. However, it is difficult to find one single index where everyone is thermally comfortable under all possible conditions.
热舒适指数 试图反映人对热舒适性的接受程度的单个指数。大量不同的热舒适指数均可适用，包括预计平均反应（PMV）、球体温度、环境温度、有效温度、修正有效温度、等效温度及操作温度。但很难找到能在一切可能状况下使所有人均感到热舒适的一个单一指数。

comfort zone A range of air temperatures, mean radiant temperatures, humidities, and air velocities where the majority of people achieve thermal comfort.
舒适区 使大多数人感到热舒适的空气温度、平均辐射温度、湿度及风速的一个范围。

commencement of work Start of a construction project. For contractual purposes, this is often the date when the contractor has possession of the site.
开工 工程开始进行。从合同角度考虑，通常把承包商入驻施工现场的日期称为开工日期。

commissioning The process of running the building services systems installed within a building for the first time and performing various tests and checks to ensure that they operate correctly.
调试 第一次运行建筑设备系统的过程，通过各种测试和检查来确保设备是正常运行的。

common bond A brick masonry wall pattern in which a header course of brick is laid after every five or six stretcher courses.
五/六顺一丁砌法 一种砖墙砌法，每隔五或六皮顺砖砌一皮丁砖。

common brick A general-purpose brick that is not considered decorative. Although it may be used in exposed positions it is not considered to be aesthetically pleasing.
普通砖 标准烧结砖，不作装饰用途。虽然普通砖也有用于外露位置的，但并无美观效果。

common brickwork Brickwork that does not need to be properly finished or faced as it will not be exposed. Brickwork to be covered in plaster and cladding is usually common. Bricks that are laid quickly, cut roughly, and not pointed fall under this category. Brickwork that is exposed, pointed, and neatly laid is called *facing brickwork.
裸砖墙 因为裸砖墙是不外露的，所以并不用对裸砖墙的表面进行适当处理。通常会在裸砖墙表面抹灰或安装外墙挂板。裸砖墙铺砌速度快、加工粗糙且铺砌后无须勾缝。外露、勾缝且整齐铺砌的砖墙称为饰面砖墙。

common joist (common rafter, common stud) One of a number of joists, rafters, or studs of uniform size placed at regular intervals within a construction.
通用搁栅（通用椽条、通用竖龙骨） 建筑内一系列均匀分布的统一尺寸的搁栅、椽条或竖龙骨。

common trench A trench that contains multiple services. Mains electricity, gas, street lighting, and telecommunications are usually laid at different levels and are covered by sand and warning tape so that future excavations can avoid or locate the services safely.
市政管沟 可以容纳多种公用设施管线的槽沟。包括电力、燃气、路灯和电信管线等，这些市政管线按不同高度铺设，铺设后再用砂子回填并留有警示带，方便以后开挖时避开或定位这些公用设施管线。

communication pipe The cold water service pipe that runs from the water main to a stop valve that is usually located next to the property boundary.
连接管 连接市政供水干管与截止阀（通常临接建筑红线）的一段冷水管。

compaction A rolling, ramming, or vibration process to make particles form a density configuration. Used in soil and concrete to expel air voids; the result is a decrease in volume but an increase in density.
压实 使颗粒形成更致密结构的辊压、夯实或捣振过程。用于土壤和混凝土以便消除气隙；压实的结果是体积减小但密度增加。

compaction factor test A test to establish the workability of fresh concrete. It is considered more accurate than the *slump test, but the apparatus is quite cumbersome and is therefore used more frequently in a laboratory than on site. The apparatus consists of two hoppers above a cylinder: the top hopper is gently filled with freshly mixed concrete so that no compaction occurs, the door is released at the bottom of the top hopper, and the concrete falls freely into the lower smaller hopper. The door is then released on the lower hopper and the concrete falls into the cylinder. Excess concrete is wiped clear and the density of the concrete in the cylinder is then calculated. This value is then divided by the density of the concrete of a fully compacted cylinder. This value is termed the **compaction factor**. The density of the fully compacted cylinder is obtained by compacting concrete in the cylinder in four equal layers.
压实系数测试 一种用于确定新浇混凝土和易性的测试，被视为比坍落度试验更准确，但所用装置很笨重，因此较多用在实验室中而非现场。装置包括置于柱体上的两个漏斗：在上漏斗中缓慢地装满新拌混凝土使其不发生压实，打开上漏斗底部的门，使混凝土自由落入下方较小的漏斗中，随后打开下漏斗的门使混凝土落入柱体内，把多余的混凝土擦去，然后计量柱体内混凝土的密度。用该密度值除以充分压实柱状混凝土的密度，得到的值即为**压实系数**。充分压实柱体的密度是通过把混凝土按四等分层压实为柱体而得到的。

comparator An instrument for comparing the properties of an object, system, or finish with standard properties.

比较仪 用于比较某一物体、系统的性能或对标准性能的满足程度的仪器。

compartment 1. A contained unit. 2. A fire-protected room or floor used to conceal and limit the spread of fire in the building. The compartment room has walls, floors, ceilings, and self-closing doors built to a specified fire resistance. The Building Regulations limit the size of compartment rooms and the travel distances to the nearest fire escape. Depending on the type of building, size, and the number of people using the building, the fire resistance will vary. The fire resistance of walls, floors, and doors are specified in terms of the time a component will resist the passage of fire. For compartment rooms a minimum size of door openings and escape routes are specified depending on the number of people who use the building. Approved Document B of the * Building Regulations provides guidance on the compartmentation.
1. 隔间 一个独立的空间。**2. 防火分区** 楼宇内用作阻止和延缓火势蔓延的房间或楼层。防火隔间的墙体、地板、天花和自闭门都是符合规定的耐火等级的。英国建筑规范会对防火隔间的尺寸以及防火隔间与最近逃生出口的距离进行规定。耐火等级取决于建筑物的类型、规模以及住户的数量。墙体、地板和门的耐火等级是用时间来表示的，即这些建筑构件能阻止火势蔓延的最长时间。防火隔间门洞的最小尺寸是有明文规定的，主要取决于大楼住户的数量。英国建筑规范文件 B 中就有关于划分防火分区的指示。

compass (magnetic compass) A device that has a magnetized pointer, which swings to always point to Magnetic North. Used as a navigational aid or in surveys such as a **compass * traverse**, where horizontal angles are determined by * magnetic bearings.
指南针（磁罗盘） 磁针总是指向磁北的仪器。可以作导航辅助或测量用，如罗盘仪导线，可通过磁方位角来测量水平角。

compatibility The suitability of a substance to be used in connection with another substance. For example, the compatibility of paint or plaster finishes to the base coat, so that it adheres correctly without forming defects such as * pinholing or * crawling.
兼容性 某一物质与其相邻物质的相容性。例如：面漆或抹面砂浆与底漆的兼容性，只有面漆或抹面砂浆与底漆是相互兼容的，它们之间才能黏结牢固，而不会出现诸如针孔或咬底这样的油漆缺陷。

compensating errors Mistakes that would cancel each other out, with the end result being accurate.
误差补偿 将误差相互抵消从而使最终结果准确。

compensation Payment or recompense for something by one party that has an effect on the other. Contracts often state what type of events will entitle a party to payment. **Compensable delays** are delays that result from specific actions or defaults specified in the contract. Where contracts specify the type of delays that will result in additional payment or extension of time, any other action would not result in payment under the contract. Also where contracts state events that are compensable, only those **compensable events** will result in additional payment. The New Engineering Contract (Engineering and Construction Contract) lists events such as variations, tests, and other similar items as compensation events.
赔偿 一方对受害的另一方的赔款或补偿。合同中通常会说明哪一类事件可让受害方获得赔偿。可补偿延误是由于合同中规定的具体事件或违约行为导致的延误。除了合同规定的延误原因可获得额外补偿或延期的之外，其他类型的延误均不能获得补偿。同样除了合同规定的可获得补偿的事件之外，其他事件均不能获得补偿。《新工程合同》（即《建设工程施工合同》）罗列了可补偿事件，如变更、试验和其他类似的事件。

competitive tendering (competitive bidding) Multiple parties bid for a project with the contract being awarded to the lowest price or the company that offers the service that is considered the best value.
竞争性招标 多方参加某一项目的投标，发包方将把合同授予报价最低或能提供最高性价比服务的公司。

completion 1. The point when all the tasks in a project have been successfully accomplished and the work is finished. 2. For contractual and payment purposes, the time when the project is sufficiently performed and is considered to have reached practical completion. When a project is considered to have reached practical completion, other events within the contract are triggered, for example, practical completion will result in either the release of * retention monies or the time period to hold retention monies commences. All standard forms of construction contracts have a provision for stating the date when all the work is expected to be finished; this is called the **completion date.**

1. 竣工 项目中所有的施工都已圆满完成。**2. 实际竣工日期** 出于合同目的或付款目的，把项目完成建造之日定为实际竣工日期。当项目被认定竣工后，将触发合同内的其他事件，如发包人应归还质量保证金给承包人；或开始计算发包人预留质量保证金的期限。所有标准施工合同都对项目完工日期（即"竣工日期"）有明文规定。

component A small self-contained part of a building, constructed from a number of smaller parts, and designed to perform a particular function, for example, a window or a pump.

建筑构件 建筑物中一部分小的独立构件，是由一系列更小零件构成的，每一独立构件都能发挥某一特定功能，如：窗或泵。

composite beam A beam constructed from two or more materials in order to improve its load-bearing properties, for example, a timber beam clad in steel.

组合梁 由两种或多种不同材料结合而成的梁，旨在提高梁的承载能力；如：木质梁外包钢板。

composite board (sandwich panel) A cladding panel, usually faced in metal, that contains a core of insulation material.

夹芯板 由两层金属面板和隔热内芯组成的外墙板。

composite construction (sandwich construction, mixed construction) A component constructed from two or more materials in order to improve its properties.

组合结构 由两种或多种材料组成，旨在提高结构性能的构件。

composite decking 1. Decking used to construct a * composite floor. 2. Decking manufactured from two or more materials.

1. 压型钢板 构成组合楼板的压型钢板。**2. 组合楼板** 由两种或多种材料组成的楼层板。

composite floor A floor that consists of a profiled metal deck spanning across the structure that is covered with concrete. Commonly used in multi-storey dwellings, particularly when the speed of construction is important.

组合楼板 由横跨整个结构的压型钢板及浇筑在其四周的混凝土组成的楼板。通常用于多层建筑，特别是工期比较紧的建筑。

composite material A mixture of different types of materials. One of the most commonly used composite materials used globally and also in the construction industry is * reinforced concrete (made from steel and concrete). Most composites are made up of just two materials, however, more are possible. It is also possible for a single generic material to be a composite material, e. g. * dual-phase steel that combines a ductile and lower strength phase with a hard high strength brittle phase.

复合材料 由多种不同性质的材料组合而成。在全世界的建筑行业中应用最广泛的复合材料是钢筋混凝土（由钢筋和混凝土组成）。大部分复合材料只由两种材料组成，但是也有由两种以上的材料组合而成的。复合材料也可以是一类材料，如：双相钢，就组合了高韧性低强度相体和高强度脆性相体。

compound beam (compound plated beam, flitch beam, flitched beam) A * composite timber beam comprising a steel plate sandwiched between two pieces of timber that are joined together at regular intervals.

钢木复合梁 由两根木板及中间的钢板按照一定间距拴接而成的梁。

compound girder (built girder, sandwich girder) The main supporting * compound beam.

钢木复合主梁 起主要支撑作用的钢木复合梁。

compressed air A specific form of air under greater than atmospheric pressure. Compressed air is usually utilized to power a mechanical device or to provide a portable supply of oxygen.

压缩空气 处在高于大气压的压力下的一种特殊形式的空气。压缩空气常被用作机械设备的动力或被用于提供便携式供氧。

compression The effect of squeezing, i. e. particles within a material being pushed closer together. A **compression** * **boom**, * **chord**, or * **flange** is one that is subjected to compression, as opposed to * tension. **Compression failure** occurs in something (e. g. a column) that has failed to take the amount of compressive forces imposed on it.
压缩 压聚在一起的作用，即把材料中的颗粒更为紧密地压在一起。**受压杆**、**受压弦**或**承压法兰**即为受到压缩力之物，与拉伸力相反。当某物（例如柱子）不能承受施加于其上的压缩力时就会发生**承压破坏**。

compression index (*Cc*) The slope of the linear portion of a curve of void ratio vs logarithmic effective pressure, using data obtained from a * consolidation test.
压缩指数（*Cc*） 孔隙率与对数有效压力之比的曲线的线性部分的斜率，所用数据来自固结试验。

compression wave (compressional wave) A mechanical longitudinal wave that propagates along or parallel to the direction of travel, e. g. a sound wave in air.
压缩波（压力波） 沿着或平行于行进方向而传播的机械纵波，例如空气中的声波。另请参考“*P* 波”词条。

compressive strength The ability of a material to withstand * compression, defined as a value quoted in terms of N/m^2.
抗压强度 材料承受压力的能力，抗压强度单位为 N/m^2。

compressor A mechanical device used to increase the pressure of a gas. Used in air-conditioning systems, heat pumps, and to produce compressed air. Various types of compressor are available, including axial flow, centrifugal, lobe, reciprocating, and rotary vane compressors.
压缩机 提升气体压力的一种机械设备。压缩机被用于空调系统和热泵系统，用来制造压缩空气。压缩机有下面几种，包括：轴流式压缩机、离心式压缩机、螺杆式压缩机、往复式压缩机和旋转叶片式压缩机。

compriband A generic name given to a wide range of joint-sealing materials (tapes, strips, and profiles) designed to accommodate movement. Used in a wide range of internal and external applications such as sealing windows and doors, sealing air-conditioning and ventilation ducts, and sealing movement joints in brick walls.
密封带/密封条 是一系列密封材料（密封带、密封条等）的通用名，用于缓冲位移。其应用非常广泛，可用于密封门窗、空调风管和砖墙上的伸缩缝。

computer-aided project management Use of software capable of producing fully resourced networks, resource plans, * Gantt charts, and resources histograms. The software uses algorithms to analyze the tasks, resources available, and fixed time activities to calculate the shortest and most cost-effective plan of work.
计算机辅助项目管理 利用软件建立全面可靠的网络、项目资源计划表、甘特图和资源柱状图。软件将利用算法分析作业、可用资源、限时活动来计算出用时最短、最经济的施工计划。

concentrated load (point load) A * load that acts or is applied over a very small area.
集中荷载（点荷载） 作用在非常小的面积上的荷载。

conchoidal facture A facture surface shaped like a shell, which is either concave or convex. Used to describe the irregular broken surface of minerals and rocks.
贝壳状断口 类似于贝壳（既可以是凹面也可以是凸面）的断口面，用于形容矿物和岩石的不规则断裂表面。

conciliation A dispute resolution process where an independent party acts as an intermediary between two parties attempting to find an arrangement that will overcome a dispute.
调解 一种争议解决过程，其中由独立方作为双方之间的中间人，尝试找到能克服争议的安排。

concrete A commonly utilized construction material composed of cement and aggregates. Concrete is used in many civil engineering applications such as pavements, structures, foundations, motorways, roads, and bridges due to its excellent compressive strength (up to $100\ N/mm^2$). The raw ingredients include * sand, * cement, and * aggregates.

混凝土 一种被广泛应用的施工材料，由水泥和骨料混合而成。因为混凝土抗压性能（抗压强度达到 100 N/mm^2）很好，所以混凝土被广泛应用于土木工程，如：路面、结构、基础、公路、道路和桥梁施工。混凝土组成材料包括：砂子、水泥和骨料。

concurrency Activities running at the same time. Where delays resulting from both client and contractor are running at the same time, the contract is considered to be in a period of **concurrent delay**. Where the design and manufacturing overlap and run at the same time this is called **concurrent engineering**.
并行 是指多个活动在同一时间发生。在同一时间内由业主和承包商引起的延误，称为**共同延误**。**并行工程**指设计和生产交替进行、同时进行。

condensation The process in which a gas cools down and changes into a liquid. In the case of moisture-laden air, if the temperature of the air is reduced below the * dew point temperature, the air can no longer retain the original quantity of water vapour, so water will be deposited as condensation. *See also* * SURFACE CONDENSATION and * INTERSTITIAL CONDENSATION.
凝结 气体冷却并变为液体的过程。如潮湿的空气，当空气温度降至露点温度以下时，空气就不再能够保留原有数量的水蒸气，水将沉积形成凝结。另请参考“表面冷凝”和“缝隙冷凝”词条。

conditions of contract (general conditions of contract, conditions of engagement) The legally binding terms and conditions that parties agree upon and which govern the expected actions of each party. Default of the terms by one party will usually entitle the other party to compensation or repudiation. The * clauses and * terms that make up the conditions can be found in the main body of the contract, usually placed between the recitals and the appendix.
合同条件（合同通用条件、聘用条件） 经当事方同意的管辖各当事方行动的具有法律约束力的条款和条件。一方对条款的违约通常将使另一方有权获得赔偿或得以拒付。组成条件的条文和条款将列在合同的正文中，通常置于序言和附录之间。

conduction The process by which heat flows along or through a material or from one material to another by being in direct contact. The rate at which conduction occurs varies considerably according to the substance and its state. Metals (e. g. copper) are good conductors. Gases are poor conductors. It is one of the three main heat transfer mechanisms. *See also* CONVECTION and RADIATION.
传导 热量沿着或通过一种材料流动，或通过直接接触从一种材料传至另一种材料的过程。传导的速度依据物质及其状态的不同而大有不同。金属（例如铜）是良导体，气体是不良导体。传导是三大主要传热机制之一。另请参考“对流”和“辐射”词条。

conductivity The ability of a material to transmit electricity. *See also* THERMAL CONDUCTIVITY.
导电性 材料导电的能力。另请参考“热导率”词条。

conductor A material that allows heat, electricity, light, or sound to pass along or through it. The **conductance** of a material is the degree at which it conducts heat, electricity, light, or sound.
导体 能够传导热、电、光或声音的材料。材料的传导性指的是材料传导热、电、光或声音的能力。

conduit A small covered channel, usually metal or plastic, used to distribute pipework and cables around buildings.
线槽 小型有盖的金属或塑料导管，可在建筑物内作管道和电缆布线用。

cone penetrometer The apparatus used in a **cone penetration test** to determine the * liquid limit of soil. The apparatus consists of a polished stainless steel cone, with a mass of 8kg having an angle of 30° and a dial gauge calibrated to 0. 1 mm. In the cone penetration test the cone is allowed to penetrate a soil sample, at a range of different moisture contents, confined within a tin 55 mm in diameter and 40 mm high. Penetration values are plotted against moisture content values and a best-fit line is drawn. The liquid limit of the sample is taken as the moisture content corresponding to 20 mm penetration.
圆锥贯入仪 圆锥贯入试验用仪器，可用于检测土壤的液限。圆锥贯入仪包括抛光不锈钢圆锥头，其锤重为 8kg，圆锥角 30°，刻度盘精度为 0. 1 mm。在圆锥贯入试验中，圆锥头可以贯入不同含水率的土样，土样应放在直径 55 mm、高 40mm 的锡盒内。画出贯入阻力随水分含量变化的最佳拟合线。把贯入深度为 20mm 时的含水率定为土样液限。

configuration management A process which identifies how each part of a project fits together and makes clear links with those responsible for each activity. By having clear links between tasks and those responsible, it is easier to manage change, those parties that need to make changes, and those affected by the change.
配置管理 确定一个项目的各部分如何组合并使各项活动的负责人与活动之间形成清晰联系的过程。通过使任务与其负责人之间形成清晰的联系，更便于对变化、需做出变化的有关方及受变化影响的有关方进行管理。

confined aquifer (artesian aquifer) An *aquifer whose water table lies above its upper boundary, which could either be above ground level or above an overlying strata of low permeability. *See also* ARTESIAN WELL.
承压含水层（自流水层） 水位高于其上边界的含水层，该水位可能高于地面或高于具有较低渗透性的上覆地层。另请参考“自流井”词条。

confined compression test *See* TRIAXIAL TEST.
侧限压缩试验 参考“三轴压缩试验”词条。

conflict resolution Informal, formal, and legally binding processes used to resolve disputes. *See also* ADJUDICATION, MEDIATION, CONCILIATION, AND ARBITRATION.
冲突解决 非正式、正式及具有法律约束力的过程，用于解决争议。另请参考“审裁”、“调停”、“调解”及“仲裁”词条。

conformity mark A manufacturer's stamp, tag, identifier, or trademark that signifies a standard or certification status.
合格标志 生产厂家的盖章、标签、标识或商标，证明产品符合标准或已通过合格认证。

congestion Excessive volumes of traffic or people making movement difficult.
堵塞 人流量或车流量过多，导致交通困难。

connector 1. A device that enables one item to be joined to another. Often enables the items to be joined together and taken apart when required. 2. (US) A highway that connects one highway to another.
1. 连接器 连接两件东西的器具。需要时可通过拆卸连接器来断开连接。**2. （美国）高速公路联络线** 连接两条高速公路的高速公路。

conservation The efficient and careful use of existing natural and human resources such as energy, the environment, and buildings, so that they are preserved for future generations.
保育 对现有自然资源和人类资源的保护，包括各种能源、环境和建筑，以备后代使用。

conservation area An area, usually of architectural, historical, or natural significance, that is designated by the local authority in order to ensure that it is protected or enhanced. Building and development work within the area is usually tightly controlled.
保护区 通常指有建筑、历史或自然价值的一块区域，这块区域是地方当局指定的，旨在对其进行保护和改善。保护区内通常严格控制建筑和开发项目。

conservatory 1. A room attached to a building that comprises mainly glazed walls and a transparent or translucent roof. Also known as a **sunroom**. 2. A glass constructed building used to grow plants.
1. 阳光房 与某一建筑物相连接的小玻璃房，包括玻璃墙和透明或半透明的屋顶。也称为“日光室”。**2. 温室** 植物栽培用的玻璃房。

consideration The part of a contract that has something of worth. For a contract there needs to be an offer, acceptance, and something performed or something of worth, e. g. agreeing to undertake a task for payment or buy something is the consideration.
对价 合同中具有一定价值的组成部分。对于一份合同而言，需要有要约、接受、需要履行的某事或有价值的某物，例如为了获得支付而承担某任务或购买某物即为对价。

consistency (consistency index, consistency limits) *See* ATTERBERG LIMITS.

稠度（稠度指数、稠度界限） 参考“阿太堡界限”词条。

consolidated-drained test A *triaxial test where free drainage of the sample is allowed during both the consolidation and axial loading stage. Loading is carried out at a slow rate so that no increase in pore pressure occurs. *Effective stress parameters (c'_d and φ'_d) are obtained from such a test and typically used for long-term stability analysis.
固结排水试验 三轴压缩试验，试样在固结和轴向加荷阶段允许排水。加荷速度不宜过快，这样孔隙水压力才不会增加。有效应力参数（c'_d and φ'_d）均可从固结排水试验中获得，并用于长期稳定性分析。

consolidated-undrained test A *triaxial test where free drainage of the sample is allowed (usually for 24 hours) under cell pressure, to allow the sample to consolidate or to become saturated. Drainage is prevented and pore pressure readings (u) are taken during axial loading. *Total stress parameters (Ccu and φcu) and *effective stress parameters (‘c’ and ‘φ’) are obtained. Typically, such parameters are used to analyze problems regarding sudden change in load after an initial stable period, e. g. rapid drawdown of water behind a dam, or where effective stress analysis is required, e. g. slope stability.
固结不排水试验 三轴压缩试验，试样在施加围压作用下排水（通常为24小时），使试样固结或饱和。在轴向加荷阶段是不允许排水的，应在此阶段读取孔隙水压力值（u）。由此可测得总应力参数（Ccu和φcu）和有效应力参数（‘c’和‘φ’）。通常，这些参数是用来分析初始周期后的荷载突变的有关问题，如坝体后的水位急速下降；或需要有效应力分析时，如边坡稳定性。

consolidation Decrease in volume of voids (expulsion of air or water) in a soil that is subjected to *compression. In a fully saturated soil the resulting increase in pressure is first taken by the pore water and is then gradually transferred to the soil particles over a period of time. The rate at which this transfer of load occurs depends upon the *permeability of the soil. In a high permeable soil (e. g. silty clay) the transfer of pressure from the pore water to the soil skeleton will occur very quickly, thus consolidation will be rapid. In a low permeable soil (e. g. clay) the transfer of pressure from the pore water to the soil skeleton will be slow, thus consolidation would occur over many years.
固结 压缩空气或水分逐渐排出的土壤中孔隙减少（的过程）。当饱和土受压后，其附加压力先由孔隙水压力承担，一段时间后再逐渐转移到土壤颗粒上。附加压力的转移速度取决于土壤的渗透性。在高渗透性土壤中（如：粉质黏土），附加压力从孔隙水转移到土壤骨架的速度非常快，因此固结速度也很快。在低渗透性土壤中（如：黏土），附加压力从孔隙水转移到土壤骨架的速度会很慢，因此土壤固结可能需要很多年。

consortium A group of people or organizations that come together for a specific task, project, or programme.
联营体 为具体任务、项目或计划而集结的一组人员或组织。

constant-head permeability test A test to determine the *coefficient of permeability (k) of highly permeable soils such as sands and gravels (i. e. $k > 10^{-4}$ m/s). Water from a constant-head supply flows through a soil sample until steady flow conditions prevail. The head loss (Dh) across the samples is then obtained by determining the difference between two manometers: one connected to the top of the sample and the other at the bottom. The head loss is divided by the length (L) of the sample between the manometers (device for measuring pressure) to obtain the *hydraulic gradient (i). The rate of flow (q) is obtained by measuring the quantity of water (Q) collected in a measuring cylinder over a measured time (t). Given that the cross-sectional area (A) of the sample is known, the coefficient of permeability can then be calculated from Darcy's equation:

$$k = q/Ai$$

常水头渗透试验 检测高渗透性土壤（如：沙砾，即$k > 10^{-4}$ m/s的土壤）的渗透系数（k）的试验方法。在常水头实验法中，水以常水头流经土样，直至水处于稳定流状态。可通过计算两个压力计的压力差来确定土样的水头损失（Dh）。一个压力计接在土样顶部，另一个则接在土样底部。水力梯度（i）=等于水头损失/试样高度（L）。渗流量（q）=量筒的总渗透水量（Q）/测试时间（t）。已知试样截面积（A），就可通过达西公式计算出土的渗透系数：

$$k = q/Ai$$

constant velocity grit channel A long narrow channel, parabolic in cross-section, that is designed to give constant *velocity with varying depth of flow. They are used in sewage treatment plants to allow grit particles to settle out. The maximum depth of the channel is calculated on the basis of a design particle which will settle at 0.02 m/s when the velocity is slowed to 0.3 m/s. This particle must reach the bottom within the length of the channel.
平流式沉砂池 抛物线截面的窄长形池，主要作用是确保不同水深的水是匀速流动的。污水处理厂用以沉淀和去除砂粒。其最大深度取决于颗粒的粒径，当污水流速为 0.3 m/s 时，砂粒应以 0.02m/s 沉淀。砂粒应在沉砂池段内沉淀到底部。

construction The process of creating or altering a building, structure, or object.
施工 建设或改造某一建筑物、构筑物或物件的过程。

construction contracts Standard forms of construction contract created specifically for building and civil engineering projects. The term is also used in the Housing Grants Construction and Regeneration Act 1996 meaning an agreement for carrying out construction operations.
施工合同/协议 专为建筑工程和土木工程制定的标准格式施工合同。英国在 1996 年发布的《住房补贴、建设和重建法案》中也用"construction contracts"一词来表示施工协议。

Construction Designand Management Regulations (CDM regs) A health and safety regulation that specifies health and safety responsibilities and management duties that must be performed by the client, designers, and contractors.
《建筑设计管理条例》(CDM regs) 健康与安全条例，要求业主、设计师和承包商应履行健康与安全责任和管理责任。

Construction Industry Training Board (CITB) Organization that organizes and subsidizes training and skills development for the construction industry. Construction companies pay training levies to the CITB, now largely replaced by Construction Skills. The companies can then draw from the organization subsidies for training employees.
建造业训练委员会(CITB) 为建筑业提供技能培训和培训补贴的机构。建筑公司先支付培训费给建造业训练委员会（现在逐渐被行业技能委员会取代）。然后建筑公司再从建造业训练委员会那里获取培训补贴。

construction joint A deliberately made joint in concrete that is formed when one area of concrete is laid at a different time to another. The joint should provide a good bond between the areas of concrete and should not allow movement.
施工缝 指的是先后浇筑的两块混凝土之间所形成的接缝。施工缝的设置应能保证两块混凝土之间的黏结力且不允许位移出现。

construction load The loads imposed on an element during construction.
施工荷载 施工过程中施加在某一建筑构件上的荷载。

construction management (construction management contracts) The management of a building project where one organization takes a fixed fee or percentage of the project value to organize and manage the designers, contractors, and subcontractors.
施工管理（施工管理合同） 工程项目管理，业主雇佣专业的工程管理机构来管理设计师、承包商和分包商的工作。管理机构将收取固定的一笔费用或项目合同价格的百分之几作为工程管理费。

construction manager The person who organizes and leads the on-site building process. Different companies and organizations may have different meanings for this, but the term is also interchangeable with *construction project manager, *site manager, and *site agent.
项目经理 组织和管理现场施工的人员。不同的公司和机构对项目经理的定义可能是不同的，本词与施工项目经理、现场经理和工地总管这几个词可以互换使用。

construction process A physical item of work that is undertaken during *construction.
施工工序 施工过程中某实体工作的实施。

constructive dismissal The process where an employee is wrongfully forced out of their job or position or is placed in a position that makes their job untenable. The illegal process means that an employee can bring an action against the employer for wrongful dismissal and should be entitled to compensation.
构定解雇（推定解雇） 雇员被非法解雇或被安排到不合适的岗位上工作。在这种情形下，雇员可以对雇主的非法解雇提起诉讼并要求获得相应的赔偿。

C

consultant A professional person providing expert advice in a given subject area. For example, a **consultant engineer** is a chartered engineer who advises the client on certain or all aspects of a project such as the design, suitability, and/or construction. They can be self-employed or work for a consulting firm.
顾问 在某一特定领域提供专业意见的专业人员。例如，顾问工程师，是就工程项目的某些方面或所有方面（如：设计、可行性和/或施工）为业主提供专业建议的特许工程师。顾问可以是直接受聘于业主的或顾问公司委派的。

consumer *See* CLIENT.
用户 参考“业主”词条。

consumer unit A box located inside a dwelling containing a *mains isolation switch and a series of *fuses or miniature *circuit breakers. It should be located as close as possible to the *cable tail intake on the internal wall.
用户配电箱 用户住处内部的配电箱，用户配电箱里边装有一个电源切断开关和一系列熔断器或微型断路器。用户配电箱应尽可能设在电缆尾线穿墙处的附近。

consumptive use The amount of water lost through evaporation, taken up by vegetation, or other natural processes.
消耗水量 通过蒸发、被植物吸收或其他天然过程而损失的水量。

contact bed *See* TRICKLING FILTER.
滤床 参考“滴滤池”词条。

contact pressure The distribution of the load beneath a foundation slab. In sand the actual contact pressure decreases from the centre to the edge, while in clays the contact pressure is higher at the edge than the centre.
接触压力 基础板作用于其下部的荷载分布。如果基础板下部是沙子，则接触压力将从中间向边缘逐渐减小，而如果基础板下部是黏土，则边缘处的接触压力要大于中间的接触压力。

contaminated land Section 78a (2) of Part IIA of the Environment Protection Act 1990 presents a legal definition of contaminated land as: ‘any such land which appears to the local authority in whose area it is situated to be in such a condition, by reason of substances in, on or under the land that (a) significant harm is being caused or there is a significant possibility of such harm being caused; or (b) pollution of controlled waters is being, or is likely to be caused’.
受污染土地 英国在1990年发布的《环境保护法》中，在IIA部分的78a（2）条对受污染土地作了如下定义：因为土壤中有污染物存在，导致地方当局所在的土地（a）明显受损害或很大可能产生明显的损害；或（b）对任何受管制水域造成或可能造成损害。这样的土地都称作受污染土地。

contemporaneous documents All of the paper-based documents that can act as evidence of events and transactions. In delay and disruption claims all such evidence shows the events and consequences relating to relevant events and the resulting delay and disruption.
同期资料/同期文档/同期证明文件 可以用作事件或交易的证据的所有纸质文件。在就工程延误申请索赔时，所有这些证据都能证明是关联事件的发生导致工程被延误或停滞。

contiguous bored piles *See* PILE.
钻孔灌注排桩 参考“桩”词条。

contingency Something that is set aside in case of events that have a likelihood of occurring but are not planned into the contract in the usual way, or an allowance made for unexpected events. A **contingency plan** identifies a specific course of action should the need arise; this may be for an event that may occur or an unexpected event. A **contingency sum** is an allowance of money for certain events that may occur.

或有事件/不可预见事件 为了具有发生的可能性但在通常情况下并未在合同中列入的事件而预留的某物，或为突发事件而做出的拨备。**应急计划**即列出如出现需求时的具体行动步骤；这可能是针对可能发生的事件或突发事件。**应急款**即针对可能发生的某些事件的款项拨备。

Continuing Professional Development (CPD) The process of updating one's knowledge and skills during a professional career. It is a requirement of all professional bodies that the members keep their knowledge and skills up to date and maintain records of their attendance on training courses.
持续职业发展（CPD） 在个人的职业生涯中提高知识和技能的过程。一切专业机构均要求其成员保持最新的知识和技能，并保持参加培训课程。

continuous flight auger (CFA) *See* AUGER.
螺旋钻孔（CFA） 参考“螺旋钻”词条。

continuous grading *See* GAP-GRADED AGGREGATE.
连续级配骨料 参考“间断级配骨料”词条。

continuously reinforced concrete pavement A *pavement made from Portland cement, which does not have any transverse expansion or contraction joints, but has continuous longitudinal steel reinforcement. Transverse cracking occurs, but these cracks are held tightly together by the reinforcement.
连续配筋混凝土路面 面层内配置纵向连续钢筋，但横向不设胀缝或缩缝的波特兰水泥混凝土路面。横向可能会出现细小的裂缝，但钢筋会把这些裂缝紧紧地拉结在一起。

contour line A line on a *map or plan joining points of equal height above datum, usually *ordnance datum but sometimes a local datum.
等高线 地图或平面图上把具有基准点（通常为水准零点但有时为地方基准）以上相等高度的各点连接起来的线。

contract The agreement between two or more parties comprising an offer, acceptance, and consideration. The contract may be a standard form of agreement between the *contractor and *client, such as a *construction contract or maybe a simple contract formed between the client and contractor.
合同 当事双方或多方之间设立的协议，包括要约、承诺和对价。合同可以是业主和承包商订立的标准协议，如施工合同；也可以是业主和承包商订立的简易合同。

contract administrator The person who oversees the construction project on behalf of the client, ensuring that the process, terms, and agreement are followed in accordance with the agreement.
合同管理人 代表业主监督整个工程项目的人员，合同管理人应确保工程施工过程、条款及协议均符合合同的规定。

contract bills The contractual documents that identify the rates, prices and quantities, and materials that should be delivered.
工程量清单 合同文件，其中列出单价、总价、工程量和应交付材料等项目。

contract documents All of the drawings, details, bills, terms, and conditions that make up the information submitted for tender and used to agree the contract.
合同文件 包括图纸、详细资料、工程量清单、条款及条件等投标及签约用资料。

contract drawings The illustrations, working details, and other graphics used to support the tender and contract documents at the point when the contract was signed and agreed.
合同图纸 构成投标资料及签署合同时的图解、施工详图和其他图表等。

contracted weir A *weir that does not extend the full width of the channel and thus has side and end contractions. Used for measuring the flow in open channels such as streams and rivers. *See also* SUPPRESSED WEIR.
有侧收缩堰 堰宽比渠道宽度要小，因此在侧面和末端收缩的堰，用于计量溪流及河流等明渠中的水流。另请参考“无侧收缩堰”词条。

contract hire An agreement to hire plant and equipment from a supplier for a fixed term.

设备租赁合同 承包人与机械设备供应商签订的定期租赁协议。

contraction joint (shrinkage joint) A joint within a concrete structure or slab allowing shrinkage to occur on drying.

缩缝 在混凝土结构或混凝土板上设置的收缩缝，用以预留混凝土干燥收缩。

C

contractor A person or organization that agrees to undertake the works.

承包商 指同意实施工程项目的人或机构。

contractor's obligations The terms, conditions, and warranties that the builder agrees to undertake for the client. The builder can assign benefits but cannot assign any contractual burdens (things the contractor agrees to do for the client) without express (written) agreement.

承包商的义务 建筑商同意为业主承担的条款、条件和保修义务。建筑商可以自行分配利润，但在未经业主（书面）同意情况下不得转让其合同义务（承包商同意为业主做的事）。

contract programme 1. The plan of work contained within the package submitted with the tender documents and agreed upon. The start, finish, and critical milestones are normally the only fixed parameters of the plan of works. Unless specified within the contract documents it is reasonable to expect other items or work to change as the project unfolds, weather conditions vary, and change orders are issued. 2. The live plan of work that is referred to during the contract. The plan of work will vary and develop slightly, making use of float, accommodating changes, and adjusting tasks to ensure full utilization of project resources.

1. 施工计划 与标书一起打包提交并据此签署的施工计划。通常，施工计划都会包括起始时间、结束时间和重要里程碑这三项。除非合同文件另有规定，在项目施工过程中是可以变更其他项目或工程的，如：天气变化或业主发出变更通知。**2. 动态施工计划** 工程施工过程中采用的动态控制计划。施工计划随时会有些许的变化和进展，利用浮动时间并根据变更调整和改善作业来确保项目资源得到充分的利用。

contract sum (contract price) The monetary figure fixed and agreed upon in the contract. The contract value can also have the same meaning or might have a different connotation depending on whether it is the expected profit, loss, or *earned value that is being discussed.

合同金额（合同价格） 合同中固定和约定的价格。合同价值有时也指合同价格，但有时也有其他含义，这取决于说的是预期利润、损失或挣值。

contract time The period of the project fixed and agreed in the contract. The period is measured from when the contractor is expected to enter the land to establish site set-up, to the date when the contractor hands over the completed project.

合同期限 合同中规定的工期。工期是从承包商进驻项目开始算起的，直至承包商完成并移交整个项目为止。

contra proferentem A rule used to decide the meaning of a term where it could have two or more different meanings. Where an ambiguous term with two or more meanings arises, the courts can construe the meaning such that the party against whom the term is used is allowed the more favourable definition. The party putting forward the claim would not get the benefit of the interpretation that they were implying and relying on.

不利解释原则/疑义利益解释原则 一条规则用于决定可能具有两个或多个不同条款的含义。如发生了条款模糊，具有两个或多个含义的情况，则法院可按照对该条款所适用的当事方更有利的定义来理解条款的含义。解释合同条款不会向有利于在条款中提出主张的一方进行解释。

contributory negligence Where the actions of an individual have increased their own risk of injury and may have added to it. In such situations, the injured party is still able to claim against the party that was negligent; however, the claim may be proportionally reduced to take into account the injured party's actions that negligently exposed themselves to and increased the risk of injury. Contributory negligence is governed by the Law Reform (Contributory Negligence) Act 1945.

共同过失 如某个人的行动已提高其自身受伤害的风险及可能已增大该风险，在此情况下，受伤害方仍可向有过失的当事方索赔；但在考虑到受伤害方有过失使其自身面临风险及增大受伤害风险的情况，该索赔可能会按相应比例被削减。共同过失受《1945年法律改革（共同过失）法案》管辖。

control gear The electrical and electronic equipment used to control discharge lighting, such as fluorescent, sodium, mercury, and metal halide lamps. Usually includes a starter and *ballast, but may include other items such as a power factor correction capacitor and dimming equipment.
灯控制器 用于控制放电灯（如：荧光灯、钠灯、汞灯和金属卤化物灯）的电气电子设备。通常包括一个启动器和一个镇流器；但有时也包括其他一些配件，如：功率因数校正电容器和调光器。

controlled water Any watercourse such as a canal, river, stream, including underground water. It is an offence to cause pollution to any of these.
受管制水域 任何水道，如运河，河流，小溪，包括地下水。使受管制水域受到污染是犯法的。

controlled waste In the UK, all waste (commercial, household, or industrial) that requires a licence for its disposal or treatment.
受管制废弃物 在英国，所有废物（商业废物、家庭废物或工业废物）的废弃或处理前应持有许可证。

controlled waste regulations Regulations governing controlled waste in the UK.
受管制废弃物的管理法规 与英国受管制废弃物相关的管理法规。

Control of Substances Hazardous to Health Regulations (COSHH) The regulations that ensure all materials are described so that the risks associated with handling and using them are clear to those who are likely to come into contact with these hazards.
危害健康物质的控制条例（COSHH） 该条例将确保所有材料均得以标注，使有可能面临有关危险的人均清楚关于处置和使用该材料的风险。

controls Electrical or mechanical devices capable of operating items of equipment according to predefined conditions.
控制装置 机电控制装置，可按预选设定的条件运行设备。

control valve *See* DISCHARGE VALVE.
控制阀 参考“排水阀”词条。

convection The process by which heat is transferred in a gas or liquid. As the gas or liquid is heated, it expands and becomes less dense. It then rises and is replaced with colder, more dense gas or liquid which in turn is then heated and rises. This results in a continuous flow or convection current from one area to another. It is one of the three main heat transfer mechanisms. *See also* CONDUCTION and RADIATION.
对流 气体或液体中的传热过程。当气体或液体被加热时会膨胀且密度降低，随后将上升且其位置将被较冷、较致密的气体或液体所取代，后者继而也被加热并上升，这将导致从一个区域向另一个区域的持续流动或对流。对流是三大主要传热机制之一。另请参考“传导”和“辐射”词条。

convector (convector heater) A space heating device that heats the space using *convection.
对流式空间加热器 用对流的方式加热空间的一种空间加热器。

conversion The process of cutting up tree trunks into sections prior to *seasoning. The aim of conversion is to maximize output in financial rather than volume terms.
木材切割 在风干之前把树干切为木段的过程。切割的目的在于获得最大的经济产出而非体积产出。

conveyor A mechanical device used to transport materials from one place to another. Usually comprises a continuous wide rubber belt supported on rollers.
输送机 用于把材料从一个地方输送到另一个地方的机械设备。通常包括一条较宽的橡胶制连续输送带，带有滚筒支撑。

cooling tower A device in which circulating air is used to cool warm water by *evaporation. Used to dispose of waste heat from the cooling coils of air-conditioning chillers and remove heat from coolant water in electricity generating plants.
冷却塔 利用循环空气来达到蒸发散热降低水温的装置。冷却塔可用来散去空调冷水机组冷却盘管产生的余热，或带走发电厂冷却水产生的热量。

coordinates A set of numbers that details the exact position of something with reference to a set of axes. *See also* CARTESIAN COORDINATES.
坐标 详细说明某物相对于一组轴线的准确位置的一组数字。另请参考“笛卡尔坐标”词条。

C

coordination The act of aligning tasks, activities, and resources so that they are managed effectively. **Coordination meetings** are used to bring together all of the contractors, subcontractors, and designers necessary to ensure that tasks and activities can be organized so that they are properly integrated together and managed effectively.
协调 安排好各项作业、活动和各种资源，确保它们的高效运行。协调会议是指召集所有承包商、分包商和设计师来开会，确保各项作业和活动的安排是合理的、紧密配合的，这样整个工程的施工才能有效运行。

coping A protective covering placed on top of a wall or parapet to prevent rainwater penetration. Usually projects slightly from the face of the wall.
压顶 在围墙或女儿墙顶部安装，以防止雨水渗入墙体的保护材料。通常，压顶会凸出墙面一些。

copper (copper alloys) A non-ferrous metal, chemical formula Cu. Copper is used in buildings for plumbing pipes (due to its excellent corrosion resistance) and fittings, and sometimes in sheet form for roofing. Due to its electrical conductance and malleability, copper is the predominant material used in electrical wiring in buildings and dwellings. Copper can be alloyed with zinc to form * brass or with * tin to form * bronze. These alloys, especially brass, find many applications in plumbing and electrical fittings. Pure copper has a density of 8, 900 kg/m^3.
铜（铜合金） 一种有色金属，化学式为 Cu。铜在建筑物中用作给排水管（由于其优秀的抗腐蚀性）和管件，有时以铜板形式用于屋顶。由于其导电性和延展性，铜是建筑物和住宅中的电气布线所用的主要材料。铜可与锌构成黄铜合金或与锡构成青铜合金，此类合金（尤其是黄铜）大量应用于给排水和电气装配。纯铜的密度为 8900 千克/立方米。

corbel A section of brickwork, masonry, timber, or other material that projects outwards from the face of a wall to support a load above.
牛腿 从墙面上凸出来用于支撑上部荷载的砖、石、木材或其他材料做成的构件。

corbelling Layers of brickwork or other materials that successively project out from the face of a wall. *See also* CORBEL.
叠涩 用砖或其他材料层层堆叠向墙外挑出。另请参考“牛腿”词条。

corduroy road A road formed from logs tied together at each end. Used for crossing muddy or swampy ground.
木排道 使用从两端捆在一起的原木组成的道路，用于跨越泥泞或沼泽地面。

core 1. The central part of something. 2. A cylinder column of material, typically soil or concrete, obtained during a sampling. A **core barrel** is a steel tube with a coring bit (i. e. a cutting bit with pieces of tungsten carbide or industrial diamonds set in a soft metal), which is used to obtain an undisturbed sample of strata during a site investigation. *See* SOIL SAMPLER. A **core box** is a container with dividers, used to keep cores from a borehole. A **core catcher** is a steel spring in a soil sampler that is used to retain a sand sample. A **core cutter** is an attachment that can be fitted to a core barrel to break and grip the core so that it can be extracted from the ground for examination. A **core drill** is a water-cooled industrial diamond-bitted power-drill tool, used to extract rock cores from the ground. A **core test** is a crushing test undertaken on a concrete core sample.
1. 核心 物体的中心部分。**2. 芯** 圆柱形物质，一般指土壤芯样或混凝土芯样，是从钻孔取芯而来的。**岩心管**是与取芯钻头（如装在软金属上的碳化钨或工业钻石切割钻头）连接的一段钢管。在土地勘测过程中，岩心管可用于在地层中取不受动扰的土样。参考“取土器”词条。**岩芯箱**是用来储存芯样的带分隔条的容器。**岩芯爪**是取土器里边的不锈钢弹性爪，可用于夹住土壤芯样。**岩芯钻头**是固定到岩心管上的用于打碎岩土的配件，这样才可以从土壤中钻取岩芯，以备检验。**岩芯钻机**是工业级水冷式金刚石钻头电动钻机，可用于从地基中钻取岩芯。**取芯检验**是混凝土芯样的压碎试验。

cornice 1. An internal decorative moulding used at the junction between the wall and the ceiling. 2. An external decorative moulding used over openings or at the top of an external wall.
1. 天花装饰线条 安装在天花和墙面相交处的室内装饰线条。**2. 门/窗套线/檐口线脚** 在门洞/窗洞上方的装饰线条或外墙顶部的装饰线条。

corridor An enclosed passageway inside a building used to access rooms that lead off it.
走廊 建筑物内有顶的通道，走廊可以通向各个房间。

corrosion The most common form of degradation mechanism for materials, especially metals.
腐蚀 材料的退化机制的最常见形式，尤其是对金属而言。

corrugated sheeting A sinusoidal-shaped sheeting material used for cladding, and available in a range of materials including aluminium, steel, GRP, and PVC.
波纹板 正弦曲线型薄板材料，可用作外墙板，有一系列材质供选择，包括：铝、钢、玻璃钢和 PVC。

cost breakdown The separation of items, resources, tasks, and projects and their associated costs and values. The ability to separate costs provides greater financial control. Comparisons of estimated against actual cost for individual items ensure that those items that result in loss and those that make a profit can be identified.
成本明细 对单项、资源、任务和项目以及其相关成本和价值的细分。把成本细分的能力可带来较大的财务控制力。各单项的估计成本与实际成本的比较能确保将亏损项和盈利项区分开来。

costing The calculation of the monetary worth of work by measuring the materials and labour used and multiplying the quantities by the rate agreed. The actual cost can be compared with the amount fixed in the contract documents.
成本核算 通过计量所用材料和劳务并把数量乘以议定费率，对工程的货币价值的计算。可用实际成本与合同文件中所列的金额作对比。

cost-plus contracts Projects where the price of the work is covered and a rate is agreed to cover the service or profit.
成本保利合同 涵盖了工程价格，并以约定对服务或利润的涵盖费率的项目合同。

cost-plus-fixed-fee contract The cost of the work performed by the contractor is covered by the client and a set amount of profit and contractor's overheads are agreed.
成本加固定费用合同 承包商发生的工程成本由业主负担，且利润和承包商管理费用的固定金额已约定的合同。

cost-plus-percentage contract The cost of the work is covered and a percentage rate for overheads and profit is calculated against the total amount of work completed. Such remuneration schemes offer little incentive for work to be undertaken quickly and efficiently, the more resources used the greater the cost and profit. Work of this nature should be performed with close supervision.
成本加成合同 涵盖工程成本，且按照已完成工程的总量计算管理费和利润的百分比率的合同。此类报酬方案对于快速和有效施工的激励不高，所用的资源越多，成本和利润就越大。对于此类性质的工程应予密切监督。

cost-reimbursement contract Contracts that are completed and the price of the work is paid as the work is completed, profit and overheads may be covered by a fixed figure or percentage of the value of the work paid. Such contracts are normally only performed where there is insufficient time to measure work and prepare bills.
成本补偿合同 已完成且按照工作完成情况支付工程价格，利润和管理费按固定数额或百分比涵盖于已支付的工程价值中的合同。一般仅在没有足够时间计量工程和编制清单的情况下才执行此类合同。

Coulomb's earth pressure theory A concept (developed in 1776) used in retaining wall analysis for

*cohensionless soils in which a failure wedge develops behind a retaining wall; the failure plane rises at an angle from the base of the wall, *See also* RANKINE THEORY.
库仑土压力理论 用于挡土墙分析的概念（于 1776 年提出），适用于在挡土墙后形成破坏楔体的无黏性土；破坏平面按照与墙壁基础的角度抬升，参考“朗肯理论”词条。

C

Coulomb's equation An equation developed by Charles Augustin Coulomb (1736—1806) that defines the shear strength (τ_f) of soil to the normal stress (σ_n) on a failure plane as illustrated below. Where c is *cohesion and ϕ is *angle of shearing resistance, the equation can be represented in terms of *total and *effective stress parameters.
库仑公式 查尔斯·奥古斯汀·库仑（1736 年—1806 年）发明的公式，用于界定如下所示的破坏面上对于正应力（σ_n）的土壤剪切强度（τ_f）。其中 c 是黏附力而 φ 是剪切阻力的角度，该公式可通过总应力参数和有效应力参数表示。

Council of Engineering Institutions (CEI) An association founded in 1965 for the engineering institutions, but was replaced in 1983 by the *Engineering Council.
英国工程学会（CEI） 一个面向工程机构的协会，创立于 1965 年，但已在 1983 年被英国工程委员会取代。

counter ceiling *See* FALSE CEILING.
假吊顶 参考“假天花”词条。

counterclaim The defending party to a claim serves a claim for damages against the claimant.
反诉 一项索赔的被诉方向索赔方提出损害索赔。

counterjib The rear part of a crane's *jib that carries the counterweight.
平衡臂 起重机吊臂的带有配重的后部。

counter-offer Following an initial offer by one party, an alternative or different offer is made by the other party. For a contract to be made there must be consideration and an offer and acceptance. If an offer to do something is made and the other party adds or alters the offer, this is not a contract, but it is a counter-offer and the party undertaking the work must agree to the alterations for the contract to be complete.
还价/还盘 在一方首先报价后，另一方做出的替代或不同报价。对于将要达成的合同而言，必存在对价和报价及接纳。如与某事项有关的报价被做出而另一方对此报价予以增添或更改，则此并不成为一份合同而仅是一个还价，收到还价的当事方须同意该更改才能使合同达成。

countersink 1. A shallow enlarged hole drilled in a material to enable the screw head to sit flush with the surrounding surface. 2. A tool used to produce a countersink.
1. 埋头孔/沉头孔 在材料表面钻的一个又浅又大的孔，埋入螺丝后螺丝头与相邻表面齐平。**2. 埋头钻/锪钻** 用来加工埋头孔的工具。

couple Two horizontal forces that are equal in magnitude but opposite in direction, which causes a turning effect.
力偶 大小相等但方向相反的两个水平力，将导致转动效应。

coupler (coupling) 1. A pipe fitting that enables two pipes to be joined together. 2. A device that enables scaffolding poles to be connected together.
1. 管接头 连接两根管道的管件。**2. 扣件** 连接脚手架杆件的装置。

course A horizontal layer of masonry, one unit in height, including any bedding material. Also used to describe horizontal layers of materials such as tiles or damp-proofing. Materials that are laid in courses are said to be **coursed**, for instance, **coursed brickwork**, **coursed blockwork**, etc.
层/皮 水平砖石层，一层砖的厚度，加上灰缝的厚度。也可表示水平的材料层，如瓦或防水层。按层铺设材料就称为砌，如：砌砖墙、砌块墙。

cove A concave-shaped moulding placed at the junction between a ceiling and a wall.
(天花板下的) 凹圆线脚 安装在天花板与墙面相交处的凹圆线脚。

cover 1. A lid or piece of material that can be placed over a surface, either to protect or to hide it. 2. The thickness of concrete that is required to be placed over steel reinforcement to protect it.
1. 盖 盖子或表面覆盖材料，可起到保护或隐蔽的作用。**2. 钢筋保护层厚度** 用于覆盖保护钢筋的混凝土的厚度。

covering Materials used on walls, floors, and roofs to hide the structure.
覆盖材料 用来覆盖墙面、地面或屋面以隐蔽结构的材料。

covering-up The process of covering over previous work.
隐蔽 将前一道工程覆盖、掩盖起来的过程。

cover meter A non-destructive test instrument, used to check the position and direction of reinforcement in concrete.
钢筋保护层厚度测定仪/钢筋仪 用于检测混凝土中的钢筋位置及分布的非破坏性测试仪器。

cowl A device, usually terracotta or aluminium, that is placed at the top of a chimney to prevent down-draughts, assist ventilation, and reduce rain penetration and bird entry. Available in a wide range of shapes and sizes and can be static or revolving (wind-driven).
烟囱帽 安装在烟囱顶部的装置，通常为陶土或铝材，可以防止出现烟囱倒灌现象、辅助排气、防止雨水渗入烟囱或小鸟掉进烟囱。有各种各样的形状和尺寸的烟囱帽可供选择，如固定式烟囱帽或旋转式烟囱帽。

crab A small portable winch.
活动绞车 小型可移动式绞车。

craft A skilled occupation or trade, for instance, bricklaying, carpentry, joinery, painting, plastering, or plumbing. The person who undertakes a particular occupation or trade is known as a **craftsperson** or **craft operative**. A **craft foreman** supervises the work undertaken by the craft operatives.
工艺 需要手工技巧的职业或专业，如：砌砖、粗木工、细木工、油漆、抹灰或水管安装。从事这种特殊职业或专业的人就称为**工匠或技工**。**技术工长**负责监督技工的工作。

cramp 1. A hand-operated tool comprising two jaws and an adjustable screw that is used to hold pieces of wood together while they are being glued. 2. A metal strap attached to a door or window-frame that is used to fix the frame into an opening.
1. 木工夹紧器 木板涂完胶水后，用来拉紧多块木板的手动操作工具，包括两个夹具和一个可调节螺杆。**2. 门/窗框上的固定片** 门框或窗框上的金属拉结片，用于把门框或窗框固定到门洞或窗洞上。

crawler track The tracks used on construction and earth-moving equipment.
履带传动 用于建筑和挖土设备的传动系统。

crawling A defect in paintwork where the paint tends to pull away (crawl) and bead on the surface. Caused by a lack of adhesion to the substrate
咬底 一种油漆缺陷，油漆有从表面脱落的趋势，表面开始形成小水珠。这是油漆与基层黏结不牢固导致的。

crawl space A shallow space under the ground floor of a house or in the loft just under the roof that is not large enough to allow a person to stand up. Used to gain access to services, such as ductwork, electrical cables, and pipework.
管线空间 房屋底层下面或屋顶阁楼里面供风管、电线和水管维护用的较矮的空间（空间较矮，人不能起来）。

crawlway A duct containing services that is large enough for a person to crawl through.
爬行检修道 容纳设备的管道，里面足够大到人可以在里边爬行。

crazing A defect in paint, plaster, or concrete where a random series of tiny hairline cracks appear on the finished surface.

微裂 一种油漆、灰浆或混凝土表面的缺陷，表现为一些不规则细小的发丝裂纹。

creasing One or more courses of tiles that project out from the face of a wall. Used as a coping and at window sills to project water away from the wall.
压顶砖 安装在砌筑墙顶部且凸出墙面的一层或多层砖。可作墙体或窗台压顶用材料，防止雨水渗入墙体。

creep Permanent deformation of materials that takes place usually at elevated temperatures, especially in metals. Creep in concrete occurs at ambient temperature.
蠕变 材料（特别是金属材料）的永久性变形，蠕变现象通常在高温情况下发生。混凝土在室温条件下发生的变形称为蠕变。

crest The top of something, e. g. the **crest of a wave** is the top of a wave or the **crest sight distance** is used in the design of carriageways where the rate of change of the summit is designed to accommodate adequate sight for overtaking.
顶部/峰值 某物的顶部，例如**波峰**即为波的顶部，而**顶视距离**则用于车道的设计中，在设计最高点的变化率时计入足够的超车视野。

crib Steel or timber members laid at right angle to each other, used to spread a foundation load. *See also* GRILLAGE. A **crib dam** is a retaining structure formed from stacked stones or *precast concrete units. A **crib wall** is a series of pens made from prefabricated timber, precast concrete, or steel filled with granular material. It acts like a gravity wall but can withstand large differential settlements. **Cribwork** is a series of timber cells filled with concrete and are sunk to form bridge foundations.
笼 由横向和纵向钢条或木条成直角堆叠而成，可用于分散基础荷载。另请参考“格排基础”词条。**木笼填石坝**一种挡水建筑物，木笼里边堆叠着石块或预制混凝土砌块。**框格式挡土墙**是由一系列预制木条、混凝土或钢条组成的内填颗粒材料的框格式建筑物。其作用与重力墙类似，但能承受不均匀沉降。**木笼基础**是用木材堆叠而成的木笼，木笼里边填有混凝土，沉入水底形成桥梁基础。

Crimp and Bruges' formula Used to calculate flow in sewers:

$$v = 56m^{0.67}\sqrt{i}$$

where v = velocity (m/sec), m = hydraulic mean depth (m) and i = the slope of the sewer. It is now considered that the equation yields conservative estimates (of the order of 20%) for the value of v when compared to experimental data. *See also* BARNES FORMULA.
柯瑞普和布鲁日公式 用于计算下水道中的水流：

$$v = 56m^{0.67}\sqrt{i}$$

其中 v = 速度（米/秒），m = 平均水深（米）及 i = 下水道斜率。目前认为，与试验数据相比，公式得出的结果是对 v 值的保守估计（大约20%的程度）。另请参考“巴尔纳斯水流公式”词条。

crippling load (buckling load) The load at which a column starts *buckling.
屈曲荷载 柱子开始屈曲的临界荷载。

critical Something that has extreme importance at a certain state. **Critical angle** is when complete reflection of a ray of light from a surface occurs. **Critical circle** is a failure slip circle in a slope stability analysis with the lowest factor of safety. **Critical height** is the height to which a *cohesive soil will stand without support. **Critical slope** is the maximum angle a slope can stand without deformation. **Critical velocity** is when the *Froude number equals one for flow in an open channel or when laminar flow changes to turbulent flow in a pressure pipe. *See also* BELANGER'S CRITICAL VELOCITY and KENNDEY'S CRITICAL VELOCITY.
临界的/关键的 某事物在某一状况下具有极大的重要性。**临界角**即为当光线在表面发生全反射时的角度。**临界圆**即为在坡面稳定性分析中安全系数最低的破坏滑弧。**临界高度**即黏性土壤在无支撑的情况下可维持的高度。**临界坡度**即为斜坡在不变形的情况下可维持的最大角度。**临界速度**即为在明渠中水流的弗劳德数等于1或压力管道中层流变为湍流时的速度。另请参考“伯朗格临界速度”和“肯尼迪临界速度”词条。

critical path The longest path of events that has no float between them, which dictates the total duration of

the project. Through critical path analysis logical links are established between task and a network of events. Tasks that do not fall on the critical path have float and potential for deviation without affecting the total duration of the project; these tasks are not critical.
关键路径 项目中最长的路径，上面没有浮动时间，其时间决定了整个项目的工期。通过关键路径法分析，可以建立任务之间的逻辑关系。不在关键路径上的任务是可以有浮动时间的，即有潜在的偏差而不影响整个项目的工期；而这样的任务就不是关键任务。

critical path method (CPM) [critical path analysis (CPA), critical path scheduling] Logical links between tasks are established and the start, finish, duration, and float of each task is determined. The links and activities are analyzed to determine which tasks are critical and have no float and those where some changes in start times can take place. Float is the ability of an event to slip backwards without affecting other events. Events that can change, be brought forwards, or move backwards, are non-critical. Slippage in non-critical events does not affect the overall completion time, unless excessive slippage makes them critical to the delivery. The critical path analysis determines the earliest start time, latest start time, earliest finish, and latest finish of each task. Where the earliest finish and latest finish are the same, the events are critical. Those events that have different earliest and latest start and finish times have a certain amount of float and are not critical.
关键路径法（CPM）[关键路径分析（CPA），关键路径进度计划] 建立任务之间的逻辑关系，确定每个任务的开始时间、结束时间、工期和浮动时间。可通过分析活动之间的关系来确定哪些活动是关键活动、哪些活动的浮动时间为零，或哪些活动的开始时间是可以往后推迟的。自由浮动时间指在不影响其他后续活动的最早开始时间的情况下本活动可以推迟的时间。可以变更（包括提前或推迟）开始时间的活动不是关键活动。非关键活动的延误不会影响整个项目的工期，除非过度的延误让这些活动成为关键活动。关键路径法可以确定每个任务的最早开始时间、最迟开始时间、最早结束时间和最迟结束时间。如果活动的最早结束时间和最迟结束时间是同一时间，则活动为关键活动。如果活动的最早开始时间和最迟开始时间、最早结束时间和最迟结束时间不是同一时间，则说明有自由浮动的时间，也就是说活动是非关键活动。

critical state (critical state theory) A condition at which phases of substances are identical or in equilibrium. For example, in soil mechanics when soil is sheared without incurring any further changes in density or stress level.
临界状态（临界状态理论） 物质的相处于相同或均衡情况下的状态，例如，在土壤力学中，临界状态即为土壤受到剪切但在密度或应力水平方面尚未发生任何进一步变化的状态。

cross-bracing Diagonal bracing or tie rods that resist shear forces from wind loads etc. to provide internal stability/stiffness in roof trusses and rectangular frames.
剪刀撑 可抵抗风荷载等作用下的剪力的斜撑或拉杆，保证屋架和矩形构架的内部稳定性/刚性。

cross-cut A cut made with a saw in a piece of wood at right angles to the *grain.
横切 锯切方向与木纹垂直的切削。

crossfall A slight gradient on a footpath or a carriageway to prevent ponding of surface water.
横坡度 人行道或车行道上可避免道路表面积水的横向坡度。

cross-hairs Fine lines that cross at right angles (in the shape of a + sign) in a telescope or theodolite, which are used to precisely align the horizontal and vertical lines of sight.
十字准线 望远镜或经纬仪中成直角（按 + 号形状）交叉的细线，用于精确地对齐视野的水平线和垂直线。

crossing Applying coats of paint with brush or roller strokes at right angles to the direction of the previous coat to ensure a smooth finish.
（横竖）交叉刷涂法 用毛刷或滚筒刷与先前一组刷涂成直角方向涂刷涂料，以确保涂膜光滑均匀。

cross-machine direction (x-machine direction) The direction perpendicular to the *machine direction that determines the width of something, for example, a geotextile. Typically, this is determined by the width of the bed of the machine.

横向 与机器运转方向垂直的方向，决定物体的宽度，如土工布的宽度方向就是横向。通常情况下与机床宽度一致。

cross-section A view of a component or building as if the item was cut in order to provide an internal perspective.
剖面图 某一构件或建筑物的视图，是通过对有关图形按照一定剖切方向所展示的内部构造视图。

cross-wall A structural internal wall that intersects the external wall at 90 °.
横墙 与外墙垂直的承重内墙。

cross-welt A type of joint used in sheet-metal roofing.
咬口 金属屋面相互固定连接的一种构造。

crown 1. The upper-most part of a building. 2. The middle of a cambered road.
1. 建筑物顶部 2. 拱形路面的中间部分

crown course The top *course of masonry on a wall. In brick walls, the crown course often comprises a brick on its edge.
顶层 砌体结构中的最顶层。最顶层的砖通常是侧砌的。

crown plate A purlin that sits on the top of the *crown posts in a *crown post roof to support the *collars. Also known as **a collar purlin** or **collar plate**.
中柱枋 屋顶桁架结构中位于桁架中柱上方用来支撑平顶搁栅的檩条。也被称为“collor purtin”或“coller plate”。

crown-post The vertical member in a *crown-post roof that spans from the *tie beam to the *crown plate.
桁架中柱 屋顶桁架结构中连接系梁和中柱枋的立柱。

crown-post roof A traditional timber *roof truss where a *crown-post spans from the centre of the tie beam to the *crown plate, which in turn supports the *collars.
桁架屋顶 一种传统的木桁架屋顶，在木桁架结构中，桁架中柱横跨系梁和支撑平顶搁栅的顶板。

crown rafter The central *common rafter in a *hipped end roof.
主椽 在坡屋顶结构中三角形坡面中间的通用椽条。

crown-strut A vertical timber roof truss member that spans from the *tie beam to the *collar.
顶拉杆 木桁架屋顶结构中连接系梁和平顶搁栅的支撑立柱。

cryogeology A term used with general reference to frozen/freezing the ground. The prefix cryo-is derived from the Greek word krous, meaning ‘icy cold’.
冻土学 研究冻土的学科。前缀 cryo-是从希腊词 krous 派生而来的，是“冰冷”意思。

crystalline Any material that has an ordered internal atomic structure.
晶体材料 内部原子结构呈现有序排列的任何材料。

crystallinity A measure of how *crystalline a material is.
结晶度 用于反映聚合物中结晶相含量的指标。

crystallization A process whereby the internal structure of a material becomes more ordered. Crystallization is an act of forming crystals or bodies during solidification, the crystals are normally symmetrically arranged.
结晶 材料内部结构变得更加有序的过程。结晶是物质凝固形成晶体的过程，晶体通常是对称排列的。

crystal structure How atoms or ions are arranged or spaced inside a material. For example, most metals have either a *body-centred (bcc) or a *face-centred cubic (fcc) structure.
晶体结构 原子或离子在某种材料中的排列或空间分布方式。例如，大多数金属都是体心立方结构（bcc）或面心立方结构（fcc）。

cubicle A small enclosure, i. e. a changing cubicle or a shower cubicle.

隔间 小隔间，如更衣室或淋浴房隔间。

Culmann line A graphically constructed curved line, which represents values of earth pressure for randomly chosen failure planes behind a retaining wall.
库尔曼线 一条以图形方式创建的曲线，这条曲线表示机选择的破裂面（挡土墙后面）土压力的大小。

culpable delay Part of the delay for which the client is responsible. A delay may be a result of changes and delays by the contractor, client, or designer. Where a delay is a result of actions taken by more than one party, the amount that results from the client's actions needs to be determined.
不可原谅的延期 业主负有部分责任的延期。延期可能是承包商、业主或设计师不能按时完成任务或需要变更工程引起的。如果延期是多方原因造成的，则应单独计算因业主原因导致延期的时间和损失的费用。

culvert A covered *channel used to convey a watercourse below ground, mainly under roads and railways.
涵洞 在路基下设置的排水孔道，通常设在公路或铁路下方。

curling The uplifting of the edges of a laminate or wooden floor due to the upper surface having a greater rate of shrinkage than the lower surface.
翘边 强化木地板边缘翘起，因上表面比下表面收缩速度快造成。

curtailment The bending of reinforcement bars at the end of the bar to ensure greater anchorage within the concrete.
钢筋弯钩 压弯钢筋的末端，以提高筋在混凝土里的锚固力。

curtaining A defect in painting that occurs on vertical surfaces where there has been excessive movement in the paintfilm, resulting in it sagging into thick lines, caused by the uneven application of the paint.
流挂 一种油漆缺陷，在垂直面上，漆膜出现大面积淌挂下垂的现象，这是由于油漆不均匀导致的。

curtain wall A non-load-bearing wall used to clad a building. Usually comprises a secondary framework that supports cladding panels or sheets of glass. Frequently used to clad non-domestic buildings.
幕墙 建筑物不承重的外墙护围。通常由外墙板或玻璃面板和后面的支撑结构组成。通常用于非住宅建筑。

curtilage The enclosed area of land occupied by a dwelling, its outbuildings, and garden.
宅地 包括住宅的庭园和附属建筑物的围闭土地。

curvature 1. The amount to which something curves, e. g. the curvature of the Earth. 2. The rate of change of a curve, e. g. in a circle, curvature is the reciprocal of radius.
1. 曲率 东西的弯曲程度，如地球曲率。**2. 曲线的曲率** 如圆的曲率等于圆的半径的倒数。

cusec The rate of flow given in one cubic foot per second.
立方英尺/秒 流量单位。

cushioned flooring A durable and tough floor covering comprising a vinyl or rubber flooring mat topped with an acrylic coating.
弹性地板 一种耐磨的地板覆盖材料，包括乙烯基地垫或橡胶地垫，通常地垫上会刷丙烯酸涂料。

cut *See* CUTTING.
开挖 参考“开挖”词条。

cut and fill A process of excavating earth (cut) and placing it (fill) to form the vertical alignment of road, railway, or other forms of earthwork.

开挖和回填 开挖土方并把开挖的土用来回填公路、铁路的边坡或用于其他形式的土方工程。

cut and fit *See* SCRIBED JOINT
裁切和拼接 参考“拼缝扣合对接”词条。

C

cut brick *See* ROUGH CUTTING and FAIR CUTTING.
切砖 参考“粗切”和“精切”词条。

cut corner A 90° corner that has been cut to form an angle of 45°.
切角 把一个 90°的角切成 45°的角。

cut nail A nail of constant thickness that has been cut out of a steel plate.
方钉 用钢板加工成的一样规格的钉子。

cut-off A construction formed below the ground to minimize water seepage into an excavation, such as a **cut-off trench** (or **cut-off wall**), which is filled with (made from) impervious material. The depth below excavation to the cut-off is termed the **cut-off depth**.
截水槽/截水墙 在地基中挖槽，并用透水性很小的材料填充，以防止地下水渗入基坑的防渗设施。基坑到截水槽的深度就是截水深度。

cutout 1. An automatic device that disconnects a circuit, e. g. a *circuit breaker. 2. A section that has been removed from a building or wall after the original construction has finished.
1. 可以自动切断电流的装置，如断路器。2. 原有建筑竣工后从某一栋建筑或某一堵墙上拆除下来的某一设施。

cut stone *See* DIMENSION STONE.
规格石 参考“规格石”词条。

cut string (open string) At one side of the staircase the *string is cut out so that the *treads and *risers profile can be viewed from the side, i. e. where a side of staircase is open rather than adjacent to a wall.
明步式楼梯 楼梯一侧的斜梁顶部被切掉，踏步和竖板外露；也就是说该侧楼梯是外露的而不是紧靠着墙的。

cutter block A steel block housing two or more knives that is fitted to woodcarving machines. It rotates at high speeds to shape or smooth timber.
铣刀 与木工铣床装在一起的双刃或多刃不锈钢铣刀。铣刀可以通过高速旋转来雕刻或抛光木材。

cutting A permanent excavation with given side slopes to allow the formation of a railway or road.
开挖 基坑开挖并回填道路边坡，开出一条铁路或公路。

cutting-in A process of painting a clean edge at the perimeter of a painted area, for example where ceilings meet walls or where walls intersect.
精确涂装 精确粉刷待涂漆区域的四周边缘，如墙面与天花交接的区域或墙面交界处。

cutting list A list of steel bars (detailing diameters and length) from which reinforcement is ordered.
下料单 钢筋下料单（详细说明钢筋直径和长度）。

cutwater (starling) The pointed projection at the base of a bridge pier, shaped to allow the water to part around the base.
（桥墩的）分水角 设于桥墩底部使水流分流的尖端分水桩。

cyanide A highly hazardous compound which is a form of hydrogen cyanide. Cyanide is extremely poisonous (with fatal consequences if consumed by human beings) in the form of compounds potassium cyanide and sodium cyanide.
氰化物 指带有氰基的剧毒化合物。具有强烈毒性的氰化物，如氰化钾和氰化钠（如果不慎服用这些氰化物将会产生致命性的后果）。

cyclic loading A loading effect (that represents a sine wave) which is repeated in cycles and used in *fatigue tests.
循环加载 循环操作的一种加载试验（波形为正弦波），循环加载可用于疲劳试验。

cylinder test involves casting concrete in a steel or cast-iron cylindrical mould; the cast concrete is then tested to destruction by compression. It can be tested with the top and bottom surfaces of the cylinder against the

loading platens to determine compressive strength. Alternatively, the cylinder can be placed with its horizontal axis between the platens to determine its tensile strength, referred to as the splitting tension test.
混凝土圆柱体抗压强度试验 先把混凝土浇筑到钢或铸铁圆筒施压型模具里面，然后通过压缩对混凝土进行破坏性测试。测试时可以让上下压板施压混凝土圆柱体的顶部和底部表面，以此来确定混凝土圆柱体的抗压强度。也可以让混凝土圆柱体水平置于上下压板之间来测试其抗压强度，具体要求参照拉伸断裂强力试验。

dabs Small amounts of soft adhesive material used for fixing sheet or board materials to walls.
胶点 少量的软性胶，用于把板材固定到墙上的。

dado 1. The lower part of an internal wall in a room, decorated differently from the rest of the wall. For instance, it may be painted differently or be clad in wood panelling. 2. A rectangular groove cut across the grain of a piece of timber to allow one piece to connect to another.
1. 护墙板 装饰层与墙的其余部分不同的内墙下部。如内墙下部被涂成另一种颜色或被木板包覆。
2. 横切榫槽 木材上与木纹垂直的方形榫槽，方便两块木材的拼接。

dado capping *See* DADO RAIL.
护墙板线条 参考“护墙板线条”词条。

dado rail (dado capping, chair rail, rail chair) A protective moulding that runs around the internal walls of a room at chair-back height. Also used for decorative purposes.
护墙板线条 沿着内墙椅背高的位置安装的一种兼具保护性和装饰性的线条。

dado trunking Surface-mounted *trunking that runs along the internal walls of a room at *dado-rail height.
明装线槽 设在护墙板线条位置的明装布线槽。

dairy Farm building for milking cows.
奶牛场 饲养奶牛的农场。

dam A barrier of concrete or earth that is constructed to retain water, typically to form a *reservoir.
水坝 混凝土或砂土筑成的拦水建筑物，一般会形成水库。

damage Physical destruction that reduces the usefulness, value, or function of something. The act results in damage, reducing the value of a service or product. The effect is usually caused by the actions of a person or the effects of the environment.
损害 物理性破坏，会影响东西的使用、价值或性能。造成损害的行为可使某项服务或某件产品贬值。损害往往是由人为因素或环境因素引起的。

damages Losses suffered as a result of failure to deliver the contract as specified. For example, where a party has failed to complete their part of the contract on time, the additional costs incurred as a result of the delay could be claimed.
损害赔偿 一方未能按规定履行合同，导致对方损失的。如遭受一方未能按时完成合同，则另一方可以要求其赔偿延误导致的额外损失。

damages for delay (US penalty clause) The loss and expense associated with a delay.
误期赔偿 （美国 误期处罚条款）与工程误期相关的损失和费用。

damp (dampness) A state of matter containing moisture, usually higher than the equilibrium amount.
潮湿 物体含有比平衡状态下更多水分的状况。

damp course *See* DAMP-PROOF COURSE.
防潮层 参考“防潮层”词条。

damping A process of reducing the effects of an oscillation or vibration in a system.
减震 减少系统振荡或震动效应的过程。

damp-proof course (damp course, dpc, DPC) A horizontal strip of impervious material that is built into a wall and is designed to prevent moisture penetration by capillary action. DPCs should be laid a minimum of 150mm above ground level. A variety of materials can be used for a DPC including slate, lead, bitumen-polymer, and polyethylene. The most commonly used materials are those that are flexible (**flexible damp-proof course**). Care must be taken when installing a **flexible damp-proof course** to avoid puncturing the material. DPCs can also be retrofitted into existing walls by injecting a suitable chemical, such as silicone, into the wall at regular intervals.
防潮层 为了防止水分通过毛细管作用渗透进砖墙而设置的防渗材料层。防潮层至少要高于室外地坪150mm。有很多材料都可以用作防潮层材料，如板岩、铅、聚合物改性沥青和聚乙烯。这其中用得最多的是柔性防潮材料。安装柔性防潮层时要十分小心，避免刺穿防潮材料。也可以按一定间隔向砖墙里注入适当的化学剂（如硅胶）来起到防潮层的作用。

damp-proof course brick A brick used to construct a *damp-proof course.
防潮层用砖 可以用作防潮层的砖。

damp proofing To prevent damp by, for example, using a *damp-proof course or a *damp-proof membrane.
防潮 例如通过设置防潮层或铺设防水膜来防止受潮。

damp-proof membrane (dpm, DPM) An impervious layer of material, usually 1200 gauge polythene sheet, that is built into a solid ground-bearing concrete floor and designed to prevent moisture penetration by capillary action. Should lap with the damp-proof course in the wall. *See also* DAMP-PROOF COURSE.
防水膜 一种不透水材料层，通常为1200ga的聚乙烯薄膜，铺设在地面承重实心混凝土楼板下方，以防止水分通过毛细管作用渗透进地板。防水膜应与砖墙结构中的防潮层叠合使用。另请参考“防潮层”词条。

darby (darby float) A large twin-handled aluminium float used by plasterers to level plaster.
刮尺 抹灰工用来抹平灰浆用的双手柄铝质大抹子。

dart valve A valve, on the bottom of a *bailer, that employs a pin-like mechanism, which closes on the up-stroke and opens on the down-stoke.
捞砂筒单流阀 捞砂筒底部的阀，可以通过控制针型装置在上行冲程时关闭和在下行冲程时开启。

dashed finish A type of render finish used on external walls to shed water. Can comprise either a dry finish, such as *pebbledash or a wet finish, such as *roughcast.
装饰砂浆饰面 一种可防水的外墙装饰砂浆饰面。装饰砂浆饰面可分为两类，一种为干饰面，如石碴类饰面；另一种为湿饰面，如灰浆类饰面。

dashpot A device to dampen vibration. It consists of a piston enclosed in a fluid-filled cylinder.
缓冲器 一种减震装置。缓冲器液压缸缸筒内装有一个活塞。

data cabling Collective name given to computer cables or optical fibres that are bound together.
数据电缆 计算机电缆或光纤电缆的总称。

datum A fixed point of reference in surveys, upon which other measurements or calculations are based. *See also* BENCHMARK AND ORDNANCE BENCHMARK.
基准点 工程测量时作为其他测量或计算标准的原点。另请参考“基准”词条。

day *See* DAYLIGHT SIZE.
采光面积 参考“采光面积”词条。

daylight factor The ratio of the interior illuminance at a particular point within a space divided by the simultaneous horizontal unobstructed exterior illuminance.
采光系数 室内某点照度与同时间无遮蔽室外水平面照度的比值。

Measured as a percentage (%). It is a combination of three separate components:
采光系数用百分比表示。采光系数是由三个分量组成：
The *sky component 天空直射光
The *externally reflected component 室外反射光
The *internally reflected component 室内反射光

daylight prediction (daylight analysis) A method of estimating the amount of daylight within a space.
采光测算 （采光分析）一种估算室内采光量的方法。

daylight protractor A calculation method developed by the *BRE to predict daylight factors.
采光系数分度器 英国建筑研究院发明的一种估算采光系数的方法。

daylight size (**daylight width, sight size, sight width**) The total area of a window opening that admits daylight.
采光面积 窗口可透光的面积。

days The glazed elements of a window.
窗户玻璃 窗户玻璃构件。

daywork (dayworks, US force account) Payment agreement to undertake work for the price of the labour and materials, with a percentage to cover overheads and profit. The method is generally used for unforeseen events or variations where the work was not specified in detail within the contract, or due to the circumstances were not covered. Where work is undertaken using this method, the daywork sheets, describing the work done, material used, and the labour, are signed off at the end of the day, or period of work, by the *resident engineer, *clerk of works, or *client representative.
计日工（美国 零星用工） 一种付款协议，按约定的综合单价来施工，该单价应包括劳务费、材料费及承包人的管理费和利润。计日工常用于不可预见的事件或变更的工作（合同中没有明确规定的工作）。如果是采用这种计价方式，则承包人应在每天下班后或计日工作完成后提交计日工明细表（计日工明细表应包括完成的工作、使用的材料以及劳务）给驻场工程师、监理或发包人代表签字。

dB (A) A method of measuring the loudness of a sound which is weighted according to the frequency response of a human's ear. *See also* DECIBEL.
分贝 一种测量声音的相对响度的单位，分贝是根据人耳对声音的频率响应来计量的。另请参考“分贝”词条。

dead 1. A material that is no longer suitable for use. 2. Disconnected from a distribution system or supply, for instance, the electrical circuit or telephone line is dead. 3. Does not contain an electrical charge, for instance, a dead battery.
1. 无用的材料 2. 不通电的、无信号的 如电路断路或电话断线。**3. 没电** 如电池电量耗尽。

dead bolt (deadbolt) The part of a dead lock that engages into the door jamb when the key is turned.
方舌 方舌门锁的一个部件，用钥匙关门时，锁舌会弹出来，卡到门边梃预留的孔里。

dead end (US false exit) A street or corridor with only one means of gaining entry and exit.
死胡同 指只有一个出入口的胡同或通道。

dead leg A length of pipe within a hot-water system where the water only circulates when it is drawn off.
盲管段 热水系统中的一节管子，正常状况下水不会流经这节管子，只有放水时例外。

deadlight *See* FIXED LIGHT.
固定窗扇 参考“固定窗扇”词条。

dead load The weight of a structure and any permanent fixtures and fittings. *See also* LIVE LOAD.
恒载 建筑结构本身的自重和任何永久构件的重量。另请参考“活载”词条。

dead lock (deadlock) A type of security lock incorporating a *dead bolt that can only be operated using a key.
方舌门锁 一种带方舌的安全锁，只有用钥匙才能操纵方舌。

dead shore A supporting timber or prop positioned directly underneath *needles to support the weight of a wall. Where a wall needs to be temporarily supported while work is carried out below the wall, beams are inserted into and through the walls, props are then positioned under the beams (needles) so that the weight of the wall can be carried and secured, enabling work below to be undertaken.
顶撑 位于支撑墙重的穿墙梁下方的支撑木条或支柱。如果要在墙下施工，则必须对墙实施临时支撑，包括往墙里边打入穿墙梁，在穿墙梁下方设置支撑立柱，以确保墙支撑稳定及施工安全。

death watch beetle (Xestobium rufovillosum) A wood-boring beetle that burrows deeply into the sapwood, which makes the beetle difficult to kill. The beetle tends to be found in decaying hardwoods, but will bore into adjacent softwoods. The entrance and exit holes are approximately 3 mm in diameter. The beetle is known for the ticking or knocking noise that it makes as it bores.
报死虫（红毛窃蠹） 一种蛀虫，深深地钻进白木质，这样人们就很难杀死这些蛀虫。报死虫一般只出现在旧硬木，但也会在相邻的软木上钻孔。这些蛀孔的直径约3mm。报死虫钻孔时发出的声音像钟表的嘀嗒声。

debris-collection fan (protection fan) A temporary canopy projecting from the face of a building that is designed to protect pedestrians by collecting any falling debris.
防坠棚（防护棚） 突出建筑外墙的临时保护棚，可以保护行人不会受到高空坠物的伤害。

decametre (US decameter) A unit of length equal to 10 meters.
公丈 长度单位，相当于10米。

decanting 1. A method of pouring fluid from one container to another without disturbing the sediment. 2. To transfer from one place to another. 3. A method of decompression.
1. 滗析 在不搅混沉淀物的情况下，把液体从一个容器注入另一个容器的方法。**2. 迁移** 从一个位置移到另一个位置。**3. 覆压减压** 一种减压方法。

decay 1. The weakening of timber that has or is being attacked by fungus (fungal decay, rot). Both *wet rot and *dry rot are types of decay, both requiring moisture to be present in the timber. Well-seasoned timbers have a natural resistance to decay, preservative treatments are also available to reduce the potential of the timber being attacked. 2. The natural deterioration of something.
1. 腐烂 木材受真菌破坏而发霉腐烂（真菌腐朽）。湿腐和褐腐均为腐烂的形式，两种都是因为木材中有水分引起的。风干木材具有天然的耐腐性，而防腐处理也可以降低木材受到腐烂的风险。**2. 指物体自然退化、变质**。

dechlorination The removal of chlorine from water; for example, by the use of granular activated carbon filters.
脱氯 去除水中的氯离子；如利用活性炭颗粒滤池进行脱氯处理。

deci- Prefix denoting a tenth.
十分之一 前缀，表示“十分之一”。

decibel (dB) A logarithmic scale used to measure the intensity of a *sound or a sound-pressure *level.
分贝 表示声音或声压级强度的单位。

deck 1. An outdoor platform, usually constructed above the ground. 2. The structural component of a floor or roof.
1. 露台 露天平台，通常是架空的。**2. 承板** 地面或屋面的结构部件。

decking 1. The material used to form the finished surface over a deck. 2. The profiled metal sheeting used

to construct in-situ concrete slabs.
1. 铺面 露台地面材料。**2. 楼承板** 现浇混凝土楼板下方的压型钢板。

decorator Skilled worker who applies finishes such as paint, varnish, wallpaper, and stippled finishes.
装修工 从事饰面装修（如油漆、清漆、墙纸和拉毛等饰面施工）的技术工人。

deed An agreement or *contract.
契约 协议或合同。

deeping The process of cutting timber to the required thickness by sawing through the thickest section of the timber. The opposite of **flatting**.
深切割 通过锯切木材最厚的部分来使木材达到要求的厚度。Flatting 的反义词。

deep-plan building A building which has a high ratio of internal floor area to external wall area, resulting in large areas of floor space that are greater than 4 m away from the external walls. Such buildings can be difficult to light and ventilate naturally, so are commonly electrically lit and mechanically ventilated or *air-conditioned.
大平层建筑 室内楼面面积与室外外墙面积比率高的建筑物，该建筑内大部分楼面距离外墙至少 4 米。这样的建筑采光和自然通风都不好，通常要采用电气照明和空调通风。

deep-seal trap A U-shaped anti-siphon *trap having a seal of approximately 75 mm or more.
反虹吸存水弯 U 型反虹吸存水弯，水封深度可以达到 75mm 或以上。

default Failure to undertake an agreed action or contractual duty.
违约 合同当事人违反合同义务的行为。

defect Something that renders an object less than specification, not as required, and not perfect. Defects include scratches to surfaces, poor workmanship, products that are not within tolerance or specification.
缺陷 导致某物达不到规范要求的地方或不够完备的地方。缺陷包括表面刮花、做工差、产品偏差超出允许范围或产品不符合规格要求。

defect action sheet Papers produced by the BRE that list and describe recognized problems, often in houses.
房屋缺陷参考手册 由英国建筑研究院出版的，说明房屋存在各种缺陷的手册。

defect correction period The time allowed to make good or correct all defects listed.
缺陷修复期 对罗列的所有缺陷给予修复的期限。

defective work Work that does not comply with standards, regulation, or specification. The work may be a result of incorrect design, poor workmanship, error, or damage to the work.
缺陷工程 不符合标准、法规或规范要求的工程。缺陷工程可能是因为错误设计、做工差、施工有误或受到破坏导致。

defect structure A reference relating to the irregularities in the structure of a metallic or ceramic compound. These include the presence of vacancies and interstitials at a microscopic level.
缺陷结构 金属或陶瓷复合材料的结构不规则。包括微观角度下看到的空隙结构。

deflection The deformation from the horizontal due to loading.
挠曲 荷载作用下引起的弯曲变形。

degradation 1. The breakdown of chemical compounds into simpler compounds. 2. The erosion of the Earth's surface by *weathering. 3. The loss of quality or performance of something.
1. 降解 把化合物分解成更简单的化合物的过程。**2. 剥蚀** 地球表面因风化作用受到的剥蚀。**3. 退化** 质量或性能退化。

degree 1. The extent or amount of something. 2. An educational qualification. 3. A unit of angular measurement quoted between 0 °and 360 °. 4. A unit of temperature measurement as defined on the Celsius or Fahrenheit scale. **Degree Celsius** (previously degree centigrade) is the *SI unit of temperature at which water freezes at 0 ℃ and boils at 100 ℃, under normal atmospheric conditions. **Degree Fahrenheit** is referred to as

the non-metric scale at which water freezes at 32 ℉ and boils at 212 ℉, under normal atmospheric conditions.
1. 程度。**2. 学位**。**3. 角度单位** 如0°～360°。**4. 温度单位** 测量温度的单位，如摄氏度或华氏度。摄氏度：国际单位、标准大气压下，水结冰的温度为0℃，水的沸点为100℃。华氏度：非国际单位，在标准大气压下，水结冰的温度为32 ℉，水的沸点为212 ℉。

D

degree-day The number of degrees in a day that the average temperature is above or below a particular reference value, known as the base temperature. In the UK, the base temperature is 15.5 ℃. Degree-days are cumulative so can be added together over periods of time, for instance a month or a year. They are used to estimate the amount of heating or cooling that a building is likely to require.
度日数 一天内平均温度高于或低于基准温度值的度数。在英国，基准温度值为15.5°C。度日数是可以随着时间累积的，如一个月或一年。度日数可以用来计算建筑物采暖或制冷所需的能量。

degree of compaction A measure of a soil's compactness. *See also* VOID RATIO.
压实度 土壤的密实度。另请参考“孔隙比”词条。

dehumidifier A mechanical device that reduces the moisture content of air either mechanically using a refrigerant, or chemically using a dessicant. Used in some types of *air-conditioning systems.
除湿器 使用制冷剂或干燥剂来减少空气中水分含量的机械装置，是某些空调系统的一个部分。

delamination A type of failure whereby one of the layers separate from the object.
剥离 一种缺陷，物体表层的分离。

delay The period that the project or events take place after the contractual or predetermined completion date.
延误时间 指在合同约定或预定的竣工日期后完成的某一项目或事项所用的时间。

delay damages Loss incurred as a result of a *delay.
误期赔偿费 由于工期延误导致的损失。

deluge sprinkler A type of *sprinkler that always remains in the open position and when activated enables large amounts of water to flow through it to extinguish the fire. Used in areas that contain hazardous materials.
雨淋喷头 一种开式喷头，只要触发启动，雨淋喷头就会大量喷水灭火。雨淋喷头适用于危险材料的区域。

demand factor *See* DIVERSITY FACTOR.
需要系数 参考“量因”词条。

demolition The act of taking apart, disassembling, breaking down, and removing buildings, structures, and works. The act of taking down buildings has become much more sophisticated, as health and safety regulations require the structure to be safely dismantled and the materials safely disposed. Recycling of waste is now a regulated process demanding the segregation and reuse of materials.
拆除 拆掉建筑物、构筑物和建筑工程。根据健康与安全法规的规定，拆除建筑物时一定要注意安全且要确保材料得到安全处理，这就使拆除建筑物的工作变得越来越复杂。废料回收（包括材料分类和重新利用）也是健康与安全法规要求执行的。

demountable partition A prefabricated unit used to form a wall that is capable of being erected, taken down, and repositioned. Demountable partitions offer greater flexibility for office and building space.
可拆卸隔断墙 一种预制的可以现场拼装、拆卸和移动的活动隔断单元。可拆卸隔断墙可以给办公和建筑空间提供更多的灵活性。

denitrification The conversion of nitrate to nitrogen gas. Used in wastewater treatment plants.
脱硝 硝酸盐转化成氮气的过程。多用于废水处理。

densification To increase the *density, the number of soil particles per unit volume, of an in-situ soil by the use of any form of ground improvement technique.
土壤压实 借助各种地基处理方法来增强原位土的密实度（单位体积内土壤颗粒的数目）。

density Mass (m) per unit volume (V) for an object measured as kg/m^3.

密度 物质每单位体积内的质量。密度的单位为 kg/m^3。

depth The distance from the top to bottom or from the front to back of something.
深度 某物从顶部到底部的距离或从正面到背面的距离。

derelict A building or structure in a state of disrepair or run-down condition that renders it beyond practical use. Such properties may be * demolished, * recycled, or * refurbished.
荒废的 建筑物年久失修，需修葺才能使用。可以拆除、回收或翻新这些建筑物。

derrick crane A simple crane with controls at the bottom of a vertical mast, used for lifting and moving objects.
桅杆起重机 一种简单的起重机，控制装置设在垂直桅杆底部，桅杆起重机可以用来提升和移动重物。

description Definitive statement of the product or service referred to in the contract to ensure the desired standard is achieved. It is important that sufficient detail is described to ensure that the required service and standard is met. The description serves two key purposes: it defines the work set that is to be completed; and if the nature, quality, standard, or type of work that has been carried out is questioned, it provides a benchmark statement against which the work can be compared.
说明 合同中有关产品或服务的规定性说明，以确保产品或服务达到规定的标准。要想确保服务或产品达到标准要求，应尽可能提供足够的细节说明。产品或服务说明有两个主要作用：一个是界定了工作范围；另一个是提供了一个作比较的标准，特别是对工作的性质、质量、标准或类型有疑问时。

design A model, sketch, drawing, outline, description, or specification used to create the vision of that which is to be created—a working item, product, building, or structure. Design can be considered as: 1. Concept design: information that conveys the general idea or vision. 2. Working design: information developed by pulling together details and specifications from the various trades and professionals building and integrating the content so that all of the information can function and fit together without gaps. The developing design is often broken down into stages with different stages being completed as the information from different groups or professionals is brought together. 3. Final design: possesses sufficient detail to ensure that a fully functional product can be created.
设计 是有关工程项目、产品、建筑物或构筑物的模型、草图、详图、轮廓、说明或规格。可以从以下三点来看待设计：1. 概念设计：传达总体理念或意见的信息。2. 施工设计：通过收集和整合各个专业的施工细节和规范制定的信息，确保所有信息都是紧密联系的。施工设计通常会划分为几个不同的阶段来完成。3. 最终设计：搜集足够的细节来确保产品是完全按照要求运行的。

design and build contract Where the contractor is contracted to undertake both the design and construction phases of the project.
设计和施工合同 规定承包商必须按要求履行工程设计和施工任务的合同。

design-build contract (design and build contract, package deal, turnkey project) A complete contract that includes development, coordination and delivery of design, and all assembly aspects of a project. The contracting party is considered wholly responsible for taking whatever concept, design, and specification information that is given to them and developing it into a fully functional structure or building. As both the design and construction are contained within a single contract, the * contractor is exposed to greater risk than is carried for contracts that separate design and construction.
设计—施工合同（设计和施工合同，统包合同，交钥匙合同） 一揽子合同，包括开发、协调、设计、安装工程的各个方面。物承包商应负责总体规划、设计、施工计划的制定，直至把一个随时可以使用的建筑物交给业主。由于这份合同里边包含了设计和施工两个方面，因此相对于独立的设计合同、施工合同，承包商要承担更多的风险。

design code A document that provides rules for design. Such documents may or may not be statutory documents.
设计规范 提供设计准则的文件。这些文件不一定是法定文件。

Design Engineer A * Chartered Engineer who is qualified to undertake the design of a structure.
设计工程师 负责结构设计的特许工程师。

design leader The professional who heads up the * design process ensuring the information is taken from concept to the final design, and ensuring the information brought together is properly integrated and coordinated.
设计总监 对设计工作起带头作用的专业设计师，设计总监应跟踪从概念设计到最终设计的一切信息，确保这些信息是适当协调和整合的。

desludging The removal of sludge from an item of equipment.
清除污泥 清除某一设备上的污泥。

desuperheater 1. A device that reduces the temperature of superheated steam to its saturation temperature. 2. A type of heat exchanger that extracts the waste heat from the compressor in an air-conditioning system or heat pump, and uses the waste heat to pre-heat water.
1. 减温器 把过热蒸汽温度降至其饱和温度的装置。**2. 换热器** 从空调系统或热泵压缩机中提取废热能，并用这些热能来预热水的装置。

detached A building, dwelling, or structure that is separate from other buildings.
独立建筑物 指与其他建筑物不相连的建筑物、住宅或构筑物。

detail Drawing or written specification that conveys more intricate information about the assembly, arrangement, or components to be used. By showing more detail, more information is given, and a greater understanding should be achieved.
详图或规范详情 有关组装、排列或构件的详细信息的图纸或文字规范。通过提供更多细节和信息，施工人员就可以更好地了解施工要求。

detail drawings Drawings that show arrangements of components and provide supporting information that ensures the assemblies can be achieved.
详图 显示构件构造细节的图纸，建筑详图会提供详细的辅助信息来确保正确安装。

details 1. Drawings that show the * detail. 2. Intricate information.
1. 详图 显示构件构造细节的图纸。**2. 详细信息**。

determination (termination) The end of a contract or agreement before all of the work has been completed. The end of a contract may be brought into force by either the employer or the contractor, for breach of contract, to the extent that it is considered reasonable to bring the contract to an end and seek damages for the breach. Suspension of works may allow the parties to continue to work together once matters have been agreed, whereas determination is final.
合同终止 合同一方在所有工程竣工前终止合同或协议。如果业主或承包商任何一方严重违约，则另一方有权要求终止合同并寻求违约赔偿。一旦双方协商并解决好纠纷，工程便可以继续；反之，终止合同将成为最终决定。

deterioration The reduction in the quality, value, or strength of something.
退化 （质量、价值或强度）逐渐下降。

detritus Fragments of rock that have been produced due to erosion.
岩屑 岩石由于风蚀或水蚀产生的碎片。

deviatoric stress (differential stress) A condition where the stress components are not the same in all directions, e.g. in a * triaxial test on soil.
偏应力（差应力） 各个方向上的应力分量是不一样的，如土壤的三轴剪切试验。

devil float (devilling float, nail float) A wooden hand * float containing nail points that project outwards from each corner of the float. Used by plasterers to scratch the surface of the plaster to provide a * key for the next coat. This process is known as **devilling**.
带钉抹子 木质抹子，在抹子的各个角落带有凸出的钉子。抹灰工常使用带钉抹子来对抹灰面进行拉毛处理，使抹灰面与下一层材料能够更好地黏结。这一过程就叫拉毛。

dewater To removal water, for example, to lower *groundwater by pumping to allow excavation to be undertake in dry conditions.
排水 如用水泵抽水来降低地下水位，以确保基坑开挖工作能在干燥条件下展开。

diagram A schematic drawing of something.
示意图 物体的简单图示。

dial gauge A mechanical instrument used to measure small displacements of a plunger by means of a pointer moving round a circular scale.
千分表 一种机械测量仪器，通过指针在刻度盘上的旋转运动来读取测量轴的细微位移。

diamond break stiffening A very slight fold, running from corner to corner, in a panel of sheet metal, which reduces the tendency of the panel to vibrate, preventing a drumming noise.
楞线加固 非常轻微的折纹，在金属板上由一个角延续到另一个角，可减低面板震动防止产生噪音。

diamond saw A tool that has industrial diamonds embedded along its cutting edge.
金刚石锯片 刀头内部嵌有工业金刚石颗粒的一种切割工具。

diamond washer A curved washer used with a hook bolt to secure corrugated roof sheet. It can be manufactured from aluminium, galvanized steel, or plastic.
菱形垫圈 一种弧形垫圈，与钩头螺栓一起使用，用来固定压型屋面板。弧形垫圈可以是用铝材、镀锌钢或塑料加工的。

diaphragm A taut thin membrane. It is can be used in a diaphragm pump, which is a reciprocal pump that has a flexible membrane. It can also be used in a diaphragm tank, which is a closed container having an upper and lower chamber separated by a diaphragm—the upper chamber containing air or nitrogen, the lower water.
隔膜 一张拉紧的薄膜。可用于隔膜泵，隔膜泵是带有柔性膜的往复式泵。也可用于隔膜式气压罐，隔膜式气压罐是一个密闭容器，隔膜把水室和气室完全隔开，隔膜上部为含空气或氮气的气室，下部为水室。

diaphragm wall A retaining wall that has been formed within an excavated trench in the ground. As the trench is excavated it is filled with *bentonite slurry to prevent collapse. A reinforcement cage is then inserted into the trench. Concrete is placed via a *tremie at the bottom of the trench. Uses include basement walls, earth-retaining walls, water storage tanks, and cut-off walls.
地下连续墙 基坑内的挡土墙。基槽挖好后，用膨润土泥浆护壁。清槽后，在槽内吊放钢筋笼。然后用导管法灌筑水下混凝土。用途：地下室墙、挡土墙、储水箱和截水墙。

die 1. A metal form used as a permanent mould for die-casting. A die is a tool responsible for cutting the shape of the label out of the material. 2. A threaded cutting block that is used to cut threads into the outer surface of bars and pipes. *See also* TAP, which is used to cut threads into the inner surfaces of pipes or holes drilled into metal.
1. 模 压铸加工用的永久性金属模。是切模、塑模的工具。**2. 圆板牙** 螺纹切削工具，可以在钢筋和管道的外表面切出螺纹形状。另请参考“丝锥”词条，一种加工内螺纹的刀具，如加工管道内螺纹或金属里边钻螺纹孔。

dielectric Any material that is electrically insulating.
绝缘体 任何经过绝缘处理的材料。

differs (key changes, variations) The amount of alternative keys that can be made for a given lock. *See also* KEY CHANGES.
密匙量 能打开一种特定锁的不同钥匙的总数量。另请参考“锁具密匙量”词条。

diffuser 1. A grille attached to an air supply duct that distributes the air in a particular direction. 2. A baffle placed in front of a luminaire to distribute the light evenly and minimize glare.
1. 散流器 送风管末端的格栅风口，向特定方向送风。**2. 灯具扩散板** 安装在灯具前面的扩散板，可以使灯光更均匀地分布并减少眩光。

digging The breaking up and loosening of ground by hand, tools, or machine. The excavation using a hand, tools, and machines, such as a * spade or * digger may be considered different from the loosening of the ground or spoil.
挖掘 用手、工具或机械破土和松土，用手、铲子或挖掘机等工具和机械开挖基坑是不一样的。

dilatometer An instrument used to measure the expansion of a liquid.
膨胀计 测量液体膨胀性的仪器。

diluting receiver (laboratory receiver) A container on a waste drain from a laboratory sink that helps to reduce the concentration of chemicals entering a sewage system.
实验室化学剂稀释接收装置 安装在实验室洗涤槽下面的一个接收装置，可以在化学药品进入排污系统前稀释掉一部分的浓度。

dimension 1. The size of something in terms of length, width, or height. 2. A property defining a physical quantity, e. g. mass or time.
1. 尺寸 如长度、宽度或高度。**2. 量纲** 是表示物理量的性质（类别），如质量或时间等。

dimensional coordination The size of various building components to ensure consistency in design and build of modular systems, e. g. a standard door size and opening.
尺寸协调 不同建筑构件的尺寸协调，以确保建筑设计和建筑模数的协调统一。如门和门洞尺寸协调。

dimensional stability Materials and products with little movement when exposed to different moisture contents and temperatures. The materials are stable in size, and do not suffer movement or * creep in different environments.
尺寸稳定性 材料和产品在不同的湿度和温度下只有些许的变化。材料的尺寸稳定，在不同的环境中不产生变化或徐变。

dimension stone (cut stone) Natural stone that is cut to shape, normally to form rectangular blocks.
规格石 由天然石加工成一定形状的石块，通常是矩形。

diminished stile Door stile that is narrower at the top allowing greater space for windows.
不等宽门窗竖梃 门竖梃由下往上宽度逐渐变小，给顶部的玻璃预留更大的空间。

diminishing courses Stone courses laid with the larger courses at the bottom and smaller courses towards the top of the wall.
宽度依次递减的石砖层 底部的石砖层最宽，然后沿墙逐渐向上层递减。

diminishing stile *See* DIMINISHED STILE.
不等宽门窗竖梃 参考“不等宽门窗竖梃”词条。

DIN (Deutsches Institut fü r Normung) German standards institute.
德国标准协会 德国标准协会

direct cold-water supply A cold-water system where all of the appliances are supplied with cold water directly from the mains.
直接供水（冷水）系统 冷水供应系统的一种，所有冷水龙头的水都是直接从市政水管输送过来的。

direct current (DC) An electric current that flows continuously in one direction. *See also* ALTERNATING CURRENT.
直流电 向同一方向流动的电流。另请参考“交流电”词条。

direct cylinder A cylinder used for storing hot water in a direct hot-water system. *See also* INDIRECT HOT WATER CYLINDER.
直接加热储水箱 直接加热热水的供应系统的热水储水箱。另请参考“间接加热热水储水箱”词条。

direct hot-water system A hot-water system where the water that is heated is the same water that is drawn off at the hot water taps. Not suitable for hot-water central heating and only used in soft water areas. *See also* IN-

DIRECT HOT WATER SYSTEM.
直接加热热水供应系统 用加热设备里的水作为热水龙头的水源的热水供应系统。该系统不适用于中央加热热水且只适用于软水区域。另请参考“间接加热式热水供应系统”词条。

direct labour Workers who are employed through the company rather than being *subcontracted or hired through an agency for short periods of time or for single projects. Labour that is employed directly is considered to be ‘on the books’, receiving the full benefits of being a full-time or part-time employee who has a contract directly with the employer. Being employed directly may result in additional benefits with regard to bonus pay, welfare, and holiday entitlement.
自有劳务人员 公司直接雇佣，而不是通过中介机构分包的或派遣的短期服务的或某一项目的劳工。公司直接雇佣的劳工都是有备案的，这些直接与公司签订合同的全职员工或兼职员工可以享受到公司的全部福利。公司直接雇佣的劳工可以享受到奖金、社保和带薪假期等福利。

direct reading tacheometer A *tacheometer fitted with a scale to enable horizontal distance and difference in level to be obtained without the need to measure vertical angles.
电子快速测距仪 带有刻度盘的快速测距仪，无需测量竖直角就可以计算水平距离和高差。

direct stress (normal stress) Tensile or compressive stress without bending or shear.
正应力（法向应力） 不包括弯曲应力或剪应力拉应力或压应力力或抗压应力。

direct supply *See* DIRECT COLD-WATER SUPPLY.
直接供水 参考“直接供水（冷水）系统”词条。

direct transmission of sound Sound (airborne or impact) that is transmitted directly through a separating wall or floor. *See also* FLANKING TRANSMISSION.
直接传声 空气传声或撞击传声，直接透过隔墙或地板传播的声音。另请参考“侧向传声”词条。

dirt-depreciation factor (US) Lux or light loss factor.
灯具尘埃减能系数（美国） 光损失系数。

dirty money Extra money awarded for having to work in difficult or messy conditions.
补偿津贴 付给在困难或肮脏工作条件下工作的工人的额外报酬。

disabled facilities (handicapped facilities) Adaptations and equipment within a building that enable the building to accommodate disabled needs. As part of increasing accessibility, stairs, ramps, toilets, signage, and hearing loops are provided to ensure people with a disability are better able to use the building.
残疾人设施 为满足残疾人的需要在建设工程中配套建设的服务设施。包括：无障碍通道、电（楼）梯、坡道、洗手间、盲文标识、助听器等，以确保残障人士更好地使用建筑里边的各种设施。

disc grinder Hand-held or fixed machinery with a rotation disk that can be used to smooth, grind, and cut metal and ceramic materials, such as concrete, stone, tiles, etc. Different discs are used, depending on the material that is to be cut.
角磨机 角磨机是一种利用高速旋转的砂轮来切削和打磨金属与陶瓷等材料（如混凝土、石材、瓷砖等）材料的手提或固定式的电动工具。加工不同材料使用的砂轮是不同的。

discharge Something that is emitted from a pipe, duct, or conduit, or something that permeates away from its source.
放出/排放 从水管、风管或电管中放出，或某物从源头扩散开来。

discharge lamp A light source emitted from an electric discharge passing through a tube containing mercury or sodium vapour or argon, krypton, or neon gas.
放电灯 通过放电管（放电管里含有水银/钠蒸气/氩、氪或氖气等气体）放电将电能转换为光的一种电光源。

discharge pipe A *branch that carries waste from a sanitary appliance to the *soil stack.
排水管 指把卫生器具里的废水排到污水立管的支管。

discharge valve (control valve, regulating valve) A valve used to control the flow of a fluid through pipe-

work.
排水阀（控制阀，调节阀） 指用于控制管道内液体流量的阀门。

disconnecting trap (disconnector, intercepting trap) A water trap on a drainage system that connects a rainwater pipe to a sewer—used to prevent foul smells being released into the environment from the sewer.
防臭弯管 连接雨水管和污水系统的存水弯，可防止污水管道系统中的臭味散布入环境中。

D

discontinuous construction (isolation) Structure or assemblies with breaks that separate construction to reduce the transmission of vibration, sound, and thermal energy.
不连续结构 有间隔的结构或组件，这些设计是为了达到抗震、隔声和保温的效果。

disc sander Hand-held machine with a rotating shaft onto which sanding discs can be attached. The sanding discs are made to different grades for different materials.
盘式砂光机 转轴上附有砂轮的手持式研磨机。不同材料使用的砂轮目数不同。

disc tumbler lock A type of cylinder lock that contains rotating disc retainers.
盘簧锁 圆筒锁的一种，带可转动的圆片锁芯。

disinfection A tertiary stage in a *water treatment process, where chlorination or ultraviolet light is used to remove pathogens (disease-causing bacteria, viruses, and protoza) so that it is fit for human consumption.
消毒 杀菌 水处理的第三个步骤，通过加氯处理或使用紫外线来去除病原体（致病细菌、病毒和原生动物），以确保水是可以饮用的。

displacement pile (driven pile) A *pile that has been pushed into the ground, displacing the soil rather than excavating the soil.
挤土桩（打入桩） 桩打入土里，取代土体的位置而不是将土挖掘出来。

displacement pump A pump that draws fluid in then pushes it out, such as a *lift-and-force pump, reciprocating pump, or diaphragm pump.
容积式泵 反复吸排液体的泵，如手压泵、往复泵或隔膜泵。

disruption Events that occur that cause events to be delayed and/or to run out of sequence.
中断/扰乱 事件被延误或被扰乱。

dissolve 1. To be dispersed in a liquid. 2. To break down and disappear.
1. 固体溶解 在液体中散开。**2. 消失、消散**。

distance piece A wedge fitted into a glazing rebate to prevent displacement.
玻璃垫块 嵌在玻璃槽里的玻璃垫片，可防止窗扇移位。

distribution board (distribution panel, distribution fuse-board, sub-board) An electrical board that subdivides the feed from the mains into the various electrical circuits contained within the building. Contains a number of fuses or circuit breakers that protect each electrical circuit.
配电箱（配电柜，熔断器板） 把输电干线分配到大楼内各条电路的电箱。配电箱里边装有保险丝或断路器等，保护每条线路。

distribution pipe A pipe that is used to move water or steam from one place to another.
输送管 把水或蒸汽从一个位置输送到另一个位置的管道。

district heating A system that utilizes a central heat source to heat a number of buildings within a town, city, or district. The heat from the central heat source (usually in the form of hot water) is distributed to the buildings via a series of underground pipes.
区域集中供热 从城市中央热源向全市、全镇或其中某一地区的用户供应用热的系统。从城市集中的热源（通常以热水为介质）经一系列地下供热管网送到每一栋建筑。

ditch A long narrow channel that has been formed in the ground, typically for drainage and irrigation purposes.
沟 地面上长而窄的渠道，通常为排水和灌溉用途。

diversity factor (demand factor, use factor) The sum of the individual maximum demands in an electrical distribution system divided by the total maximum demand of the system, usually always greater than one.
需量因数 配电系统中独立用户或单个设施的最大需量总和与系统总设施负荷之比，这个数值通常大于1。

diverter A three-way valve to direct/split the flow.
分流阀 导流/分流三通阀。

divided responsibility Duty or obligation shared between two or more parties. Duties need to be clearly stated to manage work and to avoid disputes.
责任分担 责任或义务由两方或两方以上承担。需清晰划分责任以便管理工作和避免纠纷。

division wall 1. A load-bearing wall that is used to subdivide a space into rooms. 2. A boundary wall between two properties, which is not part of either property. 3. A fire wall.
1. 隔墙 将建筑物内部空间分隔成房间的承重墙。**2. 边界墙** 两处物业之间的边界墙，边界墙不属于任何一处物业。**3. 防火隔墙**

do-and-charge contract Cost reimbursement contract.
成本补偿合同。

dock An area where ships and boats moor, load, and unload, which can be cut off from the rise and fall of the tide by dock gates.
码头 专供轮船或渡船停泊、货物装卸的建筑物，可通过关闭船坞闸门来避免受到涨潮、落潮的影响。

docking saw A type of circular saw mounted on a stand that cuts the timber by lowering the saw blade onto it.
截料锯 固定在锯台上的一种圆盘锯，可通过下压锯片来切割木材。

documents Contract documents, papers, drawings, and bills referred to in the contract.
文件 合同内有关的文件、图纸和标书等。

dog bolt *See* HINGE BOLT.
防盗铰链 参考“铰链防盗钉”词条。

dogleg stair (dog-legged stair) Stair with two flights between a floor where the stair returns on itself.
平行双跑楼梯 层间有两跑楼梯，梯段折回到原来起步的方向。

dolly *See* DRIVING CAP.
桩帽 参考“桩帽”词条。

dolphin A pile or group of piles used for mooring ships or as a *fender to protect a *dock.
码头系缆桩 单个或一组用来停泊船只的桩，也可以起保护码头的作用。

dome A structure that has a hemispherical roof.
穹顶 有半球形屋顶的构筑物。

dome-cover screw (domed top screw) A screw where the head has a small threaded hole into which a domed cover is inserted.
盖形螺丝 螺丝头小螺孔上带有圆形盖子的螺丝。

dome-head screw A screw with a dome-shaped head.
圆头螺丝 螺丝头部为半圆形的螺丝。

domelight A dome-shaped rooflight.
半球的形天窗 一半球形的天窗。

domestic hot water (DHW) Hot water supplied to a dwelling for washing, bathing, etc.
生活热水 供应给住户作清洗或沐浴用的热水。

domestic sub-contractor A subcontractor who is selected by the main contractor, being different from a *nominated subcontract who is specified in the contract documents.
一般分包商 承包商雇佣的分包商，区别于载明于合同中由业主指定的分包商。

domical grating Wire mesh or grating shaped in a dome, which acts as a strainer over the end of a rainwater outlet.
（雨水斗）半球形格栅 半球状钢丝网或格栅装置，在雨水斗末端作过滤作用。

door assembly *See* DOOR SET.
门组件 参考“门组件”词条。

door blank (blank door) A door-sized recess in a wall either for aesthetic reasons or to enable the simpler insertion of a door at a later date.
假门（装饰门） 墙面上门形状大小的凹位，可起装饰作用或为日后安装方便门准备的一个门洞。

door buck *See* DOOR CASING.
门套线 参考“门套线”词条。

door buffer A device attached to a door to allow it to close softly and prevent banging.
门缓冲器 固定在门上的缓冲装置，可以让门慢慢闭合，以避免门砰然关上。

door casing The *architrave around a door.
门套线 门周边的线条。

door closer (door check) A spring-operated mechanical device that closes a door.
闭门器 能使门自动关上的弹簧操作装置。

door control A system for controlling entry through a door.
门禁系统 控制出入门的系统。

door face The visible portion of a *door leaf.
门扇表面 门扇可视部分。

doorframe The frame that surrounds a door and on which the door is hung. It comprises a head and two jambs. Mainly used to hang external doors or doors in solid partitions. The frame itself has sufficient strength to support the weight of the door.
门框 围着门的框架，通常支撑着门扇。包括门道两旁和顶上的边梃和上槛。主要用于悬挂室外门和实心隔墙上的门。门框自身有足够的强度可以支撑门的重量。

door furniture (door hardware) The collective name given to all the items of ironmongery that are fixed to the door. For instance, door handle, *escutcheon, locks, etc.
门五金配件 固定在门上的一切五金配件的总称。如门把手，锁眼盖、门锁等。

door handle A device located on the inside and/or outside of a door used to open it.
门把手 设在门内侧和外侧用于开启门的装置。

door hardware *See* DOOR FURNITURE.
门五金配件 参考“门五金配件”词条。

door head (head member) The horizontal top member of a *doorframe or *door lining.
门框上槛 门框或筒子板顶部的水平构件。

door holder A device that is used to hold a door in the open position.
门挡 保持门打开的装置。

door jamb The vertical side members of a *doorframe or *door lining.
门边梃 门框或筒子板两侧的垂直构件。

doorkit All of the individual components used to make a *door set.
套装门 构成成套门的所有独立构件。

doorknob A knob-shaped * door handle.
球形门把手 一球形的门把手。

door latch 1. A spring-operated door lock, operated either by key or by turning the door handle or knob. 2. A door fastener comprising a metal bar that is either slid across or lowered to prevent the door from opening.
1. 弹簧锁 可以通过钥匙或旋转门把手来操作的弹簧门锁。**2. 门闩** 可通过平移或下移金属插销来防止门被打开。

door leaf The opening portion of a door.
门扇 门可自由开关的部分。

door lining (US door buck) The surround that covers the * reveal around an internal door and on which the door is hung, usually of thinner cross-section than a * doorframe. The lining does not normally have enough strength to support the weight of the door on its own, so relies on additional support from the surrounding wall or partition.
筒子板 覆盖室内门洞表面以及安装门扇的材料，通常横截面比门框要薄。一般情况下，筒子板自身是没有足够的强度来单独支撑门的，筒子板要依赖相邻的墙体或隔墙来提供额外的支撑。

door lock A * lock for a door.
门锁 用于门上的锁。

doormat A mat placed in front of or just inside an external door that is used to wipe dirt from shoes.
门垫 放在入户门前或门口用来刮除鞋子上泥尘的垫子。

doormat frame A frame sunk in the floor just inside an external door to hold a doormat.
门垫框架 嵌装在入户门处的地面上用来放门垫的框架。

door opener An electronic device for automatically opening a door.
电动开门器 可以自动打开门的电气装置。

door post A door jamb.
门侧柱 门的边梃。

door pull A handle used to pull open a door.
门拉手 用来打开门的把手。

door rail The horizontal member that subdivides a door.
门扇冒头 分割门扇的水平构件。

door schedule A list, usually in tabular form, containing information about all the doors that will be used on a project. It usually contains details of their size, specifications, location, etc.
门一览表 有关项目所有拟使用的门的信息表。这些信息包括门的尺寸、规格、位置等。

door set (doorset, door unit) A pre-assembled door and frame that can be installed on-site immediately. Comprises a door hung on its frame, linings, architraves, and door furniture.
成套门（整装门） 可以在现场随时安装的预先组装好的门和门框。包括挂在门框上的门、筒子板、门套线和门五金配件。

door sill *See* DOOR THRESHOLD.
门槛 参考“门槛”词条。

door stile The vertical side members of a door.
门竖梃 门扇竖直方向最边上的两根料。

door stop 1. Strips of wood attached to the head and/or jambs of a doorframe or lining. Used to prevent the door leaf moving through the frame and past the closed position. 2. A device that is attached to the floor, wall, or skirting board to prevent the door opening past a particular position. Used to protect the skirting board and wall from damage caused by the door or the door handle.
1. 门框挡条 固定在门框或筒子板上面的上槛和/或边梃上的木条，用于定位门扇关闭位置，避免门扇越过门框。**2. 门止/门挡** 固定在地板上、墙上或踢脚板上保持门打开位置的装置。可以保护踢脚线和墙免受门或门把手的撞击。

door switch A switch used to open or close a door.
开关门装置 用来开启或关闭门的开关。

D

door threshold *See* THRESHOLD.
门槛 参考“门槛”词条。

door unit *See* DOOR SET.
整装门 参考“成套门”词条。

doorway width The distance between the jambs of a *doorframe.
门洞宽度 指两门框边梃之间的距离。

dormer (dormer window) A vertical window that projects out from a sloping roof.
老虎窗（屋顶窗） 在斜屋面上凸出的窗。

dormer cheek The vertical side of a dormer window.
老虎窗边柱 老虎窗两侧的垂直构件。

dormer ventilator (dormer vent) A dormer-shaped roof ventilator designed to be installed on a sloping roof.
斜屋顶通风器 安装在斜屋面上的类似老虎窗形状的屋顶通风器。

dot and dab The process of attaching plasterboard to a wall using dots of plasterboard adhesive.
黏结石膏打点 用一点一点的石膏板黏结剂把石膏板固定到墙上的过程。

dote (doat) Dots, spots, and speckles on timber that indicate early signs of decay.
木材腐朽 木材上显示腐朽迹象的斑点、条纹。

dots and screeds Mounds, spots, and strips of screed that are levelled to the correct height and placed at regular intervals so that infill screed can be laid and levelled between.
打点冲筋 在规定的间隔处抹上一定厚度的砂浆（砂浆为一抹、一点或一条条的形状），以确保灰筋之间的砂浆是平整的。

double-acting hinge A hinge that enables a door to open in both directions.
双向铰链 允许门双向开启的铰链。

double-action door A door that can open in both directions.
双向开启门 可往内两边方向打开的门。

double connector A short pipe connector containing a thread at each end.
双接头 一种较短的管接头，每一头都有螺纹。

double door A pair of doors.
双扇门 一对门。

double-eaves course (doubling course) A double row of tiles at the eaves of a roof, the lower tiles are half-length to tie in with the longer tiles that overlap.
双层屋檐瓦 安装在屋檐上的双层瓦，上层瓦会盖住下层瓦一半的长度。

double-faced door A door where each face of the door is decorated differently.
双面门 内外两侧门面的饰面效果截然不同的门。

double glazing Glazing comprising two panes of glass that are separated by a cavity. The cavity can be filled with air or an inert gas such as argon or krypton.
双层中空玻璃 两片玻璃中间有空腔隔着。空腔里边可以填充空气或惰性气体（如氩或氪等）。

double-handed lock *See* REVERSIBLE LOCK.
双向锁 参考“双向锁”词条。

double-hung sash window A sash window where the sash slides vertically.
双悬上下推拉窗 窗扇可以上下移动的推拉窗。

double-inlet fan (double-width fan, double suction blower) A *centrifugal fan where air can enter from both sides.
双进风式离心风机 两边都可以进风的离心风机。

double insulation A method of insulating an electrical appliance that does not require it to have an earth connection.
双重绝缘 是一种电器绝缘方法，电器经过双重绝缘处理后便不需要再做接地连接。

double-leaf separating wall A cavity wall where the two separate leaves of the wall are not connected together in any way. Used to provide sound insulation between spaces. *See also* DOUBLE PARTITION.
双层墙 夹心墙，两堵墙是分开的且不以任何方式连接。双层墙可以起到隔声的作用。另请参考"空心墙"词条。

double lining Two layers of lining paper hung on a wall, one horizontally and the other vertically, to provide a smoother wall surface.
双层底衬纸 粘贴在墙上的双层底衬纸，一层水平粘贴，另一层竖直粘贴，以提供更加平整的墙面。

double-lock seam [double-lock welt, double welt, (Scotland) cling] A *single-lock seam with an additional fold. Provides greater strength and weather resistance than a single-lock seam. Used to join sheets in sheet-metal roofing.
双咬口 在单咬口的基础上再加一个咬合连接。比起单咬口连接方式，双咬口可以提供更大的强度和更好的耐候性。双咬口是金属屋面板的接合方式。

double margin door A single door that is divided into two halves to give the appearance of a double door.
双窗单扇门 由中梃分隔成看起来像双扇门的单扇门。

double partition A partition wall comprising two separate leaves that are not connected together in any way. Used to provide sound insulation between spaces. *See also* DOUBLE-LEAF SEPARATING WALL.
空心墙 中间有空腔且不以任何方式连接的双层砌体墙。空心墙可以起到隔声的作用。另请参考"双层墙"词条。

double-rebated doorframe A *doorframe that contains a rebate on both sides.
双槽口门框 门框两侧各有一个槽口。

double Roman tile Single lap clay tile 420 × 360 × 16mm, laps onto two half tiles making two waterways or channels. The edges of the tile are normally lapped 75 mm.
连锁瓦 单搭接陶土瓦，尺寸为420 mm × 360 mm × 16 mm。上一层瓦叠在下一层瓦后应有2个排水道。连锁瓦的搭接长度通常为75 mm。

double-suction blower *See* DOUBLE-INLET FAN.
双进风式离心风机 参考"双进风式离心风机"词条。

double time Paid at twice the normal rate for work outside normal working hours or work carried out in difficult conditions.
双倍工资 非正常工作时间或困难条件下工作应得的双倍工资。

double triangle tie Galvanized or stainless steel wall tie with two triangles formed in the wire at each end to provide a bond in the mortar course. The tie is used to link two separate skins of *cavity wall together.
三角钢丝拉结 镀锌或不锈钢拉结件，两头被拧成三角形置于灰缝中。拉结钢丝可以把空心砌块墙的两堵墙拉结在一起。

double window A window that contains two glazed sashes with an air space in between.
双层玻璃窗 中间有一层空隙的两层玻璃窗。

dovetail (joint) A fan-shaped joint used to join the corners of boxes and drawers, the tenons are cut in a fan

shape and the same shape is cut out of the end of the timber to be joined. The thicker end of the dovetail is positioned at the far end of the tenon, so that when the two parts are joined together, the shape of the dovetail prevents pullout.
燕尾榫接合 燕尾形接合，可用于木质盒子或木质抽屉的拼接。榫头做成燕尾形，并在另一根木材的端部裁切出跟榫头相匹配的榫槽。燕尾榫头端较宽，可以使两个工件紧紧接合，避免在受力时脱开。

D

dovetail cramp A cramp used for stonework shaped in a dovetail.
燕尾码 燕尾型石材挂件。

dowel A round wooden peg inserted into holes to join to pieces of timber together. The holes are drilled so that the tight fit causes friction enabling the two pieces of timber to be joined together. The joint may also be glued.
木榫/木销子 圆柱形木钉，可以插入洞中将木材拼合在一起。在木材上钻孔使其与木销紧接，靠材料的摩擦力将两块木材固定在一起。也可在相接处涂上胶水。

dowel (dowel pin) A wooden or plastic headless peg, sometimes with ridges down the side, that fits into aligned holes on two separate pieces of wood or metal to hold them together.
销子（定位销） 木质或塑料无头销钉，有时边上带有斜纹，可通过两块木材或金属上对齐的孔将其固定在一起。

dowel bar A smooth reinforcement rod that is placed across a contraction joint in large ground slabs to transfer the load from one slab to the next while accommodating axis expansion.
传力杆 指的是沿大面积水泥混凝土路面板伸缩缝布置的圆钢筋，其作用是在两块路面板之间传递行车荷载以抵消轴向膨胀。

doweller A machine for drilling *dowel holes.
木工开孔机 在木制品上开榫孔的机器。

dowelling The process of drilling dowel holes by using a *doweller and inserting *dowels into the holes.
开榫孔 用木工开孔机加工榫孔并把木榫插进榫孔的过程。

downcomer A pipe or passage in which fluids flow downwards.
下水管 向下排水的管道或管段。

down conductor The cable or rod that runs vertically down the outside of a building to transfer any lightning discharge safely to the ground. Usually made of copper.
引下线 沿建筑物外墙直立敷设的电缆或钢棒，可将避雷针接收的雷电流安全引向地面。通常情况下引下线是铜质的。

downlighter A *luminaire designed to direct light downwards.
筒灯 光线下射式照明灯具。

downpipe *See* RAINWATER PIPE.
落水管 参考“雨水管”词条。

downstand An edge that has been folded down.
肋形结构 边缘向下折弯的构件。

downstand beam A beam that extends beneath a floor slab.
肋形梁 楼板下的梁。

downtime A period when it is not possible to work due to adverse conditions or equipment failure.
停工期 因恶劣条件或设备故障导致无法工作的时间。

downwards construction *See* TOP-DOWN CONSTRUCTION.
逆作法施工 参考“ 逆作法施工”词条。

dozer *See* BULLDOZER.
推土机 参考“推土机”词条。

dpc（DPC） *See* DAMP-PROOF COURSE.
防潮层 参考“ 防潮层”词条。

dpm（DPM） SeeDAMP-PROOF MEMBRANE.
防水膜 参考“防水膜”词条。

draft（US） *See* DRAUGHT.
寒冷气流（美国） 参考“寒冷气流”词条。

drafted margin A uniform border chiselled around the edges of the face of a stone.
石材凿边 石材表面四边开凿的统一的线条。

dragline A tracked or wheel excavation machine, where the bucket is operated by ropes. The bucket is cast out from the boom; excavations can be undertaken below the level of the tracks and under water. The excavation and dumping can be widely separated.
拉铲挖掘机 铲斗靠前方钢丝绳牵引的履带式或轮式挖掘机。铲斗由吊杆甩出，拉铲挖掘机可以在停机面以下或水下挖土。拉铲挖掘机的挖土与卸土半径可以很大。

drain A channel or pipe (drainpipe) that carries surface and foul water.
排水渠/下水道 排除地表水和污水的渠道或管道。

drainage A process to convey fluids or gases to an egress point.
排水/排气 把液体或气体排到出口的过程。

drainboard A drainer, used for allowing cutlery, plates, cups, and dinner services to dry after being washed.
沥水板 用于餐具、盘子、杯子等洗后放上去沥干的去水板。

drain chute A specially designed *drainpipe that makes *rodding easy.
排水槽 一节特制的排水管，可使管道疏通工作变得更容易。

drain clearing *See* RODDING.
下水道清理 参考“下水道清理”词条。

drain cock（drain valve, drain plug, drain stopper） A device for controlling (e. g. stopping or draining) the flow of a liquid in a *drain.
排水旋塞（排水阀、排水螺塞） 可以调节（如关闭或打开）下水管里边液体流量的装置。

drained test *See* TRIAXIAL TEST.
排水剪切试验 参考“ 排水剪切试验”词条。

drainer（drainboard, draining board） A sloping surface adjacent to a kitchen sink that enables water to drain into the sink.
去水板（沥水板） 厨房洗涤槽旁边的一个倾斜表面，水会通过这个倾斜表面排到洗涤槽里。

drainlayer（pipe layer） A tradesperson who lays drains.
下水管安装工人（管道工） 负责安装下水管的技工。

drainline A series of pipes that are adjoined to form a *drain.
排水管线 组成下水道系统的一系列排水管。

drainpipe *See* DRAIN.
下水管 参考“下水管”词条。

drain rods A series of flexible rods that are screwed together and pushed down a *drain for cleaning purposes or to remove a blockage. *See also* RODDING.
清洁和疏通管道用推杆 清洁和疏通管道用推杆，推杆由几节柔性管拧接在一起。另请参考“下水道清理”词条。

drain shoe A drainage fitting that has an access cover and is connected to a *down drain.

浴缸去水 一个与下水管连接的带塞子的排水管件。

drain test A *hydraulic test on a newly laid drain to test the water tightness of each drainage run before backfilling.
管道水压试验 在回填前对新敷设的排水管进行水压测试，测试每个排水管段的水密性。

draught (US draft) The flow of cold air into a space that causes discomfort.
寒冷气流 流进某一空间令人不舒服的寒冷气流。

draught bead A bead placed on the sill of a sash window to prevent draughts at the sill.
密封条 安装在推拉窗窗框四周的密封条，可以保证窗扇的气密性。

draught stop (US draft stop) *See* FIRE STOP.
防火材料 参考“防火材料”词条。

draught strip (draught stripping) A strip of compressible material fitted between the opening leaf of a door and the frame, or a window casement and frame, to prevent draughts.
密封条 安装在活动门扇和门框之间或在窗扇和窗框之间的弹性密封条，可以保证门窗的气密性。

draw bolt *See* BARREL BOLT.
圆形插销 参考“ 圆形插销”词条。

drawbore pinning (drawboring) When the hole in the *tenon is slightly offset from the hole in the *mortise, causing the pin to be hammered into place through the holes so that it pulls the pieces tightly together.
榫头加销 当榫头与榫眼位置有轻微偏差时，可以在榫头与榫眼上钻木销孔，通过木销钉使两块木板紧紧固定在一起。

drawbridge A bridge that is hinged at one or both ends (in the latter case the deck will be split in the middle), such that the deck can be raised and lowered to allow, for example, ships to pass beneath.
吊桥 一端或两端用铰链固定的桥（如果是后者则桥板将从中间分开），可通过绳子拉起和降下吊桥，以允许船只从下面通过。

draw cable [draw wire, (US) fish tape] A thin wire that is used to pull other wires or cables through a cable duct or conduit. It can be inserted during construction or blown through using compressed air.
牵引绳 一根细长的钢丝，用来牵引其他电线或电缆穿过电管。可在电缆穿管时人工插入牵引绳，也可以通过空气压缩机把牵引绳吹进并穿过电管。

drawdown The lowering of the groundwater due to pumping from a well or surrounding wells.
水位下降 由于从某个井或邻近的井里抽水导致地下水水位下降。

draw-in box (pull box) An access point into ducting, cable tuns, or pipework that allows services to be pulled through the service tun; forms part of the draw-in system.
过线盒（拉线盒） 电管、电缆分接盒，过线盒上有开孔，电缆或电管可以从开孔穿过；过线盒是电缆引入系统的一部分。

drawing An illustration of something, e. g. a cross-section of a bridge.
图纸 某物的图示。如桥的剖面图。

drawings Normally refers to the working details used to construct the building or structure.
施工详图 通常指建筑或构筑物施工用的节点图。

drawings and details Images and supporting information, such as specifications and descriptions, that are used to inform the assembly of the building, structure, or works.
图纸和详细说明 包括一些图片和辅助说明信息（如规范和说明），这些都是用来说明建筑物、构筑物或分部工程的安装细节的。

drawing symbols Recognized shapes, used in drawings to depict components.
图例 图纸上用来表示某一建筑构件的公认符号。

draw knife Steel blade used to smooth and shape the surface of wood.
木工刮刀 用来修整和刮平木料表面的不锈钢刮刀。

draw-off pipe A pipe that enables a fluid to be removed from a system.
排水管 排空系统内液体的管道。

draw-off tap A tap that controls the amount of fluid that is removed from a system. A water tap is a type of draw-off tap.
放水龙头 可以控制出水量大小的龙头。水龙头就属于放水龙头。

draw-off temperature The temperature of the hot water that is drawn off from a hot-water storage cylinder.
出水温度 从热水储水箱流出来的热水的温度。

drencher sprinkler *See* DELUGE SPRINKLER.
雨淋喷头 参考“雨淋喷头”词条。

dress 1. To plane, smooth, and finish timber. 2. To smooth or cut stone to its finished dimensions. 3. To cut and fold lead or other malleable roofing materials, so that they are embedded into the mortar joints of walls, and are overlapped and folded to prevent water penetration. The folded lead often provides a decorative finish to the roof.
1. 刨平，刨光 刨平、刨光木材。**2. 打磨** 打磨或切割石材至规定尺寸。**3. 折叠** 切割和加工铅片或其他可锻屋面材料，确保加工后的屋面板接缝是完全咬合的，以杜绝屋面漏水隐患。压型铅质屋面板有装饰的效果。

dressed dimension (neat size) The size of stone, masonry, or timber once it has been cut or shaped to the desired dimensions.
成品尺寸 石材、砌块或木材完成加工后的尺寸。

dressed stone *Stone that has been cut or shaped to the desired dimensions.
料石 按规定尺寸切割或打磨而成的石材。

dressed timber *Timber that has been cut or shaped to the desired dimensions.
刨光木材 按规定尺寸刨切而成的木材。

dressing Shaping and cutting materials to their finished dimensions.
加工 把材料加工成预期的尺寸和形状。

dressing compound 1. Hot or cold bituminous liquid poured or levelled over roofing felt to provide an adhesive surface onto which limestone chippings can be scattered. The chippings provide some protection against degradation caused by the sun. 2. Levelling compound.
1. 沥青混合料 浇筑或在屋面油毡卷材上找平的热拌沥青或冷拌沥青混合料，这种沥青表面黏性良好，可以在上面摊铺石灰石碎石。石灰石碎石可以保护下层材料不受太阳暴晒破坏。**2. 自流平砂浆**

drift bolt A steel pin that is driven into a timber hole with a smaller diameter than the shaft diameter of the pin.
销子/销钉 用来钻木销孔的不锈钢销钉，销钉的轴径比木销孔的直径大。

drill A tool for boring (drilling) holes. The drill bit, a long pointed piece of metal, is the boring part of the drill.
电钻 一种钻孔工具。钻头，一根带尖端的金属杆，电钻的钻孔部件。

D-ring flexible 1. D-shaped sealing ring with one flat surface that prevents the ring rolling when it is slid over the male end of a pipe (*spigot). Once the ring is in place over the spigot, the pipe is pushed into the larger female pipe (socket). The rounded part of the ring makes a water-and air-tight seal by connecting with the socket. 2. Elasticated ring with a D-shape in cross section allowing the rounded edge of the ring to run and connect with rounded grooves in pulleys and wheels.
1. D型垫圈 D型的密封圈，平面端套在管道公头上（插头端）可防止滑动。当垫圈放置在插头端后，将管道插入套管母头里面（承口端）。垫圈的圆面端与承口相接起密封作用。**2. 密封圈** 横截面为D型的弹性圈，圆面端与滑轮和滚轮上的圆形槽相接并滑动。

drinking fountain A device that provides water for drinking.
直饮水机/自动饮水机 提供饮用水的机器设备。

drinking water (potable water) Water that is suitable for human consumption.
饮用水 可供人饮用的水。

drip 1. Part of a product or component that is shaped down towards the ground to encourage water to drip from it. 2. The front-lipped edge of the windowsill. 3. The 'V' bent or formed in the middle of a stainless steel, galvanized steel, or plastic wall tie that prevents water passing along the tie and crossing the cavity.
1. 滴水构件 头朝向地面，水可以沿着这个构件流下来。**2. 滴水线** 外窗台板下边的一条鹰嘴型线条。**3. 拉结件V型部分** 不锈钢、镀锌钢或塑料材质的空心墙拉结件中间的V型部分，可以防止水沿着拉结件通过空心墙。

drip edge The lipped edge of a windowsill or doorsill that encourages the water to run and drip away from the building.
滴水线 设置在窗台板或门槛下边的一条鹰嘴型线条，可以防止水沿窗台板或门槛流到建筑结构内。

drip-free paint *See* NON-DRIP PAINT.
非流淌漆 参考“非流淌漆”词条。

dripping eaves The edge or eaves of a roof that has no gutter to collect rainwater.
檐口滴水线 因为没有天沟来收集雨水而在屋顶边缘设置的滴水线。

drip tray Tray used to catch drips from condensing water from combustion, also known as a condensate pan.
滴水盘/接水盘 用来承接冷凝水的盘，因此又叫冷凝水盘。

driven pile A pre-cast *pile or pile casing (for a cast-in-situ pile) that has been pushed into the ground and displaces the soil around it; *See also* BORED PILE.
打入桩/沉入桩 打（压）入土中使周围土体侧移的预制桩或（灌注桩）桩套管。另请参考“钻孔灌注桩”词条。

drive screw (screw nail) A square section nail that has a twisted shank, giving the nail a screw effect. The nail, which is 2 mm or greater in cross section, is driven into wood in the same way as a conventional nail, but is more difficult to extract.
螺丝钉 金属杆上带螺纹的钉子，使钉子具有螺丝的效果。跟普通钉子相比，螺丝钉的横截面公称直径要大2mm以上，螺丝钉打入木材的方式跟普通钉子无异，但想拔出来就困难些。

driving cap A cover that is placed over the top of a pile to protect it from impact damage when it is being driven (hammered) into the ground.
桩帽 焊接在桩头位置的桩尖，打桩时可以避免桩头被破坏。

driving rain (wind-driven rain) Rain that is carried along by the wind and blown onto the envelope of a building.
大风雨（风夹雨） 伴有风的大雨，被大风吹到建筑围护结构上的雨水。

drop apron (drip edge) A strip of metal fixed at the edge, eaves, or gutter to act as a drip at the edge of metal-sheet roofing.
滴水线/滴水条 安装在金属屋面的屋檐或天沟位置的金属条，引导雨水顺着滴水线下沿滴落。

drop manhole *Manhole where the inlet pipe is at a considerably higher height than the outlet pipe; a drop pipe is used to take the inlet down to the *invert level of the *channel in the manhole.
跌水井 进水管高程远高于出水管高程时，用竖管将进水高程降到检修井渠道的水平高程。

droppings Materials such as mortar, snots, and other debris that have fallen down the inside face of a cavity. To ensure optimum efficiency of the cavity and ensure insulation can be placed so that it is in direct contact with the masonry, cavities must be free from debris and clean. Debris within the cavity causes cold bridges and pre-

vents insulation butting up to surfaces. The droppings within the cavity reduce thermal efficiency, increase sound transmittance, and can allow water to cross the cavity.
墙空腔杂物 从空腔内壁掉下来的砂浆、砂浆块和其他垃圾。为确保空心砌块墙能发挥最佳的性能以及保温隔热材料能直接与砌块接触，空腔内必须是干净的、无任何垃圾的。如果有垃圾，则可能会导致冷桥，且导致保温隔热材料不能紧紧平贴墙面。空腔内的垃圾将大大影响隔热、隔音性能，还会导致空腔透水。

drop system A hot-water heating system that is fed from a sub-tank at the top of the system.
上行下给式供暖系统 一种供热系统，热媒从建筑物顶层的热水箱分送到各层。

drove Scottish term for a * boaster
石工扁凿 苏格兰语中“石工扁凿”的意思。

drunken saw (wobble saw) A * circular saw where the saw blades contain slanting flanges resulting in the blade not being set perpendicular to the drive shaft. Used for cutting a groove wider than the thickness of the saw blade.
摇摆式锯 锯片上有倾斜法兰的圆锯，所以叶片与锯轴并不垂直。可用于开比叶片厚度要宽的槽。

dry 1. Paint after it has dried, either touch dry or dust dry. 2. Dry construction, a prefabricated building constructed on-site with no wet trades.
1. 油漆干了 包括指触干或不粘尘干。**2. 干法施工** 用预制件施工而无须弄湿现场。

dry area An area not expected to be exposed to water or moisture, being the opposite of a * wet area.
干区 不暴露于水或潮湿环境的区域，与湿区相反。
dry concrete A type of concrete having a low proportion of water so that the plastic mixture is relatively stiff; this type of concrete is suitable for use in dry locations, and is especially advantageous where large masses are poured and compacted on sloping surfaces.
干硬性混凝土 含水率很低的一种混凝土，这样混凝土拌和物就比较干稠。这种混凝土多用于干燥区域，特别适用于大面积斜面的浇筑和压实。

dry hip tile A hip tile that does not require bedding in mortar.
干铺脊瓦 不用砂浆进行黏结固定的脊瓦。

drying (drying-out, drying of screeds, drying of mortar) An operation in which a liquid, usually water, is removed from a wet solid. In construction this usually refers to the process of water evaporation (water loss) from a screed or mortar.
干燥（砂浆失水） 从潮湿的固体物质中去除水分的过程。在建筑上，通常指砂浆水分的蒸发。

drying shrinkage The contraction of plaster, cement paste, mortar, or concrete caused by loss of moisture.
干燥收缩 灰泥、水泥浆、砂浆或混凝土由于失水出现收缩。

dry joint A joint in masonry that does not use mortar.
干砌 砌体之间的缝隙不用砂浆而直接砌筑。

dry lining (US dry-wall) An internal wall finish comprising plasterboard sheets that are fixed to * dabs of plasterboard adhesive or timber battens.
石膏板隔墙 一种室内饰面墙，把石膏板固定到石膏板黏结剂或钉板条上。

dry masonry Masonry without mortar. In other words, blocks are held together without the presence of mortar.
干砌墙 砌体之间的缝隙没有砂浆。也就是说，不用砂浆进行黏结的砌块墙。

dry mix A mixture of mortar or concrete that contains little water in relation to its other components.
干混 只加少许水借助搅拌对各组分进行混合的工艺。

dry partition Partition made from prefabricated components, so wet trades, such as wet plaster, in-situ concrete, or masonry are avoided, keeping the process fast and dry. Traditionally, dry partitions are constructed from a timber frame and plasterboard, but can be constructed with steel frames with cellular cardboard cores,

and finished with various prefabricated panels.
石膏板隔墙 一种预制的隔断单元，无须使用灰泥、现浇混凝土等湿操作或砌块，安装过程快速、方便且可以避免潮湿。传统的石膏板隔墙都是由木质框架和石膏板组成的，但现在的石膏板隔墙也可以是轻钢龙骨框架加蜂窝墙板，再铺上各种预制饰面板。

dry riser (US dry standpipe) The main dry vertical pipe in a fire protection system that is used by the fire brigade to distribute water to the various floors within a building if a fire develops.
干式消防立管 消防系统干式立管，一旦发生火灾，消防员就可以利用干式消防立管把水分配到建筑物各楼层。

dry rot (Pula lacrymans) A fungal attack on timber that occurs in wood with high moisture content, normally around 22% moisture content.
木材褐腐 木材的一种真菌性病害，多出现在含水量高（通常含水量在22%左右）的木材中。

dry sprinkler A * sprinkler where the water is supplied from the * dry riser.
干式洒水喷头 由干式消防立管供水的洒水喷头。

dry stone walling Walls constructed of stone without the use of mortar. This was the main technique of construction during the prehistoric era and still prevails in some regions today. Extremely high-quality and solid walls can be built with the careful selection and bedding of the stones.
石材干砌墙 不用砂浆，直接把石材铺码在一起的一种墙。干砌墙是史前时代很流行的一种施工方法，即使在现在，有一些地区还是沿用这一种砌墙方法。只要选对石材且砌筑方法得当，是可以砌出高质量的坚固墙体的。

dry verge tile A verge tile fixed with clips rather than mortar or adhesive.
干铺式檐瓦 用金属挂件固定而不是用砂浆或黏结剂来固定的檐瓦。

dry wall (dry stone wall) *See* DRY STONE WALLING.
石材干砌墙 参考“石材干砌墙”词条。

DSC (differential scanning calorimetry) An item of apparatus used to determine the * glass transition temperature (Tg) and melting point for polymers.
差示扫描量热仪 用来测定聚合物的玻璃化转变温度（Tg）和熔点的仪器。

dual carriageway A road with a least two lanes in each direction, separated by a central reservation barrier or grass verge.
双程分隔车道 每个方向上至少有2个车道的双向行驶的车道，中间设有隔离带或绿化带。

dual-duct system A full air-conditioning system where two separate supplies of air are provided to all the * zones within the building. The temperature of each zone is altered by mixing appropriate amounts of hot and cold air in a thermostatically controlled mixing box, which is usually located in a false ceiling. In a basic dual-duct system, the temperature and humidity of the air are controlled by the central plant, so there may be variations in humidity between zones. In its constant volume form, this system will often mix air that has been heated with air that has been cooled. Although it can provide multi-zone control of temperature in a building, it has high space requirements, and high capital and running costs, so is no longer installed.
双风管空调系统 有两条送风管的全空气空调系统，可以送风给建筑物里的各个分区。一条风管送冷风，另一条风管送热风，冷风和热风在各区域的送风口前的混合箱（混合箱通常安装在假吊顶上）内按不同比例混合，达到各自要求的送风状态，再进入各区域。普通双风管空调系统中，空气温度和湿度是由中央空调控制的，因此各分区的空气湿度可能有差别。在定风量模式下，双风管空调系统通常都是混合已经加热和冷却过的空气。虽然双风管空调系统可以控制整栋大楼里边各个分区的温度，但是它对空间的要求高、投资高、运转费也高，所以现在很少使用。

dual-flush cistern A water-efficient cistern that contains two buttons for flushing. One button gives a full flush and the other button gives a half-flush.
双按式冲厕水箱 一种节水冲厕水箱，水箱上有2个冲水按钮，一个按钮是最大排水量的，另一个按钮则只有一半的水量。

dual phase steel A type of steel consisting of two distinct phases, allowing a compromise between strength and ductility.
双相钢 含两种不同相体的钢材，可达到强度和延性的良好配合。

dual system *See* TWO-PIPE SYSTEM.
双立管排水系统 参考“双立管排水系统”词条。

dubbing-out The filling up of holes and deformations in walls before skimming with plaster.
抹灰基层处理 在抹灰前先用灰泥填补墙上的孔洞或坑洼，使墙面平整。

duckbill nail A nail with a chisel point.
扁平尖钉 钉尖是凿子形状的钉子。

duckboard The use of wooden boards to provide a temporary solution, e. g. a walkway over muddy ground or a floor over joists.
工地排板 用木板作临时用途，如铺在泥泞地上的临时通道，或是龙骨上铺地板。

duckfoot bend (rest bend) A 90° bend (vertical to horizontal) in a pipe that is supported by a flanged base.
鸭脚弯头 由托座支撑着的一个90°（垂直到水平）管道弯头。

duct A channel, tube, or pipe through which something flows or passes.
管道 物质流过或通过的渠道、管子或管道。

duct cover A steel plate, or mesh, or a concrete slab that fits flush over the end of a *duct.
管道盖板 线槽盖板或风道末端的格栅风口或排水沟混凝土盖板。

ducted flue (appliance ventilation duct) A flue that supplies fresh air to a *room-sealed appliance for combustion.
锅炉进气管（设施通风管） 供应新风给密封式燃具以辅助其燃烧的管道。

ductile fracture A type of failure/fracture in ductile materials involving extensive plastic deformation.
延性断裂 延性材料的一种断口/断裂，包括塑性变形。

ductile iron A type of cast iron with increased ductility, sometimes known as nodular iron.
延性铸铁 延展性很高的铸铁材料，也叫作球墨铸铁。

ductile to brittle transition The temperature at which a material undergoes a change from ductile to brittle behaviour.
韧性（延性）-脆性转变温度 材料从韧性状态过渡到脆性状态的温度。

ductility The ability of a material to undergo *plastic deformation.
延性 材料的塑性变形能力。

ductwork The pipes used to move air around a building. Usually metal or plastic and can be flexible (**flexible ductwork**) or rigid (**rigid ductwork**).
风管 用于空气输送的管道系统。按材质，风管可分为金属风管或塑料风管；按刚度，又可分为柔性风管或刚性风管。

due date Date when the building is expected to be complete.
竣工日期 建筑物的预期竣工日期。

due time Expected time required and contractually bound to complete the project.
工期 计划的或合同约定完成项目所需的期限。

dummy activity An activity built into a network that is used to show a start, finish, or key point in the programme; it need have no properties, and time and resources do not need to be associated with the activity but can be. Start and finish activities are normally dummy activities.
虚工作 在网络图中显示计划内开始、结束或关键点的工序；虚工作并非实质工作，亦无时间和资源关联。开始和结束工作通常为虚工作。

dummy frame A temporary structure that is used to provide the correct spacing for a door set when a brick wall is being constructed.

临时门框 在砌砖门洞时，用一个临时的门框来表示门的尺寸。

dump truck A heavy, sometimes articulated, lorry that is used to haul material such as earth, aggregate, and rock over long distances. It has an open back bed that can tilt up to dump its load.

自卸货车 重型（铰链式）载货汽车。适用于土方、骨料、砂石的远距离运输工作。其货箱是开敞式的，可自动倾卸货物。

dumper A four-wheeled vehicle that is used on construction sites to haul material and tools over short distances. The driver sits at the back of the machine overlooking a front-tipping hopper.

前翻斗车 是工地上常用的短途输送物料的四轮车辆。驾驶座在翻斗车的后部，司机坐在驾驶座上可以看到前面的料斗。

dumpy level An optical surveying instrument with a telescope that only moves in the vertical direction, used in levelling.

定镜水准仪 一种光学测量仪器，其望远镜只在竖直方向上移动，可用来测量高程。

dungeon Medieval cells used to imprison and torture people.

地牢 中世纪的监狱，是用来关押和拷问犯人的地方。

dunnage Packaging material used to protect a ship's cargo.

集装箱充气袋 可保护海运集装箱中货物的包装材料。

duo-pitched roof A roof with two slopes that meet at a central ridge.

双坡屋顶 由两个倾斜面相互交接的屋顶，顶部的水平交线称正脊。

duplex apartment A modern term for a * maisonette.

复式住宅 以前称之为“maisonette”。

duplex control Lift control system that sends one car to a floor when requested, even though there are two or more lifts.

并联控制 每次只运行一部电梯到呼叫楼层的电梯控制系统，即使有 2 部或以上的电梯。

duplication of plant The doubling-up of plant so that one piece of plant is on standby in the event of the other piece of equipment failing or requiring maintenance.

备用设备 准备两台设备，一旦一台设备出现故障或需要维护时，另一台设备可以继续运行。

durability The ability of a material to resist wear, and loss of material through continual use. Durability of a material tells us how long a material will last in service. This is one of the most important properties an engineer must consider before selecting a material for any application.

耐久度 材料在持续使用情况下抗磨损和耗损的能力。材料的耐久度能告诉我们其使用寿命，这是工程师选择使用材料时需要考虑的重要因素之一。

duration The time taken to complete an activity or project.

工期 完成某个活动或项目所用的时间。

Dutch bond A brickwork bond in which the * stretchers (long side facing outwards) are in line in every other course.

一顺一丁式砌筑（荷兰式砌筑） 每隔一层的顺砖（长边面向外面）都是对齐排列的石砌筑法。

Dutch door (US) *See* STABLE DOOR.

两扇门 参考“两扇门”词条。

Dutchman A piece of timber used to cover up a defect or an error.

用来修补缺陷的木料。

duty of care The degree or standard of consideration that should be given to third parties, members of the public, and contracting parties. There is a duty of care to ensure other parties are not exposed to risks to their health or risks of injury when carrying out contracts. Employers have a duty of care to ensure that the health and welfare of their workers is considered when undertaking their work.
照顾责任 给予第三方、公众和合作方应有程度的考虑。确保对方在履行合同时免受伤亡危险。雇主也有责任考虑其工人工作时的健康和福利待遇。

dwang (Scotland) A horizontal member that is fitted between wall studs, floor joists, or rafters to provide rigidity. Usually timber, but can be aluminium or steel. *See also* NOGGING and STRUTTING.
水平加固条（苏格兰） 立柱、木搁栅或木椽之间的水平加固条。通常是木质的，但也有铝材或不锈钢材质的。另请参考“水平加固条”和“支撑杆”词条。

dwarf wall 1. A low external wall used as a border for gardens. 2. A low internal wall that does not extend all the way up to the ceiling. If the wall is a partition wall, it is known as a **dwarf partition**. 3. A low wall used in suspended timber floor construction to support the floor joists. Also known as a **dwarf supporting wall** or **sleeper wall**.
1. 室外矮墙 用作花园围墙的矮外墙。**2. 室内矮墙** 没有延伸至天花的内墙。如果是隔断墙，则称为“矮隔断”。**3. 地垄墙** 是指房屋底层空铺木地板下的承重矮墙。也称为“**dwarf supporting wall**”或“**sleeper wall**”。

dwelling Building or property where people reside.
住宅 人们居住的房屋或房产。

dye A substance used to colour a liquid or solid element, e. g. Paints.
染料 是能使液体或固体材料着色的物质，如油漆。

dynamic Relating to a physical change over time. **Dynamic consolidation** is the compaction of soil by repeatedly dropping a heavy weight over an area of land. **Dynamic factor** is an impact factor taking into account the effects of a sudden load to that of a static load. **Dynamic load** is an on-off loading regime, such as from the effects of wind. **Dynamic penetration test** is a standard penetration test where a rod is hammered into the ground. **Dynamic response** relates to continuous time-dependent movement, such as vibrations. **Dynamic viscosity** is the resistance to flow when a layer of fluid moves over an adjacent layer at a given speed.
动态 指事物随时间产生物理变化的情况。**强夯法**用重锤重复夯击某一片地来达到夯实土壤的目的。**动载荷系数**突加载荷与额定载荷的比值。**动载荷**受到如风力等变化因素影响下的载荷。**动力触探试验**是一种标准的贯入试验，利用重锤将探杆打入土中。**动力响应**在如振动等的作用下结构随时间变化产生的位移。**动力黏度**液体以一定速度从相邻流体层流过时产生的内摩擦力。

dynamite A powerful explosive used for blasting rock.
炸药 用于岩体爆破的烈性炸药。

E *See* MODULUS OF ELASTICITY.
弹性模量 参考“弹性模量”词条。

early warning meeting Some standard construction contracts, such as the *Engineering and Construction Contract (previously the *New Engineering Contract), require each party to call for an early warning meeting as soon as a problem is recognized. The meetings have been used to help overcome potential problems, reduc-

ing the potential for disputes. Such meetings are recognized as a proactive method of dispute management.
预警会议 一些标准施工合同，如《建设工程施工合同》（前称为《新工程合同》），会要求发包方和承包人在问题出现时尽快召开预警会议。预警会议有助于于解决潜在的问题和纠纷。预警会议是主动解决纠纷的有效方法。

ear protectors A pair of ear covers or muffs, normally connected to an adjustable headband, that are used to protect the ears from loud noise.
护耳器 保护人的听觉免受强烈噪声损伤的耳罩，耳罩通常与可调节耳机头环连接。

E

earth (US ground) 1. The ground or topsoil. 2. An electrical connection between an electrical device and the earth. The earth is assumed to have a voltage of zero.
1. 地面 地面或表土。**2. 地线** 连接电气设备和大地的导线。通常认为地线的电压为0。

earth bar A copper or brass bar that provides a common earthing point for electrical installations.
接地排 为电气设备提供接地点的铜排。

earth bond (earth bonding) The connection of all metal objects within a building, such as metallic water pipes, to earth, in order to provide protection against shock from the electrical system.
接地 建筑内一切金属装置（如金属水管）的接地。接地的作用主要是防止人身遭受电击。

earth building (earth wall construction) External walls constructed from *rammed earth, *adobe, or *cob.
生土建筑 用夯土、土坯或草泥垛为材料修筑的外墙。

earth conductor (earthing conductor) An insulated conductor that connects an item of electrical equipment, such as a *consumer unit, to the *earth electrode.
接地线 连接电气设备（如用户配电箱）到接地体上的绝缘导线。

earth dam A water-retaining structure (such as to retain water in a reservoir), formed from compacted earth material. Typically, it has an impervious clay core surrounded by sand, gravel, or rock, with the rock being placed on the upstream side (termed rip-rap) to prevent erosion from wave movement.
土坝 用压实土筑成的挡水建筑物（如水库土坝）。一般指由砂、砾石或石块围绕不透水的黏土坝心构成的坝，石块通常设在上游坡面（“抛石”），以防止河岸受水流冲刷。

earthed concentric wiring A mineral insulated cable where the conductor is totally surrounded by an earthed metallic copper sheath.
同轴接地电缆 导线由接地铜护套隔离的矿物绝缘电缆。

earth electrode (earth termination) A conductor that provides an electrical connection directly to earth. Usually takes the form of a rod, stake, or plate.
接地体 提供直接接地的金属导体。通常包括接地棒和接地排。

earthenware *See* CERAMIC.
陶器 参考“陶瓷”词条。

earthing The process of connecting electrical equipment to earth to provide protection against shock from the electrical equipment.
接地 电气设备经由导体与大地相连。接地的作用主要是防止人身遭受电击。

earthing lead The conductor that connects a device to the *earth electrode.
接地线 连接电气设备与接地体的导线。

earth-leakage circuit-breaker (ELCB) A *circuit breaker that switches off the power when a current is detected leaking to earth.
漏电断路器 当检测到接地极上有电流经过时，漏电断路器就会切断电源。

earth pressure The pressure exerted by the mass of the soil under gravity. The **lateral earth pressure** is the sideways pressure from the soil, which will act on a retaining structure (wall). The *active earth pressure and

* passive earth pressure are the minimum and maximum values of lateral earth pressures respectively.
土压力 大量土体在重力作用下产生的压力。侧向压力是土体对挡土建筑（墙）侧面作用的压力。主动土压力和被动土压力分别为侧向压力的最小值和最大值。

earth termination *See* EARTH ELECTRODE.
接地体 参考“接地体”词条。

earthworks 1. Engineering work associated with the movement and processing of quantities of soil, e. g. excavation, backfilling, compaction, and grading. 2. The process of moving large quantities of earth and soil. 3. A structure made from soil, for instance an embankment or a berm.
1. 土方工程 与土方处理和运输相关的工程。如：基坑（槽）开挖、基坑回填、土方压实和场地平整等。**2. 土方运输** 大量的土体运输。**3. 土石方建筑物** 用土体砌成的构筑物，如路堤或护堤。

earthworks support The supports used to retain soil during excavations or earthworks.
基坑支护 就是土方开挖过程中用来支撑土体的支护结构。

ease To loosen, free.
松开 使松散，松弛。

eased arris An * arris that has been rounded off.
圆棱 被磨圆的棱角。

easements A right held by a person to use land that belongs to another person for a specific or restricted purpose. The power or agreement is often used in civil engineering works, such as right to access services, to use land adjacent to a carriageway to undertake necessary works, the right to discharge water onto neighbouring property, right to services, etc. The specific acts are usually covered under an Act of Parliament, express grant under deed, express reservation when land is sold, or by prescription.
地役权 某人拥有的因特定目的使用他人土地的权力。该权力或协议常见于土建工程，例如使用邻近通道进行必要工作的权力，向邻近物业排水的权力，使用设施的权力等。

easing To lessen the pressure, slacken, allow movement, or reduce tension, to make room for works or access such as * easement of land.
使放松/使通过 减轻压力，使松弛，松开，缓和紧张状态，或为工程施工留出空间或通道等（如地役权）。

eaves (eave) The part of a pitched or flat roof that projects over the external wall. Comprises a * fascia and a * soffit. Protects the external wall from rain.
屋檐 坡屋顶或平屋顶凸出外墙的部分，包括挑檐立板和挑檐底板。挑檐可以保护外墙免受雨水打湿。

eaves board *See* TILTING FILLET.
檐口板 参考“披水条”词条。

eaves course (US starting course) The first * course of tiles at the * eaves of a roof.
檐口瓦 屋檐上的第一层瓦。

eaves drip (roof drip) The end of the * eaves of a roof where water drips off.
檐口滴水线 檐口末端的线条，雨水在这条线外就会跌落。

eaves flashing (drip edge flashing) A * flashing installed at the * eaves of a roof to prevent water penetration.
檐口泛水板 安装在檐口的泛水板，可以防止雨水渗入。

eaves gutter A gutter installed at the * eaves of a roof.
檐沟 安装在屋檐边的集水沟。

eaves lining *See* SOFFIT BOARD.
挑檐底板 参考“挑檐底板”词条。

eaves overhang The portion of a roof that overhangs the external wall at the *eaves.
出檐 屋檐伸出外墙之外的部分。

eaves plate A *wall plate.
檐口垫 一种梁垫。

eaves soffit The underside of an *eaves overhang. It may be lined with a *soffit board or left open.
挑檐底面 挑檐的下端面。可以在挑檐的底面安装一排挑檐底板或者什么也不装就空着。

eaves tile A tile used to form the *eaves course, it is usually shorter than the other roof tiles.
檐口瓦 屋檐第一层屋面瓦，通常比其他屋面瓦要短。

eaves vent (eaves ventilator) A ventilator installed at the *eaves to ventilate a roof.
屋檐通风口 安装在屋檐底面的通风口，可用于屋面的通风。

EC *See* EUROCODE.
欧洲标准 参考"欧洲标准"词条。

eccentric Away from the centre. An **eccentric load** will create a bending moment.
偏心的 偏离几何中心的。**偏心荷载**偏心荷载会引起弯曲力矩。

ecoduct A wildlife crossing, such as a channel under a highway for badgers, small mammals, amphibians, insects, spiders, etc. to pass through safely.
野生动物通道 野生动物通道，如设置在公路下方供野生动物（獾、小型哺乳动物、两栖动物、昆虫、蜘蛛等）过往的安全通道。

ECR glass A type of glass with excellent high temperature properties and resistance to corrosion.
ECR 玻璃 一种具有良好耐温性和耐腐蚀性的材料。

edge 1. The perimeter of a surface or object. 2. A sharp line created by the intersection of two surfaces or objects.
1. 边缘 物体或表面的周边部分。**2. 棱（边）** 两个平面或两个物体相交的线。

edge beam *See* RING BEAM.
边梁 参考" 边梁"词条。

edge bedding *See* FACE BEDDING.
（石材）侧砌 参考"侧砌"词条。

edge fixity The structural stability given to a suspended floor slab by an end edge beam or wall.
边缘稳定性 因为有边梁或墙体的支撑，从而确保了楼板的结构稳定性。

edge form *Formwork used at the edge of a concrete slab.
边缘模板 混凝土板边缘的支护模板。

edge grain (comb grain, vertical grain) The vertical section of grain which can be seen in wood that is *quarter-sawn.
径切纹理 径切木材中可见的垂直纹理部分。

edge joint 1. A joint along the edges of two materials. 2. A joint between two materials that has been made in the direction of the grain.
1. 接缝 两个构件接合部形成的缝。**2. 边边对接** 两个构件按照纹路的方向拼接。

edge nailing *See* SECRET NAILING.
暗钉法 参考"暗钉法"词条。

edge tool A tool used for cutting, made of specially hardened steel. Saws, chisels, knives, etc. can all be described as edge tools.
有刃的工具 用来切割的工具，淬硬钢材质。锯子、凿子、刀等类似工具都可以称作有刃的工具。

edging strip A thin strip of material, usually timber, used to cover the edge of the facing of a *flush door.

封边条 平面门封边用的条形材料，一般是木质的。

edging tool (edger) 1. A flat concreting trowel with one long edge turned down to make a radius that runs along the length of the tool. The turn-down radius is used to finish the edge of concrete, forming a bevelled edge on external corners. A bevel is often formed on the edge of concrete stairs to take off the hard 90° corner that would otherwise be formed. The bevel also reduces potential chipping that would occur at the corner of sharp concrete forms. 2. A spade-like tool with a semicircular head used for trimming the edge of grass lawn and turf.
1. 阳角抹子 一条边向下弯曲的混凝土抹子。向下弯曲的部分可用来修整混凝土边缘，在阳角处做出倒角边的形状。通常会对混凝土楼梯的边缘做倒角边处理，否则就会是坚硬的直角。倒角边还可以避免混凝土的边角出现破损。**2. 草坪修边铲** 用来修剪草坪边缘的半圆形铲头工具。

Edison screw cap A threaded fitting used to attach a bulb to a light fitting.
爱迪生螺旋灯头 用来把灯泡固定到灯座上的螺纹配件。

EDM *See* ELECTRONIC DISTANCE MEASUREMENT.
电子测距仪 参考“电子测距仪”词条。

effective angle of internal friction The shear strength value of internal friction in terms of * effective stress.
有效内摩擦角 有效应力内部摩擦抗剪强度指标。

efflorescence A white mark on walls (bricks) caused by the evaporation of water.
泛碱/起碱 水分蒸发后在砖墙上留下白色晶体。

effluent Liquid waste discharged from a sewage system or an industrial plant.
废水 从下水道或工业厂房流出的废水。

effusion The flow of gas through a very small aperture.
泄露 气体从很小的孔隙中流出。

egg-crate ceiling A * cell ceiling.
格栅吊顶 参考“格栅吊顶”词条。

eggshell 1. Paint with a finish that gives the very low sheen similar to that of an egg, it's slight shine or gloss makes it suitable for walls, furniture, and other internal finishes. 2. Shell structures are sometimes referred to as eggshell construction. The curved shell-like structures provide multiple paths for the stresses to be distributed. Although the structures are often thin, they are very strong and resilient, much the same as an eggshell.
1. 哑光漆 漆面光泽度很低，跟蛋壳的光泽度差不多，哑光漆适用于家具、墙面和其他内部饰面。
2. 壳体结构 又称为“蛋壳结构”。曲面的壳体结构可以把受到的压力均匀地分散到物体的各个部分。虽然壳体结构通常很薄，但却不乏牢固性和弹性，这一点跟蛋壳很像。

E-glass Also known as electrical glass, commonly used in buildings.
电工玻璃 也称作“electrical glass”，广泛应用于建筑行业。

egress 1. In property law, this is the right of a person to leave a property. 2. The movement of people or substances from an internal to an external environment. In the event of a fire, people within a property should be able to evacuate to a place of safety. A safe egress route must be provided to support evacuation procedures.
1. 外出权 在产权法中，外出权是指土地权利人享有使用某一出口的权利。**2. 安全出口/安全疏散** 人员或物品从室内区域转移到户外区域。例如，发生火灾时，大楼里边的人员通过安全出口疏散到户外安全的地方。应设置安全通道来协助紧急疏散。

EIA *See* ENVIRONMENTAL IMPACT ASSESSMENT.
环境影响评价 参考“环境影响评价”词条。

elastic modulus *See* MODULES OF ELASTICITY.
弹性模量 参考“弹性模量”词条。

elastomeric sealants A type of * sealant made from a specific group of polymers called * elastomers. Sealants are used to seal gaps or joints between materials or building elements, usually to prevent the ingress of water. Elastomeric sealants are considerably more expensive than other types of sealants, however the extra cost is offset by the superior durability. These sealants have the advantage of being much more resilient than mastics, together with the ability to withstand much greater joint movements (in the region of 20%). The anticipated life is about 25 years in the case of polysulphide sealants. Elastomeric sealants bond well to metals, brick, glass, and to many plastics. The maximum depth of sealant is usually about 15mm; elastomeric sealants do not change in volume when stressed.

弹性密封胶 是用一组特殊的聚合物（弹性体）制成的密封胶。密封胶可以用来填补材料或建筑构件之间的缝隙，以防止水分入侵。弹性密封胶比其他类型的密封胶要贵，但是这高出的费用跟弹性密封胶良好的耐久性也是成正比的。弹性密封胶比胶粘剂具有更高的弹性，可经受更大的接缝位移(20%范围内)。多硫化物密封胶的预期使用寿命约为25年。弹性密封胶适用于金属、砖、玻璃和大部分塑料构件。密封胶最大黏结深度通常为15 mm左右；弹性密封胶即使在外力作用下也不会产生变形。

E

elbow (elbow joint, knee) An angled pipe fitting, usually at 90°.

弯头 成一定角度的管件。90°弯头是最常用的。

elbow board (elbow lining) *See* WINDOW BOARD.

窗台板 参考“窗台板”词条。

electrical resistance strain gauge A device used to measure the amount of *strain of an object. A *Wheatstone bridge is used to measure the change in electrical resistance of aflat coil of very fine wire that is glued to the surface of the object.

电阻应变仪 测量物体应变的仪器。惠斯通电桥是一种可以测量（粘在物体表面的）细线扁平线圈电阻变化的仪器。

electrical riser A vertical shaft or duct that contains electrical cables and services. Used in multi-storey buildings.

电气竖井 也称电缆竖井，是高层或多层建筑内用于布放垂直干线电缆的通道。

electrical services (electrical engineering, electrical installation) A term that encompasses all of the electrical services found within a building, such as lighting, small power, alarm systems, etc.

电气设施（电气工程、电气装置） 指建筑物内所有电气设备的组合，包括照明系统、小型供电系统和消防报警系统等。

electric boiler A boiler that is powered by electricity. Two main types of domestic electric boiler are available; dry-core and wet storage. Dry-core boilers have a core of high-capacity thermal bricks, that are heated electrically, and the heat emitted from the bricks is used to heat pipes that are filled with water. Wet-storage boilers comprise a series of immersion heaters that are used to heat a large volume of stored water.

电锅炉/电热锅炉 用电进行加热的锅炉。生活电锅炉主要有两类，一类是干式锅炉，一类是湿式锅炉。干式锅炉的炉膛是用高强耐火砖砌成的，是用电能进行加热，而耐火砖所发出的热则是用来加热装有水的管道的。湿式锅炉则装有一系列浸入式加热器，是用来加热大量的储存水。

electric drill A hand-held electrically powered tool used to drill holes in materials. Can be mains or battery powered.

电钻 利用电做动力的手持式钻孔工具。电钻可以是有线的，也可以是电池供电的。

electric heating (electric resistance heating) The process of providing space or water heating by passing electricity through some form of resistance, such as a coil, panel, or wire.

电阻加热 是指利用电流流过电阻（如线圈、电板或电线）产生的热能对空间或水进行加热。

electrician A qualified tradesperson employed to install, repair, and maintain electrical circuits, plant, and equipment. In the US an electrician may be referred to as an inside wireman or outside lineman. The inside wireman works on building and plant with the outside lineman installing utility cables for the national distribution of electricity.

电工 从事电路、电气安装和修护的合格技工。在美国，电工通常分为室内装修电工和市政电网电工。室内装修电工主要负责楼宇或工厂内部电气的安装，而市政电网电工主要负责市政电网电缆线路的安装。

electric motor A motor that is driven by electricity which it converts to mechanical work.
电动机 是把电能转换成机械能的发动机。

electric panel heater A type of space heater that uses an electrical resistance to heat a panel.
壁挂式电暖器 小型取暖器的一种，直接用电阻对电板进行加热。

electric screwdriver (screwgun) A hand-held electrically powered tool used to tighten or remove screws. Usually battery powered.
电动起子（电批） 手持电动式拧紧或旋松螺钉的工具。通常是电池供电的。

electric shock An electric current passing through the body. It causes nerve stimulation and muscle contraction. Currents around 100 mA are fatal.
触电 电流通过人体。触电会导致人的神经受到刺激和肌肉收缩。电流到 100 毫安时可致命。

electric-storage floor heating Electric *underfloor heating.
电地暖 参考“电地暖”词条。

electric striking plate A *striking plate where the locking mechanism can be opened remotely by applying an electric current.
阴极锁片 可通过通电远程开启阴极锁的锁紧机构的一种锁舌片。

electric tools Portable tools that are electrically powered from the mains or a battery.
电动工具 利用电（电源供电或电池供电）做动力的手持式工具。

electric water heater A device that heats water using electricity. Can either be a hot water cylinder or an instantaneous water heater.
电热水器 指以电作为能源进行加热的热水器。电热水器可分为储水式和即热式。

electrodeposition Deposits on a plate due to an electric current. The following result from electrodeposition: electroplating, electroforming, electrorefining, and electrotwinning.
电沉积 通过电流沉积在板上的过程。以下结果会导致电沉积：电镀、电铸、电解精炼和电解冶炼。

electrolier (US) A *pendant fitting.
枝形吊灯（美国） 参考“枝形吊灯”词条。

electromagnetic cover meter *See* COVER METER.
钢筋保护层厚度测定仪 参考“钢筋保护层厚度测定仪”词条。

electronic distance measurement (EDM) A surveying instrument that uses an infrared or laser beam to measure the distance. The instrument transmits the beam to a reflector, which reflects the beam back to the instrument. The difference between transmitted and received signal is converted into a distance.
电子测距仪 用红外线或激光光束来测量距离的测量仪器。电子测距仪在工作时向反射器射出一束很细的激光，反射器再把光束反射回电子测距仪。最后将激光束发射和接收信号之间的差异转换成距离。

electro-osmosis A ground improvement process that involves passing an electric current through the ground so that groundwater migrates to the *cathode. Used in silty soils to lower the groundwater level or to divert the groundwater flow away from excavations.
电渗 一种地基处理方法，电流通过地基时，地下水会流向阴极位置。电渗可用于粉粒土地区降低地下水位或基坑降水。

electrostatic Electricity that is stationary.
静电 是一种处于静止状态的电荷。

electrostatic filter A filter that uses an electric charge to attract particles.

静电过滤器 利用电荷捕集微粒的空气过滤器。

element A major part of a building that has its own functional requirements, such as walls, floors, roofs, windows, doors, stairs, and services. Constructed from various components and materials.
建筑构件 指构成建筑物的各个不同功能要素。如：墙体、楼面、屋面、窗、门、楼梯等。建筑构件由不同的零件和材料组成。

elemental bills of quantities Descriptions of the building works that are organized by dividing the quantities into building elements. The labour and material quantities are rolled up into the main building elements with less detail than standard * bills of quantities, saving time on the detail of estimation.
分部分项工程量清单 按照建筑构件的格式编制的工程量清单。人工和材料费是包含在主要建筑构件中的，分部分项工程量清单的细节信息比标准工程量清单要少，这也是为了节省估算的时间。

E

elevated gravity tank A water storage tank located above ground level, usually on a roof, where the water is distributed via gravity.
重力水箱/高位水箱 高于地平面的储水箱，通常设置在屋面上，可以利用重力自流供水。

elevated road (elevated railway) Located above ground level on an embankment or supported by columns. Reasons for elevation could be to prevent flooding or to allow passage of vehicles/pedestrians beneath.
高架路（高架铁路） 将道路架空建设，下边有路堤或柱子支撑。之所以建高架路是为了避免道路被淹，或者是行人和车辆也可以从高架路下面通过。

elevation 1. The vertical distance between a particular point and sea level. 2. A two-dimensional drawing of the side of a three-dimensional object as seen from the front, back, left, or right. Commonly used to show the exterior of a building.
1. 标高 是某一点相对于海平面的竖向高度。**2. 立面图** 三维物体其中一面的二维图，例如其正面、背面、左面或右面。通常用来表示建筑物的外墙。

elevator (US) A * lift.
电梯（美国） 参考“电梯 ”词条。

elevonics (US) The electronic controls for a lift.
电子控制装置（美国） 电梯电子控制装置。

ell An extension to a building that is normally at right angles to the main building, producing a L-shape.
侧楼 楼房延伸出来通常与主楼垂直（成 L 形）的部分。

elm A hardwood, which is dull brown in colour, coming from deciduous and semi-deciduous elm trees. Deciduous trees shed their leaves seasonally and semi-deciduous trees shed their leaves for short periods as new foliage is growing. Elm suffers from warping if not seasoned properly; it should only be used in consistent environments and not ones which experience changes in moisture content. The wood has a twisted grain that makes it harder to split than oak, although it has other characteristics that make it weaker than oak. It is often used as a * veneer and for wood-block flooring.
榆木 硬木，暗棕色，落叶乔木或半落叶乔木。每年秋冬季节或干旱季节落叶乔木的叶子会全部脱落，而半落叶乔木只是短期内叶子会脱落，很快新的树叶又会长出来了。如果未能适当烘干，榆木将会出现翘曲。榆木只能用在稳定的环境中，而不能用于时而潮湿时而干燥的环境中。榆木有羽状木纹，比橡木更不易开裂。但是榆木也有不如橡木的特性。榆木通常用作饰面板和块状拼花木地板。

elongation 1. The lengthening of something. 2. The angle between the sun and a celestial body.
1.（某物）伸长或延长。2. 距角 指天体离太阳的角距离。

embankment A compacted earth structure, which is typically trapezium in cross-section. Used to retain water, i. e. acting as a dam or preventing flooding, or can be used to carry an * elevated road or railway.
路堤 压实土建筑物，通常其横截面形状为梯形。路堤有截水的作用，如作堤坝或防止洪水；路堤还可以用作高架公路或铁路的支撑。

embedded wall A retaining wall (such as a sheet pile) that is driven into the ground to produce * passive

earth pressure on the front of the wall.
打入土中的挡土墙 打入土中的挡土墙（如板桩墙），会对墙背施加被动土压力。

embossed A raised image or design on a material.
浮雕 材料上凸起的图案或设计。

embossed carpet Decorative pattern formed on carpet by cutting the strands of fabric at different lengths.
浮雕地毯 通过切割不同长度的纤维组成图案的装饰地毯。

emergency exit indicator lighting *See* SELF-ILLUMINATING EXIT SIGN.
紧急出口指示灯 参考"紧急出口指示灯"词条。

emergency shutdown A button, usually red in colour, that when activated in an emergency, cuts the power or stops items of machinery. *See also* MUSHROOM-HEADED PUSH BUTTON.
紧急关闭按钮 通常为红色按钮，紧急情况下启动该按钮可以切断电源或中断设备功能。另请参考"蘑菇头按钮"词条。

emissivity A measure of a surface's ability to emit radiation. It is calculated by comparing the amount of energy radiated by a surface to the amount a black body would radiate at the same temperature.
辐射率 辐射率是衡量物体表面以辐射的形式释放能量相对强弱的能力。物体的辐射率等于物体在一定温度下辐射的能量与同一温度下黑体辐射能量之比。

empirical Based on observation and experimental testing rather than theory.
经验主义 以实验或观察为根据，而非借助于理论。

employer (client) The person or organization that pays for the project and for whom the building or structure is being provided. The employer is not necessarily the body that will use the building once constructed, although it will often own the land on which the structure is being built.
业主 支付工程建设项目的投资人或投资机构。虽然业主通常是物业土地的所有权人，但建成后业主不一定是物业的使用者。

employer's agent (employer's representative) The representative who acts on behalf of the client, administering the contract and from whom the construction professionals take instruction and report to. Under the standard forms of contract different terms are used. *JCT WCD (With Contractors Design) refer to the employer's agent, and JCT 98 refers to the employer's representative.
业主代表 代表业主执行合同管理的工作，负责为承包商的施工提供指示及审查承包商提交的报告。标准合同中对于"业主代表"一词有不同的说法。英国合同审定联合会制定的标准合同（附承包商设计）中用 employer's agent 来表示"业主代表"，而 JCT 98 合同条件中则用 employer's representative 来表示"业主代表"。

employer's requirements Includes performance requirements, service agreements, contract brief, functional requirements, and express instructions provided under contract.
业主方的要求 包括履约规定、施工协议、合同摘要、功能要求及业主按照合同规定给出的一些指示。

emulsion (emulsion paints) A liquid of usually two immiscible components, e. g. Paints.
乳胶（乳胶漆） 通常为不混溶的双组分液体，如乳胶漆。

EN *See* EUROPEAN STANDARD.
欧洲标准 参考"欧洲标准"词条。

enamel paints A special type of paint based on *polyurethane or *alkyd resins to give very durable, impact-resistant, easily cleaned, hard gloss surfaces. Colours are usually bright. Such paints are suitable for machinery and plant in interior and exterior locations.
磁漆 以聚氨酯树脂或醇酸树脂为基料制成的一种特殊类型的油漆。主要特点是：耐磨、抗冲击、易清洁和具有良好的光泽。磁漆的颜色通常都比较鲜艳。磁漆适用于室内或室外区域机械设备表面的涂装。

encasement The process of covering or enclosing an element with a fire-resistant material in order to increase the level of *fire protection.

防火包裹 用防火材料包裹建筑构件，以提高其防火等级。

encasing 1. Encasement. 2. Enclosing elements or components, such as columns or pipework.

1. 包裹 2. 包裹（柱子或管道等）的建筑构件。

encastre support A supporting member (beam, column, etc) that is fixed, and prevents all movement.

端部固定 用于支撑的建筑构件（如梁和柱等），被固定以避免发生任何位移。

E

enclosed stage (closing in) The completion of the building envelope sufficient to allow finishing trades and internal works to begin. *Fast-track building methods often mean that internal building works start well before the external envelope is fully watertight. Traditionally, to eliminate any risk of damage to the *finishes from the elements, the external envelope would have been complete before the finishing trades commenced.

建筑封顶 指建筑外部围护结构基本完成，建筑进入内部装饰的阶段。快速跟进施工通常指的是在外部围护结构还未做好防水工作前就开始室内工作。通常情况下，为了避免对室内装饰材料造成损坏，要等到外部围护结构全部完成后才开始室内装饰工作。

enclosed stair *See* PROTECTED STAIR.

封闭楼梯间 参考“封闭楼梯间”词条。

enclosure 1. An enclosed space. 2. The external weatherproof layer of a building. 3. A box, cabinet, or cupboard that houses electrical equipment.

1. 密闭空间。2. 建筑物防风雨保护层。3. 接线盒、配电箱、配电柜。

encroachment Beyond specified limits, e. g. where property enters the land owned by another.

侵犯（他人土地） 越过边界线，如建筑物跨过他人的土地。

encrustation (incrustation) Mineral, chemical, or other deposits left in a pipe, vessel, or other equipment by the liquids that they convey.

结垢 管道、容器或其他设备在运输液体时残留的矿物质、化学物质或其他沉淀。

end bearing pile A pile that distributes its load to the strata at the base of the pile; *See also* FRICTION PILE.

端承桩 桩端进入岩石层并把荷载分布给岩石层的桩。另请参考“摩擦桩”词条。

end event The final point or task on a project network. As it is the final milestone, the last thing to complete, it will form part of the *critical path.

结束事件 在项目进度计划网络图上的尾节点或结束活动。由于是最后的里程碑，所以是组成关键路径的一部分。

end grain The exposed grain on a piece of timber that has been cut perpendicular to the length of the log.

端面木纹 沿着木头端面方向（与木头长度方向垂直的方向）切削得到的木纹。

end joint The joint that is formed when the ends of two components have been joined together.

端接/对接 两个构件端部的对头接合。

endlap (end lap) The amount of overlap at an *endlap joint.

搭接量 端部搭接的重叠部分。

endlap joint (end lap joint) A joint where the ends of two materials are halved in thickness and then lapped.

端部搭接 切削两块材料端部的厚度到一半，再进行搭接。

endoscope An optical instrument that can be inserted into pipes and orifices to inspect the internal conditions.

管道内窥检测仪 一种光学仪器，把管道内窥检测仪（摄像头）放入管道和孔口内部，可以检查管道内部的情况。

endothermic A chemical reaction that absorbs heat; *See also* EXOTHERMIC.
吸热反应 吸收热量的化学反应。另请参考“放热反应”词条。

endrin A toxic pesticide, which is harmful to human health when found in domestic water supplies.
异狄氏剂 一种杀虫剂，掺入异狄氏剂的饮用水对人的健康有危害。

energy The capacity to do work. *See also* KINETIC ENERGY, POTENTIAL ENERGY, and STRAIN ENERGY.
能量/能力 完成一项工作所需的能动力。另请参考“动能”“势能”和“应变能”词条。

engaged column (applied column, semi-detached column) A load-bearing column that is partially set within a wall.
附墙柱 部分埋在墙体里边，部分凸出墙面的承重柱。

engine A machine that converts energy into power or motion. An engine can be driven electrically, hydraulically, by compressed air, steam, or internal combustion.
发动机 一种能够把其他形式的能转化为机械能的机器。包括电动机、水力发动机、空气压缩机、蒸汽机和内燃机。

engineer A professional with applied mathematical knowledge or technical skills in either *civil, mechanical, structural, electrical, or marine engineering that enables the person to design or work with the tools of the profession. Civil engineers work on roads, sewers, and infrastructure; structural engineers design buildings and structures; electrical and mechanical engineers work with electrics, plant, and equipment; and marine engineers work with structures and plant based in the sea or that which is part of other watercourses.
工程师 在土木工程、机械工程、结构工程、电气工程或海洋工程方面具有应用数学知识或专业技术能力的人员。有了这些专业知识和技术能力，工程师就可以用专业的软件或工具进行设计或施工等活动。土木工程师负责道路、排水和基础设施的工作；结构工程师负责建筑物和构造物的设计；机电工程师负责电气设备和机械设备的工作；海洋工程师负责海上或属于河道部分的构筑物和设备的工作。

engineered brick (US) (UK engineering brick) *See* BRICK.
高强抗蚀砖 参考“砖”词条。

engineering The use of science to design things.
工程学 是用科学的原理来设计物体的过程。

Engineering and Construction Contract [New Engineering Contract (NEC)] Standard form of construction and civil engineering contract published by the Institute of Civil Engineers. The contract has attempted to remove traditional barriers. There are a number of *terms and *conditions which encourage greater collaboration especially in the areas of dispute or problem resolution. Identifying and attempting to resolve problems through proactive planning and early warning meetings are two ways that encourage the team to own the problem and attempt to mitigate the effects.
工程施工合同[新工程合同（NEC）] 土木工程师协会出版的标准土木工程施工合同。新工程合同旨在清除传统施工的一些障碍。新工程合同加入了一些条款和条件，旨在鼓励发包人和承包商更好、更广泛的协作，特别是在解决纠纷或问题方面的协作。主动规划及预警会议是工程队伍发现和解决问题的两种主要方式。

Engineering Council A regulatory body that holds a national register of Chartered Engineers (CEng), Incorporated Engineers (IEng), Engineering Technicians (EngTech), and Information and Communications Technology Technicians (ICTTech). It sets and maintains internationally recognized standards to ensure professional competence and ethics.
英国工程委员会 负责特许工程师（CEng）、技术工程师（IEng）、工程技师（EngTech）和信息通信技术技师（ICTTech）的注册和认证的管理机构。英国工程委员会的职责是：制定国际通用标准来保证工程师和技师的专业能力和职业道德。

engineering geology A branch of geology related to *engineering, e.g. the design of a foundation related to

the underlying strata (geological) conditions.
工程地质学 地质学的一个分支，是研究与工程建筑等活动有关的地质问题的学科。如设计地基的时候，应考虑到下伏地层地质上的条件。

engineering installations Collective name given to the electrical and mechanical services within a building.
工程安装设施 建筑物内部机电设施的总称。

engineering wood A mixture of different timber products to produce a generic timber material, e. g. Plywood.
复合木材 几种木质产品复合而成的普通木材，如胶合板。

English bond (Old English bond, Dutch bond) A type of brickwork bond comprising alternating courses of * headers and * stretchers.
英式砌法（传统英式砌法，荷兰式砌法） 一种砌砖方式，丁砖和顺砖交错排列。

English garden wall bond The type of brickwall bond has three courses of stretchers between every course of headers.
三顺一丁砌墙法 每层丁砖之间有三层顺砖的砌砖法。

enhanced greenhouse effect *See also* NATURAL GREENHOUSE EFFECT.
增强的温室效应 请参考"温室效应"词条。

en-suite bathroom A bathroom that is attached to a bedroom.
套间卫生间 卧室里边的独立卫生间。

ensure To do to one's best ability.
确保 尽全力做好某事。

entire completion The satisfactory conclusion of all works.
完全竣工 所有的工程都圆满完成。

entire contract The contract should be considered in its entirety, a term cannot be read on its own when it forms part of a contract; it must be considered against other terms within the contract which could mean that the term itself has a different meaning.
不可分合同 合同应看成一个整体，不能脱离整个合同背景来解读某一合同条款，而应结合其他合同条款一起理解，避免以偏概全。

entrapped air Irregular and undesirable air voids in concrete. Can be caused by poor mix design, segregation during placing, and inadequate vibration; *See also* AIR-ENTRAINED CONCRETE.
混凝土孔洞 混凝土材料中出现不规则或不符合要求的孔隙，可能是由于混凝土配合比有问题、浇筑过程中出现偏析或振实不充分引起的。另请参考"加气混凝土"词条。

entry 1. An entrance to a building. 2. A point where a component, such as a gas pipe or an electrical cable, enters a building.
1. 入口 建筑物入口。**2. 入户点** 建筑构件的入户点，如天然气管道或电缆的入户点。

entry phone A phone that is used to gain entry to a particular space. Usually used in conjunction with an * electric striking plate so that remote entry can be gained.
门禁电话机 设在门口供进入某一区域使用的电话分机。通常门禁电话机与阴极锁是配套使用的，阴极锁可以远程将门打开。

entry system An electronic system used to gain entry to a particular space. Various systems are available including card-operated systems, code-operated systems, and entry-phone systems.
门禁系统 用来控制出入某一区域的电子系统。门禁系统有下列几种：卡片识别、密码识别和电话门禁系统。

envelope 1. The external enclosure of a building, often the part of the building that resists the weather. A **building envelope** is the exterior surface of a building that separates the inside of the building from the outside.

2. The cover or surround that encases an element.
1. 围护结构 建筑围护结构，能够有效抵御不利环境的影响。建筑围护结构是区分室内和室外的分隔界限。**2. 外壳，覆盖物** 构件外壳、覆盖物。

environment Our physical surroundings such as people, building, structures, land, water, atmosphere, climate, sound, smell, and taste.
环境 包括周围的人、建筑物、构筑物、土地、水、大气、气候、声音、气味和味道。

environmental impact assessment (EIA) An investigation to determine the potential effects (both positive and negative) on the environment resulting from an existing or proposed development.
环境影响评价 对现有开发活动或拟定开发活动给环境质量带来的影响（包括积极影响和消极影响）进行评价。

environmental temperature A *comfort index that takes into account air temperature and mean radiant temperature.
环境温度 一种热舒适指数，是人对空气温度和平均辐射温度的主观反应。

EPDM *See* ETHYLENE PROPYLENE DIENE MONOMER RUBBER.
三元乙丙橡胶 参考“三元乙丙橡胶”词条。

epoxy A *polymeric material used in adhesives and pipeline coatings for corrosion protection.
环氧树脂 一种聚合物材料，可用作胶粘剂或管道涂料，有耐腐蚀的作用。

epoxy paints Special types of paint based on *epoxy polymers which are highly resistant to abrasion and spillages of oils, detergents, or dilute aqueous chemicals. Often applied to concrete, stone, metal, or wood in heavily trafficked workshops and factories.
环氧漆 以环氧聚合物为基体制成的特殊涂料。环氧漆的主要优点是：耐磨损，不易粘油、耐洗涤剂和耐化学稀释液。环氧漆广泛应用于车间和工厂等人流量较多的场所的混凝土、石材、金属或木材上面。

equal The same in size, quantity, value, or standard.
同等的 同等尺寸、数量、价值或标准的。

equilibrium moisture content The point at which the moisture content of a material is in *equilibrium at a given temperature and humidity.
平衡含水率 指在一定温度和湿度条件下，材料（木材）湿度最终达到稳定的含水率。

equilibrium phase In metallurgy, the state of a system where the phase characteristics remain constant over indefinite time periods. When a material is in equilibrium, it is considered to be stable, i. e. its properties will not change unless it is subjected to a change in external conditions i. e. force, temperature, ambience, etc.
相平衡 在冶金学方面，当一个多相系统中各相的性质和数量均不随时间变化时，称此系统处于相平衡。材料处于相平衡时是稳定的，也就是说材料的性能是不变的，除非有外部条件（如外力、温度、环境等）刺激。

equipment 1. Building services systems. 2. Any items used to undertake a task.
1. 设备 建筑设备系统。**2. 器材** 完成某一任务所需的道具。

equipment noise The noise emitted from building services systems.
设备噪声 建筑设备系统发出的噪音。

equivalent 1. Items or services of the same standard or value. 2. Legal language or terms that have the same meaning. In different countries or contracts certain terms are considered to have the same meaning or similar meaning ‘near equivalents’.
1. 同等 有相同标准或价值的产品或设施。**2. 等效的** 有相同含义的法律语言或术语。在不同国家或不同合同中，有些不同表达的术语的含义都被看作是相同的或近似的。

equivalent temperature A *comfort index that takes into account air movement, dry-bulb temperature, and mean radiant temperature.

等效温度 热舒适指数，是人对不同气流速度、干球温度和平均辐射温度的环境的主观反映。

erection The process of raising, positioning, constructing, and fixing building * elements or * components into their final position.
建造/安装 把建筑构件搭起、固定、建造和安装在最终规定位置上。

erg Unit used to measure energy or work. It is equal to work done by a force of one * dye moving through 1 cm in the line of action of the force.
尔格 功和能的单位。是一达因的力使物体在力的方向上移动一厘米所做的功。

E

erosion A physical or chemical process of wearing away, for example, through water, wind, ice.
腐蚀/侵蚀 指物质因物理或化学作用而逐渐消损破坏，如受潮或风化。

erratic 1. Irregular, has no consistency. 2. Rock that has been carried by a glacier and deposited a distance from its source once the glacier has melted.
1. 不规则的，不稳定的 2. 漂砾 冰川融化时，被冰川带到别处的石块。

error A mistake, something that should not exist.
错误 不正确，不应存在的事物。

escalation of contract prices (US) Fixed price contract that allows for an increase in the contract price if certain events or price fluctuations take place.
追加合同价格 如有特殊事件或价格波动发生，可以在合同固定价的基础上增加合同价格。

escalator A set of moving stairs between two floors. The stairs are attached to a continuously circulating belt. Commonly found in shops where they are used to move people quickly from one level to another.
自动扶梯 两层楼之间的自动行人电梯。电梯固定在连续运行的运输带上。通常可在商场里边看到，自动扶梯可以快速地把行人从一个楼层输送到另一个楼层。

escape chute A fabric tube used as a vertical escape route from upper floors of a building (not used in the UK). The tube is normally angled and twisted by a helper on the ground to prevent anyone entering the chute and falling straight to the ground. Once a person enters the tube, it is slowly untwisted allowing a person to steadily descend to the ground.
逃生滑槽/逃生滑道 一种从建筑物顶层逃离到地面的竖直布质柔性管（不适用于英国）。逃生滑道通常由地面人员协助弄好角度并扭几圈，避免逃生人员直接从逃生滑道入口直线掉到地面。一旦逃生人员进入逃生滑道，地面协助人员再慢慢往反方向扭几圈，以便逃生人员顺利地滑到地面。

escape route The method of egress out of a building in the event of a fire or emergency.
疏散通道 发生火灾或其他紧急情况时，用以离开建筑物的专用通道。

escape stair (fire escape, fire stair) Flight of steps in a building used as the main exit from the building in the event of fire or other emergency.
疏散楼梯（消防楼梯） 在发生火灾或其他紧急情况时用来疏散人群的主要楼梯通道。

escutcheon (key plate) A thin metal plate surrounding a keyhole. Used for decoration and to protect the surrounding door.
锁眼盖 安装在锁眼周围的薄金属盖板。既起到装饰的作用又可以保护锁眼周围的区域。

espagnolette bolt (shutter bolt) A long vertical bolt used to secure casements, doors, or shutters. Operated by turning a handle halfway along the length of the bolt.
天地杆插销 用来固定平开窗、门或遮板的竖直长插销。可以通过旋转插销中部的手柄来控制门窗的开关。

essence of the contract Something that is at the heart of the contract or aspect that goes to the root of the contract, a core term, or condition of contract.
核心条款 合同的核心或涉及基本价格的方面，合同的主要条款或条件。

estate 1. Property and land, belonging to an organization or individual. 2. A person's net worth, their enti-

tlements, obligations, and net assets.
1. 房产/地产 个人或机构拥有的房产和土地。**2. 个人财产** 个人的资产净值、津贴、债务和净资产等。

estimate Price given for works. If an estimate is not to be considered a legally binding offer of a price for service or works, it is prudent for the party suggesting the probable price to state so. In construction *bills of quantities are used to fix the approximate costs of the works. Changes to labour costs, and/or variations in the standard, quality, and quantity of work may alter the final cost.
估价 给出工程的价格。如果对某项工程的估价是不受法律约束的报价，则招标人应谨慎给出大概的价格并说明是非约束性报价。在工程方面，工程量清单是建设工程计价的依据。但是如果劳力成本变更或工程的标准、质量和数量变化，都会影响最终的报价的。

estimating The practice of pricing works—fixing the probable cost and expected profit. The *estimate is fixed by *taking off the works, measuring the work from the drawings, descriptions, and quantities, and projecting future labour and material costs.
预算 对建设工程的大致成本和预期利润事先加以计算的过程。预算一般包括下列几个步骤：搜集有关的工程描述，根据图纸、工程说明等计算工程量，预计未来的劳力和材料成本。

estimator The person who works out the probable costs and fixes the estimate. The estimator is often a quantity surveyor.
预算员 计算出工程的大致成本并给出估价的专业人员。预算员通常为工料测量师。

etching A surface treatment applied to metals to reveal the morphology of grains and microstructure.
蚀刻 金属表面处理工艺，蚀刻可以显示出纹理和微观结构的形态。

ethics The study of moral conduct. **Engineering ethics** is regulated by the *Engineering Council and relates to what is right and wrong with regards to carrying out the requested work, e. g. it is unethical to falsify gathering data.
伦理学 道德方面的研究。由英国工程委员会规范的，探讨的是跟工程相关的对错问题，如伪造数据是缺乏职业道德的表现。

ethylene propylene diene monomer rubber An elastomeric polymer also called EPDM and EPM. EPDM is widely used in construction applications due to its retention of properties (strength) at elevated temperatures.
三元乙丙橡胶 一种弹性聚合物，也称为 EPDM 橡胶或 EPM 橡胶。由于其优异的耐热性能，三元乙丙橡胶被广泛应用于建筑行业。

ethylene vinyl acetate copolymers (EVA) A polymer with high density and toughness. EVA is primarily used in construction as hot melt and heat seal adhesives.
乙烯-醋酸乙烯酯共聚物（EVA） 高密度、高强度聚合物。EVA 主要用于建筑领域，作热熔密封胶用。

Eur. Ing. The professional title for an engineer registered with the Fédération Européenne d'Associations Nationales d'Ingénieurs (FEANI). The federation brings together national engineering associations from European countries.
Eur. Ing. 认证 “欧洲工程师协会联盟”（FEANI）颁发的“Eur. Ing.”（“欧洲工程师”）的头衔。“欧洲工程师协会联盟”是欧盟各成员国的工程师协会联合成立的。

Eurocode (EC) The European quality standard published and controlled by the *European Committee for Standardization. The code fixes a minimum standard of quality that must be achieved. In many cases Eurocodes are replacing *British Standards. The introduction of Eurocodes helps to ensure greater consistency across Europe, making it easier to specify goods and services in other European countries.
欧洲规范 欧洲标准化委员会出版并监控的欧洲质量标准。欧洲规范确立了产品和设施的最低质量标准。在许多情况下都可以用欧洲规范来代替英国标准。推行欧洲规范是为了在其他欧洲国家中统一标准，便于更好地识别商品和服务。

European Committee for Standardization (CEN) Comité Européen de normalisation, the French body

that coordinates Eurocodes and European standards. The organization works with other member states to establish and control standards. The common European standard should help with standardization across Europe, allowing free trade, and reducing international boundaries.
欧洲标准化委员会 在法国成立的对欧洲规范和欧洲标准起到协调作用的机构，欧洲标准化委员会的宗旨在于促进成员国之间的标准化协作，制定和控制本地区需要的欧洲标准。通用欧洲标准应能对自由贸易起到促进作用，打破国际的贸易壁垒。

European Standard [EuroNorm (EN)] A standard that has national recognition by all *European Committee for Standardization states.
欧洲标准 获得欧洲标准化委员会所有成员国认可的标准。

EVA *See* ETHYLENE VINYL ACETATE COPOLYMERS.
乙烯-醋酸乙烯酯共聚物 参考“乙烯-醋酸乙烯酯共聚物”词条。

evaporation (evaporate) The process by which a liquid is converted into a gas.
蒸发 物质从液态转化为气态的相变过程。

event Task or activity on a project network or plan.
事件 项目进度计划网络图上的任务或活动。

evidence The records, artefacts, documents, and statements providing the material that illuminates a situation, which can then be used to support a claim.
证据 用来说明某一情况的材料，包括记录、人造物、文件和声明等，可作申请索赔用。

examination of site The inspection or survey of a site. The examination can take the form of a *site reconnaissance, *walk-over survey, *desk-top study, and *soil investigation.
现场考察 项目环境的考察或调研。考察的形式包括：现场踏勘、资料研究和土地勘测。

excavation 1. The process of digging holes in the ground. 2. A hole dug in the ground.
1. 开挖 在地面挖孔。**2. 基坑** 在地面开挖的土坑。

excavator A machine or person that digs an *excavation.
挖掘机/挖土者 实施开挖作业的机械或人。

excepted risks (employer's risks) The risks that are *expressly excluded from the contractor's responsibilities; these are within the risks that the employer has agreed to.
除外风险（业主应承担的风险） 合同中明确规定的承包商可以免责的风险；业主同意承担这些风险。

exceptionally adverse weather Weather conditions that impede or restrict construction activities.
极端恶劣天气 影响或阻碍施工的天气条件。

excess current *See* OVERCURRENT.
过电流 参考“过电流”词条。

excess excavation An *excavation that is deeper than required.
过度开挖 基坑的开挖深度比规定的要深。

excess voltage *See* VOLTAGE OVERLOAD.
过电压 参考“过电压”词条。

execute work Direction to commence an activity or task.
执行工作 开展一项活动或任务。

execution The undertaking of an event, activity, or task.
执行，履行 执行一项事件、活动或任务。

exfiltration The process where air exits a building through cracks, gaps, and other unintentional openings in the building envelope. It is driven by the wind and stack effect.

空气渗出 空气从裂缝、缝隙或围护结构上其他非人为预留的洞口逃出建筑物的过程。空气渗出是风压和烟囱效应的结果。

ex gratia Claims and payments are made without giving any legal liability, obligation, responsibility, or position. Payments of this nature are considered to have been paid voluntarily.
抚恤款 不附加任何法律责任或义务的付款或补偿。抚恤款通常被认为是当事人自愿支付的。

exhaust 1. Combustion gases that are ejected from a device, such as a boiler. 2. The air that is extracted from a room or building.
1. 废气 从设备（如：锅炉）里排出的燃烧废气。**2. 排气** 从建筑物内部或房间里排出的空气。

exhaust shaft A vertical duct used to extract air from a room or building.
风井 房间或建筑中用于排气的竖井。

exhaust system 1. Various parts of an appliance that discharges combustion gases to the outside. 2. An * extract system.
1. 排气装置 把燃烧废气排到室外的各个部件。**2. 排气通风系统**

exit 1. An opening or passage that enables occupants to leave a building. 2. A protected passage that is used as a means of escape in the event of a fire.
1. 出口 供住户离开建筑物的出口或通道。**2. 安全出口** 发生火灾时的逃生通道。

exit door An outwards opening door that connects the exit to outside.
出口门 连接出口通道与室外的外开门。

exit sign An illuminated sign that identifies the location of an exit.
出口标志 显示出口位置的发光标识。

expanded clay (bloated clay, expanded shale, expanded slate) A material made from common brick clays by grinding, screening, and then feeding through a gas burner at about 1482℃, thus changing the ferric oxide to ferrous oxide and causing the formation of bubbles.
黏土陶粒（膨胀陶粒、页岩陶粒） 黏土陶粒是以普通的黏土为主要原料，经粉碎、过筛、高温烧胀（烧结温度约1482℃）而成，因此氧化铁将转变成氧化亚铁，而且会有气泡形成。

expanded polystyrene (EPS, PS foam) *See* POLYSTYRENE.
发泡聚苯乙烯 参考“聚苯乙烯”词条。

expanded PVC (PVC foam) *See* POLYVINYL CHLORIDE.
泡沫聚氯乙烯板（PVC 发泡板） 参考“聚氯乙烯”词条。

expanding bit (expansion bit) A drill bit that can be adjusted to drill holes of differing diameter.
扩孔钻头 可调节钻头，可以钻出不同孔径的孔。

expanding plug A plug containing a rubber ring that is used to seal a pipe. Used for gas purging and drain testing.
膨胀堵头 带有橡胶圈的管塞。多用于换气排气测试。

expansion A process which causes an increase in size. *See also* THERMAL EXPANSION.
膨胀 尺寸变大或增长的过程。另请参考“热膨胀”词条。

expansion cistern (expansion tank) *See* FEED CISTERN.
膨胀水箱 参考“给水箱”词条。

expansion joint (EJ) A joint between two components that allows a degree of movement due to expansion.
伸缩缝 两个构件之间为了容许一定程度上因膨胀产生的位移而设置的结构缝。

expansion pipe (vent pipe) A pipe that runs from the hot water cylinder to the feed and expansion cistern in a vented system. Used to discharge any boiling hot water or steam that may be produced if the cylinder thermostat fails.

膨胀管（通气管） 暖通系统中，连接热水箱和膨胀水箱的管，用来排泄因水箱温控器故障而产生的沸腾热水或蒸汽。

expansion sleeve *See* SLEEVE.
膨胀套管 参考“管道套管”词条。

expansion strip (expansion tape, edge isolating strip) A strip of material used to form an *expansion joint.
伸缩缝止水带 覆盖伸缩缝的条形材料。

E

expansion tank *See* FEED CISTERN.
膨胀水箱 参考“给水箱”词条。

expansion vessel A small tank that allows the expansion of hot water in a sealed central heating system.
膨胀水箱 集中供热系统中允许热水膨胀的小水箱。

expense The economic cost incurred to undertake an operation, service, item of work, or task. In some contracts, the term expense will be used in reference to the contractor's delay and disruption claim.
费用 完成一项操作、服务、工作或任务所需的经济成本。在一些合同中，“expense”有时也指业主向承包商索赔的因承包商原因导致工程延期的费用。

experimental 1. Relating to scientific testing. 2. Based on an experiment. 3. Relating to something novel and not yet tested.
1. 与科学实验相关的。2. 以实验为基础的。3. 实验性事物 新奇的、有待实验的事物。

explosive fixing *See* SHOTFIRED FIXING.
螺钉枪固定 参考“螺钉枪固定”词条。

export To take material off-site.
材料出场 把材料运出现场。

exposed aggregate finish Concrete with coarse stones and rock exposed at the surface. Prior to the concrete maturing, the exposed surface of the concrete has a retarding agent applied so that the fine aggregate does not set and can be removed by spraying with water; alternatively the fine material can be brushed away. The removal of the fine aggregate at the surface of the concrete exposes the coarse aggregate.
露骨料混凝土饰面 表面暴露粗骨料的混凝土。在混凝土硬化前，先在混凝土外露表面刷一层缓凝剂，这样细骨料就不会凝固，可以用喷水的方式清掉细骨料，也可以用刷子清掉细骨料。清掉混凝土表面的细骨料后粗骨料就暴露出来了。

exposure 1. Exposing a building or materials to the elements of weather, such as wind, rain, snow, and sunlight. 2. A method used to describe the orientation of a building, e. g. the dwelling has a southern exposure.
1. 暴露 建筑物、建筑材料暴露在风、雨、雪或日照环境中。**2. 朝向** 房屋朝向，如房屋朝南。

ex-situ Away from its original location, e. g. ex-situ remediation techniques relates to cleaning the soil away from its *in-situ location.
易地/迁地 搬离原来的位置，如异位修复：将受污染的土壤从受污染区域转移到邻近地点，对其中的污染物进行治理的方法。

extended price (extension) The price or cost that has been calculated in a *bill of quantities by multiplying the measured quantities by the unit rate.
总价 工程量清单中由单价与工程量的相乘计算出来的价格或成本。

extension 1. The addition of extra space to an existing building. 2. An agreed increase in time or money to undertake work.
1. 扩建 扩大现有建筑的空间。**2. 施工延期或增加费用。**

extension bolt (monkey-tail bolt) An elongated vertical bolt that is attached to a window casement or door

leaf. When operated, the bolt slides into a socket on the head or sill of the window or door.
门窗长插销（卷尾插销） 固定在平开窗或门扇上的加长垂直插销。插销可以滑入门窗上槛、门槛、窗台板预留的洞口内。

extension ladder A ladder whose length can be extended. Either comprises two or more sections that slide together for storage and are slid apart and locked in position to increase the length of the ladder, or telescopic sections that can be extended and locked into position.
伸缩梯 长度可以拉伸的梯子。不管伸缩梯有几节，都可以收拢在一起存放，要用时再拉伸开来并在需要的位置锁定以增加梯子的使用长度，或用伸缩节拉伸并固定位置。

extension of preliminaries Addition to works generally described in the preliminaries section of the contract. The preliminaries are distributed as they occur, where they are sufficiently described and itemized, or where the descriptions are very general, the sum of money allocated may be divided equally amongst the interim certificates or monthly valuations. If there is a delay or an extension of time, there is no automatic claim to adjust the monthly figure for the preliminaries, the contractor must prove loss by reference to records and evidence. The preliminary items of a contract are for the work that has been described in general terms.
开办费超支 合同中开办费部分所述的工程出现增量。开办费按照发生的活动来计取，如果已经在清单中详细说明并分项，或者描述得非常笼统，则拨付费用按照中期支付或者按月均分。如果工程出现延误或延期，则无法通过索赔调整每月固定拨付的额度，这时候承包商需出示记录及证据去索赔损失。合同中开办费的项目应为通用条款中描述的工程内容。

extension of time The agreement to extend the contract duration by a specified time, administered by the client's agent or representative, under the terms of the contract. Under the contract, additional time may be awarded as a result of client variations, exceptionally inclement or adverse weather, or terms that are expressly agreed within the contract provisions that deal with *variations and extension of time.
延期 业主代表根据合同规定同意延长合约时间至某一时间。按照合同规定，如果是业主要求变更工程或因出现极端恶劣天气而导致的工程延期，或在合同条款中明文规定可以变更及延长工期的条款，则业主应适当地延长工期。

external angle A special brick that is angled part-way along its length to enable the brickwork to turn corners of less than 90°.
外角转角砖 一种特制转角砖，沿其长边转角（转角小于90°）。

external glazing Glazing that is located on the external faces of a building.
外墙玻璃 建筑外立面玻璃。

external insulation Insulation that is applied to the external faces of a building.
外墙保温 安装在建筑外立面的保温材料。

external leaf The outer leaf of a cavity wall.
外叶墙 夹心墙的外叶墙。

externally reflected component The light received on an internal surface directly by reflection from buildings, the ground, and obstructions outside a room. Used in conjunction with the *sky component and the internally *reflected component to calculate the daylight factor.
外部反射组分 室内表面直接接收到室外建筑物、地面和障碍物反射过来的光。同跟天空组分和内部反射组分一起计算出采光系数。

external plumbing Plumbing fixtures and fittings that are located outside the building, for instance, rainwater goods.
室外给排水配件 设在室外的给排水配件，如雨水排放件。

external rendering *Rendering that is applied to external areas and surfaces.
外墙抹灰 涂抹在外墙区域及表面的砂浆。

external wall A wall that has one face located outside.

外墙 面向室外的墙体称为外墙。

external works Construction work that is undertaken outside.
室外工程 在户外完成的工程。

extinguisher Something that ends or removes something else, such as a * fire extinguisher to put out a flame.
熄灭器 用于结束或移除某样东西的设备，例如灭火器扑灭火焰。

extra (extra work) Additional work which was not included within the original contract bills. The additional work is ordered under * variation order by the client's agent or representative.
额外工程（附加工程） 原合同内工程量清单并没有提及的工程。额外工程是由业主代表通过下发变更通知确认的。

extract air Air that is removed from a space.
排气 把空气从一个空间里排出。

extract fan (extractor fan, exhaust fan) A fan used to extract air from a space.
排气扇 用来排出空间内空气的风扇。

extractor hood A device located above a cooker that extracts moisture and cooking smells.
抽油烟机 设在炉灶上方，能将烹饪过程中产生的水气和油烟抽走，排出室外。

extract system (exhaust system) A type of mechanical ventilation system that uses grilles, ductwork, and fans to extract air from a space.
排气系统 一种机械通风系统，使用风口、风管和风机把空间内的空气排出室外。

extrados The convex outer curve of an arch.
外拱线 拱的外缘线。

extra-low voltage Electric current that is less than 50 volts AC.
超低压 小于50伏的交流电电压。

extra-over Additional sum of money allowed for items of work, normally used where the work has become slightly different from that agreed or described in the contract documents. For example, where excavation has become more difficult due to different ground conditions being experienced, e. g. bedrock where excavation through clay was described. Part of the original term and price for the excavation can be used with the addition of an allowance to provide for the difference.
另外收费 指用于支付工程变更的一笔额外资金，通常指工作内容与合同中规定的有些微差异的变更。如先前合同说的是黏土，但真正开挖时却发现是岩石，这就增加了基坑开挖难度。原合同中关于开挖的条款和报价部分可用于申请该变更的额外补偿。

extrapolate 1. To estimate a value, which lies outside a set of known values that is based on a trend. 2. To use facts as a starting point and conclude about something that is unknown.
1. 估价 在已知的价格体系以外根据趋势推断出某样东西的价值。**2. 推断** 以事实为出发点对某件未知事物进行推理、判断。

extras (extra work) Additional task or service required by the client that is not described in the * contract documents. The work would be ordered by the architect, engineer, or other client representative and treated as a * variation and priced in accordance with the contract documents. For a contractor to be paid for the work there must be a provision within the contract, and the variation order must be administered by the nominated representative.
额外工程（附加工程） 原合同文件中并没有提及的工程。额外工程是建筑师、工程师或其他业主代表根据合同规定作为工程变更加上并给出报价。合同中必须有工程变更的规定以便承包商能收到该工程款，变更通知须由指定的业主代表发出。

extruded brick A type of brick produced by the extrusion process. *See also* BRICK.
挤压砖 用挤压工艺生产的一种砖。另请参考“砖”词条。

extruded gasket A preformed gasket that has been formed by *extrusion.
挤出型密封条 挤压成型的预制密封条。

extruded section (extrusion) A component, usually metal or plastic, that has been formed by *extrusion.
挤压型材 金属或塑料通过挤出工艺制成的构件。

extrusion The process of forming something from a semi-soft material by pushing it through something, e. g. the process of forming igneous rock when magma is pushed through cracks in the earth's crust.
挤出 半软状态下的材料从某物中间穿过后成型的过程，例如岩浆从地壳的裂缝挤出形成火成岩的过程。

eye 1. The middle of a building component or element. 2. A small hole in a metal object.
1. 建筑构件的中间部分。2. 金属制品的孔眼。

eye bolt A bolt in which the head of the nut is replaced by a steel loop, such that when it has been screwed into place the loop can by used for lifting.
活节螺栓，孔眼螺栓 孔眼螺栓的头部是一个钢孔眼（而非螺帽），当螺杆拧进建筑结构时，孔眼就可以用来挂东西。

eyebrow (eyebrow dormer) A curved *dormer window in the shape of an eyebrow that contains no sides.
波形老虎窗 形状似眉毛的波形老虎窗，无侧框。

eye protection *Personal protective equipment in the form of glasses, goggles, and face masks (particularly for welding) essentially used so that vision is not impaired due to the activity being undertaken.
护目用具 眼镜、护目镜、面罩（特别是焊接面罩）等能够保护作业人员眼睛不受影响的个人防护用具。

fabric 1. Materials from which a building is constructed. 2. A textile structure produced by weaving or knitting techniques. *See also* GEOTEXTILE.
1. 建材 建筑施工材料。**2. 织物** 用机织或针织生产的织物。另请参考“土工织物”词条。

fabrication The construction or manufacture of something, which involves a number of consecutive steps or procedures.
制作/加工 制作、生产、建造某物，包括一系列连续的步骤或程序。

fabridam An air-or water-filled tube made from neoprene, laminated rubber, and nylon, anchored to a concrete plinth and used as a inflatable dam. It is most suitable for applications where the width-to-length ratio is high, such as irrigation systems, tidal barriers, etc.
橡胶坝 用氯丁橡胶、叠层橡胶和尼龙制成的锚固于混凝土底板上的充气管或充水管，通过充排管路用水（气）将其充胀形成袋式挡水坝。橡胶坝非常适用宽长比很大的构筑物，如灌溉系统、挡潮闸等。

facade The external *face of a building, usually the front.
外立面 建筑物的外立面，通常指正立面。

facade panel A prefabricated cladding panel designed to be installed on the facade of a building.
外墙板 安装在建筑外立面上的预制墙板。

facade retention The process of retaining an existing facade while the other elements of the building (roof, floors, and internal walls) are removed and replaced.
外墙支护 在拆除并置换建筑物其他构件（如屋顶、楼板和内墙）时，应对现有外墙进行适当支护。

face 1. The front of a building or wall. 2. The exposed external surface of a material that has the best appearance. 3. The working surface of a tool, such as the face of an axe.
1. 建筑物的正立面；正面墙。2. 材料外露的最佳表面。3. 工具的面，如斧头砍削面。

face brick *See* FACING BRICK.
饰面砖 参考“饰面砖”词条。

faced wall A wall clad in a *facing.
饰面墙 墙面覆盖层。

face edge (face side) The exposed edge of a material that has the best appearance.
正面 材料美观的外露表面。

face joint A joint in the *face of a material or object.
表面构造缝 设在材料或物体表面的构造缝。

facelift (face lift) Improving the external appearance or face of a building without undertaking major changes to the existing structure.
表面翻新 改善建筑（表面）外观，对现有结构并无大的改动。

face mark A temporary mark placed on a material to identify the face of the material.
材料标记 材料表面上的一个临时标记，可以让使用者识别哪个面是材料的表面。

face shovel (crowd shovel, forward shovel) An excavation machine that has a rope or hydraulically operated bucket. It removes soil from the base of excavations in a direction away from the machine; *See also* BACKHOE.
正铲挖掘机 带钢绳或带液压铲斗的挖掘机。正铲挖掘机把基坑里的土铲到远离挖掘机的方向。另请参考“反铲挖掘机”词条。

facing 1. Material used to cover a less decorative material to protect or decorate a building. 2. The process of smoothing or finishing the surface of a material. 3. The visible portion of a *door leaf.
1. 饰面材料 用来覆盖、保护或装饰建筑表面的材料。**2. 表面处理** 整平或磨光材料表面。**3. 贴面** 门扇可视部分。

facing brick (US face brick) Bricks with an aesthetically pleasing face, used where the brick is exposed and can be seen. Facing bricks are used to clad buildings to produce a desirable finish.
饰面砖 用于见光区域且表面美观的砖。饰面砖常用于覆层外墙以达到预期的装饰效果。

facing hammer *See* BUSH HAMMER.
凿毛锤 参考“凿毛锤”词条。

failure When maximum load or serviceability is exceeded, which can result in a structure or item suddenly collapsing or being unable to perform its requested task.
破坏 所施应力超出建筑构件的所能承受的最大载荷，导致建筑物突然崩塌或无法发挥预期功能。

fair cutting Cutting *fair-faced material.
精切 加工清水砌体材料。

fair face (fair faced) The neatly built surface of a material that is on show. Common materials used include brickwork (**fair-faced brickwork**), blockwork (**fair-faced blockwork**), and concrete (**fair-faced concrete**).
清水砌体饰面 材料的表面平整光滑，不加其他装饰。常见清水砌体材料包括：清水砖墙、清水砌块和清水混凝土。

fall A slope used to prevent water ponding, e. g. in gutters, on flat roofs, and in pipework.

排水坡度 为了防止积水而在天沟、平屋顶和管道上设置一个排水坡度。

falls Stability failures of steep (nearly vertical-sided) slopes. These occur in coastal regions, e. g. by the wave action undercutting a rock cliff.
边坡失稳 陡坡（直立边坡）失稳。沿海地区常会出现边坡失稳的情况，如波浪的冲击导致岩质边坡失稳。

false body Due to the high viscosity of non-drip paint, it maintains a shape when undisturbed. The shape that is formed by stirring or applying the pressure of a brush. The thixotropic characteristic, holding the shape into which it is moved, means that it does not readily drip.
假稠 由于高附着力油漆的黏度高，在不搅拌的情况下会保持一定的形状，搅拌或用刷子试压可以改变形状。由于其触变的特性停止搅拌后又会保持其形状，不容易流挂。

false ceiling (dropped ceiling) A secondary ceiling that has been installed below the existing ceiling resulting in a reduction in *ceiling height. Generally not a *suspended ceiling.
假天花（假吊顶） 安装在现有天花下面的第二层天花，假天花将导致室内净高降低。假天花通常不是悬吊式天花。

false exit A dead end, corridor, or passage that does not lead out of the building.
死胡同 不通往建筑外的走廊或通道。

false leaders The supporting framework (legs) of a pile-driving rig.
导桩架 打桩机的支撑框架（支腿）。

false tenon A separate tenon. Usually used to replace a tenon that has failed.
栽榫 单独的榫。通常用于更换坏掉的榫头。

falsework A temporary scaffolding structure, used to support *formwork and the construction of arch bridges.
膺架 用来支撑建筑模板和拱形桥结构的临时脚手架。

fan An electrically powered mechanical device that uses a series of blades to move air.
风扇/风机 依靠电力驱动叶片排送气体的机械。

fan coil unit An *air-conditioning unit comprising two coils that are supplied with hot and cold water from a central *boiler and *refrigeration plant. A fan built into the unit recirculates heated or cooled air to the space. Fresh air from a separate system can be introduced directly to the space, or may be ducted to the inlet of the unit. Air *filtration can be incorporated within the fan coil units. Fan coil units are usually located within a *suspended ceiling or are floor-mounted and are commonly used in multi-zone buildings, such as hotels and cellular offices.
风机盘管机组 由热水盘管、冷水盘管和冷热源供应系统组成的空调系统，热水盘管与锅炉相接，冷水盘管与中央空调相接。风机盘管机组装有一台风机，不断循环所在房间的空气，使空气通过冷水或热水盘管后被冷却或加热。其他系统的新风可以直接送入室内，也可以通过风管送到风机盘管机组的进风口。空气过滤器也可以安装在风机盘管机组内部。风机盘管机组通常安装在悬挂式吊顶上或直接落地安装，多用于有较多分区的建筑，如酒店和独立式办公室。

fan convector A type of space heater that uses an electric fan to draw cool air in at the bottom, passes it over a heating element or coil, and then blows the heated air out into the room. *See also* CONVECTOR.
风扇式对流空间加热器 一种空间加热器，用电风扇把冷空气吸进加热器下方的进气口，冷空气经过加热元件或加热线圈后被加热，从出气口流出，进入房间。另请参考“对流式加热器”词条。

Franki pile A reinforced concrete cast in-situ pile, with an enlarged base, which has been formed by a bottom-driving displacement and concrete-ramming technique.
法兰基灌注桩 用锤击沉桩和夯扩桩施工工艺打入的钢筋混凝土灌注桩（桩基扩大）。

fanlight (transom window) A small window located above a door to admit daylight and provide ventilation. Traditionally, semi-circular in shape with a number of glazing bars radiating out from the centre of the semi-circle resembling an open fan.

腰头窗（亮子，气窗） 在门上方的小窗，为辅助采光和通风之用。传统上为半圆形窗，中间由几根玻璃窗格条向外辐射，就像一把打开的扇子。

fan pressurization test A simple and widely used technique to determine the * airtightness of a building. The technique involves sealing a portable variable speed fan into an external doorway, using an adjustable doorframe and panel. A fan speed controller is then used to pressurize and/or depressurize the building. The airflow rate that is required to maintain a number of particular pressure differences across the building envelope is measured and recorded. The leakier the building, the greater the air flow required to maintain a given pressure differential. In the UK, a pressure differential of 50 Pa is used.
风扇增压法测试 测试建筑物气密性能的试验方法，这种方法比较简单，因此被广泛使用。风扇增压法测试所需工具包括一台变速风机、用薄膜覆盖的门框。先把薄膜覆盖的门框封住前门，风机就设在薄膜的开口。然后用风扇调速器来对建筑物进行加压和/或减压。在保证建筑物内部有一定气压差的前提下，测量和记录空气流速。建筑物渗漏率越小，就越应该增大空气流速以维持一定的气压差。在英国，风扇增压法测试使用的标准气压差为 50 Pa。

F

fan truss A timber * roof truss that contains vertical and inclined struts.
扇形木屋架 一种由木材制成的桁架式屋盖构建，包括垂直支撑与斜撑。

fan vault A fan-shaped * vault.
扇形拱顶 一扇形的拱顶。

fascia (fascia board, eaves facia) Vertical board placed at the * eaves of a * roof, and providing a neat finish to the edge of the roof, covering the edge of the * rafters. The facia board also provides a suitable surface onto which the gutter can be fixed.
挑檐立板 安装在屋顶檐口位置并盖住椽边缘的装饰竖板。挑檐立板还可以作为天沟的支撑面。

fastener Device, fitting, or unit used to secure two or more objects together.
紧固件 是连接两个或两个以上物体的装置、配件。

fast to light A colour that does not degrade in light.
不变色的，耐光的 在光照条件下不变色的。

fast track (fast-track procedures) Construction projects programmed so that they are completed in the shortest possible time; the building works commence before the building's detailed drawings are totally complete. As soon as sufficient design information is produced, the on-site construction starts with further design information being produced so that subsequent elements of the building can be procured and built. The overlapping of design and construction reduces the overall project duration.
快速跟进方式 一种工程管理方式，目的是在尽可能短的工期内完成项目。在建筑物的施工详图尚未全部完成之前就开始施工。一旦掌握足够的设计信息，就可以展开现场施工，相关设计工作可以同步进行，以便及时采购和安装后续的建筑构件。边设计边施工的方法可以缩短整体工期。

fat board Slang term for mortar board used to carry mortar for pointing.
托灰板（俚语） 勾缝时的托灰用板。

faucet A small water tap, the type used in a domestic household sink.
水龙头 安装在洗涤槽上方的小水龙头。

feasibility study Investigation undertaken to determine whether a site or building is economically viable or functionally suitable for a specific development.
可行性研究 是对工程项目的现场调研，以确定项目或建筑的经济或功能合理性是否值得开发。

feather edge board (feather edge) A tapered board used for cladding and fencing.
薄边板 一种用于外墙和围墙的楔边板。

feather edge rule (featheredge rule) A tapered straight-edged metal rule used to smooth, straighten, and level plaster and render.
刮尺 用于整平内墙和外墙砂浆的楔形边钢尺。

feathering The process of reducing a material from one thickness to another. Used in plastering to smooth out lumps in walls.
削薄 把材料的厚度削薄的过程。如抹平墙上的砂浆块。

feed The gradual supply of something to ensure that it remains operational, e. g. water to a boiler, electricity to equipment, etc.
供给 稳定供应原料给某设备，以确保其正常运行。如供水给锅炉，供电给电气设备等。

feed cistern A cold-water storage *cistern that supplies cold water to a boiler or hot-water cylinder. In indirect hot-water systems, a vent pipe may be required to feed into the cistern to accommodate any expansion in the hot-water system. Cisterns that incorporate expansion are known as **feed and expansion cisterns**.
给水箱 冷水储水箱，供给冷水给锅炉或热水箱。如果是间接加热式热水供应系统，则应接一根通气管（膨胀管）到水箱，以容纳系统中水因加热膨胀所增加的体积。装有膨胀管的给水箱也叫膨胀水箱。

felt *See* BITUMEN.
毡 参考“沥青”词条。

felt and gravel roof Flat roof with weather protection formed with a waterproof bitumen layer covered in gravel to protect it from the sun and to help prevent it lifting off the roof surface during strong winds.
油毡撒绿豆砂屋面 平屋顶上采用油毡为主要防水层材料，然后在防水层上铺设一层绿豆砂作为保护层。这层绿豆砂的作用是阻挡阳光照射沥青，保护防水卷材不被强风吹走。

felt nail Fixing nail with a large head to hold roofing felt in place. The large head prevents the felt pulling through or ripping over the nail. *See also* CLOUT NAIL.
油毡钉 用来固定屋面防水油毡的大头钉。大头钉能确保油毡不易被掀起或穿透。另请参考“大帽钉”词条。

feltwork The process of using bonded layers of *bitumen felt.
油毡铺设 一层层铺设沥青油毡的过程。

fence An enclosure around something, e. g. an area of land, to act as a barrier. **Fence posts** are the main vertical members sunk or concreted into the ground to support the fence.
围墙 用来围合某一区域（如一片土地）的隔断结构。围墙柱子是打入地里边的垂直构件，可起到支撑围墙的作用。

fender A guard or cushion that provides protection for something against impact. A **fender wall** is a low wall that carries the hearth slab of a fireplace. A **fender pile** is a pile driven into the seabed to provide protection to the berth from impacts of mooring ships.
防护物，缓冲装置 保护某件东西不受冲击。**壁炉围栏**是围着壁炉底座的矮栏。**防冲桩**是在码头岸壁前打入海床的直立桩，可以保护岸壁不受停船的冲击。

fenestral A *window.
窗 一种窗。

fenestration A term used in architecture referring to the way in which the windows have been arranged on the outside of a building.
窗户布置 建筑学术语，用来指建筑外墙窗户的布置方式。

ferrocement A thin layer of *Portland cement and sand that contains a steel wire mesh. Used to form curved sheets, such as hulls for boats, canoes, water tanks, and sculptures.
钢丝网水泥 在波特兰水泥砂浆薄层里铺设钢丝网。可用于加工弧形板，如船只、独木舟、水箱和雕塑的外壳。

ferrule 1. A metal cap or ring to protect the end of a shaft, e. g. to prevent the end of a wooden pole from splitting. 2. A metal cylinder used to join pipes.
1. 金属箍 杆状物顶端的金属帽或箍，如木杆顶端的金属箍可以保护木杆不易开裂。**2. 管箍** 用来连接两根管子的金属短管。

festooned cable A flexible electrical cable that is supported, at intervals, on an overhead bogie, allowing the cable to concertina back on itself.
拖令电缆 固定在起重机吊架上的柔性电缆，每隔一定间隔电缆都有支撑，拖令电缆可以自由折叠伸缩。

fetch A distance the wind or a wave travels without interruption.
风区长度 风浪未受阻扰情况下所走的距离。

fettle 1. The finishing of any trade work. Final adjustments, corrections, and finishing to complete the work.
修整，修复 指任何工种的表面处理。包括最终的调整、修复和表面处理。

F

ffl *See* FINISHED FLOOR LEVEL.
竣工地板标高 参考“竣工地板标高”词条。

fibre A long slender thread of filament of natural or synthetic origin from which yarns are spun to make a fabric structure, such as a *geotextile.
纤维 用于纺织织物结构中天然或人工合成的细长丝线，例如土工织物。

fibreboard (fibre building board) A timber product comprised of wood fibres, e. g. medium-density fibreboard (MDF).
纤维板 由木质纤维制成的木材产品。如中密度纤维板（MDF）。

fibre cement board A high-quality calcium, cement fibreboard with additives.
纤维水泥板 以高质量钙质为原料，加入添加剂制成的水泥纤维板。

fibreglass Material made from extremely fine fibres of glass. The fibres are embedded in a matrix, e. g. in polymers, better known as glass reinforced plastic (GRP). The resultant material has improved strength and toughness as the inclusion of these fibres helps accommodate the stress imposed. Furthermore, materials and composites with fibreglass reinforcements have excellent strength: weight ratios.
玻璃纤维 用极细的玻璃纤维制成的材料。玻璃纤维埋置于聚合物基体组成的复合材料就叫玻璃纤维增强塑料或玻璃钢（GRP）。玻璃钢材料优点是强度高、韧性高，因为玻璃钢里边的纤维可以帮助分散应力。另外，加入玻璃纤维增强材料的复合材料都具有优良的比强度。

fibre reinforced concrete Concrete with fibres embedded in the cement matrix, also known as FRC. Usually polymer fibres are used, however, steel fibres are also used. The addition of these fibres results in an enhancement of the mechanical properties, i. e. strength and toughness.
纤维增强混凝土 在水泥基体中掺入各种纤维材料而形成的混凝土，也称为“FRC”。通常会加入聚合物纤维，然而有时也会使用钢纤维。加入这些纤维可以增强混凝土的机械性能，如强度和韧性等。

fibre saturation point The moisture content at which the strength of timber remains constant—around 30%. Below this value, as the timber dries, it becomes stronger and shrinks.
木材纤维饱和点 木材强度在含水率为30%左右时保持稳定，低于该值时，随着木材变干会变强和收缩。

fibrescope (borescope) A flexible *endoscope that uses fibre-optic technology to view inaccessible places, e. g. to inspect cavity ties in a cavity wall.
纤维内窥镜 一种屈曲自如的光学纤维内窥镜，能对难以到达的位置进行检查，如检查空心墙里边的拉结件。

field 1. An area of land used for agricultural or playing purposes. 2. An area rich in natural resources that can be mined. 3. A subject discipline area—speciality. 4. An area of force, e. g. magnetic field.
1. 场地 一块用作农业或娱乐用途的土地。**2. 产地，矿区** 某种可开采自然资源的产地、矿区。**3. 领域** 专业领域。**4. 场力存在的范围** 如：磁场。

field book A notebook to record measurements and observations while on site.

工地巡检记录簿 在工地现场巡检和测量时所用的记录本。

field drain A series of interconnecting porous or perforated pipes, which have been buried in the ground to remove groundwater so that the surrounding ground is less boggy or waterlogged.
工地排水 预埋在地下的一系列相互连接的多孔渗水管，这些多孔渗水管是用来排泄地下水的，这样周围地表就不会积水。

fielded (fielded panel) A panel with a raised central surface where the edge is rebated, recessed, or shaped away from the surface.
凸面镶板 中间有一个凸起表面的镶板，凸面的边缘是凹进去的。

field order A site order given by an engineer in the US.
工程变更单 在美国指由工程师颁发的工程变更令。

field settling test A test to determine the cleanness of fine aggregate/sand. This type of aggregate should be free from dust, clay, silt, and organic materials, because such impurities reduce the bond between the cement and aggregate, hence reducing the net strength of the hardened concrete or mortar. The test involves placing 50 ml of 1% solution of common salt water in a 250 ml measuring cylinder. Sand is added slowly until the 100 ml mark is reached. Further solution is then added to bring the level to 150 ml. The cylinder is then shaken and allowed to settle for 3 hours. The thickness of the silt layer is then measured and expressed as a percentage of the sand layer. If this value is less than 10%, the sand is classed as acceptable. If it exceeds 10%, it cannot be classed as failed as this test only gives an approximate estimate, thus further laboratory testing would be required.
集料沉淀试验 用来检测细集料/细砂清洁度的试验。这种类型的集料应该是不含粉尘、黏土、粉土和有机材料的，因为这些杂质会降低水泥和集料之间的黏结度，并且会进一步降低硬化混凝土或硬化砂浆的强度。把 50 ml 浓度为 1% 的普通盐水溶液倒进 250ml 的量筒。接着再慢慢加入砂子直到溶液达到 100ml 的位置。然后再往量筒里边倒入 50ml 的溶液使溶液达到 150ml。摇晃量筒并将其静置 3 个小时。最后测量淤泥层的厚度比用百分比表示。如果淤泥层厚度小于 10%，则说明砂子是合格的。而如果淤泥层厚度大于 10%，也不代表砂子不合格，因为这个沉淀测试只是一个大致测试，还须进一步送往实验室测试。

field splice A structural steel *splice that is bolted on on-site.
现场拼接 在施工现场用螺栓拼接两块结构钢板。

field superintendent (US) Engineer in the US who often acts as the on-site project manager. This person may also manage the design process.
总监理工程师（美国） 主持项目现场监理工作的工程师。总监理工程师还可能负责设计方面的管理工作。

figure 1. A geometrical two-or three-dimensional shape. 2. A *drawing or *diagram. 3. A symbol representing something, e. g. a number. 4. The natural markings and colour of timber.
1. 图形 二维或三维几何图形。**2. 图解** 图纸或示意图。**3. 符号** 表示某种意义的符号，如数字。**4. 花纹** 木材天然的纹理和颜色。

figured dimension Dimensions on a drawing where the drawing or sketch is not to scale.
尺寸数字 图纸（非按比例绘制）上用数字标注的尺寸。

fill Material such as sand, gravel, or stone used in *earthworks to raise the level of, for example, an embankment to the desired level.
（土石方）填料 土方工程填筑施工用的材料，包括砂子、砾石或石块，以抬高路堤等到需要的高度。

filler A specific type of additive utilized to increase the bulk density/volume of an object or material.
填料 用来增加物体或材料密度/体积的特制添加剂。

filler beam floor (filler joist floor, filler concrete slab) An element that fits between the main component,

which can also be made of a different material. For example, smaller reinforced concrete beams spanning between main steel beams.
填充梁板 填充在楼板主要支撑构件之间的混凝土板，也可以用其他材料填充。如钢主梁之间的钢筋混凝土梁。

fillet 1. A triangular-shaped piece of material inserted into the angle between two materials to strengthen the joint. 2. A timber moulding used to cover the joint between two members.
1. 三角形加固条 插入两个构件相交处的三角形材料，以增强结合力。**2. 嵌缝条** 盖住两个建筑构件之间的接缝的木线条。

F

fillet chisel A tool used by a mason for working stone.
石工尖凿 石工加工石材的工具。

fillet saw A small hand-saw.
小手锯 一种小型手锯。

fillet weld A triangular weld that can join two pieces of metal at right angles to each other.
角焊 焊接后两工件相互垂直形成90°夹角的一种焊接方法。

filling knife A *stopping knife that has a thin, flexible blade, used to apply **fillers.**
油灰刀 用来填缝的刮刀，其刀片薄且有韧性。

filter A straining device to remove larger particles. A **filter bed** is a layer of sand, gravel, or any other pervious material used to remove sewage or other impurities from waste-water. A **filter drain** is a *French drain. **Filter material** is washed and graded sand, gravel, or any other pervious material.
过滤器 清除较大颗粒的过滤装置。**滤料层** 是一层砂子、砾石或其他透水材料，滤料层可以阻止污泥或其他杂质进入废水系统。**滤水暗沟** 一种盲沟。**滤料** 精选级配砂、砾石或其他透水材料。

filtrate Filtered material.
滤液 已被过滤的物质。

final account The total costs of the development based on the work that has been measured and undertaken. Any variations or fluctuations that are allowed under the contract would be included in this total. The work is normally measured and recorded by the contractor's *surveyor and checked and agreed by the client's *quantity surveyor.
竣工结算 开发项目根据已测量和已完成的工程计算出来的总造价。任何合同内允许的变更或价格浮动均已计算在内。通常该工作由承包方的测量员计算并记录，然后交由业主方工料测量师复核并审批。

final certificate Document issued by the client's representative, traditionally the *architect. The paperwork acts as a contractual trigger stating that the work is complete and entitles the contractor to payment of the *final account and part of the retention monies.
竣工验收证书 业主代表（通常是建筑师）颁发的证明工程已全部竣工的文件。竣工验收证书是承包商收到竣工结算尾款和部分质量保证金的凭证。

final completion The point when the end of the defects liability period is reached, all defects have been made good, and the final certificate is issued, releasing the second part of the retention monies.
最终竣工 缺陷责任期期满、所有缺陷都已被修复、业主已颁发竣工证书并归还剩下的质量保证金给承包商。

final fixing *See* SECONDARY FIXING.
最终固定 参考“次级固定”词条。

final grade (US) *See* FORMATION LEVEL.
平整面标高（美国） 参考“平整面标高”词条。

final inspection An inspection of a building prior to completion.

竣工验收 在工程竣工前对建筑物进行的检验。

financial control Checking the cost of the actual works against that budgeted and priced *bills of quantities; includes the forecasting, reporting, and setting new targets where necessary. Through the examination of unit costs and elements that are costing more than anticipated, plans can be produced to ensure that the future work, labour, materials, and services do not exceed the budget.
财务控制 对照实际工作量的费用与工程量清单中的预算和报价；方法包括预测、报告和有需要时设定新的目标。通过核查单价和超支的项目，可以制订计划以确保将来的工程、劳动力、材料和设备成本不会超支。

fin drain A flexible three-dimensional *geocomposite drain, consisting of a polypropylene drainage core surrounded by a nonwoven *geotextile, which acts as a *filter, and prevents the drain becoming blocked.
土工复合排水网 柔性三维土工排水板，由聚丙烯排水网芯双面粘接无纺土工布组成，可起到过滤器的作用，防止排水系统被堵塞。

fine aggregate Material used in concrete, usually in the form of sand. Fine aggregates are usually small in size, typically smaller than 4 mm in diameter.
细集料 制作混凝土用料，通常指砂子。细集料的粒径通常都很小，一般都小于4 mm。

finger joint A type of glued timber joint comprising a series of interlocking zig-zag shaped tapered fingers.
木材指接 两根木材的端部做成一系列的齿形（斜锥状指形榫），然后涂胶接合。

finger plate (fingerplate, doorplate, pushplate) A protective plate, usually decorative, fixed to the face of an internal door to prevent finger marks.
门推板 室内门表面上防止被手指染污的防护板，通常还有装饰的作用。

finial A decorative element used at the top of a gable, newel post, or fence post.
尖顶饰 山形墙、楼梯栏杆支柱或栅栏柱顶端的装饰配件。

fining off The process of applying a finish coat to the external *render.
面漆施工 把面漆涂抹在外墙砂浆层上的过程。

finish The external coat or surface of material. The exposed surface could be the material's natural appearance or be enhanced by applying another material, or polishing the surface.
饰面 材料的面层或表面。饰面可以是材料的天然表面或通过另外涂抹一层材料以及对表面进行抛光处理获得。

finished floor level (ffl) The final level of a finished floor.
竣工地板标高 地板装修完成后的最终标高。

finisher Tradesperson who trowels or tamps the surface of concrete to give it the final finish.
涂装工 用抹子磨平和压实混凝土表面以达到预期饰面的技工。

finish hardware (US) Builder's hardware that is seen.
装修五金件（美国） 看得见的五金配件。

finishing coat (fining coat, setting coat, skimming coat, white coat) Usually the final layer of paint applied to provide a durable and decorative layer of the required colour. Typically, the gloss, silk, or matt (eggshell) finishes are based on oils and alkyd resins, although water-borne products are increasingly becoming available. Waterborne gloss finishes are more moisture permeable than the traditional solvent-borne hard glosses. Waterborne glosses have the advantage of quick drying without giving off solvent odour, and generally they do not yellow on ageing. The matt and silk finishes are usually vinyl or acrylic emulsions.
饰面层 通常指最后一层油漆，这一层油漆可以提高构件的耐久性并起到装饰的作用。尽管水性漆越来越多，但通常情况下，亮漆、丝绸漆或哑光漆都是溶剂型油漆（以油脂和醇酸树脂为分散介质）。水性光泽涂料比传统的溶剂型高光泽涂料透湿性能更好。水性光泽涂料的优点是：快干且没有散发出溶剂的怪味，不会因为时间久而变黄。哑光漆和丝绸漆通常为乙烯乳胶漆或丙烯酸乳胶漆。

finishing schedule List of surface coats and the programme of works necessary to complete the project and en-

sure that all of the walls, ceilings, floors, doors, and windows have the desired appearance. The paint, varnishes, wallpaper, and other fittings are listed against each room. The order in which the operations are to be completed is also listed.
饰面计划表 面漆一览表以及墙面、吊顶、楼板、门窗的饰面施工计划。涂装计划表上应列出每个房间需要的油漆、清漆、墙纸和其他装修配件，同时还应列出施工的先后顺序。

finishing trades Painters, joiners, decorators, plasterers, carpet and tile fitters, tillers, and other trades necessary to give the building and the rooms the final appearance.
装饰专业 包括油漆工、细木工、装修工、抹灰工、地毯和瓷砖安装工、园林工和其他跟建筑物和房间的饰面施工相关的专业。

F

finish nail (US) *See* BULLET-HEAD NAIL.
无头钉（美国） 参考"无头钉"词条。

fin wall A wall that projects outwards at a right angle to the structure.
翼缘墙 与建筑结构垂直的凸出的墙。

fire 1. The process of combustion. 2. The process of baking materials in a kiln. 3. A space-heating device located within a fireplace.
1. 燃烧 燃烧的过程。**2. 窑干** 把材料放进窑里烘烤。**3. 取暖器** 设在壁炉里边的取暖器。

fire alarm A mechanical, electromechanical, or electronic device that is activated by fire or smoke to warn the occupants of a building that an unwanted fire has broken out.
火灾警报器 一种机械、机电、电子装置，一旦检测到火苗或烟雾，火灾警报器就会发出报警信号提醒住户有火灾发生。

fire alarm indicator A panel that indicates where in a building a fire has broken out.
火灾显示盘/器 显示大楼火灾位置的显示盘。

fire back 1. A wall at the rear of a *fireplace. 2. A decorative cast-iron plate inserted at the rear of a *fireplace to protect the masonry wall. This also stores heat from the fire, which is radiated back to the space once the fire has died down.
1. 壁炉背墙。2. 安装在壁炉背墙上的保护砖墙用铸铁装饰板 这块铸铁板还起到储热的作用，当壁炉里边的火变弱时，这些储存的热量就会辐射到炉膛里。

fire barrier A strip of non-combustible material that is inserted into a construction to restrict the movement of smoke and flames.
阻燃材料 安装在建筑结构里边的条形阻燃材料，可阻止火苗和烟雾蔓延。

fire block (US) A *fire barrier used inside wooden walls and floors.
防火材料（美国） 安装在木隔墙和木地板里边的阻燃材料。

fire booster A pump used to increase the pressure of the water within a *fire riser.
消防增压泵 用于增加消防立管水压的泵。

firebox The part of a *fireplace or the space inside a boiler where the fuel is burned.
燃烧室（炉膛） 壁炉或锅炉中燃料燃烧的空间。

fire break (firebreak) A gap between buildings, a fire-resisting wall between *compartments, or a strip of land where the vegetation has been cleared. Used to stop or slow down the spread of fire.
防火隔墙 用来分隔防火分区的墙，这些墙是用防火材料砌成的。（森林）防火隔离带在植被之间设置的空旷地带。防火隔墙和森林防火隔离带都能防止火灾扩大蔓延。

fire brick A brick made for use in burning stoves and fireplaces. Although they are normally used for a base, they can be used for surrounds and hearths.
耐火砖 火炉和壁炉用砖。虽然耐火砖多用作底座，但它们也可以作炉膛用砖。

fire cell *See* COMPARTMENT.

防火分区 参考“防火分区”词条。

fire certificate Under the Fire Precautions Act all non-domestic buildings must be inspected by a fire officer to check that the building complies with the Act. The certificate confirms that the building is compliant.
消防安全合格证 消防监督员将根据《消防法》的规定检查一切非住宅建筑，看其是否符合《消防法》的要求。符合要求的建筑将可以获得消防安全合格证。

fire check door A fire-resistant door where its period of integrity is less than its period of stability. *See* FIRE-RESISTING GLASS.
防火门 耐火完整性次于耐火稳定性的耐火门。参考“防火玻璃”词条。

fire compartmentation (compartmentation) The sealing of a room, corridor, apartment, or floor with fire-resisting materials to ensure that the occupants within that room or dwelling are protected from smoke andfire for a given period of time. Fire-resisting doors, floors, walls, and ceilings are specified to give fire protection from 30 minutes up to 4 hours.
划分防火分区 用防火材料来分隔建筑里各个防火分区（如房间、走廊、住宅或楼层），以确保房间或住宅里的住户在一段时间内不会受烟雾和火灾伤害。防火门、防火地板、防火墙和防火吊顶的耐火性能应介于半个小时到 4 个小时时间。

F

fire damper A fire-resisting damper installed within the ductwork of an air-conditioning or ventilation system to prevent the spread of fire and smoke.
防火阀/防火风门 安装在通风、空调系统风管上起隔烟阻火作用的阀门。

fire detector An electrical device that is designed to detect the presence of a fire and set off an alarm.
火灾探测器 探测是否有火灾发生并触发报警的电子设备。

fire door (firebreak door) A door, including the frame, which is designed to provide fire resistance for specific periods of time depending on their grading, for instance, a FD30 is designed to give 30 minutes of resistance.
防火门 在一定时间内能满足耐火性能要求的门（包括门框）。如 FD30 表示门的耐火时间为 30 分钟。

fire engineering (fire protection engineering, fire safety engineering) The use of science and engineering principles, such as measurement and calculation, to provide fire safety in buildings.
消防工程学 用科学和工程学的原理（如检测和计算）来为建筑提供消防服务的学科。

fire escape (fire exit) An egress point, stairwell, or pathway to allow people to exit a building in the safest, quickest route if there is a fire. *See also* ESCAPE STAIR.
安全出口 发生火灾时快速疏散人群用的安全通道，包括安全出口、消防楼梯或其他安全通道。参考“逃生楼梯”词条。

fire extinguisher (portable fire extinguisher) A cylindrical metal container containing foam or other extinguishable liquid that can be sprayed onto a fire. A **fire hydrant** allows a fixed connection to a continuous water supply via a valve such that a **fire-hose** can be attached by fire-fighters. A **fire-hose reel** is a roll of pipe that is often positioned adjacent to a fire hydrant. A **fire-extinguishing system** is the fixed equipment within a building, such as hoses, hydrants, and a sprinklers system, used to extinguish a fire.
灭火器（便携式灭火器） 装有泡沫或其他阻燃液体的圆柱形金属容器，可喷在火上。消火栓：与供水系统连接且供消防人员连接消防水带用的一种固定式消防设施。消防水带卷盘：放置在消火栓附近的一卷软管。消防系统：安装在建筑物内的一系列固定式灭火设备组成，如消防水带、消火栓和自动喷淋系统。

fire-fighting shaft A fire-protected stairwell, and sometimes a lift shaft, in tall buildings that allows fire-fighters, together with their equipment, to again access to the floor on which the fire is situated. The lift will have its own back-up power supply.
消防井道 高层建筑中的防火楼梯间或电梯井，可供消防人员带着消防设备到达起火楼层。电梯应有备用电源供电。

fire floor (firebreak floor) A fire-resistant floor.
防火层 用耐火材料围护起来的楼层。

fire grading *See* FIRE RATING.
耐火等级 参考“耐火等级”词条。

fire hazard Something that is liable to catch fire.
火灾隐患 易燃的物品。

fire inspection The examination of a non-domestic building by the fire officer to check that the building has all the required measures necessary to comply with the Fire Precautions Act. If compliant, a * fire certificate will be issued.
消防安全检查 非住宅建筑的消防安全检查。消防监督员将检查非住宅建筑里是否有英国《消防法》要求的一切必要灭火设施。如有，将可获得消防安全合格证。

fire lift A lift installed within a protective shaft that is designed to be used by fire-fighters in the event of a fire.
消防电梯 发生火灾时供消防人员使用的电梯，电梯设在封闭电梯井内部。

fire load density The amount of potentially combustible material within a structure. It is used to predict the * fire severity, which is measured in megajoules (MJ) per square metre. Anything up to 1135 MJ/m2 would be classed as having a low fire load density and would be applicable to domestic dwellings. Moderate and high fire load density values would be 1135—2270 MJ/m2 and 2270—4540 MJ/m2, respectively, and could be associated with various types of warehouses.
火灾荷载密度 建筑内可容纳易燃物的数量。用于预测火灾烈度，计量单位为兆焦（MJ）/平方米。低于 1135 MJ/m2 的属于低密度火灾荷载，适用于住宅用途。中密度火灾荷载为 1135 ～2270 MJ/m2，高密度火灾荷载为 2270 ～4540 MJ/m2，适合作不同类型的仓库。

fire lobby A * lobby that leads to a fire lift or forms part of an escape route from a building in the event of a fire.
消防前室 消防电梯的前室，发生火灾时，可用作消防疏散通道的一部分。

fire main A horizontal or vertical fixed pipe within a structure used to supply water to a * fire-extinguishing system. It is normally made from galvanized steel and colour-coded red.
消防管道 建筑物里边用来输送消防用水的固定式水平管道或立管。消防管道通常是镀锌钢材质且喷有红色油漆。

fireman's switch A * fire switch.
消防员专用开关 消防电梯开关。

fire modelling The use of computer programs to determine how a structure will perform in a fire and to assess the potential consequences.
火灾模拟 利用电脑程序检测构筑物在火灾情况下的表现并对可能产生的后果进行评估。

fire officer (fire prevention officer) Senior member of the fire service with duties to inspect buildings, issue * fire certificates, and provide advice on necessary changes. Employed by the fire service, the officer's work is to enforce the fire prevention laws and issue fire certificates for buildings that are compliant.
消防监督员 消防局高级管理员，主要职责是：检查建筑物的消防安全问题，颁发消防安全合格证，提供必要的整改建议。除此之外，消防监督员还要认真贯彻英国的《消防法》，为符合英国《消防法》要求的建筑颁发消防安全合格证。

fire performance (fire behaviour) A general term used to describe how a structure or a material performs when subjected to fire.
耐火性能 用来描述构筑物或材料在火灾情况下的特性。

fireplace A point within a building that has been designed to contain a fire or heating element for a building. Where open combustion is to take place, as in coal, wood, or gas-fuelled fires, a * chimney and * flue is re-

quired to remove the fumes and smoke. The fireplace is normally set on an incombustible * hearth.
壁炉 在室内生火取暖的设备。开放式燃烧壁炉以煤炭、木材或天然气为能源，内部上通烟囱和烟道可以排掉燃烧产生的烟雾。壁炉通常安装在耐火底座上。

fire point The location in a building where fire-extinguishing equipment is kept.
消防器材分布点 建筑物内的消防器材存放点。

fire precautions Measures taken to avoid the risk and effects of fire, both to people and property.
防火措施 有利于避免火灾的产生或防止火灾对人和财产造成伤害的措施。

fire prevention Taking measures to avoid or reduce the risk of a fire occurring.
防火措施 避免火灾发生或减小火灾影响的措施。

fireproof *See* FIRE RESISTANCE.
耐火性能 参考"耐火性能"词条。

fire propagation index A measure or rating of the heat index from a surface material or wall lining when it is subjected to a fire test or furnace test.
火焰传播指数 评估面材或墙衬在防火测试或燃烧炉试验中的热度指数。

fire protection Steps taken to protect materials and buildings from the effects of fire, heat, and smoke. The measures are taken so that the building is sufficiently resistant to the effects of fire to provide sufficient time to safely evacuate a building in the event of a fire. The materials will also limit the combustibility of the building, offering protection to the building, which can also be advantageous in reducing insurance premiums.
消防措施 保护材料和建筑物免受火灾、热气和烟雾破坏的措施。这些措施应能确保建筑物有足够的能力来抵御火灾，为逃生人员的安全疏散提供足够的时间。所用材料必须是耐火的，避免建筑物受损，造成经济损失。

fire rating The amount of time various building elements (*See* FIRE DOOR) have the ability to resist fire. It is determined in accordance with BS 476, which assesses absence of collapse, flame penetration, and excessive temperature rise on the cool face to determine a fire rating.
防火等级 建筑构件（参考"防火门"词条）的耐火极限，即耐火时间。按照 BS 476 标准的要求对建筑构件实施耐火测试，通过检测是否有坍塌、火焰穿透和建筑构件冷却表面是否过度升温来确定防火等级。

fire reserve Water that is specifically kept (stored in a tank) and used for fire-fighting.
消防储备水 储存在水箱里的灭火用水。

fire resistance (fireproof) The ability not to catch fire or to burn so slowly that it will cause minimum impact, e. g. a * fire door.
耐火性能 阻止燃烧或降低燃烧速度的性能，可最大限度减小火灾的破坏，如防火门。

fire resisting glass Glass that holds its form, does not fall away from the frame, and maintains a barrier in the event of a fire. The glass may crack but it will maintain sufficient integrity to ensure that it prevents the passage of smoke and fire for a given time.
防火玻璃 在火灾情况下能保持完整性和稳定性，没有从框架上脱落，且能起到阻火作用的玻璃。玻璃可能会有裂纹，但还是能够维持足够的完整性，并在相当一段时间内控制火势的蔓延或隔烟。

fire retardant An element used as a deterrent to combustion. Fire retardants either prevent or delay the commencement of ignition/combustion. Examples include intumescent coatings on steel structures.
阻燃剂 遏制燃烧的一种助剂。阻燃剂可以使建筑构件不容易着火和着火速度变慢。例如在钢结构上涂抹膨胀型涂料。

fire riser Water pipes travelling up and down a building that deliver large quantities of water for use by the fire service in the event of a fire. Both dry and wet risers can be used. A dry riser provides an empty pipe run which can be connected to the water supply by opening a valve, whereas a wet riser is kept permanently charged with water so that it is available for immediate use in the event of a fire.

消防立管 发生火灾时，用来上下输送大量消防用水的立管。包括干式消防立管和湿式消防立管。干式消防立管平时是没有水的，发生火灾时再打开阀门进行充水灭火；而湿式消防立管则任何时候都是有水的，以便于发生火灾时可以即时灭火。

fire safety sign Signage required to make users aware of building laws, fire fighting equipment, fire exits, and procedures necessary in the event of a fire.
消防安全标志 用来提醒用户与消防有关的安全信息，包括消防法律、消防设备、安全出口以及一些必要的火灾应急措施。

fire shutter Vertical and horizontal roller doors and walls that provide additional fire resistance by acting in a similar way to that of a *fire door. Can be used to seal lift shaft doors, shop fronts, and other openings.
防火卷帘门 上卷和侧卷防火卷帘门，作用和防火门相似。防火卷帘门可用于封闭电梯间门、店面和其他洞口处。

fire stair A set of stairs that can be used as a means of escape in the event of a fire.
消防楼梯 发生火灾时，用作逃生通道的楼梯。

fire stop (fire stopping) Non-combustible material used to fill gaps and voids in order to prevent fire passing through the void. A fire stop may be a solid block of non-combustible material used below *raised floors, a blanket or non-combustible quilt used above suspended ceilings, or it can be a strip of material that expands when heated to close gaps around services and doors.
防火封堵材料 用来填充建筑缝隙或空隙的不燃材料，以避免火势通过空隙蔓延。防火封堵材料包括：架空地板下方的耐火实心材料、吊顶天花上方的耐火毡以及门防火条（当发生火灾时防火膨胀密封条自动膨胀，封堵门缝缝隙）。

fire stop sleeve (pipe closer) A collar that fits around a pipe or duct at the point where the duct passes through a wall; in the event that a fire burns through the pipe, the intumescent foam within the collar expands and seals the pipe. Where walls are compartment walls offering a specified fire resistance, all openings within the wall should be sealed with a fire stop.
阻火圈 安装在管道穿墙处的套管。一旦火苗窜入管道内部，套管里边的发泡胶就会自动膨胀并密封管道。如果是有特殊要求的防火隔墙，应在所有墙壁开口处安装防火封堵材料。

fire-suppression system *See* FIRE-EXTINGUISHING SYSTEM.
灭火系统 参考“灭火系统”词条。

fire switch (fire lift priority switch, fireman's switch) A switch that enables fire-fighters to operate a *fire lift.
消防电梯开关（消防员专用开关） 供消防员操作消防电梯用的开关。

fire testing Calculations, experiments, and tests undertaken to determine the fire resistance and grading of building elements.
防火测试 通过计算、试验和测试来检测建筑构件的耐火等级。

fire tower (US) A compartmentalized or protected stairway.
封闭楼梯间（美国） 用防火材料分隔开来或围起来的楼梯间。

fire tube A long steel tube in a boiler that carries the hot combustion gases.
火管 锅炉中高温燃气流经的长钢管。

fire vent A damper located in a roof that is opened for *fire venting.
消防排烟风机 设在屋顶的排烟风机，可用于排掉火灾烟气。

fire venting The process of enabling the hot gases and smoke from a fire to escape from a building using the *fire vents.
排烟 通过消防排烟风机把高温热气和烟排到室外的过程。

fire wall (firebreak wall, division wall) A fire-resistant wall.
防火隔墙 防火隔墙

firing A process whereby an objece is heated to elevated temperatures with the objecetive of increasing density and/or strength, e. g. ceramics. A process of bakingceramics in a kiln to partly or fully vitrify (harden/glaze) them.
烧结 为增强物体密度和/或强度加热至高温的过程，如陶瓷。在窑内烘烤的过程可部分或全部玻化（硬化/釉化）陶瓷。

firm clay (firm silt) A material that does not remould easily.
硬质黏土 很难塑造的黏土。

firmer chisel A sturdy general-purpose *chisel designed to be used by hand and not to be hit with a hammer or mallet. Used by carpenters, joiners, and wood carvers. Also known as a forming chisel.
木工凿 木工通用的坚固耐用的凿子，可以直接用木工凿来刨木材，而不用结合锤子一起使用。木工凿适用于粗木工、细木工和木雕。也称为“forming chisel”。

F

first fix (1st fix, first fixing) The installation of various items within a building after the primary elements of the building have been constructed (floors, walls, and roof); these include electrical cables, pattress boxes, plumbing, ductwork, and partitions.
初装修 建筑物基本构件（楼板、墙体和屋顶）竣工后在建筑物里边安装各种配件，这些配件包括：电缆、插座盒、给排水管道、风管和隔板。

first floor The floor immediately above ground level.
二楼 在一楼上面的楼层。

first storey The space between the first floor and the floor or roof above.
二层 二楼楼面至上部楼板底面或屋顶底面之间的空间。

fishplate A *splice bar.
鱼尾板 一种拼接板。

fishtail fixing lug (fang, tang) A fishtail-shaped *lug that is either cast into concrete or built into masonry.
鱼尾固定耳铁 预埋在混凝土或砖石结构里的鱼尾型拉结件。

fish tape *See* DRAW CABLE.
牵引绳 参考“牵引绳”词条。

fissured clay A cohesive soil that contains a network of cracks, which particularly open up when the material is dry.
裂隙粘土 有很多裂隙的粘土，干燥时收缩开裂。

fissured surface A flat surface that contains shallow irregular cracks, used for decorative processes, or to absorb sound.
裂纹面 有各种不规则浅裂纹的平面，可起到装饰或吸声的作用。

fitment A factory-made element that is installed and can be removed from buildings, such as baths, basins, and cupboards. *See also* FITTINGS.
家具或卫生配件 工厂预制的可现场拆装的构件，如浴缸、洗脸盆和橱柜。另请参考“木管配件”词条。

fit-out (fitting-out, US outfitting) The process of installing finishes such as suspended ceilings, floor coverings, and partitioning systems.
装饰 安装饰面材料的过程，如安装吊顶天花、地板材料和隔断系统。

fitted bolt A bolt that has a small clearance in a hole, and is used to carry high loads.
紧配螺栓 通孔直径很小的螺栓，用于荷载较大的场合。

fitted carpet A carpet that is fixed to the floor and covers the floor completely.
满铺地毯 覆盖整个地板的地毯。

fitter A person who assembles machines or maintains machinery.
装配工 负责组装设备和维护设备的人员。

fittings 1. A term used to refer to various small items of pipework, such as a bend, coupling, or tee. 2. Items within a building that can be removed without causing damage.
1. 水管配件 各种水管配件，如弯头、管接头或三通。**2. 非关键部件** 拆除后不会造成破坏的建筑配件。

fit-up Formwork that can be reused, typically it is used repeatedly to form sections of a concrete wall, column, beam, etc.
临时支撑模板 可以重复使用的模板，特别是混凝土墙、混凝土柱子和混凝土梁等的模板。

fixed beam A beam that is restrained from rotating and moving vertically. A beam with a **fixed end** cannot move or rotate. A **fixed end moment** is set up at the restraint end because the beam cannot rotate.
固端梁 在垂直方向上不能移动或转动的梁。因为梁不可以转动，因此在荷载作用下固定端位置都将承受弯矩。

fixed-form paving train (fixed-form paver) A machine that is used to lay, compact, and finish a concrete road surface.
混凝土铺路机 用来浇筑、压实和压光混凝土路面的机械。

fixed-in The process of inserting components into the construction after it is complete. *See also* BUILT-IN.
后置埋 在已竣工工程结构上埋置连接件的过程。另请参考“预埋”词条。

fixed light (dead light) A window that does not open.
固定窗 不能开启的窗。

fixed-price contract Contracts that help secure the final price of the contract by either agreeing to a *lump sum, or schedule of rates and prices, or by a measure and value process. Clients generally prefer such arrangements, since much of the financial risk is transferred to the contractor. Generally, fixed-price contracts have positive benefits in terms of fast completion times and known price, as the contractor wants to complete the work as quickly as possible in order to reduce overheads and labour. However, without good supervision quality can suffer.
固定总价合同 通过包干价、固定单价或其他计量和估价方式确定工程最终报价的合同。因为财务风险转嫁到承包商身上所以业主通常欢迎该方式。固定总价合同有较快完工和报价已知的优点，因为承包商希望尽快完工以减少管理费和劳动力。但如果没有完善的监管，工程质量会下降。

fixed retaining wall A *retaining wall that is held in place at the top and bottom, such as a basement wall.
顶端和底部固定的挡土墙，如地下室墙。

fixer A tradesperson who fixes elements and components to the building on-site.
安装工 负责在项目现场安装建筑构件的技工。

fixing 1. A device used to hold something in place. 2. The process of attaching, adjusting, or repairing something.
1. 紧固件 用于紧固连接的配件。**2. 安装** 固定、调整或修复的过程。

fixing brick *See* NOG.
加固用木砖 参考“同厂”词条。

fixing channel A channel that enables the location of a fixing to be adjusted. It can be surface-mounted or embedded into a floor, wall, or ceiling.
固定槽 允许紧固件作位置调整的槽钢。固定槽可以是明装式的，也可以嵌装在楼板、墙体或天花结构里边。

fixing device A *fixing.
紧固件 一紧固件。

fixing fillet (fixing slip, pad, pallet, slip) A thin piece of wood inserted into the mortar joint in masonry construction to act as a fixing.
木楔 预埋在砖石结构砂浆缝里边的紧固用薄木衬条。

fixing gun *See* CARTRIDGE TOOL.
打钉枪 参考“打钉枪”词条。

fixing moment (fixed-end moment) The moment created at the support because the beam has been prevented from rotating.
固端弯矩 因为梁不可以转动，因此在支撑位置（固定端位置）将产生弯矩。

fixing strap A long narrow piece of material that is placed around an object to fasten it in place.
固定带/扎带 缠绕某物体并使其固定的扁长型材料。

fixing strip A long narrow piece of a material that an object is attached to.
衬条 使物体固定在其上面的长条形材料。

fixture 1. Any item that is fixed to a building or land, and would cause damage if removed; for instance, plumbing, electrical cables, trees, fences, etc. 2. A *luminaire.
1. 固定式装置 固定到建筑物或土地上的装置，拆除后会造成结构破坏；如给排水管道、电缆、树木和围栏等。**2. 灯具**

flag (flagstone) A rectangular piece of material used for paving.
板石 长方形铺路石板。

flame detector A type of fire detector that uses infra-red and/or ultra-violet sensors to detect the presence of flames.
火焰探测器 一种火灾探测器，用红外线或紫外线探测器来探测是否有火源。

flame retardant A material or object that does not ignite or burn easily.
阻燃剂 不燃或不易燃烧的材料或物体。

flame-retardant paints A special type of paint that emits non-combustible gases when subject to fire, the usual active ingredient being antimony oxide. Normally, combustible substrates such as plywood and particleboard can have their fire resistance increased substantially with the application of an appropriate protective paint.
阻燃涂料/防火涂料/防火漆 燃烧时会排放不可燃气体的特种涂料，通常其活性成分为氧化锑。一般情况下可燃性基材（如胶合板和刨花板）在涂抹适当的防火涂料后，其耐火性能都会大大提升。

flamespread (surface spread of flame) A measure of a material's ability to burn and spread flames.
火焰传播指数（火焰表面蔓延） 测试火焰在材料表面蔓延的速度。

flammable A term used to describe materials that easily ignite and will burn rapidly.
可燃的 用来描述易燃且燃烧速度很快的物体。

flammable liquids A liquid that easily ignites or burns rapidly.
可燃液体 易燃且燃烧速度很快的液体。

flange A projecting piece for fixing or strengthening, e. g. a flat plate on the end of a pipe used for fixing, or the top and bottom parts of an *I-beam for strengthening.
1. 法兰 连接或紧固用凸缘，如管子与管子相互连接用的板式平焊法兰。**2. 工字钢翼缘** 工字钢顶部和底部的增强刚度用部件。

flank The side of something such as the side of a carriageway, the hard shoulder, or the *intrados of an arch. A **flank wall** is a side wall of a building.
侧面 物体的侧面，一侧，如行车道、硬路肩或桥拱/门拱内圈的一侧。**侧墙** 建筑物的侧面墙。

flanking transmission (flanking sound) The indirect transmission of sound through a separating wall or floor via elements that are adjacent to the wall or floor.

侧向传声 通过隔墙或地板中间的连接构件间接传播的声音。

flanking window (wing) 1. A side window, adjacent to a door or main window, with its sill at the same level as the door or main window.
边窗 门口或大窗两侧的小窗，通常边窗的标高和门或大窗的标高是一致的。

flap trap A drainage fitting that contains a hinged plate, which only allows flow in one direction to prevent backflow.
止回阀 带铰链机构的排水配件，可以防止介质倒流。

flared column A column that widens outwards at the top.
喇叭口柱子 顶端部分加宽的柱子。

flash The process of protecting a joint from the weather using thin continuous pieces of metal. The pieces of metal are known as a **flashing**.
泛水 用连续的金属薄板包住接缝处以防风雨。这些金属薄板就称为"泛水板"。

flashing A strip of impervious material to protect roof joints and angles from the ingress of water. It can be made from lead, aluminium, galvanized steel, or bitumen felt.
泛水板 防止屋面与墙面接缝处或墙角渗水的条形不透水材料。泛水板可以是下列材料制成的，包括铅、铝、镀锌钢或沥青毡。

flat arch (French arch, jack arch, straight arch) An arch above a window or door that is almost completely horizontal.
平拱 门窗上方接近完全水平的拱。

flat-bottomed rail A rail with a rounded top and flanged bottom.
铁路钢轨 轨头呈圆形，轨底很宽的钢轨。

flat cost Cost of the materials and labour without overhead or profit.
净成本 材料和劳务费，不包括管理费和利润。

flat glass A reference to rolled glass, plate glass, float glass, and sheet glass.
平板玻璃 参考压延玻璃、平板玻璃和浮法玻璃。

flat interlocking tiles A flat roof tile that interlocks with an adjacent tile.
平瓦 与相邻屋面瓦水平咬接的屋面瓦。

flat joint (flush joint) A mortar joint where the mortar is finished flush with the masonry.
平灰缝 与砌体结构表面齐平的灰缝。

flat jointed brickwork Brickwork constructed using a *flat joint.
勾平缝砖墙 勾平缝的砖墙。

flat-pin plug A three-pin *plug that contains rectangular flat pins.
扁脚插头 三极扁脚插头。

flat pointing Pointing masonry to achieve a flat joint.
勾平缝 在砌体墙面勾出平整的灰缝。

flat roof A roof that has a pitch of less than 10°. Flat roofs should not be completely horizontal but should have a pitch that enables water to drain off the roof. *See also* PITCHED ROOF.
平屋面 坡度小于10°的屋面。平屋面并不是绝对的平，也是有坡度的，这样才能排出屋面积水。另请参考"坡屋面"词条。

flat slab A concrete slab reinforced in two or more directions, generally with drop panels at supports, but without beams or girders.
无梁楼盖 由两个或以上方向加固的混凝土楼板，通常支撑处有托板柱帽，但没有梁或主梁。

flaunching A cement mortar fillet used at the top of a chimney stack to prevent rain penetration.

烟囱泛水 屋面烟囱根部的水泥砂浆边，有防止雨水渗入作用。

Flemish bond A type of brickwork bond formed by laying alternating *headers and *stretchers within the same course.
梅花丁砌法 在同一层砖里顺砖与丁砖交替铺砌的砌砖方式。

Flemish garden wall bond A type of brickwork bond based on the *Flemish bond, that is formed by laying one header for every three stretchers on the same course. On even numbered courses, the headers are positioned under the centre of the middle stretcher on the odd numbered course.
荷兰花园墙式砌法 基于梅花丁砌法的变体，即同一层砖中砌筑三块顺砖再砌筑一块丁砖。偶数层中的丁砖正好位于奇数层三块顺砖中间那块顺砖中心位置下方。

fleuron General term for any carved floral decoration.
花形图案的装饰 任何花形图案装饰的总称。

fleur-de-lis An iris or lily-shaped ornament often used as a *finial.
百合花饰 鸢尾花或百合花形状的花饰，常用作尖顶装饰物。

flex The flexible cable between an electrical appliance or device and the plug that is inserted into the electrical socket.
软电线 电器或设备与插头之间的软电缆。

flexible damp-proof course *See* DAMP-PROOF COURSE.
柔性防潮层 参考“防潮层”词条。

flexible ductwork *See* DUCTWORK.
柔性风管 参考“风管”词条。

flexible membrane An impervious plastic or rubberized sheet.
柔性防水膜 不透水塑料膜或橡胶膜。

flexible metal conduit A spiral-shaped steel or aluminium *conduit that can be bent easily and moved around.
金属软管 波纹状钢管或铝管，易于弯曲。

flexible metal roofing *See* SUPPORTED SHEET-METAL ROOFING.
金属板材屋面 参考“金属板材屋面”词条。

flexible mounting *See* ANTI-VIBRATION MOUNTING.
减振垫 参考“减振垫”词条。

flexible pavement A *pavement where the surfacing and base materials are bound with bituminous binder, e. g. Dense Bitumen Macadam (DBM); Hot Rolled Asphalt (HRA); Dense Tar Macadam (DTM); and Heavy Duty Macadam (HDM).
柔性路面 面层和基层用沥青结合料联结的路面，如密级配沥青碎石路面（DBM）、热压式沥青混凝土路面（HRA）、密级配煤沥青碎石混合料路面（DTM）和重型沥青碎石路面（HDM）。

flexible pipe A pipe that allows a degree of movement; used in plumbing applications, allowing a water supply to be connected to the base of a tap with ease.
软管 允许一定程度的弯曲的给排水管，可用软管来连接供水管与水龙头底座。

flexible wall A wall that acts as a cantilever along its vertical axis.
柔性挡土墙 在垂直方向作为悬臂结构的墙体。

flexure rigidity The resistance to bending.
抗弯刚度 物体抵抗弯曲变形的能力。

flight A set of steps between each landing.
梯段 楼梯平台之间的一段楼梯。

flint wall A wall constructed using flint.
燧石墙 用燧石砌成的墙。

flitch A large piece of timber cut length ways from a tree, which is ready for processing into planks.
料板 沿树木纵向切割的大块木材，木板加工用料。

float 1. A tool for floating the surface of plaster or a screed. 2. A shallow sloping drain pipe suspended from a floor soffit or housed within a floor duct, used to carry wastewater from sanitary fittings. 3. To reset on or move over the surface of a liquid. 4. A *ball valve. 5. In scheduling programmes, using the *critical path method, it is the spare time to undertake certain activities. It is calculated from the difference between the earliest start and the latest time finish minus the duration of time assigned to undertake the activity.
1. 抹子 用来抹平灰浆表面的工具。**2. 坡度较小的排水管道** 坡度较小的排水管道，通常悬挂在楼板底面或安装在楼板线槽内里面，供卫生配件排水用。**3. 漂浮** 在液体的表面停留或移动。**4. 浮球阀** **5. 总浮动时间** 在以关键路径法为分析基础的项目进度计划中，总浮动时间指的是某活动可以推迟的时间。总浮动时间等于最早开始与最迟完成时间之间的差值减去执行该活动的所需时间。

float glass A high quality, flat glass sheet with smooth surfaces, manufactured by floating molten glass on a bed of molten metal at a temperature high enough for the glass to flow smoothly.
浮法玻璃 一种高质量，表面平滑的玻璃平板，通过将熔融的玻璃平稳地浮在足够高温的熔化金属槽上面制成。

floating A process of compacting the surface of plaster or a screed by scouring it with a *float to leave a rough surface.
抹底层砂浆 用抹子压实石膏砂浆或底层砂浆表面并进行拉毛处理，使表面变粗。

floating crane [semi-submersible crane vessel (SSCV)] A *crane fixed onto a barge or pontoon.
浮吊 固定在驳船上或起重船上的吊机。

floating floor A concrete screed or floorboards positioned on top of a floor structure with insulating material between them to reduce the transmission of noise.
浮动地板 混凝土面层或企口板与底层地板之间的一层隔音材料的地板。

floating foundation A reinforced concrete raft foundation designed to be buoyant; used to support structures built over silty and muddy ground conditions.
板式基础 钢筋混凝土筑成的浮式筏板基础，板式基础是用于支撑粉粒土和淤泥质软土地基的上部结构的。

floating pipeline Long lengths of interconnecting pipes that float, with the use of buoyancy aids, on the surface of water. Used to convey water, oil, gas, or other petroleum products.
漂浮式管路 借助浮具漂浮在水面上的一系列相互连接的管道。漂浮式管路可用于输送水、石油、天然气或其他石油产品。

float switch A device to detect the level of a liquid surface in a tank. It can be used to trigger a pump when the level gets too low or to raise an alarm if the level becomes too high.
浮球开关 是一种用来检测水箱中液位情况的装置。当水箱中液位过低时，浮球开关就会启动水泵进行抽水；当水箱中液位过高时，浮球开关就会触发报警。

float valve Avalve in which the flow is regulated by the position of a moving ball. Typically found in toilet cisterns. When the cistern empties, the ball moves down with the water level and causes the valve to open. As the cistern fills, the ball rises to close the valve.
浮球阀 一种阀门，其中液体的流动受到一个移动的球体所控制。浮球阀常用于厕所水箱。当水箱排空时，球体随水位向下移动并使阀门开启。当水箱充满时，球体上升关以闭阀门。

flood channel An open-air channel that is normally dry, but when excessive rain has occurred, it fills to allow water to be discharged to a safe egress point, thus preventing areas from flooding. Flood routing uses a series of dams, flood channels, storage basins, etc, to reduce the impact of flooding on an area.

泄洪渠 户外排水渠，通常情况下是没有水经过的，只有雨量过多时疏洪渠才会发挥作用，把水排到指定出口，以避免地面被淹。疏洪是利用一系列防洪设施，如水坝、疏洪渠、蓄水池等来减小洪水的冲击和影响。

floor The lower part of a room in a building—the part upon which people walk.
地板 房间里供人行走的部分。

floor area ratio (US) *See* PLOT RATIO.
容积率（美国） 参考“容积率”词条。

floor blocks *See* RIB AND SUSPENDED FLOOR.
密肋填充块楼板 参考“密肋填充块楼板”词条。

floor boarding (timber flooring) A *floor made of *floorboards.
木地板 由企口板铺设而成的地板。

floorboards A wooden floor covering, normally with tongued and grooved edges, that is fixed directly to the floor joists.
企口板 直接安装在地搁栅上的一种木地板材料，通常一侧加工成凸字形，一侧加工成凹字形。

floor centre (telescopic centre) A small extendable steel beam, used to provide *falsework support to the reinforced-concrete floor slab's *formwork.
模板支架可伸缩钢梁 用来支撑钢筋混凝土楼板模板的支架可伸缩钢梁。

floor check A *door closer with a water *rebate.
地弹簧 一种液压式闭门器。

floor chisel A tool used to remove *floorboards.
木板凿 撬木地板用的（扁）凿子。

floor clamp (flooring clamp, floor cramp, flooring cramp) A *clamp used to hold tongue and grooved floorboards together tightly while they are being laid.
木地板拉紧器 安装木地板时，用来拉紧企口板的工具。

floor clip (bulldog clip, US sleeper clip) A U-connector that is inserted into concrete floor slabs before the concrete is set to provide a fixing point for flooring battens.
地板卡扣 在混凝土凝固前埋入混凝土楼板里边的用来固定木条的U型卡扣。

floor coverings (floor finish, floor finishing) A decorative hard-wearing finish applied to the floor level, such as carpet, laminate wood, etc. Sometimes sound-reducing and moisture-resistant.
地板材料 安装在地面上的耐磨、装饰材料，如地毯、复合木地板等。有些地板材料还有隔音和防潮的作用。

floor framing The framework that provides the support to the floor, e. g. *joists and *strutting.
地板骨架 对地板起支撑作用的骨架，如地龙骨和其他支撑系统。

floor grinder (concrete grinder) A machine used to provide a finish to the floor covering once it has been laid, e. g. grinding marble or stone that has been set in mortar to give a polished mosaic effect, known as terrazzo.
地板打磨机 用来磨光已安装地板表面的机器，如打磨已混入水泥黏结料的大理石或石材地板，使地板看起来有光亮马赛克的效果，俗称“水磨石”。

floor guide A runner for a sliding door located on the floor.
推拉门下轨 推拉门在地面的滑轨。

floor heating *See* UNDERFLOOR HEATING.
地暖 参考“地暖”词条。

flooring The material that the floor is finished with; *See also* FLOOR COVERINGS.

地板材料 地面装饰材料；参考“地板材料”词条。

flooring saw (floor saw) A diamond-bladed saw used to cut concrete floors.
混凝土切割机 用来切割混凝土地板的金刚石锯机。

flooring tile (floor tile) A thin regularly shaped *floor covering laid side-by-side on an adhesive bed and grouted between. Floor tiles can be manufactured from ceramic, concrete, rubber, asphalt, etc.
地板砖（地面砖） 紧贴着铺设在砂浆层上的厚度较小、形状规则的地板材料，砖与砖之间要进行勾缝处理。地板砖可以是用下列材料制成的，包括陶瓷、混凝土、橡胶、沥青等。

floor joist A *joist that provides structural support to a floor.
搁栅 为楼（地）板提供结构支撑的龙骨。

F

floor outlet An electrical socket that is flush with the floor.
地插 与地面齐平的插座。

floor plan A *drawing detailing the layout for a room, noting the position of the walls, wall thicknesses, etc.
楼层平面图 用来显示（某一楼层的）房间的布置、墙体的位置和厚度等信息的图纸。

floor plate *See* ANCHOR PLATE.
地板扣件 参考“地板扣件”词条。

floor scabbler (floor scabber) A machine used to 'scrape' floors, e. g. to remove existing coatings.
地面打磨机 打磨地面的机器，如旧油漆的打磨处理。

floor slab A ground or suspended reinforced concrete FLOOR SLAB.
钢筋混凝土板 钢筋混凝土地板或楼板。

floor spring (spring hinge) A pivot with a spring sunk in the floor to act as a *door closer. *See also* FLOOR CHECK.
地弹簧 嵌装在地板里面的带弹簧转轴，作闭门器用。另请参考“地弹簧”词条。

floor stop A *door stop that has been fixed to the floor.
门止/门挡 固定在地板上的门止。

floor strutting Stiffening between *floor joists.
搁栅加劲杆 搁栅之间的加固条。

floor trap *See* GULLY.
地漏 参考“地漏”词条。

floor void 1. A hole made through a floor to allow pipework, cables, etc to pass. 2. A hollow space within a floor slab.
1. 楼板预留洞 供管道、电缆等穿过。**2. 楼板中间的空隙**

flue A duct or passage that is used to vent the products of combustion from an appliance or fire.
烟道 用于排除设备或燃烧产生的烟雾的管道或管段。

flue block (flue brick, chimney block) A special type of block used to construct a *flue.
烟道用耐火砖（烟囱用耐火砖） 用来砌筑烟道的特种砖。

flue isolator An automatic damper located in the flue that closes when the boiler switches off. Used to prevent heat loss up the flue.
烟道挡板 设在烟道里边的自动调节阀，锅炉关掉时，自动调节阀会自动关闭，防止热量逃出烟道。

flue liner (flue lining) A heat-and fire-resistant material used to line the interior of a chimney and transport the products of combustion. Usually made of stainless steel, fire clay, or terracotta pipe.
烟道内衬 安装在烟囱内部的耐热和防火材料，可用于排放燃烧产生的烟气。烟道内衬通常包括不锈钢管、耐火黏土砖管或陶管。

flue pipe A pipe that connects a fuel-burning appliance to a flue. Usually made of stainless steel.

燃料设备排烟管 连接燃料设备与烟囱的管道。通常是不锈钢材质的。

flue terminal A type of *cowl that is placed on the end of a *flue to help discharge the combustion products and prevent rain penetration.
烟道末端的通风帽 设置在烟道末端的通风帽，可以协助排烟并防止雨水灌入烟囱。

fluorescence A lighting (optical) phenomenon usually emanating from cold bodies; thus sometimes known as cold light.
荧光 一种发光（可见光）现象，通常指冷光源发出的光，因此有时也把荧光称为冷光。

fluorescent lamp (fluorescent tube, fluorescent tubular lamp) A tubular glass envelope filled with argon and low-pressure mercury gas and coated internally in phosphor that produces visible light through *fluorescence when an electric current is passed through it. Two main types are available: **tubular fluorescent lamps** and **compact fluorescent lamps (CFLs)**. CFLs are three to five times more efficient than *incandescent lamps, with up to 10 times the lamp life. Full-size tubular fluorescent lamps are five to six times as efficient and may last up to 20 times as long. However, the energy savings from CFLs may not be as large as claimed, because most CFLs are much brighter seen from the side than from the ends, while a typical incandescent light is almost equally bright from all directions.
荧光灯（荧光灯管） 充有氩气和低压汞气且内部镀荧光粉的管状玻璃灯，通电后荧光粉发出可见光。常见的荧光灯有两种：直管型荧光灯和紧凑型荧光灯（CFLs）。紧凑型荧光灯的发光效率约为白炽灯的 3 ～5 倍，使用寿命则是白炽灯的 10 倍左右。标准尺寸直管型荧光灯发光效率为白炽灯的 5 ～6 倍，使用寿命则是白炽灯的 20 倍左右。但是紧凑型荧光灯的节能效率并没有标榜的那么高，因为大部分紧凑型荧光灯从侧面看要比从两头看要亮很多，而标准的白炽灯从各个方向看亮度几乎都是一样的。

fluorescent luminaire A type of *luminaire that produces light using one or more *fluorescent tubes.
荧光灯具 一种包含一根或多根荧光管的灯具。

flush 1. To make water flow through something, e. g. a toilet, so as to dispose of its contents. 2. Something that is completely level with its surrounding, e. g. flush door, flush eaves, flush joint, flush panel, flush pipe, flush soffit, etc.
1. 冲洗 用水冲洗，如冲马桶。**2. 与……齐平** 与相邻表面完全齐平，如平面门、嵌装挑檐立板、平灰缝、嵌入式面板、嵌装管道、平面天棚等。

flushing cistern A *cistern that holds water that is used for flushing a *WC.
冲厕水箱 用来存水冲洗马桶的储水箱。

flushing mechanism A device used to release the water contained within a *flushing cistern to flush a *WC. Usually comprises a button on the top of a cistern or a handle on the side of the cistern.
马桶冲水装置 用来释放冲厕水箱里边的水的装置，通常包括冲厕水箱顶部的冲水按钮或冲厕水箱一侧的冲水手柄。

flushing trough (trough cistern) A long rectangular *cistern that serves several *WCs.
旧式冲水箱 同时服务于几个马桶的长方形水箱。

flush valve (flushing valve, US flushometer) A valve located at the bottom of the *cistern that separates the water in the cistern from the bowl and regulates the flow of water when the toilet is flushed.
马桶冲水阀 安装在冲厕水箱内部底面的阀门，阀门关闭时水箱里的水就不会流向马桶，冲水阀可以控制水流。

fluting (flutes) Concave vertical grooves set or cut into columns and pillars.
罗马柱凹槽 柱子上的竖直凹槽。

flux A liquid used to clean oxides and facilitate wetting of metals with solder.
助焊剂 用于清除氧化物并促进焊接的液体。

flyer A rectangular tread in a straight flight of steps.

楼梯梯级 一段楼梯中的矩形踏板。

flying Flemish bond A brickwork bond in which the bricks are laid headers (short side) or 3/4 bat then stretchers (long side) such that in surrounding courses the headers are directly central to the stretchers.
梅花丁砌法 每皮砖中丁砖或3/4砖和顺砖相隔，丁砖位于周围顺砖的正中间。

flying form (flying formwork) * Formwork that is so large it is moved by a crane.
飞模 模板太大，要用吊机提升。

flying scaffold A * scaffolding structure that is hung from the building to give access to the outside walls of tall buildings.
悬空脚手架 从建筑上悬挑下来的脚手架，以供高层建筑外墙通行用。

flying shore (flier, horizontal shore) A temporary horizontal support used above ground between two objects, e. g. If a terraced house is demolished, this can be used to span the gap and support the two neighbouring houses.
水平支撑 两个物体之间的临时离地水平支撑。如果排屋中有一间房屋要拆除，可以用水平支撑来支撑相邻两间屋子。

fly wire A fine plastic or metal wire used for * insect screens.
防虫网 防虫用的细铁丝网或塑料网。

foam-backed rubber flooring (sponge-backed rubber flooring) Rubber flooring that has a layer of foam or sponge adhered to the back of it. The adhered layer acts as an underlay, increases the resilience of the flooring, and can help reduce impact noise.
泡沫垫橡胶地板（海绵垫橡胶地板） 底下粘有一层泡沫或海绵的橡胶地板。这一层泡沫或海绵是地板的衬垫，既可以增强地板弹性，又可以帮助降噪。

foam inlet A socket that fire-fighting equipment can be inserted into in order to provide a supply of foam.
消防泡沫液接口 消防设备接口，接入后便可喷洒泡沫。

folded floor A floor that has been laid by springing the boards into place rather than using a * floor clamp.
平扣地板 用弹簧片固定而非用拉紧器固定的地板。

folding door A door that is opened by folding it back in sections. Generally used internally where there is limited space. Different types of folding door are available, such as a bi-fold door and a multi-folding door.
折叠门 通过折叠开启的门。通常用作有限空间的室内门。有不同种类的折叠门供选择，如双扇折叠门和多扇折叠门。

folding partition A * partition that folds back on itself allowing a large room to be divided into two smaller separate rooms.
折叠式隔断 可以折叠收缩的隔断，可把一个大房间分隔成两个小房间。

folding wedges Two pieces of wedge-shaped timber or metal (triangular in section) which are used for bracing and securing building components. When the triangular wedges are pointed in the opposite direction, they can be slid over each other to increase their overall thickness. The wedges are placed into gaps and adjusted increasing their thickness, thus trapping and securing objects.
双木楔 两块楔形（横截面为三角形）的木块或金属块，用于撑紧建筑构件。当两块木楔尖端相对放置时，可以互相滑动增加他们的厚度。木楔可放置在空隙中，通过调整厚度以楔紧物体。

foliage 1. Leaves or any leafage. 2. Castings or carvings that resemble a leaf or leaves.
1. 叶子。2. 叶状的铸件或雕刻品。

foliated Carved leaf decoration.
叶状饰 雕刻的叶状饰品。

follower (long dolly) A length of block used to transmit the load from a hammer to a pile head. It is used when the hammer cannot be fitted directly onto the pile head. *See also* DRIVING CAP.

长替打 能够把桩锤的荷载传递到桩头上的长替打木。桩锤不能直接套在桩头上时应使用长替打。另请参考“桩帽”词条。

folly A building or structure often with no other useful function than to attract the attention of people passing.
装饰性建筑 除了吸引路人注意力没有其他实际功能的建筑物或构筑物。

foot bolt A *barrel bolt at the foot of a door.
地插销 门底部的圆插销。

foot cut The horizontal cut in a *birdsmouth.
角口平切 承接角口的水平切割。

footing (spread footing) The foundation or base of a structure.
基础 构筑物的基础。

footing fabric (trench mesh) Steel-reinforced fabric used in a *footing.
基础钢丝网 基础结构里的钢丝网。

footprints (pipe tongs, US combination pliers) An adjustable wrench used for gripping pipes.
快速扳手 夹管用可调扳手。

foot screws (levelling screws, plate screws) The three screws connecting the *tribrach of a *theodolite or level to the tripod plate. Used to level the instrument.
校平用螺丝 连接经纬仪或水准仪三角基座和三脚架的三颗螺丝，作校平仪器用。

foot valve A *check valve positioned towards the lower end of a length of pipe.
底阀 开向管道下端的止回阀。

footway (US sidewalk) A path for people to walk along, typically located at the side of a road.
人行道 设在道路一侧供行人通行的小路。

FOPS (falling object protective structure) An enclosed space for an operator to work, e. g. the cab on a hydraulic excavator is a FOPS as it is designed to provide protection to the driver from any falling objects.
落物保护结构 机械操作员工作的一个密闭空间，如液压挖掘机驾驶舱就是一个落物保护结构，可以保护挖掘机司机不会受到任何落物的伤害。

force-action mixer A *mixer used for *lean concrete mixes.
贫混凝土搅拌机

forced air heating system *See* WARM AIR HEATING SYSTEM.
热风采暖系统 参考“热风采暖系统”词条。

forced circulation *See* MECHANICAL CIRCULATION.
强制循环 参考“机械循环”词条。

forced draught (mechanical draught) A *draught in a flue that is fan-assisted.
机械通风 利用通风机把烟排出烟道。

forced drying The process of applying a small amount of heat (up to 65℃) to paint to speed up drying.
强制干燥（人工干燥） 通过轻微加热（到65℃）的方式来加速油漆干燥的过程。

force majeure Legal language used in standard contracts to describe events or circumstances which are unforeseen and prevent the contract being fulfilled, such as war, crime, strikes, or ‘acts of god’. The force majeure term acts to free the parties to the contract from obligation or liability when the unforeseen extraordinary event or circumstances occur. The force majeure term would not cover aspects described in the contract or mentioned as these are foreseen. Force majeure does not cover negligence nor weather that is typical e. g. something that is expected within a five-year cycle.
不可抗力 标准合同中的法律用语，用于描述妨碍合同执行的不可预见的事件或环境，例如战争、犯罪、罢工或天灾。不可抗力条款在不可预见的极端事件或环境发生时可免除各方对合同的责任或义务。不可抗力条款不可以涵盖合同中已列明的或可以预见的事件。不可抗力不适用于过失或典型的天气状况，例如五年一遇的某些天气。

force pump A *pump that can deliver fluid to a great height above its own datum.
压力泵 把液体输送到远高于其基准面很多的位置的水泵。

foreman Person in charge of site operatives or a gang of tradespeople. Often it is a person who has progressed through a trade background, and has considerable experience with skills that allows them to organize labour and plant, order materials, and ensure activities on-site are organized and controlled.
工长 现场作业人员或专业施工人员的主管。通常工长都是从某一专业做起的，积累了足够多的经验和专业技能后，便有能力管理劳动力和设备，采购材料，并确保现场施工有序进行。

foreman bricklayer Person in charge of a bricklaying gang. To organize a specialist trade, the person is often a bricklayer with considerable experience. Duties would include organizing labour, organizing and checking the setting out of brickwork, and ensuring sufficient materials are available for the bricklayers.
瓦工工长 瓦工班组的主管人员。瓦工工长通常是有丰富经验的瓦工，这样他才能管理一支专业的队伍。瓦工工长职责包括安排劳动力、组织和检查砌砖工程的放线工作、确保施工材料充足。

foreman carpenter Person supervising and in charge of carpenters and *joiners. This person organizes the groups of joiners ensuring there is sufficient labour, materials, and plant to perform operations. To undertake the tasks, the foreman has considerable skill in the specialist trade.
木工工长 粗木工和细木工班组的主管人员。木工工长应分配好木工的工作，确保施工劳动力、材料和设备充足。为了完成好木工工长的职责，木工工长必须在木工方面有超高的技能。

foreman plasterer Person supervising a gang or group of plasters. This person ensures there is sufficient labour, materials, and plant to perform operations. To undertake the tasks, the foreman has considerable skill in the specialist trade.
抹灰工工长 抹灰工班组的主管人员。抹灰工工长应确保施工劳动力、材料和设备充足。为了完成好工长的职责，抹灰工工长必须在抹灰专业有超高的技能。

forepole (forepoling board, spile) A method used in tunnelling to support the ground.
超前支护 隧道工程的一种支护方法。

forked tenon A tenon (recess) is cut centrally in the end of a rail or piece of wood to slot over another piece of wood.
叉形榫 在木板端头的中间位置切割出来的榫，可与另一块木材榫接。

fork-lift truck A motorized vehicle with protruding forks that move vertically. Used for lifting, stacking, and moving equipment and materials, particularly those stacked on a pallet.
叉车 带起重货叉的机动车。叉车可用于设备和材料（特别是托盘货物）的装卸、堆垛和运输。

form An item of formwork.
模板 一件模板。

formation level The finished level of the ground before concreting takes place.
平整面标高 场地平整后未浇筑混凝土前的标高。

former 1. A device used to shape something, e. g. to bend pipes. 2. An item used to create or form a hole or voids in something, such as the use of polystyrene blocks to form a void in concrete.
1. 塑形工具 如弯管器。**2. 预留孔模具** 在物体中开孔或钻出空隙的工具，如在混凝土板里边预装发泡聚苯板，在混凝土结构中形成空隙。

form lining The facing of *formwork.
模板衬垫 模板饰面加工材料。

form of tender Document that provides the project name, tender price, duration, completion date, and other relevant details, such as the proposed schedule, and is signed by the contractor.
投标函 提供项目名称、投标价格、工期、完工时间以及其他相关细节的函，例如建议进度表等的文件，并由承包商签署。

formply * Plywood used as lining for formwork.
胶合板模板 混凝土模板用胶合板。

formwork (casing, shuttering) A mould normally made of timber used to hold freshly placed and compacted concrete until it hardens. A formwork foreman is an experienced tradesperson who oversees the construction of formwork on large projects. A formworker or formwork carpenter is a tradesperson who erects formwork.
模板 用来支护新浇筑和新压实混凝土直至混凝土固化后才拆掉的模板，通常是木质的。模板工长：有过大型项目模板施工管理经验的专业人员。模板工：从事模板搭建的专业人员。

fortress A building constructed to act as a major means of defence such as a castle, ministry of defence building, or other modern-day stronghold.
堡垒，要塞 主要起防御功能的建筑，如城堡、国防部大楼或其他现代堡垒。

forum 1. An open space within a building. 2. An open space surrounded by public buildings. 3. A group of people gathered together for a specific purpose, to discuss a specific issue, or with a specific agenda.
1. 建筑物内一块空旷的场地。2. 由公共建筑物围起来的一块空旷的场地。**3. 论坛** 一群人为了某一目标集结在一起就某个话题或议程表进行讨论。

forward shovel *See* FACE SHOVEL.
正铲挖掘机 参考“正铲挖掘机”词条。

foul air (vitiated air) Stale air that is no longer suitable for breathing.
污浊空气 不适于呼吸的混浊空气。

foul air flue A * flue that is used to transport foul air from one place to another.
通气管/排气管 用来排出污浊空气的管道。

foundation The underlying support of a structure. It is normally made from concrete, which is cast below ground level to provide a critical interface between the structure and the ground beneath it. Foundations should safely distribute the load from the structure into the surrounding ground without exceeding * bearing capacity or causing excessive settlement. The three main types of shallow foundations include * strip, * pad, and * raft—a * pile would be classed as a deep foundation.
基础 建筑物的底部支撑。通常是由混凝土在地表以下浇筑而成，是建筑物和下面地基之间关键的界面。基础应能安全地将建筑物的荷载分布到周围的地基上，不会超过承载能力或造成过度沉降。浅基础的三种主要类型包括条形基础，独立基础和筏形基础，而桩基础应归类于深基础。

foundation bolt (holding down bolt) A bolt that has been cast into a concrete foundation, which is to be used to anchor the * base plate of a steel column to the foundation.
地脚螺栓 预埋在混凝土基础里边的螺栓，可用于锚固钢柱底板到混凝土基础上。

four-in-one-bucket A multipurpose bucket that can be fitted to hydraulic excavators, which can be used to perform the function of digging, dozing, spreading, and grading.
四合一铲斗 固定在液压挖掘机上的多功能铲斗，可以执行挖土、推土、摊铺和场地平整操作。

four-pipe system A heating or air-conditioning system that has separate flow and return pipes.
四管制系统 有独立供、回水管道的暖通空调系统。

foxtail wedging Wooden wedges, such as * folding wedges, when used in pairs.
紧榫楔 木楔，如成对出现的双木楔。

foyer An entrance hall or reception room in a building used as a gathering place for guests who enter or exit the building.
门厅 进门大厅或会客室，访客进出建筑物的一个聚集地。

fracture A sudden break or crack.
折断 突然断裂或破裂。

framed door (ledged, braced, and matchboard door) A door with structural components at the side, top, bottom, and middle, which hold the panels square and true. The structure also ensures the door can be fixed within the frame.
框架门扇 （直拼工型撑门企口板门）两侧、顶部、底部和中间都有结构构件支撑的门，保证门板方正不变形。这种结构还能保证门牢牢地套在门框里。

framed partition (stud partition, stud wall) A walling unit with its own structure or frame.
骨架隔墙（龙骨隔墙） 自带龙架的隔墙单元。

frame house (US timber-frame construction) *See* TIMBER FRAME CONSTRUCTION.
木结构建筑 参考“木结构建筑”词条。

F

frame saw A saw with several blades in one frame, used to cut several pieces of wood in one pass. Such saws can have diamond blades that are capable of cutting stone or slate.
框锯 将多根锯条张紧在锯框上的锯子，可一次性切割几块木材。框锯可装上金刚石锯片来切割石材或石板。

framing gun A large *nail gun, used to insert nails of at least 100 mm long.
长直钉枪 大钉枪，适用于长度≥100mm 的钉子。

Francis turbine *See* REACTION TURBINE.
弗朗西斯式水轮机 参考“混流式水轮机”词条。

franking A haunch or step that is cut into a tenon to reduce its width.
半闭口榫头 在榫头上切削出跌级的形状以削减其宽度。

free board The level difference between the height of something (e. g. a dam) and normal water level (e. g. in the reservoir).
出水高度 某物（如大坝）与正常水位（如水库）的高度差。

free cooling The use of cold air from outside the building to act as a chiller to cool equipment and people within the building.
自然冷却 利用室外冷风来进行室内冷却。

free face An exposed face, a term typically used in rock blasting.
自由面 特指被爆破的岩石的外露表面。

free haul The distance excavated material can be moved without additional costs; *See also* MASS HAUL DIAGRAM.
免费运距 在不增加额外费用的情况下，被开挖材料可以运输的距离；参考“土方调配图”词条。

freestanding (free standing, self-supporting unit) Element of a building, object, or structure that has the necessary strength and form to maintain an upright position without additional support.
独立式的 建筑物、物体或构筑物的构件有足够的强度和必要的形状来维持直立的状态，而无须依靠其他支撑。

freeway (US) A high-speed *highway, such as a motorway.
高速公路（美国） 快速公路，例如机动车道。

freezer An electrical appliance designed to store goods below 0℃.
冰冻箱 利用低温（<0 ℃）来保存食品的电器。

freeze-thaw cycle If water is present in cracks within various materials (e. g. the road surface), it will expand on freezing by 9%. Repeat cycles of freezing will cause the crack to grow, forming a pothole.
冻融循环 当水渗入到不同材料（如路面）的裂缝中，结冰时体积会膨胀 9%。重复的结冰会令裂缝扩大，产生坑槽。

freight elevator (US trunk lift) *See* SERVICE LIFT.
货梯 参考“货梯”词条。

French casement window (French door) Glazed external double doors that open out onto a garden, patio, or balcony.
落地窗（落地双扇玻璃门） 与花园、露台或阳台连着的双扇玻璃门。

French drain A perforated pipe that is laid at a gradient at the bottom of a trench and surrounded by free-draining material, such as gravel. Used, for example, at the side of highways to drain surface water.
盲沟 以一定坡度在排水沟底部敷设穿孔管，周围再填充如砾石之类的透水材料。如在公路一侧设暗沟来排除地表水。

French roof *See* MANSARD ROOF.
复折式屋顶 参考“复折式屋顶”词条。

French tile Single lap tiles imported from France. Also used as a generic term for other imported single lap tiles.
法式屋面瓦 从法国进口的单搭接屋面瓦。也可用来指代其他进口单搭接屋面瓦。

fresco Decorative painting on fresh plaster or lime mortar. The pigment is absorbed into the wet plaster. Once the plaster has dried, the pigment is fixed into the plaster, and the decorative finish remains.
湿壁画 在新鲜灰泥或石灰砂浆上面作的画。矿物颜料应被灰泥层充分吸入。灰泥变干后，颜料就与灰泥墙永久地融合在一起，不易脱落。

fresh-air filter A filter designed to remove dust and particles from the fresh outside air as it enters an *air-conditioning system.
新风过滤器 将室外新鲜空气中的灰尘和颗粒过滤掉再输送进空调系统的过滤器。

fresh-air inlet The vent that allows fresh air to enter into an *air-conditioning system.
新风入口 允许新鲜空气进入空调系统的风口。

fresh-air rate The percentage of air that is supplied to a space or item of equipment that contains fresh air. Minimum fresh-air rates are contained within CIBSE Guide A.
新风量 从室外引入室内或某一台新风设备的新鲜空气含量。英国暖通设计手册内有最少新风量的规定。

fresh concrete (wet concrete) Concrete that has just been mixed and has not yet started to set, thus it can be poured and compacted into *formwork, etc.
新拌混凝土 刚刚拌好但尚未凝结的混凝土，可以往混凝土模板里浇筑新拌混凝土并进行压实。

fret A geometrical decorative pattern of vertical and horizontal lines.
回纹饰 由横竖线条组成一定几何图形的装饰图案。

fretsaw A *coping saw, but with a deeper U-shaped frame.
深弓锯 U 型锯框更深的线锯。

fretting (ravelling) 1. The wearing away of two surfaces, when rubbed together. 2. The loss of aggregate from a road surface.
1. 摩擦腐蚀 两个表面因摩擦而导致缺损。**2. 路面松散** 路面有集料损失。

frictional soils *See* COHESIONLESS SOIL.
非黏性土 参考“非黏性土”词条。

friction catch (friction latch) A door spring or ball door fastener.
门夹扣 弹力门碰或门碰珠。

friction head Energy loss due to friction in a pipe.
摩擦压头损失 管道内部摩擦导致的机械能损失。

friction pile A *pile that transmits the load from the structure to the soil along the length of the pile; *See also* END-BEARING PILE.
摩擦桩 把结构荷载传递到桩侧土的桩。另请参考“端承桩”词条。

frieze 1. A horizontal band between the *architrave and *cornice in Classical architecture. 2. A long decorative horizontal relief or painting.
1. 饰带 楣梁和檐口线脚之间的水平装饰线条。2. 一条长长的水平浮雕或油漆饰带。

frieze panel The panels above the *frieze rail on a panelled door or panelling.
上镶板 镶板门上冒头上方的镶板。

frieze rail The uppermost horizontal intermediate rail on a panelled door or panelling.
镶板门上冒头 镶板门最上方的横档。

F

frog A recess or indent that is pressed, moulded, or cut into the face of a brick when the brick is being manufactured or hand made. Pressing a brick to form an indent is often used to help compress the clay and fill the mould. The indent also reduces the volume of clay, which helps with the firing process and makes the brick more economical to produce.
砖面凹槽 生产砖头时，用机器或人工在砖头表面模压或切割出凹槽。通常可通过在砖面模压出凹槽来压实黏土，也便于黏土放进砖模。同时，砖面凹槽的设计还减小了黏土的体积，加速烧结速度并节省生产成本。

frogged A brick or building block which has a *frog (indent or recess) in one or more of the faces.
刻槽砖 表面上有一个或多个凹槽的砖或砌块。

front The primary face of an object that is meant to be seen or seen first. The side of a building, brick, or wall that is to be the most exposed to public viewing.
正面 物体的正面，即第一眼看到的面。建筑物、砖块或墙体最先进入人们视野的一面。

frontage The primary face of a building that faces the road. *See* FRONT.
临街面 建筑物正对着道路的一面。参考“正面”词条。

frontage line The line of a building that runs parallel to the road, often the same as the *building line, which is the line imposed by the planning authority, and beyond which no building works can project (there are some exceptions).
建筑红线 与道路平行的一条建筑线，通常也称建筑控制线，是城市规划局设定的一条建筑控制线，任何建筑物都不得超过这条线（也有例外）。

front hearth The stone, concrete, or other incombustible platform such as a tiled hearth, that sits at the front of a fireplace and projects into the room. The hearth is often decorative and serves as a feature and functional item.
壁炉底座（壁炉地台） 石材、混凝土或其他不可燃材质制成的位于壁炉前面的凸出平台，如砖砌底座。通常，壁炉底座都兼具装饰和功能的作用。

front lintel A supporting beam which is exposed and sits over a main wall opening at the front of a building.
正面过梁 设在建筑物正面外墙洞口上方的外露支撑梁。

frost resistant brick A type of brick resistant to freeze-thaw failure.
抗冻砖/防冻砖 能够抵抗冻融循环而不被破坏的特种砖。

froststat A type of thermostat, which will start a heating system to prevent the water in the system from freezing.
防霜冻调温器 一种温控器，可自动启动加热系统，避免系统中的水被冻结。

Fuller's earth A type of clay material that is highly absorbent of oil and grease, used in filtering liquids.
漂白土 具有高吸附率的一种黏土材料，具有从油和油脂里吸附杂质的能力，可用于净化液体。

full height The distance from the floor to the ceiling. A full height unit goes from the floor to the ceiling.
室内净高 楼面或地面至上部天花之间的垂直距离。

full-length bath A standard 1700 mm long by 700 mm wide bath.
标准尺寸浴缸 1700 mm（长）×700 mm（宽）的浴缸。

full-length pipe The pipe length as manufactured.
标准长度管道 工厂加工好的一定规格长度的管道。

full-scale drawing A drawing that has been produced to exactly the same size as the object it represents.
原尺寸图纸 照物体实际尺寸画的图。

full-way valve A valve that does not reduce the bore of the pipe.
全径阀门 阀门内径尺寸与管道内径尺寸相同的阀门。

fully fixed Something that has no degree of freedom (rotation), e. g. a beam.
完全固定的 某物不能够自由转动，如梁。

fully quarter-sawn timber A board that has been sawn in such a way to reduce warping.
径切木材 以这种方式切割木板可以减少木板翘曲。

fully supported metal-sheet roofing Sheet-metal roofing system with a supporting frame.
全支撑金属板材屋面 带支撑结构的金属板屋面系统。

fume Gas, smoke, or vapour that has an unpleasant smell.
烟气 气味难闻的烟气。

functionalism Design that concentrates mainly on function, ensuring that the building or structure fulfils its functions.
功能主义 注重功能的设计，确保建筑物或构筑物能发挥其功能的设计。

furnish and install (US) A contract for the supply and fixing of items.
供应及安装合同（美国） 设备供应及安装合同。

furniture 1. Items such as chair, table, cupboard, and drawers that furnish a room. 2. The hardware, fixtures, fittings, and decorative items fixed to a door.
1. 家具 装饰房间的家具如椅子、桌子、橱柜和抽屉。**2. 门五金配件** 安装在门上的五金件、配件和装饰件。

furred up A build-up of *scale in pipework. May result in a restricted flow or a blockage through the pipework.
生水垢 管道内部结水垢。水垢可能会导致水流不畅通或管道被堵塞。

furring (encrustation) *See* SCALE.
结垢 参考“结垢”词条。

fuse A safety device that automatically prevents electricity flowing through an *electrical circuit. Usually comprises a thin piece of wire (**fuse wire**) or metal strip (**fuse strip**) that is designed to melt when the currentflowing through it exceeds its rated capacity. A **rewirable fuse** or **semi-enclosed fuse** is a fuse where the wire or metal strip can be replaced. A **cartridge fuse** is one that is enclosed within a tube. The fuse wire cannot be replaced in a cartridge fuse. *See also* CIRCUIT BREAKER.
保险丝 熔断器能自动阻止电流通过电路的保护装置。通常由一条细丝（熔断丝）或金属片（熔断片）组成，能在电流超过额定限度时熔断。可重接保险丝或半封闭式熔断器是一种可以更换熔断丝或熔断片的熔断器。熔丝管的熔丝密封在管内，并且不能更换。另请参考“断路器”词条。

fuse board A series of fuses that are mounted together.
熔断器板 把一系列保险丝安装在一起的一种电流保护器。

fuse box *See* CONSUMER UNIT.
用户配电箱 参考“用户配电箱”词条。

fused switch A switch where a fuse makes the contact.
保险开关 用保险丝连接的开关。

fusible plug A plug that contains a replaceable fuse.
带保险丝插头 带保险丝插头，保险丝是可更换的。

G

gabion Wire or plastic mesh crates or boxes filled with large stone blocks linked and stacked on top of each other to form a free-draining retaining wall. As water can naturally filter through the boxes and stone, they do not have to resist any hydrostatic pressure. It acts like a *gravity wall, but can withstand large differential settlements.
石笼网箱 装满大石块的铁丝网或塑料网箱，石块堆叠放置，砌成一堵透水良好的挡土墙。因为水可以直接透过网箱和石块滤出，因此可以大大减小水对墙体的压力。石笼网挡土墙就如同一堵重力式挡土墙，但可以承受更大的沉降差。

gable The triangular portion of the end wall of a building between the sloping sections of a *pitched roof.
山尖墙 坡屋面中两个斜坡之间的三角形部分的墙体。

gable board Face boarding placed along the edge of tiles that would be exposed at the gable end of the roof. The board covers the tiles providing a neat finish to the roof. *See also* BARGE BOARD.
山墙封檐板 沿着山墙屋面瓦边缘安装的面板。封檐板可以盖住屋面瓦，增加屋面美感。另请参考“博风板”词条。

gable coping Finishing stone that sits on the top of a gable wall. Provides weather protection to the wall, preventing water percolating down the inside and outside faces of the masonry. It is the *coping at the top of a gable wall.
山墙压顶石 设在山墙顶部的压顶石。可保护山墙不受雨水侵蚀，避免雨水渗透进砌体。

gable end The end wall of a building that has a *pitched roof.
山墙 坡屋顶建筑的侧墙。

gable post A short post, often decorative, at the apex of a gable. Used for fixing the *barge boards.
山墙柱 设在山墙顶端的一根装饰小柱。可用来固定封檐板。

gable roof (gabled roof) A roof that contains one or more *gables.
人字形屋顶 有一面或多面山墙的屋顶。

gable springer (footstone, skew block, skew corbel) A projecting portion of wall at the base of a *gable. Usually constructed in brick, tiles, or concrete.
山尖墙基石 山尖墙底部的凸出部分。通常是由砖头、瓷砖或混凝土砌块砌成的。

gablet The small *gable wall found on a *gambrel roof.
小山墙 复折式屋顶的小山墙。

gable wall A wall that has a *gable.
山墙 带山尖墙的墙体。

gaboon (gaboon mahogany, okume) Lightweight commercial wood from Africa.
加斑木，加蓬红木 产自非洲的一种轻质商品材。

gallet Small pieces of stone or pebbles used to fill the joints in a *rubble wall. The process of inserting the small pieces of stone or pebbles into the joints is known as galleting.
碎石 用于填充毛石墙缝隙的碎石或鹅卵石。把碎石或鹅卵石嵌入石墙缝隙的过程就称为填石。

galvanize To coat a metal surface with *zinc.

镀锌 在金属表面镀一层锌。

gambrel roof 1. A pitched roof that has different slopes on each side of the ridge, and at the *gable. 2. A *mansard roof (US).
1. 复折式屋顶，双重斜坡屋顶 在正脊两侧及两侧山墙上各有两个坡面的屋顶。**2.** 美国英语称之为"mansard roof"。

gang A group of operatives; skilled workers who perform as a team to complete the work.
班组 一组作业人员；以小组为单位进行施工的一群技术工人。

ganger The leader of a *gang.
工长 一个班组的领导。

gang form Formwork that is linked together and remains as one unit during stripping and reassembly when on site.
组合式模板 组合在一起的成套模板，可在现场整体拆卸和吊装。

gang nail *See* NAIL PLATE.
钉板 参考"钉板"词条。

gang rate The unit price for a gang of workers, normally charged by the hour. Rates may vary if additional plant and specialist labour are required.
施工班组人工单价 施工班组人工单价，通常是按小时计价的。如果需要增加设备或技术工人，则单价也会随之增加。

gang saw A frame saw with several blades.
排锯 有多根锯条的框锯。

gangway A temporary walkway or ramp on a building site formed from planks of wood.
临时通道 在工地现场用木板搭的临时通道或坡道。

gantry A spanning framework structure. Used to span carriageways and rail tracks to display information; carry travelling cranes (gantry crane), *bogies, and *derricks at a raised level in an industrial workplace; carry pipework, or provide a walkway at an elevated level.
龙门架 一种横跨式框架结构。可用跨公路和跨铁路的龙门架来显示信息；在厂房内，可把龙门吊、轨道小车和井架提升到想要的高度；同时，龙门架还可在高处承载管道或提供人行通道。

gap An opening between two objects.
间隙 两个物体之间的空隙。

gap-filling glue A glue used to join surfaces that cannot be closely fitted together.
填缝胶 用来填充两个相邻表面的缝隙的胶粘剂。

gap-graded aggregate An *aggregate with a particle-distribution that is grouped in specific sizes, e. g. having just large and small particles to provide specific properties; *See also* UNIFORMLY GRADED AGGREGATE and OPEN-GRADED AGGREGATE.
间断级配骨料 骨料由某些特定粒径的颗粒组成，例如只由大粒径和小粒径颗粒组成以获得特别的性能；另请参考"连续级配骨料"和"开级配骨料"词条。

garbage Domestic waste such as unwanted food.
垃圾 生活垃圾，如废弃食物。

garbage disposal sink (sink grinder) A type of *waste disposal unit that has a grinder fitted directly to the sink outlet.
食物残渣处理器 一种废弃食物处理器，在洗涤槽出水口位置接有一个粉碎机。

gargoyle A *rainwater spout for a roof in the shape of a grotesque person or animal.
滴水嘴兽 怪异人形或动物形的排水口。

garnet hinge A T-shaped hinge.
T 型铰链 一种 T 型铰链。

garnet paper A finishing and polishing paper covered with powdered garnet abrasive.
石榴石砂纸 打磨和抛光用砂纸，纸上粘有石榴石粉。

garret An attic.
阁楼 阁楼。

gas A term commonly used to refer to * natural gas.
天然气 用来指代天然气的术语。

gas concrete (aerated concrete) Lightweight concrete produced by developing voids by means of gas generated within the unhardened mix (usually from the action of cement alkalies on aluminum powder—used as an admixture).
加气混凝土 在还未固化的混合料里边掺加发气剂［通常是利用铝粉（外加剂）与碱性水泥水的化学反应］而制成的轻质、多孔混凝土。

gas detector A safety device that can detect the presence of various different gases.
气体检测仪 可以检测多种气体在环境中是否存在的安全装置。

gas fitter Appropriately registered installer of gas fittings and appliances. In the UK fitters need to be approved installers.
燃气装配工 燃气配件和燃气装置的注册装配工。在英国，装配工都必须是通过认证的。

gasket A flexible material used to seal the gap between two objects to prevent * air leakage and water penetration.
密封垫圈 用来密封两个物体之间的空隙、防止空气泄漏或水分渗入的柔性材料。

gas metal arc welding (GMAW) A type of welding process involving heating metals with an arc.
熔化极气体保护电弧焊 一种用电弧加热金属的电焊过程。

gas pliers Wrench or pliers that are designed to grip tubular objects such as pipes and tubes.
夹管钳 用于夹住管状物体（如内径管和外径管）的扳手或钳子。

gas tungsten arc welding (GTAW) A type of welding process using a tungsten electrode, also known as tungsten inert gas (TIG) welding.
钨极惰性气体保护焊 一种用钨作电极的电焊过程，亦称为“TIG”焊。

gate valve (sluice valve) A type of stop valve that uses a sliding disc or gate to stop the flow of fluid in a pipe. It cannot be used to regulate the flow of a fluid in a pipe.
闸阀 一种利用滑动闸板来切断管道中水流的截止阀。闸阀不可用于调节水流。

gauge 1. A device used to measure a particular quantity of a material. 2. The thickness of sheet metal or wire. 3. The process of mixing specific proportions of materials together. 4. The spacing between the tile battens on a pitched roof.
1. 计量器 用于测量材料量值的装置。**2.（铁板等的）厚度，（电线、钢丝等的）直径。3. 掺和** 按一定比例把几种材料放在一起混合的过程。**4. 坡屋顶挂瓦条之间的距离**。

gauge box Container with set dimensions used to either measure or check the volume or quantity of a material.
量斗 标有尺寸的可用于测量或检查材料体积或数量的容器。

gauged arch An arch constructed from * gauged brickwork. Various types of gauged arch are available including a **flat gauged arch**, a **segmented gauged arch**, **semi-circular gauged arch**, and a **Venetian gauged arch**.
磨砖拱 用磨砖砌成的拱。磨砖拱有这几种类型，包括磨砖平拱、磨砖弧拱、磨砖半圆拱和磨砖威尼斯尖拱。

gauged brickwork (gauged mortar) A brick wall, usually decorative, with very narrow mortar joints.
磨砖墙 灰缝很细的一种装饰性砖墙。

gauged stuff (gauged lime plaster) A type of lime plaster containing plaster of Paris to reduce cracking and shrinkage.
罩面石膏灰 含有熟石灰的一种石灰泥，可降低开裂和收缩的风险。

gauge pot Similar to a *gauge box, but a cylindrical container with set dimensions used to either measure or check the volume or quantity of a material.
量杯 与量斗类似，但容器是圆柱形的，量杯上面标有尺寸，可用于测量或检查材料体积或数量。

gauge rod Pole or length of timber cut to the height of something, for example, a door, floor etc. The rod can be used repeatedly to check height and measurements.
皮数杆 上面划有某一建筑构件（如门洞、楼板）高度位置的一种木制标杆。可以用皮数杆来检查各种竖向尺寸和竖向标高。

gauging board 1. Flat board or surface used for mixing mortar or plaster. 2. Board with increments marked on for measuring materials placed on it.
1. 搅拌用平板 砂浆或灰泥搅拌用平板。**2. 计量用板** 上面标有尺寸的计量用板，用于计量放置在上面的材料。

gauging box *See* GAUGE BOX AND BATCH BOX.
量斗 参考“量斗”词条。

gauging trowel A round-tipped triangular trowel used for transferring plaster and applying it to a wall; due to the regular size of the trowel it can also be used to measure out quantities of plaster.
圆头砌砖刀 圆头三角形砌砖刀，可从灰桶里铲起灰浆并抹到墙上；同时，由于此类砌砖刀都是有固定尺寸的，因此还可以用它来计量灰浆的量。

gauging water Water of a known quantity and consistency, being free from impurities, that is used for mixing with plaster, mortar, or other substances.
拌合用水 灰泥、砂浆或其他物质拌合用的不含杂质、具有一定量和一定稠度的水。

gaul A small depression in the final coat of plaster or render.
空鼓 内墙或外墙砂浆面层上的小凹陷。

gavel A *gable.
山尖墙 山尖墙。

G-cramp Clamp in the shape of a letter ‘G’ or ‘C’, fitted with a clamping screw and used to secure two pieces of material together temporarily.
G 字夹 G 字形或 C 字形夹具，带旋转丝杆，可用于临时固定两件材料。

gel coat An epoxy-based coating used as a surface finish on composite materials.
凝胶漆 涂抹在复合材料表面的一种环氧类涂料。

general contractor Organization that undertakes a range of construction and building work, the term may be used interchangeably with main contractor.
总承包商 执行一系列施工活动的组织，“general contractor ” 和 “main contractor” 都是指总承包商。

general foreman Person in charge of a group of tradesmen, which could be made up of mixed trades such as bricklayers, joiners, ground-workers, etc.
总工长 一组技术工人的主管，技术工人可以是由不同工种组成的，如瓦工、细木工或土建施工员等。

general location plans Drawings showing the proposed buildings and roads in relation to boundaries, other property, and other features. The drawings give a general position and orientation.
总平面图 表示拟建建筑物和道路与周围环境、建筑或其他设施的布局的图纸。总平面图可以给出拟建建筑物的大概位置和朝向。

general operative Labourer who can be used to assist with many trades rather than one specific trade or skill.
普工 可以协助很多工种作业的工人。

geocomposite Typically a * geosynthetic consisting of two or more materials such that the properties of each material is exploited in the final product.
土工复合材料 由两种或以上材料组成的土工合成材料，合成物能综合各种材料的特性。

geogrid A specific type of * geosynthetic, formed from plain sheets of extruded polymer, which are punched and drawn. The drawing process simultaneously extends the punched holes into the required aperture size and induces molecular orientation. These structures are ideal for soil reinforcement applications.
土工格栅 一种土工合成材料，由高分子聚合物挤压成片状并拉伸和冲孔。在拉伸的同时冲固定尺寸的孔，并改变分子取向。该结构适用于土壤固定。

G

geomatrix A three-dimensional grid-like structure used to prevent soil erosion while vegetation becomes established.
土工网垫 一种三维网状结构，可以在植被成长时保护土壤免受侵蚀。

geomembrane A thermoplastic * geosynthetic, which is designed to be impermeable. Used to prevent the flow of water or * leachate in ground engineered constructions.
土工膜 一种防渗透的土工复合材料。用于防止地面工程建筑物渗水。

geometric stair A curved staircase, normally without a newel post.
螺旋楼梯 弧形楼梯，通常没有中心支柱。

geotextile A * geosynethtic (or vegetable fibre) product formed by a woven, nonwoven, or knitted process. **Woven geotextiles** are produced by the interlacing of yarns to leave a finished material that has discernible warp and weft. The main applications include reinforcement, erosion control, drainage, and separation. **Nonwoven geotextiles** are needle-punched or thermal/chemically bonded structures predominantly unilized for filtration, drainage, and erosion control—those for the latter typically manufactured from vegetable fibres, e. g. **Geojute®**. **Knitted geotextiles** involve the looping of the weft or warp yarn to produce a grid structure or to encapsulate inlay yarns for directionally stryctured fabrics, particularly for reinforcing applications.
土工织物 由纺织的、无纺的或针织加工而成的土工合成材料（或织物纤维）。**纺织土工织物**是由纱线交织而成，表面有明显的经纬线。主要用于加固、防腐蚀、排水和分隔。**无纺土工织物**是针刺或热/化学黏合而成，主要用于过滤、排水和防腐蚀——后者主要由植物纤维制成，如 **Geojute**。**针织土工织物**包括经纱或纬纱织成的线圈，构成网格结构或内嵌纱线以定向编织纤维，特别是作加固用途的。

German siding (drop siding, novelty siding) A type of * weatherboarding comprised of a concave top edge and a rebate at the bottom edge.
披叠板 一种护墙板，顶边为凹形结构，底边为槽口结构。

giant form Large section of wall or floor formwork, lifted by cranes.
大模板 墙体或楼板的大尺寸模板，可用吊机进行吊装。

Gibbs surround Architrave around a door with large blocks at the corners and sometimes a keystone set under a large triangular feature.
吉布斯饰边 在门的饰边上采用较大的石块，有时，还包括拱心石和其上方的山花。

gilding A type of decoration involving applying a metallic leaf to solid surfaces.
镀金 一种装饰工艺，指在器物的固体表面上镀一层金属箔。

gimlet A small hand-tool for drilling holes.
螺丝锥 小型手动打孔器。

gimlet point A sharp penetrating tip of a screw, used to help penetrate into the surface of a material.
螺纹钻头 锋利的螺纹钻头，可以更轻易地钻进材料表面。

girder A main supporting beam, usually of steel or concrete, but may also be timber.
主梁/大梁 起主要支撑作用的梁，通常有钢梁或混凝土梁，但也有木梁。

girder bracket A bracket used to connect components to a girder.
主梁支架 用来把构件固定到主梁上的支架。

girder truss A *trussed rafter designed to support heavy loads, such as other trusses.
桁架梁 用于支撑较大荷载（如其他桁架）的屋架梁。

girt A horizontal structural member that spans between columns or posts in framed construction, and is used to support cladding.
连系梁 在框架结构中横跨两根柱子的水平支撑构件，用于支承外墙板。

girt board A timber *girt.
木质连系梁 木质连系梁。

girth 1. The circumference of a tree trunk. 2. A *girt.
1. 树围 树干的周长。**2. 围梁** 一种连系梁。

gland A watertight seal around a shaft. The seal can be made using an *olive or *packing. *See also* GLAND NUT.
填料压盖 转轴密封结构。可以用黄铜密封圈或填料进行压紧密封。另请参考“填料压盖螺母”词条。

gland nut (packing nut) A nut that is placed around a *gland. As the nut is tightened it compresses the *packing, producing a watertight seal.
填料压盖螺母 安装在填料压盖上的螺母。通过拧紧螺母，便可压紧填料，以确保密封性。

glare Any excessive brightness that causes annoyance, a distraction, or reduces vision. It is caused by one part of a room being excessivly bright in relation to the rest. Glare can be caused by daylight, by artificial lighting, or a combination of both. There are two different types of glare: disability glare and discomfort glare. Disability glare is where the glare impairs vision, but does not necessarily cause discomfort. Discomfort glare is where the glare causes discomfort, but does not necessarily impair vision. Disability and discomfort glare can occur simultaneously or separately.
眩光 任何过量的亮光，能引起不适、干扰或降低视力。是由于房间内某部分比其他地方相比过于光亮。眩光可由日光、人造光或两者组合而成。有两种眩光：失能眩光和不适眩光。失能眩光是眩光损害视力，但不一定引起不舒适。不适眩光是眩光引起不舒适，但不一定损害视力。失能和不适眩光可同时发生或分别发生。

glass A transparent inorganic material based on silica, used primarily in construction for glazing applications.
玻璃 以二氧化硅为基底的透明无机材料，主要用于建筑中的玻璃装嵌。
glass block (glass brick) A brick made from glass. An architectural element made from glass.
玻璃砖 用玻璃料制成的砖。用玻璃料制成的建筑构件。

glass concrete A type of concrete comprising of glass (usually recycled and crushed).
玻璃混凝土 用玻璃（通常是回收玻璃或碎玻璃）作骨料制成的一种混凝土。

glass cutter Tool used for cutting flat glass.
玻璃刀 用于切割平板玻璃的工具。

glass door 1. A *panelled door where all the panels are made of glass. 2. A frameless door where the *leaf comprises one sheet of glass.
1. 玻璃门 门芯板为玻璃的镶板门。**2. 无框玻璃门** 门扇（玻璃）周边没有固结的边框。

glassfibre Fibre manufactured as a continuous filament from molten glass, normally used for reinforcement due to its high tensile strength.
玻璃纤维 由熔化的玻璃拉成连续长丝组成的纤维，由于其抗拉强度高通常作为加固材料。

glasshouse A * greenhouse.
温室 温室。

glasspaper A paper with powdered glass on the surface, used for smoothing other surfaces.
玻璃砂纸 在原纸上胶着玻璃砂制成，可用于磨光其他物体的表面。

glass size *See* GLAZING SIZE.
玻璃板实际尺寸 参考“玻璃板实际尺寸”词条。

glass stop A * GLAZING BEAD.
玻璃压条 参考“玻璃压条”词条。

glass wool (glass silk) A material comprised of glass fibres used for insulating purposes.
玻璃棉 含有玻璃纤维的保温隔热用材料。

G

glaze A thin, smooth, shiny coating.
釉面层 一层薄薄的、光滑的、亮泽的面层。

glazed door A door that incorporates some glass.
玻璃门 有玻璃的门。

glazed tile A type of tile made from clay, usually with a glossy finish for moisture resistance.
釉面砖 用黏土烧制的一种砖，通常是有光泽的、防渗的。

glazier Skilled tradesperson who cuts, installs, and fits glass.
玻璃装配工 负责玻璃切割和安装的技术工人。

glazier's chisel Knife with a squared-off blade for applying putty.
油灰刀 刮腻子刀，刀片是扁的。

glazier's putty Putty used to secure the glass in timber windows and seal in the unit.
玻璃腻子 把玻璃固定到木质窗框上的泥子，有防渗的作用。

glazing 1. Transparent or translucent sheets of glass or plastic that are installed within a frame. 2. The process of fixing transparent or translucent materials into a frame.
1. 安装到框架里的透明或半透明玻璃板或塑料薄板。2. 把透明或半透明材料固定到框架里边的过程。

glazing bar A metal or wooden bar that subdivides and supports * glazing within a frame.
玻璃窗格条 分隔和支撑玻璃的金属条或木条。

glazing bead A small moulding applied around a window frame to hold the glazing in place.
玻璃压条 安装于窗框四周对玻璃起固定用的压条。

glazing block Small plastic blocks or packers used to hold glazing centrally within the frame.
玻璃垫块 对玻璃起固定作用的小塑料垫块。

glazing fillet A * glazing bead.
玻璃压条

glazing putty A type of putty that holds glass panels in place and provides resistance to moisture penetration.
玻璃腻子 把玻璃片固定到窗框上的泥子，有防渗的作用。

glazing rebate An L-shaped recess in a window-or doorframe for fixing glazing.
玻璃固定槽 窗框或门框上的L型凹槽，作固定玻璃用。

glazing size (glass size) The actual dimensions of a pane of glass or other glazing material. The visible dimensions are likely to be smaller than these dimensions.
玻璃板实际尺寸 普通玻璃板或其他玻璃板材的实际尺寸。实际尺寸要比可视尺寸大。

glazing sprig (glazier's point, brad) A small headless nail used to secure glazing in place while the * putty

sets.
玻璃窗楔子 玻璃腻子固化过程中，用来把玻璃固定到窗框上的小无头钉。

glory hole *See* SPILLWAY.
泄洪道 参考“泄洪道”词条。

gloss (gloss paint) A surface shine or lustre applied in paint form that has the ability to reflect light in a specular direction. The ability and optical property of gloss paint to reflect light has led to different definitions based on the percentage of light returned and dispersed. At one end of the continuum is flat paint with low reflection, to gloss with a high reflection; common names used in the continuum include flat (low reflection) sheen, eggshell, semi-gloss, and gloss (high reflection).
亮光（亮光漆） 在表面上应用的光泽漆，可以以镜面方式反射光线。根据亮光漆反射和漫射光线的程度，区分为不同的定义。从一端低反射的平光，过渡到高反射的亮光；常用的名称有：平光（低反射），蛋壳光，半光，和亮光（高反射）。

glue A material used for adhesion or bonding objects together.
胶水/胶粘剂 把两个物体黏结在一起的材料。

glue block *See* ANGLE BLOCK.
木三角条 加固用直角三角形木块 参考“斜垫块”词条。

glued-laminated timber (glulam) A timber product comprised of several layers glued together.
胶合木 多层胶合木材产品。

glue line The thickness of a joint that has been glued.
胶层厚度 粘贴缝的厚度。

gluing clamp Adjustable clamp used for securing two or more objects tightly together while the glue sets.
木工夹具 胶水固化过程中，用来固定两块或多块木板的可调节拉紧器。

GMAW *See* GAS METAL ARC WELDING.
GMAW 参考“熔化极气体保护电弧焊”词条。

goal Ultimate project target.
项目的最终目标 最终的项目目标。

goat's foot clip A clip in the shape of a goat's foot used on steel reinforcement bars.
钢筋保护层垫块 用于固定钢筋位置的山羊脚状垫块。

goggles (safety goggles) Eye protection used when workers are exposed to the risk of small flying objects or working with hazardous chemicals that could damage eyes. Hazardous tasks when eye protection should be used include: using cutting or grinding tools; working in particularly dusty environments; or working with chemicals.
护目镜 施工人员为了防止眼睛受飞溅物或危险化学物品的伤害而佩戴的眼镜。在执行下列危险作业时须佩戴护目镜：切割或研磨作业；粉尘作业；或化学操作。

going A staircase measurement of the horizontal distance between the front of one step (also called the *nosing) to the front of the next step. This measurement can be different from the tread, which measures the total length of horizontal platform that is available for the user to step on. The going can be shorter than the tread as the nosings can overlap the step below.
踏步宽度 相邻两踏步前缘线之间的水平距离。踏步宽度跟踏步面宽度是不同的，踏步面宽度指的是整个踏步面的宽度。踏步宽度要比踏步面宽度小，因为上一级踏步的前缘会凸出下一级踏步一小部分。

going rod A measurement stick, cut to the length of a *going used to ensure the length of each going is consistent.
踏步定距标杆 切割成跟踏步宽度一样长的一根标杆，可确保踏步宽度的一致性。

gondola *See* CRADLE.

吊船 参考“吊篮”词条。

good ground (good bearing ground) Ground that has suitable *bearing capacity to support foundations without requiring excessive excavation.

承载力良好的地基 地基承载力良好，无须挖很深的地基就足以支撑基础。

good practice (current trade practice, site practice) The accepted method of building, with the standard and quality being satisfactory to all requirements and regulations.

良好的工程质量管理规范 工程验收规范，确保工程质量和标准是满足一切相关要求和规定的。

goods lift *See* SERVICE LIFT.

货梯 参考“货梯”词条。

G

gooseneck A 180° pipefitting, in the shape of a goose's neck, used to connect a communication pipe to the water main or a gas service pipe to the gas service main. Used to overcome differential settlement.

180 度弯管 鹅颈型 180°弯管，连接给水干管和连接管的管件，也可用于连接燃气干管和燃气进户管。鹅颈型 180°弯管可克服不均匀沉降的问题。

gouge A long curved bladed chisel used for hollowing out timber, and cutting into rebate when turning wood.

半圆凿 可在木材上挖空和剔槽的弧形刃凿子。

grab rail A securely mounted handle or rail fixed to the wall at the side of a toilet or bath to assist the elderly and those with limited mobility.

扶手 牢牢固定在马桶或浴缸一侧墙上的扶手，可以供老年人和行动不便的人使用。

grade 1. Stated quality of a material, that meets with specified strengths and standards. 2. The flat *formation level of a building from which all building works commence.

1. 等级 材料的规定质量，规定的强度和标准。**2. 参考基准面** 建筑的施工基面。

grader Typically a six-wheeled earth-moving machine used for grading/levelling the formation level of a site. A wide shallow blade (the mouldboard) is moulded below the vehicle, whose height and angle can be adjusted. The front wheels on the vehicle lean to resist side forces.

平地机 用于平整现场地面的土方机械，通常有六个轮子。在平地机（回转环）下装有又宽又浅的刮刀，刮刀的高度和角度都是可以调节的。可以通过倾斜平地机的前轮来抵消机器受到的侧向力。

gradient A measure of the rate of incline or decline on a slope. Expressed as a ratio, such as 1∶5.

坡度 斜坡的倾斜程度。通常用坡度比来表示坡度，如坡度为 1∶5。

grading The levelling off of ground to provide a flat surface.

场地平整 平整地面，使地面达到水平状态。

graffiti Drawings and writing on buildings, walls, and other surfaces that have been undertaken without the owner's consent. Can be minimized by using *anti-graffiti treatments.

涂鸦 未经业主同意，私自在建筑物墙上或其他表面上作画和写字。可通过在建筑表面上刷防涂鸦涂料来减小这种影响。

grain A part of a metal, timber, or ceramic material representing an individual crystal. Usually grains are visible with an optical aid, for example, a microscope.

纹理 金属、木材或陶瓷材料表面的独特晶体结构。通常，在光学仪器（如显微镜）辅助下是可以观察到材料的纹理的。

graining The process of painting the non-wood surface of a material to simulate the grain of wood.

木纹漆施工 在非木质表面上刷木纹漆，模仿木纹的效果。

grand master key The key that unlocks all doors on suited locks. The grand master key opens all locks that belong to the same group, with each lock also having its own unique key.

总钥匙 可以开启所有同一款锁的钥匙。总钥匙可以开启同一组别的所有锁，但这一组别的每个锁又有独立的钥匙。

granite An igneous rock composed of feldspar, mica, and quartz.
花岗岩 由长石、云母和石英构成的火成岩。

grate (grating) A frame of crisscrossed interlocking bars that cover an outlet or opening. This allows fluids, gases, and small objects to pass through the outlet or opening.
格栅/格栅板 扁钢和扭钢交叉排列制成的开口或洞口盖板。水、气体或其他微小物体可通过格栅流进或流出。

grated waste A waste pipe that has a grate over the inlet or outlet.
格栅地漏 在排水管上方的进水口或排水口位置安装一个格栅盖板。

gravel board A horizontal board fixed between fence posts and below the fencing panels at ground level. Used to protect the fencing panels from moisture and retain any soil or gravel at ground level.
木制围栏下方的长石板 安装在围栏柱子之间、围栏板下方的水平石板。可避免水分渗入围栏板，还可保持地表土壤，避免土壤流失。

gravity circulation (thermosiphon) The circulation of hot water through a system by *convection.
重力循环 热水通过对流在系统内循环。

gravity dam (gravity retaining wall) A dam or wall whose self-weight is so great as to resist sliding and overturning. Typically constructed of mass concrete; *See also* EMBEDDED WALL.
重力坝（重力式挡土墙） 依靠自身重力来维持稳定的坝体或墙体。通常是用大体积混凝土砌成的。另请参考“打入土中的挡土墙”词条。

gravity flow The movement of a fluid due to the effect of gravity. *See also* GRAVITY PIPE.
重力流 液体在自身重力作用下流动。另请参考“重力流管道”。

gravity pipe A pipe designed for *gravity flow.
重力流管道 液体依靠自身重力作用流动的管道。

green 1. A park or open area. 2. A term used to describe buildings or products that have low environmental impact.
1. 公园或露天场地 2. 绿色建筑/绿色产品 对环境有利的建筑或产品。

green belt An area of park, farmland, or undeveloped land around an urban area.
绿化带 城市里边的公园、农田或未开发的土地。

green concrete (green mortar) Concrete that has set, but has not thoroughly hardened. It will not support any significant load without deformation.
新浇混凝土 已开始凝结但还未安全固化的混凝土。新浇混凝土不可以支撑任何实质的荷载，避免出现变形。

greenfield site A site that has not been previously developed, is free from contaminates, and is normally in a rural setting; *See also* BROWNFIELD SITE.
未开发区 之前未被开发过的区域，没受到污染，通常为乡村地带。参考“工业地带”词条。

greenhouse A building with glass walls and roof that is used to grow plants.
温室 有玻璃墙和屋顶，用于种植物的建筑。

green timber Timber that has not been seasoned.
生材 未经干燥的木材。

grees Steps or a staircase.
一段楼梯 楼梯或梯级。

greywater Wastewater from domestic activities such as washing machines, dishwashers, showers, but not from toilets, which is classified as *sewage.
洗涤水 家庭中如洗衣机、洗碗机、沐浴产生的废水，但不包括厕所水（归类为生活污水）。

GRG (glassfibre-reinforced gypsum) A composite of high-strength alpha gypsum reinforced with glass fibres; used as an alternative to plaster castings because of its lightweight, superior strength and ease of installation.
玻璃纤维增强石膏 高强α型半水石膏加上玻璃纤维增强的复合物；因其轻质、高强度和易安装，可取代普通石膏制品。

grid 1. The establishment of points of reference on-site to mark out a network from which other coordinates and lines can be set. 2. Distribution network for electricity, gas, and communications across the country. 3. Horizontal and vertical reference lines marked at regular intervals on drawings.
1. 基准点 在现场标记的参考点，可供制作坐标和基准线用。**2.** 国家内供应电、燃气和通讯的线网。**3. 图纸基准线** 在图纸上以一定间隔标记的水平和垂直的参考线。

grid line A line on a drawing that is part of a larger *grid. Each line is given a unique label, e.g. A1, A2, C4, with the letters and numbers relating to different points on the drawing's horizontal and vertical axes. On scale drawings, the lines are spaced at known distances apart, the grid lines can then be transferred to the site and used to determine offsets to the building structure and features.
坐标定位轴线 属于图纸上基准线的一部分。每条线都有一个独立的编号，按照图纸上水平和垂直的坐标标上字母和数字，如 A1，A2，C4。在等比例图纸上，线以一定的间隔分布，轴线可在现场上标记，用于确定建筑构件的坐标。

grid plan A drawing where the grid lines are set so that they cut through the structural frame. The frame is set at the same regular intervals as the grid.
坐标定位图 图纸上的定位轴线已确定并标示于结构框架上。框架与坐标定位一样按一定的间隔标记。

grillage A foundation or support comprising a crisscross framework of beams, used to spread heavy loads over a larger area.
格排基础 由十字交叉梁构成的用于在较大区域传递较大荷载的基础或支撑结构。

grille (grill) 1. A ventilation *grate used to supply air to or extract air from a space. 2. A perforated screen used for decorative or security purposes.
1. 格栅风口 用于房间的送风或排风的通风格栅。**2. 镂空格栅或防盗门金属穿孔板**。

grinder A power tool with an abrasive cutting wheel, which rotates at high speed and is used for shaping and cutting materials. Different grinders and discs are available for cutting steel, concrete, stone, and plastic.
角磨机 利用高速旋转的砂轮对材料进行切削和打磨加工的电动工具。针对不同的材料（钢铁、混凝土、石材和塑料）可选择不同的角磨机和砂轮进行切削。

grinding 1. A process of rubbing the surface of a material in order to wear away a layer, such as to sharpen an object. 2. A process of crushing a material to obtain finer particles.
1. 磨光 研磨材料表面，磨掉一层，如把东西磨利。**2. 磨碎** 把材料磨成细小的颗粒。

grinding disc An thin abrasive cutting wheel used with a grinder or angle grinder.
砂轮 切削用薄片砂轮，与角磨机配套使用。

grinding wheel An abrasive wheel used with a grinder.
砂轮 与角磨机配套使用的砂轮。

grinning through A defect in painting where the undercoat can be seen through the finishing coat of paint.
透底 一种油漆缺陷，透过面漆就可以直接看到底漆。

gripper strip (carpet gripper) A thin metal or wooden strip with metal teeth used to secure the edges of a fitted carpet.
地毯钉条 带钉子的金属条或木条，用来固定满铺地毯边缘的。

grit A hard coarse-grained siliceous sandstone.
粗砂岩 硬质粗粒硅质砂岩。

grit arrestor A device that can pre-clean the supply of air or flue gases by removing grit, sand, and particulates.
捕尘器 空气或烟气净化设备，可过滤掉沙砾、沙子和微粒。

grommet A metal, plastic, or rubber ring inserted into a hole that a cable or pipe passes through. Used to strengthen or seal the hole.
索环/扣眼 穿线孔或管道预留洞的金属、塑料或橡胶环。可起到加固和密封的作用。

groove A long narrow slot in a material.
凹槽 材料上又长又窄的开槽。

groover A hand tool for making grooves in concrete, used to help control the location of possible cracks.
伸缩缝切割刀 在混凝土面切缝的手工工具，可防止混凝土开裂。

gross feature (US characteristics) Visual features of timber.
木材可视特征 木材可视特征。

gross load The total load on the underlying soil, without taking into account the soil removed during excavation; *See also* NET LOAD.
总荷载 下层土壤所承受的总荷载，不包括开挖时移走的土方；另请参考"净荷载"词条。

ground 1. The material on top of the earth's surface. 2. In electricity, an earth connection. 3. A background surface.
1. 地表材料。2. 电气接地。3. 背景，底子。

ground coat The base coat of paint that is applied prior to the *glaze coat or *graining.
底漆 在刷釉面漆或木纹漆前刷的一层底层油漆。

ground duct (trench duct) A duct set within the ground that houses cables and pipes.
埋地式电缆排管 埋地敷设的一组电缆保护管。

ground failure Instability of the ground, e. g. exceeding bearing capacity, formation of landslips, etc.
地面塌陷 如超过承载能力，产生滑坡等。

ground floor The floor of a building that is closest to the external ground level.
底层/一楼 与户外地面标高最接近的一层。

ground-floor plan A *plan of the ground floor of a building.
一楼平面图 某一建筑物的一楼平面图。

ground freezing A method of freezing groundwater so that excavation can take place. A freezing agent, such as calcium chloride, brine, or liquid nitrogen is pumped through pipes in the ground.
冻结法施工 使用冻结地下水的方法开挖土方。制冷剂如氯化钙、盐水或液氮通过管道泵入地下。

ground investigations Site investigation to determine the characteristics of the soil, nature of the ground, and site features.
土地勘测 现场调研，以检测土壤的特性、土地的性质和现场环境等。

ground level (GL) The elevation of the ground above a *datum, usually sea level.
地面标高 水准面（通常指海平面）以上的高程。

ground plate (sole plate) *See* SOLE PLATE.
（脚手架）底板 参考"底板"词条。

grounds (groundwork) Timber battens fixed to walls to provide a flat surface onto which plaster board can be fixed by nails.
木衬条 固定在墙上的打底用木条，可在木条上钉石膏板。

ground shaping *Landscaping of the ground.
地面景观布置 地面景观布置。

ground slab (US slab-on-grade) A concrete slab that is supported by the ground beneath.
底层地板 由地面支撑的混凝土板。

ground storey The level of a building that is above the ground and below the first floor.
一楼 地面至二楼楼板底面的空间。

ground terminal A terminal that connects an electrical device or lightning conductor to the earth.
接地端子 将电气设备或避雷针与大地作连接的端子。

groundwork (earthworks, work in the ground) Any work that takes place in the ground, includes excavation, laying of services, foundations, kerbs, paths, and roads.
土方工程 指任何地基工程，包括土方开挖、管道埋地敷设、基础工程、路缘石、小路和道路施工。

grouped columns More than one column located on the same pedestal.
群柱 固定在同一基座上的多根柱子。

group supervisory control Lift control system used to synchronize lifts so that they arrive at destinations allowing passengers to transfer to other levels, or return to a position that allows the cars to be moved quickly and effectively to the desired destination when called.
电梯群控系统 一种电梯控制系统，可集中控制多台电梯，以确保电梯能达到指定楼层输送乘客，或返回指定位置待命以便呼梯时能快速到达指定楼层。

grout A liquid cement mortar used to embed rebar in masonry walls, connect sections of pre-cast concrete, fill voids, and seal joints.
灰浆 水泥砂浆，适用于覆盖砖墙上的钢筋、预制混凝土砌块的黏结、填充缝隙和灌缝。

grouting A process of injecting *grout into ground (soil or rock) or concrete to improve strength, stability, and impermeability characteristics, by filling and binding cracks, fissures, and voids.
灌浆 把灰浆灌入地基（土壤或岩石）裂隙或混凝土接缝中，通过填充或密封裂缝、裂隙或空隙来提高被灌地层或建筑物的强度、稳定性和抗渗性。

growth ring *See* ANNUAL RING.
年轮 参考“年轮”词条。

groyne A piled wall built to prevent erosion from the actions of the sea or a river.
防波堤 为阻断海水或河水的冲击而修建的桩墙。

GRP (glass fibre-reinforced plastic) A polymer strengthened by the addition of glass fibres. GRPs are renowned for their exceptional toughness.
玻璃钢（玻璃纤维增强塑料） 通过加入玻璃纤维强化的高分子聚合物，特别的坚韧。

grub screw A screw or bolt with no head, driven into the threaded hole using an Allen key. The screw is driven in so that it embeds below the surface, and covers can be inserted so that there is limited evidence of the fixing.
平头螺丝 无头的螺丝或螺栓，可用内六角扳手拧进螺孔。拧进螺孔后，无头螺丝就隐藏在表面下，然后可以塞上螺丝盖，这样就没有固定的痕迹。

guard A device that is used to provide protection against accidents, such as the guard on a circular saw, or a guard for an open fire.
防护装置 提供保护、避免发生事故的装置，如圆盘锯防护罩，或壁炉围栏。

guard bead (corner bead) 1. An *angle bead. 2. A vertical moulding used to protect corners.
1. 护角网。2. 护角条 保护转角位的垂直线条。

guard board *See* TOE BOARD.
安全挡板 参考“挡脚板”词条。

guard rail (safety rail) A temporary rail installed during building work to prevent workers from falling.
护栏（安全栏杆） 施工临时护栏，避免工人从高空坠落。

guide coat A thin coat of paint applied to a surface, particularly one that has just been filled, to highlight any bumps and imperfections.
标记漆 涂在表面特别是刚涂装完表面的一层薄漆，用于标记刮碰和瑕疵的地方。

guide price A price given to indicate the perceived sale price and value.
指导价 对某商品的感知销售价值所制定的价格。

guide rails T-shaped vertical steel rails located on the inside of a lift shaft. Used to guide the lift car and the counterweight.
导轨 安装在电梯井道内部的垂直 T 型钢导轨，可保证轿厢和对重沿其作上下运动。

guide shoes (runners) Devices used on elevators to guide the lift car and counterweight along the *guide rails.
导靴 可保证轿厢和对重沿导轨作上下运动的电梯装置。

gully (gulley, floor drain) 1. A narrow channel formed by running water. 2. A type of underground drainage fitting that receives the discharge from rainwater or waste pipes.
1. 水沟 水流冲成的窄沟。**2. 雨水口地漏** 管道排水系统汇集雨水或废水的配件。

gunstock stile *See* DIMINISHING STILE.
不等宽门窗竖梃 参考"不等宽门窗竖梃"词条。

gusset piece A small triangular piece of sheet metal used at an internal corner to make the corner watertight. It can be lead welded or soldered.
角撑板 设在内角处起固定和防水作用的三角形金属板。可用铅焊或低温焊。

gusset plate *See* NAIL PLATE.
钉板 参考"钉板"词条。

gutter (guttering) 1. A narrow, usually semi-circular channel at the eaves of a roof used to convey rainwater. 2. A narrow channel at the side of a road used to convey surface water.
1. 天沟 设在屋檐处的用来排除雨水的半圆形窄沟槽。**2. 排水沟** 设在路边的用来排除地表水的窄沟槽。

gutter bearer A short piece of timber that supports the *layer boards of a *box gutter.
挑板支撑 支撑 U 型天沟竖向挑板的一块短木。

gutter bed Sheet metal inserted behind a gutter at the eaves to prevent any rainwater that may overflow from the gutter entering into the wall.
天沟挡水板 安装在天沟后面的金属薄板，可防止雨水溢流渗入墙体。

gutter board *See* LAYER BOARD.
天沟竖向挑板 参考"天沟竖向挑板"词条。

gutter bolt A metal bolt used to attach an eaves gutter to a *fascia board.
天沟螺栓 把天沟固定到挑檐立板上的螺栓。

gutter end *See* STOP END.
天沟封盖 参考"天沟封盖"词条。

gutter plate A beam used to support a sheet metal eaves gutter.
天沟梁 支撑金属天沟的梁。

gypsum A white mineral (hydrated calcium sulphate) used to produce cements and plasters.
石膏 一种白色矿物质（主要化学成分为硫酸钙），可用于制作水泥和灰泥。

gypsum plaster A type of *plaster based on *gypsum.
石膏料浆 以石膏为原料制作而成的一种灰浆。

gypsum plasterboard A structure comprising gypsum plaster sandwiched between sheets of paper, and used

commonly in interior walls and ceilings.
纸面石膏板 以石膏料浆为夹芯，两面用纸做护面而成的一种板材。广泛用于内隔墙和天花。

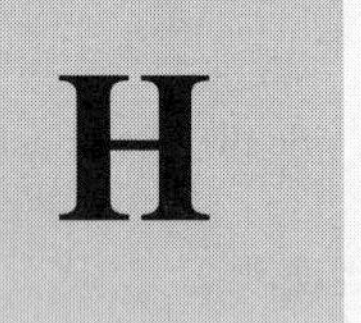

H

habitable room A room used for living in, excludes kitchens, toilets, bathrooms, and other similar areas.
居住空间 供居住的空间，不包括厨房、卫生间、浴室和其他类似区域。

HAC (high alumina cement) A type of cement that hardens at a rapid rate.
高铝水泥 能快速硬化的一种水泥。

hacking 1. Creating a keyed surface, by making indentations using a comb hammer. 2. The digging out or removal of existing putty.
1. 打毛处理 用澳式石工锤在石材表面上做打毛处理。**2. 清除旧腻子**。

hacking knife Blade used to remove old putty from window and doorframes.
油灰刀 用来刮掉窗框或门框上的旧腻子的刀。

hacking out Removing putty from a window or doorframe where the window pane has been broken or damaged. Once the glass and putty have been removed, the window frame can be reglazed.
清除玻璃腻子 门窗玻璃板碎掉时，需要用油灰刀来刮掉窗框或门框上的旧腻子。等玻璃腻子清掉后，就可以重新在窗框上安装玻璃了。

hacksaw A saw for cutting steel. The cutting blade is stretched and held in a frame. The releasable clamps enable the saw blade to be removed and replaced when the teeth are broken or worn.
钢锯 可切断钢制品的锯子。锯条张紧在锯架上。锯齿磨损时，可通过拧松拉紧螺母来卸下锯条。

haft The handle of a small hand tool.
柄 手工工具的把手。

ha-ha (sunken fence) A ditch used to contain or prevent people and animals from crossing a boundary; it has the advantage over a fence that it does not obscure views.
哈哈墙 一道壕沟，可阻止人或动物跨过园林的边界；比起围墙，壕沟的优点在于不阻碍人们的视野。

hair Animal hair used as natural binder in lime plaster. The animal hair acts as a fine reinforcing agent reducing cracking in the plaster.
动物毛 动物毛，用作石灰泥的天然黏结剂。动物毛可以作为绝好的增强剂，防止灰泥开裂。

hair cracking (hairline) Fine cracks in the surface of plaster, concrete, screed, and other ceramic materials, caused by shrinkage during hydration, thermal movement of different materials below the substrate, or slight structural movement. The most common cause of hairline cracking is drying shrinkage.
毛细裂纹 灰泥、混凝土、砂浆和其他陶瓷材料的表面出现的细小裂纹，这是由于基层下不同材料的水化反应、热运动或结构位移引起的。导致毛细裂纹的最普遍原因是材料的干燥收缩。

half bat (snap header) A brick cut through the middle to create a half brick. The brick is cut halfway along its width, making a half bat different from a *queen closer or *half header which is cut along its length.
半砖 沿着砖的宽度从中间把砖切成两半，这与竖半砖是不同的，竖半砖是沿着砖的长度把砖从中间切成两半。

half bond (stretcher bond) Brick courses are staggered so that the joint on each course sits directly above the centre of the brick below.
全顺式砌法 每一皮砖错缝搭接，上皮砖的竖缝坐中于下皮顺砖。

half-brick wall A brick wall made of a single skin of brickwork, where the width of the wall is equal to half the length of a brick, less a 10 mm joint.
半砖墙 单层砖墙，墙厚等于砖的长度的一半，灰缝厚度小于10mm。

half header A brick cut in half along its length, also known as a *queen closer, used next to a *quoin to complete the bond.
竖半砖 沿着砖的长度把砖从中间切成两半，也称为“queen closer”，砌在墙角砖旁边。

half-hour fire door (FD 30) A door designed to resist the passage of fire for a minimum period of 30 minutes. Fire doors are normally covered in timber with a core made up of less combustible material.
丙级防火门 耐火极限大于等于30分钟的门。通常防火门的门面为木材，门芯由难燃材料制成。

H

half landing A level platform formed between two flights of stairs, it allows users to rest and change direction. The platform is positioned so that the flights of stairs run into the platform adjacent to each other, enabling the total stair length to be reduced. Users travel down or up the stairs, turn 180° and travel down or up the other *flight. The landing should be at least the depth of the width of one flight and the width of the two flights that it joins.
中间平台 两楼梯梯段之间的水平板，可供使用者休息或楼梯转折。在梯段之间设置中间平台的目的是为了缩减楼梯的总长度。使用者爬楼梯或下楼梯时，必须经过一个180°的平台后再继续上/下另外一段楼梯。中间平台的深度应大于等于其中一段梯段的宽度，宽度应大于等于两段楼梯的宽度相加。

half-lap joint Method of joining timber by recessing two pieces of timber to half their thickness and lapping them together. If joined together end-to-end the surfaces of the timber are flush and when joined at a 90° angle, only half the end grain timber is exposed.
(木材) 半槽搭接 一种木材搭接工艺，分别在两块木材上开槽，切掉木材一半的厚度，然后再将两块木材搭接固定。可以是端部水平搭接，也可以是端部成90°直角搭接，只有端部的一半露在外面。

half pitch roof A pitched roof where the rise is half the span.
50%坡屋面 高跨比为1：2的坡屋顶。

half round A pipe cut in two lengthways, often used as a channel.
半圆槽 沿着管道直径把一根管道切成两半，通常作排水沟管道用。

half round bead (US astragal) A half round moulding.
半圆形装饰线条 半圆形装饰线条。

half round channel A pipe cut along its length into two parts. The channels can be bedded in concrete surrounds to form a watercourse in a manhole.
半圆形排水槽 沿着管道直径把管道切成两部分。半圆形排水槽可以埋在混凝土里，作为人孔里边的一截排水道。

half-round screw A screw with a dome head, also known as a button-headed screw.
半圆头螺丝 头部为半圆形的螺丝，也称作“button-headed screw”。

half-round veneer Strips of timber cut from the edge of wood that has been rotated and rounded on a lathe. The radius of the veneer can be altered by the depth of the cut and the radius of the timber on the lathe.
半圆旋切薄木/单板 原木边缘靠着车床削刀旋转而制成的木条。可通过调整切削深度和旋切半径来改变薄木的半径。

half space landing *See* HALF LANDING.
中间平台 参考“中间平台”词条。

half span roof A roof with a * pitch in one direction, also known as a single pitch or mono pitch. Similarly, a lean-to roof is a half span roof, although the roof abuts another building at its highest point.

单坡屋顶 只有一个坡面的屋顶，也称为"single pitch"或"mono pitch"。类似地，lean-to roof 也是单坡屋顶，尽管 lean-to roof 最高点是紧靠另一栋建筑的。

half timbered A building where the structural element is formed of timber and brickwork; lath and plaster, wattle and daub, or other materials are used to infill the gaps between the structure.

砖木结构 由砖、木结构砌成的建筑；半木结构建筑的墙体可采用板条抹灰隔墙、抹灰篱笆墙或用其他材料砌筑。

hall 1. A hallway, passage, or corridor within a building, that leads to the other areas or rooms within a building. 2. A central open area or grand room used to greet and entertain guests.

1. 门厅，过道，走廊 建筑物内的门厅、过道或走廊，可以通向建筑物内的其他区域或房间。**2. 宴会厅** 一块公共区域或一个大房间，接待和娱乐客人用。

H

halogen lamp (tungsten-halogen lamp) A type of high-pressure incandescent lamp that is filled with halogen. It produces a bright white light, has excellent colour rendering properties, is dimmable, and has a high efficacy, but it does operate at a much higher temperature than conventional incandescent lamps.

卤素灯（卤钨灯） 填充气体内含有部分卤族元素的一种高压白炽灯。卤素灯发出的光是亮白的、光源显色性好、可调节亮度且具有高光效。卤素灯的工作温度要比传统的白炽灯高很多。

halving (halved joint) The joining of timber by recessing two pieces of timber to half the thickness of the original piece and joining the matched joints together. If joined together end-to-end, the surfaces of the timber are flush and when joined at a 90° angle only half the end grain timber is exposed.

（木材）半槽搭接 一种木材搭接工艺，分别在两块木材上开槽，切掉木材一半的厚度，然后再将两块木材搭接固定。如果是端部水平搭接则木材表面是平滑的，也可以是端部成90°直角搭接，这样则只有一半的端部是外露的。

hammer Tool used for striking items, such as nails which are driven into timber or for hitting resistant objects so that they are moved into position, or are dislodged from their original position.

锤子 锤击工具，如用锤子把钉子钉进木材里，用锤子固定容易松动的东西或拆卸某样东西使其离开原位。

hammerbeam (hammer-beam) A beam used to tie the rafters of a roof together, but does not extend across the full width of a roof as a tie beam would.

悬臂托梁 用于连接固定屋顶椽条的短梁，与下弦杆不同的是：托臂梁并没有横跨整个屋架。

hammerbeam roof (hammer-beam roof) A roof with a hammer beam tie. A hammer beam tie joins the diagonal roof members together, providing triangulation to the roof structure.

悬臂托梁屋顶 有悬臂托梁的屋顶。悬臂托梁有连接固定屋顶椽条的作用，为屋架提供三角支撑。

hammer-dressed stone (hammer-faced stone) Stone that is shaped and finished using a hammer.

锤琢石 经石工锤打磨的石材。

hammer drill A power tool with a tungsten drill bit that vibrates up and down as it rotates. The movement helps to smash through the concrete as the rotating drill also bores into the surface.

锤钻 带钨合金钻头的可上下旋转钻孔的电动工具。上下快速旋转可捣碎混凝土表面。

hammerhead crane A tower crane with a horizontal * jib and counter jib.

锤头塔式起重机 带起重臂和平衡臂的塔式起重机。

hammer-headed key A hammer-head-shaped tenon joint used to secure members in curved joinery. The hammer-headed tenons slot into recesses with corresponding profiles. The hammer-head prevents the joint being pulled apart. The joints are wedged in together.

锤头销 一种锤头型榫销，用于固定弯曲的构件。将锤头型榫头插入对应形状的榫槽，可防止构件被拉开，然后用楔子紧固。

hand drill (wheelbrace) A hand-powered drill, in which a lever rotates a cog that drives a drill. The drill is useful for working in confined space, making slow, precise, and careful movements with delicate materials.
手摇钻 用手摇柄转动齿轮钻孔的手工钻。手摇钻适用于在密闭空间内部作业，但是钻速低不适用于钻较硬的材质。另外，手摇钻更容易控制钻孔位置。

hand float A plasterer's flat surfaced tool for laying on and smoothing the finishing coat of plaster. Measuring 300 mm × 100 mm, it is made from timber or light plastic.
抹子 抹灰工涂抹和抹平灰泥用的工具。尺寸为：300 mm × 100 mm，有木制的和塑料的。

handhole 1. A hole in carcassing for latching onto. 2. A hole in an object through which a hand can pass.
1. 闩孔 为上闩而在构件上开的孔。**2. 手孔** 物体上能让手穿过的孔。

handmade brick A brick made by hand, either hand-cut, moulded, or pressed into shape.
手工砖 手工制作的砖块，包括手工切割、模制或压制成形。

handover (handing over) At final completion, the building is passed over to the client, when it is inspected by the client's representative and possession is returned to the client. When the building is handed over, the *defects liability period starts; this is the period in which the building owner is expected to report defects, which the contractor is obliged to return and rectify. At the end of the defects liability period, any retention monies withheld by the client should be released if the defects have been rectified. Any defects that occur after the defects liability period should still be put right by the contractor if they are the result of poor or defective workmanship.
移交 在最终完工后，建筑物移交给业主，由业主代表检验并将所有权归还给业主。当建筑物移交后，缺陷责任期开始；在该期限内，住户可以上报缺陷而承包商有责任整改。在缺陷责任期结束后，被业主扣留的保证金在缺陷整改好的情况下可以返还。在缺陷责任期后，如有因承包商原因造成的缺陷，责任仍归承包商。

handpull A recessed handle for opening and closing sliding windows.
嵌入式拉手 用于打开和关闭推拉窗的嵌入式拉手。

handrail A rail positioned at waist height or slightly higher to provide support when moving up and down stairs, the rail may also provide security and safety at the edge of platforms and landings. There are different standard dimensions and regulations for handrails that are to be used to assist people who have greater difficulty using stairs. Stairs that serve a specific use and do not need to accommodate older people, disabled users, and children can be built to less demanding standards, although handrails and stairs should be designed to accommodate all users.
扶手 设在腰部位置或比腰部位置更高一些的协助行人上下楼梯用的扶手，扶手还起到坠落保护的作用，避免行人从楼梯平台边沿坠落。有多种规格尺寸和标准的扶手供行动不便的人使用。某些楼梯只针对特定人群使用，特定人群中并不包括老人、残障人士或小孩，像这种楼梯扶手要求就不是很高，尽管楼梯和扶手应要满足各类用户的需要。

handrail bolt (joint bolt) A bolt that has threads at both ends for securing two lengths of handrail timber together.
扶手螺丝 用于连接固定两截木质扶手的螺丝，两头皆有螺纹。

handrail screw *See* HANDRAIL BOLT.
扶手螺丝 参考“扶手螺丝”词条。

handrail scroll A scroll or spiral at the end of a handrail.
扶手涡形起步弯 楼梯扶手末端的涡形弯头。

handsaw A saw that can be operated manually.
手板锯 可以手动切割的锯子。

hand tool (small tool) Piece of equipment that can be held and operated by hand, and may be manual or power driven.

手工工具 用手握持和操作的工具，也可以是手持式电动工具。

handwheel (wheelhead) A handle shaped as a wheel that sits on top of *valves and is used to open and close the valves.

阀门手轮 安装在阀门顶部的轮形手柄，可用来开启和关闭阀门。

hang 1. To fix a door or window into its frame using hinges or mounts. 2. To fix a painting, picture, clock, or ornament to a wall. 3. To fix wallpaper, boards, or tile finishes to a wall or ceiling.

1. 用铰链或支架把门/窗固定到门框/窗框上。2. 挂在墙上悬挂画作、照片、时钟或装饰品。3. 在墙上贴墙纸或在墙上/天花上安装饰面板或吊顶板。

hanger A tie, chain, strap, threaded bar, or fixing that is used to secure fittings, fixtures, mechanical units, and suspended ceilings to the structure. The length of the hanger needs to be adjustable.

吊件 用于把水管、灯具、机械设备和龙骨吊顶固定到其上部结构的五件配件，包括扎带、吊链、管箍、螺纹吊杆或挂件。吊件的长度是可以调节的。

H

hanging To *hang or mount objects on walls.

悬挂 把某物挂在或固定在墙上。

hanging device The fitting or ironmongery needed to *hang a door or window.

挂件 安装门窗所需要的五金配件。

hanging gutter A gutter fixed to the ends of rafters on a roof or the eaves of the roof.

天沟 安装在檐口位置的排水沟。

hanging jamb (hanging post) The side of a doorframe to which the door or window is fixed, the sides of the *jamb that the hinges are fixed to.

带铰链的门/窗边梃 固定门窗的一侧的边框，带铰链那边的门/窗边梃。

hanging sash A *sash window.

上下推拉窗 一种推拉窗。

hanging stile The side of the stile to which a window is hung.

挂窗竖梃 窗户悬挂那边的竖梃。

hardboard A type of fibreboard, which is an engineered wood product; also called **high density fibreboard**. Hardboards are very popular in construction and provide a cheaper option in comparison with plywood.

硬质纤维板 一种复合木材，也称为“高密度纤维板”。硬质纤维板广泛应用于建筑中，且比起胶合板，硬质纤维板更加经济。

hard burnt (well burnt) The description of clay bricks, tiles, and other ceramics that have been fired to the point of vitrification. Bricks which are made of good-quality clay and are well fired have high compressive strength and are durable.

烧透的 黏土砖、瓷砖或其他陶瓷材料烧结到玻化温度。优质黏土烧结的砖具有较高的强度和耐磨性。

hardcore Crushed limestone or sandstone aggregate used to make compacted hard surfaces. Compacted hardcore is used as the base layers in road construction.

碎石垫层 用石灰岩碎石或砂石骨料碾压而成的硬质表面。碾压碎石垫层可用作道路施工的基层。

hard dry Paint that has dried to a solid hard surface, has lost its rubbery feel, and is not easily marked with a fingernail.

硬干 油漆已完全干燥，漆膜表面坚硬、无橡胶触感，且涂膜上不会残留指纹。

hardener An element used specifically to increase the *hardness properties of a material. Different hardeners exist for different types of materials.

固化剂 增强材料硬度的一种化学剂。不同的材料要选用不同的固化剂。

harden hardware Ironmongery that has been case-hardened.

硬化处理五金 表面硬化处理后的五金件。

hard hat (hardhat, safety helmet) *Personal protective equipment in the form of a helmet, manufactured out of resilient plastic, to prevent injury to the head from falling objects. It is now compulsory for a hard hat to be worn on all construction sites.
安全帽 个人防护设备，用弹性塑料制成的头盔，可防止高空坠物伤害头部。进入施工现场时必须佩戴安全帽。

hard landscaping The finishing of external areas in brick, block or clay paving, concrete, bitumen, tarmac, stone or chippings, or anything that creates a firm surface suitable for walking on or driving vehicles over.
硬质景观 室外区域的砖、砌块或黏土砖路面，混凝土、沥青、柏油碎石、块石或碎石路面，或其他可供行人和车辆通过的硬质路面。

hard plaster (hard finish) Plasters that are resistant to knocks and scuffs, such as rendering, squash court plaster, and renovation plaster.
硬质石膏 耐敲击和磨损的石膏，如外墙石膏、壁球室墙面石膏和翻新用石膏。

hard plating An abrasion-resistant metal coating achieved by electroplating the surface of iron or steel with chromium plating.
硬镀铬 通过在钢或铁表面镀铬层，以达到耐磨的效果。

hard solder A solder that has copper within it, which raises the melting point.
硬焊料 焊料含铜，可提高熔点。

hardstanding A paved, concreted, or stabilized area that is suitable for storing materials or parking vehicles.
硬化区域 可供储藏材料或停车的铺平、浇混凝土或固化的区域。

hard stopping Paint formed from a powder mixed with water or oil that is used to fill small holes or defects.
补洞涂料 由粉末加水或油混合而成的一种涂料，可用于填补小洞或修复其他缺陷。

hard surface *See* HARD LANDSCAPING.
硬质景观 参考“硬质景观”词条。

hardwood Timber or wood that originates from broad-leaved dicotyledonous trees—not based on the hardness of the material.
硬木 取自阔叶树的木材，而不是以材料的硬度来区分或命名。

harl Render with stones cast in the surface to provide a rough finish. The finish is porous enabling the surface to both absorb moisture and allow moisture to escape.
粗灰泥打底 灰泥和碎石混合覆盖表面，形成一个粗糙的饰面。由于饰面是多孔的，这样墙面就可以吸收和排出水分。

harsh mix A concrete mix with a low percentage of fine aggregate.
干硬性混凝土拌合物 细集料比例很小的混凝土拌合物。

hasp and staple A slotted latch that drops over a large raised staple to provide a method of locking doors, gates, and box lids. The hinged latch is secured by positioning a peg or lock through the staple once the latch is closed.
搭扣锁 开槽搭扣压到一个凸起的锁环上，以实现锁紧的一种方法，适用于各种门和箱柜。带铰链的搭扣扣到锁环上时门就锁上了。

hatch An opening in a wall, roof, or ceiling through which objects can be passed. A loft hatch is used to gain access from the habitable rooms to the roof space.
检修口 墙面、屋面或吊顶上的可供进出的开口。阁楼入口是指从居住空间进入阁楼的开口。

haul The movement or transportation of materials by using *haulage plants, e.g. *dump trucks and *bulldozers.
搬运 通过运输车辆（如自卸货车和推土机）来搬运材料。

haulage The transportation of goods and materials.
搬运 货物和材料的运输。

haunch A triangular brace used to add strength to a frame at the junction of roof members or the junction between columns and beams, or columns and roof beams. The thickening helps create stiffness to resist bending at stress points.
牛腿 设在屋架构件相交处或柱子与横梁相交处的加固用三角支撑。加腋部分可以在受力点通过增加其刚度来抗弯。

haunched tenon A * tenon that has a thicker section at its base than it does at the tip.
加腋榫 根部比头部厚的榫。

hauncheon The thick or wide section of a * haunched tenon.
榫腋脚 加腋榫加厚或加宽的部分。

H

haunching The build-up of concrete behind or against a drain or kerb used to hold it securely in place. The concrete sits on top or at the side of the * bedding sloping up to the side of the * kerb, * edging, or * drain.
加腋 在下水道或路缘石背面或正面浇筑混凝土以把它牢固固定在位置上。混凝土浇筑于基座的顶部或侧边，向路缘石、边缘或下水道一侧倾斜。

hawk (mortar board) A board used for holding wet plaster and mortar, providing a temporary place to hold the material while it is being applied. The board is usually 300 mm square with a handle enabling the operative to hold the board and apply the material at the same time.
灰浆托板（托灰板） 用于托住湿的灰泥及灰浆的板块，在材料涂抹时可临时托住。该板块通常是 300 mm^2，带手柄，可在托着板的同时涂抹材料。

HAZAN Hazard analysis is the use of numerical methods to assess how often a hazard will manifest itself, and the consequences of the hazard for people, process, and plant.
危害分析 危害分析是使用计算方法来分析危险发生的频率，以及对人、进程及设备带来的后果。

hazard An area, object, or situation that presents a risk to those who work in the proximity of the situation or object.
风险 会给在该处境或物体邻近工作的人员带来风险的一片区域、一个物体或一种处境。

Hazen-Williams formula An empirical expression that relates the velocity of water in a pipe or open channel to the physical properties of the pipe/channel, causing a pressure drop due to friction.
海森—威廉公式 一种实证表达式，把管道或明渠内水速与管道/渠道的物理特性联系起来，得出由于摩擦而造成的压降。

head 1. The horizontal top member of a door or * window frame. 2. The striking part of a * hammer. 3. A measure of the pressure exerted by a column of water, usually measured in feet. 4. The vertical distance between a * header tank or * storage cistern, and the outlet below (measured in metres).
1. 上槛 门或窗框的水平顶部构件。**2. 锤头** 锤子的敲打部位。**3. 水位差、水头** 用于测量水柱的施加压力，常以英尺为单位。**4. 高差** 蓄水池或水箱与较低处排水口之间的垂直距离（单位为米）。

head board 1. A board that was positioned on a labourer's head and used to carry bricks or other building materials. 2. A cushioned board at the top of a bed.
1. 头顶板 放置在劳工头顶上的一块板，用于搬运砖块或其他建筑材料。**2. 床头板** 床头的一块加垫板。

head casing (US) The * architrave that is used at the * head of a door.
门楣（美国） 用于门顶部的门框线。

header A brick or block with its narrowest facer exposed with the length of the brick, the stretcher face, penetrating into the wall. Traditionally, bricks that penetrate into the walls were used to link the two skins of brickwork together forming a solid single brick wall. With the introduction of cavities and insulation, header bricks rarely penetrate across the cavity as this would form a * cold bridge.

丁面 砖块或砌块的最窄面即丁面外露，而砖的长边及顺面穿入墙壁。一般来说，穿入墙壁的砖块是用于连接墙的两面，从而成为一面坚固的砖墙。采用空腔及安装保温层时，砖块的丁面很少会穿过空腔，因为会引起冷桥现象。

header bond (heading bond) Courses of brick laid end on with only the header exposed, used for curved walls and foundation courses.
全丁砌筑 在砌砖过程中，使砖的两端相对，只露出丁面，这种砌法适用于弧形墙及地基建造。

header tank (head tank) A raised tank that contains cold water. Used in central heating systems, hot water systems, and cold water supply systems. Also known as a **feed cistern**, a **feed and expansion cistern**, and a **cold water cistern**.
高位箱 装着冷水的位于高处的水箱。用于中央加热系统、热水系统及冷水供应系统。也可叫作给水箱、供水及膨胀水箱和冷水箱。

head flashing The lead work around an opening in a roof that creates a gutter to remove rainwater from the area around the opening, shedding it to the roof surface.
天沟 建筑物屋面开口周围的排水沟槽，用于把开口周围的雨水排放至屋面。

H

head guard A damp-proof course positioned over the top of the lintel to a door or window opening.
门窗头泛水 位于门或窗过梁顶部的防水层。

heading course Layer or course of header bricks, this could sit on top of a stretcher bond as in *English bond or sit on top of another course of headers as in *header bond.
丁砌层 整层或整皮的丁砖，可以顺丁砌筑方式砌于顺面砖上；或以全丁砖砌筑的方式砌于另一层丁砖上。

heading joint A joint made to bond two pieces of timber together end-to-end, such as a *finger joint, *butt joint, or *half-lap joint.
顶头接缝 用在端对端连接两块木材的接合处，例如"指形结合""对接"或"半搭接"。

head joint (cross-joint) A vertical joint also known as a **perpendicular joint** (perpendicular end) in brickwork and blockwork.
竖缝（十字缝头） 砌砖或砌块的一竖缝，也被称作垂直缝（垂直端）。

headlap The lap produced when the bottom of one roof or wall tile sits over the top of the other.
端搭接 屋面或壁砖末端相互重叠的部分叫作搭接。

head lining 1. A lining positioned around the top of a doorframe. 2. A *damp-proof course positioned over the top of a door or window opening *head guard.
1. 顶沿贴边 门框顶部周围的衬里；**2. 顶防潮层** 门或窗顶部的防水层，也就是门窗顶泛水。

head loss The drop in pressure between two points on a pipe or a duct.
水头损失 管道上两点之间的压力差。

head of water The pressure exerted by water on an object.
水压 由水施加的压力，即水的压力。

headroom (headway) The space allowed between the floor and ceiling or any obstruction that hangs below the ceiling. In stair construction and design, it is the gap between the *pitch line and ceiling, or obstructions below the ceiling.
净空（净空高度） 地板与天花或悬挂于天花下面障碍物之间预留的空间。在楼梯建造与设计中，则指的是坡度线与天花或天花下面障碍物之间的空隙。

headworks Fittings used to supply hot and cold water to a bath.
淋浴开关 提供洗澡用的冷热水的开关装置。

health and safety The discipline concerning safety, health, and welfare of people, particular in the work place.

健康与安全 关于人们的安全、健康及福利等方面的法律规定，尤其是在工作场所的。

heart The centre of a tree.
树心 树干的中心部分。

heart bond A type of brickwork bond where two *headers meet in the centre of the wall and the joint between them is covered by another header.
中心砌筑 一种砌砖方式，两个砖块丁面在墙中心相接，它们之间的接缝由另一丁面覆盖。

heart centre (pith) The wood at the centre of a tree that mainly consists of *parenchyma. *See also* HEARTWOOD.
髓心（树芯） 位于树中心的木材，主要是由柔组织组成。另请参考“芯材”词条。

hearth The base of a fireplace. It may extend out from the fireplace into the room.
壁炉底座 壁炉底部。可从壁炉延伸进房间。

heartwood Wood around the centre of a tree or cut log that has finished growing; it is more durable than the newer sap wood. The wood at the very centre, *heart centre, of the tree may not be as strong and durable.
（树木的）芯材 树心周围的木材或已经停止生长的木段，比新长的边材更耐用。树木正中心的木材可能没芯材牢固及耐用。

heat 1. Energy caused by the movement of atoms and molecules that can be transferred from one place to another by *conduction, *convection, or *radiation. 2. The perception of warmth. 3. The process of increasing the temperature within a space.
1. 热能 由于原子及分子移动而产生的能量，可通过传导、对流及辐射从一处传递至另一处。**2. 热量** 热度的感知。**3. 加热** 在一空间范围内升温的过程。

heat bridge *See* THERMAL BRIDGE.
热桥 参考“热桥”词条。

heat capacity *See* THERMAL CAPACITY.
热容 参考“热容量”词条。

heat detector A device that measures the temperature within a space and reacts when a particular temperature is reached. Used as a *fire detector.
热探测器 用于测量一空间范围内的温度，并在达到一定温度时作出反应。可用作火警探测器。

heat diode *See* THERMAL DIODE.
热二极管 参考“热敏二极管”词条。

heated floor (heated screed) Heating pipes laid within the screed or under the surface of the floor, also called *underfloor heating.
地暖 位于砂浆层内或地板底的加热管道，也称作地板下供暖。

heater A device that provides heat.
发热器 提供热量的装置。

heat exchanger A device that can transfer heat from one medium to another without allowing them to come into contact with one another.
热交换器 一个可把热量从一个媒介转移至另一个媒介，而无须两者相互接触的装置。

heat fusing (fusion welding, polyfusing) *See* THERMOFUSION WELDING.
热熔（熔焊、聚合物热熔） 参考“热熔焊接”词条。

heat gun (heat welding gun) A hand-held electrical device that emits a jet of hot air from a nozzle. Used in *thermofusion welding and paint stripping.
加热枪（热焊枪） 一个便携式电气装置，可从喷嘴喷射热气。用于热熔焊接及脱漆。

heating 1. The process of increasing the temperature within a space or an object. **2. Heating, ventilation**

and air conditioning (HVAC) A single integrated system that provides the heating, ventilation, and air conditioning within a building.
1. 加热 一个空间或一个物体内升温的过程。**2. 暖通空调(HVAC)** 建筑内一个可提供暖气、通风及空气调节的单个综合系统。

heating and ventilation engineer A person who is trained to design, install, and maintain heating and ventilation systems.
供暖与通风工程师 一个受过设计、安装及维护供暖与通风系统知识培训的人。

heating battery A type of *heat exchanger that uses hot water pipes to heat the air within a duct.
加热盘管 一种使用热水管道加热风道内空气的热交换器。

heating coil 1. Wire or pipes arranged in a coil that is used for *heating. 2. A **heating battery**.
1. 加热线圈 盘成线圈的线材或管道,用于加热。**2. 加热盘管**。

heating element An electrically heated wire coil.
加热元件 一种电加热的线盘。

heating medium A heat-transfer fluid, such as a coolant, used in central-heating systems.
热媒 一种传热液体,如冷却剂,用在集中供暖系统中。

heating panel *See* PANEL HEATING.
加热板 参考"板式辐射取暖"词条。

heat load The amount of heat that needs to be removed or added to a space to maintain the required internal temperature.
热负荷 为维持一空间内所需的内部温度而排出或增加的总热量。

heat loss A measure of the amount of heat that is transferred through the various elements of building fabric, such as windows, doors, floors, roofs, and walls, due to the exfiltration of warm air.
热损耗 由于暖气渗出,通过各种建筑结构元素(如窗户、门、地板、天花及墙壁)转移的热量。

heat of hydration The amount of heat produced from *hydration.
水合热 水合作用产生的热量。

heat pipe A relatively simple and efficient type of heat transfer device. Consists of a sealed metal rod or pipe that is partially filled with a fluid. Heat is transferred by vapourizing the fluid at one end and then condensing the fluid at the other.
热管 一种相对简单及有效的传热装置。由一个装有部分液体的密封金属杆或管道组成。通过在一端蒸发液体,然后在另一端冷凝成液体来转移热量。

heat pump An electrical device that moves heat from a low temperature source and upgrades it to a higher, more useful temperature, using the *vapour compression cycle. It comprises a compressor (usually driven by an electric motor), a circuit containing a refrigerant, an expansion valve, a condenser coil, and an evaporator coil. Heat pumps can be used to heat and/or cool a building. Refrigerators and air conditioners are common applications of heat pumps that operate only in the cooling mode.
热泵 一种利用蒸汽压缩循环原理把低温处的热量提升至一个更高、更实用的温度的电气装备。热泵一般由一个压缩机(一般由一个电动机驱动)、带制冷机的电路、一个膨胀阀、一个冷凝盘管及一个蒸发器盘管组成。热泵可用于加热或冷却一个建筑物。冰箱与空调是热泵的常见应用,只是仅以冷却模式应用。

heat recovery The process of transferring heat that would otherwise be wasted from one medium to another using a heat exchanger.
热回收 使用一热交换器将一媒介至另一媒介浪费掉的热量回收的过程。

heat-recovery wheel *See* THERMAL WHEEL.
转轮式热回收 参考"转轮式热交换器"词条。

H

heat-resisting glass A type of glass with excellent resistance to thermal shock.
耐热玻璃 一种具有极好耐热震性的玻璃。

heat-resisting paint Usually an aluminium paint having a lustrous metallic finish, and resistant to temperatures up to 230 ～260℃. A dry film thickness of 15 microns (μm) is typical.
耐热漆 常指一种有光泽金属面的铝粉漆，可耐 230 ～260℃的高温。通常为 15 微米厚的干膜。

heat-transfer fluid (coolant, heating medium) A fluid that is designed to transfer heat from one medium to another.
传热液体（冷却剂、热媒） 一种可把热量从一媒介转移到另一媒介的液体。

heat-transmission value *See* THERMAL CONDUCTIVITY.
导热系数 参考"热导率"词条。

heat welding gun *See* HEAT GUN.
热焊枪 参考"热风枪"词条。

H

heave The upward movement (expansion) of soil due to a reduction in effective overburden pressure or an increase in water pressure; *See also* SETTLEMENT.
隆起 由于有效上覆层压力降低或水压增加而造成土壤发生上升运动（膨胀），另请参考"沉降"词条。

heavy-body impact test A door strength test. A 30 kg sand bag is swung to impact the door, adjacent to the lock, three times, at increasing standard height. The test is repeated on the other side of the door.
重物撞击测试 门强度的测试，用一个重 30kg 的沙袋撞击门，靠近门锁的部位，以递增的标准高度撞击三次，在门的另一侧重复此测试。

heavy industry A business that requires a large amount of space to operate heavy equipment.
重工业 一种需要大量空间来操作重型设备的行业。

heavy metal A toxic metal with a relative density of 5. 0 g/cc or greater, examples include lead, mercury, copper, and cadmium.
重金属 密度达 5. 0 g/cc 或更大的有毒金属，如铅、汞、铜与镉。

heavy protection Aggregate, chippings, asphalt, or mortar laid over roofing to prevent uplift and protect the material from the sun.
重物保护层 铺放于屋顶的集料、碎屑、沥青或灰浆，用于防止屋顶凸起，并起到保护材料不受太阳照射的作用。

hecto- A prefix denoting 100 times.
表示"百" 前缀，表示"100 倍"。

heel The part of a structure, normally a beam or roofing rafter that rests on a support.
柱脚 结构的一部分，通常指搁在一支柱上的横梁或屋顶椽条。

heel strap A galvanized steel strap, produced in a U-shape, which is used to hold down roof trusses and tie beams.
托梁连接件 镀锌钢板，U 型，用于固定桁架及系梁。

height board/rod Strip of timber or rod that is cut to a known length, e. g. the height of a door, window, or floor, and is used to produce a consistent measure of the element of construction.
标尺板/棒 切割成一定长度的木材板块/木棒，如门、窗或地板的高度，用于统一测量建筑构件。

helical hinge A sprung hinge, with two coiled springs, that allows doors to be opening and self-closing in both directions. Used for lightweight swing doors.
旋转铰链 一种装弹簧的铰链，带两个螺旋弹簧，可使门从两个方向打开或自动关闭。适用于轻质双向门。

helical reinforcement Steel reinforcing bars that are used as cross-links in a circular column.

螺旋箍筋 环形柱中用作交叉连接的钢筋。

helical stair A spiral stair.
螺旋式楼梯 一种旋转楼梯。

helices Spiral ornamentation.
螺旋饰品 螺旋装饰品。

heliograph 1. A communication signalling device (such as a mirror) that reflects the light from the sun and can be used to send Morse code messages in terms of flashes. 2. An apparatus to take pictures of the sun. 3. A surveying instrument used at a station to reflect sunlight, such that a point can be sighted up to 100 km away. 4. An instrument to determine the intensity of the sun's rays.
1. 日光反射信号器 一种通讯信号装置（如一面反射镜），可反射太阳光，并可借助闪光灯发送莫尔斯电码信息。**2. 太阳摄影机** 可拍摄太阳的一种设备。**3. 阳光反射器** 位于测站的一种测量仪器，可反射太阳光，且可在100km处观测到该点。**4. 日照计** 可测定太阳射线密度的一个仪器。

H

helix A spiral line that follows the path of a cylinder at a constant angle.
螺旋线 以固定角度沿着圆柱体路径的螺旋状线条。

helmet A hard protective covering worn on the head to prevent injury. *See also* HARD HAT.
头盔 戴在头上的一坚固保护套，可防止头部受伤。另请参考“安全帽”词条。

helper A worker's mate or labourer.
帮手 工人的助手或劳工。

helve A handle of a pick axe or other similar heavy striking tool that is lifted and thrown using two hands.
斧柄 鹤嘴锄或其他类似的重敲击工具的手柄，可用两只手提起及投掷。

hemihydrate plaster Gypsum plaster produced by heating gypsum so that it loses 75% of its chemically bound water: $CaSO_4 \cdot 2H_2O$ becomes $CaSO_4 \cdot \frac{1}{2}H_2O$. *See also* HEMIHYDRITE.
半水化合物石膏 通过加热石膏而得的石膏灰泥，这样可失去它75%的化学结合水：$CaSO_4 \cdot 2H_2O$ 变成 $CaSO_4 \cdot \frac{1}{2}H_2O$。另请参考“半水化合物”词条。

hemihydrite Another term for hemihydrated plaster. Gypsum plaster that has been steadily heated to 150～170℃.
半水化合物 “半水化合物石膏”的另一术语。已经稳定加热至150～170℃的石膏灰泥。

hemisphere Half a sphere, particularly in relation to the earth, e. g. northern hemisphere, but also relates to other celestial bodies.
半球 球体的一半，尤其是指地球，如北半球，但也可指其他星球。

hemp (Cannabis sativa) An annual plant having a stem diameter of 4～20 mm and a stem length of 4.5～5 m containing a *bast fibre. Cultivated throughout Russia, Italy, China, Yugoslavia, Romania, Hungary, Poland, France, Netherlands, UK, and Australia. Requires mild climates with high humidity and annual rainfall >700 mm. Uses include: ropes, marine cordage, ships' sails, carpets, rugs, paper, livestock bedding, and drugs as a result of the narcotic THC content of certain varieties (30 varieties in all).
大麻（大麻植物） 一年生植物，茎径为4～20mm，茎长为4.5～5m，含有韧皮纤维。俄罗斯、意大利、中国、南斯拉夫、罗马尼亚、匈牙利、波兰、法国、荷兰、英国及澳大利亚均有栽培，适合生长于高湿度的温和气候，且年降雨量要求大于700 mm。用途有：绳索、船用绳索、船帆、地毯、毯子、纸、家畜草垫和药物，某些品种（总共30种）的含量的药物含有麻醉作用THC成分。

heptagon A seven-sided shape, with all seven sides and seven angles being equal.
七角形 七边形，七条边的边长和七个角的角度都相等。

hermetic Sealing an object to ensure airtightness.

密封的 密封一个物体以确保是密封状态的。

hermetic seal An edge seal that is airtight. Also refers to a barrier to prevent water vapour entering a cavity.
密封 密封封边，也可指防止水蒸气进入空隙的屏障。

hermitage A small dwelling in a secluded location.
隐居处 位于隐蔽处的一个小住所。

herringbone (herringbone pattern) Bricks, pavers, or timber placed in a diagonal pattern with units moving towards a point and placed at 90°to each other. The pattern is supposed to be similar to that produced by fish bones.
鱼骨状的（人字形） 砖块、铺路材料或木材呈对角线型排列，且向一个点移动，成直角。该模型与鱼骨形状相近。

herringbone drain A surface drain filled with granular material (stones) that has been laid in a herringbone fashion. *See also* CHEVRON DRAIN.
鱼骨式排水渠 填满颗粒材料（石块）的地面排水沟，以鱼骨型铺砌。另请参考"人字形排水沟"词条。

herringbone strutting Timber struts placed at angles to provide bracing to the floor beams. The strutting runs from the bottom of one beam to the top of the adjacent beam. The struts cross each other making the beams stable. As the struts cross, a *herringbone pattern is produced.
人字形支撑 成一定角度放置的木材支柱，为楼板梁提供支撑。支撑从一条梁的底部延伸至相邻梁的顶部。支撑相互交叉以稳固梁。支撑交叉从而形成了人字形状。

hertz (Hz) SI unit of frequency that is equal to one cycle per second.
赫兹（Hz） 频率的国际单位，即每秒一周期。

hessian An open-weave fabric, made from jute or hemp fibres, used to produce sacks, also sheets for protecting newly laid brickwork from frost attack. The fabric can also be laid within the ground to provide a ground reinforcement, and can also be introduced into plastic products to provide reinforcement.
粗麻布 一种稀松组织织物，由黄麻或大麻纤维制成，用于制造保护新铺砌砖块免受霜冻的麻袋及麻布。织物铺放在地下可以加固地基，加进到塑料制品里可以提升其牢固性。

hew To shape dressing stones or square timbers with an axe or hatchet.
砍 使用斧头或短柄斧开凿装饰石材、砍方木的动作。

hexastyle A porch or portico that has six columns at the front.
六柱式门廊 前面有六根柱子的走廊或门廊。

HGV (heavy goods vehicle) A large vehicle used for transporting goods.
重型货车（HGV） 用于运输货物的大型车。

H-hinge A parliament hinge with two legs that extend away from the knuckle. The heavy H-shaped hinge is used for carrying large doors and gates.
H 型铰链 一种工字型铰链，两只脚从轴承向外延伸。重型 H 型铰链用于承载大门。

hickey A tool used for bending steel.
弯管器 用于折弯钢铁的工具。

high alumina cement To develop a concrete with an early strength that is much higher than ordinary Portland cement (OPC), aluminium oxide is used instead of clay. In the longer term, the cement has been found to be unstable so it is now rarely used.
高铝水泥 可使混凝土的初期强度比普通硅酸盐水泥（OPC）的大，其使用的是氧化铝而不是黏土。但从长远来看，此混凝土成分不稳定，所以现在很少使用。

high bay lighting A type of lighting system mainly used in industrial buildings that have a high floor to ceiling height. Usually comprises a series of high intensity discharge lamps or fluorescent fixtures.

工矿灯 一种照明系统，主要应用于较高天花板的工业建筑物中。通常包括一系列高强度气体放电灯或荧光照明设备。

high-calcium lime (fat lime, non-hydraulic lime, rich lime, white chalk lime, white lime) A lime that contains mostly calcium oxide or calcium hydroxide, and not over 5% magnesium oxide or hydroxide.
高钙灰（也称富石灰、非水石灰、优质石灰、白粉灰、白灰土） 主要成分为生石灰或氢氧化钙的石灰，且含有不超过5%的氧化镁或氢氧化物。

high-carbon steel A steel alloy containing between 0.6 and 2.0% carbon by weight; usually high-carbon steel alloys contain 0.6% ~ 1.0% carbon. The advantage of higher carbon is increased *ultimate tensile strength (UTS), however high-carbon steels have lower ductility in comparison with low or medium carbon steels. Higher carbon steels also have increased *hardness (resistance to surface indentation).
高碳钢 一种碳含量（按重量计算）在0.6%至2.0%之间的合金钢；高碳钢合金一般含0.6–1.0%碳。因碳含量高会增加极限抗拉强度（UTS），越高碳钢强度（抗表面压痕）越大，然而与中低碳钢相比，高碳钢延展性较低。

H

high-efficiency particulate air filter (HEPA filter) A high-efficiency air filter that removes at least 99.7% of the airborne particles that have a diameter of 0.3 microns (mm) or more.
高效微粒空气过滤器（HEPA 过滤器） 一种高效空气过滤器，可清除至少99.7%的粒径为0.3微米或更大的大气尘粒。

highlighting Emphasizing parts of decoration by making areas lighter in colour.
凸显 通过使某些区域在颜色上更明亮，从而起到装饰的效果。

high pressure hot water system (HPHW) A type of *central heating system that circulates the hot water through small bore pipes at high temperature and pressure, usually above 200℃ and 16 bar. Used in industrial and commercial buildings.
高压热水采暖系统（HPHW） 一种集中供暖系统，以高温及高压（通常为200摄氏度及压力16巴以上）循环小口径管道内热水。用于工业与商业建筑。

high pressure system A term commonly used to refer to a *high pressure hot water system or a *high velocity system.
高压系统 常指高压热水采暖系统或高速系统。

high rise (high rise building) Buildings that are greater than eight storeys are often said to be high rise buildings.
高层建筑（高层建筑物） 指楼层超过八层的建筑。

high spot A part of a surface or substrate that is higher than the general profile or surface.
高处 比一般外形或表面高的部分表面或基底。

high-strength concrete Concrete with strengths in excess of 40 kN/mm^2. Concretes with additives such as microsilica, superplasticisers, and other strengthening materials can be produced that have strengths in excess of 100 kN/mm^2.
高强度混凝土 强度超过40 kN/mm^{2}的混凝土。添加硅粉、减水剂及其他强化材料等添加剂的混凝土强度可超过100 kN/mm^2。

high tensile steel A steel alloy with a high *ultimate tensile strength (UTS), the tensile strength is so high that in comparison, high tensile steels may have up to ten times the tensile strength of wood, and more than twice that of mild steel. Normally high carbon steels have increased UTS, however, other special treatments can also impart this property.
高抗拉钢 一种有极限抗拉强度（UTS）的合金钢，抗拉强度相当高，相比之下，高抗拉钢的强度是木材的十倍，是低碳钢的两倍以上。一般来说，高碳钢有更强的极限抗拉强度（UTS），并且，其他特殊处理也会增加此性能。

high velocity system A type of air-conditioning system that delivers conditioned air to the space at high ve-

locity using small diameter ductwork.
高速系统 一种可通过小直径管道来高速传送已调节好温湿度空气的空调系统。

high voltage A cable that operates with a voltage that is greater than 600 volts.
高压 电缆运作电压大于600伏特。

highway A main public road, particularly one connecting large towns or cities. HIGHWAY ENGINEERING is the discipline that deals with the planning, design, construction, and maintenance of highways.
公路 指公共道路，特别是连接大城镇或城市的道路。公路工程则指的是关于公路的规划、设计、建造及维护。

hinge A pinned joint between two elements that enables one of the elements to pivot relative to the other. Used to hang doors to their frames. Many different types of hinge exist, including butt hinges, back-flap hinges, butterfly hinges, concealed hinges, continuous hinges, cleaning hinges, flush hinges, friction hinges, H-hinges, parliament hinges, strap hinges, and T-hinges.
铰链 两构件之间的铰接，其中一构件环绕枢轴做相对于另一构件的转动运动。一般用于把门悬挂于门框上。有很多不同类型的铰链，包括普通铰链、轻型铰链、蝶形铰链、暗式铰链、钢琴铰链、四连杆铰链、子母铰链、摩擦铰链、H形铰链、工字形铰链、条形铰链及T形铰链。

H

hinge bolt (dog bolt) A type of *security hinge where a bolt pin is fitted into the hinged side of the door.
铰接防盗螺栓 一种防盗铰链，销栓可嵌入门的铰链侧。

hinge-bound door A door in which the recess for the hinge has been cut too deep.
铰链错位门 安装铰链的凹口切得过深的门。

hip (hip ridge) The upper angle created by the intersection of the two sloping roofs. The rainwater flows away from the hip, down the two faces of the roof.
斜脊 由于两斜屋面交接处所形成的较高角。水流从斜脊往下流向屋顶两面。

hip capping The material, felt, or flashing that is placed over the *hip.
斜脊盖 覆盖在斜脊上的毡制品或挡水板。

hip hook (hip iron) A strip of metal fixed to the *hip rafter that extends under the lowest hip tile, providing support for the tile, and preventing it slipping from the roof. The metal is usually visible and may be angled, twisted, or rolled to provide an ornamental feature.
戗脊挂瓦钩 安装于四坡屋顶面坡椽的金属条，延伸至在最低的屋脊盖瓦底下，并支撑屋脊盖瓦，防止盖瓦从屋面滑落。该金属一般是可见的，且可能是成角度的、扭曲的或成卷的，具有装饰性。

hip knob A decorative point (finial) fixed on top of the ridge where it intersects with the *hip or *gable.
戗脊端饰 安装在斜脊或山形墙相交的屋脊上的装饰性尖点（尖顶饰）。

hipped end (hipped end roof, hip) The pitched triangular end of a hipped roof. A hipped end roof is different from a *gable ended roof, which would just be left with the two standard pitched roofs meeting the wall of the dwelling at the *ridge and *eaves, forming a triangular section of external wall.
四坡顶山墙端（四坡屋顶、斜脊） 四坡屋顶的人字形三角端。四坡屋顶不同于山墙端屋顶，四坡屋顶是两面标准斜屋顶与房屋的墙于屋脊及屋檐处相交，从而形成一个三角形的外墙。

hipped-gable roof A *hipped and gable end roof. The pitched roof has a hip at one end, that extends partway down the roof where it is cut short by a gable end, also called a **jerkin roof**.
人字屋顶 一个有斜脊及山墙端的屋顶。斜屋顶的每端有一个斜脊，部分延伸至屋顶底下山墙端中断的位置，也叫**“山墙尖呈斜坡的两坡式屋顶”**。

hipped roof (hip roof) A pitched roof with a *hipped end.
四坡屋顶 有一个四坡顶山墙端的斜屋顶。

hippodrome A building that was traditionally used for chariot racing and staging equestrian events, and in Victorian times was a name commonly associated with music halls.
竞技场 传统上用于举办战车赛或马术比赛的建筑；在英国维多利亚时代，常指音乐厅。

hip rafter (angle rafter, angle ridge) The rafter that is used to create the structure of the roof *hip, with *jack rafters resting on the central hip rafter.
四坡屋顶面坡椽(角椽) 用于构建屋顶的斜脊结构,使用短椽支撑中间的屋面坡椽。

hip rib A curved *hip rafter used to create domed roofs.
弯椽 用于构建穹顶的一弧形椽。

hip roll A turned length of wood with straightened sides and a V-cut in the under side of the timber to cover the hip joint. The rounded length of timber is referred to as a roll.
半圆脊帽 一边为直边的半圆木材,且木材底侧为V型切口,以覆盖斜脊接缝。圆木材的一边被称为"卷"。

hip tiles Curved, angular, *bonnet or V-profile tiles that are used to cover the *hip of the roof. The tiles are nailed, dowelled, or bedded in mortar, the lowest tile is supported and held in position by the hip iron.
斜脊盖瓦 弧形、有角的、帽状或V型瓦,用于覆盖斜脊。盖瓦的安装是用钉钉牢、用暗销钉合或嵌入灰浆。使用斜脊挂瓦钩支撑并固定最低的盖瓦。

hip truss A trussed rafter that is used to create the *hip of a roof.
斜脊桁架 用于构建屋面斜脊的桁架椽。

histogram A statistical bar graph to denote distribution of data. Vertical rectangles of different heights are drawn up the y-axis to represent corresponding frequencies.
柱形图 条形统计图,可表示数据分布情况。不同高度的垂直矩形,以y轴表示相应的频率。

hit and miss fencing Boarded fence with gaps placed between the panels.
木格栅围栏 面板之间有间隙的木板栅栏。

H-method A method used in finite element analysis to increase the accuracy of the solution by increasing the number of elements.
H方法 有限元分析中所采用的一种方法,可通过增加单元数来提高计算精确度。

hoarding A secure tall fence, often closely boarded, that is used to prevent the public gaining access to the construction site. Many of the fences have open areas or viewing windows that allow people to see the construction, while ensuring that access is prevented to keep the public safe. The hoarding should be particularly resistant to the passage of children. Safety notices should be displayed on the boundary of the site, and direction should be given to the entrance and site offices. Passage onto the site should be secured and controlled at all times.
临时围墙 一防盗高墙,通常是封闭式木板,用于防止公众进入到施工现场。大多数围墙设有开放区或观察窗,在确保防止进入及保证公众安全的同时可使人们看到施工。临时围墙应特别预防小孩子通过。在现场边界贴上安全告示,并指示入口及现场办公室方向。应始终保护及控制现场通道。

hob 1. An electric or gas device built into a kitchen worktop, and used for cooking. It will contain the controls and the burners or electric plates. Also known as a **cooktop**. 2. A shelf at the rear or side of a fireplace used to keep food warm.
1. 炉盘 安装于厨房橱柜操作面的电气或煤气装置,用于烹饪;含控制装置及炉灶或电加热板,也被称为"炉灶"。**2. 铁架** 位于壁炉后面或侧边的架子,用于保温食物。

hod A V-shaped trough attached to a pole that is used to carry bricks or mortar.
砖斗或灰浆桶 附属于柱上的一个V形槽,用于承载砖块或灰浆。

hod carrier (hodman) Bricklayer's labourer who traditionally carried a *hod of bricks and distributed the bricks around scaffolds and adjacent to the bays where the bricks were to be used.
瓦工力工(小工) 瓦工的劳工,通常搬着一砖斗并把砖块发放到脚手架周围或将要用砖区域附近。

hoe excavator A wheeled tracked excavator with its bucket operated by a two-part hydraulic arm that digs in a downward action. *See also* BACKHOE.
反铲挖掘机 轮履式挖掘机,挖掘机斗由一个两部分组成的液压臂操作,向下进行挖掘。另请参考

“反向挖土机”词条。

hog Bending upwards, something that is concave, e. g. an arch. A **hogging bending moment** is a bending moment with tension along the top and compression along the bottom; *See also* SAG.
拱起 向上弯起，凹形物，如拱形。拱起弯矩指的是顶部受拉而底部受压所产生的弯矩；另请参考“下垂”词条。

hoggin Fine aggregate of ballast used as infill material.
粗砂 用作填充料的碎石细骨料。

hogging A cambered surface, a beam that bends upwards towards its centre.
中拱 一拱形表面，一横梁朝其中心方向向上弯起。

hogsback A curved ridge tile.
拱形瓦 一弧形屋脊瓦。

hoist Equipment used for lifting building materials to the level needed.
起重机 用于把建筑材料提升至所需水平高度的设备。

hoisting Lifting or lowering building materials or equipment using a *crane or *hoist.
起重 使用吊车或起重机升降建筑材料或设备的操作。

hoistway *Lift shaft.
井道 电梯井。

holderbat A galvanized steel fixing, with a dovetail end that is built into a masonry joint, and a ring on the other end for securing pipes.
吊码喉箍 镀锌钢配件，一端为楔形榫头，可并入到砌砖缝，另一端则带一圆环用于固定管道。

holding-down bolt Long steel bolt that is set in concrete foundations to hold fast the *base plate of columns.
地脚螺栓 长的钢螺栓，安装于混凝土基座以固定柱子基板。

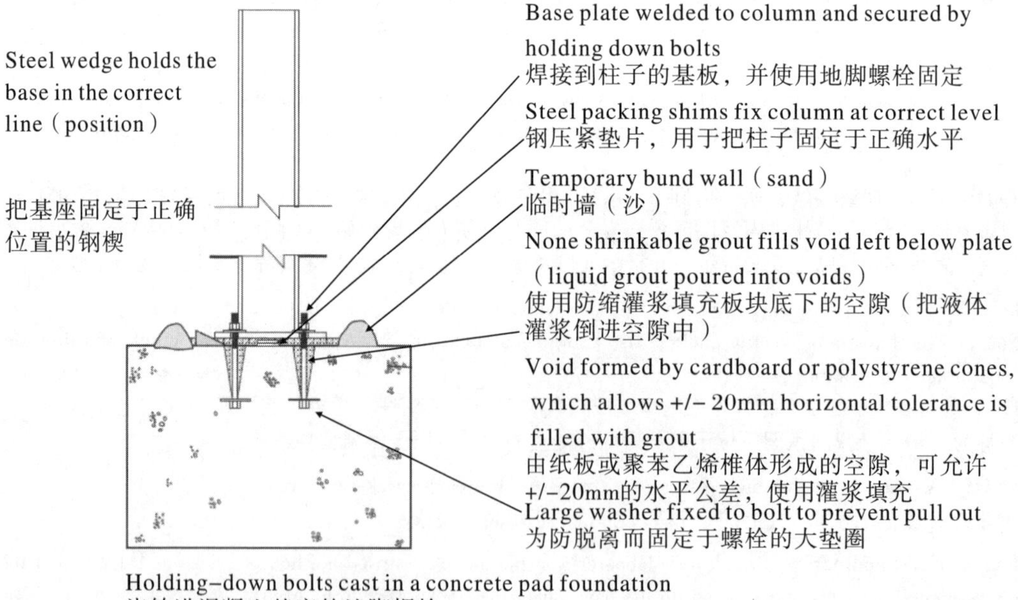

Steel wedge holds the base in the correct line (position)
把基座固定于正确位置的钢楔

Base plate welded to column and secured by holding down bolts
焊接到柱子的基板，并使用地脚螺栓固定

Steel packing shims fix column at correct level
钢压紧垫片，用于把柱子固定于正确水平

Temporary bund wall (sand)
临时墙（沙）

None shrinkable grout fills void left below plate (liquid grout poured into voids)
使用防缩灌浆填充板块底下的空隙（把液体灌浆倒进空隙中）

Void formed by cardboard or polystyrene cones, which allows +/- 20mm horizontal tolerance is filled with grout
由纸板或聚苯乙烯椎体形成的空隙，可允许+/-20mm的水平公差，使用灌浆填充

Large washer fixed to bolt to prevent pull out
为防脱离而固定于螺栓的大垫圈

Holding-down bolts cast in a concrete pad foundation
浇铸进混凝土基座的地脚螺栓

hole saw (annular bit, tubular saw) A tool used to cut circular holes through wood and plastics.
孔锯（环形钻头、开孔锯） 用在木材及塑料上切割圆孔的工具。

hollow A sunken or recessed surface.
空洞 一个凹陷表面。

hollow block A concrete or clay block with a cavity within the block or a cellular core.
空心块 砌块或多孔芯中间有一空洞的混凝土块或粘土块。

hollow brick Cellular brick, normally considered hollow when it has greater than 25% voids within it.
空心砖 多孔砖中心有超过25%是空洞，一般被认为是空心砖。

hollow clay block (US structural clay tile) Block with voids or void within it.
空心黏土砖（美国 空心黏土砖） 中心有空洞的砖块。

hollow clay tile (hollow pot) Tiles used for creating drains, vents, underfloor ducts, or other passages within the tile.
空心黏土瓦（空心块） 用于建造排水沟、通风孔、地板下管道或其他砖瓦内通道的瓦。

hollow concrete block A concrete block with a cellular core.
空心混凝土砌块 有多孔芯的混凝土砌块。

hollow core door A *door that has a hollow core.
空心门 有一中空心的门。

hollow floor A hollow pot floor or a suspended floor.
空心地板 空心楼面或悬垂楼板。

hollow partition A partition with a cellular core.
空心隔间 带多孔芯的隔间。

hollow roll A method of forming a water-resistant joint between two pieces of sheet metal, used on roofing or as a cladding material. The joint is usually used to joint the two sheets running down the slope of the roof. The edges of the sheet metal are lifted, positioned together, and rolled to form a cylinder.
立边咬合 在两块金属片之间形成一防水接缝的方法，用于屋面或用作覆盖材料。一般是用作连接沿着屋面坡度下延伸的两块金属片的接口。提起金属片的边缘，放置在一起然后滚制成一圆柱。

hollow wall A *cavity wall construction, traditionally constructed with an air gap within the cavity; it is now becoming more common to totally fill the cavity with insulation.
空心墙 夹心墙构造，空洞中有一气隙，现在一般都是使用保温层完全填满空心。

homogeneous Of the same consistency throughout.
同质的 始终是相同一致性的。

honeycomb bond A *half-brick wall built with a *stretcher bond, with gaps between the stretchers.
蜂窝式砌筑 顺砖式砌合的半砖墙，顺砖之间有空隙。

honeycomb door A *door that has a cellular core.
蜂窝芯门 有一蜂窝型芯的门。

honeycomb fire damper A honeycomb-shaped intumescent *fire damper. The intumescent material expands at a given temperature, preventing the passage of fire and smoke.
蜂巢式防火挡板 蜂巢形状的膨胀型防火挡板。膨胀型材料会在特定温度时扩张，防止火焰及烟雾通过。

honeycombing Gaps in concrete caused by loss of fine aggregate, due to failure to compact concrete properly or over-vibration, causing segregation of the concrete's aggregates. Where concrete is cast against formwork, gaps within the formwork may allow the fine aggregates to escape and leave voids between the coarse aggregate.
形成蜂窝状 由于未能适当碾压或过度振动混凝土，导致混凝土骨料分离、细骨料耗损，从而使混凝土中出现空隙。如果把混凝土浇筑于模架上，模架内的空隙会使细骨料逸出，并在粗骨料之间留下空洞。

honeycomb structure A structure consisting of a collection of six-sided cells, similar to that found in a bee hive.
蜂窝状结构 包含许多六边单元的结构，类似于在蜂巢箱所看见的结构。

honeycomb wall *See* HONEYCOMB BOND.
蜂窝状墙 参考“蜂窝式砌筑”词条。

hood (cooker hood) A mechanical device found above a cooker that is used to extract the vapour and smells from cooking.
抽油烟机 位于炉灶上方的机械装置，可用于抽走烹饪时的蒸气与气味。

hood mould (drip cap, hood moulding, label cap) An external moulding that projects from a wall above an opening to throw off rainwater.
雨篷 从一开孔上方墙上突出的一室外线条，可用于引流雨水。

hook bolt A bolt with a threaded straight end that is used to hold steel profile sheet roofing material, and with a J-shaped end that hooks around roof * purlins, securing the sheet material to the roof structure.
钩头螺栓 一端为螺纹端而另一端为 J 型端的螺栓；螺纹端可用于固定钢板的屋面材料，J 型端可钩住屋面的桁条周围，把屋面材料固定于屋面结构上。

hook bolt lock A lock built into a mortise in a door stile; the hook locks into the * jamb of the door when closed.
钩锁 内嵌于门竖梃内榫眼的锁；当关门时，钩锁会锁进门边梃。

hook curtailment A reinforcement bar with the end bent into a hook shape.
钢筋钩头 一端弯成挂钩型的钢筋。

Hooke's law A relationship that states the extension of a spring is directly proportional to the load that is applied to it. This applies to many materials such as steel up to its elastic limit. *See also* MODULUS OF ELASTICITY.
胡克定律 一条说明弹簧的伸展与其所承载的荷载成正比关系的公式。可应用于许多材料，如达到弹性极限的钢筋。另请参考“弹性模数”词条。

hook height The maximum height a hook on a crane can be raised; this is also the maximum lifting height of the crane.
吊钩提升高度 吊车上吊钩能被提升的最大高度，也就是吊车的最大提升高度。

hook intake (cable sleeve, turn-down) A duct or pipe through which cable can be added; the duct has a downturn end to prevent rainwater penetrating into the duct.
电缆套管 电缆可穿过的管道；管道有一向下翻的端，可防止雨水渗入管道。

hook time The time planned for each specialist group or subcontractor to have dedicated access to the crane.
吊车使用时间 专门安排给每个专业组或分包商使用吊车的时间。

hoop iron Reinforcement occasionally inserted in the bed joints of walls to add strength to the wall.
带钢 隔一段距离嵌入墙平缝以加固墙的钢筋。

hopper 1. An open container used for the storage of materials. 2. The wide square opening to a funnel.
1. 料斗 存放物料用的敞式容器。**2. 斗槽** 漏斗的方形开口。

hopper light (hopper window) A window that has a bottom-hung inward-opening casement.
内开下悬窗（下悬窗） 悬挂式底部并向室内方向打开的窗户。

horizon 1. A line where the land or sea meets the sky in the distance. 2. Circle on a celestial sphere, such as a plane through the centre of the earth. 3. A distinct layer or band of rock or soil.
1. 地平线 远方陆地或海洋与天空相接的一条线。**2. 地平圈** 天球上的圆环，如穿过地球中心的平面。**3. 岩层** 一个有分明层次的岩石或土壤层。

horizontal Parallel to the horizon.

地平线的 与地平线有关的。

horizontal shore A shore that is braced on a horizontal plane; it supports one vertical surface by bracing itself off another vertical surface. *See also* FLYING SHORE.
水平支撑 水平面支撑的支柱；通过倚靠在另一垂直面上来支撑一垂直面。另请参考“横撑”词条。

horn The extended jambs or head of a doorframe that provide additional strength to the mortise and tenon joints during transport. Once the frames are delivered to the site, the horns can be cut off when they are ready to be built in or inserted into the wall.
凸出保护 门框边梃或门楣的延伸，可在运输过程中为榫卯接头提供额外的强度。门框送到现场后便切割掉角状物，以便装入或嵌入墙内。

horsed mould A *running mould.
石膏线模具 现场制作石膏线的模具。

horsepower (hp) A unit of power equivalent to 550 foot-pounds per second or 746 watts.
马力（hp） 功率单位，550 英尺磅力每秒或 746 瓦特。

horseshoe tunnel A tunnel that has a cross-section in the shape of a horseshoe, typically used when tunnelling through rock.
马蹄形通道 横截面为马蹄形的通道，一般存在于挖岩石通道的情况。

hose A flexible tube used to convey liquids and gases.
软管 用来传送液体及气体的挠性管。

hose reel A flexible reel of tube used for delivering water to the place needed.
软管卷盘 用来把水输送到所需地方的挠性盘管。

hose union tap (hose cock) A tap with a threaded outlet to which can be fastened the threaded connector of a hose.
软管接头（软管龙头） 有一螺纹出口的水龙头，可把软管的螺纹接头固定在水龙头上。

hospital door A large flush door used in hospitals.
医院平面门 医院用的大平面门。

hospital sink A large sink, often stainless steel, used in hospitals.
医用水槽 医院用的大水槽，一般为不锈钢。

hospital window An inward-opening bottom-hung casement window.
医院窗户 向室内开启的下悬式窗户。

hot-air stripping Heated air that is used to soften paintwork, making it easier to remove. Heated paint will soften and often bubble off the surface as it expands and loses its bond.
热气脱漆 用热气软化油漆，可更容易去除油漆。加热的油漆会软化，并随着其膨胀而脱离表面，失去黏结性。

hot bonding compound A bituminous material that is heated to increase fluidity and used as adhesive between sheets of felt and flat roof substrates.
热黏结剂 一种沥青材料，可通过加热来增加流动性，并可用作毡板与平屋面基底之间的黏合剂。

hot coil *See* HEATING COIL.
热盘管 参考“加热盘管”词条。

hot cupboard A static or portable unit that is used to keep cooked food at a regulated temperature.
保温柜 一种固定或便携式装置，可用于把熟食保持在一固定温度。

hot isostatic pressing A technique for compacting powder at high temperature and constant pressure.
热等静压 在高温恒压条件下压制粉末的一项技术。

hot pressing A technique used to compact particles at high temperatures and varying pressures to form ceram-

ic materials.
热压制 在高温、不同压力条件下压制颗粒以形成陶瓷材料的一项技术。

hot water system A system that supplies hot water to a building.
热水系统 供应热水至一建筑的系统。

hot work Work that uses naked flames or glowing hot equipment. The work could be dangerous if chemicals or other flammable materials are in the vicinity. Hot work is normally associated with welding, and bitumen and asphalt work on roofs and roads.
热工作业 使用明火或发热加热设备的工作。如果附近有化学品或其他易燃物，那这工作可能是危险的。热工作业通常与屋面或道路的焊接、柏油沥青工作关联。

hot working A technique used to alter the properties of a metal by subjecting the material to different heat treatments, thus changing its chemistry.
热加工 通过使材料受到不同的热处理来改变金属性质，从而改变它的化学性质的技术。

hot-work permit A permit to undertake * hot work in a specified area at specified times. Where hot work is considered a risk it is necessary to control the operation and activities around the work to ensure the risks are reduced. To ensure the necessary risks have been assessed and effective measures have been taken, a hot-work permit is requested, and the manager or safety officer responsible will issue the permit once appropriate measures have been taken. The hot-work permit helps to avoid fires, explosions, or other dangerous events in areas of high risk, such as chemical or material storage sites.
热作许可证 在一个特定区域特定时间进行热工作业所需的许可证。如热作被认为是个风险，那么有必要控制热作周围的操作和活动，以确保降低风险。为保证对风险进行了必要的评估，并已采取有效措施，要求取得热作许可；一旦采取了适当的措施，管理者或所负责的安全员将发放热作许可证。热作许可证有助于避免高风险区如化学或材料储存场所发生火灾、爆炸或其他危险事件。

house A standard description for a small two- or three-storey dwelling, normally used as a home for individuals or families.
房屋 对一幢小的两或三层住宅的标准描述，通常用作个人或家庭的一个住所。

house breaking Demolition of houses and other small dwellings.
房屋拆除 拆除房屋及其他小住所的工作。

housed joint (let in) A rebate, recess, or mortise that lets in the end of another piece of timber such as a * tenon or the end of a timber. As the timber is held within another piece of timber, it is said to be housed, or let in.
闭口榫 连接另一块木材一端的卯口或榫眼，如木材的凸榫或尾端。由于木材是嵌入另一块木材之中，也称为闭口。

housed string The side of a staircase that encases the treads and sits against the wall, also known as a **close string**.
嵌入式楼梯梁 梯梁包住楼梯踏步并靠墙的楼梯一侧，也被称为暗步楼梯。

housewrapping Wrapping a house with a breather membrane, may also double up as the air barrier to ensure air tightness. In timber frame buildings the breather membrane is often fitted to individual panels before they arrive on-site, however, the membrane will still need to be lapped and sealed at the panel joins to make an airtight structure.
房建用透气膜 使用一块透气膜包裹房屋，也可折叠起来形成气密层以确保气密性。至于木构架建筑，透气膜一般在到达现场之前就包裹于单独板块上，仍需在面板接缝上重叠及密封透气膜以使结构密封。

housing 1. A protective covering. 2. A number of dwellings. 3. A * housed joint.
1. 防护层 2. 房屋群 若干住宅。**3. 镶嵌接头**

hovel A structure that has an open front and is used to house cattle.

茅棚 有一前开口的结构，通常供养殖牲畜。

Howe truss A roof truss that has both vertical and diagonal members, distinguished by the diagonal members at each side of the truss sloping up towards the centre so that they are in compression—the reverse of a *Pratt truss.
豪式桁架 有垂直及斜构件的屋架，特性是桁架每侧的斜构件向上中心位置倾斜，使其处于受压状态——跟普拉特桁架相反。

HPHW *See* HIGH PRESSURE HOT WATER SYSTEM.
HPHW 参考“高压热水系统”词条。

HSE *See* HEALTH AND SAFETY EXECUTIVE.
HSE 参考“健康与安全执行局”词条。

H-section A *universal section that has a cross-section in the shape of an ‘H’, such as an H-column or an H-pile.
H 形型材 一种通用型材，其横截面为 H 形，如 H 形柱或 H 形钢桩。

hub (US) The enlarged bell-shaped end of a pipe into which another pipe is joined.
承口（美国） 管道扩大的钟形端，可嵌入另一连接管。

hubless *See* UNSOCKETED PIPE.
无承接口管道 参考“无承接口管道”词条。

human error An unintentional mistake made by a human that has a negative consequence.
人为误差 由人造成的无心的但有负面效果的错误。

humidifier A mechanical device that is used to increase the moisture content (humidity) of air.
加湿器 用于增加空气水分含量（湿度）的机械装置。

humidifier fever A flu-like illness that results in a fever, chills, and a headache. Although the exact cause is unknown, it has been linked to the bacteria and fungi found in humidifiers.
湿热病 类似流感的疾病，会引起发热、发冷及头痛。尽管还不清楚确切起因，但与在加湿器中发现的细菌及真菌有关。

humidity A measure of the amount of water vapour in the air.
湿度 空气中水蒸气总量的测量。

humidity controller *See* HYGROSTAT.
湿度调节器 参考“恒湿器”词条。

humus Dark brown organic component of topsoil, which has been produced from decaying vegetable and animal matter.
腐殖土 深褐色有机表层土，由腐烂蔬菜及动物有机物质生成。

hung Any fixture or fitting that has been attached to a wall, door jamb, ceiling, or roof of a building. The fitting of *doors, opening *windows, *curtains, and *wallpaper are described as hung, once fitted.
悬挂件 悬挂于墙上、门边梃、天花板或建筑物屋顶的固件或配件。门、窗户、窗帘和墙纸上安装的配件在安装后被称为悬挂件。

hush latch A form of *striking plate that allows the latch bolt to close and open under gentle pressure without the need to withdraw the latch bolt.
平钩门闩 一种锁舌片，可使锁舌在轻压下便关闭或开启，无须拉开锁舌。

HVAC *See* HEATING (VENTILATION AND AIR CONDITIONING).
暖通空调 参考“暖通空调”词条。

hybrid Containing two different elements. A **hybrid car** is a vehicle that can run on both electricity and petrol.

混合的 包含两种不同元素的。"混合动力车"指的是可通过电力及燃油运行的交通工具。

hydrant A vertical pipe and valve that protrudes up from the ground. Enables water to be taken directly from the water main.
消防栓 从地面伸上来的一垂直管道与阀门，能直接从总水管取水。

hydrated lime (calcium hydroxide) A caustic substance produced by heating limestone, which is widely utilized in construction. Hydrated lime is one of the primary constituents for * mortars.
熟石灰（氢氧化钙） 通过加热石灰岩产生的一种腐蚀性物质，广泛应用于建筑业中。熟石灰是灰浆的一种基本成分。

hydration When used in connection with concrete, mortar, render, or plaster it refers to the chemical reaction of a substance that combines it with water.
水化作用 指的是结合混凝土、灰浆、外墙抹灰砂浆或内墙抹灰砂浆使用时，一种物质与水结合时发生的化学反应。

H

hydraulic Relating to the flow of fluids.
水力的 与液体流动相关的。

hydraulic breaker A mechanically operated demolition drill with the spring and return action driven by compressed air. Smaller units can be hand-held, although the vibration from the drill and return can lead to conditions such as white finger (poor circulation in the digits on the hand). Larger breakers are often fitted to the * backhoe of excavators.
液压破碎机 一种机械操作的拆卸用钻，带压缩空气驱动的弹簧与回位。较小的装置可为手提式的，但是钻的振动及回位会导致白指（手指血液循环不良）。较大的破碎器可安装在挖掘机的反向铲上。

hydraulic door closer A door closer with a hydraulic damper to prevent the door slamming shut. Equally, rams and pistons can be used to assist opening and closing of doors that are mechanically rather than manually operated.
液压闭门器 带液压缓冲器的闭门器，可阻止门砰地关上。同样的，可使用油缸及活塞帮助机械开启及关闭门而不是手动操作。

hydraulic equipment Mechanical equipment that is worked using fluid (hydraulic oil) which is pumped into rams to drive motors or move objects. Hydraulic rams usually have a drive and return valve making sure the units can be opened and closed.
液压设备 使用泵送进油缸的流体（液压油）驱动的机械设备，可驱动电机或移动物体。液压缸通常具有一个驱动器和返回阀以确保可打开和关闭装置。

hydraulic excavator An excavator (such as a * backhoe), with the shovel being operated by a hydraulic system.
液压挖掘机 一种挖掘机（如反向铲），其铁铲是通过液压系统操作的。

hydraulic fill Material that has been excavated by high-powered water jets and transported in water through a pipe, then placed and drained to form an embankment.
水力充填 通过大功率水射流挖掘的材料，并通过管道在水中运输，然后安放并排出以形成一堤坝。

hydraulic fracture A * fracture in rock produced by fluid pressure. An artificial hydraulic fracture can also be made into rock strata to increase the permeability, when mining for gas or oil.
水力裂缝 由液体压力产生的岩体裂缝。开采气体或石油时，为提高渗透率，可在岩层内人工制造水力裂缝。

hydraulic hammer A tool that can be placed on the * jib of an * excavator.
液压锤 一种可以放置在挖掘机悬臂上的工具。

hydraulic jump When the flow in an open channel suddenly changes from high to low velocity, there is a corresponding rise in the liquid surface. The rise causes eddies and turbulence to form with accompanying energy lost, with some of the initial kinetic energy being converted into potential energy.

水跃 明渠内流体突然从高速变低速时，液表上会出现相应的升高。上升引起涡流和湍流，伴随着能量损失，一些初始动能转换成势能。

hydraulic lift A *lift where the car is moved between floors by a hydraulic piston.
液压电梯 由液压活塞移动的楼层之间的电梯轿厢。

hydraulics The discipline that relates to fluid flow, particularly in rivers, water supply, and drainage networks, as well as irrigation schemes.
水力学 一类涉及液体流动，特别是在河流、供水、排水网路以及灌溉方案的学科。

hydraulic test A test to ensure water/air tightness—tested to a value just above the design pressure.
液压试验 一个确保水密/气密性的测试——测试值应高于设计压力。

hydrodynamic drag The resistance an object will experience when moved through a liquid.
流体动力阻力 当一物体穿过一液体时受到的阻力。

hydrodynamics The study of fluid dynamics that is concerned with water flow, such as over weirs, through pipes, and along channels.
流体力学 流体动力学的研究，与水流有关，如堰上、管道内或沿着渠道的水。

hydroelectric The use of flowing water to turn a turbine, which is attached to a generator to create electricity.
水力发电的 使用流水带动安装在发电机上的涡轮机来发电。

hydrophilic Having an affinity for water, the opposite is hydrophobic, a fear of water.
亲水的 对水有亲和力，反义词是“疏水性的”，怕水的。

hydropower The use of flowing water to produce electrical or mechanical power.
水力发电 使用流水来产生电或动力。

hydrostatic Relating to fluid at rest. **Hydrostatic pressure (hydrostatic head)** is the pressure exerted on an object from a column of fluid at atmospheric pressure and at a given height above the object. Also known as **gravitational pressure**.
流体静力学的 与静止状态的流体相关。液体静压力（静压头）指的是在大气压力及物体上方一定高度条件下，一液柱对该物体施加的压力。也称为**重力压力**。

hygrometer An instrument that is used to measure the amount of *relative humidity within a space.
湿度计 用于测量一空间内相对湿度总量的仪器。

hygrostat A device that is used to control *relative humidity.
恒湿器 用于控制相对湿度的装置。

hyperbolic paraboloid roof A roof formed from a sweeping curved shape that is made up from straight lines running between sides of a rectangle. The shapes resemble the curves that you get if you bend a piece of paper; in construction such shapes can be achieved by spanning cables, planks of wood (glue-laminated beams), or using reinforced concrete.
双曲抛物面屋顶 由一连续曲线形状形成的屋面，由沿着一矩形两面之间的直线组成。形状就像弯曲一张纸所得到的曲线；在施工中，可通过斜拉索、木板（胶合叠板梁）或使用钢筋混凝土来获得该形状。

I

I (intensity) 1. A measure of the average amount of power emitted by sound, solar radiation, or a light source. Measured in watts per square metre (W/m^2). *See also* LUMINOUS INTENSITY. 2. A measure of the brightness of an object. 3. The symbol used for * electric current.
1. 发射强度 可测量的由声音、太阳辐射或光源发射的能量的平均量。测量单位为瓦/每平方米（W/m^2）。另请参考“发光强度”词条。**2. 亮度** 一个物体亮度的测量。**3. 电流符号** 电流使用的符号。

I-beam *See* I-SECTION.
工字钢 参考“工字型型材”词条。

ice Frozen water; occurs when water cools to below 0℃ (32°F) at standard atmospheric pressure. As it freezes, water expands by 9% and can cause damage to roads creating potholes, uplift of shallow foundations, etc.
冰 冻结水，当水在标准大气压力条件下冷却至0℃（32°F）以下时会出现。水结冰时，体积会膨胀9%，可损坏道路，出现凹坑、浅基隆起等情况。

ICE *See* INSTITUTION OF CIVIL ENGINEERS.
ICE 参考“土木工程师学会”词条。

ignitability The ability of a material to ignite.
可燃性 一物质燃烧的能力。

ignition The process of lighting a material so that it begins to combust.
点燃 点燃一物质让它开始燃烧的过程。

ignition temperature The lowest temperature at which a material will ignite.
燃点 一物质会燃烧起来的最低温度。

illuminance (*E*) The * luminous flux density, or amount of light, incident on a surface. Measured in lux, where 1 lux = 1 lm/m^2. Also known as the illumination value or illumination level.
照度（*E*） 指的是射入一表面的光通量密度或光量。单位为“勒克斯”，1 勒克斯 = 1 lm/m^2。也可称为“照度值”或“照明水平”。

illuminated push A push button that * illuminates when operated.
光照按钮 一按钮，开启时会照明。

illumination Lighting an area or object, usually artificially.
照明 照亮一区域或物体，通常是人为的。

imbrex A curved roof tile laid over the joints between the tegula to provide a waterproof roof covering. Used in ancient Greek and Roman architecture. *See also* TEGULA.
槽瓦 铺放于沟瓦之间接缝上的弧形屋瓦，可防水。常用在古希腊及罗马式建筑上。另请参考“平瓦”词条。

imbricated Overlapped, layered, or woven in a regular pattern.
叠瓦 成规律模式的重叠、分层或交织。

Imhoff cone A graduated one-litre conical vessel that is used to determine the amount of settled solids in a given time. It gives an indication of the volume of solids that can be removed from wastewater by settling in sed-

imentation tanks, clarifiers, or ponds.
英霍夫式锥形管 一个有刻度的一公升锥形容器，可用于确定指定时间内的沉积固体数量。可指示从废水中移除沉淀于沉淀池、澄清池或池塘内固体的体积。

Imhoff tank A two-storey tank that receives and processes raw sewage. The upper part of the chamber allows sedimentation to take place. Solids slide down the inclined surface of the upper chamber and enter the lower chamber. Anaerobic digestion of the sludge then takes place in the lower chamber. The lower chamber is vented and requires the digested sludge to be emptied periodically.
英霍夫式沉淀池 一个两层的水槽，可接收并处理原污水。容器的上部分可用于沉淀。固体沿着上段水室的倾斜面滑下并进入下段水室。沉淀物于下段水室发生厌氧消化。排放下段水室并要求定期清空消化污泥。

immersion heater A thermostatically controlled electric element installed within a tank or cylinder that is used to heat a liquid. Commonly found in hot water cylinders.
浸没式加热器 一个安装在水槽或水箱内用于加热液体的恒温控制的电气元件。常见于热水箱中。

immiscible Solutions or materials that do not mix together—unmixable.
不融合的 不会混合在一起的溶液或材料——不相容的。

impact sound (impact noise) Sound that is generated by something impacting upon the structure of the building. Typical sources include footsteps, slammed doors and windows, noisy pipes, and vibrating machinery.
撞击声（撞击噪声） 由于某物撞到建筑结构而发出的声音。一般声源有脚步声、使劲关门窗声、有噪音的管道及机械的振动声。

impact strength The strength of a material or structure to withstand stock loading.
冲击强度 一材料或结构可承受负载的强度。

impact test A standardized test that is used to determine the brittleness of a material. *See* CHARPY TEST.
冲击试验 可用于确定一物质脆性的标准化测试。另请参考“简支梁冲击试验（恰贝试验）”词条。

impeller A rotor that transmits motion in a centrifugal or rotary pump, turbine, compressor, or fan.
叶轮 可在离心泵或旋转泵、涡轮、压缩机或风机中传递运动的转子。

imperial units Units of measurements that relate to foot, pound, and gallon. Superseded in the UK by the *SI units.
英制单位 计量单位，如英尺、英镑及加仑，英国才采用的，正被国际标准单位取代。

imperishable Does not decay.
不朽的 不会腐烂的。

impermeable Does not allow a liquid or a gas to pass through. *See also* PERMEABLE and IMPERVIOUS.
不可渗透的 不允许液体或气体穿过。另请参考“可渗透的”与“不能渗透的”词条。

impervious Does not allow a liquid to pass through. *See also* PERMEABLE AND IMPERMEABLE.
不能渗透的 不允许液体穿过。另请参考“可渗透的”与“不可渗透的”词条。

impingement filter (viscous impingement filter) A filter that removes particles by forcing them to strike a sticky filter medium.
撞击滤尘器（黏滞撞击滤尘器） 通过强制使微粒撞击一黏性滤材从而清除掉微粒的一种过滤器。

imposed load *See* LIVE LOAD.
外加负荷 参考“活荷载”词条。

improved nail (ringed shank nail) Fixing nail with raised rings around the shank to increase friction and improve its ability to hold itself within the wood.
加强钉（螺纹钉） 钉杆上有凸起环的钉子，从而增加摩擦并提高其固定于木材内的能力。

improvement line A position from which development will take place in the future, e.g. a point from which

buildings will be built and land developed.
建筑扩建线 将来要开发的位置，如建筑物的建造位置及土地的开发位置。

improvement notice An order issued by the *Health and Safety Executive under Health and Safety legislation which enforces organizations to take corrective action to avoid serious risk to health. The notice is served when Health and Safety law has been broken and states a period required for improvement.
整改通知 由健康与安全局根据健康与安全法发行的一则通知，强制机构采取纠正措施避免对健康带来严重威胁。如违反健康与安全法，将会发出此通知书，要求在一定时间内整改。

inactive leaf The less frequently used opening leaf on a double door. *See also* ACTIVE LEAF.
固定门扇 双扇门中不经常打开的其中一门扇。另请参考“活动门扇”词条。

incandescent lamp A type of lamp that produces light by passing an electric current through a thin tungsten wire (filament), which heats up. Inexpensive and easily dimmed but has low efficacy, short lifespan, and relatively high running costs.
白炽灯 一种灯的类型，电流通过电灯泡里的细钨丝（灯丝）使其发热而发光。便宜，容易变暗淡，但效能低、寿命短，运行成本相对高。

I

incentive A factor that is used to provide motivation to increase performance or do something.
诱因 用于给予动力以提高性能或做某事的一因素。

incentive scheme A financial reward system that provides extra money for specific and targeted performance.
奖励方案 为特定及针对性表现而给予额外金额的一经济奖励体系。

incineration (soil incineration) High temperatures (up to 1200°C) are used to volatilize and combust organic compounds in contaminated soils, particularly those contaminated with explosives and chlorinated hydrocarbons such as *PCBs. Auxiliary fuels are used to initiate and sustain combustion. Air pollution control systems are employed to remove gases produced. Other combustion residues usually require treatment/disposal. Such gases and combustion residues may be highly toxic, so effective air-pollution control systems are essential. Combustion residues may not be suitable for re-use and may still require disposal to landfill. Contaminated soils may be treated on-site using mobile plant or taken off-site to a static plant.
焚烧（土壤焚烧） 高温（高达1200℃）挥发及燃烧受污染土壤（特别是那些受到爆炸物及氯化烃类如多氯联苯污染的）中的有机化合物。使用辅助燃料开始并维持燃烧。采用空气污染控制系统来清除所产生的气体，并需处理/清理其他燃烧残渣。此类气体及燃烧残渣有可能是剧毒的，所以必须采用有效的空气污染控制系统。残渣燃烧有可能不适合循环使用，仍需要填埋处理。使用可移动式处理厂于现场处理受污染土壤或运至场外的固定处理厂进行处理。

incise To carve or cut into an object using a sharp implement.
切割 使用一锋利的工具切开或切断一物体。

inclement weather (unfavourable weather) Any type of weather, such as rain, snow, high winds, etc. that slows down or prevents work being undertaken on a building site.
恶劣天气（灾害性天气） 任何阻碍或延误工地工作的天气，如雨、雪、强风等。

inclinator (US) An inclined domestic stairlift.
椅式升降梯（美国） 一种家用的斜挂式座椅电梯。

inclined shore *See* RAKING SHORE.
斜撑 参考“斜撑”词条。

inclinometer *See* CLINOMETER.
倾斜计 参考“倾斜仪”词条。

inclusion 1. Something that is embedded or trapped into something else, e. g. a gas or liquid trapped within a rock stratum. 2. The relationship between two sets such that the second set is a subset of the first.
1. 融入 嵌入或陷入别的东西里面，如陷入岩层内的气体或液体。**2. 包含** 两组之间的关系，第二组是第一组的子集。

incombustible (non-combustible) Not able to be burnt.
不燃性的（不易燃的） 不可燃烧的。

Incorporated Engineer An engineer who is registered with the Engineering Council, UK, and has gained the post-nominal letters 'IEng'. According to the Engineering Council, UK, 'Incorporated Engineers maintain and manage applications of current and developing technology, and may undertake engineering design, development, manufacture, construction, and operation.' In doing so they carry out similar, but less conceptual and detailed work than a *Chartered Engineer.
注册工程师 已于英国工程委员会注册的工程师，并获得有'IEng'字样的勋衔。根据英国工程委员会要求，注册工程师需维持并掌握应用当前及未来发展的技术，并从事工程设计、开发、制造、施工与操作工作。在这种情况下，他们才可执行类似的，但相对特许工程师而言，少些概念上及精细的工作。

increaser A pipe that tapers from one diameter to another allowing pipes of different diameters to be joined together. The pipe is usually fixed with the pipe increasing in size along the direction of the flow.
异径接头（大小头） 从一直径逐渐变为另一直径并可接入该直径管道的锥形管道。沿着水流方向，通常使用较大尺寸的管道安装于该接头上。

I

indent A groove or recess that is cut in a material or component.
锯齿形缺口 在材料或部件上所切割的凹槽或凹处。

indented joint A joint that is formed from two matching indents.
锯齿状接缝 由两相配的锯齿形缺口形成的接缝。

indenting 1. The process of cutting an *indent into a material or component. 2. The grooves, tracks, or marks left in flooring from heavy loads and traffic.
1. 压痕加工 在材料或部件上切割锯齿形缺口的过程。**2. 压痕** 由于重载及通行而在地板上留下的凹槽、痕迹或标记。

indenture A written contract with two or more parties, typically used for an apprentice to serve a master.
契约 两方或多方之间签订的书面合同，常用于指学徒拜师的协议。

independent float Time before or after an activity that provides some flexibility over when a task can be started or finished, without affecting subsequent tasks. Tasks with float can be started as soon as a preceding task finishes or the task can be started later, within the specified float time, without affecting the start or finish date of any subsequent tasks.
独立时差 一行为发生前或后的时间，当一任务可以开始或完成时，可提供一些灵活性，不影响后续任务。有浮动时间的任务可在上一任务结束时立刻开始，或该任务在特定浮动时间内稍后开始，不影响任何后续任务的开始或完成日期。

independent scaffolding Tubular support system, with two sets of parallel *standards providing vertical support independent of the building, that allows operatives to work at heights. The support system is made of individual poles and clips that are fastened together to suit the structure that they are tied into. All of the vertical support is provided by the vertical poles in the scaffolding, horizontal support is provided by *braces and ties into the building. The ties are normally fixed into window openings or around columns.
双排脚手架 管状支撑系统，有两组并列支柱，可提供独立于建筑的立式支撑，允许操作人员在高处施工。此支撑系统由单独的杆及扣件组成，杆与扣件顺着要拉结的结构固定在一起。由脚手架的立杆提供所有的垂直支撑，而水平支撑则由固定于建筑上的支架与拉结提供。拉结点一般设于窗户开口或柱子周围。

indicating bolt A type of *privacy latch that when engaged indicates whether the door is locked or unlocked or if the room is occupied or vacant.
带指示门锁 一种安全门闩，用于指示门是否上锁或房间内是否有人。

indicator A device that shows the status of something. It may take the form of a meter, a gauge, or a light (indicator light).

指示器 一种用于指示某物状态的装置。可为仪表、计量器或指示灯（灯光指示器）。

indicator panel (annunciator) A large panel comprising a number of different indicators.
指示盘（信号器） 由许多不同指示器组成的一个大仪表板。

indigenous Originating or belonging to a particular region or country.
本土的 起源于或属于一特定地区或国家。

indirect Not direct, not in a straight line, or not immediate.
间接的 非直接的，或不在同一直线上，或非即时的。

indirect cold water system A cold water system where all of the appliances, except the cold water drinking outlets, are supplied with cold water indirectly from a cold water storage tank. All of the cold water drinking outlets are supplied with water directly from the mains.
间接冷水系统 除饮用冷水出口外，冷水系统的其他所有设备的冷水均由冷水储存箱间接供应。所有饮用冷水出口的水将由总管直接供应。

indirect hot water cylinder A cylinder where the water is heated indirectly by passing hot water from the boiler through a coil within the cylinder. The water from the boiler does not mix with the water within the cylinder. Also known as a CALORIFIER.
间接热水箱 通过让锅炉的热水流过线圈再而流至水箱的方式间接加热水。锅炉的热水并不会与水箱的水相混合。也称为“水加热器”。

indirect hot water system A hot water system where the water that is heated by the boiler is not the water that is drawn off at the taps. Instead the water that is heated by the boiler, or other heat source, passes through a coil inserted into the hot water cylinder, where it heats the cold water that is fed directly into the cylinder. This heated water is then drawn off when the hot water taps are activated.
间接热水系统 热水系统的水是通过锅炉加热，并非是水龙头流出的水。相反，通过锅炉或其他热源加热的水，流过安装于热水箱内的盘管，从而加热冷水，然后直接流入水箱。打开热水龙头时便会流出加热的水。

indirect lighting A method of lighting a space where the light has been reflected and diffused by a ceiling or wall, rather than falling on the space directly.
间接照明 通过反射灯光及天花或墙扩散的方法而不是直接照亮一空间。

indoor pool A swimming pool that is located indoors.
室内游泳池 位于室内的游泳池。

induced siphonage The removal of water that forms a seal in a trap due to suction; it is caused by the pressure generated from water flowing through other parts of the drainage system.
诱导虹吸作用 利用吸力作用通过一密封弯管移除水；虹吸是水流流过其他部分排水系统而产生压力所引起。

inductance (L) The property of an electric circuit or device that relates the electromotive force to the current flowing through it or near it.
电感（L） 一电路或装置的性能，电动势与所流过它或在它附近电流的相互作用关系。

induction 1. A progress of introducing information to someone at the start of something, e. g. a structured programme someone attends at the start of a new job. 2. Conclusion based on logic. 3. The production of electric or magnetic forces in a circuit by being in close proximity to (but not touching) an electric or magnetic field.
1. 入门培训 在开始的时候向他人介绍信息的过程，如某人在开展一份新工作时参加的一个结构化规划。**2. 归纳推理** 基于逻辑的推论。**3. 电磁感应** 电路接近（但不接触）电场或磁场时电力或磁力的产生。

induction unit A type of air-conditioning unit where high velocity air is injected through nozzles to induce the circulation of room air over a coil to which heating or cooling is applied. These tend to have relatively high fan

power and may result in a noise nuisance.
诱导器 一种空调机组，其高速气流由喷嘴喷出，并通过加热或冷却盘管引起室内空气流通。一般需要有相对较高的风机功率并可能会引起噪音。

industrialized building Systemized construction using prefabricated building modules that are manufactured off-site for quick assembly on-site. Off-site manufacture reduces the time spent on-site constructing the building. When there is a high degree of standardization, the economics and speed of factory building methods are considerably better than traditional bespoke methods.
工业化建筑 使用工地外制造但在现场快速安装的预制安装建筑模块的系统化建造。工地外制造可减少现场建筑施工所用的时间。如果标准化程度高，相对于传统的订制方法，经济及快速的工厂建造方法更加好。

inelastic Not stretchy or easily changed—not able to return back to its original shape after deformation; *See also* PLASTIC.
无弹性的 不可伸展的或不可轻易改变的——变形后不可回到原来形状，另请参考“可塑的”词条。

infilling 1. Units or panels of material placed between the structural frame to increase stiffness or provide weather protection. 2. Material placed within the cavity to improve fire resistance or thermal insulation.
1. 填充物 放置于结构框架中组件或材料，用于增加刚性或耐候性。**2. 空隙填料** 空隙中放置的材料，用于提高防火性或隔热。

I

infill wall Non-loadbearing walling units, brickwork, or blockwork positioned between the structural frame.
填充墙 位于结构框架之间的非承重墙、砌砖或砌块墙。

infiltration 1. The permeation of a liquid through a substance by filtration, such as rainwater entering into a soil's groundwater; *See also* RUN-OFF. 2. The process where air enters a building through cracks, gaps, and other unintentional openings in the building envelope. It is driven by the wind and stack effect.
1. 渗透 液体渗透过一物质，如雨水进入土壤地下水，另请参考“流掉”词条。**2. 漏风** 空气通过裂缝、空隙及建筑围墙等其他非计划内的开口而进入建筑。由风及烟囱效应引起的。

infinite (∞) Not measurable—having no limits.
无限的 不可测的——没有限制。

inflammable A substance/material that is highly susceptible to ignition, describing the ease of ignition. Inflammable and flammable have the same meaning, inflammable may be misinterpreted as meaning not flammable, as this could be confusing, it is often suggested that the word flammable is used.
易燃的 高度容易着火、易点燃的物质/材料。“inflammable”与“flammable”意思都是“易燃的”，“inflammable”可能会被误解为“不可燃的”，容易混淆，所以建议使用单词“flammable”。

inflated structure *See* AIR INFLATED STRUCTURE.
充气结构 参考“充气结构”词条。

inflection A change in curvature, e. g. from convex to concave.
弯曲 弧线的变化，如凸面体变成凹面体。

inflow The action of flowing in, for example, the point where fresh water flows into a lake.
流入 流入的作用，如淡水流入湖泊。

influent A fluid entering a system, for example, where a stream enters a lake, or where a stream loses its flow by recharging the groundwater, i. e. it does not have a *base flow.
流入的 液体进入一系统，如溪流进入一湖泊，或溪流补给地下水从而流失，即是溪流没有了基流。

infrared The portion of the invisible electromagnetic spectrum that consists of radiation with wavelengths between about 750 nm and 1 mm.
红外线 不可见电磁谱的部分，波长在 750 nm 与 1 mm 之间的辐射。

infrastructure The physical public systems, services, and facilities of a country that are necessary for society and economic activity. These include buildings, roads, bridges, and utilities (electricity, gas, water, sewers,

and telecommunications).
基础设施 一个国家的社会及经济活动所需的实体公共系统、服务与设施。包括建筑物、道路、桥梁与设施（电、汽、供水、下水道及通信）。

inglenook A seated area built into a recess next to a fireplace.
炉边座 靠近壁炉凹处的座位。

ingo (**ingoing** Scotland) A window or door * reveal.
门窗边框（苏格兰"ingoing"） 门窗侧。

ingo plate (Scotland) A * reveal lining.
门窗边框板（苏格兰） 门窗侧衬里。

inherent Something that exists as a permanent or existing feature, for example, * knots are an inherent defect in timber.
固有的 永久性存在的东西，如节疤是木材的固有缺陷。

initial ground levels The level of the natural ground before any construction operations have taken place.
原始地面标高 在任何施工开始前的天然地面标高。

I

initial rate of absorption The * absorption rate.
原吸收率 吸收速率。

initial set The very early stages of concrete maturity where bonds between the cementious materials have started to form; concrete should not be worked once initial bonds have started to form.
初凝 混凝土成熟度的初级阶段，凝结材料之间的黏结才刚开始形成；一旦初级黏结开始形成后，混凝土不应再进行浇筑。

initial surface absorption test (**ISAT**) A test to determine the * porosity of concrete as defined in BS 1881 part 5. The test employs a plastic cap, which is sealed to the concrete surface. The cap has a water area of 5000 mm^2 and a head of water of 200 mm. The absorption of water is measured by observing the movement of water in a connecting capillary tube over afixed time period.
初始表面吸水性测试（ISAT） 如 BS 1881 第 5 部分中定义的用于判断混凝土孔隙率的测试。该测试需用到一个塑料盖，密封于混凝土面。塑料盖的水域面积为 5 000 mm^2 以及一个 200mm 的水柱高度。通过观察于固定的时间段内在连接的毛细管内水的运动而测量水的吸收性。

injection A method of introducing a liquid to something under pressure, e. g. spraying fuel into an engine.
注入 在压力下把一液体引入某物的方法，如把燃料喷入引擎中。

injection damp course (**chemical injection damp course**) A damp-proof course that is formed by injecting a chemical into a wall at regular intervals under pressure. Various chemicals can be used to form the damp course including silicone resins, aluminium stearate, or methyl siliconate. Used where there is no existing damp-proof course (old properties), or if the existing damp-proof course is no longer functioning correctly.
注射法防潮层（化学剂注射防潮层） 在压力下把化学品注入墙内而形成的一防潮层。可使用各种化学品来形成防潮层，包括硅树脂、硬脂酸铝或聚甲基硅酮。所用的地方本来没有防潮层（旧建筑），或原有的防潮层已经不再起作用。

inlaid parquet Parquet * flooring that is set flush within a decorative border.
镶木地板 于装饰边范围内平齐安装的木地板。

inlay The process of decorating the surface of a material or component by inserting another material into prepared indentations. The result is a decorative flush finish.
镶嵌 通过把另一材料嵌入准备好的缺口来装饰一材料或组件表面的过程。装饰完后表面是齐平的。

inlet 1. A narrow indentation in a coastline or lake. 2. A narrow stretch of water between adjacent islands. 3. An opening through which a liquid or gas passes to enter another device.
1. 水湾 海岸线或湖泊狭窄进水口。**2. 海峡** 相邻岛屿之间的狭窄水域。**3. 入口** 液体或气体流过从而进入到另一设备的开口。

innings Land that has be reclaimed from the sea or other waterlogged areas.
围垦地 从海洋或其他水涝区开垦的土地。

innovation To develop something new and originally, rebuild or modify, to create something different from the old that represents a noteworthy development or change to the item or process.
创新 开发一些原创的新事物，重建或修改，把原来旧的创造成新的，可表现为一个显著的开发或对该项目或过程的改变。

input system A type of mechanical ventilation system that supplies fresh air to a building using a fan. In dwellings, if fresh air from the roof space is supplied via a small fan, it is known as **positive input ventilation**.
输入系统 一种机械通风系统，可使用风机把新鲜空气供应至建筑。在住宅处，如果屋顶空间的新鲜空气是通过小风机供应，那么又叫作“正压通风”。

insect screen (fly wire screen) A fine mesh screen used to prevent insects entering a building.
纱窗（防蚊蝇纱窗） 用于防止昆虫进入一建筑的精细筛网。

inset The process of placing a material or component within the boundary of another. For example, a kitchen sink will be inset within the kitchen worktop.
嵌入 把一材料或组件放入另一个范围内的过程。如把厨房洗涤槽嵌入厨房操作面。

inside glazing Term used to describe external *glazing that has been installed from the inside of the building. *See also* OUTSIDE GLAZING.
室内安装玻璃 该术语用于描述从建筑内部安装的外部玻璃。另请参考“室外安装玻璃”词条。

in-situ In its original position. Refers to components or elements of a building that are formed on-site in their final position, for example in-situ concrete.
原位 在其初始位置。指的是在现场其最终位置安装的建筑的组件或元件，如现场浇筑混凝土。

inspection The visual checking of work to ensure that operations have been carried out properly and established standards have been satisfied. Various professionals and authorities will be required to check the works as part of their job. Local authorities, clerks of works, architects, and the client's representative may all carry out visual inspections.
检查 肉眼检查以确保正确执行操作并按照既定标准。要求各种专业人士及权威机构来检查工作。地方当局、工程监督、建筑师以及业主代表均可进行肉眼检查。

inspection certificate Written record that the works have been checked and have achieved the specified standard. The Building Authority will issue such certificates following inspections of drains and other works.
检验证书 书面记录，证明已检查工程且工程符合规定的标准。在检查完排水沟及其他工程后，营建处会颁发的证书。

inspection chamber A chamber with a removable cover that enables access to be gained to an underground drain for inspection and maintenance. Also known as a **manhole**.
检查井 带活动盖的小房室，可进入到地下检查及维修下水道。也称为检修孔。

inspection cover The removable, usually flush-fitting cover over an *inspection chamber.
检查井盖 检查井上平齐安装的可移动盖。

inspection door A door or panel located in a wall, floor, or ceiling that can be opened or removed to enable access to be gained to an installation or services. Also known as an **access door**.
检查口 位于墙壁、地板或天花内的一扇门或一板块，可打开或移动以进入设备处。也可称为检修门。

inspection fitting (inspection eye) *See* ACCESS COVER.
检查口（检查孔） 参考“检修盖”词条。

inspection junction A short section of drainpipe with a removable cover that runs at a 45° angle from the main underground drain to the surface. Used to insert a drain rod. Also known as a **rodding point**.

立管检查口 一小段排水管，带一活动盖，在地下总排水道与地面之间成 45 度角。用于插入管道疏通棒。也称为**清扫口**。

inspection notice Notification given to the local authority to state that an area of works is complete or exposed, and is available for inspection. Certain building stages require the building inspector or nominated authority to inspect and certify that the works are satisfactory.
检查通知 发给地方当局的通知书，声明可接受检查的竣工区域。特定的分段建筑需建筑检查员或指定当局进行检查并证明工程符合要求。

inspector Person employed to check the quality of the work.
检查员 受聘去检查工程质量的人员。

instability The condition of being unstable or not in *equilibrium.
不稳定性 不稳定或不在平衡状态。

installation 1. The act of installing equipment. 2. A large building or facility, for example, a chemical installation.
1. 安装 安装设备这一行为。**2. 装置** 大型建筑或设施，如化学装置。

I

instantaneous hot water heater (single point heater) A device designed to produce hot water instantaneously only when it is required. Usually only supplies hot water to a single point. Can be gas-fired or electric.
即热式热水器（单点出热水器） 设计用于在需要时可即时供应热水的设备。一般只向一处提供热水。可用燃气或电气加热。

instant-start tube (rapid-start tube) A type of fluorescent lamp that can be switched on instantly. It incorporates a ballast that has a continuous input high enough to start an arc through the tube instantly.
瞬时启动荧光灯（快速启动荧光灯） 一种荧光灯，可立即启动。含有一镇流器，有足够高的连续电压以立即通过灯管启动电弧。

institution An important professional or public body, or organization; *See* INSTITUTION OF CIVIL ENGINEERING.
机构 一个重要的专业或公共机构或组织，另请参考“土木工程学会”词条。

Institution of Civil Engineering (ICE) An international membership organization, founded in 1818, which promotes and advances civil engineering around the world. The purpose is to qualify professionals engaged in civil engineering, exchange knowledge and Institution of Electrical Engineers best practice, and promote their contribution to society. There are around 80,000 members worldwide.
土木工程学会（ICE） 一国际会员组织，于 1818 年成立，该组织旨在推动及提升全世界的土木工程发展。目的是使从事土木工程的专业人员具备资格、交流知识及最佳实践并促使他们为社会做出贡献。目前世界各地大约有 80 000 名会员。

Institution of Electrical Engineers (IEE) Now called the Institution of Engineering and Technology, this is a leading professional society for engineering and technology, providing a global network to facilitate knowledge exchange and promote science, engineering, and technology.
电气工程师学会（IEE） 现在被称为工程技术学会，是工程与技术的领先专业团体，提供全球网络以方便知识交流并促进科学、设计与技术发展。

instructions to tenderers (US notice to bidders) Direction and guidance contained within the bills of quantities on how to include prices for materials, labour, and other items of work described.
投标人须知（美国投标人注意事项） 工程量清单中所含的关于如何给材料、劳力及其他所述工程项报价的指示及引导。

instrument 1. A piece of equipment or tool used to aid performance. 2. A legal document such as a statutory instrument.
1. 仪器或工具 有助于施工的设备或工具。**2. 文件** 一种法定文件。

insulate The process of reducing the rate of heat transfer, sound transmission, or the flow of electric current.

隔热/隔音/绝缘 降低传热、声音传播或电流流动率的过程。

insulator (electric) A non-metallic substance that has very low electrical conductivity at room temperature.
绝缘体（电气） 一非金属物质，在室温下导电性极低。

intact clay A clay with no visible fissures; *See also* FISSURED CLAY.
原状黏土 无可见裂缝的黏土；另请参考“裂隙黏土”词条。

intake An opening or structure through which fluid passes to enter a system, e. g. water intake to a treatment plant.
进口 液体流进一系统所通过的开口或结构，如进水口至污水净化厂。

intake unit (house service cutout, fuse link) A device that contains the service fuse and connects the incoming electric service cable to the cables that in turn connect to the electricity meter. Usually located within the electricity meter box.
保险装置（室内配线断路器、熔断连杆） 含有分户保险丝的装置，并把输入电力供电电线连接至依次连接电表的电线。一般位于电表箱内。

integral waterproofing The process of making concrete waterproof by adding the waterproofing component to the cement or the water.
整体防水 通过把防水组分添加到水泥或水中而制造混凝土防水层的过程。

integrity In *fire resistance, the length of time that a component will remain structurally intact in a fire without failing.
完好性 在火灾时，组件可保持结构的完整性而不受损的时间长度。

intelligent building (smart building) A building in which the control systems for the building, such as heating, lighting, ventilation, security systems, etc. are capable of automatically adapting to changing external conditions.
智能楼宇 带控制系统如供暖、照明、通风、安全系统等的建筑，可自动适应外部条件的变化。

intelligent fire detector (smart fire detector) A fire detector that monitors a range of parameters in order to respond more quickly to the presence of a fire.
智能火警探测器 为迅速对火灾作出反应而监控一系列参数的火警探测器。

intelligent vehicle highway system (IVHS) A system that allows the interaction of vehicles, highway, and people to be monitored. It can be used to improve safety, reduce wear, and improve transportation times.
智能车辆公路系统（IVHS） 可监控车辆、公路和行人交互的系统。可用于提高安全性、降低磨损并提高运输效率。

intensifier A device that increases the amount of something, e. g. the strength of a signal, the quality of an image, the pressure over the source pressure.
增强器 一种可以增加某物（如信号强度、图像质量及压力源的压力）总额的装置。

interactive Two-way responsive communication (for example, between user and computer).
交互式的 双向响应的通信（如用户与电脑两者之间）。

intercepting trap (interceptor) A *trap that prevents unwanted material entering a drain.
拦截器 一个可阻止多余物质进入排水道的分隔器。

interceptor sewer A large sewer system that is designed to direct dry weather flow to the treatment plant and discharge wet weather flow into a receiving river when the sewer capacity is exceeded.
污水截流渠 一种大型污水管道系统，设计用于把旱流污水引向污水处理厂；当降雨期流量超过截流管容量时排入河道。

interchange A major road junction where vehicles pass without stopping, by means of slip roads, bridges, and underpasses.
立体交叉道 主干道交叉处，车辆可走岔道、桥梁、下穿交叉道通过，不需停下。

intercom A device that allows communication between different parts of a building.
内部通话系统 可允许建筑内不同位置之间进行通讯的装置。

interface 1. The point at which two or more components of a building meet. 2. Where building operations come together, and different contractors or professionals have to coordinate and integrate their activities to ensure a joined-up service.
1. 交界 建筑的两个或更多组件相遇的点。**2. 施工协调** 建筑施工一块进行的地方，为保证设施相容，不同承包商或专业人员需协调及整合各方的施工。

interim certificate Document that certifies the value of the works, normally issued on a monthly basis.
中期付款通知书 证明工程价值的文件，通常是每个月发布。

interim payment Instalments paid to the contractor or partial payment of the contract sum for works performed. Such payments are usually made on a monthly basis with retention money held back to ensure continuation of work, and that any defects can be made good before full payment is made.
中期付款 向承包商支付的分期款项，或是对竣工工程的合同金额部分付款。这些款项一般是每月支付，并扣留保证金，以确保工程的持续进行，在支付全部款项之前，必须修复所有的缺陷。

interim valuation Calculation of the quantity of work performed and the value of that work so that an *interim payment can be made.
中期付款计算书 对所执行工程量及工程价值的估算，以便支付分期付款。

interior adhesive Substance used to stick fabric and building components together. The glue is of reasonable durability but not suitable for external use.
室内用胶粘剂 用于把织物与建筑部件粘贴在一起的物质。该胶粘剂应为较耐用的但不适合室外使用。

interlock 1. The joining of two components together such that they are interconnected to one another. 2. A mechanical device or arrangement of controls that prevents a device functioning unless other devices are functioning in a particular way. For example, a boiler interlock prevents the boiler from firing unless other devices (thermostats, programmers, time-switches, and TRVs) indicate that there is a demand for heat.
1. 联结 两个组件互相联结起来。**2. 连锁装置** 一个机械装置或是控制装置的布局，可阻止一设备运行，除非其他设备以一特定方式运行。如一个锅炉的连锁装置可防止锅炉点火，除非其他设备（恒温控制器、程序编制器、定时开关与温控阀）指示需要加热。

interlocking joint A joint where a projection in one component connects into a groove on another component.
榫接缝 一组件的突出部位嵌入另一组件凹槽所出现的接缝。

intermediate joist A common *joist that runs from one wall to another.
平顶搁栅 从一面墙延伸至另一面墙的普通托梁。

intermediate rafter A common *rafter that runs at right angles from the wall plate to the ridge. *Jack rafters, which do not run the full slope of the roof, are not included within this classification.
中间椽 从梁垫呈直角延伸至屋脊的普通椽木。短椽，由于不是沿着整个屋面斜坡延伸，所以不属于中间椽。

intermediate rail A horizontal *rail located between the top and bottom rail of a door.
中冒头 位于门顶部与底部冒头之间的水平冒头。

internal angle The internal corner of a room.
阴角 房间的内角。

internal diameter (ID, bore) The diameter of the inside of a pipe.
内径（ID，钻孔） 管道内直径。

internal door A door located inside a building.
内门 位于建筑内的门。

internal dormer A vertical *dormer window that does not project above the slope of the roof.
室内天窗 一立式屋顶窗，不会延伸至屋顶斜坡上面。

internal glazing Glazing that is located inside a building. *See also* EXTERNAL GLAZING.
室内玻璃 位于建筑内的玻璃，另请参考“外墙玻璃”词条。

internal hazard A fire *hazard that is located inside a building.
室内危害 建筑内的火灾。

internal leaf The internal leaf of a cavity wall. *See also* EXTERNAL LEAF.
内叶墙 夹心墙的内层。另请参考“外叶墙”词条。

internally reflected component (IRC) The light received on an internal surface that is reflected from the surfaces inside a room. *See also* EXTERNALLY REFLECTED COMPONENT AND SKY COMPONENT.
内部反射组分（IRC） 由房间内表面反射到一室内表面上的光。另请参考“外部反射组分”与“天空组分”词条。

internal pipework (internal plumbing) Pipework located inside a building.
室内管道工程（室内给排水） 位于建筑内的管道工程。

I

International Building Code (IBC) A model building code developed by the International Code Council (ICC) and used throughout the US. The code deals with regulations in regards to design and construction, structural stability, and health and safety.
国际建筑规范（IBC） 由国际规范委员会编制的一模型建筑规范，并在美国通用。此规范涉及关于设计与施工、结构稳定性、健康与安全的条例。

International Standards Performance requirement established and set by the *International Standards Organization (ISO). The ISO works with the national quality standards organizations, such as the British Standards Institute, the European Committee for Standardization (CEN), etc. to coordinate and set international standards.
国际标准 由国际标准组织（ISO）建立及制定的性能要求。ISO 与国家质量标准组织共事，如英国标准协会、欧洲标准委员会（CEN）等，共同协调并制定国际标准。

International Standards Organization (ISO) The organization that has the world's largest body of published standards. Based in Geneva, it coordinates international standards by working with the national standards organizations, enabling greater consensus between standards and operations.
国际标准组织（ISO） 该组织是拥有世界上最多已发行标准的组织。总部位于日内瓦，通过与国家标准组织共事而协调国际标准，使得标准与执行达成一致。

interpaver A paving brick that *interlocks with the other paving bricks and blocks.
连锁砖 与其他铺路砖及砌块互联的一种铺路砖。

intersection 1. The point where two lines meet one another. 2. The point where two roads meet one another.
1. 交叉点 两条线相遇的点。**2. 十字路口** 两条路相遇的路口。

interstitial condensation Condensation that occurs within and/or between the individual layers of the building envelope, when the temperature of some part of the building envelope equals or drops below the *dew point temperature. May occur on the surfaces of materials within a structure, particularly on the warm side of relatively vapour-resistance layers, within the material when the dew point and structural temperatures coincide throughout the material, or on more than one surface in a structure. This is because moisture may evaporate from one surface and recondense on a colder one.
缝隙冷凝 当部分建筑围护结构温度等于或低于露点温度时，建筑围护结构单层内部或两层之间出现的冷凝。有可能出现在一结构内的材料表面上，特别是防潮层相对暖和的一面；当整个材料结构温度与露点温度一致时，冷凝会发生在材料内部或结构内的至少一表面上。这是由于水分从一表面蒸发并重新凝结于另一较低温的表面。

interstitial level A level located halfway between two floors of a building that houses mechanical services,

such as air conditioning or mechanical ventilation equipment.
夹层楼 一建筑内两层楼中间的一层，用于放置机械设备，如空调或机械通风设备。

intertie A horizontal intermediate member used in framed construction to help strengthen and stiffen the vertical members.
水平系杆 用在框架结构上的一水平中间构件，用于加固及固定垂直构件。

intruder alarm system *See* **BURGLAR ALARM.**
入侵报警系统 参考“防盗报警器”词条。

intumescent coatings A special type of protective coating, applied as a paint, offering fire protection to structural steel components. Intumescent coatings are typically 1 or 2 mm in thickness without any noticeable visual effect.
发泡涂层 一种特殊的防护涂层，可用作涂漆，为结构钢组件提供防火性能。发泡涂层厚度一般是 1 至 2mm，肉眼一般看不出。

invar A non-ferrous alloy based on nickel and steel utilized for its low coefficient of thermal expansion.
殷钢 镍和钢基的有色合金，特点是热膨胀系数小。

I

inverted roof (protected membrane roofing, US inverted roof membrane assembly) A type of flat roof where the insulation is applied on top of the weatherproof covering. An earth-sheltered (or turfed) roof is an extreme example of such a roof.
倒置式屋面（薄膜防护层屋面，美国倒置式屋面薄膜装配） 一种平屋顶，其保温层安装在防水层上。掩土屋面（或草皮屋面）就是这种倒置式屋顶的一个极端例子。

invertor An electrical device that converts *direct current to *alternating current.
逆变器 一电气设备，可把直流电转换成交流电。

investment casting A type of casting process used to mould metal alloys; also known as **lost wax** and **precision casting**.
熔模铸造 浇铸金属合金采用的一种铸造法；可称为失蜡铸造和精密铸造。

invited bidder (US) *See* SELECTED TENDER.
特邀投标人（美国） 参考“选定投标人”词条。

iron fairy A small wheeled crane.
小型吊车 一小型轮式起重机。

ironmongery A term used to refer to metal window and door fixtures and fittings, such as handles, locks, hinges, catches, etc. Originally, these would have all been made from iron.
五金件 指的是门窗金属配件及固件，如把手、锁具、铰链、拉手等。起初，这些配件全是由铁制成的。

irregular paving Irregular sizes and shapes of a material that has been laid as *paving.
不规则铺路 把一种不规则尺寸与形状的材料当作铺路材料铺放。

irreversible Impossible to progress backwards, i. e. to its original state, shape, or place.
不可逆的 不可倒退的，即不可回到初始的状态、形状或位置。

irrigation The supply of water via channels and pipes, particularly to enable crops and plants to grow.
灌溉 通过水槽及管道供水，特别是为促进庄稼及植物生长。

I-section A rolled steel section with a cross-section in the shape of an ‘I’. *See also* UNIVERSAL SECTION.
工字型型材 横截面为“工”字形的轧制型钢。另请参考“通用型材”词条。

island 1. A piece of land surrounded by water. 2. A small area which is surrounded by something else, e. g. a traffic island.
1. 岛屿 被水围绕着的一片土地。**2. 岛** 被一些其他东西围绕着的一小区域，如交通岛。

iso- Prefix meaning of equal.
相同 表示“相同、相等”之意的前缀。

isobar A line on a weather map connecting points of equal atmospheric pressure.
等压线 天气图上大气压相等各点的连线。

isochromatic Of equal colour.
等色的 相同颜色的。

isochrone A line on a map that joins points of equal time.
等时线 地图上时间相同点的连线。

isoclinic line A line on a geological map joining points that have the same magnetic dip.
等磁倾线 地质图上磁倾角相等各点的连线。

isohel A line on a map that represents the same average number of hours of sunshine in a course of a year.
等日照线 地图上表示一年当中日照平均时数相等各点的连线。

isohyet A line on a map that represents the same average amount of rainfall in the course of a year.
等降雨量线 地图上表示一年当中平均降雨量相同各点的连线。

isolated ceiling A type of false ceiling that is suspended from the floor above on special hangers or clips that are designed to limit the amount of sound transmission. Used to improve sound insulation between floors.
隔音天花板 一种从上方楼板悬挂于特殊吊架或固件上的假天花板，设计用于限制音量传播，改善楼板之间的隔音效果。

isolated column A column that is located in a different position away from the majority of the other columns.
独立柱 位于不同位置的一根柱子，远离大多数其他柱子。

isolated footing A foundation, such as a *pad, that is not connected to other foundations.
独立基础 一种地基，如阶形基础，不与其他地基相连。

isolating membrane A separating barrier used in flat asphalt roofing to allow asphalt to expand and contract without being affected by the different thermal movement of the roof structure. Breaks and holes in the barrier allow a certain amount of adhesion of the asphalt to the roof. The resulting adhesion and friction ensure that the asphalt does not contract too much in cold weather nor slide off the roof structure.
隔离膜 沥青平屋面上使用的一隔离屏障，可使沥青在不受到屋面结构不同热位移影响条件下发生膨胀及收缩。屏障上的裂缝及洞可允许一定数量的沥青黏附到屋面。所产生的黏附与摩擦力可确保沥青不会在寒冷天气中过多收缩或脱离屋面结构。

isolating strip Length of expandable and compressible material that forms a break in the structure, allowing independent movement and preventing cracking.
隔离带 一段可膨胀及可压缩材料，可在结构上形成一断裂，允许独立位移和防止开裂。

isolating valve (stop valve) A mechanical device inserted within pipe lengths to enable the water supply to be completely closed off.
隔离阀（截止阀） 嵌入管道内的一机械装置，可完全关闭供水。

isolation Separate from something—remote.
隔离 与某些东西分隔——疏远。

isolator An electrical device that, when activated, ensures that a circuit cannot become live.
隔离器 一电气装置，启动时可确保电路断电。

isopleth A line on a map connecting points of equal value, e. g. a contour line.
等值线 地图上表示数值相等各点的连线，如等高线。

isoseismal A line connecting points of equal earthquake intensity.
等震线 地震强度相同各点的连线。

isostasy A state that denotes the equilibrium condition within the earth's crust.
地壳均衡 表示地壳内平衡状态。

isotach A line on a weather map connecting points of equal wind speed.
等风速线 天气图上表示风速相等各点的连线。

isotherm A line on a weather map connecting points of equal temperature.
等温线 天气图上温度相同各点的连线。

isthmus A narrow strip of land, surrounded by water, that connects two larger areas of land.
地峡 被水围绕的一个狭长地带，连接两块较大的陆地。

IStructE The Institution of Structural Engineers is a professional institution that has around 23, 000 members worldwide. It promotes professional standards in structural engineering and public safety within the *built environment. Chartered members of this institution gain post-nominal letters 'MIStructE'.
结构工程师学会 结构工程师学会是一个拥有世界上约23000名会员的专业机构。该协会推广建筑环境中关于结构工程及公共安全的专业标准。其特许会员获有“结构工程师”的勋衔。

jack A mechanical device for lifting or moving heavy objects by applying a force through a screw thread or hydraulic cylinder.
千斤顶 一种机械装置，通过对螺纹或液压缸施加压力来升起或移动重型物体。

jacket The insulated outer cover around a hot water cylinder.
护套 热水箱周围的一层隔热外罩。

jacket platform An offshore platform supported by a tubular steel welded framework located on driven pile foundations in the seabed.
导管架平台 由位于海床打入桩桩基上钢管焊接框架支撑的海上平台。

jack hammer (pneumatic drill, US rock drill) A hand-held pneumatic drill, used for breaking up and splitting hard density material, such as concrete, rocks, and pavements—typically used in highway works to break up the surface layers of a pavement.
手持式凿岩机（风钻、美国凿岩机） 手持式风钻，用于破裂及拆分坚硬的材料，如混凝土、岩石及道路——一般用在道路工程中，用于打碎道路的表层。

jack-leg cabin A small portable building with adjustable steel legs. Commonly used as accommodation on-site.
简易活动房 一种有可调节钢支架的小型移动式建筑。一般用作施工现场的住处。

jack pile (jacked pile) A short pile used in *underpinning, where the pile is driven into the ground by jacking against the structure.
千斤顶桩 基础托换中所用的短桩，通过用千斤顶支承结构从而把桩打入地面。

jack plane A hand-held plane used to shape wood and provide a smooth surface.
粗刨 用于塑形木材并使其表面平滑的手持式刨子。

jack rafter A short *rafter that spans from the *eaves to the *hip rafter in a hipped roof, or from the *valley rafter to the *ridge in a valley roof.

小椽 从屋檐跨越至四坡屋顶的屋顶面坡椽，或屋顶斜沟至其屋脊上的一短椽。

jack shore (back shore) Part of a raking shore that adds stability to the structure.
可调节支撑（顶撑） 斜撑的一部分，可使结构更稳固。

jamb The vertical members of a door or window which are adjacent to the wall; also used to describe the side/s of an opening in a wall.
门窗边梃 门窗的垂直构件，毗邻墙；也可指墙内开洞口的侧边。

jamb form Removable form used inside formwork to enable an opening for the window or door to be formed inside a concrete wall. A profile of a timber door or window is inserted on the inside face of the formwork so that when the concrete is poured a space in the concrete wall remains for the door or window. *See* BOXING OUT.
门窗洞口侧模 模板内使用的可拆除模板，可在混凝土墙内为门窗的建造预留洞口。把门窗木型材嵌入框架的内表面，从而在浇铸混凝土时可在混凝土墙内为门窗预留空间。另请参考“预留洞口”词条。

jamb lining Timber inserted on the face of the *jamb to improve the finish of the door or window.
边梃内衬 在边梃表面嵌入的木材，可改善门窗饰面。

Japanese saw A hand-held *saw that cuts through material on the pull rather than the push stroke.
日本手锯 手持式锯子，通过拉力而不是推力来切割材料。

jaw breaker A machine that breaks rock.
颚式碎石机 一种破碎岩石的机械。

JCB A proprietary name for a *backhoe excavator.
杰西博 “反铲挖掘机”的专有名称。

jemmy (crowbar, pry bar) A short bar about 400 mm long used to prise or lever two objects apart.
撬棍（铁撬、撬杆） 一根约长400mm的短杆，用于把两个物体撬开。

jerry builder A builder who produced poor-quality buildings and cut corners during the First World War were given this name. It is still associated with poor workmanship.
偷工减料的施工员 在第一次世界大战期间为建造质量差的建筑且走捷径的施工员取的名字，现在用于指工程质量低劣。

Jersey barrier (Jersey wall) A 3-ft high reinforced concrete *barrier used on highways to separate lanes. It has a distinct shape that tapers outward towards the bottom, and is designed to minimize damage and direct vehicles back into the direction of flow of traffic in the event of a collision.
新泽西护栏 公路上用于隔开车道的高3英尺的钢筋混凝土护栏。其形状独特，朝底部向外形成椎体，设计用于降低损坏，并在发生碰撞时导引车辆弹回行驶方向。

jet A thin stream of liquid or gas which has been forced at high velocity out of a vessel or nozzle.
喷射 细流液体或气体在高速条件下从一导管或喷嘴中喷射出来。

jet freezing Method of freezing pipes and ground to temporarily halt the flow of water.
喷射式冻结 为暂时停止水流而冷冻管道及地面的方法。

jetty 1. Bank or structure that projects into deep water enabling people to access boats. 2. A projection from a building beyond the face of the main wall.
1. 突堤式码头 延伸至深水中的堤岸或结构，方便人上船。**2.（建筑物的）突出部分** 从建筑物承重墙伸出来的突出物。

jib A crane arm that is used to lift and move objects from one place to another.
起重臂 起重机的吊臂，可把物体从一地方提起并移动至另一地方。

jib door (gib door) A door that is concealed to match the surrounding wall.
隐门（与墙面齐平门） 暗式门，与周围墙面相配。

jig 1. A clamp made to a set profile so that wood can be glued or moulded to a desired shape. 2. A template made so that material can be repeatedly marked or cut to the same shape.
1. 夹具 固定式夹具，以便木材被胶水粘住或塑造成预定的形状。**2. 模具** 可重复标记材料或把材料切割成相同形状的一个模板。

jig saw A power saw with a thin protruding blade used for cutting curves in timber.
曲线锯 一动力锯，露出一片薄的刀片，用于切割出木材上的弧形。

Jim Crow A hand-operated tool, used for bending rails. The tool is shaped in a 'U' or a 'V' to grip the rails, which is connected to a thread rod to provide leverage. The grip can be mounted on rollers to enable continuous bending to be undertaken.
弯轨器 一手动操作的工具，用于弯曲导轨。工具为"U"或"V"形以便夹紧导轨，并连接螺纹杆以提供杠杆作用。手柄安装在轧辊上以进行连续弯曲。

jobber (handyman, builder's mate) A semi-skilled or experienced operative who can undertake most remedial and repair work on a building project, such as plastering, bricklaying, pointing, joinery, plumbing, and minor electrical works.
散工（杂务工、施工员助手） 一名半熟练或熟练的技工，可从事建筑项目上的大多数修复及修理工作，如抹灰、砌砖、勾缝、细木作、给排水与一些简单的电气工程。

J

job site The building or construction site.
工地 建筑或施工现场。

job specification Description of the building project, which might be accompanied by material, labour, and plant required.
工作规范 建筑项目的描述，包括所需材料、劳动力与设备。

joggle Up and down shape produced by cutting trapezoidal shapes next to each other.
折曲 通过切割互相挨着的梯形而形成一上一下形状。

joggle joint (stop end and key) Joint formed at the end of a concrete pour to provide a key for the next pour. A piece of trapezoidal shaped timber is fixed to the *formwork stop end so that a rebate will be formed in the concrete face when the concrete is poured against it.
混凝土企口接缝 在混凝土浇筑块一端上所形成的企口，可为下一个浇筑块提供榫头。在模板封端安装一片梯形木材，浇筑混凝土时便可在混凝土面板上形成一个槽口。

joiner A skilled tradesperson who works with wood, traditionally at a bench; however, the term is now generally applied to all skilled operatives who work with timber on building projects.
细木工 传统上是指在工作台上从事木工作业的一名技术工人。然而，现在一般指的是所有从事建筑工程上木材工作的技术操作人员。

joiner's gauge A tool used for scribing straight lines in timber.
画线规 用于在木材上划直线的工具。

joiner's hammer (Warrington hammer) A hammer with a double head—one with a rounded face and the other with a wedge-shaped head (cross peen head).
木工锤（尖头锤） 一个双头锤子——一头为圆面，另一头为楔形头（横头锤锤头）。

joiner's labourer (carpenter's mate) Operative with general knowledge of woodworking equipment and tools such that they can assist with *joinery tasks.
木匠小工（木匠助手） 掌握木工设备及工具常识的技工，可协助细木工工作。

joinery (finish carpentry) Working with timber for furniture, buildings, and structures, and includes assembling, jointing, fitting, and finishing timber.
细木工（木工修整） 家具、建筑与结构用木材的工作，包括装配、接合、安装与装饰木材。

joint The junction between two components or elements.
接缝 两组件或元件之间的接合。

joint backing *See* BACKUP STRIP.
接缝背衬 参考"背衬条"词条。

joint bolt A bolt used to join two components together.
接头螺栓 用于把两个组件连接在一块的螺栓。

joint cement (joint filler, joint finish) A plaster-like material that is used in plasterboard dry lining to make joints and cover over nail or screw heads.
填缝水泥（接缝填料、填缝涂料） 石膏状的材料，用在石膏板干内衬中以形成接缝，并遮住钉子或螺钉头。

joint cover Strip of wood, bead, or wood roll used to hide a gap.
接缝盖条 用于掩盖空隙的木条、饰条或木卷材。

jointing Filling and pointing masonry joints in one operation.
填缝 砌石时填缝和勾缝一并完成。

jointing compound Malleable putty-like substance that is smoothed around threaded pipework connections to make a water- or gas-tight seal. The seal is formed as one pipe is screwed or located in another, and the substance fills the threads and gaps. Also used to seal other plumbing equipment and appliances. Used in place of *jointing tape. *See* JOINT CEMENT.
填缝剂 抹于螺纹管道连接处周围的韧性油灰状物质，可防水或气密的。当一管道用丝扣或其他方式连接于另一管道时需加以密封，用该油灰状物质填充螺纹及空隙；还可用于密封其他的给排水设备与装置。可代替"生料带"，另请参考"填缝水泥"词条。

jointing fluid Liquid used when making a solvent-welded joint.
粘接胶 胶粘接缝时所用的液体。

jointing mortar Grout used for sealing and finishing tiles.
填缝灰浆 用于密封及涂饰瓷砖的薄浆。

jointing strip A strip of material used to join two materials.
连接带 用于连接两种材料的条状材料。

jointing tape 1. Thin PTFE (polytetrafluoroethylene) tape used around threaded pipework connections to make the joint watertight. Used in place of *jointing compound. 2. A multi-purpose tape used in plasterboard dry-lining to join boards, reinforce the joint between boards, patch boards, and provide protection to the corners of boards.
1. 生料带 螺纹管道连接处周围的薄聚四氟乙烯（PTFE）胶带，可使接缝防水。可代替"填缝剂"。**2. 填缝带** 用在石膏板干内衬中的一种多功能胶带，可以接合板块、加固板块之间的连接、修补板块并为板块边角提供保护层。

jointing tool 1. A tool used to join two components. 2. A hand-held tool used to produce a *bucket-handle joint in brickwork. 3. The collective name given to a wide range of hand-held tools used for jointing plasterboard dry-lining and brickwork, such as jointing knives, edging trowels, taping knives, and so on.
1. 连接工具 用于接合两组件的工具。**2. 勾缝工具** 在砖缝上做成一半月形接缝的手持式工具。**3. 填缝工具** 用于接合石膏板干内衬及砌砖的范围广泛的手持式工具的总称，如填缝刀、修边泥刀、镶边刀等。

jointless flooring A flooring laid as one continuous surface, rather than a number of individual components that are joined together.
无缝地板 铺设的地板为一个连续表面，而不是把若干单独的组件拼凑在一块。

joint reinforcement 1. Steel reinforcement used in movement and construction joints. 2. Steel embedded into horizontal brickwork beds to provide extra tensile strength.
1. 接缝配筋 变形缝及施工缝内使用的钢筋。**2. 拉结筋** 为取得额外的抗拉强度而在水平砌砖处埋入的钢铁。

joint ring (rubber ring, sealing ring) Ring of rubber that slides onto the end of clay, plastic, and steel pipes which provides a seal when one pipe is pushed inside another. Push-fit pipes can be sealed with rings of rubber with O profiles ('O' rings), D profiles ('D' rings), and rings that fit on the end of the spigot (lip rings).
垫圈（橡胶圈、密封环） 放在黏土、塑料及钢管末端上的橡胶圈，当把一管道推入另一管道时可密封接口。可使用O与D型橡胶环密封推入接合管道，橡胶环安装于插口末端上（唇形环）。

joint runner (pouring rope) Asbestos rope or similar fire resisting rope placed around the outside of a cast iron pipe to act as a guide when forming a lead-caulked joint.
填缝油麻丝 用于铸铁管外周的石棉绳或类似的耐火绳，在做灌铅嵌缝时可起导引作用。

joint tape Adhesive tape, usually 50 mm wide to cover the joints between plaster boards so that they can be easily covered with a light plastered finish.
接缝胶带 胶带，通常是50mm宽，用于盖住石膏板之间的接缝，以便使用轻质石灰饰面覆盖石膏板。

joist A beam that supports the floor.
地龙骨/地梁 支撑地板的骨架。

J

joist anchor A tie that fixes the wall to the beam to provide lateral support to the walls, restraining the walls, and preventing bowing.
墙锚固 为向墙提供横向支承、约束墙并防止弯曲而把墙固定于横梁上用的连接件。

joist hanger Steel bracket fixed to the internal skin of the wall to fix and secure the floor *joists. The brackets can be fitted with nail holes, which if used, tie the wall to the floor joists, ensuring the walls have greater lateral stability.
托梁吊件 安装于墙内壁的钢支架，用于安装与固定地板托梁。可使用钉孔安装支架，如使用钉孔，可把墙联系于地板托梁，确保墙板有较大的横向稳定性。

joist trimmer (grinder bracket) A special *joist hanger that allows one joist to be fitted at right angles to another joist. The bracket allows one joist to carry another. It is used to form the structural opening for a stairway in timber upper floors.
托梁支撑（支架） 一特殊托梁吊件，用于把一托梁以直角安装到另一托梁上。该支架可使托梁去支承另一托梁。可用于为上层木楼板楼梯构成一结构开口。

joule (J) The international unit of energy.
焦耳（J） 表示能量的国际单位。

journeyman A craftsperson.
熟练工人 工匠。

judas A peep hole in a door that allows the occupants to see those who are calling on them.
窥视孔 门上的一窥视孔，居住者可通过其看到拜访的人。

jumbo brick (US) A brick that has larger dimensions than standard bricks.
大型砖（美国） 尺寸比标准砖大的砖块。

jump form Concrete formwork where panels are struck and removed from the bottom, where concrete has previously been cast and lifted to the top, ready for the next pour.
爬升模板 之前已浇筑混凝土的混凝土模板的板块脱离墙面并从底部移出，然后提升至顶部，准备用于下一次浇筑。

jumping jack A piece of plant used to consolidate ground.
夯土机 用于夯实地面的一件设备。

junction 1. The point where two components or elements join together. 2. The intersection of two or more roads.
1. 接合点 两组件或元件连接在一块的点。**2. 十字路口** 两条或多条道路的交集。

junction box A box that contains the connections between various electrical circuits.
接线盒 一个容纳各种电路连接的盒子。

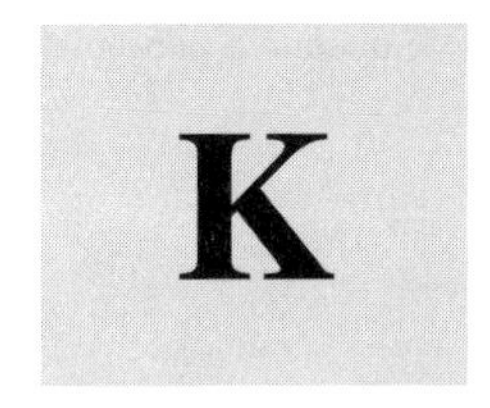

keel moulding A curved decorative moulding shaped like the keel of a boat.
弧形线条 形状像船的龙骨的一个弯曲装饰线条。

keep The main tower of a castle, often used as a dungeon.
城堡主楼 城堡的主塔，常用作地牢。

keeping the gauge (keeping to gauge) A term used in bricklaying where the thickness of the bed joints is maintained over a set area.
砌砖保持齐平 砌砖用术语，水平灰缝厚度保持在一固定范围。

keeping the perpends (keeping the perps) A term used in bricklaying where the perpends are maintained in a vertical line.
砌砖竖缝保持竖直 砌砖用术语，竖缝应保持在一直线上。

Kelly (Kelly bar) A hollow bar above a drilling rig used to drive the drilling stem.
伸缩钻杆 钻机上用于驱动钻柱的一空心杆。

Kennedy's critical velocity The flow of a fluid in an open channel at which it will neither pick up nor deposit silt.
肯尼迪临界速度 明渠内液体以既不冲走也不沉淀粉土的速度流动。

kentledge Heavy material (e. g. large blocks, scrap metal, water tanks, etc) used to provide weight/stability. Applications include a counterweight on tower cranes, *ballast on a ship, and a reaction point for a jack or sheet pile loading test.
压载物 用于增加重量/稳定性的重物料（如大型砌块、废金属、水箱等）。可用于塔式起重机上提供平衡力、压舱用、千斤顶的支点或钢板桩荷载测试。

kerb Raised concrete edging between a road and the pavement.
路缘石 公路与人行道之间凸起的混凝土边缘。

kerf 1. The groove left in a material that is being cut by a saw. 2. The width of the groove caused by cutting with a saw.
1. 锯口 锯切割材料留下的槽。**2. 锯缝** 锯切割时形成的槽缝。

kerfed beam A beam that has been curved by *kerfing.
刨槽折弯梁 通过槽折来弯曲的横梁。

kerfing The process of curving a material by applying a series of *kerfs.
刨槽折弯 通过运用一系列锯口来弯曲材料的过程。

kerosene A colourless flammable oil, refined from crude oil, used mainly for heating, cooking, and lighting.
煤油 一种无色易燃油，从原油中精炼，用于加热、烹饪及照明。

key 1. A device, usually metal, that is inserted into a lock to open or close it. 2. A written explanation of the various abbreviations and graphical symbols used on a drawing. 3. A rough surface that provides a bond for another material, such as paint, plaster, or render.

1. 钥匙 一种设备，通常是金属的，用于插入一锁具开启或关闭锁。**2. 图例** 用在图纸上的对各种缩写词及图解符号的书面说明。**3. 毛面** 可黏合另一种材料（如涂漆、石膏或灰泥）的一粗糙表面。

key changes (key combinations, differs) The total number of keys that can be made to operate a particular type of lock.
密匙量 可用于操作一套特定锁具的钥匙总数。

key drop *See* KEYHOLE COVER.
钥匙孔盖片 参考“锁眼盖”词条。

keyed 1. Held in position using a key. 2. Making a surface rough enough for bonding. *See* KEY.
1. 锁着的 使用钥匙固定其位置。**2. 毛化处理** 使表面足够粗糙以便黏合，另请参考“毛面”词条。

keyed beam A notched lap jointed beam.
榫插梁 一锯齿状的搭接梁。

keyed construction joint *See* JOGGLE JOINT.
企口施工缝 参考“企口接缝”词条。

keyed joint 1. A notched joint between two materials or components. 2. A joint in brick or blockwork that has been raked out to provide a key for wet plaster.
1. 齿槽连接 两种材料或组件之间的锯齿状连接。**2. 拉毛面** 砖块或砌块中的刮缝，为湿石膏提供毛面。

K

keyed mortise and tenon *See* TUSK TENON.
齿槽榫卯 参考“多齿榫”词条。

key event An important or critical point in a plan. *See also* MILESTONE.
关键节点 一个计划中的一重要或关键点。另请参考“里程碑”词条。

keyhole cover (key drop) A flap, usually round, that is attached to the escutcheon and covers the keyhole on the inside and/or outside of a lock.
锁眼盖（钥匙孔盖片） 一活板，通常为圆形，安装于锁眼盖上，用于覆盖锁具内侧及/或外侧的锁眼。

keying-in The process of joining a new masonry wall to an existing wall.
新旧墙搭接 把一面新的砌石墙连接到一现有墙上的过程。

keying mix A mix applied to a surface to enable another material to bond to it.
糙面灰 应用于表面使其可黏合另一材料的混合物。

key plan A plan showing the location of the development.
总平面图 显示开发区位置的一平面图。

key plate *See* ESCUTCHEON.
钥匙孔盖 参考“锁眼盖”词条。

key saw (compass saw) A hand-held saw, with a small blade that tapers to a point, used for cutting tight curves. The narrow blade can be used to cut out holes in the centre of a piece of timber. A large hole is first drilled into the timber. The saw can then be inserted to cut away from the hole.
鸡尾锯 一手持式锯，带一逐渐缩窄成细尖的锯片，用于切割小半径曲线。狭窄锯片可在木材中心内切割出孔，首先在木材上钻出一个大的孔，然后再把锯插入孔内进行切割。

key schedule A list of all of the keys and their related rooms, cupboards, and other lockable units. The list is organized into key suites and associated master keys.
钥匙一览表 一个关于所有钥匙及其对应房间、橱柜与其他可锁装置的列表。该列表被组织成成套钥匙及相关的万能钥匙。

keystone The central wedge-shaped stone at the top of a masonry arch.
拱心石 砌块拱门顶部的中心楔形石块。

key suite A set of related keys listed on a key schedule. The suite of keys may be grouped into floors of a building, particular offices, or rooms.
成套钥匙 钥匙一览表上列举的一套关联钥匙。成套钥匙可按建筑楼层、特定办公室或房间钥匙分组。

keyway miller A powered hand-held mechanical device for *chasing out grooves and rebates in brick and block walls so that service ducts can be set in the walls.
墙体开槽机 一种电动手持式机械装置，用于在砖块或砌块墙上刻出凹槽及槽口，以使管道可嵌入墙内。

kicker The bottom section of a column that is constructed first, to support the shuttering and/or reinforcement for the remainder of the column.
柱基 首先建造的柱子底部部分，可支撑模板及/或柱子剩余部分的钢筋。

kicker formwork (kicker frame) The *formwork used to construct the *kicker of a column.
柱基模板（柱基框架） 用于建造柱子柱基的模板。

kick plate (kicking plate) A metal plate attached to the bottom of a door to protect it against being kicked.
踢脚板 安装于门底部的一金属板，用于保护门以免被踢。

K

killed steel Steel that is treated with a strong deoxidizing agent, such as silicon or aluminium, which reduces the oxygen content to a level where no reaction occurs between carbon and oxygen during the solidification process.
镇静钢 使用强脱氧剂（如硅或铝）处理过的钢，可把氧含量降低至在碳与氧固化过程中两者之间不发生反应的水平。

kiln An oven or furnace that is used to burn, dry, or fire items such as bricks, ceramics, glass, and timber.
窑 一烤炉或熔炉，用于焚烧、干燥或火烧物品，如砖块、陶瓷、玻璃及木材。

kiln seasoning (kiln drying) A drying process predominantly utilized for reducing the moisture content in timber. The process is carried out in a closed chamber, providing maximum control of air circulation, humidity, and temperature. Drying can be carefully regulated so that shrinkage occurs with the minimum of degradation problems (cracking and splitting). One advantage of kiln seasoning is that lower moisture contents are possible than can be achieved using *air seasoning. Other advantages include rapidity of turnaround, adaptability, and precision. It is the only way to season timber intended for interior use, where required moisture contents may be as low as 10% or less. It is necessary to regulate kiln drying to suit the particular circumstances, i. e. different species and different sizes of timber require drying at different rates.
窑时效处理（窑干） 指主要用于降低木材水分含量的一个干燥过程。在一个密封室内进行处理，提供对空气循环、湿度与温度的最大程度控制。应小心控制干燥程度，尽可能减少收缩时发生的退化问题（破裂与裂开）。相比自然干燥，窑时效处理的一个优势是，可使含水量降到更低。其他优势包括，可迅速转向、适应性强及精确度高。这是干燥指定室内用木材的唯一方式，因为该木材所需的含水量应低至10%或更低。必须控制窑干燥使其适应特定环境，即不同品种及不同尺寸的木材要求不同等级的干燥。

kilogramme (kg) The international unit of mass. Equivalent to 1,000 grammes.
千克（kg） 国际质量单位，等于1 000克。

kilojoule (kJ) The international unit of energy. Equivalent to 1,000 joules. *See also* JOULE.
千焦耳（kJ） 功的国际单位，等于1 000焦耳。另请参考“焦耳”词条。

kilometre (km) The international unit of length. Equivalent to 1,000 metres. *See also* METRES.
千米（km） 长度的国际单位，等于1 000米。另请参考“米”词条。

kilonewton (kN) The international unit of force. Equivalent to 1,000 newtons. *See also* NEWTONS.

千牛顿（kN） 力的国际单位，等于1 000牛顿。另请参考“牛顿”词条。

kilowatt hour（kWh） The basic unit of electrical energy, equivalent to 1 joule per second. *See also* JOULE.

千瓦时（kWh） 电能的基本单位，等于每秒一焦耳。另请参考“焦耳”词条。

king closer A brick closer with a diagonal slope cut from half-way up the * header face to half-way along the * stretcher face. Closer bricks are used to make up the corners, close reveals, and fill in the gaps and joints, often used in and around window and door openings. The brick may also be called a three-quarter brick as three-quarters of the brick remains.

砍角砖 一种封口砖，在砖丁面中间与顺面中间呈斜坡形切割开。封口砖用于构成边角、窗侧并填入空隙与接缝，常用在门窗孔内及周围。该类砖块又称为“四分之三砖块”，因为只剩下一整块砖块的四分之三。

king post The vertical strut in a traditional * king post truss.

桁架中柱 传统中柱桁架中的垂直支柱。

king post truss A traditional timber roof truss consisting of two rafters and a tie beam with a central vertical post (a king post), running from the centre of the tie beam to the apex, and two diagonal struts. Traditional roof construction has now largely been replaced by prefabricated roof trusses.

中柱桁架 一传统木材屋架，由两根上弦杆以及一根连梁、两根斜撑构成；其中，连梁上有一根从其中间延伸至顶点的中心垂直杆（桁架中柱）。大部分传统的屋顶构造现在已经被预制屋架代替了。

K

kite winder A triangular- (kite) shaped stair tread (step) used to turn corners without the need for a landing.

三角踏步楼梯 一三角（风筝）形状的楼梯踏板（台阶），用于拐角处，从而无须楼梯平台。

kit home A dwelling constructed on-site from prefabricated parts assembled in a factory.

预制件组装房屋 使用工厂组装的预制件在现场建造住宅。

kit joinery Prefabricated joinery components that are assembled or installed on-site, such as door sets, shelving units, windows, etc.

预制件组装细木工制品 现场装配或安装的预制细木工组件，如门、架子、窗户等。

knapping hammer A masonry hammer for shaping building stones.

碎石锤 用于塑形建筑石材的凿石锤。

knee The convex bend in a handrail found at the top of a flight of stairs where the handrail enters the newel post. *See also* KNEELING.

扶手上弯头 一段楼梯顶部栏杆内的凸状弯管，即位于栏杆连接楼梯栏杆支柱的地方。另请参考“扶手下弯头”词条。

knee brace (corner brace) A short support that runs diagonally between the post and beam to strengthen the joint.

斜撑（角撑） 支柱与横梁之间斜安装的一短支撑，用于加固接缝。

knee elbow A right-angled pipework bend.

直角弯管 一直角管道弯头。

kneeler (skew, skew table) The angled masonry on a * gable wall that carries the * gable coping.

斜交石 山墙上成角度的砌块，承载着山墙压顶.

kneeling The concave bend in a handrail found at the bottom of a flight of stairs where the handrail enters the newel post. *See also* KNEE.

扶手下弯头 一段楼梯底部栏杆内的凸状弯管，即位于栏杆连接楼梯栏杆支柱的地方。另请参考“扶手上弯头”词条。

knob A round-shaped handle on a door that is used to open or close it.
球形把手 门上的圆形把手，用于开启或关闭门。

knob set A pair of internal and external door * knobs.
成套门把手 门内外的一对球形把手。

knocked down Building components that have been cut and shaped ready to be assembled on-site.
拆装组件 已切割成形并准备在现场装配的建筑构件。

knockings A defect in the surface of concrete caused by chips to the exposed faces.
夹渣 由于碎木片残留在表面而造成的混凝土面的一种缺陷。

knocking up (retempering) Reworking mortar or concrete that has already started to set. The mortar or concrete may have to be forcefully moved around or have extra water added to make it workable. In practice, concrete or mortar that has started to set should not be reworked then used, as the final strength of the mortar will be significantly reduced.
混凝土再搅拌 重新加工已经开始凝固的灰浆或混凝土。需用力搅拌灰浆或混凝土或加水使其可搅拌。实际上，已经开始凝固的混凝土或灰浆，在使用之后不可重新加工，因为会大大降低灰浆的最终强度。

knockout Parts of a unit that are made so that they can be pushed or knocked out if needed, for example electrical boxes with perforated parts so that they can be pushed out to allow wires and cables in and out.
敲落盖 一个装置的部件，需要时可推开或敲落，例如，带穿孔部件的接线盒，可推出来以使电线或电缆进出。

K

knoll A small natural rounded hill or mound.
小山丘 一座小的天然圆形山或土墩。

Knoop hardness test An indentation test for ascertaining a material's resistance to surface indentation or scratching. All hardness tests measure how easily diamond can produce an indentation on a material. The **Knoop hardness method** uses calibrated machines to force a rhombic-based pyramidal diamond indenter, having specified edge angles, into the surface of the test material. By measuring the indentation depth, the hardness value of the material can be determined.
努氏硬度测试 一个用于确定材料对表面压痕或划痕的抵抗力的压痕测试。所有硬度测试都是通过测量金刚石在材料上产生一压痕的难易程度。努氏硬度测试使用校准机器把一个带特定棱角的棱锥体金刚石压头推进测试材料表面。通过测量压痕深度以确定材料的硬度值。

knot The circular dense grain that appears in cuts of timber. The change in grain is the result of branches growing out from the main trunk of the wood. Knots reduce the strength of the wood.
木材节疤 木材切口的圆形稠密木纹。木纹的变化是由于树枝从树的主干长出来。木材节疤会减弱木材的强度。

knot area ratio (KAR) The KAR is a part of the visual grading process of timber to identify the number of knots in an area of wood. Timber is graded based on the number of defects it has and is given a grade identifying its potential structural use.
节疤面积比（KAR） KAR 是木材目观定级过程的一部分，可确定一定区域内木材的节疤数量。根据木材的缺陷量来定级木材，并给予木材一个可标记其潜在结构用途的等级。

knot brush A thick paint brush in a round shape for painting walls.
圆头刷 用于涂刷墙的圆形浓厚涂漆刷。

knuckle The cylindrical holes in a hinge through which the pin passes.
销孔 铰链内的可让销子穿过的圆柱孔。

knuckle bend A sharp turn in a pipe.
转向弯管 管道内的突转弯头。

knuckle joint A sharp bend or joint between the two different slopes of a Mansard roof.

转向接头 复折式屋顶上两个不同斜坡之间的突转弯头或接头。

***K* value (K-value, k-value, k value)** *See* THERMAL CONDUCTIVITY.
***K* 值** 参考“导热系数”词条。

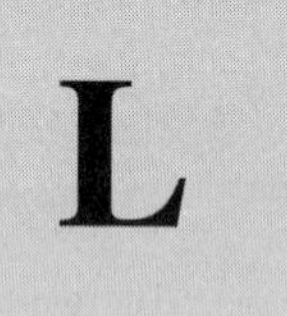

label A drip or mould providing a projection out from the wall over the top of an opening.
窗檐 从一开口顶部上方的墙伸出来的一个滴水槽或滴水线。

laboratory A room equipped with testing equipment, where scientific testing is carried out or taught.
实验室 一个装备着测试设备的房间，可进行或教授科学测试。

labour The human resource associated with a task or job.
劳动力 与一项任务或工作相联系的人力资源。

labourer A person employed to undertake and assist with building work, but does not have a skill, trade, or profession.
劳工 被雇佣去从事及协助建筑工程但无技能、行业或专业知识的人员。

labour only Description of a worker who is not a direct employee, but is contracted to work on a subcontract basis. The term 'labour' only implies that the person is contracted to complete the work, however, the materials and equipment are to be supplied by the employer or main contractor.
包清工 非正式员工，但以转包合同形式被雇佣工作。术语“劳动力”仅指的是签约人员需完成这项工程，但是是由业主或总包商提供材料与设备。

labyrinth A complex interconnected series of maze-like paths or tunnels.
迷宫式通道 一系列错综复杂的迷宫式路径或地下道。

laced valley A *valley created using slates or tiles where the courses run from one roof slope to the next. They do not contain a *valley gutter. Also known as a **swept valley**.
搭瓦斜沟 斜沟用板岩或瓦从一个斜屋面到另一个斜屋面搭接而成。不需要排水斜沟。也称为 **swept valley**。

lacer A reinforcing bar that is tied at right angles to the direction of the main reinforcement to create a mesh.
拉筋 朝主钢筋方向呈直角系住的钢筋，以形成一钢筋网。

lacing 1. The interweaving of roof slates at an intersection to form a *laced valley. 2. Tying together the horizontal members in formwork to improve their stability.
1. 交织 在屋面交汇处板岩相互交织以形成一搭瓦斜沟。**2. 对拉** 把模板的水平构件绑在一块以提高其稳定性。

lack of cover Inadequate or no concrete around steel reinforcement. For reinforced concrete to act as a matrix structure there must be an adequate level of concrete surrounding the reinforcement. Where the concrete and steel are properly bonded together and positioned, the tensile forces can be transferred to the steel and the compressive forces to the concrete. A layer of concrete over the top of reinforcement also provides protection against corrosion and the effects of fire. Exposed reinforcement is not afforded any protection and does not form a proper bond with the concrete.
露筋 钢筋周围无足够的或没有混凝土。充当矩阵结构的钢筋混凝土，钢筋周围必须有充足的混凝土。混凝土与钢筋适当粘结并固定，拉力可转移至钢筋而压力则转移至混凝土。钢筋顶上的一层混凝土同样可提供防腐蚀保护及防火。外露的钢筋无任何保护层，不能与混凝土适当粘结。

ladder A piece of equipment with horizontal rungs encased by two vertical styles that is used to facilitate vertical movement. Ladders can be movable and used for jobs such as cleaning windows and accessing heights for maintenance purposes and inspections, or can be permanently secured to a structure to provide safe, easy access whenever required.
梯子 一件设备，带装在两垂直支柱内的水平梯级，用于帮助垂直移动。梯子可移动，并适用于清洁窗户、通向高处进行维修及检查等工作，或可永久固定在一结构上以便在需要时可用作安全方便的通道。

lag bolt (coach bolt) A heavy-duty fixing with a large square head. As the bolt is screwed into the wood, the square head imbeds itself into the surface of the timber, preventing it unscrewing.
方头螺栓 一种大方头重型固件。由于螺栓是螺旋进入木材内，所以方头也会嵌入进木材的表面，以防松开。

lagging 1. Thermal insulation used to wrap around hot surfaces, such as pipes, boilers, and hot water tanks to prevent heat loss. Also used to prevent heat gain and condensation on cold pipes and surfaces. 2. A timber frame used to support the sides when constructing an arch.
1. 绝缘层 环绕在热表面如管道、锅炉及热水箱上的防止热损耗的隔热层。也可用在冷管道及表面上防止吸热及冷凝。**2. 支拱木架** 建造拱形时用于支撑侧面的木构架。

laminate (decorative laminate, laminated plastic) An object consisting of a number of layers (laminae) or plies, bonded together. Laminates may be symmetrical or asymmetrical. A symmetrical laminate is one having a mirror plane lying in the plane of the laminate at its mid thickness.
层压板（装饰层压板、层压塑料） 一种由许多层（薄层）板或板层结合在一起的物体。层压板可以是对称的或非对称的。对称层压板是一板层镜像对称于中面的层压板。

L

laminated glass A special type of glass that is a combination of layers, and provides a generic material with improved toughness utilized in structures such as balconies, shop fronts, roofs, and the like.
夹层玻璃 一种特殊玻璃，是多层玻璃的结合，是一种带改良韧性的通用材料，应用于阳台、商店正面、屋面等结构中。

lamination The process of creating *laminated glass.
层压 制造夹层玻璃的过程。

lamp 1. An electrical device used to provide artificial light. 2. A lightbulb.
1. 照射器 一个用于提供人造光的电气设备。**2. 灯泡**。

lamp base (lamp cap, lamp holder) The portion of a lightbulb that provides support for the bulb and is used to connect it into place. A variety of bases are available, such as metal, plastic, or glass that incorporate screw, bayonet, pin, or wedge-shaped fittings.
灯座（灯头） 灯泡的一组成部分，可支撑灯泡并可用于把灯泡连接到安装位置上。有各种各样的灯座，如金属、塑料或玻璃的等，带螺丝钉、卡销、销子或楔形配件。

lancet Architectural style that makes use of shapes that resemble a lance.
尖顶拱 一种建筑风格，其形状像一柳叶刀。

land drain *See* FIELD DRAIN.
地面排水沟 参考“工地排水”词条。

landfill The disposal of waste material by burial.
垃圾填埋池 通过掩埋处理废物。

landing 1. Flat platform at the top and bottom of a flight of stairs, escalator, or a ramp. Allows people to stop, pass, rest, and change direction. A landing between two flights of stairs is called an **intermediate landing**. 2. A place where a lift comes to rest, and allows passengers to safely enter or depart from the carriage.
1. 楼梯平台 一段楼梯、自动扶梯或斜坡顶部及底部的平台。可允许人们停止、经过、休息及转向。

两段楼梯之间的平台叫作“中间平台”。**2. 电梯站台** 电梯停止移动时的所在位置，可允许乘客安全进入或离开电梯轿厢。

landing button The switch (button) used to summon a lift. In multi-storey buildings where there is opportunity to ascend or descend, there will be two buttons to indicate the desired direction of travel.
层站按钮 用于召唤一电梯的按钮。在多层建筑物中，如果需要上升或下降，那么一般会有两个按钮来指示所需的运行方向。

landing valve A * valve attached to the rising main that enables connections to be made to the water supply for fire-fighting purposes. Usually located in the fire cabinet on each floor of a tall building.
室内消火栓 安装于上行水管的一种阀，可为消防用水提供连接。一般位于高层建筑每一层的消防栓箱内。

landscaped office An open office where the furniture, screens, equipment, plant, and decorations can be moved around to the desired layout.
庭院式办公室 一种开放式办公室，家具、屏风、设备、植物及装饰物可按照预定的布局随意移动。

landscaping The reshaping or restructuring of the external built and natural environment, includes the alteration of existing land, plants, footpaths, roads etc. to provide improved aesthetics.
景观美化 外部建筑与自然环境的重新修整或重建，包括改变现有土地、植物、人行道、道路等，使之看上去更美观。

landslide (landslip) A downwards mass movement of soil or rock, which occurs when the disturbing forces (weight of soil and effects of gravity) exceed the resisting forces (shear strength of the soil) along a potential slip plane. This can be as a result of unsuitable geometry, a change in the groundwater regime, presence of weak planes, bands, or layers, progressive deformation, increase in effective slope height, or additional imposed loading.
滑坡（山崩） 大量土壤或岩石向下移动。当潜在滑动面的扰动力（土壤重量及地心引力影响）大于阻力（土壤的抗剪强度）时就会发生滑坡，也有可能是因为几何结构不相配、地下水体系变化、存在不牢固滑面、地带或面层、渐进变形、有效斜坡高度的增加或额外的外加荷载。

L

land survey A topographical survey of the land to document various points by distances and angles. It is undertaken by a **land surveyor**, who is a member of the * Royal Institution of Chartered Surveyors (RICS).
土地勘测 土地的地形测量，记录按距离和角度的不同点的数据。由土地测量师进行，该测量师需为英国皇家特许测量师协会（RICS）的会员。

lantern 1. A small square, octagonal, or dome-shaped structure found at the top of a roof or dome. Used to admit daylight, provide ventilation, and as a lookout. 2. A portable lighting device comprising a light source encased within a transparent or translucent cover.
1. 通风天窗 屋顶或圆屋顶上的一个小方形、八角形或圆形的结构。用于采光、通风并可观望外面。
2. 灯笼 一个便携式照明装置，其光源放置于一个透明或半透明的护壳内。

lap (overlap, lapping) 1. The distance that one material overlaps another. 2. A weather-tight joint used in profiled sheeting and roof tiling. 3. The thickened film of paint that results when one coat of paint overlaps another.
1. 搭接长度 一材料与另一材料重叠的距离。**2. 搭接接缝** 压型钢板与屋瓦上使用的防风雨接合面。**3. 油漆重叠** 由于一层油漆与另一层相重叠而形成的加厚油漆层。

lap joint (lapped joint) A joint formed between two components, where one component overlaps and joins with the other.
搭接接头 两组件之间的接缝，一组件与另一组件重叠并结合的位置。

lapped tenons Two * tenons that overlap one another within a mortise.
搭接榫 两凸榫在榫眼内搭接在一起。

lapping A process used to achieve a smooth polished surface.

抛光 实现光滑抛光面的过程。

larder A storeroom for food; historically, it was a room used for the storage of meat.
食品室 食物储藏室，以前是用于存放肉的房间。

large-panel system A precast concrete panel cladding system where the panels span the full length between the structural columns of the building.
大模板建造法 一种预制混凝土墙围护体系，墙板横跨建筑结构柱之间的全长。

larrying (US larrying up) A method of constructing brick walling by sliding bricks into position, then filling the joints with a fluid mortar that is poured between the joints.
薄层砂浆砌筑法 砖墙的建造法，把砖块滑移至适当位置上，然后在接缝之间倒入流态灰浆进行填缝。

laser An acronym for light amplification by stimulated emission of radiation. Lasers are a coherent source of light, usually in the form of straight rays.
激光 全称是“辐射受激发射光放大”。激光是光的相干光源，通常为直射线的形式。

laser soldering A type of soldering process using laser light.
激光焊接 使用激光的焊接过程。

lashing The tying of cable or rope around the top of a ladder to ensure that it is secured in position.
系索 绕着梯子顶部绑缆绳或绳索以确保其固定在位置上。

latch A lock or catch with a sloped or bevelled tongue that is secured into a mortise as a door is pushed shut. The tongue is bevelled so that as it strikes the doorframe, the slope of the tongue pushes the catch back into the door. The bevel enables it to slide over the frame until it is over the mortise where a spring pushes the catch securely into place. The tongue is removed from the mortise and the door opened by twisting a handle that is connected to the tongue; this releases the catch allowing the door to be opened.
门闩（锁舌） 带倾斜或斜角锁舌的锁或插销，当门关上时锁或插销将插入榫眼内。由于锁舌是斜角的，所以当它碰撞到门框时，锁舌的斜度会把插销推入门内。斜角可使其滑过门框到榫眼处，弹簧把插销牢固推进去。扭转连接锁舌的把手时，锁舌从榫眼中移出，便可松开插销从而打开门。

latchet A piece of metal that is used to secure or hold something in place. Generally lachets or tingles are made from folded metal strips that are fixed over the corner of a panel that it is holding. Part of the tingle is fixed to the substructure while the other part laps over the panel that it is securing in place. *See also* TINGLE.
门扣 用于固定某物的一片金属。一般来说，门扣或夹片是由折叠金属条制成，安装在其所固定面板的角上。一部分的夹片是安装在底部结构上，另一部分则重叠在其所固定的面板上。另请参考“夹片”词条。

lateral Sideways or at the side, such as **lateral movement**, referring to sideways movement rather than vertical or downwards movement.
横向的 向侧面的或在一边的，如横向运动，指的是“侧向运动”而不是垂直或向下运动。

latest date In a logical or precedent network, the latest time an activity or event can be started or finished (latest start time or latest finish time) without delaying subsequent activities and the planned completion date.
最迟日期 在一个逻辑或前导网络图中，一项活动或时间的最迟日期可指最迟启动时间或最迟结束时间，不耽搁后续活动以及设定的完成日期。

latex A polymeric material with rubber-like mechanical properties. Latex refers generically to a stable dispersion (emulsion) of polymer microparticles in an aqueous medium. Latex may be natural or synthetic.
乳胶 有着类似橡胶的机械性能的一种聚合材料。乳胶一般指的是聚合物微粒在水介质中的稳定散布（乳化）。乳胶可以是天然的或合成的。

lattice 1. At a microscopic level in metals, the space arrangement of atoms in a crystal. 2. A crisscross framework.
1. 晶格 在金属的微观水平上，晶体原子的空间布局。**2. 交叉格架** 十字交叉框架。

L

lattice window A window containing small diamond-shaped panes of glass that are fixed in position using strips of metal.
花格窗 由小菱形玻璃格组成的窗，使用金属片固定。

laundry A place or room for the washing, cleaning, and drying of clothes and linen.
洗衣房 清洗及晾干衣服与布草的地方或房间。

lavatory 1. A room where the toilet, or WC (water closet), or urinal is installed. 2. Another name for a toilet or the receptacle for foul water.
1. 盥洗室 安装厕所或冲水马桶或小便池的房间。**2. 厕所** 洗手间或污水池的另一名字。

lay bar A horizontal *glazing bar.
窗棂横条 一水平玻璃窗格条。

layer board (lear board, gutter board) The board that forms the base onto which the felt or sheet metal for a *box gutter is laid.
天沟托板 用于铺放匣形水槽的毛毡或金属薄片的基板板块。

laying-off Applying very fine, light strokes to a surface containing wet paint to remove any brush or roller marks, and provide a smooth surface.
油漆涂匀 使用精细轻质的笔刷去除油漆未干表面上的刷痕或辊痕，并取得一光滑面。

laying-on trowel A plasterer's trowel, which is square in shape and used to place, smooth, and finish plaster on a wall.
涂抹泥刀 泥水匠用的泥刀，形状为方形，用于把灰泥涂在墙上，并抹平、修整墙上灰泥。

L

laying trowel A bricklayer's trowel.
泥刀 瓦工用的泥刀。

laying-up (laminating) To *laminate.
层压 进行层压。

lay light (laylight) A horizontal window in a ceiling (ceiling light) or a roof (roof light).
顶棚采光（平顶窗） 天花上（顶灯）或屋顶上（采光天窗）的水平窗。

layout The plan positioning of a building. Layout drawings are produced that show the planned position of the building, walls, furniture, and other items that can be seen on a plan.
布局 一建筑的布置规划。布置图显示建筑的设计位置，也可在平面图上看到墙壁、家具及其他构件。

lay panel A panel that is longer horizontally than vertically.
横镶板 水平长度比垂直长度更长的板块。

leach To remove something slowly, typically relating to liquids. Leachate is the liquid from the decomposing waste in a *landfill.
过滤 缓慢地移除某物，一般与液体相关。滤出液指的是从垃圾填埋堆分解废物中渗出来的液体。

lead-caulked joint (lead joint) A type of joint used in cast-iron pipework where molten lead is poured in to seal and join the pipes.
灌铅嵌缝（青铅接口） 铸铁管道的一种接缝，把熔铅倒入以密封及接合管道。

lead-clothed glazing bar (lead glazing bar) A steel *glazing bar covered in lead.
包铅玻璃窗格条 铅覆盖的钢玻璃窗格条。

leaded light (lead glazing) A decorative window comprising a number of small glass panes, usually rectangular or diamond shaped, that are held together with strips of lead. *Stained-glass windows are an example of leaded lights.
花饰铅条窗（花窗玻璃） 一个由许多小玻璃窗格组成的装饰窗，通常是矩形或菱形，使用铅条固定在一块。彩色玻璃窗就是花饰铅条窗的其中一种。

leader head (US) *See* RAINWATER HEAD.
雨水斗（美国） 参考“雨水斗”词条。

lead flat *See* LEAD ROOF.
铅皮平屋顶 参考“铅皮屋顶”词条。

lead glazing *See* LEADED LIGHT.
花窗玻璃 参考“花饰铅条窗”词条。

lead glazing bar *See* KEAD-CLOTHED GLAZING BAR.
铅玻璃窗格条 参考“包铅玻璃格条”词条。

leading hand Person in charge of a group of tradespeople; a *charge hand.
领班 一群专业人员中的负责人，即负责人。

lead joint *See* LEAD-CAULKED JOINT.
铅接 参考“灌铅嵌缝”词条。

lead plug A small hole drilled into a wall that is filled with lead and used as a fixing for a screw, nail, or other fastener.
铅塞 使用铅填堵在墙内钻出的小孔，并用于固定螺丝、钉子或其他固件。

lead roof (lead flat) A flat roof covered in sheet lead.
铅皮屋顶 使用铅皮覆盖的平屋顶。

lead sheath A covering of lead around a steel bar. An example is a *lead-clothed glazing bar.
铅皮 钢筋周围包的铅层，例如“包铅玻璃窗格条”。

L

lead slate (leadslate) A lead flashing that provides a weatherproof joint where a pipe, such as a flue or soil stack, penetrates a roof.
铅皮管脚泛水 铅防水板，用于管道如烟道或排水立管穿透屋面处，提供防水接合。

lead time The time between the placing of an order and the delivery of materials, plant, units, or equipment to site. It is essential to be aware of how long it takes between placing an order and the item arriving on-site. Where changes are made during the construction phase and items are changed which have long lead times, considerable delays to the project may be experienced. Lead times can be affected by global demand, reduction in the availability of raw materials, and supply chain economics.
交付周期 材料、设备、装置或设施从下单至交货到现场的时间。有必要知道从订货到把货送到现场的所需时长。在施工期间，如果交付周期长的项目出现变更，会导致项目出现较长的延误。交付周期会受到全球需求、原料供应减少、供应链经济的影响。

leaf 1. A *door leaf. 2. One of the two main layers of a cavity wall. The main internal layer is known as the inner leaf, and the main outer layer is known as the **external leaf**.
1. 门扇。2. 叶墙 夹心墙两叶墙的其中一层。在内部的称为“内层”，外部的叫作“外叶墙”。

leaf frame (leaf structure, framing) The structural components used to construct a *door leaf, namely the *door rails, *door stiles, and the *facings.
门扇框（门扉结构、框架） 用于建造一扇门扉的结构构件，也就是门扇冒头、门竖梃及贴面。

lean lime Lime with impurities such as magnesium and silica compounds contained within it.
贫石灰 含有如镁、硅化合物等杂质的石灰。

lean-to (lean-to roof, half-span roof) A mono-pitched roof that is supported at its highest point by a wall that is higher than the top of the roof.
披屋（披屋顶） 由高于屋顶的墙在其最高点处支撑的单坡屋顶。

leasehold Where the owner has bought the right to occupy or use a property or land for a given period of time; this is different to a freehold, where ownership is purchased outright.
租赁 业主已经购买了该房产或土地的一段时期的占有权或使用权；跟永久业权有所不同，永久业权指的是完全购买了所有权。

leat A channel that diverts water to a watermill to provide power.
水渠 把水转接至水磨从而产生能量的渠道。

ledge 1. A horizontal platform. 2. The horizontal platform at the foot of a window. 3. The horizontal timber used to hold the panels in a matchboard door together.
1. 平台 一个水平平台。**2. 窗台** 窗户下部的水平平台。3. 横档 用于把企口板门内的板块固定在一起的横木。

ledged and braced door (matchboarded door, batten door) *See* MATCHBOARDING DOOR.
直拼 Z 形撑门（企口板门、板条门） 参考“企口镶板门”词条。

ledger 1. A horizontal beam fixed to a wall that supports the ends of the floor joists. 2. A horizontal scaffolding pole used to support *putlogs.
1. 檩条 用于支撑楼板托梁端并固定于墙的水平梁。**2. 横杆** 用于支撑脚手架跳板横木的水平支撑。

Legionnaires' disease A severe and potentially fatal form of pneumonia that is caused by the bacteria Legionella pneumophila. It is contracted by inhaling tiny water droplets contaminated with the bacteria. The bacteria is commonly found in rivers and ponds, but can also grow in other water systems such as centralized air-conditioning systems, cooling towers, evaporative condensers, hot water systems, ornamental fountains, room air humidifiers, showers, and whirlpool spas. The Legionella bacteria grows in water at temperatures from 20℃ to 50℃ and is killed by high temperatures.
退伍军人症 由嗜肺军团菌引发的一种严重且有可能致命的肺炎。是由于吸入受细菌污染的微小水滴而感染。该细菌一般存在于河流及池塘，但也会在其他供水系统中生长，如中央空调系统、冷却塔、蒸发冷凝器、热水系统、装饰性喷泉、室内空气加湿器、淋浴设备及漩涡水疗。军团菌在水中的滋生温度为20℃至50℃，可通过高温杀灭。

lengthening joint (heading joint) A joint made between two materials that are joined end to end.
加长接头（端接） 两材料两端相接的接头。

let in (sunk in) Fixed flush with the surface.
埋头（沉头） 与表面齐平的安装。

levée An embankment adjacent to a watercourse. An **artificial levée** is built to prevent flooding; a **natural levée** is formed from sediment during flooding.
堤坝 毗邻水道的一个路堤。人造堤坝用于防止洪灾，而天然堤是在洪水发生时的沉淀物形成的。

level 1. An instrument used to provide a level line of sight so that the relative levels of the ground, parts of the structure, and buildings can be determined. There are different types of level: automatic level, tilting level, and laser level. 2. A straight edge with a clear glass tube, containing liquid and an air bubble. The tube is set parallel with the straight edge, so that when the bubble floats in the centre of the tube, the straight edge is horizontal. 3. When the ground is graded or moved by bulldozers so that it is horizontal and flat.
1. 水准仪 用于提供水平视线的仪器，从而确定地面、结构部件与建筑的相对水平。有不同种类的水准仪：自动水准仪、微倾水准仪、激光水准仪。**2. 水平尺** 带一含液体与一个气泡的透明玻璃管的直尺。该玻璃管与尺边平行，所以当气泡浮动于玻璃管中心时，则表明直尺是水平的。**3. 场地平整** 使用推土机推平或移动土地以使地面平整。

levelled finish A relatively smooth flat finish achieved using a float or trowel.
平坦表面 使用抹子或泥刀所取得的相对平滑表面。

levelling 1. Moving material around so that it is flattened. 2. Moving earth or other material around so that it is horizontal. 3. The act of taking readings using a tilting or automatic level to determine the height of objects and land, or to work out the profile of the land.
1. 整平 不断移动材料使其平整。**2. 调平** 不断移动土壤或其他材料以使其水平。**3. 水平测量**

使用微倾水准仪或自动水准仪读取数据以确定物体及土地的高度或制定出土地的轮廓。

levelling rule A large straight edge, normally at least 3 m long, with a spirit level set in it, used for levelling screeds.
水平刮尺 一把大的直尺，通常至少3米长，内装一气泡式水准仪，用于整平砂浆层。

level-tube A tool to determine if an object is parallel to the ground, it has a transparent cylinder containing liquid with an air bubble trapped in it.
水准管 用于测量一物体是否与地面平行的工具，带一装有液体的透明圆筒，并且有一个气泡在里面。

lever 1. A simple machine comprising of a rigid bar that is able to pivot about a fixed point. It is capable of multiplying the force that can be applied to an object. 2. A flat metal tumbler in a lever lock.
1. 杠杆 由一个刚性杆构成的简单机械，以一个固定点为支点。可增大应用于一物体上的力。**2. 锁杆** 杆锁内的一个扁平金属杆锁簧。

lever arm The distance a force rotates about. *See also* MOMENT.
杠杆臂 力旋转的距离。另请参考“力矩”词条。

lever boards Adjustable wooden *louvres.
百叶窗板 可调节的木百叶窗。

lever handle A horizontal lever attached to a door that operates a *latch. An alternative to a door knob.
杠杆手柄 固定于门的水平杠杆，可控制门闩，是门把手的替代选择。

lever lock A lock in which the key lifts a number of levers to operate the lock.
杆锁 该锁的钥匙通过升降若干操控杆来控制锁。

lever mixer (lever-operated thermostatic mixing valve) A mixer tap operated by a single lever. The water flow can be controlled by lifting the lever up or down, and the water temperature can be controlled by moving the lever left or right.
提拉杆式龙头（杠杆操作的恒温混合阀门） 由单杆操控的冷热水混合龙头。可通过升降提拉杆来控制水流，左右移动提拉杆来控制水温。

lewis A mechanical device that fits into a dovetail hole and expands. Used for lifting stone or concrete blocks.
吊楔 一个嵌入楔形榫头孔并膨胀的机械设备。用于吊起石块或混凝土块。

liability The consequences of failing to deliver legal obligations under contract or statutory instrument. Persons or organizations have legal responsibilities when operating under a contract or within the legislation that applies to operations undertaken; they will be liable for actions that breach any term or regulation.
责任 未能履行合同或法定文件规定的法律义务所产生的后果。当个人或组织执行合同时或所履行法律中具有的法律责任；他们需为其违反规定或条款的行为负责。

lid A cover placed on top of a vessel, bin, or lift shaft.
盖子 容器、箱子或电梯井道顶部的盖子。

lift (US elevator) A mechanical device that uses a *lift car to transport goods and people vertically from one floor to another.
电梯（美国 升降机） 利用电梯轿厢把货物及人从一楼层垂直运送至另一层的机械设备。

lift and force pump A *displacement pump where the reciprocating action of the piston within the cylinder imparts movement to the water.
压力抽水泵 泵缸内活塞往复式活动从而使水移动的容积式泵。

lift car The enclosed part of a lift that moves vertically from one floor to another.
电梯轿厢 电梯的封闭部分，可从一楼层垂直运行至另一层。

lift controls (US elevonics) The electronic controls used by passengers to control and operate a lift.

电梯控制装置 由乘客用于控制及操作电梯的电子控制装置。

lifting beam A straight steel beam or frame that is attached to a crane's hook, and has a number of fixings hanging from the length of the beam to attach onto the object being lifted. The beam is used for lifting long objects such as beams, wall panels, cladding panels, and so on.
吊梁 附属于起重机吊钩上一直钢梁或框架，有若干附件悬挂在钢梁上并附于被吊升的物体上。钢梁用于提升长的物体，如梁、墙板、外墙板等等。

lifting equipment Chains, hooks, slings, clips, and plant used by cranes in order to hoist, elevate, or lift items from the floor.
起重设备 用于把物体从地面吊起、提升或升起的链条、吊钩、吊索、夹具及设备。

lifting eye A steel ring used to secure clips, hooks, and slings when performing lifting operations with a crane or other lifting devices.
吊环 使用起重机或其他起重装置进行起重操作时，用于固定夹具、吊钩及吊索的钢环。

lifting frame A piece of equipment that is attached to a crane's lifting hook to assist the lifting of long or wide objects. The frame is often built to lift certain objects, such as cladding panels, large beams, columns, formwork, and so on.
吊装架 附属于起重机吊钩上的一件设备，用于协助吊起长的或宽的物体。吊架一般是用于起重特定的物体，如外墙板、大梁、柱子与模架等。

lift machine room (lift motor room, LMR) The *plantroom that contains all the equipment required to operate the lift. Usually located at the top of the *lift shaft.
电梯机房（电梯电机房，LMR） 含有所有用于操作电梯的设备的机房。通常位于电梯井顶部。

lift-off hinge (loose butt hinge) A hinge where the top and bottom leaf can be slid apart. When the hinge is fixed to a door, the door can be simply lifted on and off the part of the hinge that is fixed to the door casing. When the hinge is coupled together, the hinge operates as normal.
可脱卸铰链（抽芯铰链） 铰链的上下两叶可滑动并分离。当铰链固定于门上时，可轻易把门抬起并从固定于门框上铰链的叶片中抽离出来。当铰链连接时，可正常操作铰链。

lift pit (runby pit) The space at the bottom of the lift shaft that extends below the lowest floor level. In hydraulic lifts the pit will house the pump and the ram.
电梯底坑 位于电梯井底部的空间，一直延伸至最低楼层以下。液压电梯的底坑中放置着泵与油缸。

lift safety gear Safety equipment that is designed to ensure that the lift operates safely.
电梯安全装置 设计用于确保电梯安全操作的安全设备。

lift shaft (lift well, US hoistway) A vertical shaft within a building that houses a *lift. Usually incorporates a *lift machine room at the top of the lift shaft and a *lift pit at the bottom.
电梯井（电梯井道，美国 井道） 位于一建筑内放置电梯的竖井。一般在电梯井顶部有电梯机房，底部则是电梯底坑。

lift shaft module A prefabricated portion of a *lift shaft.
电梯井模块 电梯井的预制构件。

ligature 1. Reinforcement binders fixed around the main bars of reinforcement to form a cage. *See* STIRRUP. 2. Slang term used to describe a wire, band, or other device that can close something off, for instance, restrict or close the movement of fluid along a pipe, like a tourniquet device.
1. 箍筋 绕着主钢筋安装的箍筋，以形成一笼状物，另请参考“箍筋”词条。**2. 扎带** 俗语指的是金属丝、绑带及其他可封锁东西的装备，如限制或切断管道内液体流动，类似于止血带。

light 1. A *window. If it can be opened, it is referred to as an opening light, otherwise it is known as a fixed light. 2. Electromagnetic radiation that can be detected by the human eye. 3. A source of illumination, either natural or artificial.
1. 窗户 一窗户。如果是可开启的，则为“开启窗”，否则为“固定窗”。**2. 光** 人眼可看到的电磁波。**3. 光线** 光源，可以是自然的也可为人造的。

light alloys A low-density metal such as *aluminium, *magnesium, *titanium, beryllium, or their alloys. These type of alloys are extensively utilized as they combine high strength with low specific density.
轻合金 低密度金属，如铝、镁、钛、铍，以及它们的合金。由于这些金属合金集强度高与比密度低为一体，所以被广泛应用。

light fitting *See* LUMINAIRE.
灯具 参考"照明装置"词条。

lighthouse A tall round tower building with a powerful flashing light, which is located on hazardous coastal regions to warn sailors of dangerous rocks.
灯塔 带一强力闪灯的高圆塔建筑，位于危险的沿海地区，用于警告海员出现危险岩礁。

lighting 1. The process of providing light, either artificial or natural, to a space. 2. The electrical equipment used to provide light.
1. 照明 为一空间提供光线的过程，不管是人造的还是自然的。**2. 照明设备** 用于提供灯光的电气设备。

lighting bollard A low post that protrudes from the ground and incorporates a *luminaire. *See also* LIGHTING COLUMN.
灯柱 从地面伸出的一根矮柱，带一照明装置，另请参考"灯杆"词条。

lighting column A tall post that protrudes from the ground and incorporates a *luminaire. *See also* LIGHTING BOLLARD.
灯杆 从地面伸出的一根高柱，带一照明装置，另请参考"灯柱"词条。

lighting control panel A *distribution board for all of the lighting circuits contained within a building. Can also be used to programme various lighting circuits.
照明控制面板 一建筑中所有照明电路的配电箱。可用于编程不同的照明电路。

lighting fitting *See* LUMINAIRE.
照明配件 参考"照明装置"词条。

lighting installation All of the electrical cables and equipment required to provide artificial lighting within a building.
照明设备 一建筑中用于提供人工照明的所有电缆与设备。

lighting point The point in a building where a *luminaire is to be installed.
照明点 一建筑中安装照明装置的点。

light-loss factor (LLF, US dirt-depreciation factor) A factor used to calculate the luminance of a lamp at a particular period of time, usually just before relamping. It takes into account factors such as the age of the lamp, dirt accumulation on the luminaire, the frequency and effectiveness of cleaning, maintenance regime, and the age of the ballast. Formerly known as the maintenance factor.
光损失系数（LLF，美国 尘埃减能系数） 用于计算在特定时期的灯亮度的系数，通常是在换灯之前。应考虑各种因素，如灯的年龄、灯具上的脏污积累、清洁频率与效果、维护保养制度与镇流器的年龄。曾用名"维护系数"。

lightness (value tone) The perceived intensity of light reflected by a colour.
亮度（色调值） 由一颜色反射的可感知光强。

lightning arrestor (surge protector) A device that is used to protect telecommunications and power lines from a lightning strike.
避雷器（浪涌保护器） 用于保护电讯及电线免受雷击的设备。

lightning conductor (US lightning rod) A thick strip of metal, usually copper, that runs from the highest point on a building to the ground. Protects the building from a lightning strike by transmitting any lightning that

does strike the building to the ground.
避雷针 一根粗大的金属条，通常为铜，从建筑的最高点延伸至地面。通过把雷电传递至地面从而保护建筑以免受到雷击。

lightning protection system A system installed on a building to protect it against a lightning strike. Usually comprises a series of *lightning rods, *lightning conductors, and *ground terminals.
防雷系统 安装在建筑上的一系统，用于防雷击。通常由一系列避雷针、地面终端设备组成。

lightning rod (US) *See* LIGHTNING CONDUCTOR.
避雷针（美国） 参考“避雷针”词条。

light switch A switch, usually located on a wall, that is used to operate electric lights.
照明开关 安装于墙上的用于操控电灯的一开关。

light well (lightwell, air shaft) An open space that usually extends several floors down a building that admits daylight and provides ventilation.
采光井（通风井） 建筑中向下延伸至几层楼的开敞空间，可引入光线并通风。

lime The common name for calcium oxide (CaO). Lime is most commonly used in mortar along with sand and cement. In mortar, lime has a plasticizing effect, which improves the workability. Lime is also used as hydraulic lime in mortars, specifically termed 'lime mortar'. Hydraulic lime is a useful building material both as mortar and for conservation of historical properties.
石灰 氧化钙（CaO）的常用名。石灰最常用于与沙、水泥混合的砂浆中。石灰在砂浆中起到增塑作用，可改善施工性能。石灰也可用作砂浆中的水硬性石灰，专门术语为“石灰砂浆”。跟砂浆一样，水硬性石灰也是一种有用的建筑材料，可保护历史性建筑。

L

lime putty A type of *putty made from lime (calcium carbonate or calcium oxide).
石灰膏 一种由石灰制成的油灰（碳酸钙或氧化钙）。

limestone A sedimentary rock rich in calcium carbonate. Used as dimension and aggregate stone, it is also burnt with clay to make cement. It can dissolve in rainwater to form *karst and sinkholes in the ground.
石灰岩 富含碳酸钙的沉积岩。用作规格石料及骨料石材，可与黏土一块烧制成水泥。也可溶解于雨水中以形成喀斯特地形及地下的落水洞。

limit of works The point at which one contractor's or designer's work ends and another's begins. For example, where the main mechanical and electrical subcontractor's work package ends and the supplier of specialist electrical equipment begins, or where the builder's work ends and the plumber takes over. These points are particularly important as the contract's description must ensure that the interface between works covers all the work required. Where the gap between the limit of work is vague, claims and disputes over additional work are common.
工程施工界限 承包商或设计师的一个工程结束而另一方开始的点。如机电分包商的工程结束了，然后专业电气设备供应商开始其工作，或者是建筑工工程结束，然后由水管工接手。这些点特别重要，因为合同对工程的描述必须确保工程之间的界面需覆盖整个工程。如果工程施工界限不明确，容易出现关于额外工程的索赔及争执。

limpet washer A steel or plastic washer that is bent so that it can fit over the curve of a profiled metal or composite roof sheet. The bolt simply passes through the washer with the washer sitting in contact with the profiled roof.
波纹垫圈 弯曲的钢或塑料垫圈，刚好匹配金属压型板或复合屋顶板的曲线。螺栓穿过垫圈，垫圈与屋顶压型板相接触。

line (bricklayer's line, builder's line, string line) Cord or string pulled between two points used for setting out straight lines in building works. The line can be securely fixed to profile boards, steel pins, dowels, or line pins, depending on the operation.
放线（瓦工放线，施工放线，吊线） 在两点之间拉的绳索或细绳，用于在建筑工程中定直线。该直线可牢固固定于压门板、钢钉、销钉或挂线钉，取决于操作。

lineal measure (run) The length of something.
长度测量 某物的长度。

line and level (line level) A spirit level suspended at the centre of a taut horizontal string line. With skilled use, the level can measure accurately to +/- 1mm per metre.
挂线水平仪 一拉紧水平标线中心悬挂的水准仪。熟练使用，可准确测量水平面，精确度为每米+/-1mm。

linear cover The width of a tile minus any side lap.
平瓦铺宽 屋面平瓦的宽度减去所有搭盖宽度后的长度。

linear diffuser (slot diffuser, strip diffuser) A *diffuser that comprises a number of slots through which air can be diffused.
线型出风口（条缝型散流器，条形散流器） 由若干条缝构成的散流器，空气可通过该条缝扩散。

lined eaves *Eaves that contain a *soffit board. Also known as closed eaves.
封板檐 含有一挑檐底板的屋檐。也称为“封闭檐口”。

linen chute (laundry chute) A *chute that is used to collect dirty laundry.
污衣槽（布草井） 用于收集脏衣服的斜槽。

line pins Steel pins, 80 - 100 mm long, which are inserted into mortar joints at each end of a wall and used to hold a *builder's line. The bricklayer first builds up square and level corners of the wall. The pins are then inserted into the mortar joints and a builder's string line pulled taut between them. The bricks in the panel between the corners can then be laid level between the two corners.
挂线钉 80～100 mm长的钢钉，嵌入墙两端的灰浆接缝以固定施工放线。瓦工首先建造墙的直角与水平角。然后把挂线钉嵌入灰浆接缝，在挂线钉之间紧拉一施工直线。然后两墙角之间的砖块就可水平铺设在两角之间。

line-tapped connection A direct connection that is made to an electrical cable without having to disconnect it.
电缆直连 电缆的直接连接，无须断开它。

lining 1. First cover or layer of paper, paint, or plaster that ensures a better ground for the finish. 2. The internal surround of a door or window to cover the reveal.
1. 底衬 第一层的贴纸、涂漆或灰泥，为确保给完成面预留更好的安装基底。**2. 筒子板** 门窗内侧覆盖贴面的饰边。

lining paper The first cover of paper pasted on the wall to act as a base for the wallpaper.
底衬纸 黏附于墙上的第一层纸，用作墙纸的衬底。

link 1. Ring which connects to another ring to make a chain. 2. Reinforcement bars that bend to the profile of the desired cage, spaced at regular intervals, and to which the main reinforcement is attached. The main reinforcement is normally attached on the inside of the link bars. The link bars appear to wrap around the main reinforcement.
1. 链环 环相互连接从而构成一条链条。**2. 箍筋** 弯曲成所需笼状物外形的钢筋，隔一定间隔放置于所固定的主筋上，主筋一般固定于箍筋的内侧。箍筋卷绕着主钢筋。

linoleum (lino) A floor covering made from linseed oil (linoxyl). This type of covering is highly durable and is very commonly used in hospitals and clinics.
亚麻油地毡 由亚麻籽油制成的地面覆盖层。该种覆盖层相当耐用并常用于医院及诊所中。

lintel (lintol) A horizontal beam that spans an opening to support the load from the structure above. Used above doors and windows. *See also* LINTEL DAMP-COURSE.
过梁 横跨一开口上用于支撑上方结构负载的水平梁。用在门窗上方。另请参考“过梁防潮层”词条。

lintel damp-course (combined lintel) A lintel that is shaped to act as a *damp-proof course for a cavity wall.

过梁防潮层（复合过梁） 用作夹心墙防潮层的过梁。

lipping (banding, edging strip) A thin strip of timber fixed to the edge of a *door leaf to cover the *door face, and protect the door core.

门扇封边（封边木材，边沿衬条） 固定于门扇边沿的用于覆盖门面并保护门芯的一薄木片。

lip-seal joint A joint used for push-fit pipes where the socket end of the pipe contains an inward-facing projection of rubber which seals the pipe when the spigot end of a pipe is inserted.

唇形密封接头 用于推入式安装管道的接头，该管道的承端含有一向内的凸出橡胶，可用于安装管道承插端时密封管道。

liquefaction Transforming into a liquid. Occurs in saturated cohesionless soils when the pore pressure is increased and the effective stress is reduced, resulting in overall reduction in shear stress. *See also* QUICKSAND.

液化 当孔隙压力增加而有效应力减少时，剪应力总体减少，从而发生饱和无黏性土转变为液体。另请参考“流沙”词条。

liquefied gas A gas that has been converted into a liquid by pressurizing it and/or lowering its temperature. Examples of liquefied gas include butane and propane.

液化气 通过加压气体及/或降低其温度而转化成液体。液化气有丁烷及丙烷等。

L

liquidated damages An amount of money that is stated in the contract as the sum which would be payable in the event of a specified breach; the amount must be a realistic estimate of the loss that is likely to occur as a result of the said breach. The purpose of the term is to compensate and not act as a penalty. If the sum specified is excessive, then it is a penalty, and cannot be claimed.

违约赔偿金 合同中规定用于支付特定违约行为的一笔金额。该金额必须是对由于上述违约有可能造成的实际估算损失。违约金的目的是赔偿而不是罚款。如果规定金额过多，则变成了罚款，不可进行索赔。

liquid limit (LL) The point at which the water content in soil changes it from plastic to liquid behaviour. It is an *empirical boundary defined as the *moisture content at which the soil is assumed to flow under its own weight and is measured using a *cone penetrometer or *Cassegrande cup.

液限（LL） 土壤由可塑状态到流动状态时其界限含水率。假设土壤的含水量可使其在自重条件下流动，并使用圆锥贯入仪或卡氏液限仪测量，从而实证此界限。

liquid metal embrittlement A process whereby a metal becomes more *brittle when in contact with a liquid.

液态金属致脆 金属与一液体接触从而变得易脆的过程。

liquidus A line indicating the temperature at which a metal alloy of a specific composition will start to melt.

液相线 指特定成分金属合金开始融化时温度的连线。

litigation The progressing of a dispute to a court of law. Litigation differs from adjudication in that the decision of the courts is final and binding, and evidence is given under oath; however, the court's process is slow and costly.

诉讼 法庭处理一纠纷的过程。诉讼与裁决的不同之处在于，裁决是指法院判决是决定性的，并具约束力，证词是经宣誓的；然而，法院程序是缓慢并且昂贵的。

little joiner Small pellets, dowels, or pieces of wood that are used to fill, hide, and make good holes.

小件接合物 小的芯块、销钉或一小木块，用于填充、隐藏及修复孔。

live (live wire, live conductor) The red sheathed conductor in an electrical cable that carries an electrical current.

带电（带电电线，通电导体） 电缆中带电流的红色护壳芯线。

live edge (wet edge) The wet edge of a fresh coat of paint.

湿边 刚涂的涂漆层的潮湿边缘。

live load (imposed load) A load that can be removed or replaced (for example, furniture in a building); *See also* DEAD LOADS.
活荷载（外加荷载） 可撤掉或替换的一个负载（如建筑物内的家具），另请参考“恒荷载”词条。

live main A gas or water main that is transporting gas or water under pressure.
输送干管 在压力下传输气体或水的煤气总管或总水管。

live tapping The process of connecting a pipe, such as a *communication pipe, to a *water main while it is live.
带压开孔 在管道通水时，把管道（配水管）连接至干管的过程。

load A force or weight on a structure. *See also* LIVE LOAD and DEAD LOAD.
荷载 一结构上的力或重量。另请参考“活荷载”及“恒荷载”词条。

load-bearing Carrying load, for example, a **load-bearing wall** offers support to other structural components such as floors and roofs, rather than just acting as a *partition.
承载 承受负载，如承重墙可支撑其他结构组件如地板或屋面，而不仅仅是作为隔墙。

loader shovel Tracked or wheeled earth-moving equipment, used for loading and transporting bulk materials such as soil, rock, and aggregate into dumper trucks.
装载机铲斗 履带式或轮式运土设备，用于装载及输送散状材料如土壤、岩石及骨料至自卸车。

loading 1. Something being added or carried by something else. 2. A *filler or *pigment added to paint. 3. The use of *ballast to prevent uplift. 4. A pipe filled with sand to prevent distortion during bending. 5. Forces imposed on a structure.
1. 装载 由别的东西添加或承载的某物。**2. 填料** 添加到油漆的填充料或颜料。**3. 压载物** 用于防止升起的压载物。**4. 填沙** 用沙填充管道，以防在弯曲时发生变形。**5. 荷载** 强加于一结构上的力。

loading coat (loading slab) A concrete slab that is placed on top of asphalt or other tanking materials to resist hydrostatic forces or wind pressure which may otherwise force the waterproof material from its substrata.
承载板 放置于沥青或其他防水层材料顶部的一混凝土板，用于抵抗静水压力或风压，以防防水材料脱离其基底。

load path The route a force takes through a structure, from the point of application to the point of exit; for example, a *live load could be transmitted through a floor slab, beams, and columns to the foundation, where it is then distributed into the ground.
传力路径 一结构的力从受力点传递至出口点的路径。如活荷载可通过一楼板、梁及柱子传递至基础，然后分散到地基上。

load schedule A record to show the distribution of power across a number of electrical circuits or machines.
荷载一览表 显示若干电路或机械的电力分配的记录。

load shedding (automatic load limitation) The process where the power to certain customers or circuits is cut off due to a shortage in supply.
减载（自动限载） 由于供应不足而切断特定用户或电路的电源的过程。

load smoothing (resource levelling) *See* RESOURCE SMOOTHING.
负荷平滑（资源平衡） 参考“资源平滑”词条。

loam A material that contains roughly equal proportions of sand, clay, and silt.
肥土 含有大致相同比例的沙、黏土与粉土的土壤。

lobby (vestibule) A short passage or room with entrance and exit doors at each end of the room used to control the movement of people or air. The opening and closing of the doors at each end of the room acts to seal and secure the passage of people or air. Where people are being controlled, the entrance door can be opened

and the number of people passing through can be screened and controlled before exit, normally for security purposes. The room can also be used to prevent warm air escaping, or serve as an air trap that would restrict the development of fire.
大厅（前厅） 一小段通道或房间，该房间的每端有出入口门，用于控制人或空气的流动。房间每端门的开关可密封并确保人或空气的通过。如人流受到控制，出于安全考虑，可打开入口门，在人离开之前可筛选并控制一定量通过的人。该房间还可防止暖气流出，或用作气阱，防止发生火灾。

lobster-back bend (cut and shut) A pipework bend that is formed by cutting the pipe into angled segments, which are then joined together to form the bend.
虾米腰弯头（多节弯头） 把管道切割成一段段成角度的管，然后连接在一块形成弯管。

local control A device that is capable of locally controlling remotely located equipment, such as an air-conditioning system.
就地控制 可就地控制远处设备如空调系统的一个装置。

location block Thin strips of resilient plastic that are used to locate and help hold the position of glass in a window frame.
定位块 薄的弹性塑料带，用于定位并协助固定窗框中玻璃的位置。

location drawings (location plan, key plan) Plan drawings that show the placing of the building. The drawings are supported with dimensions to enable the positioning of the building and its components on-site.
位置图（索引图） 显示建筑位置的平面图。该图纸配有维度，可定位建筑及其现场构件。

L

location schedule A list of rooms with the finishings, fittings, and equipment to be located in the rooms identified. Normally, such schedules are used in buildings with many rooms, such as apartments or hotels.
房间布局一览表 包含将安装于各房间内的饰面、配件与设备的一览表。一般来说，此类一览表用于多房间的建筑中，如公寓或酒店。

lock A device fitted to a door or window, usually key-operated, to prevent entry.
锁 安装于门窗，通常用钥匙操作，防止进入的装置。

lock block A block of timber in a hollow-core door that accommodates the *mortise for the lock, and provides a fixing for any lock furniture.
锁块 空心门内的一木块，可安放锁孔，并为所有带锁家具提供固定作用。

lock joint 1. An interlocking joint used to join two pieces of timber together at right angles. 2. A type of standing seam used in sheet metal roofing.
1. 锁合槽榫 用于呈直角连接两块木材的连锁接合。**2. 咬口接缝** 金属板屋面使用的一种直立锁边。

lock nut (locking nut) A nut that is prevented from becoming loose under load. This can be in the form of a nut being tightened onto an existing nut, or having a nut which contains a portion that deforms elastically when tightened to prevent it from coming loose.
锁紧螺母（防松螺母） 可防止在负载下变松的螺母。可以是螺母固定在现有螺母上的形式，也可以是使一螺母部分在固定时发生弹性变形，以防止变松。

lock rail (middle rail) A *door rail that accommodates the mortise for a lock.
门锁中冒（中冒头） 安放锁孔的门扇冒头。

lockset (lock set) The complete set of hardware required to lock a door. Includes the keys, handles, levers, lock mechanism, escutcheons, and so on.
锁具（成套门锁） 用于上锁门的一整套五金，包括钥匙、门把手、锁杆、锁定机构、锁眼盖等。

lockshield valve (lock shield valve, LSV) A valve located on the return side of a radiator that controls the flow through the radiator to ensure that tall radiators receive enough hot water. Used to balance the system when radiators are first installed.
暖气片调节阀（LSV） 位于暖气片回水侧的阀门，用于控制经过暖气片的水流量以确保高处的暖气片有足够的热水。当暖气片第一次安装时，可用于平衡系统。

lock stile (closing stile) A *door stile that accommodates the mortise for a lock.
门锁梃（装锁竖挺） 安放锁孔的门竖梃。

lodge 1. Dwelling set up and used by stonemasons; used as the living quarters and workshop when major medieval structures were built. 2. A small building offering accommodation for short stays. 3. A small gate house at the entrance to an estate or park.
1. 小屋 由石匠建造及使用的住宅；当建造大型中世纪建筑时，可用作生活区及工作间。**2. 临时住宿** 为短暂停留提供住宿的小建筑物。**3. 门房** 房产或公园入口处的小门房。

loft (attic) 1. The space in the roof that can be used for storage, or converted to a room in the roof. 2. A floor of a commercial building, let unfurnished.
1. 阁楼（顶楼） 屋顶空间，可用作贮藏室或改变成顶部的房间。**2. 未完工楼层** 商业建筑的一未装修楼层。

loft insulation Thermal insulation that is placed both between and over the ceiling joists in a dwelling to reduce heat loss. Can be applied in the form of a batt, a board, a quilt, or as loose-fill material.
阁楼保温层 安装于住宅天花托梁之间及上方用于减少热损失的隔热层。应用形式有棉絮、板块、被子或其他疏松填料。

loft ladder (disappearing stair) A telescopic or folding ladder that enables access to the loft.
阁楼伸缩梯（隐藏式楼梯） 可伸缩或折叠楼梯，用于通向阁楼。

log-construction A form of wall construction using tree trunks/logs, which are positioned horizontally on top of each other. The logs can be planed to give semi-circles or flat sections.
圆木构造 使用树干/圆木水平层叠安装的墙构造。圆木也可刨成半圆或平剖面。

long arm Long pole with fixing on the end that allows it to be used to manoeuvre overhead equipment, such as roller shutter doors, Velux roof lights, roof vents, or other equipment that is too high to reach.
操纵杆 于末端固定的长杆，用于操控高处设备，如卷帘门、Velux 采光天窗、屋顶通风或其他过高而够不着的设备。

long float A flat trowel operated by two people for smoothing screed and concrete.
长抹子 两人操作的平泥刀，用于抹平砂浆及混凝土。

longscrew A pipe connector with threads that run parallel.
直螺纹接头 有直螺纹的管道连接头。

longstrip roofing Method of laying long strips of metal roof covering over a structure from end to end without the need for intermediate seams.
屋面长条扣板 在结构上的一端至另一端铺设长金属条屋顶覆盖层的方法，无中间接缝。

long sweep bend *See* SLOW BEND.
长半径弯头 参考“长半径弯头”词条。

looped-pile carpet Woven carpet where the yarn is looped through the backing fabric.
圈绒地毯 纺织地毯，其纱线环形穿过底布。

loose-butt hinge *See* LIFT-OFF HINGE.
抽芯铰链 参考“可脱卸铰链”词条。

loose-fill insulation Thermal insulation material that is either blown or poured into a loft space or cavity wall. A number of loose-fill materials are available including mineral wool, cellulose fibre, mineral fibre, polyester beads, and vermiculite.
松散材料保温层 吹入或倒入阁楼空间或夹心墙内的保温材料。松散料保温有矿物棉、纤维素纤维、矿物纤维、聚酯珠子及蛭石。

loose key 1. A tongue or *tenon that is not fixed to the main timber, hence it is loose and located into a

mortise in each piece of timber that it is fixed to. 2. A large T-shaped bar with a square recessed end that is used to turn stop cocks that are below the surface of the ground, often placed in the footpath or * pavement.
1. 插条榫 非固定于主木材的一榫舌或凸榫，所以它是不牢固的，安装于其所固定的每片木材的榫眼中。**2. 水龙头开关钥匙** 一种大型的T型匙，带一凹式方形端，用于转动地下的龙头，一般位于小路或人行道。

loose material Particles of the main substrate, such as rust, scale, or flacking, or dust and debris on the surface of a material. The particles are removed by blast cleaning, scrubbing, or with a wire brush.
松散物质 基底主材的颗粒，如锈、结垢、脱皮或材料表面的尘土及杂物。可通过喷砂清理、擦洗或钢丝刷除去这些颗粒。

loose-pin butt (pin hinge) A butt hinge with a pin that can be removed, enabling the door to be quickly unhinged by removing the pin. This avoids the need to unscrew the hinge.
抽芯铰链（销铰） 带一可拆卸销的普通铰链，通过拆除销芯可快速取下门。无须拧松铰链的螺丝。

loose ring An O-ring used to seal pipes or joints. The O-ring is not fixed in place, but seals itself through its elasticity as it is stretched over the inside pipe and is compressed when the outer pipe is pushed into place.
松套法兰 用于密封管道或接缝的O型环。O型环并非是固定安装的，而是通过自身的弹性在嵌进内管道时密封，当外管推到安装位置时O型环则被压紧。

loose side (slack side) The side of a veneer that was cut away from the main timber when it was formed. A timber veneer is formed by finely stripping timber with a cutting knife. As the knife peels the thin strip of a veneer from the timber, the side of the veneer that is in contact with the knife and pulled away from the main timber is called the loose side.
旋切单板背面（松面） 从主木材上切割下来的一块单板的一面。通过使用切刀精致剥离木材从而形成一薄木片。当切刀从木材上旋切出薄木片时，与切刀接触并从主木材上脱离的一面叫作单板背面。

loose tongue (key slip tongue, slip feather, spline) An oblong strip of timber (* tenon or tongue) that fixes two pieces of timber together by locating the timber in two mortises formed in the timbers to be joined. The tenon is not fixed to the main timber, hence it is loose and located tightly into a mortise in each piece of timber that it is fixed to. *See* LOOSE KEY.
插条榫 通过把榫条安装在将要连接的两木材中的两榫眼从而连接该两木材的椭圆形木块（凸榫或榫条）。该榫并非固定于主木材，因此它是松散的，并牢固嵌入其固定的每块木材的榫眼中。另请参考“插条榫”词条。

lost-head nail A nail with a small slightly pointed head, which makes the head difficult to see when the head is driven just below the surface with a * punch.
无头钉 只有一小尖钉头的钉子，当钉子打入表面时就很难发现钉头。

loudness The perceived intensity of a sound.
响度 声音可感知的强度。

louvre (US louver) A series of horizontal slats designed to admit air, light, and sound, but exclude rain. Used in doors, windows, lights, and ventilators.
百叶式 一连续水平板条，用于引入空气、光线及声音，但不包括雨水。用于门、窗、灯及通风机上。

low-angle light A light that impinges on a surface from a low angle.
低角度灯 以一低角度照亮一表面的灯具。

low bid (US) The lowest priced tender.
最低标价 价格最低的投标价。

low block A low-level podium.
矮工作架 矮的工作平台。

lower bound theorem A structure will not collapse under an external applied load if the distribution of

*bending moments are in equilibrium and the plastic moment of resistance is not exceeded.
下限原理 如果弯矩的分布处于平衡状态且不超过塑性抵抗力矩，那么一结构受到外加荷载时将不会倒塌。

lowering wedges Formwork stays or wedges that are eased before the formwork is stripped.
拆支撑楔块 拆除模板之前先拆除的模板支撑或楔块。

lower yield point The lowest value of stress after the onset of *yield, for example, before the load starts to increase again. Particularly related to annealed carbon steels.
下屈服点 产生屈服时（如在荷载再次开始增加之前）的最小应力值。与退火热处理碳钢密切相关。

lowest tender The tender that has the lowest price out of all those that have tendered. *See also* LOW BID.
最低投标价 在所有的投标中，其价格为最低的投标。另请参考“最低标价”词条。

low pressure hot water system (LPHW) A wet central heating system that operates at atmospheric pressure and the flow temperature of the water is at a maximum of 80℃. Used in the domestic sector and some small non-domestic buildings.
低压热水系统（LPHW） 一湿式集中供暖系统，在大气压力条件下运作，水流最高温度为80℃。用于住宅区及一些小型的非住宅大厦。

low-rise (low-rise building) A building with eight storeys or less.
低层房屋 楼层不超过八层的建筑。

low-velocity system A type of *air-conditioning system that moves air around the building at low speed through large diameter ductwork.
低速空调系统 一种空调系统，通过大直径管道以低速运转建筑周围的空气。

low voltage Electric current greater than 50 volts but less than 250 volts. Usually supplied from the mains at 240 volts.
低压电 在50至250伏特之间的电流。通常由总电源以240伏特供应。

LPHW *See* LOW PERSSURE HOT WATER SYSTEM.
低压热水系统（LPHW） 参考“低压热水系统”词条。

luder bands Elongated surface markings or minor indentations in sheet metal caused by discontinuous yielding.
拉伸应力痕 由于不连续屈服引起的金属薄片上的细长表面印痕或小压痕。

luffing (derricking) Reducing and increasing the radius of a crane by raising and lowering the jib. Luffing jib cranes are used on tight sites where over-sailing rights could not be obtained. As the crane passes the side of the site where it does not have the right to pass over, the crane jib is raised preventing any infringement of the neighbour's rights.
起重机臂的上下回转（用动臂起重机移动） 通过升起或降低起重机的臂来减少或增加起重机的半径范围。动臂起重机应用于禁止跨越的狭窄现场。当起重机经过现场（无权越过该现场）一侧时，起重机吊臂将升起，避免侵犯邻居的权利。

lug 1. A projecting piece of metal or plastic used for fixing. The lug projects out from the main unit and provides a fixing that enables the unit to be secured in position. Some fixing lugs have holes for screws or bolts, or fishtail ends that enable the lug to be bedded into mortar joints. 2. A metal end fixed to a wire used to make an electrical connection. 3. A plastic spacer used between the joints of tiles to accurately position floor or wall tiles.
1. 耳板 用于固定的一片突出的金属或塑料。耳板从主件伸出并把组件固定在位置上。有些固定耳板带螺丝或螺栓孔，或鱼尾形端，可把耳板嵌入砂浆缝。**2. 接线片** 固定于电线的金属端，用于进行电气连接。**3. 瓷砖定位器** 瓷砖接缝之间的塑料卡子，可准确定位地砖或墙砖。

lug end The end of the extensions on some window frames that are built into the wall reveal. Some windows have extensions on the sills that are longer than the window opening and are built into the wall; *See also* LUG

SILL.
窗台边耳 一些窗框的延长端，嵌入墙内的窗侧。一些窗户的窗台延长端比窗口长，并嵌入墙内；另请参考“窗台边耳”词条。

lugged pipe bracket A wrap-around bracket with a *lug for fixing pipes to a wall.
耳板管箍 环绕式带耳板的支架，可把管道固定于墙上。

lug sill A window sill that is longer than the window opening such that the ends of the sill are built into the wall at the jambs.
窗台边耳 比窗口长的窗台，窗台的末端嵌入窗侧壁处的墙内。

lumen (lm) The SI unit of *luminous flux.
流明 光通量的国际单位。

lumen method A method of calculating the average illumination in a room.
流明法 计算房间内平均照度的方法。

luminaire (lighting fitting) A device that holds or contains a lamp, and controls the amount of *luminous flux emitted from a lamp. To achieve the required *illuminance within a space, the choice of lamp must be combined with the correct choice of luminaire.
照明装置（照明配件） 固定或容纳一光源的装置，并控制光源所射出光通量总量。为取得一空间内所需的照明度，选择光源的同时必须配合正确的照明装置。

luminance (*L*) A measure of the brightness of light emitted or reflected from a surface in a particular direction. Usually measured in *candelas per square metre.
亮度（*L*） 测量一表面以特定方向射出或反射的光亮度。一般以“坎德拉/每平方米”为单位。

L

luminous ceiling A suspended ceiling comprising back-lit translucent panels.
流明天花板 含有背光半透明面板的吊顶天花。

luminous flux The amount of visible energy emitted by a light source, measured in *lumens (lm).
光通量 光源射出的可见能量总量，测量单位为流明（lm）。

luminous intensity (*I*) The amount of energy emitted from a light source in a particular direction. Measured in *candelas (cd).
发光强度（*I*） 一光源以特定方向射出的能量总量。一般以“坎德拉（cd）”为单位。

lump hammer A hammer with a large heavy head used to break up objects such as concrete, drive in large pins or nails, or ease stubborn objects.
大槌 带一沉重大头的锤子，用于打碎如混凝土等物体、钉入大的钉子或敲松坚硬物体。

lump-sum contract (fixed price) A contract where the price is fixed and agreed with the client before the work starts. Cost-plus is a similar type of arrangement with most of the work being fixed price, but variations allow for work that is difficult to determine at the start of the contract.
总价包干合同（固定价格） 价格是固定的合同，且在开工之前已与客户达成协议。与成本加成定价法类似，大部分工程的价格固定，但容许合同开始之前很难确定工程部分作变更。

lux (lx) The SI unit of *illuminance. One lux is equivalent to one *lumen per square metre.
勒克斯（lx） 照度的国际单位。1 勒克斯等于 1 流明/平方米。

luxmeter A device used to measure the amount of *illuminance.
照度计 测量照度量的仪器。

macadam In 1869 the first *flexible pavement road surface was built in London using tar mixed with aggregate and became known as '**tar macadam**' after its inventor John Loudon McAdam.
碎石路面 这是于1869年在伦敦第一条使用柏油混合骨料铺设的柔性人行道路面，以其发明者 John Loudon McAdam 的名字命名为“柏油碎石路”。

machine direction The direction in which something (e.g. a *geotextile) is manufactured. It is the direction it moves through and exits the machine and this defines the length of the product. *See also* CROSS-MACHINE DIRECTION.
纵向 某物（如土工布）的生产方向。这是一物体穿过机械并从机械出来的方向，该方向决定了该产品的长度。另请参考“横向”词条。

made ground (made-up ground, built-up ground) Ground that has been formed by *fill.
人工填土地面 由填料堆填形成的地面。

maglev (magnetic levitation) A high-speed electrically operated train that uses a magnetic field to glide above the track.
磁悬浮列车（磁悬浮） 利用磁场在轨道上滑行的高速电动列车。

mahogany A type of timber commonly used in furniture applications.
桃花心木 一种常用于制作家具的木材。

main contractor (general contractor, prime contractor) The party who has the main contract with the client is responsible for the work on-site and the employment of subcontractors and specialists necessary to complete the construction work. Some contractual arrangements may result in contractors being employed directly by the client; in such cases the main contractor would be the party responsible for the majority of the construction work.
总包商（总承包人） 与客户签订主体合同的一方，并负责现场工程及雇佣需要的分包商与专业人士来完成该建筑工程。一些合同安排可能是由客户直接雇佣承包商，在这种情况下，总包商则为负责主体建筑工程的一方。

main distribution frame (building distribution frame) A place where telephone *terminals and exchange equipment is housed.
总配线架（建筑配线架） 放置电话终端及交换设备的地方。

mains voltage Low voltage power, defined as 240 volts in the UK and used, for example, to power household appliances.
电源电压 低压电力，在英国界定为240伏特，用于为家用电器供电。

maintainability The ability to keep something operational by regularly checking it and undertaking repairs as and when required.
可维修性 可通过定期检查某物并在需要时进行维修从而使其正常运作。

maintenance The work necessary to keep things operating properly and in a good state of repair. To ensure that plant and equipment are always operational, a 'planned preventive maintenance' programme should be used. To ensure that machinery operates continually, parts that wear and have a limited life span need to be checked and replaced before they cease to function; regular maintenance should ensure that the parts are always

in good working order. Other decorative items will also need to be checked and any necessary work undertaken to ensure that they are visually appealing.
维护 为保持事物正常运作并处于良好维护状态所需的工作。为确保设备一直运作，应启动“计划性定期检修”方案。为确保机械持续运作，应检查易磨损及有使用年限的配件并在它们终止运作之前更换。定期维护可确保配件处于良好的工作状态。其他装饰性物品也需检查并采取必要工作以确保其仍具有视觉吸引力。

maintenance painting Repair and repainting to ensure the finish remains visually appealing or functional if protecting the substrate from deterioration, for example, protecting steel from rusting and timber from decay.
维护涂装 维修并重新涂漆以确保完成面仍具有视觉吸引力，或为保持实用性，预防基底变质，如保护钢以防生锈，保护木材以防腐烂。

maintenance period The time that the supplier or contractor agrees to keep the property and equipment free from defects; in contractual terms this may be referred to as the defects liability period.
维修期 供应商或承包商同意保持房屋与设备无缺陷的时间段；在合约条款中又被称为保修期。

make good To repair an item so that it is in proper working order, and so that it appears as new.
修复 维修一物品从而使它可正常运转，并且看上去像新的一样。

make up To add material so that it is the full amount or finish something off so that the work is aesthetically pleasing and complete.
补足 添加材料使某物补满，或完成某物使其完整美观。

malleable How easy or difficult it is to shape or deform a material.
可锻的 一材料成形或变形的难易程度。

M

malleable cast iron White cast iron that has been subjected to different heat treatments to increase its graphite content, resulting in a relatively ductile and malleable material.
可锻铸铁 为增加其石墨含量而经受不同热处理的白口铸铁，最终成为一种相对易延展及可锻的材料。

management contract (US construction management) A standard form of contracting where the main contractor acts as the procurer and manager of the contract and sub-contract packages. The chief contractor acts as an advisor to the client, setting up and placing the contracts for the design and construction works, although this contractor never enters into a contract with the suppliers. The contracts are made directly with the client. The chief contractor organizes and manages the work packages.
管理合同（美国 施工管理） 签约合同的标准形式，总包商作为合同及分包合同工程的采购人及管理者。总承包商充当客户的顾问，为设计与建筑工程订立及发放合同，但不会与供应商签订合同。供应商直接与客户签约。总承包商组织并管理施工工程。

managing contractor The chief contractor leading the management contract.
施工总承包管理方 负责管理合同的总承包商。

mandatory Something that must be undertaken, followed, or complied with by law or any other authority.
强制的 根据法律或任何其他主管部门必须执行、遵守或服从的。

M & E (Mechanical and Electrical) Building consultants and contractors who are specialists in building services, including such activities as: *heating, cooling, *air conditioning, *plumbing, *ventilation, fire control, lifts, escalators, security, and *renewable energy. As M & E works are now a significant part of most building projects, some consultants and contractors are now taking the lead role in building projects.
机电 在建筑设施方面（如加热、制冷、空调、给排水、通风、消防、升降梯、扶手梯、安保及再生能源）作为专业人士的建筑顾问或承包商。由于机电工程现在是大多数建筑项目的重要部分，其顾问与承包商在建筑工程中起主导作用。

man door (wicket door) A small gate or a door of sufficient height and width to allow a person to enter easily. Can be a door within a larger door; the larger door could allow plant and other mechanical equipment into

a shed, however, when not in use the smaller door would allow people to enter and egress.
人行小门（边门） 有足够高度及宽度允许一人轻易进入的窄门。可以是一较大门内的一个小门，较大的门可允许设备及其他机械设备进入小屋，然而，当大门不使用时，可允许人从小门进出。

manhole *See* INSPECTION CHAMBER.
检修孔 参考“检查井”词条。

man hour The quantity of work performed by an average worker during one hour.
工时 一普通工人在一小时内所执行的工作量。

manifold A chamber or pipe with a number of openings that branche into a series of smaller pipes.
歧管 带若干开口的管道，可分支成一系列较小的管道。

man lock An air lock entrance to a pressurized working chamber, tunnel, or shaft.
人孔闸 进入一受压工作腔、隧道或竖井的气闸入口。

manometer An instrument for measuring fluid pressure, such as a *piezometer.
测压计 测量液体压力的仪器，如压强计。

manor house The house which is the main administrative centre of a country estate or manor. The houses are sometimes fortified.
庄园主宅第 田庄或庄园的主要行政中心。此类房子有时是经过加固的。

manpower The power provided by human effort; the number of people needed to physically lift something.
人力 人工作产生的力量。物理上提升某物需要的人数。

mansard 1. A roof that has two slopes on each face, with the lower at a steeper angle than the upper slope, also known as a French roof. 2. The area directly under (enclosed by) a mansard roof.
1. 孟莎式屋顶 每面均有两个斜坡的屋顶，下部坡较上部坡陡一些，也被称为法国式屋顶。**2. 复折式屋** 由孟莎式屋顶包围并位于其正下方的区域。

mansion A very large well-appointed house often with more than ten bedrooms, a ballroom, library, and set in sizable grounds. In the US, a mansion is a building with over 740m^2 of internal space set in extensive grounds.
宅邸 一栋设备完善的大房子，一般含有不少于十个卧室、一个舞厅、图书馆及大的铺砌地面。在美国，宅邸指的是内部空间超过740平方米巨大铺砌地面的建筑。

manufacture To produce something, typically on an industrial scale.
工厂生产 制造某物，通常是工业规模上。

manufactured aggregate Thermally or otherwise modified *aggregate.
已加工骨料 热处理或以其他方式改良的骨料。

manufactured gas A mixture of gases obtained by thermal decomposition of oil (such as petroleum gas), destructive distillation of coal (such as coal gas), or steam reaction.
人造煤气 通过热分解油（如石油气）、分解蒸馏煤（如煤气）或蒸汽反应而取得的油混合物。

marble A *metamorphic rock formed from carbonate minerals, such as calcite or dolomite that have recrystallized under heat and pressure. Used as a decorative stone for tiles, worktops, and sculptures. Marble cladding is the use of a thin marble slab to face items such as columns to achieve a decorative effect. Marbling is representing the appearance of the crystalline structure of marble artificially—by the use of paints.
大理石 由碳酸盐矿物如方解石或白云石在加热及压力条件下再结晶形成的变质岩。用作砖、工作台及雕像的装饰石材。大理石包层是指柱子等物品表面使用薄的大理石板以取得装饰性效果。仿大理石涂料是通过使用涂漆，呈现大理石晶体结构外观。

margin 1. The outer edge of an enclosed area. 2. Boundary limit. 3. Profit.
1. 边缘 密封区域的外缘。**2. 界限**。**3. 利润**。

margin trowel A thin rectangular *trowel.

M

边角抹子 一种薄的矩形抹子。

marker tape *See* TRACER.
警示带 参考"警示物"词条。

marking drawing A drawing showing the position of building elements, often used to show the location of prefabricated modules and components. The drawing shows the location on a grid plan and identifies the position of each component. Each component is given a reference number in relation to its grid position.
标注图纸 显示建筑构件位置的图纸，常用于显示预制模块及组件的位置。该图纸显示了坐标网上的位置并标识了每个组件的位置。每个组件均有一个与其网格位置相关的索引号。

marking gauge (joiner's gauge) A tool used to mark a line parallel to the edge of a piece of wood. The tool consists of a wooden block with a steel projecting pin, the block slides along a bar, which can be fixed at any point to define the depth of the line.
画线规（木工画线器） 用于标记一条平行于木材边缘的线的工具。该工具由一木块以及突出钢针构成，木块沿着一根棒滑动，该棒可固定于任何点，以确定线条的深度。

marking out *See* SETTING OUT.
放样 参考"放线"词条。

marl A naturally occurring mixture of clay and calcium carbonate (lime). Used as a fertilizer and to soften water.
泥灰 自然形成的泥土与碳酸钙（石灰）的混凝物。用作肥料并可软化水。

marsh An area of low-lying water-logged ground.
沼泽 低洼积水区域。

mash hammer *See* LUMP HAMMER.
大锤 参考"大槌"词条。

masking The act of concealing the existence of something by obstructing the view of it.
遮蔽 通过挡住某物来隐藏它的存在。

mason A craftsperson skilled at working with stone and brickwork. Traditionally, the term was reserved for those who could cut, chisel, and work stone; in some areas it is now extended to include specialist bricklayers.
泥瓦匠 在砌石及砌砖工程方面技巧熟练的工匠。传统来说，该术语指的是那些会切割、雕刻、从事石块的人；在一些区域，其使用范围已扩展至包括专业瓦工。

masonry A term used to refer to the construction of walls, dwellings, buildings, etc. using essentially brick, mortar, and concrete.
砌筑 指的是主要使用砖块、砂浆及混凝土建造墙、住房及其他建筑物等。

masonry cement *See* CEMENT.
砌筑水泥 参考"水泥"词条。

masonry drill *See* TUNGSTEN-TIPPED DRILL.
砖石钻 参考"钨焊接钻头"词条。

masonry fixing Any fittings used to fix an object to brickwork, for example, an expansion bolt.
砖石固件 用于把一物体固定于砌砖上的任何配件，如膨胀螺栓。

masonry nail A square twisted hardened steel nail that can be hammered into brickwork.
墙钉 可钉入砌砖的方形螺旋硬化钢钉。

masonry paint Special type of paints which are typically smooth and textured, and are suitable for application to exterior walls of brick, block, concrete, stone, or renderings. Fine cracks in the substrate can often be hidden using the sand-textured material.
外墙漆 一种平滑且有纹理的特殊涂漆，适用于砖块、砌块、混凝土、石块或抹灰的外墙。可使用砂纹型材料掩盖基底的细裂纹。

masonry primer A primer usually based on asphalt used to prepare masonry specimens for bonding with other asphalt-based materials.
砖石底漆 以沥青为基底的一层底漆，用于为砌石试块黏合其他沥青基材料做准备。

masonry unit A building block, brick, or stone.
砌体 建筑的砌块、砖块或石块。

mason's putty A lime putty to which Portland cement and stone dust have been added; especially used in ashlar work.
瓦工油灰 添加到硅酸盐水泥与岩粉中的石灰膏，特别用于琢石工程中。

mason's scaffold A free-standing *scaffold.
瓦工脚手架 是一种独立式脚手架。

mass concrete An amount of concrete, which does not contain reinforcement, but is sufficiently large to develop cracks from the heat of hydration if the amount of heat generated is not controlled.
素混凝土 不含钢筋的混凝土，如果不控制水化热作用所产生的热量，体积过大时将会出现裂缝。

mass haul diagram A graph to illustrate the volume of cut and fill in earthworks. The horizontal axis represents chainage; the vertical axis represents cut and fill—cut is shown as positive and fill as negative *ordinates.
土方调配图 一个显示土方工程中开挖及填埋量的图表。横轴指的是里程距离。纵轴则指开挖与填埋——用纵轴正值表示开挖，纵轴负值表示填埋。

mat 1. A US term for a *footing. 2. A mesh of reinforcement. 3. A *filament resin mixture used in *fibre-reinforced plastics.
1. 支柱底板 在美国是表示“基础”的一术语。**2. 钢筋网**。**3. 纤维毡** 用在纤维增强塑料中的细丝树脂混合物。

mate A labourer or helper who assists with the organization and preparation of materials for the tradesperson.
助手 协助专业人员组织及准备材料的劳工或帮手。

material (building material) The matter or substance used to make things, particularly material used to construct building, such as aggregates, bricks, blocks, steel, and plaster.
物料（建筑材料） 制作某物使用的物质，特别是用于建造建筑所用的材料，如骨料、砖块、砌块、钢筋及灰泥。

matt A non-gloss flat finish, particularly with regards to paintwork.
哑光 无光泽平面，特别与油漆作业相关。

mattock A type of pickaxe, with a long wooden handle containing a metal head that is pointed at one end and flattened (spade-like) at the other.
十字镐 一种鹤嘴锄，带一长木柄，木柄上金属头一端是尖的，另一端则为扁平（铲子形状）的。

mattress 1. A large concrete ground slab used to support plant and equipment. 2. A layer of blinding concrete to seal something. 3. A layer of geotextile weighted down with rock to prevent *scour.
1. 混凝土地面板 用于支撑设备的大块混凝土地面基板。**2. 混凝土垫层** 用于密封某物的一层基础垫层混凝土。**3. 砂石垫层** 使用岩石压住的一层土工布，可防止冲刷。

MCB (miniature circuit breaker) Automatic breaker positioned in an electrical circuit that breaks the circuit when there is an overload. The circuit can be reconnected once the fault is diagnosed and corrected.
小型断路器 安装于电路内的自动断路器，可中断出现超负荷的电路。一旦诊断并修正好故障后可重新连接电路。

MDF (medium density fibreboard) A timber product made of wood fibres utilized as a board.
中密度纤维板 使用木纤维制成的木板。

mechanical properties The *properties of a material in relation to stress and strain.

M

机械性能 一种材料涉及应力与应变的性质。

measure A procedure to determine the size (length, area, or volume) or weight of something, by using a measuring device such as a measuring tape or balance.
测量 通过使用测量仪器如卷尺或天平确定某物尺寸（长度、面积或数量）或重量的过程。

measure and value contract A standard contract where the items of work are described in the * bill of quantities. The bills are priced by the contractor and submitted as part of the tender. Once the work is completed the prices are used to calculate the value of work. The work is remeasured to check the actual quantity of work completed.
计量估价合同 于工程量清单中载有工程项的标准合同。该清单由承包商报价并作为标书的一组成部分提交。工程竣工后，使用清单上的价格进行计算工程的价值。重新计量该工程以检查实际竣工工程量。

measured item Part of works which is included within the descriptions in the * bill of quantities.
计量项 工程量清单描述中所包含的一部分工程。

measured quantities *See* MEASURED ITEM.
计量量 参考“计量项”词条。

measured separately (ms) Aspect of the works which is quantified under a different section of the * bill of quantities, due to its size or unique qualities.
单独计量 由于其尺寸或特有的质量，在工程量清单不同部分中加以定量的工程。

measurement The act of recording quantities, prices, and valuing work, the process of which is undertaken by a * quantity surveyor. The main part of the process is * taking off the quantities from the drawings so that the work can be priced in accordance with the * Standard Method of Measurement.
测量 由预算师记录数量、价格并评估工程。该过程的主要部分是根据图纸测量数量，从而按照标准计量方法给工程报价。

measuring frame *See* BATCH BOX.
量斗框 参考“量斗”词条。

measuring tape A long, narrow strip of metal or plastic that contains marks at precise intervals (in millimetres, centimetres, metres or inches, feet, yards, or a combination of various units) to enable distances or lengths to be measured.
卷尺 一条长的窄的金属或塑料带，上有精确距离间隔标记（单位有毫米、厘米、米或英寸、英尺、码或者是各种单位的结合），可用于测量距离或长度。

mechanic A skilled person who repairs, maintains, or operates machinery or engines.
机修工 一名维修、维护或操作机械或发动机的技术人员。

Mechanical and Electrical *See* M & E.
机械和电气 参考“机电”词条。

mechanical barrow *See* MOTORIZED BARROW.
机械手推车 参考“电动手推车”词条。

mechanical computer-aided engineering (MCAE) The use of computer software packages to aid mechanical design and engineering.
机械计算机辅助工程（MCAE） 使用计算机软件包协助机械设计与工程。

mechanical degradation of timber A form of degradation of timber, it is the most common type of degradation that occurs in timber when subjected to continuous loading for long periods of time. For example, after 50 years the load that a piece of timber can withstand is less than half the stress it can carry at the onset of loading. Similarly, there is a reduction in stiffness (* Young's modulus of elasticity) with time. Another form of mechanical degradation is the induction of compression failure in the cell walls of the timber. Timber overstressed in compression in the longitudinal direction forms kink bands and compression creases. Such defects can reduce the tensile strength of the wood by 10 – 15%, but the loss of toughness under impact conditions can be as high as 50%.

木材的力学性能退化 木材退化的一种形式。当木材受到长时间的持续荷载时，力学性能退化是木材的最常见退化方式。例如，在五十年后，一片木材可以承受的应力少于它一开始可承受荷载的一半。同样地，随时间的推移，硬度（杨氏弹性模数）会降低。力学性能退化的另一个形式是诱导木材细胞壁出现压缩破坏。木材的纵向受应力过度会形成扭折带及应力折痕。这些缺陷会把木的抗拉强度降低 10 – 15%，但在冲击条件下，韧性损耗高达 50%。

mechanical engineering A profession that deals with the discipline of designing, constructing, and operating machines, tools, and mechanical plant.

机械工程学 涉及设计、建造及操作机械、工具及机械设备原理的学科。

mechanical floor (mechanical level, machine levelled floor) A floor in a building that contains mechanical plant such as air-conditioning units. Some mechanical floors have to be designed to take large imposed loads from the plant and need to be perfectly level to ensure that the equipment runs smoothly.

设备层 一建筑物中用于放置机械设备如空调组件等的楼层。有些设备楼层的设计必须可承受设备的外加大荷载，且必须完全水平以确保设备平稳运作。

mechanical room A room that houses mechanical plant.

机械房 放置机械设备的房间。

mechanic's lien The right over property if payment is not made for work undertaken, the supplier of materials, labour, or professional services can stop ownership of the property being transferred if a debt is not settled.

施工留置权 未能向已执行工程支付款项时的对财产的拥有权，如果不偿还债务，材料、劳动力及专业服务的供应方可阻止财产所有权被转移。

medium 1. The material substance that something (such as heat and sound) can be transmitted through. 2. Something that is in the middle, i. e. between two extremes.

1. 媒介 可传播某物如热及声音的物质实体。**2. 中间物** 位于中间的某物，即两个极端的中间。

medium-density fibreboard *See* MDF.

中密度纤维板 参考“MDF”词条。

medium rise Buildings between four and six storeys high. Buildings tend to be defined as low rise, less than four storeys or high rise, more than six; however, buildings of four to six storeys are sometimes described as medium rise.

中层建筑 楼层为 4 至 6 层之间的建筑物。少于四层的则定义为低层建筑，而多于六层的为高层建筑；然而楼层为 4 至 6 层之间的建筑物有时被称为“中等高度建筑”。

medium voltage Electrical supply between 1 kV and 50 kV, medium voltage installations would include any builders' work, such as the laying of cables, which will host the supply.

中压 电压在 1 kV 与 50 kV 之间的供电，中压安装包括某些建筑人员的工作，如铺放电缆以作供电用途。

meeting (coordination meeting, progress meeting, management and design meeting, site meeting) The bringing together of two or more people in a formal setting to discuss project issues.

会议（协调会议、进度会议、管理与设计会议、工地会议） 把两个或更多的人集合在一个正式的场合以讨论工程问题。

megalithic A large stone that has been used to construct a structure or monument, either alone or together with other stones.

使用巨石的 用于建造一个结构或纪念碑的大石，可以是单独的或是伴有其他石块。

meltdown Devastating collapse or breaking, for instance, when a nuclear reactor melts due to overheating of the fuel rods, resulting in radiation pollution.

熔毁 灾难性崩溃或破裂，如当一个核反应堆由于燃料棒过热而熔化，从而导致辐射污染。

member 1. A structural element in a building, for example, a beam. 2. A person belonging to a specific group.
1. 构件 建筑物的一个结构部件，如横梁。**2. 会员** 属于一个特定团体中的一员。

mensuration The calculation of lengths, areas, and volumes from measurements or dimensions.
测量法 根据测量结果及尺寸计算长度、面积及数量。

mesh An object consisting of semi-permeable barriers made of connected strands of metal, fibre, or other flexible/ductile materials. Mesh is similar to a web or net in that it has many attached or woven strands.
网丝 由半渗透栅栏组成的一物体，栅栏由金属、纤维或其他柔韧材料相连而成。网丝类似于网格或网，有很多连接或交织的丝线。

metal arc welding (SMAW, MMA) A welding process utilizing an electrode.
金属电弧焊（SMAW，MMA） 一个使用电焊条的焊接过程。

metal primer Paint used as a protective and first coat on iron or other metals.
金属底漆 用作铁或其他金属保护层及底涂层的涂漆。

metal-sheathed mineral-insulated cable Cables which are used primarily as heating units and power cables. Mineral-insulated cables typically comprise one or more wire conductors contained within a bendable metal sheath.
金属护套矿物绝缘电缆 主要用作供热组件及电源线的电缆，矿物绝缘电缆一般含有一个或多个导线，包含在可弯曲的金属护套内。

metal spraying (metallization) General term applied to the spraying of one of several metals onto a metal substrate.
金属喷涂（金属喷镀） 把几种金属的其中一种喷涂到一个金属衬底上。

M

meter A device for measuring the amount of flow, e. g. a *water meter. A meter bypass allows the flow to bypass the meter such that the amount is not recorded. It can be used for fire-fighting equipment so that no charge is incurred by the owner of the property for the water used by the fire brigade.
仪表 测量流量总量的设备，如水表。旁路仪表可允许流量绕开该仪表，从而不记录该总量。可用于消防设备，那么消防队所用的水将不会对业主产生费用。

method of measurement The technique or system used for describing and measuring units of construction. *See also* STANDARD METHOD OF MEASUREMENT.
计量方法 用于描述及测量建筑组件的方法或系统。另请参考“标准计量方法”词条。

method specification Detailed description of how to complete the task.
方法规范 关于如何完成工作的详细说明。

method statement A document that identifies how a particular job or task is to be undertaken; this is particularly relevant with regard to developing a safe working practice.
施工方案 一份说明如何完成一项特定工作或任务的文件，特别指的是制定安全操作准则。

micaceous iron oxide paint A high build finishing or undercoat containing micaceous iron oxide.
云母氧化铁涂漆 一种含云母氧化铁的厚涂完成面或底漆。

MICE Membership of the *Institute of Civil Engineers, which can either be as an *Incorporated Engineer (IEng) or *Chartered Engineer (CEng) level.
土木工程师协会会员 土木工程师协会会员，也可以是注册工程师或特许工程师级别。

microbore (microbore pipework) Very small pipework diameter, typically 6, 8, 10 or 12 mm. Used in central heating systems to enable the pipework to run up through walls and under floors.
微内径（微内径管道） 非常小的管道直径，一般为6、8、10或12mm。用在集中供暖系统中，可使管道穿过墙壁及安装在地板下。

microtunnel A very small tunnel, used for inserting pipelines. *See also* PIPE JACKING.

微型隧道 一条非常小的通道，用于嵌入管道。另请参考“顶管技术”词条。

microvoid coalescence (MVC) The formation of very small and minute cracks or surface irregularities occurring in a material due to a change in chemistry.
显微空穴聚结 一材料由于发生化学变化而出现细小裂缝或表面不均匀性。

middle-third rule A rule that states there will be no tension in an unreinforced wall if the resultant force lies within the middle-third of the wall.
三等分法则 一条说明如果合力位于墙中间的三分之一处，那么无钢筋墙将无张力的法则。

MIG (metal inert gas welding) A type of welding using an inert gas, primarily used on *non-ferrous metals.
金属惰性气焊 使用一种惰性气体的焊接，主要应用于非铁金属。

mild steel A type of *steel containing no more than 2% carbon content and very few additional alloying elements. This type of steel is most widely used, especially for civil engineering (structural applications); although prone to corrosion, a protective coating (galvanizing) is used.
低碳钢 碳含量低于2%且含极少额外合金元素的钢。这种钢使用最广泛，特别是应用于土木工程(结构应用)；容易受到腐蚀，所以会使用防护涂料（镀锌）。

milestone Significant point or event in a project. Key parts of the project may need to be delivered by this date or it may mark the point of key deliverables.
里程碑 一个工程中的重要节点或事件。工程的主要部分在该规定日期内完成，或者标记为交付重要成果的节点。

milling The action of grinding to cut or smooth metal objects.
碾轧 为切割或磨平金属而进行碾磨。

M

mill scale The surface of hot rolled steel, usually flaky in texture.
翘皮 热轧钢的呈片状纹理表面。

mimic diagram Illuminated diagram showing floor plans, zones, and equipment. It may provide information on building use and equipment location.
模拟图 显示平面图、区域及设备的发光图表。可提供关于建筑物使用及设备位置的信息。

mineral-insulated cable (mineral-insulated metal-sheathed cable, mineral-insulated copper covered, MICC) Electrical conducting cables made from copper.
矿物绝缘电缆（矿物绝缘金属护套电缆，矿物绝缘铜套） 由铜制成的导电电缆。

mineral spirits A solvent based on petroleum used as a paint thinner.
矿质油漆溶剂 石油基溶剂，可用作涂漆稀释剂。

mineral wool A light fibrous material used as an insulator.
矿物棉 用作绝热体的轻质纤维材料。

mitre Of matching ends. A mitre joint has been cut with a mitre saw to produce two ends to be joined at 45°.
斜接 端边接合。使用斜切锯切割的斜接头，可使两端接合角度为45°。

mix A mixture of different materials to form something, e. g. a concrete mix contains cement, water, sand, and aggregate.
混合物 不同材料的混合，可形成某物；如混凝土混合物包含了水泥、水、砂及骨料。

mixed construction (mixed use) Buildings with different uses, e. g. residential, commercial, and retail within one property.
混合构造（混合用途） 有不同用途的建筑，如一建筑内有住宅区、商业区及零售店。

mixed liquor A mixture of sewage effluent and organic material in an aeration tank for activated sludge treatment.
混合液 活性污泥法进行污水处理的曝气池中的排放污水与有机物的混合。

mixer A device for mixing various components to produce a *mix, e. g. an *asphalt mixer.
混合器 混合不同组件而形成一种混合物的装置，如沥青混合机。

mixing water (gauging water) The water used in a concrete or mortar mix. It should be free from contamination—typically drinking water is used.
拌和水（定量水） 用于混凝土或砂浆混合物中的水，应是不受污染——一般会使用饮用水。

moat A broad ditch filled with water that surrounds a castle. Traditionally, the moat was part of the castle's defences; however, more modern buildings are surrounded by water to give an aesthetic appeal.
壕沟 围绕着一城堡的灌满水的宽沟渠。壕沟曾经用作城堡的防御。然而，很多现代建筑由水围绕是为了添加美观。

mobile concrete pump A vehicle containing a pump and a folding boom with a delivery pipe. Used to pump ready-mix concrete into hard-to-reach places.
混凝土电动泵 由一个泵及一个带输送管的折叠式悬臂构成的一个交通工具，可用于把预拌混凝土泵送入难以抵达的地方。

mobile crane A *crane with a telescopic boom that sits on a wheeled undercarriage, forming a vehicle that can be driven from site to site.
移动吊车 位于轮式底盘上的有伸缩臂的吊车，可在场地间驾驶。

mobilization costs The cost of setting up a site, providing temporary accommodation, hoardings, and temporary power to the site.
开办费 布置一现场、为现场提供临时住处、临时围墙及临时电源的费用。

mock-up A scaled or full-size model of the building, roads, component, or item of plant.
实体模型 一个按比例缩小或完整尺寸的建筑、道路、构件或设备模型。

M

modification A *variation or change to that specified or previously described.
修改 已注明或之前已提及的变更或更改。

modular construction The use of standardized building components to form a building.
模块化搭建 使用标准化构件建成一建筑。

modular ratio The ratio of the *modulus of elasticity of steel to that of concrete, with regard to reinforced concrete.
模量比 钢筋混凝土中，钢与混凝土的弹性模数比例。

modular system Building constructed out of *prefabricated components that are designed to fit together quickly and easily.
模块化系统 使用可快速方便组装的预制构件建造的建筑物。

module (building module) Prefabricated unit of a building. Can be used within a modular system or could be a single unit, e. g. bathroom pod, or plant unit, that can be inserted into a more traditional form of construction.
模块（建筑模块） 建筑物的预制构件，可用在一个模块化系统中，也可用作单一组件，如浴室或设备组件，可嵌入更为传统的建筑中。

modulus of elasticity (E) *See* YOUNG'S MODULUS.
弹性模数（E） 参考“杨氏模量”词条。

modulus of incompressibility The ratio of pressure to volume change in a soil mass.
不可压缩模数 土壤体中压力与容积变化之比。

modulus of plasticity (plastic modulus) The ratio between the plastic moment of a beam and its yield stress.
塑型模数 横梁塑性弯矩与其屈服应力之间的比。

modulus of rigidity The ratio between shear stress and shear strain; *See also* POISSON'S RATIO.

刚性模数 剪切应力与剪切应变之间的比，另请参考“泊松比”词条。

modulus of rupture Breaking strength in a non-ductile or *brittle solid as measured by bending.
断裂模数 通过弯曲一个非可塑性或脆性固体从而测量其抗断强度。

modulus of volume change *Coefficient of volume compressibility divided by one plus the initial void ratio.
容积变化模数 体积压缩率除以一加上初始孔隙比的和。

moisture Liquid, such as water vapour, held within a solid or condensed on a cool surface.
潮湿 一固体中含有的或冷却表面上凝固的液体，如水蒸气。

moisture content The amount of moisture held within something, e.g. wood or soil, divided by the weight of the dry wood or soil, expressed as a percentage. A moisture meter is a device that can measure the amount of moisture held at the surface of wood or concrete. It consists of two prongs that are pushed into the surface—a digital display then indicates the amount of moisture presented.
含水量 某物体如木或土壤中所含水分量除以干燥木材或土壤重量，用百分比表示。湿度计可测量木材或混凝土表面的含水量，湿度计有两支插针可推入表面，数字屏显示出所测的含水量。

moisture-resistant adhesive (MR) An adhesive resistant to the ingress of water or moisture.
耐水黏合剂（MR） 可防止水或水分渗入的黏合剂。

molding (US) *See* MOULD.
线条（美国） 参考“模具”词条。

mole 1. SI unit of amount of a substance, defined at molecular level. 2. A massive sea wall constructed of mass stone, used to protect a harbour. 3. A tunnelling machine.
1. 摩尔 一物质（分子水平）的量的国际单位。**2. 防波堤** 由大块石块建造的结实海堤，用于保护一海港。**3. 隧道钻机**。

molten When a material has reached the liquid stage, passed its melting point.
熔化 一物质温度高于其熔点而变成液体。

moment A force multiplied by a lever arm, producing a turning effect.
力矩 力和力臂的乘积，从而产生转动效应。

moment distribution A structural analysis method for calculating the bending moments in statically indeterminate beams and frames. Initially, every joint is fixed, so as to develop fixed-end moments. Joints are then released one at a time with the fixed-end moments being distributed to adjacent members until equilibrium is maintained.
力矩分配法 计算静不定梁与框架弯曲力矩的结构分析办法。首先，固定每个连接点从而产生固定端力矩；然后每次释放一个连接点，使固定端力矩分配至相邻构件，直至维持均衡。

moment of inertia A measure of the resistance to rotational change. It is the sum of the mass of every component in a body multiplied by the square of their distance from the axis.
转动惯量 测量对于转动状态改变的抵抗程度。它是由物体上所有构件乘以各自与轴线距离的平方再相加所得。

moment of resistance The *bending moment that can be carried by a beam or element.
抵抗力矩 可由横梁或元件承载的弯矩。

momentum The measurement of movement, the capability of progressing forwards.
动量 移动的测量，可前进的能力。

monitor 1. To regularly check or take observations. 2. An instrument for viewing data, such as a computer monitor.
1. 监视 定期检查或观察。**2. 显示器** 用于查阅数据的仪器，如电脑显示器。

monitoring well A groundwater well to allow water quality and/or level to be checked regularly.
监测井 一地下水井，可定期检查水质及/或水平面。

monk bond Brick bond laid with two *stretcher faces between *headers; this is a variation on the *Flemish bond (header and stretcher).
跳丁砖砌法 砖块的两顺面铺设在砖块的丁面之间。这是梅花丁砌法（顺丁式）的演变。

monkey A mechanism to release a drop hammer on a pile-driving rig, so that it falls freely under its own weight.
打桩锤 释放打桩钻具上一个落锤的机械，落锤是在自重作用下自由落下。

monkey tail The scroll of the handrail at the bottom of a staircase.
扶手卷尾端 楼梯底端的卷状扶手。

monkey tail bolt *See* EXTENSION BOLT.
卷尾插销 参考“门窗长插销”词条。

monolith A single great stone (often in the form of a column or obelisk).
整块石料 单块巨石（通常是用作柱子或方尖碑）。

monorail A single track railway, typically with the carriages hung from overhead.
单轨 单轨轨道，一般在顶部有悬挂的托架。

monthly certificate (interim certificate, progress certificate) A statement issued to the client recording the value of work completed to date, and recommending to the client the sum of money that should be issued to the contractor. The certificate is usually issued by the client's representative, quantity surveyor, architect, or engineer who has measured and valued the work. The certificate also states the retention, which is the amount of money that is to be withheld.
月度证书（临时证书，进度证书） 发布给业主的一份声明，记录着到证书日期为止所完工工程价值，并建议业主应向承包商支付款项数目。一般是由估价该工程的业主代表、工料测量师、建筑师或工程师出具此证书。该证明还注明了保留金金额，即扣留的金额。

M

monthly instalments (interim payment, progress payment) Money given to the contractor for the work completed, the payment usually follows the valuation and issue of the *interim certificate.
月度付款（临时付款，进度款） 按已完工工程向承包商支付的金额，一般会在估价及临时证书发布后支付该款项。

monthly statement (monthly claim) Statement of work completed and value claimed.
对账单 已完工工程及申报价值声明。

monument A structure, building, or significant item created to commemorate an event or person of note.
纪念碑 为纪念一事件或著名人士而建立的一个构体、建筑物或标志性物件。

mortar A material used to bond bricks and blocks together in the construction of dwellings and buildings. Comprised mainly of sand, cement, and lime, although *lime mortars are made of lime, sand, and water.
灰浆 在房屋与建筑物施工中，用于把砖块与砌块黏合在一块的材料。主要由砂、水泥及生石灰构成，而石灰砂浆是由生石灰、砂与水构成的。

mosaic A decorative wall or floor covering using small coloured cubes (*tesserae) stuck onto the surface.
马赛克 粘在装饰墙或地板表面上的彩色小方块（色块物）。

motive power circuit A circuit that is capable of providing a high start demand, e. g. to drive a lift.
动力电源电路 可提供高压启动需求的电力，如驱动一电梯。

motorized barrow (mechanical barrow, US buggy, concrete cart) A power-driven two-wheeled cart, used for transporting heavy goods such as fresh concrete.
电动手推车（机械手推车，美国八吉车 ，混凝土手推车） 电力驱动的双轮车，用于运输重型货物，如新拌混凝土。

motorized valve A valve that is opened and closed by an electric motor, used in central heating systems.
电动阀 使用电动机开启及关闭的阀门，用于集中供暖系统中。

motorway A road with fast-moving traffic with limited access, typically with three lanes in each direction, separated by a central reservation barrier or grass verge, with hard shoulders on the outer edges.
高速公路 供汽车快速行驶、并有限制出入口的道路，一般每个方向有三条通道，使用中央分隔带或绿化带分隔开，外缘有硬质路肩。

mould (US mold) 1. A container for giving shape to things until it has set, e.g. concrete cube moulds. 2. Fungus. 3. Soil that is rich in humus (organic matter).
1. 模具 使某物成形并凝固的容器，如混凝土方块模型。**2. 真菌**。**3. 松软沃土** 富含腐殖土（有机质）的土壤。

mouldboard A level board or curved blade on a *bulldozer or *grader used in *earthworks to push soil.
铲刀 推土机或平地机上的一水平或弯型铲刀，用于土方工程中以推动土壤。

moulded brick A type of *brick shaped in accordance to desired or intended application.
模制砖 按所需或规定应用要求形状制作的砖。

mouth An opening or entrance.
开口 洞口或入口。

movable Capable of being transported from one place to another with ease.
可移动 可轻易从一个地方运输到另一地方。

movement The change in dimension or position of something, *See also* THERMAL MOVEMENT.
移动 某物尺寸或位置的改变，另请参考“热移动”词条。

movement detector A device that detects movement due to the disturbance in low-power microwaves that it omits, used to control automatic doors, lights, etc.
移动探测器 一种可探测由于物体发出低功率微波受到干扰而发生移动的装置，可用于控制自动门、照明等。

movement joint A joint that is flexible, to allow expansion and contraction to occur.
变形缝 柔性接缝，可膨胀或收缩。

muck Excavated soil or mining waste.
废土 开挖的土方或采矿的废土。

mud Wet sticky soil.
淤泥 湿润黏稠的土。

mudsill *See* SOLE PLATE.
底梁 参考“底板”词条。

mulch A protective covering of organic material spread on soil or around plants to prevent erosion and moisture loss from the ground.
覆盖层 覆盖在土壤上或植物周围的有机物保护层，可防止受到地面腐蚀及水分损失。

Mullen burst *See* BURST STRENGTH.
破裂强度 参考“破裂强度”词条。

mullion A vertical member that divides a window.
竖梃 分割窗户的垂直构件。

multi-storey building A tall building having a number of floors.
多层建筑物 有许多楼层的高层建筑物。

municipal engineering Roads, sewers, water supply, etc. related to towns and cities.
市政工程 在城镇及城市内的道路、排水沟及供水等。

mural A picture painted directly onto the face of a wall or ceiling.
壁画 直接涂画在墙面或天花上的图画。

murder clause (US) A contract clause that unfairly shifts the responsibility to another party who should not carry the burden.
霸王条款（美国） 将责任不平等地推卸到不应承担的另一方的合同条款。

mushroom construction Reinforced concrete construction where there are no beams—the column has been enlarged at the top (**mushroom-headed column**) to take the load of the floor slab.
无梁板构造 无横梁的钢筋混凝土构造，顶部扩大的柱子（柱帽柱子）可承载楼板的荷载。

mushroom-headed push button A mushroom-shaped stop button, typically used to stop a machine in an emergency and so it is painted red.
蘑菇头按钮 蘑菇状停止按钮，一般用于在紧急情况下停止一机械，所以涂成红色。

natural draught (US natural draft) The air flow that occurs due to the *stack effect.
自然通风 由于烟囱效应引起的气流。

natural ventilation The process of ventilating a space by opening windows and doors or purpose-provided ventilation openings. The ventilation is driven by the *wind effect and the *stack effect. Natural ventilation does impose restrictions on the plan form of the building, in that all of the ventilated spaces must be within a certain maximum distance from an open window or ventilation opening. However, this spatial constraint does permit the provision of natural lighting to much of the space, and allows the occupants good views of the outside world.
自然通风 通过打开窗户及门或专门的通风口给一空间通风的过程。通过风力效应及烟囱效应进行通风。自然通风对建筑的平面类型施加约束，因此所有的通风空间必须在敞开窗户或通风口的最远通风距离内。然而，空间约束可允许自然采光遍及大部分空间，并可使住户更好地观看外界。

neat cement grout A grout made of cement and water.
净水泥浆 由水泥与水制成的灌浆。

neat size Timber or other materials cut to the sizes needed; *See also* DRESSED TIMBER.
净尺寸 切割到所需尺寸的木材或其他材料，另请参考“刨光木材”词条。

needle A support beam that is inserted through a hole in the wall and is either propped at both ends to support the wall above, or rests on the wall and supports parts of the building or scaffolding.
临时托梁 穿过墙上洞口的支撑梁，在两端支撑以支承上方的墙，或是倚靠在墙上并支撑建筑物或脚手架的部件。

needle gun Power tool with 2.5 mm diameter steel rods, grouped together at the end of a gun which are vibrated against concrete and steel. The gun is used for cleaning surfaces and to roughen the surface.
气针枪 带直径2.5mm钢条的电力工具，钢条集合在枪的尾端，枪对着混凝土及钢振动。该枪可用于清洁表面及使表面粗糙。

needle scaffold Scaffolding or platform supported by beams that penetrate the wall (needles).
小梁托撑脚手架 由穿过墙（临时托梁）的横梁支撑的脚手架或平台。

needling Inserting a beam through a wall to act as a *needle.
穿墙梁 把一横梁穿过墙内以用作临时托梁。

negative moment A sign convention to denote the direction of a * moment, typically in the anti-clockwise direction.
负力矩 一个指示弯矩方向的符号，一般是以逆时针方向。

negative pressure Any pressure that is lower than atmospheric pressure.
负压力 低于大气压力的任何压力。

negative skin friction An additional load that can develop on the * shaft of a * pile due to * consolidation of the soil adjacent to the pile, often assumed to be in the region of 10 kN/m^2.
表面负摩阻力 由于桩附近的土的巩固而使桩轴产生额外负载，一般在 10 kN/m^2范围内。

negotiated contract Legally binding agreement where part or all of the contract terms are agreed by negotiation. The general terms of the contract are agreed and in place, with further discussions expected in order to arrive at an agreement for the remaining sections of the contract. The main contract may be a standard contract with parts, such as the contract sum, being negotiated, or the formation of the contract could be a process where the whole agreement is subject to negotiation. Usually an estimate of the contract value is stated, but the actual contract sum is discussed and agreed. In a negotiated contract the client may have selected the contractor on the ability to perform the works rather than on a tender value, therefore the value of the works and the contract sum should be agreed by negotiation and inserted into the contract.
协商合同 具有法律约束力的协议，通过协商达成该合同中的部分或全部条款。一般合同条款已达成并执行，为使合同的剩余部分达成协议，将按计划作进一步讨论。主合同可为标准合同，即仅有部分条款如合同金额需要协商，又或者整个协议都需要协商，则合同的订立需过一个过程。通常来说，所述的合同价格只是一个估计值，需经商讨并协议实际的合同金额。在协商合同中，业主可能是根据承包商的能力而不是投标金额来选择其执行工程，因此工程的价值跟合同金额应通过协商达成并载入合同。

negotiation stage Following the appointment of the main contractor, the contract value or contract sum is negotiated; *See* * NEGOTIATED CONTRACT. The term can also refer to * tendering.
协商阶段 确定总包商后，与其约定协商合同价值或合同金额；参考"协商合同"。该术语也可指"投标"。

neoprene A polymeric material from the family of rubbers used primarily in construction for plumbing applications.
氯丁橡胶 一种橡胶聚合材料，常在给排水管道应用。

nephelometric turbidity unit (NTU) A standard unit to denote the amount of * turbidity (suspended solids) in afluid. A nephelometer is typically used to measure turbidity. It measures the scatter of light caused by the suspended solids, when a light source is passed through a column (small sample bottle) of cloudy water.
比浊法浊度单位（NTU） 用于指示液体中浊度（悬浮体）数量的标准单位。一般使用浊度计测量浊度。使用一光源照射柱型容器（小样品瓶）中浑浊水，然后测量悬浮体引起的散射光。

net A safety net used to prevent workers and materials from falling.
网 用于防止工人及材料跌落的安全网。

net load The load on the underlying soil from the structure and backfill; it is the difference in the loading conditions on the soil before excavation and when the structure is complete. Net values are used in settlement calculations; *See also* GROSS LOAD.
净荷载 结构及回填对基土施加的负载，与挖掘前及结构完成时土壤所处的负载条件有所不同。净荷载应用在沉降计算中，另请参考"总荷载"词条。

neutral axis A location, in the cross-section of a beam, where the bending stresses and strains are zero. In a straight symmetrical beam, before loading, the neutral axis is located at the beam's geometric centre.
中性轴 横梁横截面内弯曲应力与张力为零的位置。在一直对称梁内，在其负载之前，中性轴位于梁的几何中心。

neutral conductor The neutral wire in an electrical circuit.

中性线 电路中的中性线。

neutralization To counteract something to make it ineffective, for example, to make a liquid neither acidic nor alkaline, or to zero an electrical charge.
中和 中和某物而使其不起作用，如使一液体既不是酸性也不是碱性，或使电荷为零。

newel The vertical member of timber into which the diagonal strings of a stair can befixed. The newel can form the base of the * newel post, which can be used to carry the stair * handrail.
楼梯扶手大柱 固定于楼梯斜梁上的垂直木构件。大柱可成为楼梯两端栏杆支柱的基础，可承载楼梯扶手。

newel cap The top of a newel post, often provides a turned decorative finish to the * newel post.
楼梯扶手柱帽 楼梯两端栏杆支柱的顶部，为其提供一个雕刻装饰面。

newel drop The bottom of the upper * newel post that projects below the ceiling level, making a decorative feature.
楼梯扶手柱垂饰 从上方楼梯两端栏杆支柱往下伸出低于底面的凸出，起到装饰作用。

newel end An escalator balustrade end.
自动扶梯端 自动扶梯的端部。

newel post The vertical timber at the end of a stair flight, used to carry the strings.
楼梯两端栏杆支柱 一段楼梯两端的垂直木构件，可承载楼梯梁。

nib A small projection from a surface.
尖 表面上细小的凸起。

nibbler A hand-tool that clamps together to cut or bite holes into thin sheets of metal. The tool can be powered or used with handles similar to pliers.
电冲剪 手持工具，钳紧以在金属薄板上切割孔。该工具可以是电动的或可与类似于钳子的手柄使用。

N

niche A recess in a wall, constructed to house an ornament or vase.
壁龛 墙壁内的一凹处，用于放置装饰品或花瓶。

nickel brass A copper alloy containing zinc and a small quantity of nickel.
镍黄铜 含锌及少量镍的铜合金。

nickel sulphide A compound that occurs in some glass-making processes. If trapped the compound can swell and crack the glass.
硫化镍 在一些玻璃制造过程中出现的一种化合物。如果夹杂在玻璃中，该化合物会膨胀并使玻璃破裂。

nicker A large flat mason's chisel used for cutting a groove into stone before splitting it.
石工扁凿 在破裂石块之前，用于在石块上开槽的一大型扁平石工凿子。

nidged ashlar (nigged ashlar) Stone dressed with a pointed hammer, giving a roughened effect.
琢石 使用一尖头工具修琢的石块，使石块粗糙化。

night lock (night latch) Cylinder lock or bolt.
弹子锁（弹簧锁） 圆筒销子锁或销栓锁。

night vent (night light) *See* FANLIGHT.
气窗 参考“腰头窗”词条。

ninety-degree bend (90° bend) A * quarter bend.
直角弯管

nippers (steel-fixer's nips, tower pincers) Pair of wire cutters used to tighten and cut tie wire when fixing and binding steel reinforcement.

钳子（钢筋钳；扎线钳） 当安装及捆绑钢筋时，用于拉紧及剪断扎线的一把剪钳。

nipple 1. A rounded fixing pipe with a one-way valve used to fix a grease pump so that lubricant can be pumped into moving joints. 2. A piece of pipe less than 12" long and threaded on both ends.
1. 黄油嘴 为把润滑剂泵入活动接缝，用于安装润滑脂泵的带单向阀的一圆形固定套管。**2. 双头螺纹管** 一段少于12英寸两头有螺纹的管道。

NIST（National Institute of Standards and Technology） A US federal agency measurement standards laboratory. Known as the National Bureau of Standards（NBS）between 1901 and 1988.
美国国家标准技术研究所 一个美国联邦机构计量标准实验室。在1901年至1988年期间被称为美国国家标准局。

noble metal A type of metal renowned for its inertness, with a high resistance to chemical reactions.
贵金属 一种以其惰性出名的金属，对化学反应的抵抗力相当大。

node An intersection point; on a curve where it crosses itself, where lines converge on a chart or diagram, or where members meet on a *truss.
节点 一个交叉点，弧形上交叉的点，图表上线条汇合的点，或桁架上构件相交的点。

no-fines concrete A concrete made from coarse aggregate and cement, with no sand or fine aggregates present. It creates a lightweight concrete with voids between the coarse aggregates.
无砂混凝土 由粗骨料与水泥制成的混凝土，不含砂或细骨料。属于轻质水泥，其粗骨料之间有空隙。

nog 1. A nog brick or block, a fixing brick made from wood or other malleable material. 2. A *nogging.
1. 加固用木砖 由木材或其他可锻材料制成的加固用木砖或木块。**2. 横撑**

nogging（noggin） Short horizontal timbers used to provide bracing and stiffen up stud work. Noggings are also used to provide grounds and fixings in timber walls to enable them to support radiators, shelves, and other fixtures and fixings.
横撑 用于支撑及加固立柱的短块横木。横撑也可用作木墙内的基础及固件，以使木墙可支撑暖气片、搁物架及其他配件。

noise absorption *See* SOUND ABSORPTION.
消噪 参考“吸音”词条。

noise control on-site Noise from construction sites is monitored by the local authority. Guidance is contained within BS 5228 - 2: 2009 Code of Practice for Noise Control on Construction and Open Sites and under Section 60 of the Control of Pollution Act 1974.
现场噪音控制 施工现场的噪音由当地有关部门监控。BS 5228 - 2：2009中关于建筑和露天场地的噪声控制实施规程以及1974年污染控制法第60节中均有指引。

Noise Criteria（NC） A rating system used to design a maximum allowable noise in a given space. It uses a series of standard curves of sound pressure level against a range of frequency bands. The system is used to assess the noise produced from building services systems, such as ventilation systems, but it may be applied to other noise sources. *See also* NOISE RATING.
噪音标准（NC） 用于规划特定区域的最大可允许噪音的一个评级系统。其采用了一系列频带范围内声压级的标准曲线。该体系可用于评估建筑设备系统如通风系统所产生的噪音，也可应用于其他噪声源。另请参考“噪声标定值”词条。

noise insulation *See* SOUND INSULATION.
隔音 参考“隔音”词条。

Noise Rating（NR） The maximum noise level that can be tolerated from an item of equipment.
噪音标定值（NR） 一项设备发出的可忍受的最大噪声级。

noise reduction coefficient（NRC） A rating system that produces a single figure for the amount of noise that is absorbed by a material, object, or construction. It is calculated by taking an average of the sound absorption

coefficient at four separate frequencies: 250 Hz, 500 Hz, 1000 Hz, and 2000 Hz.
降噪系数（NRC） 为材料、物体或建筑所吸收噪声总量定义一个单一数值的评级系统。是在 250 Hz、500 Hz、1000 Hz 与 2000 Hz 这四个频率的吸声系数的算术平均值。

nominal dimension (nominal size) The sizes used to describe the approximate characteristics of the product rather than the exact dimensions. Most timber products are described in this way with the dimensions describing or naming the cross section and length that was close to the rough dimensions of the timber before it was cut, planed, or worked. The dimensions given are within a set tolerance ensuring the product resembles the description.
标称尺寸 用于描述产品近似特征而不是精确外形的尺寸。大多数木材在切割、刨平或使用之前采用这种方式来描述或指定横截面及长度的尺寸，该尺寸接近木材的大致尺寸。所给尺寸在设定容差内，以确保产品符合所描述的。

nominated Product, supplier, or contractor identified for a purpose or position.
指定的 为一用途或位置而确定的产品、供应商或承包商。

nominated sub-contractor Company selected by the client and identified in the contract to undertake a specific aspect of the works. The *sub-contractor is normally identified in the *bill of quantities during the *tender process.
指定分包商 由业主选择且合同中确定执行特定工程的公司。指定分包商一般在投标时在工程量清单中确定。

nominated supplier Company that supplies goods or specialist services selected by the client and identified as the supplier within the contract documents. The supplier is listed within the tender documents and bill of quantities. Within government contracts it is essential that the process allows for fair competition. Framework contracts, where a number of suppliers have qualified to be part of a contractual agreement, are becoming more common in government contracts.
指定供应商 由业主选择并在合同文件中确定为供应商的提供货物或专业服务的公司；已在招标文件及工程量清单上列入该供应商。这在政府合同中，确定供应商的过程必须虑及公平竞争。在框架合同中，若干供应商有资格成为合约协议一部分，这在政府合同已经是越来越普遍了。

non-bearing wall A wall that provides a partition but does not take any of the structural loads of the building. A non-load-bearing wall has to sustain its own loads, and loads of fittings placed on it, and remain rigid and stable, but does not carry any of the building loads.
非承重墙 只起到分隔作用而不承载任何建筑结构负载的墙。非承重墙必须承载其自身负载以及放置于它上面的配件的荷载，并保持牢固及稳定，但不承载任何建筑负载。

non-biodegradable Materials that do not break down, for instance, plastics.
不可生物降解的 不可分解的物质，如塑料。

non-combustibility test A test undertaken in accordance with EN ISO 1182 where a material is exposed to a temperature of 750℃ for a minimum of 30 minutes. Used to determine whether a material is combustible or not.
不燃性测试 依照 EN ISO 1182 要求执行的测试，把一材料暴露在 750℃ 温度下 30 分钟。用于确定一材料是否可燃。

non-combustible A solid material that is not capable of burning. *See also* NON-FLAMMABLE.
不可燃的 不可燃烧的一固体材料。另请参考“不易燃的”词条。

non-concussive action A tap or valve that is self-closing.
按压式延时水龙头 自动关闭的水龙头及阀门。

non-destructive testing (NDT) A process used to determine the properties of a material without inflicting any permanent damage or alteration in properties.
非破坏性试验（NDT） 用于确定一物质特性的过程，不会对物质特性造成任何永久损伤或改变其特性。

non-displacement pile *See* DISPLACEMENT PILE.
非打入桩 参考“打入桩”词条。

non-drip paint (drip-free paint) A thixotropic paint with a jelly-like consistency and used as an alternative to gloss paint.
防流挂剂（非流淌漆） 一种胶状稠度的触变漆，可代替光泽涂料。

non-ferrous A material that does not have iron as its prime constituent.
非铁的 主要组成成分非铁的一种物质。

non-ferrous metal alloy A metal alloy not based on iron. A non-ferrous metal may contain iron in small amounts, however, a non-ferrous alloy is not based on iron. Examples include aluminium, copper, lead, titanium, and zinc. In construction, non-ferrous alloys are not used as extensively as *ferrous metal alloys.
非铁金属合金 一种非基于铁的金属合金。非铁金属可能含少量铁，然而，非铁合金是不基于铁的，如铝、铜、铅、钛与锌。在建筑中，非铁金属的使用广泛程度不如含铁金属合金。

non-flammable A liquid or gas that is not capable of burning. *See also* NON-COMBUSTABLE.
不易燃的 不可燃烧的一种液体或气体。另请参考“不可燃的”词条。

non-load-bearing wall A partition wall that carries no load other than its own weight; *See also* LOAD-BEARING.
非承重墙 一面只承载自身负载的隔断墙。另请参考“承重”词条。

non-manipulative joint A simple joint made by inserting pipes into a coupling tube. The jointing tube or sleeve has rings, olives, or glands within it that are compressed against the pipes inserted into the sleeve, making a watertight joint. Different compression joints are needed for light gauge pipes, plastic pipes, and copper tubes. Many of the fittings can be unscrewed and used again
卡套式接头 一个通过把管道嵌入连接管而形成的简单接头。连接管或管套里面含密封圈、铜环或密封压盖，通过推压管道压紧，使接头不漏水。轻型管、塑料管及铜管需用到不同的压紧式接头。很多配件可旋松并重复使用。

N

non-mortise hinge (surface-fixed hinge) A butt hinge that is cut so it is only as thick as one of its flaps when closed; the hinge may be cut so the flaps fit one inside the other.
子母铰链（表面安装铰链） 切割普通铰链，使其合上时的厚度仅为一叶片的厚度；铰链的切割可使其中一叶片刚好嵌入另一叶片。

non-performance When the contractor does not do the assigned work and is in default of the contract obligations. If the default goes to the root of the contract and is fundamental to the contract performance, the client may be entitled to bring the contract to an end; *See* DETERMINATION OF THE CONTRACT.
不履行 承包商不执行指定工程且不履行合同义务。如果违反基本合同条款，业主有权终止合约；另请参考“终止合同”词条。

non-renewable energy Energy that comes from a natural resource, which cannot be renewed, for example, coal, gas, and oil.
非再生能源 不可再生的自然资源能源，如煤、天然气与石油。

non-return valve A valve that allows a fluid or gas to flow in one direction only.
单向阀 只允许液体或气体往一个方向流动的阀门。

non-rising spindle A spindle that does not move up or down when turned; the spindle moves a threaded piston which controls the flow.
非上升型阀杆 转动时不会向上或向下移动的阀杆。阀杆移动可控制流量的螺纹活塞。

non-setting glazing compounds A flexible non-setting material used to bead glazing into the frame. Used in instances where the glazing may be liable to thermal or structural movement.
非固化玻璃密封胶 用于把玻璃密封装配进框架的一种柔性非固化材料。用在玻璃可能会发生热位移或构造位移的地方。

non-slip floors, treads, or nosings Finishes designed to be slip-resistant even when wet, dusty, or greasy. The surfaces may be embossed with shapes and patterns, have recessed grooves, have abrasive materials inserted into the surface, rubber raised studs, or use other materials that are naturally slip-resistant.
防滑地板、踏板或踏板前缘 设计为即使是潮湿、多尘或油腻时也能防滑的表面。表面饰以图形及图案，有暗式凹槽，粗糙材料嵌在表面，凸起的橡胶钉，或使用其他自然防滑的材料。

non-trafficable roof A roof that is not designed to be walked on.
不上人屋面 设计为不可在上面行走的屋面。

non-vision glass *See* TRANSLUCENT.
半透明 参考“半透明”词条。

non-vision grille A grille that has overlapping horizontal louvres to limit vision through it.
不透视格栅 有重叠水平百叶的格子窗，不可透视。

normalizing (metals) For ferrous alloys (especially steels), heating up to or above a certain temperature (usually above 500℃), then air cooled. This type of treatment is very common in steels to improve their *ductility, which is vitally important for structural steel.
正火（金属） 将铁基合金（尤其是钢）加热至或超过特定温度（一般是500℃以上），然后空气冷却。这种类型的处理常见于钢金属，可提高其延展性，对结构钢来说尤其重要。

normal soluble salts A clay brick classification that means that the surface of the brick is liable to a white powdery efflorescence stain caused by soluble salts within the brick that leach out when the brick becomes wet.
泛霜砖 一种黏土砖类别，指的是由于砖块内含可溶盐，在砖块潮湿时会析出到表面的一种白色粉末状的污渍。

normal stress *See* DIRECT STRESS.
法向应力 参考“正应力”词条。

N

northlight roof *See* SAWTOOTH ROOF.
北面采光的屋顶 参考“锯齿形屋顶”词条。

nose An overhang or prominent part of a projection.
突出部分 突出物外悬或凸出的部分。

nosing (nose) The overhang or front part of each tread.
踏板前缘 每块踏板外悬或前端部分。

nosing line The theoretical projection produced if a line is strung across the tip of each nose on the treads of a flight of stairs. Also known as the **pitch line**, the angle between the horizontal and the nosing line (pitch line) is the degree of pitch.
梯段坡度线 楼梯梯段中各级踏步前缘的连线。也可被称作坡度线，水平线与坡度线之间的夹角为坡度。

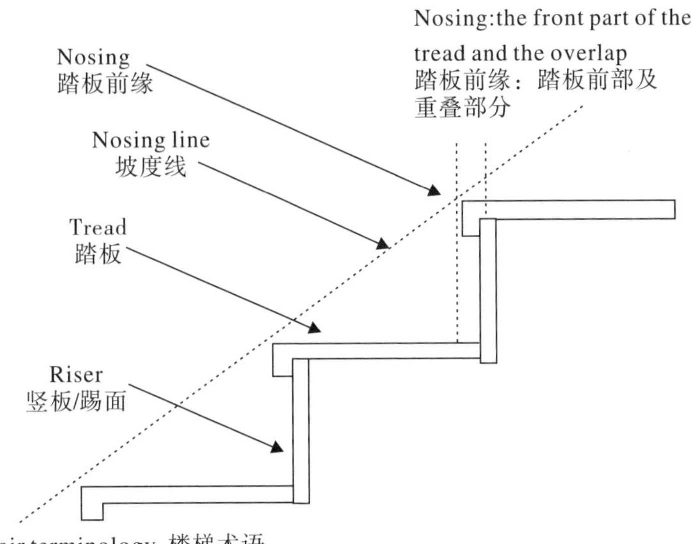

Stair terminology 楼梯术语

notch board A cut string.
凹槽侧板 经切割的楼梯侧板。

notched joint A joint made of notching sections of timber to make an interlocking joint.
凹口接合 木材凹槽部分组成的接缝，可形成连锁接合。

notched trowel A square trowel with rectangular notches cut out allowing adhesive to be pasted onto walls, floors, and ceilings. The rebates cut into the adhesive allow tiles and other fixtures to be placed and levelled. Placing tiles on adhesive without recesses proves much more difficult to level.
锯齿镘刀 带矩形槽口的方形镘刀，可把胶粘剂涂抹在墙上、地板及天花。凹口可涂抹胶粘剂以铺放及整平砖块与其他配件。把砖块铺放在无锯齿的胶粘剂上会更难整平。

notch sensitivity A material susceptible to failure due to the presence of a crack or opening.
缺口敏感性 指一材料由于存在裂缝或开口而容易受损。

notice to bidders (US) Instruction that opens a contract to tenders.
招标通知（美国） 告知投标者可进行投标的一份说明。

notice to proceed Instruction from the client or client representative to commence work.
开工通知 由业主或业主代表出具的一份表示可进场施工的告示。

noxious Harmful to life/health—something that is poisonous.
有毒的 对生命/健康有害——有毒的物质。

N-truss *See* PRATT TRUSS.
N形桁架 参考“普拉特桁架”词条。

null No value, relating to zero, invalid, or meaningless.
空值 无数值，等于零，无效或无意义。

N

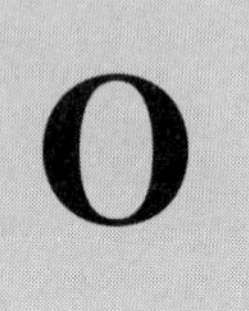

obligation What the parties to a contract agree to do for each other. A full list of the commitments and duties are contained in the conditions of contract. In a simple construction contract both the client and the contractor have obligations to each other; for example, a contractor has obligations to deliver a service and provide a building, while the client has obligations to allow the contractor access to the site and make payment for the works.
义务 签约方约定为对方做的事情。完整列表的承诺及责任均列载于合同条款中。简单的施工合同中，业主跟承包商均对对方负有责任。例如，承包商有义务提供服务并建造建筑物，而业主有义务允许承包商进入现场与支付工程款。

OBM *See* ORDNANCE DATUM.
水准基点 参考“水准零点”词条。

observation panel A glass panel incorporated within a door that enables a view to be gained.
门窗亮子 嵌入门上的一玻璃板，可供观看。

obsolescence The process of an item, building, structure, or service passing out of use so that it is redundant and no longer useful. The process can apply to old and new items, services, structures, or buildings that no longer serve a useful function.
荒废 一项项目、建筑、结构或设施废弃不用、变得多余且失去作用的过程。该过程可应用于已经失去使用功能的新旧项目、设施、结构或建筑。

occupancy The number of people that a room or building is capable of accommodating while still adhering to fire and safety regulations. The number of people a room is capable of holding is important for determining how it will fulfil evacuation requirements and meet the standards set by the building regulations.
居住率 一房间或建筑在遵循消防与安全规程的同时可容纳的人数。一房间可容纳的人数对于确定如何满足疏散要求以及满足由建筑规程规定的标准来说至关重要。

OD *See* ORDNANCE DATUM.
水准零点 参考“水准零点”词条。

odour A smell, either pleasant or unpleasant.
气味 能闻到的气体，包括香味或臭味。

offcut 1. Any piece of material that remains when something is cut down to the required size. 2. Wood remaining when a standard length of wood has been cut to the required size. These short, irregular pieces of wood are increasingly being used and recycled.
1. 下脚料 按所需尺寸进行切割时被切削掉的残余料。**2. 边料** 按所需尺寸切割掉标准长度的木材后剩余的木料，这些短的、不规则的木片日益得到广泛使用及循环利用。

offer First part of a contract, an expression of interest by a party to be bound by an obligation to another party. If there is an offer, acceptance, and some form of consideration, a binding contract is formed.
要约 一份合同的第一部分，由一方向另一方出具的意向书，使得一方对另一方负责。如果有了要约、承诺及某种形式的对价，那么可构成一个有约束力的合同。

off-form concrete The surface finish obtained when concrete is simply struck from its formwork or mould. The texture and shape of the mould provide the finish for the concrete.

清水混凝土 拆除混凝土模架或模板后的自然饰面。模板的纹理及形状即为混凝土饰面。

offset A horizontal distance measured perpendicular to the survey line to obtain the position of a specific point or object.
支距 垂直于测量线的水平距离，可确定一指定点或物体的位置。

off-site Work carried out outside the site boundary: work undertaken in a joinery shop, or work that is * prefabricated.
场外工作 在工地范围外执行的工作：在细木工工厂进行的工作，或预制件制作。

off-tamp finish (tamped finish) A textured concrete finish achieved by dabbing a straight edge of timber or metal up and down over the surface of the concrete, the tamping creates an undulating finish. The rough finish is often used for roads, ramps, and car parks to provide a skid-resistant surface.
拉毛路面 通过使用一直棱木板或金属来回轻拍混凝土表面所形成的有纹理混凝土饰面，拍打后的混凝土可形成一个波状型饰面。此粗糙饰面常用于道路、坡道及停车场，起到防滑作用。

ogee (ogee joint, OG joint) 1. A rebated joint or half-lap joint. 2. A spigot and socket joint that allows one tube to be located and connected within another. The widened neck of one end of one pipe allows the narrower end of another pipe to be housed within it. The narrowed neck is formed by the wall of the pipe being rebated to half its original thickness.
1. 搭接接合 一个搭口榫或半搭接榫。**2. 承插式连接** 可为一管道放置于另一管道内，或与另一管道连接。管道一端管颈加宽以容纳另一管道的较小管颈。通过把管壁厚度减少至原来厚度的一半从而形成较小管颈。

olive A brass ring that is slid over the narrow end of a coupling and then compressed using a nut to form a compression joint.
黄铜密封圈 在连接窄端上放置的铜环，然后使用螺母压紧以形成一压紧式接头。

on-costs The overheads—the site, head office costs, and personnel that support the company operations. All of the costs must be covered within construction costs. Costs are added to contracts on a percentage or fixed fee basis.
间接成本 用于现场、总公司开支及维持公司运营的人员的日常开支。所有这些开支必须包含在工程造价中。以一个百分比或固定费用添加到合同中。

one-and-a-half-brick wall (brick-and-a-half wall) A solid wall that has been constructed with a thickness of one and a half brick lengths (327 mm thick).
一砖半墙 一实心墙的建造厚度为一砖半长度，即厚327mm。

one-brick-wall A wall that is constructed with the thickness of one brick length.
一砖墙 墙的建造厚度为一砖长度。

on edge *See* BRICK-ON-EDGE.
侧砖 参考"侧砌砖"词条。

one-hole basin *See* SINGLE-HOLE BASIN.
单孔面盆 参考"单孔面盆"词条。

one-line diagram A single line electrical diagram.
单线图 单线电气图。

one-pipe system (single pipe, single pipe system) An above-ground drainage system that comprises a single vertical discharge stack, which conveys both soil and wastewater to the drain. An anti-siphon arrangement is required on most branch pipes to enable unrestricted layout of the appliances.
单管排污系统 一地面排水系统，含有一个单一的排放立管，可把污水跟废水排放至下水道。需要在大部分支管上安装防虹吸装置，以使设备的布置不受限制。

one-way switch A simple switch that either switches an appliance on or off.
单路开关 只可开或关一个装置的简单开关。

on/off controls An automatic controller that enables equipment to be switched on or off.
开关控制装置 一种自动控制器，可开关设备。

on-site Work carried out within the site boundaries.
现场工程 在现场范围内执行的工程。

OPC (Ordinary Portland cement) *See* CEMENT.
OPC（普通硅酸盐水泥） 参考“水泥”词条。

open assembly time The time allowed when adhesive is applied to bring two or more components together. Fixing the components after this time will result in a weaker bond.
晾置时间 黏结两个或更多组件时涂胶后至完成所需的时间。在这个时间之后再固定组件将会导致黏结不牢固。

open bidding (open tendering) Competitive tendering where the work is openly advertised and contractors are invited to submit tenders for the project. No restrictions are placed on the initial tendering process although there may be pre-qualification required to secure the work, for instance, good financial status and experience of certain types of work.
公开招标 公开对外宣传的竞争性投标，并邀请承包商投标该项目。为保障工程，可能需要进行资格预审（例如良好的财务状况以及有某些类型工程的经验），但最初的招标过程不会受到限制。

open boarding Planks of wood placed side by side with gaps between each board, often used to form a fence.
开缝铺板法 并排铺放木板，每两块木板之间有间隙，可用于构成一栅栏。

open cast mining (open cut mining, open pit mining) A method of excavating rocks and minerals at the surface of the ground where the mine is open to the atmosphere, rather than from a tunnel deep below ground.
露天开采（露天采矿） 在地面挖掘岩石跟矿物的一种方法，矿山是直接接触空气的，而不需要通过深入地下的隧道。

open-cell ceiling A suspended ceiling with a cellular opening in the finish and through the tile. Although you can see through the open cells in the ceiling, one's sight is naturally drawn to the suspended ceiling finish and distracted from the structural ceiling behind it.
格栅吊顶 饰面为格栅的吊顶。尽管可以通过格栅看到天花，但人的视线会自然被吸引到吊顶表面而不是吊顶背面的结构天花。

open cornice (open eaves) The part of the roof rafters that extend beyond the wall of the property creating an overhang. The eaves are classed as open when the soffit is not boxed in.
椽条外挑 部分屋面椽延伸至建筑墙外所形成一外悬部分。当檐底不被围起来时，这种屋檐定义为椽条外挑。

open cut An excavation that is open to the atmosphere. *See also* OPEN CAST MINING.
明挖 开敞式的挖掘，另请参考“露天采矿”词条。

open defect Any surface features of wood such as knots, shakes, splits, and worm holes that are easily recognizable.
外观缺陷 木的任何表面特性，如容易辨认的节、轮裂、裂隙及虫孔。

open drained joint (drain joint) A V-joint formed in the edge of concrete panels for drainage. The indented V is formed along the vertical edge of the panel. When adjacent concrete panels come together the two V joints form a square drainage channel. A neoprene gasket is placed in the centre of the channel to act as a baffle. Wind and water can be blown into the open joint but the baffle prevents the water passing through, allowing it to drip down the cavity and out of the open drain.
排水接缝 在水泥挂板边缘所做的一个 V 型排水接缝。沿着板块垂直边缘做一内凹 V 型。当相邻混凝土板块放在一块时，两个 V 型接缝则构成了一方形排水沟。在排水沟中心放置一氯丁橡胶垫以用作一挡条。风跟水可被吹进排水接缝但挡条会阻止水通过，使它沿空隙滴下并流出排水缝。

open eaves *See* OPEN CORNICE.

椽条外挑 参考“椽条外挑”词条。

open floor A floor in which joists are visible from the floor beneath.
明龙骨地板 下层可看到上层地板的龙骨。

open-graded aggregate An aggregate with particles of similar size that on compaction result in large air voids. *See also* GAP-GRADED AGGREGATE.
开级配骨料 相近尺寸颗粒的骨料，压实时会出现大的空隙。另请参考“间断级配骨料”词条。

opening Structural gap in a wall, roof, or floor ready to receive a window, door, or panel.
开口 墙壁、屋顶或地板上预留安装窗户、门或面板的结构孔。

opening face The side of a door or window, which when open, is furthest from the frame.
开口面 门窗的一侧，打开时离门窗框最远。

opening light A window that opens; a window which does not open is called a *fixed light.
可开式窗户 可开启的窗户，不可开启的窗户叫作“固定窗”。

open joint Gap or rebate left between components when joined together to provide a feature.
间隙接缝 组件接合时预留的间隙或企口缝，作为一特别造型。

open metal flooring A mesh type open flooring is formed by laying strips of metal on edge, or bars or expanded metal mesh on a steel frame. The metalwork provides a strong, slip-resistant floor structure with gaps between preventing dust and dirt build-up and allowing light to pass through the floor. Useful in warehouses where good vision at high level is needed.
金属格栅地板 在边缘上铺放金属条或在钢架上铺放钢筋或钢板网而形成的格栅型镂空地板。该金属件可提供一牢固、防滑地板结构，中间有间隙，可防止灰尘及尘土积聚，并可使光线透过地板。对于在高处需要充足光线的仓库相当实用。

open mortise (open mortice, slip mortise, slot mortise) Joint in wood used to fasten two pieces of timber at a 90° angle. A slot or groove is cut in the end of timber and made so that it is ready to receive the thin tenon of another timber to form a 90° angle joint.
开口贯通榫 可用于把两块木材呈 90 度固定在一块的木材接头。在木材端切割一狭槽或凹槽，可匹配另一块木材的薄榫，形成一个 90 度的接头。

open newel stair Geometrical stair without intermediate *newel posts.
无中柱螺旋梯 螺旋形楼梯，无楼梯中柱。

open plan A room, normally an office, that is just divided by furniture and screens, no dividing walls.
开敞式平面布置 指仅仅使用家具及屏风而不是分隔墙来分隔的一个房间，通常用作办公室。

open plan stair Flight of stairs without any solid risers, *open risers exist between each *tread.
镂空楼梯 不带任何实体踢板的一段楼梯，每块踏板之间为露明踢板。

open riser The vertical component of the staircase is left open between the horizontal foot *treads. Where a stair case has treads mounted in strings but no back or vertical riser, it becomes an **open rise staircase**.
镂空踢板 楼梯水平踏板之间无垂直踢板。楼梯梁上安装踏板但无背衬或垂直踢板，所以称为镂空踢板楼梯。

open roof A roof without a ceiling, exposing the trusses to those inside the building.
明造屋顶 无天花的屋面，在室内可看到桁架。

open stair A stairway which is open on one or both sides.
开敞楼梯 一侧或两侧敞开的楼梯。

open string A string cut to match the profile of the treads, a zigzag string is produced as the string is cut around the profile of each tread and riser.
明步楼梯梁 切割以匹配踏板轮廓的楼梯斜梁，沿着每个踏板及踢板周围切割楼梯斜梁时便得到一个 Z 字形的楼梯斜梁。

O

open tendering *See* OPEN BIDDING.
公开投标 参考“公开招标”词条。

open traverse A *traverse that ends at a *station whose relative position is not previously known, it is not referenced back to the first station.
支导线 导线的最终测点相对位置是未知的，但不会回到第一测点位置。

open valley An exposed valley. The tiles on the roof lead into the valley but do not cover it.
屋面明沟 外露屋面斜沟。屋面上的砖瓦通向斜沟但不遮盖它。

open-vented system A type of wet central heating system that has a feed and expansion, and is open to the atmosphere via an open-vent pipe. The water temperature in open-vented systems must be kept well below its boiling point in order to stop steam from being formed. *See also* SEALED SYSTEM.
开放式供热系统 一种湿式集中供暖系统，带一高位膨胀水箱，并通过一外露通气管与外部相连通。开放式供热系统的水温必须保持在低于它的沸点温度，以防蒸气形成。另请参考“封闭式水箱系统”词条。

open well A shallow *well that is hand-dug to depths between 1 m and 15 m, with a 1—2 m diameter.
开口式井 手动挖掘的一口浅井，深度在 1 至 15 米之间，直径为 1 至 2 米。

open-well stair A stair flight with an open well and a gap between the flight of stairs and the wall.
井式楼梯 带开口式梯井的一段楼梯，且楼梯与墙壁之间有间隙。

operative On-site worker, usually refers to someone operating plant, a labourer, or a chain person rather than a professional or skilled worker.
操作人员 现场工人，一般指的是操作设备的人、劳工或工人，而不是指专业人员或技术工人。

optical smoke detector A type of smoke detector that uses a photoelectric cell and an optical beam to detect the presence of smoke.
光电感烟探测器 使用一个光电管及一束光线来检测烟雾的一种烟雾探测器。

orangery Framed building with a large area of glass, much like a conservatory, which allows much natural light into the building.
玻璃温室 有大面积玻璃的框架建筑，跟温室很像，可允许大量自然光透射进建筑。

orbital sander A circular motorized sander with a flat surface turned in a circular motion to grindflat and smooth the surface on which it is placed.
旋转式砂光机 一圆形电动砂光机，其平整表面绕圆周方向转动，以磨平其所接触的表面。

ordnance benchmark (OBM) A permanent mark on a fixed object, denoting an elevation which has been related back to *ordnance datum. Used by surveyors to establish the initial elevation of a site.
水准基点（OBM） 一个固定物体上的永久标记，指示与水准零点相关的海拔。测量员用于确立一现场的原始海拔。

orders Instructions given to undertake work.
指令 下达执行工作的指示。

Ordinary Portland Cement (OPC) *See* CEMENT.
普通硅酸盐水泥（OPC） 参考“水泥”词条。

organic landscaping Soft landscaping of living matter: shrubs, bushes, trees, and other plants.
园林景观 植物软景观，有灌木、灌木丛、树及其他植物。

organic rendering 1. Use of masonry paint. 2. Use of grass, ivy, and other wall-climbing plants to cover a wall.
1. 有机砂浆 使用水泥漆。**2. 外墙绿化** 用草、蔓藤及其他爬壁植物覆盖墙体。

oriel window A bay window that projects only from an upper floor.
凸窗 仅从上层建筑外墙面突出的窗户。

orientation The position or direction of something.
定位 某物的位置或方向。

orifice An opening (hole) in a body from which fluid is discharged.
孔口 排放液体所流经的一机体内的孔。

original ground level The height of a site before * groundwork.
原始地面标高 挖掘地基前的现场高度。

O ring (O-ring) A rubber ring, normally mounted in a groove, used as a seal against air, water, or oil.
O 型密封圈 一种橡胶圈，一般安装在凹槽内，可用于密封空气、水或油。

ornament Statue or feature used to enhance landscape, building, or interior.
装饰物 用于丰富景观、建筑或内饰的雕像或造型。

outer hardening A treatment used in metals to increase the * hardness or resistance to surface indentation. This treatment is undertaken at a microscopic level and involves moving the * dislocations within the metal; this alters the microstructure of the metal which is tailored to increase the hardness of the material.
表面硬化 金属所采用的一种处理方式，可增加硬度或使表面耐磨。该处理在微观水平上进行，包括金属位错；可改变金属的微观结构，并增强金属硬度。

outer string One of the two diagonal components of a staircase that carries the * treads and * risers; it is the * string which is furthest away from the wall.
外侧楼梯梁 楼梯两条斜梁中离墙壁最远的一条，可承载踏板及踢板。

outfitting The construction work, fixing, finishes, and furniture needed to complete the internal aspects of a building.
装备 需用于完成一建筑内部的施工工程、设备、饰面及家具。

out for tender The period between the advertisement of the contract and the time when the tender has to be submitted; this is also the maximum period the contractor has to prepare the contract.
投标有效期 公开招标至投标提交这段时间，也是承包商准备合同的最长期限。

outgassing The removal of excess or unwanted gases from a material.
除气 除去一物质中多余或不需要的气体。

outgo The pipe from a sanitary fitting that connects it to the drain. The final part of the pipework before the gulley or sewer connection.
排出管道 从卫浴装置连接至下水道的管道。集水口或下水道接头前面管道的最后部分。

outgoing circuit The wiring between the switchboard and connection points or outlets.
输出电路 开关柜与连接点或输出口之间的电线。

out of plumb Not vertical. Usually used to describe the degree that something is not vertical. Recorded using the horizontal measurement between the top and bottom of the item from a vertical line, or the degrees difference that the centre line is from the vertical.
倾斜 不垂直。通常用来描述某物体不垂直的程度。通过记录该物体垂直线上顶部和底部的水平位置测量，或其中心线与垂直线的角度。

out of the ground Used to describe the moment when the building structure starts to emerge above ground level. The first parts of a structure that extend above the ground are normally the structural frame or the walls.
出正负零 用于描述建筑结构开始高出地面的瞬间。一结构伸出地面上的最初部分一般是结构框架或墙。

out of wind Area that is sheltered from the wind.
无风区 免受风吹的区域。

outreach arm (bracket arm, carrying arm) The arm that extends from the main column or wall to carry a light or luminaire.

伸臂式杆（支架臂，支承杆） 从主柱或墙壁延伸出来的臂杆，可用于承载光源或灯具。

outrigger 1. A beam that extends from a building to carry scaffolding. The beam removes the need for the scaffolding to extend down to the ground. The beam is propped and secured against the structural ceiling and extends out of a building opening such as a window. Scaffolding, such as flying scaffolding, can be secured to the beam. 2. Legs that extend out from a crane or excavator to provide extra stability.
1. 悬挑梁 从一建筑中伸出来的横梁，用于承载脚手架。该横梁可免使脚手架延伸至地面。横梁由结构天花支撑及固定并从一建筑开口如窗户延伸出去；而脚手架如悬挑脚手架则固定在该梁上。**2. 支腿** 从一吊车或挖掘机伸出来的撑脚，可提供额外的稳定性。

outside glazing Term used to describe external *glazing that has been installed from the outside of the building. *See also* INSIDE GLAZING.
室外安装玻璃 该术语用于描述从建筑外部安装的外部玻璃。另请参考“室内安装玻璃”词条。

outstanding works 1. Defects or work that need to be rectified. 2. Work that is yet to be completed.
1. 需返工重做工程 不及格或需返工的工程。**2. 剩余工程** 还未完工的工程。

oval wire brad Nail with an elliptical shape that is positioned with the longer dimension in line with the grain to reduce the chance of the wood splitting.
椭圆钉 椭圆形状的钉子，顺着较长的纹理钉钉子以降低木材裂开的可能性。

overcladding The process of applying an additional external cladding to a building as part of refurbishment.
建筑外包层 在建筑物上应用一层额外的外包层的过程，外包层可作为翻新的一部分。

overcloak Used in sheet metal roofing to describe the part of the sheet that overlaps the sheet beneath at a drip, roll, or seam.
彩钢板搭接部分 用在金属薄板屋面上，指的是与滴水线、绲边或接缝下方薄板重叠的金属薄板部分。

overcoating The process of applying an additional coat of paint or varnish to an object.
涂面漆 在一物体上涂上另外一层涂漆或清漆的过程。

O

over-consolidated clay A clay that has previously been subjected to a higher stress than its present-day overburden stress. The reduction in stress could have resulted from melting of ice sheets, erosion of overburden pressure, or a rise in the water table. Such a layer is more liable to swelling when wet.
超固结土 先期受到应力大于现有上覆岩层应力的土。应力降低有可能是起因于冰川融化、上覆岩层侵蚀压力或水位上升。该层土潮湿时容易膨胀。

overcurrent (excess current) The presence of a larger-than-expected current within a circuit. Usually occurs due to a *short circuit.
过载电流（过电流） 一电路中电流大于预期的，通常是因为短路。

over-fascia vent A hole cut into the top of the *fascia to allow airflow into the roof space at the eaves.
挑檐立板通风口 在挑檐立板顶部切割的一个孔，可允许气流流进屋檐处的屋顶空间。

overflow (overflow pipe) A pipe connected to a basin, bath, sink, or tank that discharges excess water when it is full. The excess water will either be discharged to a *waste pipe or externally via a *warning pipe.
溢流管 连接脸盘、浴缸、水槽或水箱的管道，在其满水时可排放过剩的水量。可把过剩水量排放至废水管或使用溢水管排放至外面。

overhand work 1. Bricklaying from the inside of the building reaching out to the external leaf or the cavity to lay bricks or blocks. 2. An awkward or complicated method of working.
1. 里脚手架砌墙法 从建筑内侧探身出外叶墙或空腔砌砖。**2. 一种怪异或复杂的施工方法**。

overhang Projection from a structure, roof, or wall that extends beyond the face of the support.
外悬部分 从一结构、屋面或墙壁延伸出支撑面外的突出部分。

overhanging eaves The lower-most edge of the roof that extends out and projects from the building, protec-

ting the face of the building.
挑檐 从建筑延伸出去的最低屋檐，可保护建筑外墙。

overhaul The haulage distance that exceeds the *free haul distance.
超运距 超出免费运距的运输距离。

overhead crane (gantry crane) A *crane that runs on a *gantry.
桥式起重机（龙门起重机） 在龙门架上运行的起重机。

overhead door *See* SWING-UP DOOR.
卷帘门 参考“上开门”词条。

overheads (administrative charges, establishment charge, on-costs) The operating costs of the company. The head office, management, administration charges, and the costs of offices that are used to operate the company and are charged proportionally to each contract, or at a flat rate per job.
日常开支（管理费，开办费，间接成本） 公司的营业成本。公司运转所用的总部、管理、行政性费用以及办公室开支，所有这些开支按比例包含在合同中，或是根据每项工作的统一费用收取。

overlap The degree that something extends over something else, the *lap. Roof tiles need to extend over each other to ensure rain does not pass through. Some work and management packages will need to extend into the next operation to ensure successful handover.
搭接 某物延伸至另一物体上方，即搭接。屋顶瓦片需互相搭接以防雨水通过。一些工程或管理范围也需延伸至下一个操作以确保可顺利移交。

overlay 1. Something that is spread over the top of something else to provide protection or enhance performance. Insulation that is spread over the top of a ceiling is often called an overlay. 2. Transparent drawing placed over the top of a plan to show the position of services or fixings to assist design coordination.
1. 覆盖 覆盖在某物上面的东西，可起到保护作用或增强性能。覆盖于天花顶的绝缘层常被称为覆盖层。**2. 蒙板** 把透明背景的图纸放置于平面图上面以显示设施或设备的位置，可有助于设计协调。

overload A load that exceeds the design load in structures and electrical circuits.
超载 超过结构及电路的设计荷载的一荷载。

O

overpanel A infill panel placed above a door to close the gap between the ceiling and the top of the door.
门顶板 放置在一门上方的填充板，可盖住天花与门顶之间的间隙。

oversail To overhang or extend over something. A crane needs to ensure there is an oversailing agreement for it to be used within the airspace of another person's property.
悬吊 悬于或延伸至某物上方。起重机在另一人的物业领空上操作时，必须确保已签订一份允许其在上空操作的协议。

oversailing course Brickwork course that extends out from the wall to act as a feature or to offer protection to the wall below.
砖墙挑出层 从墙壁延伸出来的砖层，可用作一造型墙或保护下方的墙壁。

oversite An operation, task, or feature that extends across the boundaries of the site.
现场外 延伸至现场范围外的一项操作、任务或造型。

oversite concrete Concrete placed across an excavation to seal the ground. The concrete normally covers the full area of the excavation.
混凝土垫层 铺放在一基坑上以密封地面的混凝土。该层混凝土一般会完全覆盖基坑区。

oversite work Operations that take place within the full area of the construction site or up to a boundary.
全场性施工 发生在施工现场全部范围内或达到某边界处的操作。

oversize Piece of equipment or component that has larger-than-normal dimensions but has been specifically chosen for an operation.
特大的 专门为一项操作所选的一件大于正常尺寸的设备或构件。

overspray The excess paint outside the intended area to be sprayed. When paint is applied by a spray gun it will cover an area larger than intended. Masking of surfaces and covering of areas ensures that the spray does not damage the other areas.
超范围喷涂 在预期区域外喷涂过量的涂漆。当使用喷漆枪喷涂涂漆时，所覆盖的区域会比预期的大。遮蔽表面及覆盖区域可确保喷漆不会损坏其他区域。

over-tile 1. Decorative tiling laid over other tiles, also called imbrex or Spanish tiling. 2. General tiling on top of existing tiles that have strong structural fixing to the substrate and are flat; this is not considered good practice but is sometimes undertaken in refit and refurbishment work.
1. 叠瓦 铺放于其他瓦片上面的装饰性盖瓦，也被称为波形瓦或西班牙瓦。**2. 旧瓷砖上铺新砖** 位于已有瓷砖上面的通用瓷砖，旧砖需牢固黏结在基层上且要求平整；虽然并不是推荐的方法，但有时候在修整及翻新工程中会采用。

overtime Additional working compared to standard or contracted hours. Sometimes paid at a higher rate.
加班时间 超出标准或合同工时的额外工作。有时候会以更高的单价支付。

overturning When something has the tendency to invert; retaining walls and other such structures are checked for overturning moments.
倾覆 有倾覆倾向的某物；应检查挡土墙及其他此类结构的倾覆力矩。

overvibration Vibration of concrete for longer than required leading to segregation of the aggregates and cement. Concrete is vibrated to compact it and remove air bubbles, but excessive vibration can lead to the large heavier aggregates dropping to the bottom of the concrete, and the finer aggregates being forced to the top, resulting in a defective concrete.
过度振捣 混凝土振捣时间比所要求的长，从而导致骨料与水泥离析。振捣混凝土使其密实并去除气泡，但过度振捣会使粗的较重的骨料跌到混凝土底部，而细骨料则被压到顶部，导致出现有缺陷的混凝土。

overvoltage *See* VOLTAGE OVERLOAD.
超电压 参考“电压过载”词条。

O

Owen tube A sampler used to measure the settling velocity of fine sediment.
欧文管 用于测量细颗粒泥沙沉降速度的采样器。

owner (building owner, client) The person who owns the property and who often instructs work to be carried out. The contractors and architects in building projects undertake the works for the building owner.
所有者（业主，客户） 拥有所有权并经常指示执行工作的人员。建筑工程中的承包商及建筑师为业主开展工程。

oxy-cutting Use of oxy-acetylene flame to melt metal, then once metal is glowing, an increased supply of oxygen is used to react with the molten steel, producing greater heat which cuts the steel away.
氧气切割 使用氧乙炔火焰熔化金属，然后在金属开始发热时，补给氧气，用于与熔化钢起反应，从而产生更多的热量以切割钢。

ozonation A water treatment technique. *Ozone is bubbled through the water to destroy bacteria, viruses, and traces of pesticides; it also helps to break down compounds that cause colour, taste, and odour in the water.
臭氧化 一种水处理技术。臭氧在水中冒泡以消灭细菌、病毒及杀虫剂，并有助于分解会使水发生变色、变味及臭味的化合物。

package deal A standard form of contract where both the designing and building of the project are contained within the same legally binding document. *See also* DESIGN AND BUILD CONTRACT.
一揽子协议 一种标准的合同形式，在同一份具有法律约束力的文件中包含了项目的设计与施工。另请参考“设计施工合同”词条。

packer Piece of rigid material used to lift, raise, or secure an item at the right height, level, or position. The hard resilient piece of material is placed under or between an object, ensuring it remains in the new position.
封隔器 一块刚性材料，用于把一项物品提起、提高或固定在正确高度、水平或位置上。此硬弹性的材料则放置于一物体的底下或中间，以确保其处于新位置上。

packing 1. Strip of *packers. 2. Material used to fill a void.
1. 封隔条。**2. 填充物** 用于填充一空隙的材料。

packing gland The packing and seal around the stem of a tap, the packing within the gland makes the seal watertight. The *packing nut is tightened onto the fixing, securing the seal.
填料压盖 填料并密封阀杆周围，压盖中的填料可形成不透水密封。压盖螺母拧紧于装置上，固定密封。

packing nut Gland nut used to secure a watertight seal on the tap.
压盖螺母 用于固定阀门上防水密封的压盖螺母。

pad 1. A block of stone or dense concrete placed under a concentrated point load such as a beam or column, the pad stone helps to distribute the point load over a wider area, reducing the load per unit area. 2. A foundation that is square in plan, a point load such as a column sits in the middle of the pad, and the pad foundation distributes the load to the ground. 3. A block or cushion.
1. 支承垫石 放置于集中荷载如梁或柱底下的一块石块或密实混凝土，该垫石有助于把集中荷载分散于更广的范围，从而降低每单位面积的荷载。**2. 阶形基础** 平面为方形的基座，集中荷载如柱则坐落于基础上间，而阶形基础会把荷载分散至地面。**3. 垫子** 一砌块或垫子。

padbolt A bolt that can be padlocked.
可挂锁插销 可用挂锁锁上的门闩。

pad footing (base, pad foundation) An isolated *foundation that is square or rectangle on plan and used to distribute point loads, such as columns, to the ground. Most foundations are reinforced with steel rods. The columns that rest on the pad are secured and positioned using *holding-down bolts, which are cast into the *concrete. Other foundations may be connected into the *pad. To ensure there is a good link, steel reinforcement *starter bars are used to link the reinforcement in the pad foundation to the reinforcement in other foundations. Starter bars protrude from the concrete footing, enabling the adjoining footings reinforcement to be linked to it. *Ground beams are often connected into the side of pad foundations. Ground beams carry the longitudinal loads, such as walls and floors.
阶形基础（基座） 平面为方形或矩形的一独立基础，可用于分散集中荷载，如把柱的载荷分散至地面。大多数基础会使用钢筋以加固。使用地脚螺栓来固定搁在垫层上的柱子，并使用混凝土浇筑。其他基础可浇筑入垫层内。为确保搭接良好，使用锚筋把阶形基础中的钢筋搭接到其他基础的钢筋。锚筋从混凝土基础中伸出来，使相邻基础的钢筋可与其相连。一般把基础梁连接至阶形基础一侧。基础梁承载着纵向荷载，如墙壁与楼板。

padsaw A pointed saw blade that is secured in a hand block.
鸡尾锯 固定于一手柄内的尖锯条。

padstone A grinding stone for sharpening cutting blades.
磨石 用于磨砺刀片的研磨石。

paint (painting) A method of applying a thin coating of material to the surface of another solid or structure. The paint is applied in liquid form, which then becomes solid before being put into service. Thus paints are surface coatings, marketed in liquid form, and usually suitable for site use. They serve one or more of the following purposes:
涂料（油漆） 施加一层薄层材料到另一固体或结构的表面的方法。涂料以液体形式应用，然后在投入使用之前变成固体。涂料为表面涂层，以液体形式出售，通常适合现场使用。它们具有以下一个或多个用途：

- To protect the underlying surface by excluding the atmosphere, moisture, chemicals, fungi, insects, etc.
  可保护底面免受空气、水分、化学品、菌类、昆虫等的影响。
- To provide a decorative, easily maintained surface.
  可提供装饰性、容易维护的表面。
- To provide light and/or heat reflecting surfaces.
  可提供光和/或热反射的表面。
- To give special effects, for instance, inhibitive paints for protecting metals, electrically conductive paints to provide a source of heat, condensation-resisting paints.
  可提供特殊效果，如可保护金属的防锈涂料、可提供热源的导电涂料、防结露涂料。

painter Skilled tradesperson who applies paint to the surface of materials.
油漆工 受训在材料表面涂漆的专业人员。

painter's labourer Normally an apprentice painter who assists the painter by mixing paint, ensuring materials are ready, and also assists with painting that requires less skill.
油漆小工 通过混合油漆、确保已备好材料以协助油漆匠的一名学徒，同时也会帮忙做些技能要求不高的涂漆工作。

P

painting The application of undercoat and paint to a surface to achieve the desired finish. The substrate or surface to which the paint is applied is cleaned and prepared ready to receive the undercoat and final finish.
涂漆 应用底漆及面漆以取得预期的饰面。施加涂料的底层或表面应已清洁干净并已准备好接受涂抹底漆及面漆。

paint removal The stripping of paint from a material's surface; for best results the removal of the finish should be down to the substrate. Chemicals can be used to break down the paint, which can then be removed with a *paint scraper. Alternatively, a blow torch or hot air blower can be used on oil-based paints to heat the paint, causing it to bubble and then, once soft, it can be scraped off the surface. Paint can also be removed by shot blasting.
除漆 从材料表面上去除涂漆；最佳结果是去除饰面，只剩下底层。可使用化学品分解涂漆，然后使用油漆刮刀清除。或者也可使用喷灯或热风枪来加热油基漆，使其冒泡至变软，然后把其从表面上刮掉。也可通过喷砂处理清除涂漆。

paint scraper Flat tool used to remove paint from the substrate such as wood or metal. The paint is first treated with chemicals or a hot air blower or torch. The treatment helps to lift the paint away from the surface making it easier to remove the paint with the scraper.
油漆刮刀 用于从基层（如木材或金属）去除油漆的扁平工具。首先使用化学品或热风枪或喷灯进行处理。该处理有助于把涂漆从表面脱离出来，便于使用刮刀去除。

paint stripping The removal of paint from the surface of a material. *See* PAINT REMOVAL.
脱漆 从一材料表面上去除涂漆。另请参考“除漆”词条。

paint system The prescribed application, preparation, and build-up of layers of paint to provide a robust surface finish.
涂料体系 按规定应用、准备及施工的数层涂料，可使表面坚固耐用。

paintwork The finish achieved when paint is applied to a surface.
涂漆表面 涂漆应用于表面上时所得到的饰面。

pale (pales, paling) Vertical metal or wooden struts, stakes, or boards, forming a palisade fence.
栅栏 立式金属或木支柱、桩或板块，形成一个围栏。

palisade Vertical struts driven into the ground or fixed together to form a fence.
栅栏 打入地面或固定在一起以形成一围栏的竖柱。

Palladian window A window that has three vertical sections; the top of the middle section is arched in shape and wider than the two side sections, which are horizontal along their tops.
帕拉第奥式窗 有三个立式部分的窗户，中间部分的顶部为一拱形，并且比两侧的宽，而两侧窗顶部对齐。

pallet 1. A thin strip of wood used as a fixing fillet. 2. A timber frame used to stack, store, and transport materials such as bricks, tiles, and stone. The frame is built so that it can be lifted easily by a forklift and transported on a truck.
1. 嵌条 用作固定嵌条的一块薄木条。**2. 托盘** 用于堆放、储存及运输砖块、瓷砖及石材等材料的木框，此木框的构造可使用铲车轻易抬起并使用卡车运送。

pan 1. Shallow tray used for catching liquids or debris, placed underneath an object to catch and contain the substance. 2. The bowl-shaped part of the toilet (*WC or water closet) that contains the water and receives human excreta and urine.
1. 盘 用于装载液体或杂物的浅托盘，放置于一物体下面以接住并装载物质。**2. 坐便器** 厕所（或冲水马桶）内的碗状部分，装载水及人类排泄物。

pan connector (sanitary connector, WC connector) A flexible or rigid connector, usually plastic, that connects the *WC *pan to the soil pipe.
坐便器接头 一个柔性或刚性接头，一般为塑料，把洗手间坐便器连接至污水管。

pane A framed sheet of glass installed within a window or a door.
窗格玻璃 安装在窗或门上的一带框玻璃。

panel A distinct sheet of material or sections of infill (brick, concrete, or stone) that are placed between a structural frame.
嵌板 一独立的板材或填充型材（砖块、混凝土或石块），放置于结构构架中间。

panel form Small formwork panels, 600 mm × 1200 mm, that lock together to create larger sections of permanent formwork.
散拼模板 尺寸为600 mm × 1200 mm的小块模板连接在一起以成为一块大的永久性模板。

panel heating A type of space-heating system comprising a panel that is heated using hot water pipes or, more commonly, electric conductors. The panels can be ceiling-, floor-, or wall-mounted.
板式供暖 一种含一块面板的空间加热系统，通过热水管或更为常见的电导体加热。该面板可安装于天花板、地板或墙壁。

panelled door (panel door) A door comprising panels between the *rails and *stiles.
镶板门 门冒头与门竖梃之间安装镶板的门。

panelling Regular units used to clad or line walls, ceilings, or floors, providing a modular decorative finish.
镶板 用于墙壁、天花或地板的覆盖或衬底的标准构件，可提供一模块化饰面。

panel pin Small wire nail with a small head, which makes the nail less visible when embedded in the timber.
无头钉 钉头很小的小圆钉，钉入木材时可使圆钉不太明显。

panel planer A piece of equipment that reduces lengths of timber down to a set thickness.
木工刨床 可把木材厚度减少至规定厚度的一台设备。

panel products Long units of building materials used for cladding and lining walls, ceilings, and floors.
板材 用于墙壁、天花板及地板的覆盖或衬底的长件建筑材料。

panel saw A small handsaw, with short narrow-gauge cutting teeth, used for making precise cuts.
木工锯 一把小的手锯，有短的窄距切齿，可进行精确切割。

panel wall A non-load-bearing wall made up of regular units of timber, brick, or block.
隔墙 由木材、砖块或砌块标准构件构成的一非承重墙。

pan form (waffle form) Void formerly used within in-situ concrete floors. The box-shaped *formwork creates regular rectangular indents in the soffit of the floor resembling a waffle pattern.
密肋楼板模壳 过去现场浇筑混凝土楼板时使用的空格。箱形的模架可在楼板底面形成如网格状图案的矩形凹陷。

panic bolt A device that is capable of operating the cremone bolt of a fire or emergency door from the inside.
消防推杠锁 可从里面操作消防门或紧急出口门通天插销的装置。

panic hardware *Door hardware that can be used to open a fire or emergency door from the inside when required.
推杠锁五金 需要时，用于可从里面打开消防门或紧急出口门的门用五金。

panic latch A device that is capable of operating the *latch of a fire or emergency door from the inside.
推杆门闩 可从里面操作消防门或紧急出口门门闩的装置。

pantile An elongated S-shaped clay or concrete interlocking roof tile.
筒瓦 细长S形的黏土或混凝土屋面扣瓦。

pan wash sink 1. A *sink located in a kitchen and used to wash cooking pans. 2. A *bedpan sink.
1. 锅盘洗涤槽 在厨房用于清洗锅盘的洗涤槽。**2. 便盘槽**。

paperhanger Decorator skilled at hanging wallpaper.
裱糊工人 专业贴墙纸的装修工。

P

parallel coping A *coping used to cover the sloping *parapet wall at a gable end.
女儿墙压顶 用于覆盖山墙端女儿墙斜墙的压顶。

parallel gutter A *box gutter.
U型排水沟 一匣形水槽。

parallel thread A *thread that has a constant diameter. *See also* TAPER THREAD.
圆柱螺纹 指直径恒定的螺纹。另请参考“圆锥螺纹”词条。

parameter A measurable quantity of a material, which can be altered to vary the results or outcome; *See also* PROPERTIES.
参数 一材料可测量的量，可随其变化而导致结果的变化，另请参考“性能”词条。

paramount partition Prefabricated non-load-bearing partition wall.
蜂窝隔墙 预制非承重隔墙。

parapet A low wall along the edge of a roof, balcony, or terrace. Used to protect people against a sudden drop and for decorative purposes.
女儿墙 沿着屋顶、阳台或露台周围的一矮墙，用于防止人员突然坠落，并起装饰作用。

parapet gutter A concealed gutter, rectangular in section, located behind a *parapet wall, between the wall and the roof.
女儿墙内天沟 位于女儿墙后面，墙壁与屋顶中间的一隐蔽式矩形天沟。

parge (pargeting, parging) 1. A thin coat of *plaster or *mortar that is used to coat walls. Traditionally

used as a lining material for chimney flues or to waterproof external walls. Currently used to improve the acoustic performance of masonry aggregate block party walls and to improve the * air tightness of masonry aggregate block party walls. 2. Decorative ornamental plasterwork.
1. 灰泥层 用于覆盖于墙面的一层薄灰泥或灰浆层，过去用作烟囱的衬里材料或是外墙的防水层，现在用于改善砌块分户墙的隔音性能及其气密性。**2. 灰浆涂层** 有装饰效果的抹灰涂层。

paring chisel Long narrow cutting tool used for clearing out mortise joints, the sharp bladed tool is worked by hand without the need of a mallet.
长木工凿 用于开凿榫眼的狭长切削工具，可手工操作此刀刃锋利的工具，无须借助木槌。

Parker truss A * Pratt truss that has an inclined (arch shape) top chord.
帕克式桁架 上弦倾斜（拱形）的普拉特桁架。

parliament hinge (butterfly hinge, H-hinge, shutter hinge) A hinge in the shape of the letter H.
工字型铰链（蝶形铰链，H 型铰链） 形状为 H 形的铰链。

parlour A room that was traditionally set aside for conversations.
会客室 传统上用于交谈的房间。

partially separate system A combination of the combined and separate below-ground drainage systems, where the majority of the surface water is discharged via a surface water drain to the surface water sewer. The remaining surface water, usually at the rear of the building, is discharged via the foul water drain to the foul water sewer, which flushes the foul water drain. Cheaper to install than a separate system. *See also* COMBINED SYSTEM and SEPARATE SYSTEM.
半分流制排水系统 合流制与分流制地下排水系统的组合，大部分地表水通过地表水渠排放至地表水下水道。而剩下的地表水（一般位于建筑物后面）则通过污水渠排放至污水下水道，继而流向污水管。安装半分流制排水系统比分流制排水系统便宜。另请参考“合流制排水系统”与“分流制排水系统”词条。

partition An internal wall used to separate one space from another.
隔墙 用于把一空间与另一空间隔开的内墙。

partition block A hollow clay bock used to construct * partitions.
隔墙砌块 用于建造隔墙的空心黏土砖。

P

party floor A shared dividing floor between two properties. *See also* PARTY WALL.
分户楼板 分隔开两栋房产的共用楼板。另请参考“分户墙”词条。

party wall (parting wall) A shared dividing wall between two properties. *See also* PARTY FLOOR.
分户墙 分隔开两栋房产的共用墙。另请参考“分户楼板”词条。

pass door *See* WICKET DOOR.
边门 参考“便门”词条。

passive earth pressure (Pp) The maximum horizontal stress that is exerted from the soil on a retaining wall (typically from the soil in front of the wall) as the wall moves towards the soil. *See also* ACTIVE EARTH PRESSURE and COEFFICIENT OF PASSIVE EARTH PRESSURE.
被动土压力（Pp） 当挡土墙（在外力作用下）向土壤移动时，土壤向墙（一般是来自于墙前面的土壤）施加的最大水平应力。另请参考“主动土压力”与“被动土压力系数”词条。

passive fire protection A method of fire protection where the form, layout, and fabric of the building is designed in such a way as to protect the building and its occupants from a fire. *See also* ACTIVE FIRE PROTECTION.
被动防火 一种防火方法，指建筑的形态、布局及构造的设计目的是防止建筑及其住户遭受火灾。另请参考“主动防火”词条。

passive solar heating Where the form and fabric of a building are designed in such a way as to maximize the collection of solar radiation that is received directly from the sun, the sky, and the ground. Any solar radiation

incident on the building is transmitted indoors through windows or other glazed elements, and converted into heat by absorption on opaque elements of the building. Nearly all UK buildings benefit from some passive solar heating, except where special measures have been taken to exclude solar radiation.
被动式太阳能供暖 一建筑的形态及构造设计目的是最大化采集直接从太阳、天空及地面接收的太阳辐射。照射至建筑物的任何太阳辐射将通过窗户或其他玻璃组件传输至室内，然后由建筑物的不透明组件进行吸收从而转化成热量。几乎所有的英国建筑都受益于被动式太阳能供暖，一些采取特殊措施来阻挡太阳辐射的建筑除外。

patching The process where minor defects in concrete are repaired.
修补 修复混凝土次要缺陷的过程。

patent axe (comb hammer) A hammer with nine sharp points on the striking part of the hammer head. Used for scabbling concrete to provide a good key to which new concrete will bond.
凿毛锤（打毛锤） 在锤头敲击部分有九个锋利锯齿的锤子。用于凿花混凝土以造成毛面黏结新混凝土。

patent glazing Transparent or translucent materials, usually glass, used to clad a building.
橡胶镶嵌玻璃 透明或半透明材料，一般指玻璃，可用于覆盖一建筑物。

patio Paved or flagged surface of a garden.
露台 一花园的铺砌面或石板面。

patio door A horizontally sliding external door, usually comprising two panes of glass. *See also* FRENCH DOOR.
露台玻璃门 一水平滑动外门，通常由两块窗格玻璃构成。另请参考“落地双扇玻璃门”词条。

pattern 1. A design, instruction, plan, model, or *template, which is used as a guide to make something. 2. Something that is repeated over and over, e. g. certain shapes on wallpaper, or a trend in a data series.
1. 模式 一个设计、规划、模型或模板，可作为做某事的一个指引。**2. 图案** 再三重复的某物，如墙纸上的某种形状，或数据系列中的一个趋势。

patterned glass Glass that has a pattern incorporated into one or both sides of the glass. Used for privacy and decorative purposes. Also known as **rolled** or **figured glass**.
压花玻璃 指在玻璃一面或双面融入了一个图案的玻璃。可起到保护隐私及装饰作用。也被称为“滚花玻璃”或“花纹玻璃”。

pattern staining Discolouration of plaster due to the substrate to which it is fixed. Where the plasterboard is in contact with plaster dabs, timber, or other fixing, it is usually colder or warmer at this point. If colder, due to a *cold bridge, it is more likely that condensation will form on the surface of the plaster. If the surface is warmer due to the substrate, the discolouration is often due to the dust which is trapped in air, and is more likely to circulate over the warm area of the plaster and deposit on the surface.
墙面变色 由于固定灰泥的基底的原因而褪色。在石膏板与灰泥、木材或其他固件接触的位置，此时温度一般会变得更低或更高。如果由于出现冷桥现象而温度变低，很有可能会在灰泥表面形成冷凝；而如果表面由于基底原因而温度升高，灰尘被困在空气中，很有可能在灰泥的温暖区循环并沉积在其表面上，从而褪色。

pattress box (back box, conduit box, socket box) A box used to mount electrical items, such as a switch or *socket outlet.
电线底盒（底盒，分线盒，插座盒） 用于安装电气元件的箱子，如开关或插座。

pavement Hard flat surface laid for pedestrians either as a footpath on its own or adjacent to a road used for mechanical vehicles, such as cars and lorries. Equally, pavements can be classed as any hard surface laid for use by either vehicles or foot traffic.
人行道路面 为行人铺设的硬路面，单独存在或与机动车辆（汽车或货车）通行道路相邻。同样，人行道也可视作为供车辆或行人通行的任何硬路面。

pavement lens (pavement prism) A glass block or lens installed within a pavement light.
路面采光砖 为路面采光而安装的玻璃砖或玻璃镜片。

pavement light (vault light) A window, usually constructed from glass blocks, that is located within the pavement to provide daylight to the space below.
路面采光窗（地下室采光窗） 路面上的窗户，通常由玻璃砖制成，为下方空间提供光线。

paver (paving brick, paviour) Building units, bricks, and blocks used for external hard standings.
铺路材料（铺砖，铺路料） 铺设室外硬地面使用的建筑材料、砖以及石块。

pavilion A small structure set aside for an ornamental summerhouse in the garden or a building on a cricket or sports field used for viewing the activities.
亭子 公园内起装饰作用的凉亭结构，或者在板球、运动场所中用于观看比赛的建筑物。

pavilion roof (polygonal roof) A roof that is hipped equally on all sides and is of a regular polygon shape in plan.
亭顶（多坡屋顶） 各边均等地斜接为屋脊的屋顶，平面图上为正多边形。

paving 1. The surface of a *pavement. 2. The process of laying a *pavement. 3. The material used to form the pavement, such as concrete slabs, *pavers, or stone.
1. 铺砌面 路面的表面。**2. 铺路** 铺砌路面。**3. 铺路材料** 用于铺设路面的材料，如水泥板、路砖或石头。

paving slab (flag, pre-cast flag) A large pre-cast concrete or stone *slab used for *paving.
铺路板（铺路石板，预制石板） 铺路用的大块的预制混凝土板或石板。

paviour *See* PAVER.
铺路料 参考“铺路材料”词条。

PC sum *See* PRIME COST.
总成本 参考“主要成本”词条。

pea gravel Course rounded gravel used to surround buried pipes.
豆砾石 层砌的圆石，用于围在埋管周围。

pearlite (steels) A phase found in steels and cast irons. Pearlite consists of alternating layers of alpha-ferrite and cementite (both are chemical constituents of iron).
珠光体（钢） 在钢和铸铁中出现的一个体相。珠光体中有阿尔法铁素体和渗碳体（两者均含有铁成分）的交替层。

peat Fibrous soil of organic origin, which forms boggy ground.
泥炭 由有机物生成的纤维土，可形成沼泽地。

pebble A small stone that has been rounded by the action of water, sand, or wind. **Pebbledash** is a rendered wall containing pebbles that have been pushed into the render.
鹅卵石 由于水、沙土或风的作用而被磨圆的小石头。**干粘石** 是水泥浆里压入鹅卵石的抹灰墙。

pedestal A base or support usually for a column or statue.
基座 一种垫于柱子或雕像下面的建筑支撑物或底座。

pedestal basin A *basin mounted on a *pedestal.
立柱盆 安装于底座上的洗脸盆。

pedestal WC A *WC where the bowl is mounted above the floor on a *pedestal.
厕所坐便器 安装在厕所地面上的便池基座。

pedestrian A person travelling by foot in an area used by vehicles. **pedestrian crossing** A place on a road designated for people to cross.
行人 在机动车辆通行区域步行的路人。**人行横道** 是路面上设计为供行人通行的区域。

P

pedology The study of soil science in its natural environment, concerning its properties and classification.
土壤学 对土壤自然环境的属性及分类方面的科学研究。

peelable (strippable) Wallpaper fixed with adhesive that is made to be easily removed.
可剥性（可剥去的） 用胶粘剂固定的墙纸便于移除的特性。

peeling Finish such as paint that has not properly adhered to the surface and has started to come away from the substrate.
脱皮 油漆等饰面没有紧紧黏附附在表面上而是开始从基底上剥落的现象。

peen hammer Striking tool with a rounded end, used for hammering and working metal.
圆头锤 圆头的捶打工具，用于锤击和锻造金属。

peening Working of metal using a * peen hammer.
锤平 使用圆头锤的金属（锻造）工作。

peg 1. A post fixed into the ground, marking building gridlines and levels used for setting out the building. 2. A timber dowel, normally oak, fixing nail, or galvanized steel dowel used to locate and hold a tile in place.
1. 栓 规划建筑物时固定于地下的柱子，用于标示建筑物地界和标高。**2. 木榫** 通常为橡木钉、固定钉或镀锌钢钉，用于将瓷砖固定到位。

peg tile A plain tile with a hole at the top of the tile for a peg or nail to be inserted. The peg fixes over the top of the batten and holds the tile in place.
带孔平瓦 顶部有孔可嵌入栓或钉子的平瓦，瓦条和瓦片用栓钉连接固定。

pellet (stud) A small cylindrical-shaped piece of wood that is used to cover the head of a screw that has been * countersunk into a piece of timber.
压片（螺栓） 小型圆柱状木片，用于遮盖嵌入木块中的螺钉头。

pelmet A thin horizontal panel or curtain placed above the head of a window to conceal the curtain rail and fixings.
窗帘盒 薄的水平板或遮板，装在窗户上方用以遮盖窗帘吊杆和紧固件。

P

Pelton wheel A water turbine wheel, where a jet of water impacts on blades or buckets around the perimeter of the wheel causing it to turn; *See also* REACTION TURBINE.
培尔顿水轮机 因水流冲击到水轮周围的刀片或水桶上而使其转动的一种水轮机；另请参考“反击式涡轮机”词条。

penalty (penalty clause) A condition within the contract that requires payment in excess of losses suffered. Terms that require payment in excess of loss anticipated are not legally binding in the UK.
罚款（罚款条约） 合同中要求支付超过损失的费用的条款。要求支付超过预期损失的条款在英国并不具备法律效力。

pencil arris (pencil round) A corner or angle of timber or plaster rounded to a radius of approximately 3 mm.
圆角 木材或石膏的拐角或转角，磨圆后半径约为3mm。

pencil bar A thin reinforcing bar, 6 – 8 mm in diameter.
细钢筋 直径为6 – 8 mm的细钢筋条。

pencilling The process where the mortar joints in a brick wall are painted white.
白灰勾缝 将砖墙中的灰缝涂成白色的过程。

pendant (pendant fitting, droplight) A light fitting that hangs from the ceiling.
吊灯（餐吊灯） 天花上垂挂的灯具。

pendulum A mass that swings freely under the influence of gravity from a fixed point.
悬垂物 悬挂在一个固定点处，在重力作用下能自由摇摆的物体。

penetration An opening made in a material or component.
贯穿孔 在材料或组件中的开口。

penetrometer An instrument used to undertake a penetration test. The instrument consists of a cone or plunger, which is fixed to a series of connecting rods. The penetrometer is pushed or hammered into the ground. From the amount of resistance offered from the ground to penetration, an indication of bearing capacity, shear strength, and an indication of the amount of settlement can be obtained.
贯入仪 用于进行贯入度测试的仪器。该仪器有一个圆锥体或柱塞，固定在一连串的连杆上。使用时把贯入仪压进或敲进地下，根据穿透地面的阻力可以得出承载力、抗剪强度以及沉降量的标示值。

penning gate A *sluice gate, rectangular in shape that moves upwards to open.
直升式水闸门 矩形闸门，向上移动时打开。

penstock 1. A valve or *sluice gate used to control the flow of water or the discharge of sewage. 2. A pipe or channel to supply water, typically under pressure, to something, for example, to a *hydroelectric plant.
1. 水闸 用于控制水流或污水排放的阀门或闸门。**2. 压力水管** 尤其指压力下向水电厂等处供水的管道或渠道。

pentagon A five-sided polygon, with each side of equal length.
五边形 各边长度相等的五边形。

penthouse The most prestigious apartment in an apartment block located on the uppermost floor, may occupy the whole floor and have terraces.
顶层公寓 公寓大楼中的最高级楼层，位于顶层，可占用整层并且有露台。

penthouse roof A single pitch roof which does not abut a wall.
阁楼屋顶 不与墙邻接的单坡屋顶。

peptizing agent A product that increases dispersion of a substance into colloidal form (*See* COLLOID) by depolymerization, or reducing *flocculation.
胶溶剂 可通过解聚或减少絮结产物来将一种物质形成胶体分散体（参考“胶质”词条）的物质。

percentage Proportion of a whole expressed as an amount out of a hundred, for example, half of an amount would be written down as 50%.
百分比 占整体的比例，以百分比计，如占整体一半则计作50%。

perched groundwater When the groundwater lies above the surrounding groundwater due to an isolated body of impervious soil, such as a clay lens.
上层滞水 由于不渗水土壤（黏土）的隔离体的存在，位于周围地下水上方的地下水。

percolation The slow movement of gas or water passing through a porous substance, such as rainwater percolating downwards through the soil to the water table.
渗漏 水或气体穿过可渗透物质时的缓慢活动，例如雨水从土壤向下渗透至地下水面。

percussive drilling A process used to break through rock using repeated hammering action.
冲击钻探 反复锤击打穿岩石的过程。

perforated brick A brick that incorporates a number of vertical holes. It is lighter than solid brick and has better thermal insulation properties.
多孔砖 有一些垂直孔的砖，比实心砖轻，且保温隔热性能更佳。

performance Implies the satisfactory carrying out of the works in accordance with the contract.
执行 指按照合同要求进行工作。

performance bond (completion bond) Bank guarantee provided by the contractor that secures the client with ensured remuneration in the event of the contractor's default.
履约保函（完工保证） 承包商提供的银行担保，如果承包商违约，由银行向业主提供赔偿保证的法律文件。

performance specification Description that specifies the requirements of the building, structure, and materials rather than specifying or naming the materials themselves.
性能规范 规定建筑、结构和材料要求的说明，而不是只规定或命名材料。

pergola An arch formed by a double standard wooden frame that allows plants to be trailed over it to provide a walkway in a garden.
凉棚 由一个双标准木框架架成的拱形，绿植可沿着向上长，在花园中形成人行道。

perimeter 1. A boundary line around an area. 2. The length of the boundary line around an area or shape.
1. 周界线 环绕一块区域一圈的界线。**2. 周长** 一块区域或一个图形的周界线的长度。

perimeter angle The angled trim installed around the perimeter of a suspended ceiling at the junction where it meets the wall.
边龙骨 沿着吊顶天花板周围，在与墙壁的连接处安装的成角镶边。

perimeter beam The most external beam or beams that run around the circumference of a building or structure.
圈梁 沿着一栋建筑或结构的周围最外面的一根或多根梁。

perimeter diffuser A *diffuser that is located along the perimeter of a floor or ceiling.
周界扩散器 沿着地板或天花周围安装的扩散器。

permafrost An area of land (for example the polar regions), or subsurface layer of the ground that remains permanently frozen throughout the year.
永久冻土 一整年都保持结冰状态的土地（如极地区域）或亚表土层。

permanent formwork (permanent shuttering, sacrificial form, absorptive formwork) Formwork that remains in place once the concrete has set and becomes part of the structure.
永久性模板（一次性消耗模板） 混凝土凝固后就安置到位，并且构成结构的一部分的模板。

permeability The rate at which water under pressure can flow through the interconnected voids (or pore spaces) within a material, such as soil.
渗透性 在一种材料（如土壤）内部，水在压力作用下流经连通的空隙（或孔隙）的速率。

permeameter A laboratory instrument used to determine the *coefficient of permeability of a soil sample.
渗透仪 用于确定土壤样本渗透率的一种实验室仪器。

permissible deviation (tolerance) Differences in moisture contents, weights, mixtures, ingredients, strengths, and sizes from that specified, that is considered acceptable. It is also the distance or movement from the required position that is allowed. Such allowances are necessary to accommodate inaccuracies in instruments, casting, variability of materials, and ability of humans to accurately cut, weigh, set out, and measure.
容许偏差（公差） 含水量、重量、混合物、成分、强度、尺寸、位置或位移与规定的值允许存在的偏差。(规定) 这样的允许值对于仪器的不精确度、材料的差异性、人工精确切割能力，以及称重、放线和测量（工作）都是必需的。

permissible stress The maximum allowable stress that can be specified under certain conditions in elastic design.
容许应力 弹性设计时在某种情况下可以施加的最大压力。

permit (permit to work) Form that provides authorization to undertake work in a controlled or hazardous area. The permit would normally state the area controlled, risks and hazards, equipment, and procedures that must be undertaken, persons authorized to work, periods of work allowed, period for which the permit is valid, and signature of the person authorizing the work.
许可（允许开工） 允许在受控或危险区域开工的授权文件。许可证一般包含对受控区域、风险和危险、设备以及必须采取的措施的规定，准予工作的人员、允许工作的期限、许可有效期的陈述以及授权该工程的人员的签名。

permutation An ordered arrangement of elements from a set.

排列 一套组件的有序排放。

perpend (cross-joint, perp, perp joint, US head joint) The vertical joint in masonry construction. Commonly referred to as perps.
竖缝 砖石工程中的竖向接缝。通常写作 perps。

perpendicular 1. At right angles to another object, for example, a perpendicular joint is at 90° to the horizontal mortar bed in a wall. 2. Something that stands vertical to the ground, is at 90° to the ground.
1. 垂直 与另一物体成直角，如在一堵墙中垂直接缝就是与水平的砂浆平缝成 90°角的竖直接缝。
2. 垂直式 某件竖立的物体与地面垂直，即与地面成 90°角。

perpendicular style Gothic architecture that is exaggerated by vertical straight and slender aspects such as windows and panelling.
垂直式风格 哥特式建筑的风格，通常为立式竖直的，如窗户和镶板都是细长型。

perpend stone (bond stone, parpend, perpender, perpent stone) A large, long stone that extends through a wall from the inner to the outer face. Used to bind the wall together.
穿墙石 一块从内墙面穿到外墙面，用于将墙联结到一起的大的长条石。

perps *See* *PERPEND.
竖缝 参考“竖缝”词条。

Perry-Robertson formula An empirical formula used to derive the buckling loads for long slender beams and axial loaded struts.
佩里-罗伯逊公式 用于导出细长梁和轴向载荷撑条的压曲临界荷载的实验公式。

personnel door A door to an area of a building that has restricted access, such as a staff area.
员工通道 通往建筑物限制区域（如员工区）的通道。

Perspex (Plexiglas) *See* *POLYMETHYLMETHACRYLATE.
有机玻璃（树脂玻璃） 参考“聚甲基丙烯酸甲酯（有机玻璃）”词条。

PERT (Project Evaluation Review Technique) Planning technique for analyzing logical networks to determine earliest start, latest start, earliest finish, latest finish, and float of activities to determine the *critical path.
计划评审技术 通过分析逻辑作业网确定活动开始与结束的最早、最晚（时间）以及活动浮动时间并最终确定关键路径的规划技术。

pest An unwanted organism that causes damage to livestock, crops, or humans.
害虫 会对庄稼、牲畜或人类带来损害的有害生物体。

pesticide A chemical compound (such as a herbicide or insecticide) used to kill *pests.
杀虫剂 用于杀虫的化学剂（如除草剂或敌敌畏）。

pet cock *See* AIR RELEASE VALVE.
小旋塞阀 参考“放气阀”词条。

petrol intercepting chamber (petrol interceptor, petrol intercepting trap) A trap used to remove petrol from surface water runoff.
汽油截流室（汽油截流阀，汽油截气弯管） 用于将汽油从地表水径流中移除的弯管。

phase Section of works normally linked to an aspect of the works that has distinct characteristics, for instance, a design or construction phase.
阶段 通常为与工程中有显著特征的某方面有关联的部分工程，如设计或建设阶段。

phenolic foam A foamed insulating board made from phenolic resin. Phenolic foam is light with a density of only about 48 kg/m^3, which has the advantage over many other expanded plastics foam boards in that they burn only with considerable difficulty, with very little smoke or toxic gas, and are usable at temperatures up to 150℃. The boards are not strong enough to carry the feet of a ladder, but covered with quarry tiles their per-

P

formance is greatly improved.
酚醛泡沫体 由酚醛树脂制成的发泡保温板。酚醛泡沫质轻，密度仅为48 kg/m^3，比其他发泡塑料板更有优势。酚醛泡沫不易燃，燃烧后也几乎不产生烟雾和有毒气体，并且可在高达150℃的温度下使用。酚醛板不够坚硬，不可用于放置梯脚，但是放上方砖后其性能将大大提升。

Phillips head screws Fixing device with helical thread and a crosshead rebate in which a crosshead (Phillips) screwdriver is inserted to drive the fixing into the timber.
十字螺丝 有螺旋形螺纹及十字头槽口的固定件，可使用十字螺丝刀嵌入十字槽口将螺丝拧进木头里。

phon A unit used to measure perceived loudness.
方 响度级单位。

phosphate dosing A method of adding small amounts of phosphate into the drinking water to reduce lead becoming dissolved into the water from lead pipework, which can be found within some old domestic properties.
磷酸盐添加 一种在饮用水中添加少量磷酸盐的方法，可降低铅管中溶解于水中的铅含量，在一些老式物业中可见。

photic Relating to light, particularly the upper layer (photic zone) in lakes and seas where there is sufficient penetrated light for photosynthesis to occur.
透光的 与光相关的状态，尤其指湖泊或海洋的上层（透光层）是有充足的透光可供发生光合作用的。

photochemical degradation of timber A form of deterioration of timber material caused exclusively due to the presence of the sun. Exposure to sunlight causes the colouration of the heartwood of most timbers to lighten. This degradation can be very appreciable and can be slowed down by careful application of various finishes to the wood. It is important that degraded surface layers are removed before applying any protective surface coatings, otherwise they will not adhere to the surface.
木材的光降解 专指由于太阳的照射而引起的木材降解。暴露于阳光下可造成木头芯材的着色变浅。这种降解是相当明显的，可通过进行多层表面处理来减慢该过程。进行表面处理前一定要将已降解的部分去除，否则这些涂层不会粘连到表面上。

P

photogrammetry The use of photographs to create representations, particularly the use of aerial photography to produce maps.
摄影制图法 使用照片来呈现（事物），尤指使用空中摄影制作地图。

photographic survey Images of the site, structure, or building recorded and logged with notes for future reference. The survey may be conducted prior to development taking place showing the condition of the land and surrounding buildings. Surveys can also be undertaken during the construction period to record progress and the actual construction of components.
摄影测量 标有供今后参考注释的现场、结构或建筑物的图片。可在开发前进行测量，以记录土地及周围建筑的状况。也可在施工过程中进行测量，以记录施工过程及建筑构件的设计情况。

phreatic *See* *GROUNDWATER.
地下水层 参考“地下水”词条。

pH value A number indicating if a solution is acid (1 - 6) or base (8 - 14). Pure water has a pH value of 7, which is neutral.
酸碱度 用于指示溶液酸性（1—6）或碱性（8—14）程度的数值。净水的pH值为7呈中性。

phytoremediation The use of plants to absorb or degrade contamination from soil or water.
植物修复技术 用植物吸收或降解土壤或水中污染物的技术。

piano hinge (continuous hinge) A long butt hinge that often runs along the full length of the door that is opening and closing. The hinge comes in long strips which can be cut to length.
琴式铰链（连续铰链） 通常指沿着打开或关闭过程中的门的全长延伸的长对接铰链。该铰链一般为长条形，可按需要剪成不同长度。

piano nobile Classical term for the first or main floor of a building containing the living and reception rooms.
主厅 建筑物中包含客厅和会客室的第一层或主楼层区域，传统术语。

piano wire Long thin wire that can be attached to a plumb bob or used for setting out. Wire can be pulled much tighter than string and is often considered more accurate for setting out; however, in most cases laser lines have superseded both string and wire.
测距丝 可连到铅垂线上或用于放线的细长线。因为钢丝比线拉得更紧，放线时通常更精确；但通常情况下使用激光谱线替代钢丝和线来进行测定。

piazza Open square surrounded by buildings. Many piazzas earn a reputation because of the historic or eminent buildings that surround them.
广场 有建筑环绕的空旷场地。许多广场都因历史原因或周围的著名建筑而闻名。

pick axe A double-ended pick with a point on one end and chisel head on the other, used for breaking up concrete and loose stiff materials.
鹤嘴锄 一种双头镐，其中一头为尖头，另一头为凿子，通常用于击碎混凝土或打散坚硬材料。

picked stock facings *Stock bricks that have been selected to be used as *facing bricks.
精选面砖 被选作面砖的普通砖。

picket A pointed piece of timber that is driven into the ground. Can also be used to form part of a fence (a **picket fence**).
尖桩 楔入地里的尖木头，也可用于构成栅栏（尖桩篱栅）。

pick hammer A small hammer used by roofers for making holes in slates and for driving in nails.
鹤嘴锤 屋顶工人用于在石板上凿洞或钉钉子的小锤子。

picking (stugging, Scotland **colouring)** Using a pointed pick to create a pitted effect on the surface of stone.
凿毛 用鹤嘴锄在石头表面凿，形成凹痕。

picking up Painting next to wet paint and joining without any visual signs of the two edges being painted at slightly different times, sometimes referred to joining the live edges.
修边 在未干漆的旁边喷漆，且不能明显地看出两处边沿的喷漆时间不同，有时指将不规则边连接到一起。

pictorial projection Three-dimensional drawings used for marketing purposes, includes axonometric, isometric, and oblique projections.
立体投影 营销用的三维图，包括三面投影、正等测投影和斜等测投影。

picture gallery Room in a building used for displaying pictures.
画廊 建筑物中用于展示图片的空间。

piecework (piece work) Work where the employee is paid for each unit (piece) of work that is completed.
计件工作 按照雇员完成的产品数量支付薪水。

piecing-in Repairing a damaged portion of a material or surface by inserting a replacement piece that is the same size as the damaged piece.
修补 通过嵌入与受损部位大小一致的替换件对材料或表面进行修复。

piedmont A region at the base of a mountain or mountain range.
山麓 山峰或山脉的底部区域。

piend (Scotland) A *hip.
屋脊（苏格兰） 斜脊。

pier 1. A load-bearing buttress or brickwork column between two openings. 2. A structure made out of

wood, steel, or concrete that provides a platform that extends out into the sea, river, or lake. Used as a landing stage for boats or as an attraction to walk along to admire the coastal view.
1. 桥墩 两个开口之间的承重墩或砌体柱。**2. 码头** 木头、钢材或混凝土制成的建筑，用于筑成延伸到海洋、河流或湖泊的一个平台。可用作栈桥供船停泊，或者用于吸引人们沿着栈桥欣赏海岸风景。

piezometer An instrument for measuring pressure, for instance, the compressibility of a material or the amount of fluid pressure.
压强计 用于测量压力（如材料的压缩性或流体压力总量）的仪器。

pig Brickwork that has been built out of gauge, so that one corner has one course more than the other. As the brickwork travels to the other course it often tapers up or down meaning that the brickwork is not level. Such mistakes are made when string lines are pulled to the wrong course at a corner and the brickwork is built out of line.
螺丝墙 砌砖不成直线，一个拐角比另外一个多一层。在砌一层砖时通常会高于或低于同层砌砖，即砌砖层不是水平的。这种情况通常是由于在拐角处把线绳拉到了另一层处导致砌砖不成直线。

pigeon hole wall A wall constructed using a * honeycomb bond.
镂空墙 筑成蜂巢式的墙。

pigment A powder used to colour, especially paints.
颜料 用于上色的颜料粉末，尤指油漆涂料。

pile 1. Concrete column that is driven or bored into the ground; piles are either end-bearing with the load being transferred to the end of the pile that rests on good load-bearing strata, or friction piles that transfer the loads by friction along the sides of the piles. 2. The fibres that project from a woven carpet.
1. 桩 楔入或钻入地下的水泥柱；可为端承柱，将承重转移至良好承重层上的桩端，或为摩擦桩，通过摩阻力将承重分散至桩体各边。**2. 绒毛** 在纺织地毯中突起的纤维。

pile cap A steel plate or reinforced concrete slab that is placed on top of a * pile to distribute the load from the superstructure evenly over the pile or pile group.
桩承台/桩帽 置于桩顶部的钢板或钢筋混凝土板，用于将上层建筑的压力均匀地分散于桩或桩组。

P

pile driver A machine, consisting of a hoist and a leader, used to drive (hammer) piles into the ground. Other pile-driving techniques include jacking, jetting, screwing, and vibrating.
打桩机 将桩打入（锤入）地下的机器，包含一个起重机和导架。其他的打桩技术包括顶压、水冲、螺旋和振动等。

pile helmet A temporary steel cap that is fitted to the top of a * pile to prevent damage to the pile during driving.
桩帽 安装到桩顶的临时性钢帽，用于预防打桩过程中对桩造成损害。

pile shoe A cast-iron or steel point, which can be fitted to the foot of driven piles to facilitate penetration and provide protection.
桩靴 可装到打入桩底部的铸铁或钢尖，打桩时更易穿透还能提供保护。

piling The boring or driving of concrete, steel, or timber * pile foundations into the ground.
打桩 将混凝土、钢铁或木制桩基打入或钻入地下。

pillar A non-circular free-standing vertical pier. *See also* * PIER.
墩 无须支撑物的非圆形竖直墩。另请参考“桥墩”词条。

pillar tap A tap that stands proud of the basin or bath on a pillar.
柱式水龙头 脸盘或浴盆中突出的柱状龙头。

pilot hole A small hole drilled in a material that acts as a guide for a nail or screw, or for a larger drill bit. The pilot hole prevents the material from cracking or splitting when the nail, screw, or larger drill bit is inserted.

定位孔 钻在材料上的小孔，用于引导钉子、螺丝或者较大的钻头定点。定位孔在嵌入钉子、螺丝或较大的钻头时可防止材料出现裂纹或被劈开。

pilot light A small continuous flame in an appliance, such as a boiler, used to automatically ignite a much larger burner when required.
长明火 锅炉等设备中的长燃小火，必要时可以自动点燃更大的燃烧炉。

pilot nail A nail that is driven in a material to hold it in place temporarily until the main nails are installed.
临时钉 钉进材料中的钉子，在主钉钉入前起临时固定的作用。

pin 1. A flexible joint that is held together by bolts or rivets. 2. A slender wire nail, wooden dowel, or peg.
1. 销 用螺栓或铆钉连在一起的弹性接头。**2. 细长的圆铁钉、木销钉或桩钉**。

pinch bar (case opener, claw bar, jemmy, wrecking bar) Small hexagon-shaped rod 14 – 16 mm diameter with one chisel end and a hooked chisel on the other end. A smaller version of the *crow bar, which can be used in one hand.
撬杆（开壳器，撬棍，短撬棍，起钉器） 六边形的小杆，直径 14—16 mm，一头为凿子，另一端为钩状凿。是撬杠的缩小版，可单手使用。

pinch rod Measuring and checking rod, cut to measure the height between floors, windows, or doors. Used to check heights quickly.
测量杆 切割成合适尺寸的测量和检查棒，用于测量楼层间、窗户或门的高度。用于快速测量高度。

pine Fast-growing coniferous tree that produces softwood used for structural timber components and finishes.
松树 生长迅速的针叶树。其木质为软质木材，可用作结构木料配件和装饰。

pin hinge A butt hinge with a pin that can be removed.
销铰 带销轴的对接铰链，可拆卸。

pinhole A very small hole on the surface of a material usually caused by surface imperfection or trapped air. The very small holes caused by trapped air on a film of paint are referred to as **pinholing**.
针孔 由于表面缺陷或气泡导致的材料表面的小孔。漆膜上滞留的空气引起的小孔称作针孔。

P

pin joint A connection in a structure where members can rotate with respect to each other—such joints do not transmit moments.
铰链接合 结构中构件可以互相旋转的连接，此类接合不会传递力矩。

pinnacle 1. The natural peak or top of something. 2. An ornamental turret on the top of a spire, buttress, cone, or pyramid-shaped roof.
1. 尖峰 某些东西的自然尖峰或顶部。**2. 尖顶** 尖塔、支墩、圆锥体或金字塔状屋顶上的装饰性塔楼。

pinning 1. Use of panel pins to fix pieces of timber together. 2. Use of dowels to fix timber together.
1. 钉板 用镶板钉把木块钉到一起。**2. 暗榫接合** 用销子把木块固定到一起。

pinnings Different coloured stones set in rubble walling to provide a chequered effect.
花石墙壁 毛石墙上装饰的不同颜色的石材，用以取得棋盘状的效果。

pinning up Filling the gap between an underpinning foundation and existing foundations by inserting and ramming in dry or semi-dry mortar. Non-shrinkable grout or expanding grout is often used to avoid shrinkage and settlement.
托换基础填砂浆 嵌入和夯实干的或半干的灰浆填充托换基底与现有基底之间的缝隙。通常使用防缩灌浆或膨胀灌浆来防止收缩和沉降。

pin tumbler lock A cylinder lock with spring-loaded pins offering many different key combinations, ensuring good security.
弹子锁 带有弹簧锁销的圆筒锁，可提供多种弹子组合，更加安全。

pipe A long hollow cylindrical tube, usually constructed from metal, plastic, clay, or concrete.
管子 长的中空圆管，通常由金属、塑料、黏土或混凝土制成。

pipe bracket A *bracket attached to a wall, floor, or ceiling to support a pipe.
喉码 连到墙壁、地板或天花上用于支撑管道的托架。

pipe clip A *clip used for fastening pipes to walls, floors, or ceilings.
管夹 将管道固定到墙壁、地板或天花上的夹子。

pipe closer (fire-stop sleeve) A *fire-stop that fits over a pipe where it penetrates a wall, floor, or ceiling.
阻火圈 穿透墙壁、地板或天花处管道上的阻火件。

pipe cutter Tool for cutting pipes. Pipes are rotated in a vice causing two cutting discs to cut a V-groove into the pipe. Clamps holding the cutting discs are tightened to cause the discs to cut through the pipe.
切管机 切割管道用的工具。管道在老虎钳内旋转，两片切割片在管道上切出一个 V 形槽。夹住切割片的夹子收紧固定以便切断管道。

pipe duct 1. A pipe that is used to draw cables through from one position to another. 2. A *duct that only contains pipe runs.
1. 穿线道 从一个地方到另一个地方拉电缆线所用的管子。**2. 管槽** 仅包含管路的管道。

pipe fitter Skilled person who fits water, gas, oil, and steam pipes.
管道安装工 安装水、气、油和蒸汽管道的专业人员。

pipe fitting A *fitting that is used to join pipes together. A number of different fittings are available such as *bends, *elbows, and *tees.
管件 将管道连接到一起的配件。配件有许多种类，如弯管、弯头和三通等。

pipe flashing (vent soaker) A *flashing installed around a pipe where it penetrates through a roof.
出屋面防水套管 安装在穿透屋顶的管道周围的防水板。

pipe hook 1. A hook used for lifting and moving large pipe sections. 2. A hook-shaped device, usually metal or plastic, used to support a pipe from a *rafter or a *joist.
1. 管钩吊具 用于托起和移动大型管段的钩子。**2. 管钩钉** 钩状设备，通常由金属或塑料制成，用于从椽条或托梁处支撑管道。

pipe jacking A trenchless technique for installing underground small-diameter pipelines, ducts, and culverts. From a launch pit or drive shaft, hydraulic jacks push specially designed pipes through the ground behind a shield, as excavation is taking place within the shield.
顶管技术 安装地下小口径管线、管道和电缆管道的非开挖技术。在护盾内围进行挖掘工作时，在护盾后面从一个工作井或顶进井处，用液压千斤顶把定制的管道打入地下。

pipe layer *See* *DRAIN LAYER.
管道工 参考“管道工”词条。

pipeline (pipe line) A long pipe formed from lengths of pipe joined together. Used to convey fluids or gases.
管道 将很多段管子连接到一起构成的长管道，用于输送液体或气体。

pipeline failure Mechanical and/or chemical degradation of gas and oil pipelines. In general, most pipelines are made from corrosion-resistant steels, and the failure mechanism is a combination of creep, fatigue, weld defects, and other forms of corrosion.
管道失效 输油输气管道的机械及/或化学降解退化。大多数管道是由耐腐蚀钢铁制成的，管道失效是金属变形、金属疲劳、焊接缺陷及其他形式的腐蚀的综合表现。

pipe lining A process of driving a length of steel pipe into the ground to form a shaft, which is then filled with concrete to form a *pile. The steel pipe can be open-ended or fitted with a steel shoe.
管道内衬 把一根钢管打入地下的过程，形成竖井，可在竖井内填入混凝土形成桩。钢管可以是开口的也可以是带钢制桩靴的。

pipe ramming A trenchless technique of driving a pipeline through a short distance under the ground, for example, below a road. Similar to *pipe jacking, but employs a percussion hammer rather than jacks.
夯管 一种将管子在地下（例如在道路下面）贯通一段短距离的非开挖技术，与顶管法类似，只是采用夯管锤而不是液压顶。

pipe ring A cylindrical bracket used to support a pipe. Can comprise either one or two pieces.
喉箍 用于支撑管道的圆柱形托架。可以是一件式或两件式。

pipe sleeve A short section of pipe that fits around a smaller-diameter pipe as it passes through a floor, ceiling, or wall.
管套 穿过地板、天花或墙壁时安装于较小口径管道外面的一段短管。

pipe tail 1. The open end of a section of pipe that is installed prior to the connection of additional pipework. 2. A short length of pipe that connects a sanitary fitting to a branch pipe.
1. 管端 与其他管道相连之前安装的一段开口管道。**2. 短管** 将卫浴设备与支管连到一起的一段短管。

pipework A collective term used to describe a collection of pipes and their associated **fittings.**
管道工程 描述管道及其配件的统一术语。

pipe wrap (pipewrap, wrapping tape) Collective name given to various tapes that are wound around pipes. Tapes are used to provide protection, thermal insulation, acoustics insulation, or for identification purposes.
管道包扎布 缠绕在管道周围的胶带的总称。胶带起到保护、保温、隔音或识别作用。

pipe wrench (cylinder wrench, Stillson) A wrench for gripping circular objects, threaded bars, threaded pipes, and pipes.
管扳钳 用于钳紧圆柱状物体、螺纹钢、螺纹管道及管道的扳钳。

piping 1. Instability due to seepage as a result of the pore pressure exceeding the weight of soil and water above it. It can occur on the downstream (dry) side of *dams and *cofferdams; it appears as if the soil is boiling. 2. A section of pipe. 3. Cavities in metal. 4. A high-pitched noise.
1. 管涌 孔隙压力超过其上方的水土重量导致的渗流形成的不稳定性（现象）。可发生于大坝和围堰坝的下游（干燥侧），发生时土壤如同在沸腾。**2. 一段管子**。**3. 金属内的空腔**。**4. 尖声** 声调高的噪音。

P

pit A large hole in the ground.
深坑 地上的大坑。

pitch 1. The angle of a sloping roof to the horizontal, measured in degrees. 2. A perceptual quality of a sound that is dependent upon the frequency of the sound source. 3. Name given to petroleum-derived bitumens used for waterproofing. 4. Term used to refer to the distance between items that are spaced equally apart, such as reinforcement bars in concrete or nails in wood.
1. 倾斜角 斜屋面与水平面之间的角，以度计量。**2. 音高** 由声源频率决定的感知音质。**3. 沥青** 防水用的石油衍生物的名称。**4. 间距** 用于指代间隔物体间的距离，例如混凝土中钢筋或木头中钉子之间的距离。

pitch board (gauge board, pitch block) A triangular-shaped template used when constructing stairs. The sides of the template are the same size as the rise, the going, and the pitch of the stairs.
踏步模板 垒楼梯时用的三角形模板。模板的三边边长，分别对应楼梯的垂直高度、梯段级距和斜长。

pitched roof A roof that has a pitch greater than or equal to 10°. *See also* FLAT ROOF.
斜屋顶 倾斜角大于或等于10度的屋顶。另请参考“平屋顶”词条。

pitch-fibre pipe (US bituminized fibre pipe) A type of pipe commonly used for sewers that was manufactured from wood or asbestos fibre and was heavily impregnated with pitch. No longer used.

沥青纤维管 下水道常用的一种管道，由木材和石棉纤维制成，浸有大量沥青。现已不再使用。

pitching The action of positioning a pile and runners ready for driving it into the ground.
吊桩 将单桩和滑动装置放置到位以便打桩的动作。

pitch line *See* NOSING LINE.
坡度线 参考“梯段坡度线”词条。

pitch mastic A jointless floor made from pitch with limestone or silica sand aggregate, fluid when hot, spread to a thickness of 16 – 25 mm. Oils and fats affect it less than they do asphalt, and it can have a polished or matt finish.
沥青砂胶 沥青制成的无缝地板，拌有石灰岩或硅砂骨料，热时呈流体，铺成后厚度为16—25mm。油脂对沥青胶粘剂的影响要小于对沥青的影响，并且沥青砂胶可取得抛光或哑光饰面。

pith The core of a tree, containing weak parenchyma and most of the log's defects. It is contained in boxed heart or a centre plank.
树芯 树的芯，有脆弱的软壁组织以及原木的大部分缺陷，存在于髓芯材或中心木板。

pitting corrosion A type of corrosion resulting in the formation of holes or pits on the surface of metals.
点状腐蚀 在金属表面形成洞或坑的一种腐蚀。

pivot A device used to hang swing doors or a rotating window.
转轴 用于悬挂双开弹簧门和旋转窗的装置。

pivot window A window that opens by rotating on horizontally or vertically located pivots.
旋转窗 通过旋转水平或垂直安装的转轴来开关的窗户。

placing of concrete (pouring) The process of laying wet concrete.
浇灌混凝土 浇筑湿态混凝土的过程。

plafond A ceiling or soffit.
顶棚 天花板或底部。

plain bar A smooth-surfaced steel reinforcement bar.
光面钢筋 表面光滑的钢筋条。

P

plain tile A small rectangular roofing tile, usually clay or concrete, that has a slight camber.
平瓦 小块长方形屋面瓦，通常由黏土或混凝土制成，有轻微弧度。

plan Drawing that depicts the view if the person was to look down on an object or building, it provides an outline of the horizontal plane. Plans are often provided to scale and with dimensions, and are used for setting out the building and components.
平面图 从上往下看一个物体或建筑时绘制的图纸，可看到水平面的轮廓。平面图通常会给出比例和大小，用于测定建筑和其他组成部分。

plancier piece (US) A *soffit board.
挑檐底面（美国） 挑檐底板。

plane A handtool for smoothing and levelling the surface of timber.
刨子 用于磨平木材表面的手持工具。

planed timber Wood with its surface dressed with a plane.
刨平木料 表饰面为一块刨平板的木材。

planer (rotary planer) A portable handheld power tool for smoothing, shaping, and levelling the surface of timber. Rotating cutting blades cut into the surface of the timber.
刨机（旋转式刨机） 用于磨平、塑形并将木材表面弄平的手提电动工具。由旋转的刀片切入木材表面。

planing machine (planer) A static power tool for smoothing, shaping, cutting, and levelling the surface of

timber, metal, and stone.
刨床 抛光、塑形、切割和刨平木材、金属以及石材表面的固定重型机床。

planing mill (US) A workplace (saw mill) used to cut and shape timber into planks of matchboard, floor-boards, and other *planks.
刨削车间（美国） 用于切割或将木材制成企口板、地板或绘制其他木板的车间（木材加工厂）。

plank A long, flat, solid timber board, usually installed face down in parallel rows.
木板 长的、扁平的实心木板，通常正面朝下呈平行状进行安装。

planking and strutting Braced planks laid vertically at the side of an excavation to provide temporary support.
挡土板支护 竖直地放于基坑边上的支撑板，用于提供临时支撑。

plank-on-edge floor (US) *See* *SOLID WOOD FLOOR.
毛地板（美国） 参考“实木地板”词条。

planned maintenance A regular maintenance programme for a building and its services.
维护保养计划 建筑物及其设备的定期维护计划。

planner 1. A professional responsible for the organizing and sequencing of operations on-site, who produces a programme and networks, and examines resource and cost implications. 2. Person employed by the local authority responsible for controlling the development of land.
1. 计划员 负责在现场安排操作步骤的专业人员，负责生成计划表和联络网，并负责检查资源和成本。**2. 规划师** 负责土地开发管控的当地部门的工作人员。

planning 1. Local authority controlling the development of land. 2. Organizing, scheduling, and sequencing activities.
1. 规划 当地部门控制土地开发。**2. 计划** 组织、规划和安排事项。

planning and scheduling The sequencing of works and resources for efficient project delivery.
计划调度 为高效率交付项目而做出的施工和资源安排。

plant Mechanical equipment and services. Can be mobile such as cranes, dumpers, and excavators, or can be stationary and also include buildings, such as substations and refineries.
设备 机械设备和设施。可为可移动的，如起重机、前翻斗车或挖掘机等；或固定的，如建筑物，如变电站和冶炼厂等。

planted A surface-mounted moulding bead, for example, an architrave.
明装线条 表面安装的装饰线条，例如门框线。

plant level (mechanical floor, mechanical level) A floor in a building that houses mechanical equipment, such as boilers and air-conditioning systems. *See also* *PLANTROOM.
机械设备层 建筑物中装备了如锅炉、空调系统等机械设备的楼层。另请参考“机械设备间”词条。

plantroom (US mechanical room) A room in a building that houses mechanical equipment, such as boilers and air-conditioning systems. *See also* *PLANT LEVEL.
机械设备间 建筑物中装备了如锅炉、空调系统等机械设备的房间。另请参考“机器设备层”词条。

plasma Ionized gas containing roughly equal numbers of ions and electrons. Used in welding, cutting, and metal spraying.
等离子体 包含的离子和电子数量大致相等的电离气体，用于焊接、切割和金属喷镀。

plaster base A continuous surface to which plaster will adhere.
抹灰底层 将要涂抹灰浆的连续面层。

plaster bead *See* *ANGLE BEAD.
灰泥护角 参考“护角条”词条。

plasterboard A board comprised of gypsum, fibreboard, or paper, and used in the construction of internal walls.

石膏板 包含石膏、纤维板或纸的一种板，用于建筑内墙的构筑；也称作石膏板或墙板。

plasterboard nail A galvanized or zinc-coated flat, round-headed nail used to fasten plasterboard to a wall or ceiling. *See also* PLASTERBOARD SCREW.

石膏板钉子 镀锌或有锌涂层的圆头或扁头钉子，用于把石膏板固定到墙壁或天花上。另请参考“石膏板螺丝钉”词条。

plasterboard screw (drywall screw) A screw used to fasten plasterboard to a wall or ceiling. Generally, more expensive than *plasterboard nails.

石膏板螺丝钉 用于把石膏板固定到墙壁或天花板上的螺丝钉，比石膏板钉子贵。

plasterboard trowel Plaster trowel with a thin flexible blade used to smooth over tapered joint plasterboard.

石膏板泥刀 带有一个活动薄刀片的抹泥刀，用于抹平凹缝连接的石膏板。

plaster dabs Small amounts of plaster or gypsum-based adhesive that are used to fix plasterboard to walls.

石膏板黏合剂打点 少量的灰泥或石膏基黏合剂，用于把石膏板固定到墙壁上。

plasterer Skilled worker capable of applying a smooth jointless surface of plaster.

抹灰工 能够涂一层无接缝、平滑的灰泥层的技工。

plasterer's float A large rectangular trowel used for levelling plaster.

抹灰工抹子 大的矩形泥刀，用于把灰泥抹平。

plasterer's labourer Unskilled operative used for cleaning, sealing, and preparing surfaces, mixing plaster, organizing materials, and cleaning tools. Often an apprentice or trainee *plasterer.

抹灰工助手 非技术工人，负责清理、密封和准备表面，混合灰泥、安排材料、清理工具等工作。通常为学徒或实习抹灰工。

plasterer's small tool A small tool with a flat blade at one end and spoon blade at the other, it is used to shape and finish small areas that are difficult to access with a standard float. Often used to access tight corners.

抹灰工小工具 一端为平面刀片另一端为浆勺的小工具，用于对标准抹子无法涂抹的小片区域进行塑形和修整。通常用于拐角处。

plasterer's trowel A large rectangular trowel used for levelling and smoothing plaster.

灰泥抹刀 大的矩形泥刀，用于抹平灰泥。

plastering The process of applying plaster to a ceiling or a wall.

抹灰 把灰泥抹到天花或墙壁的过程。

plastering machine Piece of plant that mixes and pumps plaster onto the walls ready to be finished and trowelled by hand.

抹灰机 混合灰泥并将灰泥喷到墙上的机器，之后可以用人手进行涂抹和修整。

plaster stop *See* STOP BEAD.

灰泥压条 参考“墙压条”词条。

plasterwork Any work undertaken using plaster.

抹灰泥工作 使用灰泥开展的任何工作。

plastic A type of polymeric material. *See also* *POLYMERS.

塑料 一种聚合材料。另请参考“聚合物”词条。

plastic analysis The analysis of a structure using *plastic theory, mainly used to analyze statistically indeterminate structures.

塑性分析 用塑性理论对一种结构进行的分析，通常用于分析超静定结构。

plastic deformation Permanent or non-recoverable deformation after release of the applied load. To effect

plastic deformation the material must be subjected to a force higher than the * yield stress.
塑性变形 释放附加荷载后导致的永久性或不可恢复性变形。要产生塑性变形，施加于材料的力必须高于屈服应力。

plastic failure A failure mode characterized by large deformation and no brittle failure.
塑性破坏 一种失效模式，特征为大变形、非脆性破坏。

plastic hinge In * plastic theory, it is a point in a structural section that has lost all stiffness, the stress in the material is at or above its yield point.
塑性铰 塑性理论中指结构截面中已经损失了所有刚度的一个点，材料所受压力处于或高于其屈服点。

plasticity The ability of a material to deform under load, without fracture, and when the load is removed the deformation remains; *See also* * ELASTICITY.
塑性 材料受压后变形的一种特征，不会折断，撤除压力后形变仍然存在；另请参考“弹性”词条。

plasticity index (PI) The range of moisture content over which the soil remains in a plastic condition.
塑性指数（PI） 塑性条件下土壤中水分含量的范围。

plasticizer A substance added to organic compounds to create a more flexible finished product.
增塑剂 添加于有机化合物中使其制成更有弹性的成品的一种物质。

plasticizer migration The loss of * plasticizer from plasticized plastics in contact with other materials, making the plastics brittle.
增塑剂渗移 与其他材料接触后塑料中增塑剂减少、塑料变脆的现象。

plastic limit The moisture content at which a fine-grained soil becomes plastic. It is the moisture content at which a 3 mm diameter thread of soil can be rolled by hand without breaking up.
塑性极限 细粒土壤变为塑性时的水分含量。直径 3 mm 的一抔土壤可以用手滚压而不会散开时的水分含量。

plastic modulus *See* * MODULUS OF PLASTICITY.
塑性模数 参考“塑性的模数”词条。

P

plastic moment (M_p) The theoretical maximum * bending moment that a structural section can resist, and at this point a * plastic hinge will form.
塑性弯矩（M_p） 理论上结构截面可以承受的最大弯矩，并且在这个点上会形成一个塑性铰。

plastics *See* * PLASTIC.
塑制品 参考“塑料”词条。

plastic theory The analysis and design of structures based on the idealization of elastic-perfectly plastic material behaviour.
塑性理论 对基于理想弹性、塑性材料的结构的分析和设计的理论。

plastisol A protective of * polyvinyl chloride coating applied to galvanized steel to provide further resistance to corrosion.
塑料溶胶 涂到镀锌钢上的聚氯乙烯保护层，提供进一步的防腐蚀保护。

plate A horizontal structural timber member, normally 100 mm × 50 mm, that is used to support studs and roof trusses, for instance, a **sole plate** or a **wall plate**.
板 平直的结构木料，通常为 100 mm × 50 mm，用于支撑立柱和房顶支架，例如底板或梁垫。

plate-bearing test An in-situ test used to determine an approximate value for the bearing capacity of the soil. A steel plate is loaded against the ground, typically at the bottom of a trial pit, until failure of soil occurs, or until a specified amount of settlement has been reached.
板承重试验 用于测定土壤承载力的近似值的一种现场测试。通常在试验坑的底部用一块钢板对土壤施压，直到土壤遭到破坏或者达到一定的沉降量。

plate cut *See* * FOOT CUT.
角口平切 参考“角口平切”词条。

plate exchanger A type of * heat exchanger comprising a series of metal plates that transfer heat between two fluids or gases. Due to the large surface area of the plates, they are more efficient than a conventional heat exchanger.
板式换热器 一种热交换器，包含在两种液体或气体之间传递热量的一组金属板。由于金属板的表面积较大，所以比传统的热交换机效率高。

plate girder A very large steel section, typically an I-section, with flanges that have been welded to the web; in older girders, rivets and bolts were used rather than welds; *See also* ROLLED SECTION.
板梁 非常大的型钢，通常为工字型型材，上下翼板焊接到腹板上；老式板梁通常采用铆钉和螺栓而非焊接；另请参考“轧制钢材”词条。

platen A smooth metal plate that loads or holds an item in place, e. g. the loading plates that are in contact with the surface of a concrete cube during a compression test.
压板 负载或支撑着一个物体，并使其处于合适位置的一块平滑的金属板，如抗压测试时与混凝土立方块表面直接接触的承载板。

platen-press A hydraulically or pneumatically operated press that presses a component between two rigid metal plates. Used to manufacture laminated board.
板式挤压机 对两块刚性金属板之间的组件进行挤压的一台液压或气动挤压机。用于制造层压板。

platform A raised level area, for example, to keep things clear of the ground, or for people to stand on so that they can be visible to the audience.
平台 高于地面的平坦区域，例如为了使东西不接触地面或者让人站在上面方便观众看到而建造的台面。

platform floor *See* * RAISED FLOOR.
平台层 参考“高架地板”词条。

platform frame A type of timber-frame construction where the structure is built up one floor at a time. Once each floor is complete, it is used as a working platform to construct the next floor. The majority of timber-frame construction in the UK is platform frame. *See also* * BALLOON FRAME.
平台型底架 一种木材框架结构，一次构筑一层。一层完工后，就用作构筑另一层的工作平台。英国大多数的木材框架结构都是平台型底架。另请参考“轻型木构架”词条。

platform roof (Scotland) A flat roof.
平台顶（苏格兰） 平屋顶。

plenum An enclosed space that is filled with air, such as the space under a raised floor or above a suspended ceiling. Often used as an air duct for an air-conditioning system.
集气室 充满空气的封闭空间，如活动地板下方或吊顶上方的空间。通常用作空调系统的风道。

plenum system A type of air-conditioning system where air from a plenum is distributed through ducts.
正压通风系统 空调系统的一种，集气室里的空气通过风道配送。

pliable conduit *See* * FLEXIBLE CONDUIT.
可弯曲导管 参考“挠性导管”词条。

plinth 1. The projecting base at the bottom of a * pedestal. 2. The projecting base of a wall. 3. The base of a cupboard. Also known as a * kick plate.
1. 柱基 基座底部突出的基座。**2. 墙基层** 墙壁的突出墙脚。**3. 橱柜的柱脚** 也称作踢板。

plinth block *See* * ARCHITRAVE BLOCK.
基底石块 参考“门头线墩子”词条。

plinth course The projecting course of masonry at the base of a wall that forms the plinth.

P

墙基层　墙脚处突出的石材层，构成墙基。

plinth return　A special brick that has an angled face along one of the long edges and one of the short edges. Used to form a right-angled corner on a plinth.
基座转角砖　沿着一条长边和一条短边上有倾斜角的异型砖，用于在基座上垒成直角。

plinth stretcher　A special brick that has an angled face along one of the long edges.
墙基顺砌砖　沿一条长边、有倾斜面的异形砖。

plot　1. A small area of land, for example, where a building is constructed. 2. To establish a graph from data points.
1. 小块地皮　例如用于建造建筑物的地方。**2. 标绘图**　用数据点构造的图形。

plot ratio (US floor area ratio)　The maximum floor area of the building compared with the land area; this limits the number of storeys that can be placed on a building.
容积率（美国　建筑容积率）　最大建筑面积与土地面积的比率；这个比率限制了一栋建筑物的楼层数。

plough (US plow)　A plane used for making a groove in timber.
槽刨　用来在木材上挖出沟槽的刨子。

ploughed and tongued joint　Two rebates or mortises formed in timber with a strip of wood (loose tongue) inserted into the rebate forming a joint.
条形榫拼接　木制的两个槽口或榫眼，由一条木头（条形榫）嵌入槽口形成拼接。

plug　1. A cylindrical device, usually plastic, that is inserted into a pre-drilled hole to hold a screw. 2. An electrical device that is inserted into a socket to provide an electrical connection. 3. A device that is inserted into the base of a bath or basin to prevent water escaping. 4. Any object that is designed to tightlyfill a hole.
1. 堵头　圆柱状装置，通常为塑料材质，用于塞住螺丝预钻孔。**2. 插头**　插入插座中提供电力连接的电气元件。**3. 排水塞**　塞在浴缸或脸盆底部的装置，防止水流失。**4. 塞子**　用于紧紧塞住孔洞的任何物体。

plug cock (plug tap)　A valve comprising a tapered plug that is used to stop the flow of a fluid or gas. Operated by turning the valve a quarter turn.
旋塞　包含一个锥形螺塞的阀门，用于关闭水流或气流。通过转动 1/4 圈来控制阀门。

plugging　Insertion of material, plastic or wood, into a hole to provide a fixing for nails or screws. As the screws or nails are inserted into the timber or plastic, the material is pushed against the surrounding material making a tight fixing.
防松填料　把织物、塑料或木头塞入洞内以固定钉子或螺丝。当螺丝或钉子钉入木材或塑料内时，防松填料能与周围材料紧紧固定在一起。

plugging chisel (plugging drill, US star drill)　A steel bar with a cross-shaped tip that is struck by a hammer to make a small hole in masonry or concrete walls.
嵌缝凿（美国　星形钻）　带十字形尖头的钢筋条，用锤子锤击，可以在大理石或混凝土墙上凿出小洞。

plug-in connector (flex cock, plug-and-socket gas connector)　A type of connector for a gas pipe that incorporates a socket.
插入式接头　插入承口的煤气管道的一种连接头。

plug-in switchgear (withdrawable switchgear)　*Switchgear that is plugged in rather than screwed into the *switchboard.
插入式开关　插入而不是拧入开关柜的开关装置。

plugmold (US)　Surface-mounted trunking providing power outlets.
插座线槽（美国）　明装线槽，提供电源输出。

P

plug tenon (spur tenon) A *stub tenon that has a four-shouldered tongue.
插销榫 有一个四肩榫舌的短榫。

plumb Vertical. Something which is plumb stands in an upright position, being exactly vertical.
垂直的 竖直的。垂直竖立的某物，完全垂直。

plumb bob (plummet) Small steel weight placed on the end of a string line to hold the line tight and vertical. Used for checking that something is plumb or checking the position of something over the ground, grid reference, or in relation to other objects.
铅锤 一条线的末端系的小块钢铁，用于保持线的紧绷和垂直。用于检查某物是否垂直，或者某物与地面、网格参照物或者其他物体的相对位置。

plumb cut (US) The cutting of a *birdsmouth on a roofing rafter where the rafter sits on the *eaves or *wall plate.
垂直切面（美国） 椽条与屋檐或梁垫相接处屋面椽条上的承接角口的切面。

plumber Skilled tradesperson who works with pipes, sanitary fittings, and lead work on roofs. As pipes used to be made out of lead, the original meaning of the word plumber, meant a person who works with lead.
水管工 与管道、卫生设施以及屋顶上的铅制工作打交道的技工。虽然如今所用管道不含铅，但是水管工原指与铅打交道的人员。

plumber's labourer (plumber's mate) Operative with skills to assist a *plumber by measuring and cutting pipes, organizing fittings and equipment, and clearing away rubbish and debris.
水管工助手 可协助水管工进行管道测量和切割、整理配件及设备，并且清理垃圾和碎屑的技工。

plumber's metalwork Sheet metal work undertaken by a plumber, includes dressing roofs, fitting flashings, and gutters.
水管工金工 水管工承担的钣金工作，包括装配屋顶、安装防水板及排水沟槽。

plumbing 1. Working with pipes, sanitary fittings, and lead work. 2. Setting something vertical and *plumb.
1. 卫生管道工程 与管道、卫生设施和铅制品相关的工作。**2. 铅垂** 把某物放置为完全垂直的状态。

P

plumbing unit Prefabricated bathroom pod or module, craned and *plumbed in position.
卫生装备组合 预制构件的卫生间组件或构件，吊装并竖直安装到位。

plumb level Spirit level fitted with two bubbles, one set for horizontal work and the other setting out vertical work and checking for *plumb.
水准仪 装有两个水准器的水准仪，一套用于水平工作，另一套用于铅垂工作及垂直检查。

plumb line String line with *plumb bob attached, which holds the line vertical, used for checking and setting out vertical work.
铅垂线 系着铅锤的线，把线拉成垂直状态用于检查及测定垂直工作。

plummet *See* PLUMB BOB.
垂球 参考“铅锤”词条。

plunger Flexible or rubber sucker stuck to the end of a wooden rod. The tool is placed over openings such as sink drains and *WCs, then pressed to cause suction and create negative pressure drawing water in the opposite direction to its normal flow. Used to unblock pipes.
马桶吸 粘到木杆一端的弹性或橡胶吸嘴。把该工具放到落水管或厕所的开口处，然后按压进行抽吸，产生负压把水从正常流向往反方向吸，用于疏通管道。

ply To join together, as by moulding or twisting.
叠合 通过铸模或缠绕连接到一起。

plymetal *Plywood covered on one or both sides with sheet metal.

金属箔胶合板 一面或两面都覆有金属薄片的胶合板。

plywood Layers of timber material glued together to form a whole structure; for additional strength and durability the timber is often formed with the grains of adjacent layers at right angles to each other.
胶合板 数层木料板胶粘在一起形成一个整体结构，为了获得更佳的韧度和耐用度，通常将两块相连的板子的纹理成直角进行胶粘。

pneumatic Relating to the use of compressed air, particularly to operate tools.
气动的 与使用压缩空气相关的，尤其用在操作工具。

pneumatic structure An air-inflated or air-supported structure.
充气结构 一种充气或气承式结构。

pneumatic tools (air tools) Equipment driven by compressed air; can be used in wet areas without risk of shock.
气动工具 由压缩空气驱动的工具，可用于潮湿区域且不会漏电。

pocket (box out, US blockout) 1. A hole that is cast in concrete using a *former as the concrete is laid. 2. A hole in a wall that is constructed to support the end of a beam.
1. 预制孔 浇筑混凝土时使用预留孔模具在混凝土中预制的洞。**2. 承梁口** 墙壁中的洞，用于支撑梁端。

pod A volumetric part of a building, such as a bathroom or a bedroom pod.
隔间 建筑物的一部分空间，例如一个卫生间或卧室隔间。

podium A section of a building at its base, which is clearly differentiated from the spaces above it by its physical form or by the type of space inside it.
群楼 建筑物底部的一部分，在外观上或内部空间上与上部建筑有明显区别。

pod urinal *See* BOWL URINAL.
小便池 参考“小便槽”词条。

point A sharp V-shaped end of an object, such as the point of a nail.
尖端 物体的锋利的尖端，呈 V 形，如钉子的尖头。

point detector A fire detector that is designed to detect the presence of a fire at a particular point within a building.
点型探测器 用于探测建筑物内某个特定点处是否有火情的火警探测器。

pointed arch An *arch formed from two curves creating a point at the apex.
尖拱 两条曲线形成的拱形，尖端处形成一个点。

pointed vault A type of ribbed vault where the ribs meet at a point.
尖顶肋拱 肋架拱顶的一种，肋架都相交于一个点。

pointing 1. The external mortar joint between individual masonry units. 2. The process where the joints between masonry units in a wall are filled with mortar after the wall has been constructed. The mortar can be coloured or finished in a number of ways to improve the appearance of the joint, or to provide greater weather protection. *See also* JOINTING and REPOINTING.
1. 灰缝 单独砌块之间的外部灰缝。**2. 勾缝** 墙壁砌成后在砌块之间的接缝中抹上灰浆的过程。可以为灰浆上色或者以不同的方式进行修饰，以美化接缝外观或者提供更佳的耐候保护。另请参考“填缝”和“重新勾缝”词条。

point of articulation A link that allows movement between two or more joined components.
活动连接点 两个或多个连接组件之间可以活动的一种连接。

Poisson distribution A probability distribution that expresses the number of random events occurring in a fixed period of time.
泊松分布 描述固定时期内随机事件发生次数的概率分布。

P

Poisson's ratio A constant that relates longitudinal strain in the direction of the load to lateral strain perpendicular to the load, for example, it defines the ratio of how a material will increase lengthwise and contract widthwise when stretched.
泊松比 描述负荷方向上纵向应变与垂直于负荷方向上横向应变关系的一个常数，例如材料受到拉伸时纵向伸长与横向收缩的比率。

polar second moment of area Defines the ability of a section (e. g. a beam) to resist torsion about an axis.
截面极惯性矩 界定截面（如梁）抵抗轴性扭转的能力。

polder Low-lying land that has been reclaimed from a body of water (sea, lake, or flood plain) by constructing *embankments or *dykes, and *dewatered. Polders have a high risk of flooding.
围垦地 通过建造堤防或护堤从一块水域（海、湖或冲积平原）开垦出的一片低洼地。围垦地极易遭受洪灾。

pole A long, usually circular in section, piece of a material.
杆 一段截面通常为圆形的长长的材料。

pole-frame construction (post-frame construction) A construction method where the building is supported on vertical poles or posts that can either be driven into the ground or supported on a slab or footing.
木框架结构 建筑物支撑于竖直杆或竿上的一种施工方法，竖直杆可被打入地下或者支撑在板子或基脚上。

pollute To cause *pollution.
污染 产生污染物。

pollution The introduction of any contaminants into an environment.
污染（作用） 环境中污染物的引入。

polyamide (nylon) A polymer containing monomers of amides. Polyamides are renowned for retaining excellent mechanical properties at elevated temperatures.
聚酰胺（尼龙） 包含单体酰胺的聚合物，以高温下仍具备极佳的力学性能而著称。

P

polycarbonate (PC) A *thermoplastic polymer material made from carbonate. PCs are commonly used as an unbreakable glass substitute especially as bullet-proof or vandal-proof windows utilized in banks, public transport stations, high-risk (federal government) buildings, etc. It has half the density of glass and transmits more light in comparison.
聚碳酸酯（PC） 碳酸盐制成的热塑性聚合物。通常用作不碎玻璃替代品，尤其用作银行、公共交通站、高风险性（联邦政府）建筑物所需的防弹防爆玻璃。其密度只有玻璃的一半，相较之下能透射更多的光。

polychlorinated biphenyls (PCBs) A polymeric material based on chlorine. PCBs have a variety of uses, most common applications are in electrical wiring and components.
多氯联苯（PCBs） 氯基聚合物材料。其用途广泛，通常用于电线和电气组件（的制造）。

polyethylene (PE) A widely utilized thermoplastic polymeric material. In construction applications it is used for interior plumbing pipes, and waterproof sheets for damp-proof applications as polyethylene is impermeable to the passage of water (waterproof). It is available in many different densities, which determines its application.
聚乙烯（PE） 应用广泛的热塑性聚合材料。由于聚乙烯的不渗水性（防水），其在建筑行业通常用作室内管道以及防潮的防水板。聚乙烯的密度有很多层级，不同的密度决定了不同的应用。

polyfusing Heat fusing of plastics.
聚乙烯熔断 塑料的热熔断。

polygonal roof *See* *PAVILION ROOF.
多坡屋顶 参考“亭顶”词条。

polygon of forces If more than three forces act through one point and are in one plane (termed coplanar

forces) they can be represented in magnitude and direction by the sides of a *polygon. If the polygon closes, the forces are in equilibrium.
力多边形 如果有三个以上的力作用于一个点且在同一平面内（共面力），则这些力可用量级表示，方向以多边形表示。如果多边形封闭，则力处于平衡状态。

polymer A material usually composed of hydrocarbon compounds with extensive applications and usage. Polymers are widely utilized in construction, and are generally low-density materials used mainly in non-load bearing applications. Approximately 20% of polymers produced in the UK go into construction. The most commonly used polymer is *polyvinyl chloride (PVC), and this material finds use as pipe materials for rainwater, waste and sewage systems, electrical cable sheathing, cladding, window frames and doors, and flooring applications. Polymers are generally classified into three categories based on their mechanical properties, they are thermoplastics, thermosets, and elastomers.
聚合物 通常由碳氢化合物构成的材料，用途广泛。聚合物广泛应用于建筑业，非承重型应用采用的通常为低密度的材料。英国生产的聚合物中大约20%被用于建筑业。最常用的聚合物为聚氯乙烯(PVC)，用于雨水、废水污水处理系统的管道、电缆包皮、外墙挂板、窗框、门和地板等。按照力学性能，聚合物通常被分成三类，即热塑性、热固性及高弹性聚合物。

polymers *See* POLYMER.
聚合物（复数） 参考“聚合物”词条。

polymethyl methacrylate (PMMA, Perspex, acrylic resin) A polymer material based on methyl methacrylate. Perspex (PMMA) is a transparent plastic, most commonly used as a shatterproof replacement for glass, especially in municipal buildings and windows in bus stops. PMMA has excellent transmission properties and toughness. In comparison with glass, it has half the density and transmits more light.
聚甲基丙烯酸甲酯（PMMA，有机玻璃，亚克力） 甲基丙烯酸甲酯基聚合材料。有机玻璃通常用作玻璃的防碎替代品，用作市政建筑及公交站的玻璃。PMMA的透光性及韧性都极佳。与玻璃相比，密度仅为其一半且透光性更好。

polypropylene (PP) A *thermoplastic polymeric material. PP has properties which are similar to low-density and high-density polyethylene. In construction it is used for sewer pipes or lavatory seats.
聚丙烯（PP） 一种热塑性聚合物。聚丙烯的属性与低密度和高密度的聚乙烯类似，建筑业中通常用作下水管或马桶座板。

polystyrene (polystyrene foam) A polymeric material based on styrene. Polystyrene foam has low thermal conductivity making it ideal for cavity-wall insulation; however, it needs protection due to its flammability.
聚苯乙烯（发泡聚苯乙烯） 苯乙烯基化合物。发泡聚苯乙烯的导热率低，非常适合用于夹心墙保温；但是由于易燃需要进行特殊保护。

polytetrafluoroethylene (PTFE, Teflon) A polymeric material with a very low coefficient of friction, PTFE is thus commonly used in containers and pipework for reactive (corrosive or hazardous) substances.
聚四氟乙烯（PTFE，特氟龙） 摩擦系数极低的聚合材料，因此通常用于制作接触易起化学反应（腐蚀性或危险性）的物质的容器或管道。

polythene *See* *POLYETHYLENE.
聚乙烯 参考“聚乙烯”词条。

polyurethane (PU) A polymeric material made from urethane, usually utilized in foam form. Polyurethane is widely used in the construction industry, particularly as thermal insulation panels in cavity wall insulation. Polyurethane-based glue is also used as an adhesive. Polyurethane-based coatings and varnishes are used in carpentry or woodworking as this results in a hard, inflexible coat that is especially popular for protecting floors. Another common use of polyurethane foams is in commercial and domestic furniture.
聚氨酯（PU） 由尿烷制成的聚合材料，通常以泡沫形式使用。聚氨酯广泛应用于建筑行业，通常用作夹心墙保温中的隔热板。聚氨酯基胶水也通常用作胶粘剂。聚氨酯基涂层和清漆广泛用于木器或木工艺中，因为会形成坚硬的刚性涂层，尤其用于保护地板。聚氨酯泡沫还可用于商业和住宅家具。

polyvinyl acetate (PVA or PVAc) A polymer material based on vinyl acetate monomer commonly used as wood glue, and in paints and industrial coatings.
聚醋酸乙烯酯（PVA 或 PVAc） 醋酸乙烯酯基聚合材料，通常用作木胶水、油漆和工业涂层等。

polyvinyl butyral (PVB) A polymer resin used to bind glass; this further improves the toughness of glass panels.
聚乙烯醇缩丁醛（PVB） 用于黏合玻璃的聚合树脂，可以进一步提高玻璃板的韧度。

polyvinyl chloride (PVC, uPVC) A common thermoplastic polymer based on vinyl used extensively in the construction industry. UPVC (unplasticized polyvinylchloride) is used as window frames, doors, and pipes in most dwellings and buildings; unplasticized refers to the fact that the material contains no *plasticizers.
聚氯乙烯（PVC, uPVC） 常见的热塑乙烯基聚合物，广泛应用于建筑业。UPVC（未增塑聚氯乙烯）用于大多数住房和建筑的窗框、门和管道。未增塑指材料中不含增塑剂。

polyvinyl fluoride (PVF) A polymeric material mainly used in flammability lowering coating applications, particularly in the construction industry. Polyvinyl fluoride is a *thermoplastic fluoropolymer and has similar properties to *PVC.
聚氟乙烯（PVF） 通常用作易燃性低的涂层，是用于建筑业的一种聚合物。聚氟乙烯为热塑含氟化合物，属性与聚氯乙烯类似。

polyvinylidene chloride (PVDC) A polymer material based on vinylidene chloride. PVdC has excellent resistance to chemical attack and is also impermeable to the passage of moisture.
聚偏二氯乙烯（PVDC） 偏二氯乙烯基聚合材料，具有极佳的耐化学腐蚀性且可以防潮。

polyvinylidene fluoride (PVDF) A polymeric material used in metal paints.
聚偏二氟乙烯（PVDF） 制金属涂漆使用的一种聚合材料。

pommel A globe-shaped ornament used at the top of a *pinnacle.
圆头 尖顶处使用的球形装饰物。

ponding The accumulation of shallow pools of water on a horizontal surface such as a flat roof or flat surface.
水洼 在平面如平屋顶或平台上积累的浅水坑。

P

pontoon A floating platform.
浮箱 浮式平台。

pony truss A bridge truss that contains a deck supported by two side trusses. It is characterized by having no lateral bracing between the top *chords of the two side trusses. This is because the side trusses are relatively low and, having lateral bracing, would prevent traffic passing beneath.
矮桁架 桥桁架，包含一座由两个侧桁架支撑的桥面。特点为两个侧桁架的上弦杆之间没有横系杆。由于侧桁架较低，如果有横系杆则会阻碍下方的交通。

poplar A type of tree from the genus Populus.
白杨 杨属的一种树。

popout (knockout) A partially cut or thin section of a component that is designed to be knocked out to provide an access hole.
敲除 敲掉一个组件中部分切除的或薄的切面，用于提供检修孔。

popping Where plaster looses and comes away from its background. Commonly occurs at the heads of plasterboard nails.
脱浆 灰泥松动并脱落。通常发生于石膏板的钉头处。

pop rivet A type of permanent fastener used to join two sheets of metal together. Comprises a hollow cylindrical shaft with a flat head and a pin that extends through the centre of the shaft and head. It is inserted into a pre-drilled hole, and a specially designed tool is then used to draw the pin through the head and shaft. The pin expands the shaft, securing it in the hole, and then is broken off, creating a 'popping' sound.
波普空心铆钉 一种把两片金属连到一起的永久性紧固件。包括一个带平头的圆柱空心轴，和一个贯穿了轴部和头的心轴。铆钉中嵌入一个预钻孔，然后使用专用工具把心轴穿过头和轴部抽出。心轴在轴内不断膨胀，把轴固定到洞内，然后就会破裂，发出“啪”的声音。

pop-up waste A drain plug at the bottom of a sink or basin that is operated by a lever.
弹出式排水塞 水池或洗脸盆底部的排水塞，用提杆进行控制。

porcelain enamel An inorganic coating bonded to metal by the fusion process.
搪瓷 通过融合黏结到金属上的一种无机涂层。

porch A covered approach to a doorway or a small roof placed over the entrance to a house, the sides of which can be closed in or left open.
门廊 带顶的通向门口的通道，或者房屋入口上方的小屋顶，两边可以是封闭的或者留有开口的。

pore A tiny hole on the surface of a material that allows the passage of a liquid or gas. These holes are present in many construction materials—timber, blocks, bricks, concrete, etc.
气孔 材料表面的一个小孔，可允许液体或气体通过。很多建筑材料中都有此类小孔，如木材、木块、砖、混凝土等。

poretreatments The process where a chemical barrier, usually silicone, is used to reduce the amount of rainwater that penetrates masonry walls. Although the treatment will reduce the penetration of rainwater into the wall, it can also inhibit the flow of water vapour out of the wall.
孔隙处理 用化学屏蔽（通常为硅酮）减少石砌墙的雨水渗透量。尽管该处理能减少墙壁的雨水渗透量，但是也会阻止水分从墙壁散发出来。

porosity The presence of holes, space, or gaps inside a solid. A *porous material is thus not fully dense. Porosity is generally undesirable as pores can act as crack nucleation points; however, a porous material is a very good insulator.
多孔性 固体中存在小孔、空间或缝隙的性质。多孔材料不是全致密的。通常多孔为有害的，因为那可能会形成裂缝起泡点；但是多孔材料是很好的隔音材料。

porous A material which is not fully dense containing holes or voids. Examples of porous building materials include *autoclaved aerated concrete, *mortar, and *bricks.
多孔材料 一种为有孔或空隙的非全致密性材料。多孔性建筑材料有蒸压充气混凝土、灰浆和砖等等。

P

porous pipe A pipe, which lets in water, used for subsoil drains.
多孔渗水管 让水流入的管道，用作地下排水管。

portable Capable of being transported (moved) with ease.
便携式 易于运输（移动）的。

portal frame A simple structural frame comprising two vertical columns and two sloping roof beams that join in the centre. Typically used where a large internal free span is required, such as in factories and warehouses.
门式刚架 简单的框架结构，包含两个垂直柱，以及两个在中心处接合的斜面顶梁。通常用于要求室内有大片开阔空间的地方，如工厂和仓库。

portcullis Historically, this was a defensive gate at the entrance to castles. The metal reinforcement grid is lowered to close and protect the castle or fortress and raised to allow people through.
城堡吊门 历史上指城堡入口处的防御门。可降落以关闭金属加固格栅对城堡和要塞进行保护，以及升高让人员通行。

Portland cement (Ordinary Portland cement) A material with adhesive and *cohesive properties capable of bonding mineral fragments (sand, bricks, stone, etc.) together. It is capable of reacting with water to give a hard, strong mass. The main ingredients of Portland cement are: calcium carbonate (from *chalk or *limestone), silica (from *clay/shale), and alumina (from clay/shale). It is manufactured by heating limestone and clay together to form *clinker rich in calcium silicates. The main stages in the manufacture process are:

硅酸盐水泥（普通硅酸盐水泥） 一种具备黏结性和黏性的材料，能够将矿物碎片（沙子、砖、石头等）黏结在一起。遇水会起反应可形成坚硬的物质。硅酸盐水泥的主要成分为：碳酸钙（来自于白垩岩或石灰岩），二氧化硅（来自黏土或板岩），氧化铝（来自黏土或板岩）。通过一起加热石灰岩和黏土形成富含硅酸钙的熟料。主要制造过程如下：

· Chalk and clay are mixed together either in a slurry (known as the wet process), or blended and transported in an air stream (known as the dry process).
将白垩岩和黏土搅拌成泥浆（称作湿法）或者在空气流中掺混并输送（称作干法）。

· The mixture then moves down a kiln undergoing a number of changes, as it becomes hotter:
——Water evaporates at 100℃.
——Carbon dioxide is given off at 850℃
——Fusion takes place at 1400℃ , *calcium silicates and *aluminates form in the resulting clinker (partly glassy, partly *crystalline material).
然后把混合物运送至干燥室进行一系列反应，随着温度升高：
——在 100℃水分开始蒸发
——在 850℃开始释放二氧化碳
——在 1 400℃开始融合，生成的熟料（部分为玻璃，部分为晶体材料）中开始形成硅酸钙和铝酸盐

· The resulting clinker is ground to form a fine powder and *gypsum is added to "control" the rate of setting of the concrete.
将生成的熟料研磨成细粉，添加石膏控制混凝土的凝结速率。

The end product is Portland cement and consists of four compounds:
1) Tricalcium silicate $3CaO \cdot SiO_2$ C_3S
2) Diacalcium silicate $2CaO \cdot SiO_2$ C_2S
3) Tricalcium Aluminate $3CaO \cdot Al_2O_3$ C_3A
4) Teracalcium alumino ferrite $4CaO \cdot Al_2O_3$ C_4AF
最终产品为硅酸盐水泥，包含如下四种成分：
1）硅酸三钙 $3CaO \cdot SiO_2$ C_3S
2）硅酸二钙 $2CaO \cdot SiO_2$ C_2S
3）铝酸三钙 $3CaO \cdot Al_2O_3$ C_3A
4）铁酸铝四钙 $4CaO \cdot Al_2O_3$ C_4AF

By varying the properties and types of these compounds and changing the fineness of the cement particles during manufacture, the four main types of Portland cements can be produced: Ordinary Portland Cement (OPC); Rapid Hardening Portland Cement (RHPC); Low-heat Portland Cement (LHPC); and Sulphate Resisting Cement. The reaction between water and the chemical compounds of cement is *exothermic (known as the *heat of hydration) and produces a largely crystalline structure referred to as cement gel. The product of these chemical reactions are calcium silicate (CSH), calcium hydroxide $Ca(OH)_2$, and calcium aluminate hydrates (CAH).
生产过程中通过更改这些化合物的属性和类型以及水泥颗粒的精细程度，可生产出四种主要的硅酸盐水泥：普通硅酸盐水泥（OPC）；快硬硅酸盐水泥（RHPC）；低热硅酸盐水泥（LHPC）以及抗硫酸盐水泥。水与水泥化学成分的反应为放热过程（称作水合热），并且产生大量的晶体结构，称作水泥凝胶。此类化学反应的产物为硅酸钙（CSH）、氢氧化钙 $Ca(OH)_2$，以及水化铝酸钙（CAH）。

Portland stone A limestone from the Isle of Portland, Dorset. Portland stone has been used in a number of famous buildings in the UK.
波特兰石 多赛特郡波特兰岛产的石灰石。英国许多著名的建筑物中都用到了波特兰石。

possession Indicates which party has control of and rights to use a site. Once a contractor is in control of the site, the contract can commence. Without possession, the contractor is prevented from starting. Once the works are complete, the contractor hands back the site and building. The control of the site is transferred back to the

owner. Even if there is no statement within the contract terms about possession of the site, it is taken that the contractor must have control (possession) of the site in sufficient time to complete the works.
支配 表明一方对某一场所拥有支配权。一旦承包商接手掌管某地，合同即开始生效。没有取得占用权，承包商不得开工。完工后，承包商应从现场及建筑物撤离。现场支配权重归业主。即使合同条款中没有现场占用权的规定，也视作承包商有足够的时间支配现场以完成工程。

post A vertical member that may be used to provide support for a structure. *See also* COLUMN.
桩/杆 可用于支撑某一结构的竖直构件。另请参考“柱子”词条。

post-and-beam construction A traditional method of timber-frame construction where a skeletal frame of vertical posts and horizontal beams supports the floors and the roof.
桩梁构造 一种传统的木框架构，用立柱和横梁龙骨框架支撑地板和房顶。

post-contract stage Building work has finished, *possession of the site has been handed back to the client, the *final account can be prepared, and the final period for retention monies starts.
后合同阶段 建筑工程完工后，现场支配权重归业主，应准备好竣工结算，另外开始计算质保期。

post-frame construction *See* POLE-FRAME CONSTRUCTION.
柱框架结构 参考“木框架结构”词条。

potable water *See* DRINKING WATER.
可饮用水 参考“饮用水”词条。

potboard The lowest shelf of a cupboard on which the pots are placed.
锅架 橱柜中最底层的架子，用于放锅。

potential difference The different voltage between two ends of an electrical path that allows voltage to move through it.
电势差 两个电路终端的电位差，允许电压通过。

pot floor A suspended floor formed with hollow clay blocks spanning between the concrete beams with a *screed or concrete topping placed on top.
空心砖楼板 混凝土梁之间用中空黏土块形成的悬挂楼板，顶部抹砂浆层或混凝土。

P

pothole 1. A hole formed in the surface of a road. 2. A deep vertical hole in rock (particular limestone), or in a river bed.
1. 坑洼 公路面上形成的洞。**2. 壶穴** 岩石（尤其是石灰岩）中或河床中的垂直深坑。

pot life The working life of glue, paint, and other chemicals stored in containers.
贮存期 胶水、油漆及其他储存在容器内的化学品的适用期。

poultice attack The corrosion of aluminium kept in contact with a wet, porous material. It can destroy aluminium foil in insulation, for example.
铝霉斑 铝由于长期接触潮湿多孔渗水的材料后发生腐蚀的现象，例如可以破坏保温铝箔。

pound (lb) An *imperial unit of mass containing 16 ounces, which is equivalent to 0.45 kg.
磅（lb） 英制质量单位，1 磅等于 16 盎司，约等于 0.45 千克。

pour The placing of concrete, normally associated with large operations involving the placing of large quantities of concrete, such as concrete walls, frames, foundations, and floors.
浇灌 混凝土浇灌，通常与浇筑大量混凝土的大型作业相关，例如混凝土墙、框架、地基、地板等。

pour and roll The pouring of hot liquid bonding compound placed prior to the rolling out of asphalt or bitumen roofing felt.
浇注与推铺 在铺设柏油和沥青油毡之前先浇灌热的液态黏结混合料。

pouring rope Non-combustible rope (asbestos rope or similar) wrapped around cast iron pipes while lead caulking is poured.
浇注绳 不易燃的绳子（石棉绳或类似的），灌铅嵌缝浇注时缠绕在铸铁管上。

powder coating A continuous film/coating formed from fine, dry powder using the electrostatic process.
粉末涂层 精细的干粉经过静电处理后形成的一层连续的膜或涂层。

power 1. Authority to act and control. 2. Low-voltage temporary electric supply. 3. Capacity for exerting physical or mechanical force. 4. In physicsthe rate of energy output from an object.
1. 权力 采取行动和控制的职权。**2. 电源** 低压临时供应的电力。**3. 动力** 运用物理或机械力的能力。**4. 功率** 物理学中指从一个物体中能量输出的速率。

power-activated tool A mechanical gun that fires nails into timber.
射钉枪 将钉子钉入木材内的机械枪。

power and lighting installation The components and services within a building that provide power for the appliances and provide artificial lighting.
电力和照明设备 建筑物内为电气设备提供电力及人工照明的组件和设备。

power float (machine trowel, rotary float, helicopter) Mechanical piece of plant with rotating blades (floats) used for smoothing concrete.
抹平机 带旋转刀片（抹板）的电动工具，用于抹平混凝土。

power panel A *distribution board that only contains the circuits that provide power to the building.
电源板 仅包含为建筑物提供电力的电路的配电板。

power point A *socket outlet.
电源插座 参考“插座”词条。

power tool Portable hand-held electrical tools used on-site, such as mechanical drills, breakers, sanders, and saws.
电动工具 现场用的便携式手持电动工具，例如电钻、粉碎机、打磨机和电锯等。

pozzolan A powdery siliceous material used to make *Portland cement. Pozzolans react with hydrated lime to form calcium silicate and silicoaluminate hydrates. Pozzolans commonly used in modern concrete construction include fly ash, which is also known as pulverized fuel ash (PFA), rice husk ash, ground granulated blast furnace slag (ggbs), silica fume, maize cob, and metakaolin (calcined clay).
火山灰 制造硅酸盐水泥用的一种粉末状硅酸材料。火山灰与氢氧化钙反应生成硅酸钙和水化硅铝酸盐。火山灰常用于现代混凝土构造物中，还包括粉煤灰，也被称作磨细粉煤灰（PFA）、谷壳灰、矿渣微粉（ggbs）、硅粉、玉米芯粉和偏高岭土（焙烧黏土）。

pozzolana *See* POZZOLAN.
火山灰 参考“火山灰”词条。

practical completion (substantial completion) Works are accepted to be sufficiently complete to hand over to the client, although minor defects noticed are still prefabricated building required to be rectified. In some contracts it is the point at which the defects liability period begins, the contractor's liability for insurance ends, plus it is the trigger for other contractual machinery to begin.
实际竣工 工程完全竣工验收后将交给业主，虽然建筑中可能还有一些次要缺陷需要修整。某些合同中实际竣工意味着维修责任期的开始，承包商的保险责任终止，同时也是其他契约机制的触发点。

Pratt truss A *truss with vertical web members in compression and diagonal web members sloping down towards the centre so that they are in tension—the reverse of a *Howe truss. It is a statically determinate structure and as such can have long spans, up to 75 m. Variations of the Pratt truss include the *Baltimore truss and *Pennsylvania truss.
普拉特桁架 有垂直受压腹杆以及斜向中心的斜杆的桁架，因此处于受力状态——与豪威式桁架相反，为静定结构，所以可有长达75米的跨度。普拉特桁架的变形有巴尔的摩桁架和宾夕法尼亚桁架。

preamble Information provided at the start of each description in the bills of quantities. The introduction will provide rules of measurement, departures from standard methods of measurement, and description of the trade.
前言 工程量清单中每项说明的开头处给出的信息，提供测量规则、偏离标准的测量方法以及专业描述。

pre-boring 1. Putting a *pilot hole in a material. 2. Drilling a hole for a pile.
1. 导向钻孔 在材料上开一个引导孔。**2. 预钻孔** 为打桩钻孔。

pre-camber A beam that has an unloaded upward deflection so that when loaded it avoids the appearance of sagging.
预拱度 梁有一个向上的预荷载，承重后可避免下沉。

precast Previously made material or structure which is transported to a site for installation.
预制件 提前制好的材料或结构，等到运送至现场后再安装。

precedence diagram (precedence network) Logical network that calculates the *earliest start and *latest start for each task, the amount of *float available, and the *critical path.
顺序图 计算每项任务最早开始和最晚开始的逻辑网络，包括可使用浮点时间数以及关键路径。

precedence of documents The hierarchy of documents and the recognition of those documents which are to be accepted as correct if two or more documents disagree. Provides a legal and contractual position as to which documents are of most importance in determining any dispute.
文件优先次序 两个或多个文件出现不一致时，文件的排序。对于解决争端时哪个文件最有决定权给出的一个法定、契约性的排序。

pre-chlorination The chlorination of drinking water prior to another treatment stage, for instance, filtration or chemical process.
预加氯 饮用水做另一项处理之前（例如过滤或化学法）进行的加氯消毒。

precipitate 1. Water in the form of condensate falling to the ground. 2. The separation of a solid from a liquid.
1. 凝结 水以冷凝形式掉在地上。**2. 沉淀** 从液体中分离出固体。

precision The exactness of detail.
精度 细节的精确性。

preconstruction process The tendering, pre-contract planning, and coordination meetings prior to construction commencing.
施工前进程 开工前的投标、合同前计划以及协调会议过程。

pre-contract stage Tendering, contract negotiations, and other meetings prior to the contract being signed and work commencing on site.
合同前阶段 签署合同及现场开工之前进行的投标、合同谈判及其他会议等阶段。

prefabricated The process of manufacturing building components or elements in a factory and then assembling them on-site.
预制的 在工厂制造建筑构件或元件，然后运送至现场组装。

prefabricated building Structure with components that have been manufactured in a factory, delivered to site, and assembled. The building may be delivered in flat-pack prefabricated modules form with panels lifted up and bolted together, or produced in full volumetric modules that have walls, roofs, and floors.
装配式建筑 组件在工厂已制造好，运送到现场后再组装而成的结构。以平板式包装的预制模块运送建筑构件，将板块立起并紧固，或者以包括墙、屋顶和地板的整体模块来制造。

prefabricated modules (modularization, prefabricated units) Components of a building that have been manufactured in a factory to be assembled on-site. The process of off-site manufacturing reduces the build time on-site.
预制模块 工厂制造现场安装的建筑组件。现场外的生产工序缩短了现场的建筑时间。

prefabricated scaffold (frame scaffold) Scaffolding system with fixing lugs and clips built into the poles and boards making the scaffolding easier to assemble.

P

预制脚手架 勾头螺栓和扣件预装于杆和板子上的脚手架系统，使脚手架更易于组装。

pre-filter (dust collector, sand filter) A *filter installed before the main filter in an air-conditioning system that is designed to remove larger particles such as dust from the fresh air.
预滤器（预除尘器、集尘器） 空调系统中安装主滤器前安装的过滤器，用于过滤掉新鲜空气中的较大颗粒（如粉尘）。

pre-finished Components that have received their final finish (paint, sealant, or stain) in the factory and are delivered to site in a finished state.
预处理 在工厂内对组件进行过表面处理（油漆、密封剂或上色），以成品的状态运送到现场。

preliminaries 1. General description of the works and list of the contractor obligations and restrictions imposed by the employer. 2. In the US mobilization costs.
1. 序言 工程的一般描述以及业主对承包商责任和限制的明细。**2. 开办费** 在美国指开办费。

preliminary work Proprietary work undertaken by others prior to works taking place on-site.
前期工作 现场动工之前其他人进行的专项工作。

pre-loading (pre-stressing, pre-tensioning) A method of introducing load, stress, or tension to a material before it is in service to counteract undesirable effects. A **pre-load** may be applied to highly compressible soil prior to the construction of a structure so that the final settlement of the structure is reduced. A **pre-stress** may be introduced to structural members of a building to offset in-service loads. A reinforcing bar may be pre-tensioned before the concrete is cast to produce a pre-stressed concrete beam—a beam that is in compression before it is loaded.
预荷载（预应力，预张拉） 投入使用前对材料施加荷载、压力或拉力以抵消不良效果的方法。构筑一个结构之前应先在高度压缩性土壤上施加预荷载，可减少建筑物的最终沉降。可在建筑物的结构组件上施加预荷载以抵消使用荷载。浇筑混凝土之前可对钢筋进行预拉伸，形成预应力混凝土梁——负载之前已受压的梁。

premixed plaster Plaster supplied factory-mixed, either wet in drums, ready-mixed, or dry in bags, and usually containing lightweight aggregate. A dry mix may be based on retarded hemihydrate gypsum or *Portland cement, so that on-site only water is added. The high thermal insulation of these plasters makes them useful in cold climates, and for fire encasement of structural steelwork. Thin-wall plaster is one type.
预拌砂浆 工厂拌好的灰泥，可以是桶装湿状、预拌的或袋装干状，通常包含轻骨料。干拌混合料以缓凝熟石膏或硅酸盐水泥为底料，现场只需添加水即可。此类灰泥具有高保温性，常用于寒冷气候以及钢结构的防火外壳。薄抹灰砂浆是其中一种。

prepainting Priming a surface prior to painting.
预涂漆 喷漆之前在表面上涂底漆。

preparation (surface preparation) The process of preparing a surface prior to it being painted or coated.
预处理（表面处理） 在涂漆或喷涂层之前对表面进行处理的过程。

prequalification In order to enter the bid process, the client may request minimum criteria that must be met, including financial stability, experience of undertaking similar work, demonstration of competence, and relevant insurance. If such factors are not met, the contractor would not be qualified to enter the tender.
资格预审 业主制定的进入投标程序需达到的最低要求，包括财务稳定性、从事类似工程的经验、竞争力展示以及相关保险。如此类因素未达到要求，则承包商没有资格进入投标。

preservation The process of maintaining and protecting buildings and natural resources.
保护 维护建筑物及自然资源的过程。

preservative An element or compound used to extend the longevity of an element or object. This usually involves the addition of a chemical to inhibit spoilage.
防腐剂 用于延长一种成分或物体寿命的成分或化合物。通常会添加一种阻碍腐败的化学成分。

pressed brick A brick that has been subjected to moulding under mechanical pressure.

压制砖 机械压力作用下模制的砖。

pressed glass Glass units, such as pavement units or glass block, that are pressed into shape.
压制玻璃 玻璃组件，如压制成型的铺路组件或玻璃砖。

pressure The force that is applied to an object over a particular area. Measured in newtons per square metre or * pascals (SI unit).
压力 特定区域内对一个物体施加的力。单位为牛每平方米或帕斯卡（国际单位制）。

pressure bulb A stress or settlement contour beneath a loaded area (foundation). The size of the bulb is proportional to the area of foundation and the magnitude of load it supports.
压力泡 承载区域（地基）下的压力或沉降等值线。压力泡的大小与地基的面积以及承重的大小成比例。

pressure circulation Circulating air or water through a system using fans or a pump.
压力循环 通过风机或泵使空气或水在系统中流通。

pressure cutout A sensor that operates a control circuit when a preset pressure is reached.
压力切断器 达到预置压力后操作控制电路的感应器。

pressure-equalized joint A type of joint used in cladding systems and windows that minimizes rainwater penetration by having a cavity within the frame or behind the cladding that is ventilated to outside air. Since the air pressure within the cavity can equalize with the external air pressure, there is no force to push the rainwater through the joint.
等压接缝 外墙挂板结构和窗户中使用的接缝，通过与外界连接的围护体后面或框架内的空腔来使雨水的渗漏量降到最小。由于空腔内部气压与外部气压相同，就不会存在压力差迫使水从接缝渗透。

pressure gun A sealing or caulking gun, used to force the fluid mixture out of a nozzle enabling easy application.
压力枪 密封或填缝枪，用于将液体混合物从喷嘴中挤出，方便敷涂。

pressure head The pressure exerted by a column of fluid above a certain point.
压头 某一点上方液体柱释放的压力。

pressure-maintenance vessel A * diaphragm tank installed on a boosted cold water supply system.
压力维持罐 升压冷水管系统上安装的一个隔膜式气压罐。

pressurization test A * fan pressurization test.
气密测试 风扇加压法测试。

pressurized escape route A fire escape route within a building that is slightly pressurized to prevent smoke entering the escape route in the event of a fire.
正压送风逃生通道 建筑物内的火灾逃生通道，进行过轻微加压防止火灾时烟雾进入逃生通道。

pressurized structure An * air-inflated or air-supported structure.
加压结构 充气式或气承式结构。

pre-standard European standard for materials and processes.
协调标准 材料和工序的欧洲标准。

pre-stressed concrete A type of concrete, pre-conditioned to improve the tensile properties of concrete. This involves using high tensile steel rods, which can absorb high tensile loads. As a result pre-stressed concrete is commonly used in buildings as floors, and also in bridges.
预应力混凝土 一种混凝土，预处理过以提升混凝土的拉伸性能。处理方式包括使用可以吸收承受高强度荷载的高强度钢筋。所以预应力混凝土通常应用于地板、大桥等建筑物中。

priced bill (priced bill of quantities) Document submitted as part of the tender which has rates fixed against the work described and measured. The descriptions and measurement of the works are normally distributed to those tendering for the contract. The contractor, within the tender, places prices and rates against the descriptions, and submits these with their tender documents. If the contractor is successful, the priced bill becomes part of the contract documents.

报价清单（工程量清单报价） 作为标书一部分递交的文件，其中对说明及测量的工程进行报价。对工程的说明和测量一般在招标时给出。承包商在标书中报出符合此类说明的价格，随标书一起递交。如果中标，承包商的报价清单构成合同文件的一部分。

price index List of prices of materials and building works published and updated at monthly intervals.

指导价 物料和建筑工程的物价清单，每月进行一次更新刊发。

price work Work based on measured rates and outputs rather than hourly rates. Prices are fixed against the work to be performed; at the end of the week the work is measured and payment is made for the amount of work performed.

计件工作 按次和产出量计费而非按小时计费的工作。按照完成的工作量进行计费；每周结束时对完成的工作量进行计量和支付。

pricking up The first of three coats of plaster applied to laths. Once dry, the surface of the plaster is scratched to form a key for the second coat. *See also* SCRATCH COAT.

粗底涂 涂到板条上的三层抹灰中的第一层。变干后在灰泥表面划痕，形成涂抹第二层中层灰泥的毛面。另请参考“底涂层”词条。

primary beam The main beam in a structure that supports the main loads and may support secondary beams. *See also* SECONDARY BEAM.

主梁 结构中承受主要荷载的梁，也可支撑次梁。另请参考“次梁”词条。

primary cell An electrochemical cell where an irreversible chemical reaction produces an electric current.

原电池 由不可逆的化学反应产生出电流的一种电化电池。

primary element A main structural or supporting element of building. The primary elements include the foundations, walls, floors, and roof. *See also* SECONDARY ELEMENT.

主要构件 建筑物中主要的结构性或支承构件。主要构件包括地基、墙壁、地板和房顶。另请参考“次级构件”词条。

P

primary fixing 1. A fixing built into the structure of a building to support a secondary fixing, for instance, a *fixing channel. *See* SECONDARY FIXING. 2. A type of fastener used to secure cladding panels or profiled metal sheeting to walls and roofs.

1. 初级固定 安装于建筑物结构中，用于支撑次级设备的固定件，如固定槽。另请参考“次级固定”词条。2. 墙板钉 用于固定墙壁或房顶的外墙挂板或压型金属板。

primary flow and return pipes The main pipes from which water flows to and from an appliance.

流出及回流主管 可供水流流向设备并从设备回流的土管。

primary lining A permanent lining, typically pre-cast or prefabricated segments, used to provide initial support in soft ground tunnelling.

初期支护 永久性支架，通常为预埋或预制的组件，用在软土隧道内以提供初期支护。

primary treatment The initial stage in a water treatment plant involving coagulation and sedimentation, followed by *secondary treatment, and *tertiary treatment.

水预处理 水处理的初级阶段，包括凝结和沉淀，后面还需进行二级和三级处理。

prime cost (PC) Description and fixed cost within the bill of quantities for work or goods which cannot be properly measured. It is often viewed as a fixed allowance for the price or the work; however, even when the price is stated, the courts have determined that, unless otherwise expressly worded in the contract, the figure is only an estimate. If the contractor is unable to obtain the goods and services within the stated figure the costs could increase. The supplier of the service or product is often nominated. Some terms of contract may provide the contractor with a percentage for managing the supplier and providing the goods.

暂估价（PC） 工程量清单内关于无法准确测量的工程或物项的说明和固定成本。通常视作报价或工程的固定酌加费用，尽管注明了价格，法院仍会判定该价格仅为估价，除非合同中有明确规定。如承包商不能在规定价格下购置物品或设施，则可以增加成本。设施和产品供应商通常是指定的。合同中某些条款可让承包商以一定比例收费去管理供应商和提供货物。

primer A component of paint which must adhere well to the substrate, seal the bare surface, offer protection against deterioration or corrosion, and provide a good base for the undercoat. To ensure good adhesion, the surface must be free from loose or degraded material.
底漆 一种油漆组分，必须牢牢贴合在基底上，可用于密封裸露面防止其劣化或腐蚀。为保证良好的黏性，（刷漆前）基底表面不得有任何散落或降解的材料。

priming An initial coat of paint applied to a surface.
底涂层 涂到表面上的第一层漆。

principal A traditional roof truss, such as a king and queen post truss; *See also* PRINCIPAL RAFTER.
三角屋架 一种传统屋架，如单柱桁架或双柱桁架；另请参考“主椽”词条。

principal rafter The main *rafter in a traditional trussed roof that provides support for the *purlins. The purlins then provide support for the *common rafters.
主椽 传统的桁架屋顶中用的主椽，为檩条提供支撑。檩条再为次要屋椽提供支撑。

principal stress The normal stress perpendicular to the plane on which the shear stress is zero. In an element of soil there are three mutually orthogonal planes where the shear stresses are zero: the largest of these is known as the **major principal stress**, and the smallest of these is known as the **minor principal stress**.
主应力 在一平面上剪应力为零时与平面垂直的正应力。土壤中有三个相互正交的平面剪应力为零；三个平面上最大的正应力称作最大主应力，最小的正应力称作最小主应力。

priority circuit An electric circuit that supplies power to essential items of equipment.
优先电路 为设备的核心组件提供电力的电路。

prison Secure building for keeping people in captivity, often used to contain those convicted of criminal acts.
监狱 用于关押犯人，有监禁设施的建筑物。

privacy latch A latch that enables a door to be locked from the inside. Used on bathroom and *WC doors.
安全门闩 可从里面把门锁住的门闩，用于浴室或卫生间的门。

P

private branch exchange (PBX) A telephone exchange that serves a building or office. It serves as the central point for all internal and external calls and connects together all of the internal telephone lines. Automated electronic systems are known as a **private automatic branch exchange (PABX)** or an **electronic private automatic branch exchange (EPABX)**.
用户交换机（PBX） 服务于一栋建筑或一个办公室的电话交换台。它是内部和外部电话的中心点，并将所有的内部电话线路连接到一起。自动电子控制系统也被称作专用自动交换机（PABX）或电子专用自动交换机（EPABX）。

privatization The transfer or sale of public utilities and buildings to private companies. The gas, water, electric, and telecoms industries in the UK used to be owned by the state, but are now owned by private companies.
私有化 将公用设施或建筑转让或出售给私有公司。在英国，煤气、水、电力和电信业过去为政府所有，现在由私有公司所有。

processed shakes *Shakes that have been modified to look as though the timber has split.
裂纹处理 加工制作的木材裂纹，看起来像木材被劈开一样。

Proctor compaction test A standard laboratory test to determine the compacted state of a soil sample, by varying the moisture content of different compacted samples. It is possible to determine the maximum compacted state (at optimum moisture content) a soil can attain.
普氏压实试验 通过改变不同压实样本的含水量来测定土壤样本压实状态的标准实验室测试。可以测出一种土壤能达到的最大压实状态（最优含水量）。

product (building product) A component, element, or material that has been designed and manufactured to be used in a building.
产品（建材产品） 被设计制造用于构筑建筑物的组件、构件或材料。

proeutectoid steel A type of *steel subjected to a specific heat treatment to influence its chemistry (micro-structure) and hence its mechanical properties. This type of treatment involves heating the steel to the *austenite phase and then cooling the material.
先共析钢 经过特定热处理的一种钢，改变其化学性（微结构）从而改变其力学性能。热处理包括加热至奥氏体相然后再对钢进行冷却。

profile The outline, edge, or vertical section through an object.
轮廓 一个物体的外形、边缘或垂直切面。

profiled sheeting (metal decking, tray decking, troughed sheet) Metal sheeting, steel, or aluminium, that is folded into a sinusoidal or trapezoidal shape in order to improve its stiffness. Used for cladding walls and roofs, and to construct a *composite floor.
压型钢板 折叠成波形曲线或梯形的金属板、钢板或铝板，用以增强硬度。多用于挂板墙、屋顶以及复合地板的构筑。

program A computer software program; a list of instructions, written in a programming language, that tells a computer to execute a given task.
程序 电脑软件程序，以编程语言写出的一串指令，指示电脑执行一项给定任务。

programme (US progress chart) Schedule or plan of work, showing the activities, start times, durations, and progress (work completed) in a horizontal bar chart. The activities are normally linked by a network of tasks. Resources may also be built into the network calculating resource demand. By distributing and levelling the resources, they may be used more effectively and efficiently.
计划表（美国 进度表） 工程时间表或计划表，以横条图标明各项活动、开始时间、持续时间和进展（完工）。各项活动通常由任务网连接。资源也应计入任务网内以计算需求量。通过分配和标准化资源，计划表会被更高效地使用。

P

programmer Person who builds and constructs the *programme.
程序员 设计并创建程序的人员。

progress certificate A monthly certificate that shows a measure of the work done to date.
进度证书 说明到目前为止已完成的工程状况的月度报告。

progress chart *Programme of works prepared by the contractor that shows the percentage of work carried out and completed for each task. Normally presented in a horizontal bar chart with the bars shaded to denote the amount of work completed.
进度表 承包商编制的工程计划，标明了每项任务实施和完工的百分比。通常以横条图表示，其中横条的阴影部分代表已完工的工程量。

progressive failure 1. Something that is failing gradually, for example, the slow movement of soil along a slip plane, failure could have occurred at one point along the plane, while further down the plane the soil has yet to fail. 2. Something that causes something else to fail (having a knock-on effect) with the consequence of the first failure being quite different to the end result of the final failure.
1. 连续坍塌 某物逐渐损坏的现象，如斜面上土壤的缓慢移动，坍塌可发生在沿着斜面上的一个点，继续往下就一定会坍塌。**2. 渐进破坏** 导致其他东西破损的某物（有撞击作用），第一次破坏的后果与最后一次破坏的后果不一样。

progress payment Monthly payments based on the amount of work completed.
工程进度款 基于完工工程量的月度付款。

progress report The progress chart and a summary report provided to show the planned work and the actual work undertaken. The report also identifies suggested activities and areas of work to be rescheduled.

进度报告 进度表和总结报告，用于说明计划的工作以及实际完工的工程。报告中还列明了建议的活动以及需要重新安排的工作区域。

prohibition notice A notice given by the factory inspector, a representative of the Health and Safety Executive, which requires dangerous activities to stop immediately or situations to be made safe. For the notice to be given, the activities are considered to be a health and safety risk to the public or workers that could result in serious injury.
停工通知 由工厂检验员（健康与安全部代表）发出的通知，告知须立即停止危险活动或将现状扭转为安全状况。通知中，被认定为对公众或工人存在健康安全风险的活动可能会造成严重损伤。

project The work or tasks to be undertaken. The process of *project management is used to deliver the *outcome.
项目 承担的工作或任务。项目管理的进程用于交付成果。

projecting hinge A hinge that allows a door to open 180°beyond an architrave or any other projection.
180 度铰链 能让一扇门呈 180°开合，越过门框线或其他突出物的铰链。

projecting scaffold A cantilevered scaffolding that is built out of one of the upper storeys of the building, the scaffolding does not extend to the ground.
悬挑脚手架 悬臂式脚手架，安装于建筑的上方楼层之上且没有延伸到地面上。

projection 1. An estimate of the rate of something. 2. Something that protrudes. 3. A drawing method to represent a three-dimensional object on a two-dimensional plane.
1. 推测 对某物比率的估计。**2. 突出物**。**3. 投影图** 一种绘图方法，在二维平面上展现三维物体。

projection plastering (mechanical plastering, spray plastering) Applying plaster via a spray plastering machine and a nozzle.
喷涂抹灰（机械抹灰） 通过喷涂砂浆机和喷嘴涂抹灰泥。

project management The management of a set of tasks, events, and resources to deliver one significant outcome. The management of multiple projects with related outcomes is often referred to as programme management.
项目管理 为取得重大成果而对一系列任务、事件和资源的管理、对相关成果的多个项目的管理通常称作项目管理。

project manager Person responsible for the management of the building *project.
项目经理 负责管理建筑项目的人员。

project network *See* NETWORK.
施工程序网络图 参考“网络图”词条。

project representative 1. *Clerk of works. 2. Person who acts on behalf of the main contractor and liaises with local residents, the general public, and other interested parties.
1. 工程员工。**2. 项目代表** 代表总承包商与当地居民、公众及相关方联系的人员。

proof stress The amount of stress needed to cause a definite small permanent extension (usually 0.2%) of a sample of material. This value is roughly equal to the *yield stress in materials that do not exhibit a definite yield point.
试验应力 材料样本上形成一个明显的永久性延伸（通常为 0.2%）所需的力。这个值约等于一种材料不会形成明显屈服点的屈服应力。

prop A temporary support that is placed under or against an object to prevent it from falling down.
支柱 放于物体下方或靠着物体放置的临时支撑，防止物体掉落。

propeller A rotating shaft fitted with spiral blades. Used to push a ship (or aircraft) through water (or air).
螺旋桨 带螺旋叶片的旋转轴。通过水（或气体）推动轮船（或飞机）。

propeller fan (axial flow fan) A fan that uses a propeller-shaped rotor.

螺旋桨风机 使用螺旋状转子的风机。

properties The inherent characteristics of a material; *See also* PARAMETERS.
属性 材料的内在特性；另请参考“参数”词条。

proportion *See* RATIO.
比例 参考“比率”词条。

protected membrane roof An inverted roof.
倒置式保温屋面 倒置屋顶。

protected opening An opening in a fire-resistant construction that is constructed from fire-resistant materials so that the fire resistance of the construction is not compromised.
防火开口 耐火结构中由防火材料构成的开口，这样就不会降低结构的防火性能。

protected route A fire-resistant stairway or corridor in a building that is used by the occupants as an escape route in case of a fire.
走火通道 建筑物中可防火的楼梯或走廊，发生火灾时供业主用作逃生通道。

protected shaft A fire-resistant vertical shaft.
保护井 防火竖井。

protected stair (enclosed stair, US fire tower) A fire-resistant stairway used as an escape route in case of a fire.
防火楼梯（封闭楼梯，美国 防火塔） 发生火灾时用作逃生通道的封闭楼梯。

protection The process of taking measures to ensure that a building, building works, building occupants, and operatives are kept safe.
保护 为确保建筑物、建筑工程、住户及操作人员安全而采取措施的过程。

protection fan *See* DEBRIS-COLLECTION FAN.
防护棚 参考“防坠棚”词条。

protection gear Electrical devices or equipment that are designed to protect large electric circuits. *See also* PROTECTIVE DEVICE.
防护装置 用于保护大型电路的电气元件或设备。另请参考“保护装置”词条。

protection of finishes The process of temporarily covering finishes during construction to prevent them from being damaged. Once the construction is complete, the coverings are removed.
饰面保护 施工时临时覆盖饰面保护使其免受损害的过程。完工后覆盖物将会被移走。

protective device Electrical device that is designed to protect small electrical circuits, such as a *miniature circuit breaker. *See also* PROTECTION GEAR.
保护装置 保护小型电路的电气设备，如小型断路器。另请参考“防护装置”词条。

protective finishes to metals Finishes, usually in the form of coatings, that are applied to metals to protect them from corrosion.
金属的保护涂层 涂到金属上防止腐蚀的饰面，通常为涂层。

protective rail A rail mounted on a wall to protect it from impact damage.
防护栏杆 安装在墙上，防止墙壁受到冲击碰撞的栏杆。

provisional sum (PS) A figure (sum) included in the tender or *bill of quantities for work that cannot be sufficiently measured at the outset of the contract; included for items that cannot be properly specified.
暂定金额（PS） 在合同初期由于无法充分测量工程而列于标书或工程量清单中的暂定金额（总额）；包括不能准确说明的事项。

pry bar (crowbar, jemmy bar, jimmy bar, prise bar) A tool used for opening gaps, separating timbers that have been previously joined together, or forcing components apart. Where two pieces of timber are nailed

or fixed together, and they can be prised apart, the bar is worked into a gap and moved backwards and forwards forcing the components apart. Where the bar has aflattened and forked end, the bar may also be used for removing nails.
撬杆 开缝用的工具，用于将原来连在一起的木材分开，或者迫使两个组件分开的工具。两块木头被钉在一起或固定在一起时，把撬杆嵌入缝隙里前后移动迫使两块木头分开。若撬杆的末端为平面叉状，则它还可用于起钉。

pseudorange The approximate (uncorrected) measured distance between a satellite and a receiver in a *global positioning system.
伪距 全球定位系统中卫星与接收器的粗略（非精确）测量距离。

psychrometer An instrument used to measure relative humidity—also known as the *wet and dry thermometer.
干湿计 用于测量相对湿度的仪器，也被称作干湿温度计。

P-trap A P-shaped *trap used on the waste outlet of a sink or toilet to prevent sewer gas entering into the building. *See also* U-BEND and S-TRAP.
P 型存水弯 水池或厕所排水用的 P 字形存水弯，防止下水道的臭味进入建筑物内。另请参考“U 型弯管”和“S 型存水弯”词条。

public service Work undertaken by someone or an organization that benefits the general public, such as the emergency services.
公共服务 由组织或个人承担的造福公众的工作，如应急服务。

puddle Clay and sand that has been mixed with water and tamped to form a nonporous material, used as waterproof lining.
黏土膏 黏土、沙子和水混合到一起再夯实形成的一种无孔材料，起到防水衬砌的作用。

puddle flange A ring, usually rubber, that is placed around a pipe that passes through an external wall or shaft in a basement. It prevents water penetrating through the wall or shaft.
胶泥圈 安装在穿过外墙或地下室机井的管道周围的环，通常为橡胶环。可防止水分渗透。

pugging A material that is placed between the joists on a timber floor to increase the floor's mass and so improve the airborne sound insulation of the floor. Materials used include mineral wool, plasterboard, and sand. If the material is laid on a board, the boards are known as **pugging boards** or **sound boarding**.
隔音层 放于木地板的龙骨之间的一种材料，用于增加地板质量从而增强地板的隔音性能。所用材料有矿物棉、石膏板和沙子。如果将材料放置于一块板上，这块板被称作隔音层板或隔音板。

pugging boards (sound boarding) *See* PUGGING.
隔音层板 参考“隔音层”词条。

pull *See* DOOR PULL.
拉手 参考“门拉手”词条。

pull box *See* DRAW-IN BOX.
拉线盒 参考“过线盒”词条。

pull-cord switch (cord switch, pull switch) A ceiling mounted switch for a light or fan that is operated by pulling on a cord that hangs from the switch. Usually found in **bathrooms**.
拉绳开关 装在天花上用于开关灯或风扇的开关，操作时拉动挂在开关上的绳。通常用在浴室。

pulley A simple type of machine comprising a grooved wheel that a rope or chain runs through. Used for lifting heavy objects.
滑轮 包含一个可通行绳索的槽轮的一种简易机械，用于提升重物。

pulley stile The jambs of sash windows that incorporate the pulleys.
滑槽 包含有滑轮组的推拉窗边梃。

pulp A soft, moist, shapeless mass of matter.
浆状 柔软潮湿无形状的物质状态。

pulverized Consisting of fine particles, normally a solid in powder form, usually prepared by grinding.
粉状 由微细的颗粒组成，一般为固体粉末状，通常为研磨制成。

pumice A material used as an abrasive.
浮石 用作磨料的一种材料。

pump A mechanical device that can move fluids and gases.
泵 能输送流体和气体的一种机械装置。

pumpability The ease with which a mix of a material, such as concrete or plaster, can be pumped from one place to another.
可泵性 一种材料混合物（如灰泥或混凝土）易于从一个地方抽到另一个地方的性质。

pumping main The main pipe that sewage is pumped through under considerable pressure.
泵水主管 大压力下泵污水时穿过的主管。

punch Pointed hand-held tool, with a sharp or flattened head, used to make indents in metal or drive the heads of nails deeper into the timber so that they are hidden. When used in metal, the pointed punch is used to create an indent in the surface of the metal. The indent created provides a temporary housing for the tip of a drill so that the metal can be drilled without the drill bit slipping over the surface. When nails are to be driven below the surface of the timber, the flat head punch is used.
钻孔机 手持式打孔工具，有尖头或平头两种，用于在金属上开口或把钉头钉入木材深处以便隐藏钉头。用在金属上时，使用尖头钻孔机在金属表面形成凹点，打出的凹点为钻头提供了一个定位口，这样钻头就不会在表面上打滑。当把钉子钉入木材时应使用平头钻孔器。

puncheon 1. A tool used for piercing or punching. 2. A short wooden vertical post.
1. 冲孔器 用于穿孔或打孔的工具。**2. 木制短立柱**。

punching shear A type of failure that can occur in reinforced concrete slabs or highly compressible soils (such as loose sand), which is due to a high localized load.
冲击剪力 由于局部压力过高，钢筋混凝土板或高压缩性土壤（例如松散砂土）中出现的一类失效。

punch list (US) A list of defective works that are required to be undertaken.
竣工查核事项表（美国） 需要承担的有缺陷工程的清单。

purchasing officer Historical title of a *buyer.
采购员 买手的旧称。

purge The process of cleaning or flushing a pipe or system.
疏通 清理或冲洗管道或系统的过程。

purlin A horizontal roof member that runs parallel to the ridge and spans between the roof trusses. Used to support the roof covering.
檩条 与屋脊呈水平状态的水平屋顶构件，横跨在屋顶架之间。用于支撑屋顶覆面。

purpose groups The division of buildings according to their main uses, for example, residential and social housing, commercial, industrial, and non-residential.
目标群体 按照主要功用对建筑群的分组，例如住宅区和公用住房、商业型、工业型以及非住宅型建筑。

push bar A horizontal bar on a door that is used to open a fire door in an emergency.
消防推杆 门上的一条水平杆，可用于在紧急情况下打开防火门。

push button A button that operates an electrical circuit when pushed. For example, the button that is pushed to call a lift.
按钮 按下时可操作电路的按钮。例如，按下按钮呼叫电梯。

push-button control A simple type of control system for a lift.
按钮控制 电梯使用的一类简单控制系统。

push-fit joint (gasketed joint, push-on joint) A simple type of pipe joint that is made by pushing a spigot into a socket. A rubber seal within the joint is compressed forming an air and watertight seal.
推入式快速接头 将内接头推入套接口的一种管道接头。压缩接头内的橡胶塞形成防水密封和气密密封。

push plate A *finger plate.
推板 推手板。

push-pull prop A type of telescopic steel prop.
伸缩撑杆 一种套筒式的钢支柱。

putlog A short horizontal scaffold pole with one flat bladed end and one tubular end. The flat end is built into the wall while the tubular end is supported on a *ledger.
脚手架跳板横木 脚手架上的一根水平短杆，一端为平面叶片，另一端为管状。平端安入墙壁内，管端支在横杆上。

putlog adaptor A flat blade that is attached to the end of a scaffold tube to convert it to a *putlog.
脚手架跳板横木接头 连在脚手架管子末端的平面叶片，以将管子转换成脚手架跳板横木。

putlog clip A clip that holds a *putlog to a *ledger.
脚手架跳板横木夹片 将脚手架跳板横木固定到横杆上的夹片。

putlog hole A hole in a masonry wall used to take the flat end of a *putlog.
脚手架架眼 砌石墙上的洞口，用于搭接脚手架跳板横木的平面端。

putlog scaffold (bricklayers' scaffold) Load-bearing scaffolding that is built in the wall as the building is constructed.
横木脚手架（瓦工脚手架） 建造建筑物时装在墙上的承重脚手架。

putties Materials mainly used for glazing purposes that are designed to act as a bedding material, as well as a filler and a sealant against rainwater. The most common type is linseed oil putty, which consists of vegetable oil (usually linseed oil) and whiting or filler (usually powdered chalk). When used, a hard skin forms by oxidation within six to eight weeks, depending on the temperature. When used in wooden frames, stiffening of the putty is aided by soaking the linseed oil into the wood; this does not happen with metal window frames. Long-term, putties fail by shrinking and cracking brought on by excessive oxidation, and as a result the putty loses its sealing and bonding properties.
腻子 装玻璃时使用的材料，用作垫底材料、填缝剂以及防水材料。最常见的是亚麻籽油腻子，由植物油（通常为亚麻籽油）和白垩粉（通常为粉末状白垩）或填料制成。使用时在六到八周内（由温度决定）经氧化后会形成硬皮。用在木制框架中时，亚麻籽油浸入木头中会加速硬化；而金属窗框架则没有上述现象。长期来看，腻子会因为过度氧化缩水而出现裂缝，从而失去密封性和黏性。

putty knife Flat knife with curved and flat edge used for applying putty to glazed doors and windows.
腻子刀 带弯边和直边的平刀，用于把腻子涂到玻璃门和玻璃窗上。

pycnometer A glass jar with a special top that is used for measuring *specific gravity of soil particles.
比重瓶 带有特殊顶部，用于测量土壤颗粒的比重的玻璃瓶。

pylon 1. A tall steel lattice structure, used for carrying high-voltage cables. 2. A tall vertical structure, which has been used historically to form a decorative entrance or approach to a building or bridge.
1. 高压线铁塔 高的钢制格状结构，用于搭载高压电线。**2. 塔门** 高的垂直结构，古代用作建筑物或桥梁的装饰性入口。

Pyran (Pyrostop) Fire-resisting glazing.
派南（音译） 皮尔金顿的一种耐火玻璃。

pyrethroid An insecticide.
拟除虫菊酯 一种杀虫剂。

pyro Reinforced and insulated electrical cable.
pyro 加强型绝缘电缆。

Q

Q-bop A version of the basic oxygen process in a furnace. A Q-bop furnace has no overhead oxygen lance; instead, the oxygen is blown in through tuyeres (pipes) at the bottom of the furnace. Q-bop is a more efficient process for making steel than other processes (more efficient than the basic oxygen process).
底吹转炉 熔炉碱性氧气吹炼法的一种。底吹转炉没有高架氧气喷枪，而是由转炉底部的鼓风口（管）吹入氧气。炼钢时采用底吹转炉比其他加工方法更高效（比碱性氧气吹炼法更高效）。

quad (quadrant moulding, quarter round) A decorative *moulding in the shape of a quarter circle, used with wooden flooring to cover the uneven gap between the floor and the *skirting board.
嵌条 形状为四分之一圆的装饰性线条，用于填平地板和踢脚板之间的缝隙。

quadrangle 1. A four-sided geometrical shape, for example, a square or a rectangle. 2. A courtyard surrounded on all four sides by tall buildings.
1. 四边形 有四条边的几何图形，如正方形或矩形。**2. 四合院** 四边都围有高的建筑物的院子。

qualification 1. An award given to show that a person or body has met a standard. The standard is normally met by achieving specific and predefined criteria. 2. A condition that must be met before the right can be gained to enter into a certain group. For example, preconditions or set criteria may need to be demonstrated before a company can tender for a specific contract. The process of checks ensures that companies have the right attributes to be included on the list of companies that will be used for future work; also called the prequalification process. Prior to a company being included on a tender list, it may be asked to demonstrate that it is financially sound, has relevant experience, and has the suitable expertise to be considered for future work.
1. 合格 表明某人或某主体达到某种标准，通常来讲达到预先制定的具体标准即视为达到标准。**2. 资格** 获得权利进入某个团体之前必须满足的条件。例如，在公司就某个具体合同投标之前必须设定好先决条件和既定标准。审核过程应确保该公司具备特定属性可被列于未来工程将采用的公司名单之上；也称作资格预审。在被列入投标名单上之前，该公司需证明其具备稳定的财务状况、相关经验以及相配的专业技术承担未来的工作的能力。

Q

quality assurance (QA) An administrative and procedural control system used to ensure that products and services are delivered to a required standard and meet the clients' and customers' needs. The system is used to standardize processes so that they are consistent and easily replicated. By recording and specifying correct procedures, other workers are able to follow the description of the process, ensuring that the end product or deliverable is consistent with that expected. The formal procedures are normally broken down into stages or parts that are checked and signed off by the person undertaking the task and by others, normally more senior, or a nominated quality assessor. The checks validate that the standard has been achieved.
质量保证（QA） 确保交付的产品和服务达到指定标准，满足客户和用户需求的一套行政性和程序性控制体系。该体系将流程标准化，确保其一致且易于重复。通过记录和指定正确程序，其他工人可以按照工序说明，确保最终交付的产品与预期一致。通常将正规的程序分成几个步骤或部分，这些步骤需由执行该项任务的人员，以及更有经验的人员或指定的质量评审员检查并签字。通过检查则确认产品达标。

quality audit The observations, assessments, and checks used to ensure that participants are following the specified procedures. The audit ensures that procedures have been defined, are in place, and are being followed.
质量审核 为确保参与者遵循了规定程序而采取的检查、评估和检验的工作。审核确保了常规程序已被明确、(实施)到位和遵循。

quality control (QC) A system of checks and tests that ensures components are produced within accepted tolerances and meet with a set criteria. Products that do not comply with the standards set will be rejected. The checks and associated procedures rigidly enforce standardization and consistency—specification and tolerance is of key importance. Where products fall outside the acceptable tolerance, they cannot be used. Products that are rejected normally trigger investigations that aim to identify the reason for the defects. If necessary, modifications are made to the process to prevent unacceptable products.
质量控制系统(QC) 确保生产的元件处于规定公差范围内，也能满足既定标准的一套检查测试系统。未达到既定标准的产品将被拒收。检查和相关程序严格控制了标准化和一致化——规格说明和公差十分重要。如果产品超出容许公差范围，则不得使用。被拒的产品通常会进行调查，目的在于确定导致缺陷的原因。必要时需修改程序防止不合格产品的出现。

quality management (QM) A management system used to ensure operations are carried out consistently, reliably, and to the reasonable satisfaction of all the parties that have an interest or stake in the services. The system makes use of quality checks, *quality control, *quality assurance, *quality management systems, and meetings and workshops to deliver an overall approach to the management of quality. Quality is managed in ways that are clear, identifiable, documented, efficient, and controlled.
质量管理(QM) 确保执行的操作的一致性、可靠性且能达到该服务的权益和所有利益相关方的要求的管理体系。该系统通过使用质量检查、质量控制、质量保证、质量管理体系、会议及工厂，形成综合的质量管理方法。以清晰、可辨认的、可记录的、高效的方法对质量进行管理和控制。

quality management system (QMS) All the administration, management, and support operations that are used in the management of quality. The systems are based on documentation and clear communication that inform processes to ensure that procedures are correctly followed so that expected standards of work are achieved. Ensuring that correct procedures are known and followed requires the implementation of training, reporting systems, and audits. Effective systems are based on good communication systems, clear lines of responsibility, efficient and clear documentation, training, auditing procedures, and a working culture that is interested in the quality of service which it delivers.
质量管理体系(QMS) 质量管理中采用的所有行政、管理和支持操作。该体系立足于通知流程的文件和清晰的沟通之上，以确保严格遵循了规程并可达到预期标准。为确保人们周知并遵照正确的规程，需要实施培训报告系统以及审计。有效体系是建立在良好的沟通体系、明确的责任、高效清晰的文件、培训、审计程序以及对其交付的服务质量有利益关系的工作文化的基础之上的。

quantitative risk analysis The analysis of possible consequences of an action using numerical and statistical methods of assessment such as the *Monte Carlo simulation technique.
定量风险分析 使用数值的或统计的评估方法(蒙特卡洛模拟法)对一种行为的可能后果进行分析。

quantitative safety assessment The use of numbers to rate the risk or hazards. Tables can be used to compile and compare the hazards and risks. Each item in the table or matrix is given a risk weighting, e. g. 1—low risk, 2—medium risk, 3—high risk.
定量安全评估 用数字对风险或危险进行评级。可使用表格对风险和危险进行编制和比较。表格或矩阵中的每一项都有一个风险加权，如1—低风险，2—中级风险，3—高风险。

quantities The amount of work, materials, labour, and services required to undertake a task. The building quantities are listed in the *bill of quantities. The bills break work down into measurable units of labour, materials, and services. Products are described using the *Standard Method of Measurement for building works (SMM), and Civil Engineering *Standard Method of Measurement (CESMM). Using standard systems of

measurement, the bills can be priced up by different parties and prices compared. The standardization helps to ensure that each party involved in the bidding process is pricing a comparable unit or volume of work.
工程量 完成一项任务需要的工程、材料、人力和服务的量。建筑物的工程量列于工程量清单中。工程量清单明细表被分成可衡量的人力、材料和服务单位。产品使用建筑工程标准计量方法、土木工程标准计量方法对产品进行描述。各方使用标准计量方法报价并可进行比较。这样可确保参与投标的各方能在一个可比较的单位或工程量的基础上给出定价。

quantity surveying The practice of a *quantity surveyor.
工料测量 工料测量师的测算。

quantity surveyor A construction professional whose work lies in the field of measurement, valuations of work, cost, and contractual advice. Work can be varied and include preparing cost plans, estimating the value of work as a project progresses, preparing final accounts, administering contracts, and assessing the value of properties. Most quantity surveyors are chartered by the *Royal Institute for Chartered Surveyors.
工料测量师 建筑行业中的专业人员，从事计量、工程估算、成本及合约咨询等相关工作。工作内容广泛，包括制订成本计划、随项目进度评估工程价值、制备最终决算、管理合约以及评估物业价值。大多数工料测量师是经过皇家特许测量师学会特许的。

quantum The quantity of something.
总额 某物的总量。

quantum meruit A reasonable sum for work provided; the term is normally used when there was no expressed agreement of price. The price relates to as much as a person deserves for the work, or as much as the item is worth. In practice, it normally relates to a payment for work, fair remuneration according to the amount and quality of work provided.
按劳计酬 对完成的工作给出合理报酬；没有明确的价格协议时通常使用这个方法。该价格是一个人做出的工作应得的报酬，或者为该事项的价值。实际操作中与工作的支付相关，按照完成工作的数量和质量给出合理的薪酬。

quarantine anchorage A docking place where vessels anchor while being granted permission (a clean bill of health) to enter the port.
检疫锚地 允许停靠等待批准（检疫健康证）进入港口的船只的地方。

quarrel A square or diamond-shaped pane of *glass within a leaded panel.
菱形玻璃 铅条玻璃内的方形或菱形玻璃板。

quarry An open-pit mine used for the extraction of rock or minerals, such as stone and other building aggregates.
采石场 开采岩石或矿物的露天矿场，例如石材和其他建筑骨料。

quarry sap The moisture that can be found in stone that is newly quarried. The moisture quickly dries out, having a case-hardening effect.
荒料水分 可在新开采的石材中发现的水分。水分会快速蒸发，形成表面硬化的效果。

quarter bend A pipe fitting that turns through 90°.
直角弯管 弯成90°的管件。

quarter point Two points on a beam set a quarter of the way in from either end. These points are used by two-point lifting equipment to minimize bending stresses in the beam during installation.
四分之一吊点 横梁上距两端距离都为梁长四分之一的两个点。两点式起重设备通过抓住这两个点，在安装时可将梁承受的弯曲应力降至最小。

quarter round (quadrant moulding) *See* QUAD.
四分之一圆 参考"嵌条"词条。

quarter-sawn The cut pattern of timber boards from a felled tree trunk (log). The log is first cut into quarters along its *axis, after which a series of cuts are made perpendicular to the tree's *growth ring. This pro-

duces a reasonably consistent grain.
径切 从伐倒的树木上切割木板的方式。首先沿着树干将圆木切成四等份，然后沿着与年轮垂直的方向进行一系列切割。这样木板的纹理可以相对一致连贯。

quarter space landing A small square *landing where the sides of the landing are equal to the width of the stairs.
正方形转角平台 小的直角楼梯平台，平台的各边长与楼梯宽度相等。

quatrefoil An architectural feature with four leaves, it is used to describe elements of a building which have four overlapping circles. Typically found at the top of gothic church windows.
四叶饰 一种有四个叶子的建筑装饰，用于描述有四个重叠圆形的建筑物的构件。通常在哥特式教堂窗户的顶部会有这种装饰。

queen closer A brick cut in half along its length to form a half *header. Used on each course of *Flemish and *English bonds to complete the pattern.
纵半砖 沿砖长切成一半的砖，形成半丁砖。在梅花丁式和英式砌法的每一皮砖都会用到。

queen post The vertical members of a *queen post truss.
双柱桁架立杆 双柱桁架的竖杆。

queen post truss A traditional timber *roof truss that has two vertical *queen posts, positioned a third of the way across the span of the *tie beam, that extend up to the sloping sides of the roof truss.
双柱桁架 传统的木料屋顶架，有两个竖直的桁架立杆，放置于连梁跨度三分之一处，向上延伸至屋顶架的斜边。

quench Rapidly cooling a material to achieve a desired property in the material, for instance, rapid quenching of low carbon steel from approximately 750℃ to room temperature achieved within a few seconds results in increased hardness.
淬火冷却 快速冷却一种材料以达到所需的属性，例如将低碳钢在几秒钟内从750℃快速淬火至室温，以增强材料硬度。

Quetta bond A type of *Flemish bond used to construct reinforced brickwork. Comprises a one-and-a-half-brick-thick Flemish bond wall where the resulting half-brick-thick vertical spaces (pockets) in the centre of the wall are filled with steel reinforcement and concrete infill.
奎塔砌法 构筑加筋砖结构所用的一种梅花丁砌法。包含一个一砖半厚的梅花丁砌合墙，其中墙壁中有半砖厚的垂直空间填充有钢筋混凝土。

quick bend A sharp or tight bend in a pipe.
短半径弯头 管道中的急转弯头。

quickset level A levelling instrument, similar to a *dumpy level, but where the three levelling screws have been replaced by a ball and socket to provide quick initial setup.
速调水准仪 与定镜水准仪类似的水准仪，但是三个水准螺旋被换成了一个球窝，可进行快速初始化设置。

quick step A decorative circular *moulding of small radius.
收边条 小半径的圆形装饰线条。

quirk A V-groove cut into timber, the edges of the groove are rounded, giving the appearance of an open book; often used as a decorative feature in timber beading.
凹槽 木料中的V型凹槽，其边缘是圆滑的，像一本打开的书；通常为木材圆边中的装饰性造型。

quirk bead Rounded edge of timber (creating a bullnose or bead) with a rebate cut directly next to the rounded edge. The rebate (*quirk) is set in from the edge of the timber and next to the bead. The quirk, in a quirk bead, often has one straight edge and the other rounded.
凹槽圆角 有一个槽口的木材圆边（形成外圆角或圆线条）。从木料边缘到圆边之间切一个槽口（凹槽）。槽口为凹形圆线脚的，通常一侧为直边，另一侧为弧形。

quoin The exterior corner of a building or a wall.
外墙角 建筑物或墙的外角。

quoin block An L-shaped block used to form the corner of a wall.
墙角砖 用于筑成墙脚的 L 字形砖块。

quoins The brick or stonework used to form the *quoin. Commonly used for decorative purposes when they project slightly from the face of the wall.
隅石 砌成外墙角的砖块或石块。出于装饰目的，通常从墙面上突出一点。

quotation (quote) The price given to undertake work.
报价 承包工程给出的价格。

R

rabbet A rebate or recess.
槽口 企口缝或凹槽。

racking back (raking back) The stepping of bricks at the ends of the wall to be laid/ constructed. The *bricklayer uses the stepped bricks as a guide for level and line to complete the wall.
斜槎 需要垒砌的墙两端的砖块台阶。瓦工以台阶式砖作为对准线以完成墙的砌筑。

racking course A layer of aggregate that is used to fill voids on the foundation layer of a road before it is surfaced.
水泥稳定碎石层 铺路面之前填满路基层上方空隙的一层骨料。

radial circuit An electrical power circuit for lights and appliances that starts at the consumer unit and finishes at the last socket or lighting point. There is no connection from the last socket or lighting point back to the consumer unit, unlike a *ring main circuit, which is connected to the consumer unit at both ends.
放射式电路 从供电箱出发，到插座或照明点结束，为灯具和电器供电的电源电路。最后的插座和照明点没有回接到供电箱，不像环状主电路的两端都是与供电箱相连。

radial-sett paving A paving set laid in semi-circular patterns, often constructed to a radius within which the paver's arm will reach.
扇形地铺 以半圆形图形铺设的路面，通常以铺砌工胳膊能够着的范围为半径进行铺砌。

radiant heating A type of space heating system that heats a space using *radiation and *convection. Usually comprises a series of pipes that are embedded in a floor, ceiling, or wall—*underfloor heating is a type of radiant heating system although emits much of its heat through convection.
辐射供热 使用辐射和对流进行供暖的一种空间加热系统。通常有一组埋于地板、天花或墙壁内的管道——地热是一种辐射供暖系统，但是是通过对流方式散热的。

radiation The process where heat is emitted from a body and is transmitted through a space as energy.
(热) 辐射 热从一个物体中发出，在空间中以能量形式传播的过程。

radiation detector *See* FLAME DETECTOR.
辐射探测器 参考"火焰检测器"词条。

radiator 1. Any heater that warms a space by emitting heat energy primarily by convection and a small amount by radiation. 2. A metal heater that comprises a series of pipes and metal fins to transfer the heat from

hot water to a space.
1. 散热器 大部分通过对流、小部分通过辐射形式进行供暖的加热器。**2. 暖气片** 金属加热器，包括一组管件和金属散热片，将热水的热气传送到一个空间。

radiator key A small key used to open the air vent located on the top of a radiator.
暖气片阀门钥匙 用于打开散热器顶部的通气孔的小钥匙。

radioactive testing A *non-destructive test using *X-rays or *gamma rays to inspect for any sub-surface flaws in concrete and welded joints.
放射性测试 使用X光或伽马射线检测混凝土和焊缝次表面裂纹的非破坏性测试。

radio-frequency heating The process of using electromagnetic radiation to heat materials.
射频加热 使用电磁辐射材料加热材料的过程。

radiography A *non-destructive test that uses *X-rays to inspect the composition of the materials. Used to locate sub-surface defects in concrete and welded joints.
射线探伤 使用X光检验材料结构的非破坏性试验。用于定位混凝土和焊缝中的次表面缺陷。

radius rod A length of timber or moulding used by plasterers to create radial patterns in wall finishes. The rod will have a nail or point at one end, which is used as the centre of the radius.
抹灰半径杆 抹灰工用于在墙面上形成放射状图案的一段木头或板条。半径杆一端有一个钉子或尖端用作半径的中心。

rafter The sloping beam that spans from the ridge to the eaves of a roof. Intermediate rafters span the full length from eaves to ridge, whereas jack rafters span part of the roof linking between hips and valleys. Also describes the sloping metal sections of a portal frame roof and used as an abbreviation for *truss rafter.
椽条 屋顶中从屋脊到屋檐上搭设的斜梁。中间椽横跨了整个屋脊到屋檐的距离，小椽横跨了斜脊和斜沟之间的部分距离。也可指门式刚架的斜梁截面，以及用作桁架椽的缩写。

rafter filling (beam filling, wind filling) The infilling between roof *rafters above the *wall plate. Bricks, blocks, or insulation are laid on top of the wall plate to fill the void between joists, reducing air flow.
椽条填充 梁垫上屋顶椽条之间的填料。梁垫上可放置砖块、砌块或保温材料以填充龙骨之间的缝隙，减少空气流动。

raft foundation Support underneath a building that transfers and disperses the building loads over a large area, often the total plan area of the building. By dispersing the load over such a large surface area, the load transferred per unit to the ground is significantly reduced. Due to its ability to reduce the loads on the ground, the raft foundation is often used on poor ground. Raft foundations are normally heavily reinforced to ensure the loads are spread over the total area.
筏式基础 建筑物下方的支撑，在大面积范围上转移并分散建筑物重量，通常分散到整个建筑规划区。将荷载分散到大面积区域上，会大大降低单位地面承受的转移荷载。由于能够降低地面的荷载，所以在劣质土地上通常采用筏式基础。筏式基础通常会采用加大配筋的方式确保荷载分散到整个区域。

ragbolt Bolt with barbs or protruding metal fixings that help lock the bolt into the substrate into which it is fixed.
棘螺栓 带倒刺或突出的金属配件的螺栓，可将螺栓锁在其固定的基底内。

ragging A paint finish created by dabbing a scrunched-up cloth, with paint on it, on the surface of a wall, or created using a roller made from loose flapping rags.
布纹漆面 通过将一团涂有油漆的布压在墙面上，或者由松散布条制成辊子形成的油漆饰面。

raglet (raggle, raglan, reglet) The rebate made in a wall or the recess in brickwork joints to receive the *flashing. The flashing is turned into the recess and secured using mortar. The recess and the inbuilt flashing prevents water getting behind the face of the flashing.
披水槽 墙上的槽口或砖缝中的凹缝，用以安装防水板。把防水板嵌入凹缝然后用灰浆固定。凹缝和

R

内嵌的防水板能防止水溅到防水板背面。

ragwork Using a rag roller or scrunched-up rag to create a pattern and texture on a wall.
涂布纹 使用卷布辊子或卷皱布在墙面上形成花纹或纹理。

rail 1. A horizontal piece of timber or steel from which objects can be hung or secured. Objects typically hung from rails include curtains, kitchen utensils, and tools, and objects secured include fence panels and boarding. 2. The top part of a stair balustrade that provides a continuous pole, which allows those using the stairs to assist movement by holding the rail. 3. The horizontal member that subdivides a doorframe. 4. A * handrail. 5. A horizontal member on a fence.
1. 横杆 木制或钢制水平杆，可以悬挂或固定物体。可悬挂的物体主要包括窗帘、厨房用具或其他工具，用于固定的物体主要有栅栏板。**2. 栏杆** 楼梯栏杆的顶部，有连续扶杆，可供使用楼梯的人扶住协助上下楼梯。**3. 冒头** 分隔门框的水平构件。**4. 扶手**。**5. 围栏横梁** 栅栏上的水平构件。

rail chair *See* DADO RAIL.
护墙板 参考“护墙板线条”词条。

railing A fence constructed from rails and vertical posts.
栏杆 由扶手和竖直杆构成的围栏。

rainwater goods Collective name given to all the components required to convey water from a roof. The components will include gutters and downpipes.
雨水管件 从屋顶上排水所需的所有组件的统称。主要包括水槽和落水管。

rainwater head (rainwater header, rainwater hopper, US conductor hopper, US leader hopper) A box-shaped container at the top of the downpipe that collects rainwater from the gutters. Usually constructed in cast-iron or lead.
雨水斗 落水管顶部的盒状容器，用于收集水槽处的雨水。通常由铸铁或铅制成。

rainwater outlet An outlet that takes rainwater from a roof and discharges it into a * gutter.
雨水排水口 雨水从屋顶排出去流入排水沟所经过的口。

rainwater pipe (RWP, downpipe, downcomer, US downspout, US conductor, US leader) *See* DOWNPIPE.
雨水管 参考“落水管”词条。

rainwater plumbing (external plumbing) The process of installing * rainwater goods.
雨水管道工程 安装雨水管件的过程。

R

rainwater shoe A short curved fitting at the bottom of a * downpipe that discharges rainwater into a * gully.
落水斜口 落水管底部的一个短的弯管，用于将雨水排放到集水沟内。

rainwater spout A duct used to discharge rainwater from a roof. *See also* GARGOYLE.
排水口 用于排放屋顶雨水的管道。另请参考“滴水嘴兽”词条。

raised and fielded panel A panel that is raised in the centre and has a sunken margin. Commonly used in panelled doors.
凸面镶板 中间突起边缘凹陷的板。通常用于镶板门。

raised countersunk head screw A wood screw with a domed head.
半沉头螺钉 有球形头的木螺丝。

raised fibres Timber fibres that stand proud of the surface. As timber dries out, the softer wood between the fibres shrink, leaving the fibres proud. This characteristic also remains after painting or varnishing; by lightly sanding and reapplying the finish, the raised appearance of the grain can be reduced.
凸起纤维 从表面隆起的木材纤维。木材变干后，纤维之间的较软木质开始缩水，纤维就会凸出来。喷漆或上清漆后仍会保持这个特征，轻微打磨和饰面处理可以改善这种情况。

raised floor A floor mounted on pedestals or battens, allowing services to be installed below the floor. *See al-*

so ACCESS FLOOR.
高架地板 安装于基座或板条上的地板，可以在地板下方安装设备。另请参考“活动地板”词条。

raising piece A packing material used to lift the unit to the correct position so that it is fixed in line and level.
垫高片 用于帮助组件固定到位并保持水平齐平的一种填充材料。

rake (batter) A sloping angle from the horizontal. Earth banks are often landscaped to a set slope; this is the rake.
斜坡 路堤上通常建一个与水平面成倾斜角的斜坡用于景观美化；这就是斜坡。

raked-out joint Mortar joints that are recessed out to approximately 10 mm in depth. The joint, which is recessed along the full length of the brickwork bed, exaggerates and enhances the appearance of the joint. To improve the water resistance of the joint, the mortar should be compressed and compacted into the joint when *pointing.
凹式灰缝 凹入深度大约为 10 mm 的灰缝。沿着砌砖层全长凹陷的接缝，扩大并加深了接缝的外表。为了增强接缝的防水性，勾缝时应将灰泥压进接缝并压实。

raking cutting 1. The grading of soil or earth at an angle. 2. The cutting back of existing brickwork to provide an exposed edge ready to receive new brickwork.
1. 斜坡修整 修整土壤或泥土，使其形成一定角度的斜坡。**2. 切削** 截短已有的砌砖，提供一个开放的边缘砌新的砖。

raking flashing The stepped cutting of flashing so that it follows the line of the roof and is rebated into the mortar joints of the abutting brickwork.
斜遮雨板 对防水板进行跌级切削，以便遮雨板对齐屋顶边缘并可嵌入邻接砌砖的灰缝中。

raking out The removing of mortar at the surface of brickwork to create a recessed or raked-out joint.
清缝 把灰泥从砌砖表面移去，形成嵌入式或凹式灰缝。

raking riser A vertical riser of a stair that is positioned at an angle to increase the tread area of a step.
斜踢面 为了增加台阶的踏面面积，以一定角度设置的楼梯垂直踢面。

raking shore (inclined shore, shoring raker) Prop or strut that is placed at an angle and secured to a wall to provide lateral stability. The rake refers to the slope or angle of the prop. Shores can be made out of steel or timber.
斜撑木 以一定角度放置的支柱或（其他）支撑物，用于固定墙壁、增强其横向稳定性。斜度指斜坡或支柱与木平面所成的角度。撑木可由钢或木材制成。

rammed-earth construction Building of walls out of raw materials, such as earth, chalk, sand, gravel, clay, and lime. The materials are normally sourced locally, making the form of construction sustainable. With the right selection of materials little energy is demanded to extract and use the materials. *Formwork is used to provide a mould for the materials, and mechanical plant may be used to compress the materials into the forms.
夯土结构 由土、白垩、沙子、砾石、黏土、石灰等原材料筑成的墙。原料通常来自当地，这样可以保证持续供应。材料选择恰当就可以少花时间去筛选使用材料。模板用于为原料提供模具，将原料压进模板时可能要用到机械设备。

ramp Inclined platform used to get from one level to another. An incline of 1 in 12 or less is often used to enable wheelchair users to travel up and down the incline safely.
坡道 从一层到另一层所用的倾斜的平台。通常采用倾斜度为 1/12 或更小的坡度，以方便轮椅使用者安全上下坡。

rampant arch An arch that has one support higher than the other.
高低拱 一个拱足比另一个高的拱形。

rampart A defensive wall.
壁垒 防御墙。

ram pump A very simple displacement pump containing two moving parts. For the pump to operate, an elevated water source is required. Water flows into the lower chamber causing the air in the upper chamber to compress. Once the pressure is sufficient, water is pushed up the outlet pipe. Various valves open and shut to control the flow of the water.
水锤泵 一种非常简单的将动能转换为压力能的机械。泵运作时，需要一流动的水流。水流进较低的泵体造成较高泵体内的气体压缩。当压力足够时，水被推动至出水管。通过开关不同的阀门控制水流。

random ashlar (random bond, irregular bond) Wall made from cut stone of different heights that prevents the stone from being laid in regular courses.
乱砌方石 用不同高度的琢石筑成的墙面，使石块堆砌得错落有致。

random courses Walling constructed from courses that differ in height.
乱砌 砌层高度不等的墙壁。

random pattern Paving or walling constructed from units of different size, that do not fall into a regular arrangement.
随机图案 由不同尺寸的组块筑成的路面或墙壁，形成的图案不会落入俗套。

random paving Arrangement of paving slabs that do not follow a regular pattern.
随机铺砌 铺路板的铺砌不遵循统一模式。

random rubble (random rubble walls) Stone of different shapes and sizes placed together without any attempt to build in regular courses.
乱砌毛石 不为取得规则图案，只是铺在一起的不同大小和形状的石块。

random slates Slates that have been split in random form and are rugged in texture.
乱砌石板 被分裂成不同形状，纹理不同的石板。

range 1. Allowable set of values or items. 2. Large cooker with double oven and hob.
1. 范围 容许的一组数值或事项。**2. 炉灶** 有两个炉头和搁架的灶。

range masonry (coursed ashlar, range work) Stones cut to equal heights so that they can be laid in regular courses.
平毛石 切成等高的可堆砌成规则砌层的石块。

ranging A method to establish a straight line (ranging line) using *ranging rods.
测距 使用测距杆确定一条直线（测距线）的方法。

R

ranging rod A pole that is used in surveying for *setting out. Typically the pole is 2 m long and is painted in alternative red and white sections of 0.5 m.
测距杆 放线测量时使用的杆。杆长通常为 2 m，涂有宽 0.5 m 的红白相间条纹。

rapid hardening Portland cement A type of *Portland cement that reaches optimum strength at an accelerated rate.
快硬硅酸盐水泥 能够快速达到最佳硬度的一种硅酸盐水泥。

rapid sand filter Used to filter water in the tertiary stage of a water treatment plant. The filter contains coarse sand and is primarily used to remove impurities that have been trapped in the floc. It is cleaned by backwashing; *See also* SLOW SAND FILTER.
快滤池 净水厂中用于过滤水的滤池。过滤池中有粗砂，主要用于去除棉絮状杂质。通过反冲洗进行清洁。另请参考“慢滤池”词条。

rasp A tool for shaping wood, which consists of a long steel bar with cutting teeth that protrude from the flat surface.
粗锉刀 锉削木材的工具，在一根长钢条的平面上有突出的锉齿。

rate 1. Output, quantity of work performed per hour. 2. Price per unit of work, or price per hour.

1. 输出 每小时完成的工作量。**2. 单价** 单位工作量的价格或每小时的价格。

rate of pay (wage rate) Payment per hour or per unit of work performed.
薪酬率 每小时或完成每单位工作的报酬。

rate of rise detector A *fire detector that responds more rapidly to an abrupt rise in temperature than a fixed-temperature heat detector.
差温式火灾探测器 与定温火灾探测器相比，对温度的突然上升反应更为迅速的一种火灾探测器。

rating flume A tank that is used to calibrate current meters and pitot tubes.
计量槽 用于校准电流计和皮托管的槽。

rat-trap bond Similar bond to a Flemish bond (alternate *header and *stretcher), but with the bricks placed on their side. The bond uses fewer bricks, making the construction method more economical. Rather than header and stretcher, when a brick is placed on its side the header end of the brick is called the rowlock and the top face of the brick, which is also exposed, is called the shiner. In a one-brick wall the two shiners are placed between the rowlocks. As the bricks are on their side there is a gap in the centre of the wall between the two shiners. The name "rat trap" is given due to the gap created within the centre of the wall.
空斗墙砌合法 与梅花丁砌法类似，但是砖是以侧边方式放置。这种砌合法用的砖更少，施工方法更经济。不同于丁砖和顺砖，当丁砖侧面竖放的时候称为眠砖，当砖的顶面外露时则称作斗砖。在单砖墙中，两块斗砖夹在眠砖之间。由于砖是侧面砌合的，所以在墙中心斗砖之间有空隙。之所以叫"空斗墙"就是因为墙心中间的空隙。

rawbolt A bolt contained within a sleeve. When inserted into a drilled hole and tightened, the bolt is drawn into the sleeve causing it to expand (tighten) against the side of the hole.
膨胀螺栓 套管内的螺栓。塞入钻孔并拧紧时，螺栓旋入套管，使套管膨胀（拧紧）并顶住孔边。

Raykin fender A *fender that has a sandwich-type construction, being made from a series of rubber layers bonded between steel plates. Used to protect docklands from the impact of mooring ships.
雷金式围护物 有夹层式结构的围护物，由夹层之间粘有橡胶层的钢板制成。用于保护码头区域免受停泊船只的撞击。

Raymond pile *See* STEP-TAPERED PILE.
雷蒙桩 参考"锥形桩"词条。

razor socket Plug point created for electric two-pronged shaver.
剃须刀插座 电动两头剃须刀的专用插座。

RC *See* REINFORCED CONCRETE.
钢筋混凝土 参考"钢筋混凝土"词条。

reaction turbine (Francis turbine) A *turbine that is driven by moving water entering a wheel round the circumference, it combines both radial and axial flow components. Principally used for electrical power production.
反击式涡轮机/混流式水轮机（弗朗西斯式水轮机） 由水流进入轮周驱动的涡轮，有径流和轴流组件。主要用于电力生产。

reactive Will produce a chemical reaction. Reactive components can be used in resins and paints to enable them to harden/set.
活性的 能产生化学反应的性质。活性成分可用于树脂和油漆中，使其硬化/固定。

readout The information obtained from a monitoring device.
读数 从显示器中获取的信息。

ready-mixed concrete (ready mix, ready mix concrete) A type of concrete that has been proportioned or batched at a plant and mixed at the plant in a static mixer or truck mixer prior to transport to site.
预拌混凝土 一种预搅拌的混凝土，在工厂中已按配比分装，在搅拌器或混凝土搅拌车中搅拌好后再运送至现场。

ready-mixed mortar A type of mortar that has been proportioned or batched at a plant and mixed at the plant in a static mixer or truck mixer prior to transport to site.
预拌灰浆 一种预搅拌的灰浆，在工厂中已按配比分装，在搅拌器或搅拌车中搅拌好后再运送至现场。

realignment The adjustment made to alter the path/route of a road or railway line.
重新定线 对一条公路或铁路的路径或路线的调整。

rebar A *reinforcing bar.
钢筋 加固（钢）条。

rebate (Scotland rabbet) 1. A cut into a material. The rebate is usually used to house another material such as a glass window. 2. A rectangular recess cut on the edge of a material.
1. 企口缝 切入材料的槽口，通常用于搭接其他材料，如玻璃窗等。**2. 矩形切口** 切在材料边上的矩形凹口。

rebated joint A joint created by forming a rebate in a material, usually formed in wood capable of receiving the end of another material.
企口接合 在一种材料上（通常为木材）形成一个榫接口，便于搭接其他材料的端部的一种接合。

rebated weatherboarding *Weatherboarding that incorporates a rebate along the thickest edge of the board.
百叶护墙板 沿着板的最厚边有一个槽口的护墙板。

rebound hammer (Schmidt hammer) A device to determine the strength of concrete non-destructively. The device consists of a spring-loaded plunger that is pressed against the surface of the concrete to load the plunger. The plunger is then released to give an indication of the surface hardness and penetration resistance; with reference to the conversion chart an indication of the strength of the concrete can be obtained.
回弹仪 以非破坏性方式确定混凝土强度的一种工具。该设备中有一个装弹簧的弹击锤，抵着混凝土面按压弹击锤使其承受荷载。松开弹击锤后会给出一个表面硬度及穿透阻力的指示值；参照换算表，就可获得混凝土强度的指示值。

receptacle (US) A *socket outlet.
插座（美国） 插座。

recess An indent in a wall or surface, sometimes built-in as a design feature, other times to provide a storage area for folding doors or other building furniture.
凹槽 墙壁或表面的凹口，某些情况下内嵌作为一种设计造型，其他情况下为折叠门或其他建筑设备提供存放空间。

R

recessed joint Brickwork mortar joint that is raked out to a consistent depth, between 5 – 10 mm below the surface of the brickwork. The joint tends to hold water that falls on the surface of the wall. Porous brick walls with recessed joints are more susceptible to frost attack than *flush mortar joints.
凹缝 刮去灰泥到统一深度的砌砖灰缝，深度大约为砌砖面下方 5 mm 到 10 mm 深。灰缝用于留存落到墙面上的水。带凹缝的多孔砖墙比平灰缝更易受霜冻攻击。

recessed luminaire A *luminaire that is fixed flush to a ceiling, wall, or other surface.
嵌入式灯具 安装后与天花板、墙壁或其他表面齐平的灯具。

recirculated air Air extracted from the conditioned space that is sent to the air-conditioning or ventilation system and mixed with *fresh air. The mixed air is then supplied to the conditioned space.
再循环空气 从空调空间抽取的空气被送到空调或通风系统里并混合新鲜空气，然后再送回至空调区域的混合。

reconfigure The ability to alter or change.
重新配置 改变或改动的能力。

recovery 1. To return to a normal state. 2. The reclamation of useful substances from waste.
1. 复原 回到正常状态。**2. 回收** 从废物中回收有用物质。

recreational accommodation Buildings constructed for pursuits of leisure: sports, theatre, cinema, concerts, and other leisure activities.
娱乐设施 为休闲目的而建的建筑：运动场、剧院、影院、音乐厅及其他休闲活动（设施）。

rectangular on plan Building that is a regular quadrilateral when viewed from above.
矩形平面 从上往下看时呈规则四边形的建筑物。

recycle To reuse something.
再循环 重复使用某物。

redevelopement To clear an area of buildings and structures so that the area can be replanned and developed as if the original development had not existed.
二次开发 清理该片区域上的建筑和结构以重新规划开发，如同未进行过第一次开发一样。

red lead A type of lead oxide, red in colour used as pigments in paints.
红铅 一种铅氧化物，用作油漆中的红色颜料。

reduced level The elevation of a point to a reference datum, e. g. *ordnance datum.
归化高程 一个点距参考基准面的高度，例如水准零点。

reducer 1. A pipe fitting used to connect a large diameter pipe to one of a smaller diameter. 2. A substance that is capable of reducing the concentration of another substance.
1. 异径管 一种将大口径管连接到小口径管上的管件。**2. 退粘剂** 能够降低另一种物质浓度的物质。

redundancy 1. Dismissal from work; the employee is no longer required. 2. Duplication or superfluous, for example, additional members within a structure that are not necessary but have been included to increase robustness and rigidity.
1. 裁员 解雇，该雇员已不再被需要。**2. 重复或冗余** 例如，结构内的附加构件，虽然不必要，但可用于增加稳健性和硬度。

reeded Convex or beaded moulds that sit together in parallel rows giving a fluted appearance.
条纹面 凸面或珠状线条装饰平行排列，形成有沟纹的外表。

reference The use of something as a source of information.
参考 使用某物作为信息来源。

reference panel (sample panel) Unit of work constructed to provide an example of the materials to be used, quality of workmanship, and the finish to be expected.
参考板（样板展板） 为所用材料、工艺质量以及预期饰面提供样板而建的工作单元。

reference specification Description of standards, codes of practice, and functions that the contractor will perform rather than describing or listing in detail the products to be used.
参考规范 对承包商需遵循标准、行为守则及职责的说明，并不对所用材料做出具体描述。

reflective glass Glass that incorporates a coating or film to reflect solar radiation from the sun.
反射玻璃 有一层反射太阳光的薄膜涂层的玻璃。

reflective insulation Insulation materials that have a reflective surface to reduce heat loss by *radiation.
反射保温 保温材料，有降低因辐射导致的热损失的反光表面。

reflector A device incorporated behind the lamp in a luminaire to redirect light back into the space.
反光板 灯具中灯背面的一种装置，用于将灯光反射回空间中。

reflow soldering A type of *soldering whereby solder fillets are formed in metallic areas.
回流焊 一种焊接方式，焊条被焊接于金属区域。

reflux valve *See* CHECK VALVE.
回流阀 参考“止回阀”词条。

refractory A heat-resistant material. The term refractory refers to the quality of a material to retain its strength at high temperatures. Refractory materials are used to make crucibles and linings for furnaces, kilns, and incinerators. The best example of a refractory material used in construction is *brick that has remarkable strength retention at elevated temperatures. There is no clearly established boundary between refractory and non-refractory materials, though a practical requirement often cited is the ability of the material to withstand temperatures above 1100℃ without softening. Refractory materials must be strong at high temperatures, resistant to thermal shock, chemically inert, and have low *thermal conductivities and coefficients of expansion.
耐火材料 耐高温材料。耐火性是指高温下依然能保持其强度的材料特性。耐火材料可用于制作坩埚以及熔炉、窑炉、焚化炉的衬里。耐火材料应用的最佳例子是砖，其在温度升高后能够极好地保持其硬度。耐火材料和非耐火材料之间没有明确界定，但实际操作中耐火材料在1100℃以上的高温下依然不会软化。耐火材料在高温下依然保持其硬度，能抵抗热冲击，保持化学惰性，具有低热导率和膨胀率。

refractory mortar A type of mortar with excellent retention of properties and performance at elevated temperatures.
耐火灰浆 高温下依然能极好地保留其性能的一种灰浆。

refrigerant A fluid that provides cooling by changing state from a fluid to a gas. Used in air-conditioning systems, heat pumps, refrigerators, and freezers.
制冷剂 从液态变成气态而冷却的液体。应用于空调系统、热泵、冰箱和冷冻机中。

refrigerated storage A storage area that is refrigerated. *See also* COLD STORE and REFRIGERATOR.
冷藏室 冷冻的储藏区域。另请参考“冷藏库”和“冰箱”词条。

refrigeration unit (US chiller, refrigeration machine) A mechanical device that produces large amounts of cooled water for *air-conditioning systems.
制冷机组 为空调系统提供大量冷却水的机械设备。

refrigerator A mechanical device that is designed to store goods at a low temperature.
冰箱 低温储藏物品的机械设备。

refuge 1. A traffic island in the middle of a road. 2. A shelter to provide protection from something, for example, a chamber cut into the side of a tunnel for pedestrians to shelter from passing traffic.
1. 安全岛 道路中间的交通岛。**2. 安全处** 提供保护的地方，如在隧道旁边开一个洞室（避车洞）保护行人免受车辆影响。

R

refurbishment (rehabilitation) The improvement of a structure or building so that it returns to being a functional property that is aesthetically pleasing and structurally sound.
翻新（复原） 对结构或建筑物的修缮，使其功能特性恢复到外观美观、结构坚固的状态。

refuse Rubbish or waste.
废弃物 垃圾或废物。

refuse chute A vertical *chute used in multi-storey buildings to dispose of rubbish.
垃圾槽 多层建筑中倾倒垃圾用的竖直溜槽。

region An area with a defined boundary, which would indicate geographical, political, or cultural characteristics.
区域 有明确边界且具备地理、政治或文化特征的地区。

register An outlet or terminal unit for air supply from an air-conditioning unit. The outlet is usually fitted with a damper and diffuser.
出风器 空调机组送风的出口或终端。出口处通常装配有一个气闸或散流器。

reglet A narrow groove chased into a wall to receive lead flashing. *See also* RAGLET.
披水槽 刻入墙内用于安装铅皮防水的窄槽。另请参考“披水槽”词条。

regrating Cleaning of stone walls and other masonry.

石面翻新 对石墙和其他砖石面的清理。

regular (regular coursed rubble, US coursed ashlar, range masonry, range work) Stone rubble that is laid in regular courses, but due to the irregular nature of the stone, the depth of separate courses varies.
平毛石层 整层垒砌的毛石，但是由于石头本身不规则，所以每层厚度都不一样。

regulating reservoir A reservoir designed to control the supply, whether to prevent the supply running dry or flooding.
调节池 用于控制供水量的蓄水池，防止过旱或过涝。

rehabilitation *See* REFURBISHMENT.
复原 参考"翻新"词条。

re-heat unit (terminal re-heat unit) An *air terminal unit that incorporates a heater to increase the supply air temperature.
再热机组 包含一个加热器的风道终端装置，用于送风加热。

reinforce To strengthen something by providing internal or external support.
加固 通过提供内部或外部支撑加固某物。

reinforced beam (reinforced column) A concrete beam or column that contains *reinforcement, typically containing a collection of *reinforcing bars that have been assembled to form a cage of reinforcement, around which concrete is poured and sets.
加固梁（加固柱） 包含钢筋的混凝土梁或柱，通常为各种钢筋条装到一起形成钢筋笼，然后在其周围浇筑混凝土使其凝固。

reinforced brickwork *Brickwork that has been reinforced by wire mesh.
配筋砌体 有钢丝网加固的砌砖结构。

reinforced concrete Also known as ferroconcrete, a type of concrete in which *reinforcement bars (rebars) or fibres have been incorporated into the concrete matrix to strengthen the material that would otherwise be brittle. Reinforced concrete is widely used in civil engineering.
钢筋混凝土 也称作钢丝网混凝土，混凝土块内混合有钢筋条或钢丝网，用于增加混凝土强度，若不添加则脆性很大。钢筋混凝土广泛应用于土木工程中。

reinforced concrete degradation Material degradation of reinforced concrete either by rusting of the steel components or freeze-thawing of the concrete matrix.
钢筋混凝土退化 钢件生锈或混凝土基体反复冻融导致的钢筋混凝土退化。

reinforced soil wall A retaining earth wall that has been constructed in layers, with a tensile inclusion (*geosyntheticor steel strips) between each layer, which are attached to the excavated face.
加筋土墙 层砌的加筋土挡土墙，各层之间夹有抗拉材料（土工材料或钢条），附在开挖面上。

reinforcement *See* REINFORCED CONCRETE.
钢筋 参考"钢筋混凝土"词条。

reinforcement bar *See* REINFORCING BAR.
钢筋条 参考"钢筋"词条。

reinforcement schedule A list with quantities of all of the reinforcement needed for a job, the list specifies the type, size, and shape of the bar, also known as the bending bar schedule.
钢筋材料计划表 一项工程所需钢筋材料的清单，列明了钢筋的种类、尺寸和形状，也称作弯钢筋计划表。

reinforcing bar (rebar) A steel bar that is placed in concrete to increase its tensile strength. The bar is normally manufactured from carbon steel and may be smooth or ridged—the latter to increase bond strength.
钢筋 加在混凝土中增强其抗拉强度的钢材。通常由碳钢制成，可为光面或带肋——后者可以增强其黏结强度。

rejointing The pointing or repointing of brickwork.
重新填缝 砌砖勾缝或重新勾缝。

relamping The process of replacing the lamps within a *luminaire.
换灯泡 更换灯具内灯泡的过程。

relaxation The reduction in tensile stress with time when a constant strain is applied.
松弛 受到持续拉伸时，拉伸应力随时间推移慢慢减弱（的现象）。

relieving arch (safety arch) An arch built into a wall above a lintel. The arch reduces the load that is imposed directly on the lintel.
卸荷拱（分载拱） 构筑于墙内的拱，位于过梁上方。该拱能减少直接施加到过梁上的载重。

religious accommodation Building used for spiritual needs or place of worship, and includes mosque, church, synagogue, cathedral, and temple.
宗教建筑 出于宗教需要而建的建筑或者做礼拜的场所，包括清真寺、教堂、犹太会堂、天主教堂和寺庙。

reloading To restore power to a circuit.
再启动 恢复电路电力。

relocatable partition Movable and demountable partition wall.
可拆卸隔断 可移动和拆卸的隔断墙。

remeasurement The measurement of the quantity of work after the work has been completed. The measurement is then presented against the list of agreed rates for calculation of the value of the work.
重新测量 完工后对工程量进行测量。测量结果会与协定的费率表进行对照去计算工程价值。

remedial work Procedures, activity, and work to improve the condition of something, make good for a specific purpose, or restore to a former condition. Remedial work to walls and floors often involves structural work; such work on the ground can include improving its stability and/or removing or sealing contamination.
补救工作 改善某物状况的程序、活动和工作，为某种目的进行修复，或者使其恢复到原有状态。对墙壁和地板的补救工作通常包含结构工作；地面上的补救工作包括增强其稳固性以及/或移除或密封污染。

remediation A technique used either to remove a contaminated source from the ground, modify contaminant pathways without necessarily removing it, or to destroy/ modify the contamination. Such techniques include: excavation and disposal, thermal treatment (desorption, incineration, and vitrification), bio-remediation, soil washing, capping, vapour extraction, and stabilization/solidification.
修复 从地面移除掉污染源、修缮而不是完全移除掉受污染路径，或者摧毁/修缮污染的一种技术。此类技术包括异地填埋、热处理（解吸附、焚化和玻璃化）、生物修复、土壤淋洗、加盖、气体排抽和稳定/凝固处理。

remote-entry system A system where the occupant of a building can permit or deny individuals access to the building by electronically activating the door.
远程门禁系统 业主通过电子控制允许或拒绝个人进入大楼的系统。

render (rendering) 1. A sand, cement, and lime or resin mix, which is applied to walls to provide a smooth or textured finish. Can be used to improve water resistance and durability of a wall. 2. Layer of sand, cement, and lime, or a resin mix that is applied to give a smooth or textured finish to a wall, and is called a rendering coat. 3. Base coat or first coat on three-coat plaster work.
1. 水泥砂浆 沙子、水泥以及石灰或胶的混合物，刷到墙上形成光滑的或有纹理的饰面，用于增强墙壁的防水性及耐用度。**2. 抹灰层** 刷到墙壁上的一层沙子、水泥以及石灰或胶的混合物，形成光滑或有纹理的饰面，称作抹灰层。**3. 底层抹灰** 三层抹灰泥中的底涂层或第一层。

renovation The improvement of the structural and aesthetic appearance of a building, or **item of furniture**, or equipment; could also include restoring the item to resemble its original appearance and quality.

翻修 改善建筑物或家具、设备的结构和（美学）外观；也包括比照原始外观和品质对其进行修复。

renovation plaster Lightweight plaster that is often used in damp conditions, it tends to be stronger and more adhesive than other plasters.
修补砂浆 通常用于潮湿环境的轻质灰泥，比其他灰泥更强韧，胶黏性更强。

repainting (maintenance painting, overpainting) Painting over existing paint, which has been prepared ready to receive the paint. All cracked and flaky paint is removed and the surface is sanded down ready to receive the new coats.
重涂 在已做过处理、可上漆的现有漆面上重新涂漆。已经移除了表面上的所有裂纹和剥落漆片，并打磨过，可重新上新漆。

repair To fix or restore something that is broken, to a good/working condition.
修理 将弄坏的某物修理或修复至良好/工作状态。

repointing Grinding or raking out of existing mortar and replacing with new mortar and pointing as required.
重新勾缝 对现有的砂浆进行碾磨或清理，替换成新的砂浆并按要求进行勾缝。

repressed brick A wire-cut brick; once it is cut, it is also pressed to create a more regular finish with true sides.
压缩砖 机制砖；切割后进行压制，使形成的饰面更规则平整。

re-sawing The sawing of large pieces of timber (flitches) into smaller pieces, during the conversion from large logs to usable sections of timber.
分锯 在将大块原木转变成可用木料时，把木料（料板）锯成更小的木块。

rescheduling The updating and reconfiguring of the project programme in order that the project deliverables are managed. The project programme will be updated with the actual progress, which once added to the project logic may affect the delivery of key activities and deliverables. If activities have slipped causing a delay to the project, extra resources may be added to reduce the time required to complete critical activities.
重新安排 更新或重新调整项目方案，以确保按时交工。按照实际进展对项目方案更新，一旦计入项目日志后会对主要活动的交付以及可交付物产生影响。如果因为漏掉某事项而拖延了项目进展，应额外投入更多资源来缩短完成主要事项所需的时间。

resealing trap A trap that allows air to pass through the water seal under negative pressure.
防虹吸存水弯 能让空气在负压下穿过被水密封处的存水弯。

reservoir 1. A lake or container for storing water. The lake can be either naturally or artificially formed. The latter is typically formed by constructing a *dam across a river. 2. Water being stored for human consumption or agricultural use.
1. 蓄水池 储存水的湖泊或容器。湖泊可为天然湖泊或人工湖泊。后者（人工湖泊）通常由横跨河流构筑的水坝形成。**2. 贮水** 供养人类或用于农业灌溉的贮藏水。

reshoring Back propping, adding additional support to formwork or other temporary works. Back propping is often used under concrete floors slabs while setting to ensure that the concrete has temporary support until it reaches full maturity.
加强支撑 背撑，对模板或其他临时工程增加附加支撑。背撑通常用于混凝土楼板下方，确保在混凝土完全硬化之前提供临时支撑。

residential Property built for people to live and sleep in.
住宅 为供人们生活休息而建的物业。

residential engineer (RE) The engineer who represents the client (client's engineer) and is based on the site so that he or she can observe and check the work. Normally the engineer will have an office on-site.
驻场工程师 代表业主驻扎于现场以便监视和检查工程的工程师（业主工程师）。通常现场应有工程师的办公室。

residual current device (RCD, residual current circuit-breaker, current-balance) A safety device that

disconnects an electric circuit if any current leakage is detected. Modern consumer units incorporate RCDs and they can also be plugged into a socket that powers appliances.
漏电断路器 如检测到任何电流泄漏即会切断电路的安全装置。现代的配电箱中带有漏电断路器，漏电断路器可插入通电插座内。

residual stress The level of *stress resultant on a material when there is no external force or load. This type of stress persists in a material even when there is no external force or load in the material. Residual stresses can be either *tensile or *compressive, and is an undesirable property as it can accelerate failure of a material.
残余应力 无外力或荷载时一种材料上的应力合力。无外力或荷载时材料中仍会存在残余应力。残余应力可为拉力或压力，是一种负面特性，可加速材料失效。

resilience How easily a material returns to its original shape after an *elastic deformation.
弹力 经过弹性形变后一种材料恢复至原来形状的容易程度。

resilient mounting *See* ANTI-VIBRATION MOUNTING.
柔性底座 参考“减震底座”词条。

resin A polymer material, usually a viscous fluid at ambient temperature.
树脂 有机聚合物，通常室温下或受热后为黏性流体。

resin flux A *resin and small amounts of organic activators in an organic *solvent.
树脂型助焊剂 内含树脂和少量有机活化剂的有机溶剂。

resistance Hindrance caused to flow, for example, the force water experiences as it flows through a pipe due to wall friction.
阻力 对流体产生的阻碍，如水流通过管道时由于壁面摩擦而受到的阻力。

resistance welding A type of *welding which utilizes an electric current.
电阻焊 运用电流的一种焊接方式。

resource smoothing (load smoothing, resource levelling) The logical rearranging of activities so that the resource use is more evenly distributed over the project or programme activities. When the project logic alone is taken into account, the resource usage will vary considerably, with resource peaks and gaps in resource use. By altering the timing of activities, without affecting the project logic, the resource use can be more evenly distributed. In programme management, where more than one project is running concurrently, the use of a single resource pool across all projects helps to ensure the most effective and continuous use of resources.
资源平滑（负荷平滑，资源平衡） 对各项活动进行合理安排，使资源可以更平均地分配于项目或计划活动中。单独考虑项目逻辑框架时资源使用的波动较大，使用时有资源的峰和谷。通过调整活动的时间节点而不影响到项目逻辑框架，可以更平均地分配资源。管理项目时如有多个项目同时进行，所有项目使用单一资源库可以保证最有效地持续使用资源。

respirator Personal protective equipment in the form of a mask, which one breathes through to filter out toxic airborne particles.
防毒面罩 一种面具形式的个人防护用品，其过滤掉有毒气体后，人可进行呼吸。

responsibility The contractual or project commitment to obligations and duties.
责任 契约性的或项目承诺的责任和义务。

responsibility matrix A table showing a list of resources, and the activities and duties that each resource has been allocated.
责任分配矩阵 显示资源清单以及资源分配的活动和责任的表格。

rest bend *See* DUCK FOOT BEND.
鸭脚弯头 参考“鸭脚弯头”词条。

restoration The repair, cleaning, and structural improvement of an item to bring it back to as close to its original condition as possible.
复原 对某一物品进行修理、清理和结构优化，尽量使其恢复至原有状态。

restraint To hold back or prevent movement, such as with a *fixed end connection.
拘束 抑制或阻止位移，例如带一个固定端连接。

restricted tendering Process that allows only those contractors that have met the prequalification criteria, are part of the framework agreement, or those that have been selected to submit bids for the contract.
限制投标 只允许符合预审标准的承包商（构成框架协议的一部分）或指定的承包商提交投标书的过程。

retaining wall Structural encasement (wall) constructed to hold back soil, water, or materials. Retaining walls are used to increase the amount of level usable building area, retaining soil at a higher level, and preventing it from encroaching into the building or another useable area.
挡土墙 建于阻挡土、水或其他材料的结构围护（墙）。挡土墙用于增加垂直方向的可用建筑面积，将土挡在高处防止其慢慢渗入建筑物或其他使用区域内。

retarder (retarding admixture, retarding additive) A chemical used to slow down a chemical reaction.
缓凝剂 用于减慢化学反应的化学物质。

retempering (knocking up) The addition of water and remixing of concrete that has started to stiffen. Retempering significantly reduces the strength of concrete.
重新混合 添加水重新搅拌已开始变硬的混凝土的过程。重新混合会大大降低混凝土的强度。

retention Sum of money held back by the client at each intermediate payment, often 5% of the *payment certificate value, with half of the sum being released back to the contractor at *practical completion, and the other half being paid to the contractor at the end of the defects liability period.
保留金 每一次中期付款业主都会保留一定金额，通常为支付证书价值的5%，实际竣工后一半的保留额将会支付给承包商，另一半在维修责任期结束后支付。

reticule (reticle) A grid of fine lines in the eyepiece of an optical instrument (telescope or microscope) to determine the scale or position of the object being viewed.
标度线 光学目镜（望远镜或显微镜）中的细线网格，用于确定被观察物体的大小或位置。

retrofit The strengthening, upgrading, or fitting of extra equipment to a building once the building is completed.
改造 完工后对建筑物进行巩固、升级或装配额外设备。

return 1. A short section of wall at right angles to the main wall. 2. A pipe that carries a fluid or gas back to an appliance.
1. 墙垛 与主墙成直角的一小段墙。**2. 回管** 将液体或气体运回至设备的管子。

R

return air Air that is extracted from a space and returned to the air-conditioning or ventilation system. *See also* RECIRCULATED AIR.
回风 从一个空间抽取并送回至空调或通风系统的空气。另请参考“再循环空气”词条。

return fill *See* BACKFILL.
回填 参考“回填”词条。

return latch (catch bolt, spring latch) The tongue of a door latch.
回弹插销 门闩的锁舌。

return pipe A pipe that carries a fluid or gas back to an item of equipment from which it originated. *See also* FLOW PIPE.
回管 将液体或气体送回原来设备的管道。

return wall A part of a wall with a change in direction.
转角墙 方向改变的一部分墙壁。

reuse of formwork Temporary forms that are used as moulds for concrete, then stripped, cleaned, and used again.

模板重复使用 用作混凝土模板的临时模板，剥离、清洁后可再使用。

reveal The area of wall at a jamb that is not covered by the frame.
门窗侧 门边梃处未被门框包住的区域。

reveal lining A finish applied to the inside of a window or door * reveal.
门窗侧板 门窗侧内面应用的饰面。

reveal pin An adjustable clamp that is used to secure a horizontal piece of scaffolding across a * reveal (a **reveal tie**). The reveal tie is used to tie scaffolding to the building.
洞口拉结夹件 一种可调节的夹子，用于固定穿过洞口的脚手架的节杆（洞口拉结）。洞口拉结用于将脚手架和建筑物拉结。

reverberation The repeated reflection of a sound from hard surfaces within a room.
混响 声音在房间硬壁上的重复反射。

reverberation time The time taken for a sound within a space to reduce to a millionth of its starting level (i. e. by 60 dB). The reverberation time of the space can be reduced by the insertion of sound absorption materials.
混响时间 空间中声音衰减至原来的百万分之一（如 60 dB）所需的时间。混响时间可通过安装吸音材料减少。

reversible lock (double-handed lock) A lock that can be installed on either jamb of a door.
双向锁（双把手门锁） 可安装于任意一边门边梃的锁。

reversible movement The expansion and contraction of materials, building elements, and building components.
可逆变形 材料、建筑组件及构件的膨胀与收缩。

revision The sequential notation attached to a document, drawing, or schedule showing whether the information is current or a previously updated document.
修正版本号 文件、图纸或清单附加的连续性注释，用于说明信息是最新的还是之前更新的。

revolving door (tambour) A cylindrical door, normally comprising three or four leaves, that rotates around a central axis. Used to reduce heat loss through the doors and reduce draughts. Usually located at the entrance to office buildings and supermarkets.
旋转门 圆柱状门，通常包括三个或四个可绕着中心轴旋转的门扇。用于减少热损失和气流。通常位于办公大楼或超市的入口。

R

rewirable fuse (semi-enclosed fuse) *See* FUSE.
可更换保险丝（半封闭式熔断器） 参考“保险丝”词条。

rework 1. To improve, make good, or put right something that has a defect. 2. The remixing of concrete that has started to hydrate. Once concrete has started to hydrate (set) it should not be disturbed, since remixing concrete breaks bonds that have been created during the initial hydration, which will not be restored. Water added to concrete once the hydration has commenced is not part of the chemical reaction and reduces the concrete's strength. Water that does not take part in the chemical reaction eventually evaporates leaving voids where water previously existed. *See* RETEMPERING.
1. 返工 改善、修复某物或将有缺陷的某物恢复正常。**2. 重新混合** 将已开始水化反应的混凝土重新搅拌。一旦混凝土开始水化（凝固）就应保持不动了，因为重新搅拌会破坏开始水化形成的粘结，是无法还原的。水化反应开始后，重新加的水并不构成化学反应的一部分，并且会减弱混凝土的强度。不参与化合反应的水分最终会蒸发掉，水原先占的位置会留下空隙。参考“重新混合”词条。

rib A longitudinal projection from the surface of a wall or ceiling; the projections can be structural beams or columns, but often they are ornamental projections adding a feature to the ceiling. The projections are often seen on the surface of vaulted ceilings at the intersection of arcs.
肋 墙面或天花面上的纵向突出物；可以是结构梁或构造柱，但通常是装饰性物体，为天花面增添某种造型。常见于拱顶天花的圆弧交叉处。

rib and block suspended floor A *floor made from strengthening beams and concrete, clay, or hollow blocks. The floor can be cast *in-situ, but mostly the floor is constructed from precast concrete beams with concrete blocks laid between them. Modern suspended floors of this nature are now using insulation board laid between the structural beams and topped off with structural *screed. Such floors are quick to assemble and construct.

密肋填充块楼板 由加固梁和混凝土、黏土或空心砖制成的楼板。楼板可在现场浇筑，但大部分是将混凝土砖放于预制混凝土梁之间制成的。现代的这种楼板将保温板放于结构梁之间，顶层浇有结构砂浆层。此类楼板易于组装和构筑。

RIBA (Royal Institute of British Architects) The professional awarding body of British architects. For a person to be called an architect in the UK they must be a member of RIBA.

皇家建筑师协会（RIBA） 英国建筑师的专业认定机构。在英国，如果想成为建筑师必须成为 RIBA 会员。

ribbed-sheet roofing Corrugated or profiled sheet roofing, the profiles add strength to the roofing material, reducing sag between the roofing spars or beams.

拱形屋顶 波纹状或压型钢板屋顶，压型钢板会增加屋顶材料的强度，减轻屋顶桁梁之间的下沉程度。

ribbon board (ledger) A horizontal beam that is fixed to a wall or onto timber studs to provide support for floor or ceiling beams to rest on.

条板（横木） 固定到墙壁或木龙骨上的水平梁，为地板或平顶梁提供支撑。

ribbon cable Flat electric cable that can be used underneath carpets or floor coverings without causing a large undulation, often used for telephone connections or distributing data.

带状电缆 可用于地毯或地板覆盖物下方且不会形成大的凸起的扁平状电缆，通常用作电话连接或数据分布线。

ribbon course A course of bricks or tiles of a different colour to the main brickwork to create an aesthetic feature. To enhance the appearance, the brickwork may project out from the main course and tiles may have a different shape or texture.

造型砖 为了造型美观而颜色与其他砌砖不一致的（瓷）砖层。为改善外观，砌砖可从主要的砌层中凸出来，瓷砖的形状或纹理也可以不同。

ribs and lagging A temporary support while tunnelling in unstable rock. Ribs are steel I-sections in the shape of the tunnel's profile. Timber lagging then spans between the ribs; however nowadays *shotcrete is sometimes used in place of timber.

拱梁和拱板 在不稳固的岩石中开挖隧道时所用的临时支护。拱梁为工字形剖面的钢材，形状与隧道截面一致。木板条横跨于拱梁之间；但现在有时会用喷浆混凝土替代木板条。

Richter scale A logarithmic scale from 1 to 10 used to measure the *magnitude of an *earthquake.

里氏震级 用于测量地震震级的对数级数，从 1 到 10。

ride A door that scrapes the floor as it is opened or closes is said to ride; it should be adjusted by trimming the underside of the door or slightly adjusting the hinges. Where possible the floor should be levelled.

蹭地 开关门时门蹭到地板称作蹭地；可通过修整门的底面或者微调铰链来矫正。必要时可整平地面。

rider shore A short shore that does not extend all of the way to the ground, but relies on a secondary longer shore for support. It is the upper part of a raking shore that is connected to the wall and back shore.

斜撑木 不是直接延伸到地面上而是支撑在一个次长横木上的短木。它是连接到墙壁和顶撑上的斜撑柱的上半部分。

ridge The top horizontal part of a roof structure where the two sloping parts of a roof meet.

屋脊 屋顶上两个斜屋面相交的屋顶结构的水平段。

ridge board (ridge piece) Horizontal timber board to which the diagonal rafters are secured.
屋脊板 用于固定斜撑椽条的水平木板。

ridge capping The covering over the top of the ridge of a roof. The ridge is often covered with capping tiles, ridge tiles, folded sheet metal such as lead or copper, or impermeable fabric such as bitumen or felt.
脊盖 屋脊顶部的覆盖物。屋脊上通常覆盖有脊盖瓦、屋脊瓦、折叠金属板（铅或铜）或者不渗水材料，如沥青或毛毡。

ridge course The uppermost course of tiles that is next to the ridge. Ridge tiles and the ridge course need to be fixed firmly in position as they are exposed, and are prone to damage in high winds if not fixed properly.
屋脊瓦层 紧挨屋脊的最上层瓦。由于屋脊瓦和屋脊瓦层暴露在外面，风速大时如果固定不到位容易被损坏，所以要将其牢牢固定。

ridge end A special tile used at the ends of the ridge to give a desired appearance.
屋脊端部 用于屋脊末端以获得预期外观的专用砖。

ridge stop A flashing to cover the intersection of a ridge that meets a wall. The flashing is sunk into the *raglet or *chase in the wall, then dressed over the wall and abutting ridge.
屋脊防水板 覆盖于屋脊与墙壁相交处的防水板。防水板嵌于墙上的披水槽或管槽内，附于墙壁与邻接的屋脊上。

ridge terminal An outlet for a soil and vent pipe or duct that exits through or close to the ridge.
屋脊端口 污水管、通气管或风道的排放孔，通过屋脊或从屋脊附近排出。

ridge tile A roof tile suitable for covering a ridge or a hip.
屋脊瓦 用于覆盖屋脊或斜脊的屋面瓦。

ridge vent A raised unit that allows air to escape from the roof space, ensuring that stagnant moist air does not build up in the roof.
屋脊通风器 凸起的组件，可使空气从屋顶消散，确保屋顶不会有污浊的湿空气堆积。

rift A gap or break—something that splits apart. A broad central steep-sided valley formed by two parallel faults is a **rift valley**. An area that has been subject to a large number of faults is a **rift zone**.
裂谷 缝隙或裂纹——某物裂开。两个平行裂层形成的宽阔峭壁中心峡谷称作大裂谷。有大量裂层的区域称作裂谷带。

R

rigid Firm or stiff.
坚硬的 坚实的或硬的。

rigid damp-proof course A course of bricks or slates that provides an impermeable barrier to rising moisture or damp.
硬质防水层 能形成阻挡上升的水汽或湿气的不透水屏障的砌砖层或石板层。

rigidity Resistance to deformation from shear forces. *See also* MODULUS OF RIGIDITY.
硬度 受到剪切力后的抗变形性。另请参考“刚性系数”词条。

rigid pavement A *pavement that is constructed with a concrete slab laid on the *sub-base rather than *asphalt or *tarmacadam; *See also* FLEXIBLE PAVEMENT.
刚性路面 底基层上铺砌混凝土板而不是沥青或柏油的路面。另请参考“柔性路面”词条。

rill 1. A small stream. 2. A groove or channel within the ground.
1. 小溪。**2. 细沟** 地面内的沟或槽。

rim latch A latch that is fixed to the surface of the door and the locating locking style is fixed on the surface of the doorframe.
外装锁闩 安装于门板上的锁闩，上闩的位置定位在门框上。

rim lock Lockable * rim latch.
外装锁 可上锁的外装锁闩。

ring A cylindrical strip of metal used to join two pipes together.
金属环 用于把两根管道连接到一起的圆柱形金属条。

ring beam (edge beam) 1. A horizontal beam placed at the top of a wall to tie the wall together. 2. A horizontal beam placed around the internal face of an enclosure to provide support for the floor. 3. A circular beam at the base of a dome.
1. 圈梁 放于墙壁顶部将墙拉结到一起的水平梁。**2. 边梁** 沿着围护墙内壁放置的水平梁，为楼板提供支撑。**3. 环梁** 圆屋顶底部的环梁。

ring main circuit (ring circuit, ring main) An electrical power circuit for lights and appliances that links a number of sockets or lighting points together. It is connected to the consumer unit at both ends.
环形主电路 灯具和电器的供电电路，把一些插座或照明点连到一起。两端都与供电箱连接。

ring-shanked nail (improved nail) Nail with raised rings on the shank that increase the hold and withdrawal resistance of the nail.
环纹钉 钉杆上有凸起螺纹的钉子，可增加钉子的抓紧力。

rip The sawing of wood in the same direction as the grain, parallel to the grain.
顺纹锯 沿着纹理方向或与纹理平行的方向锯木头。

riparian Relating to the riverbank.
河岸 河边的。

ripper A large pointed steel tool that is used to penetrate and spilt open firm ground and thinly bedded rock. The tool is fitted to the back of a bulldozer or tractor and is hydraulically forced into the ground. The bulldozer/tractor will have sufficient traction to pull the ripping tool through the ground.
松土器 尖的大型钢制工具，用于击穿并打散坚硬的土地及薄壁基岩。该工具安装于推土机或牵引机后面，通过液压打入地面。推土机或牵引机有足够的牵引力拉着松土器穿透土壤。

rip-rap Large stones placed on a riverbank or on the upstream side of a dam to prevent erosion from the wave movement.
海漫 在河堤上或水坝的上游填埋大块石头，防止水波运动导致的侵蚀。

ripsaw Saw with coarse chisel-cutting teeth with each tooth being set at alternate angles, enabling the saw to * rip wood.
粗齿锯 有粗糙的凿状切削齿的锯，相邻齿错角排列以确保能顺纹锯断木头。

R

rise The vertical distance from one * tread to the next on a * flight of * stairs. The rise of each step should be equal, and for private stairways in the UK, should be less than 220 mm high.
踏步高度 楼梯上从一个踏面到相邻踏面之间的竖直距离。梯面高度应相等。在英国，私人住宅楼梯的梯高度不得超过220mm。

riser 1. The vertical face at the back of a step. 2. A vertical service duct, pipe, or cable used to take water, gas, or electricity to an upper floor.
1. 踢面、踢板、竖板 台阶背面的竖立面。**2. 立管** 用于把水、气或电力送到上层的竖直给水管、风道或电缆。

rising damp Water that rises from the ground through masonry, timber, or concrete by capillary action causing dampness in the lower parts of a building. Rising damp leads to mould growth, creates damp conditions that lead to insect infestation, and causes the decay of adjoining materials. The use of damp-proof courses (DPCs) will largely alleviated the problem of rising damp.
潮气上返 水汽通过砖、木材或混凝土的毛细作用造成建筑物低层返潮。上返潮气会造成发霉，潮湿环境导致虫害，并使得相邻材料腐烂。使用防潮层（DPCs）可大幅改善潮气上返的状况。

rising main (rising pipe) The cold water service pipe that rises vertically from the floor of a building. It is

required to have an internal * stop valve and drain-off valve provided within it at the lowest point.
上行水管 从建筑物的地面处上行的冷水给水管。在管内的最低点处应有一个截止阀和排水阀。

rising spindle A threaded spindle that rises up and down when a tap or valve is turned on or off. *See also* NON-RISING SPINDLE.
上升型阀杆 打开或关闭龙头或阀门时上下升降的带螺纹的阀杆。另请参考“非上升型阀杆”词条。

risk assessment A procedure to assess the possible risk involved, the magnitude of the risk, and the likelihood of it occurring.
风险评估 评估可能出现的风险、风险等级以及风险发生可能性的过程。

risks The likelihood of an unwanted event occurring, such as variation, an accident, additional costs, delay, price fluctuations, etc. In a contract, items described as contractors' risk are those which the contractor takes on.
风险 发生意外事件的可能性，例如变更、意外、额外费用、延误、价格波动等。合同中作为承包商风险的条款是指承包商承担的风险。

riven slate Slate that is reduced to the desired thickness by splitting it rather than sawing it to size. Due to the separation of the layers, the slate has a natural mottled surface.
劈开面石板 被自然劈开而不是机切到指定厚度的石板。由于板层是自然劈开的，石板面有斑驳的纹路。

rivet A short metal rod with a head at one end. When the shaft of the rod is inserted through aligned holes on sheets of material, the projecting end is then flattened to form a head on the other end, thus holding the sheets of material together. Used in aircraft, steel frameworks to buildings, and in bridge construction.
铆钉 短的金属杆，一端有帽。铆钉穿透几块材料板中对齐的洞时，突起端会变平，在另一端形成钉帽，从而把材料板连到一起。常用于飞行器、建筑物钢构件和桥梁建设。

road A stabilized length of earth, typically containing a hard surface, on which vehicles can be driven.
道路 一段坚固土路，通常为硬路面，可通行车辆。

rock anchor (rock bolt) A steel cable or rod that is used to stabilize rock. A hole is initially drilled into the rock, after which the rod is inserted. An expansion device or * grouting is then used to secure the rod in place. Typically used when tunnelling through rock and on highway cuttings through rock to stabilize steep side slopes.
岩石锚杆 用于固定岩石的钢索或钢拉杆。先在岩石内钻一个洞，然后塞入钢拉杆。用膨胀件或灌浆将拉杆固定到位。通常用作在岩石中钻隧道或高速路上开凿岩石时加固陡坡上的岩石。

R

rocker shovel A mechanical loading device that is used where headroom is limited, such as in a tunnel or a mine. The shovel lifts spoil in a rocking action from the front of the machine and deposits it at the back of the machine.
铲斗式装岩机 机械装载设备，常用于净空高度有限的地方，如隧道或矿井。装岩机把机身前方的石块抓起然后转到机身后方把石块卸下来。

rock excavation Contractual term used to describe excavation that is too difficult to undertake without mechanical plant such as breakers or by means of blasting. When excavation can only be undertaken by such means, an extra cost for the dig can be claimed.
坚石开挖 用于说明没有机械设备（如碎石机）或者不爆破很难进行的开挖工程的合同术语。如果必须通过上述方式进行开挖，可以索要额外的开挖费用。

rock face Stones and rocks with the natural surface exposed and used for walling.
自然面 用于墙面的岩石的天然外露面。

rock pocket (US) Air bubbles or holes in rocks caused by loss of fines in the concrete aggregate, insufficient compaction and vibration, or using a poor concrete mix.
气泡（美国） 混凝土骨料中由于缺少细骨料、夯实振动不充分或者搅拌不佳导致的石块中生成的气泡或洞。

rock wool (rockwool) A type of insulation made from glass fibre.
岩棉 玻璃纤维制成的保温材料。

rod A measuring timber cut to a predetermined length with increments marked, used for setting out elements of construction that have consistent dimensions or spacing, for example, brickwork courses, windows openings, floor to ceiling heights, etc.
皮数杆 测量用的木杆，切割至预定长度并刻有度数，用于测定有连续长度和间隔的建筑构件，如砌砖层、窗户开口、地板至天花层高等。

rodding A method using a drain rod to unblock a pipe. A **rodding eye** is a small hatch at ground level that allows access to the drain for rodding. A **rodding point** is a section of pipe at 45°, which connects the drain to the surface of the ground so that it can be rodded.
通管 使用排水棒疏通管道的方法。通管孔是在地面上的一个小口，疏通棒可穿过这个小孔进行疏通。疏通点位于管道的45°截面处，其将排水管与地表连接起来。

roll 1. Sheeting material that has been wound around itself to form a compact cylindrical shape. 2. A semi-cylinder ornamental and partially functional shape used to round off the rim of sinks and baths; it can be totally aesthetic when used at the top of columns. 3. A joint used in sheet-metal roofing where the edges of two metal coverings are rapped over a semi-cylinder length of wood. The raised semi-cylinder joint, which runs down the slope of the roof preventing moisture entering the joint. 4. In Roman architecture, a semi-circular architectural feature found at the head of columns.
1. 卷 片状材料卷到一起形成紧实的圆柱状。**2. 旋涡状装饰** 半圆柱形的装饰性型材，围在水槽和浴缸边沿时有一定的实用功能；用在柱端时完全为了美观。**3. 绲边** 有两块金属屋板搭接在一段半圆形木头处的金属板屋顶所使用的接缝。凸起的半圆状接缝沿着屋面斜坡向下可防止潮气进入接缝内。**4. 柱顶装饰** 罗马建筑中柱子顶部的半圆柱状建筑造型。

roll-capped ridge tile A ridge tile with a roll feature added to the top of the tile.
波纹屋脊瓦 瓦面顶部呈波纹状的一种屋脊瓦。

roller An absorbent cylindrical painting tool, used to apply paint to a wall in long even strips.
辊子 有吸收性的圆柱状上漆工具，呈长条状，可用于将油漆均匀地涂到墙上。

roller door (rolling shutter) A vertically revolving door or shutter comprising a series of vertical slats that are rolled up over a horizontal roller located at the top of the opening when open and are rolled down when closed. *See also* ROLLING GRILLE.
卷帘门（卷帘闸） 垂直转动的门或百叶窗，有一组垂直板条，打开时沿着开口顶部的水平轴卷上去，关闭时滚下去。另请参考“格栅卷帘门”词条。

rolling grille A type of security grille that operates in the same way as a *roller door.
格栅卷帘门 操作方式与卷帘门相同的一种安全格栅。

Roman cement A type of cement obtained by burning septaria, found in clay deposits.
罗马水泥 发现于黏土矿中，由燃烧龟甲石得到的一种水泥。

Romanesque Medieval European architecture characterized by semi-circular and pointed arches. In England, Romanesque is better known as Norman architecture.
罗马式 中世纪的欧洲建筑，以半圆形和尖拱为特征。在英国，罗马式（建筑）通常被称作诺曼式建筑。

Roman tile A type of *single-lap roof tile. Both single and double versions are available. Single Roman tiles have one central channel while double Roman tiles have two channels.
罗马瓦 一种单搭接屋顶瓦。有单槽和双槽两种。单槽罗马瓦有一个中心槽，双槽罗马瓦有两个槽。

roof The uppermost part of a building that protects the structure from the rain and other external elements.
屋顶 建筑物顶部，保护建筑物免受雨水和其他外部因素影响。

roof abutment A part of the structure that is in direct contact with the roof.

屋面墙接 建筑物的一部分，与屋顶直接接触。

roof boarding Decking or boards placed adjacent to each other on the top of the roof enabling the roof to carry flexible sheeting material. The boarding provides a firm base on which insulation and impervious weather protection can sit.
屋面底板 屋面顶部相邻排列的盖板或板子，确保屋面能够承载柔性板材的重力。板子提供了一个坚实的基础，上方可安装保温和耐候材料。

roof cladding The external finish or material that is placed on the roof structure.
屋面层 屋顶结构上的外饰面或材料。

roof conductor A copper tape lightning conductor fixed to the highest part of the roof or parapets to provide the air terminal through which lightning can be safely conducted to the ground and earthed.
屋顶避雷针 安装在屋顶或女儿墙顶端的铜带避雷针，有一个尖端可将雷电安全地转移到地面并接地。

roof covering The external surface of the roof, normally the part that protects the structure from the rain and other external elements.
屋面覆盖层 屋顶的外表面，覆盖层通常用于保护结构免受雨水及其他外部因素的影响。

roof decking The covering of a roof using boarding or timber planks to provide a solid surface on which the flexible sheeting or liquid (asphalt) roof covering can be placed.
屋顶板 使用板子或木板以提供坚固表面的屋面覆盖层，可铺上柔性片材或流体状（沥青）覆盖面。

roof drain The outlet from a roof used to collect rainwater (* surface water) and discharge it through a * downpipe.
屋顶排水 用于收集雨水（表层水）并使其通过落水管排出的排水口。

roof drip A point of rainwater discharge at the eaves of the roof.
屋顶流水槽 屋檐上的排水点。

roofed-in stage The phase of construction where the roof is erected and provides weather protection to the workers and structure below.
屋顶吊装阶段 安装屋顶，为下方工人和结构提供保护的施工阶段。

roofer (roofing specialist) A tradesperson who constructs or repairs roofs.
屋面工 构筑或修葺屋顶的工匠。

R

roof extract unit (roof extractor) A extractor fan or series of fans that are located on the roof of a building.
屋顶排风机 建筑物屋顶上的排风扇或一组风机。

roof guard 1. A protective handrail or fence placed around the perimeter of a roof to prevent personnel falling from the roof. 2. A snowboard placed on the roof to prevent large volumes of snow sliding off the roof and injuring people below or damaging property.
1. 屋面护栏 沿屋顶周围的防护栏杆或栅栏，可防止人员从屋顶掉落。**2. 挡雪板** 放在屋顶上的挡雪板，防止大量的雪从屋顶滑落而伤到底下人员或造成财产损失。

roofing (roof covering) Uppermost part of the building used to provide a waterproof structure and protection from the weather. The weatherproof skin can be made from a membrane, slates, or sheeted material.
铺盖屋顶 用于提供防水和耐候保护的建筑物的顶部结构。耐候层可由薄膜、石板或板材制成。

roofing felt Weatherproofing sheet material that contains bitumen or asphalt.
屋面油毡 含有沥青或柏油的耐候片状材料。

roofing insulation Sheet, wool, or loose fill material that has good thermal resistivity, reducing heat flow through the roof space. Rigid insulation is often used on top of the roof rafters with the weatherproof covering being fixed over the insulation; wool and loose-fill insulation is often used between the ceiling joists providing a thermal barrier between the * roof space and main rooms within the house.

屋顶保温材料 具备良好保温性的薄板、毛料或松散的填充材料，可以减少通过屋顶的热量散失。屋面椽条的顶部通常使用硬质保温材料，防风雨层固定在保温层上方；吊顶龙骨之间通常使用毛料和松散填充料，在屋顶空间与屋子内的主要空间之间提供保温层。

roofing nail Any nail used for fixing roof tiles, battens, or roof coverings to the roof.
屋面钉 固定屋顶上的屋面瓦、板条或覆盖层所用的钉子。

roofing punch A steel rod with a sharp point used for making holes in sheeting or corrugated roofing material.
屋面冲孔机 有尖头的钢杆，用于在板材或瓦楞状屋顶材料上钻孔。

roofing screw Long threaded screw used for fixing corrugated and profiled roofing sheets.
瓦楞螺丝 用于固定瓦楞状和压型屋面板材的长螺纹螺丝。

rooflight (roof light, skylight) A window located in a roof.
天窗 屋顶上的窗子。

rooflight sheet Transparent sheet of roofing material used to allow natural light in through the roof.
天窗板 透明的板状屋面材料，可使自然光穿过屋顶。

roof space The attic or void contained within the roof structure, in many cases the attic is converted to form a room in the roof.
屋顶空间 屋顶结构内的阁楼或空间，通常会将阁楼改装成屋顶的一个房间。

roof tile (roofing tile) Clay, slate, plastic, or concrete unit in a flat form that is laid on a roof, overlapping and interlocking into other tiles to enable rainwater to be discharged from the roof.
屋面瓦 平铺在屋顶上的黏土、石板、塑料或混凝土单元，通过搭接和扣接其他瓦片使雨水排出屋面。

rooftop unit An air-conditioning, heating, or ventilation system that has been designed to be located on a roof.
屋顶空调机组 设计安装于屋顶的空调、供暖或通风系统。

roof truss Prefabricated timber roofing unit that utilizes triangulation to maintain its rigid form and transfer roof forces. The timbers are held together using pressed gangnail plates.
屋顶桁架 预制的木屋架组件，通过三角形特性保持其刚性结构稳定，并转移屋顶压力。通过冲压齿板把木材连到一起。

roof waterproofing (roof weatherproofing) The membrane or covering that prevents the ingress of rain, snow, and wind through the roof.
屋面防水层 防止雨水、雪和风穿透屋顶的膜或覆盖层。

room air conditioner A type of wall-mounted air-conditioning unit that is designed to condition only the room in which it is situated. Usually consists of a small refrigeration unit with an integral air circulation fan. Air is drawn from the room, cooled, and returned. They can be noisy, are generally not very efficient, and normally only provide comfort cooling, although filtration and heating can also be provided. Humidification is not available.
室内空调 一种壁装的空调装置，仅对安装房间进行空气调节。它通常包括一个小的制冷机组，带一体化空气流通风机，可从屋内抽取空气，将其冷却后再送回。这可能会产生噪音且不够高效。室内空调尽管也可以过滤和制暖，但通常仅提供舒适性制冷功能，也不可用于加湿。

room-heater A heating appliance that is designed to heat only the room in which it is situated.
室内暖气 设计用于对其安装房间供暖的采暖设备。

room-sealed appliance An appliance that obtains fresh air for combustion from outside and discharges the flue gases externally.
密闭式燃具 从室外获取燃烧用的新鲜空气并把烟道废气排到室外的装置。

root The section of a tenon where it widens out at the shoulders.
榫根 在榫肩部加宽的榫截面。

ropiness (ropy finish) 1. A paint surface where the brush marks have not flowed out 2. Rough or untidy finish.
1. 丝纹 刷痕未流平的漆面。**2. 粗糙或不整洁的饰面**。

rose 1. A ceiling rose is a decorative surround placed where the light fitting passes through the ceiling. 2. A * shower rose. 3. The decorative part of a door handle that the lever passes through.
1. 天花灯座 于灯具与吊顶联结处安装的装饰性环绕物。**2. 花洒头**。**3. 装饰盖** 门把手和手柄穿过处的装饰部分。

rostrum A plinth or stand for holding documents, used by presenters while making an address to an audience.
讲台 放文件的底座或讲坛，供演讲者演讲时使用。

rot Decay in wood, caused by moisture or mould growth.
腐烂 由于受潮或霉菌生长导致木头腐烂。

rotary cutting Method of spinning logs on a lathe and cutting with a blade so that a continuous veneer is cut from the log. Prior to cutting, the log is soaked in hot water making it easier to cut the veneer.
旋转切割 让圆木在机床上旋转用刀锋切割的方法，这样可从圆木上切割下连续的薄木片。切割前先用热水浸泡圆木，切割木片时会更容易。

rotary drilling A method of drilling where the drill bit moves in a rotational movement.
旋转钻 一种钻头以旋转状钻孔的方式。

rotary float A power float for levelling and smoothing concrete floors.
旋转整平机 磨平混凝土地面的电动抹平机。

rotary planer A powered planer with a rotating blade used to produce a flat smooth surface.
旋转刨床 打磨光滑表面所用的带一个旋转刀片的电动刨床。

rotary veneer A thin sheet of timber veneer produced from * rotary cutting.
旋切薄板 旋转切割形成的薄木板。

rotunda Any building that is circular on plan. The buildings are often covered by a dome.
圆形建筑 平面图上呈圆状的建筑。圆形建筑通常有圆屋顶。

R

rough arch An arch that is constructed of rectangular blocks, stone, or bricks that are not cut to the voussoir (wedge) shape.
粗制拱顶 由未切割成拱状（楔形）的矩形砌块、石块或砖筑成的拱顶。

rough ashlar Stone that is not cut neatly, having been roughly broken, hewn, or chiselled out of the quarry.
荒料 未进行整齐切割，只是从采石场中粗略地剥离、粗削或凿出来的石材。

rough axed Tiles or bricks that have been cut or shaped using an chisel-shaped hammer or axe.
斧剁 用凿状锤子或斧头切割或塑形的砖或瓦。

roughcast (slap dash, wet dash) Plaster or render that is sprayed or roughly thrown and levelled on the surface of a wall.
粗抹灰 喷到墙上或草草涂到墙上并大致磨平的内墙抹灰砂浆或外墙抹灰砂浆。

rough-cast glass Translucent, rolled sheet glass, one face of which has a slightly rippled texture.
毛玻璃 半透明的轧制玻璃板，一面有细微的波纹状纹理。

roughcast machine (plastering machine, roughcast applicator) Piece of plant used to quickly apply plaster to the surface of walls. The plaster is pumped through the machine and sprayed onto the wall. Once applied it is levelled and finished.
砂浆喷涂机 一种将砂浆快速抹到墙面上的设备。通过机器将砂浆泵出然后喷到墙壁上，涂好后就是齐平的饰面了。

rough cutting Cutting or breaking of bricks, tiles, or slates to approximate size with a *hammer, *bolster, or *trowel edge.
粗切削 用锤子、砖石凿或泥刀边将砖、瓦或石板切削至大致相等的尺寸。

rough floor A sub-floor constructed from timber to carry floorboards.
毛地面 木材筑成的底层地板，用于承接地板。

rough grounds Timber battens fixed to provide a level base upon which to nail plasterboard.
木龙骨 安装后能提供一个可以钉石膏板的水平基础的木制板条。

roughing filter A graded filter used in the pre-treatment process to separate out suspended solids.
粗滤池 分离出悬浮体的预处理中使用的级配滤池。

roughing in (roughing out) The first fix, first level, or general installation of work that will be unseen. The levelling out of plaster or the first fix in plumbing.
毛坯 后期不可见的工程初安装、第一层的或通用的安装。内墙砂浆的抹平或给排水的初安装。

roughness The level to which something is not smooth.
粗糙 某物的表面不光滑的情况。

rough string 1. A beam that runs under the centre of a flight of stairs, shaped to the same profile as the stairs, also known as the carriage of a stair. 2. The sides of a stair, otherwise known as strings, that are shaped to the profile of the stairs rather than having parallel edges. The stair *string is the diagonal beam that is used to carry the *treads and *risers of a stairway.
1. 楼梯踏步梁 一段楼梯中心下方的一段梁，形状与楼梯横剖面一致，也称作踏步梁。**2. 楼梯侧面** 也称作楼梯斜梁，形状与楼梯截面一致而不是平行边。楼梯斜梁是承接楼梯踏面和踢面的斜梁。

rough-terrain fork-lift Truck designed for lifting pallets and packages, and carrying them over building sites and other rough terrain. The trucks are either two- or four-wheel drive with tractor tyres.
越野叉车 用于抬升托板和组件并送至建筑现场及其他地形不平区域的叉车，可以是两轮或四轮驱动并带农用轮胎。

rough work Trade work that will not be seen so does not have to be aesthetically pleasing. In brickwork this would be *common brickwork that is not seen once the finishes are applied; in plasterwork it is the first coat of plaster used to dub out or build up the plaster to provide a level surface to work on; in joinery it is the carcassing or the first cut of timber that provides the approximate shape before finishing; and in plumbing it is the first fix work that will be hidden behind finishes.
粗工 由于不会被看见所以无须注意美观的工作。砌砖工程中指涂抹饰面后不会被看到的砌砖；抹灰工作中指用于刮平表面或填满以形成平整的工作面的第一层灰浆；细木工中是指木构架或粗切削的木材，在终饰前提供一个大致形状；给排水中指初安装固定工作，会被隐藏于饰面之后。

roundel A circular-shaped object.
圆形饰物 圆形物体。

rout To cut out a groove in wood, stone, or metal.
刻槽 在木头、石头或金属上刻出槽。

router Tool for cutting grooves in timber using a rotating cutter.
刨槽机 使用旋转切割刀在木材上割槽的工具。

rowlock cavity wall (all-rowlock wall) A wall constructed in *rat-trap bond.
竖砌砖空心墙（立砌墙） 以空斗墙砌合构筑的墙壁。

rowlock course A course of bricks constructed with gaps between using *rat-trap bond.
竖砌砖砌层 砖与砖之间的空隙采用空斗墙砌合的砌砖层。

Royal Institute of British Architects *See* RIBA.

R

英国皇家建筑师协会 参考“RIBA”词条。

rubbed arch A flat arch formed of bricks that are cut and shaped (rubbed) to provide angles that extend from the *key stone and hold the arch in place. The mortar joints are very thin to ensure good contact and compression between the bricks.
磨面拱顶 砖砌平拱，经过切削和塑形（打磨）形成一个从拱顶石开始延伸的角度并使拱顶处于合适位置。灰缝很细，可确保砖与砖之间有良好的黏结和压缩。

rubbed bricks (rubbers) Bricks made of a soft clay, often without a frog, that are easily cut, ground, or sanded into a desired shape.
磨砖 软黏土制成的砖，通常没有砖面凹槽，容易切割、打磨或研磨至预期形状。

rubbed finish Concrete rubbed down with an abrasive carborundum stone to create a very smooth finish.
磨光面 用粗糙的金刚砂打磨混凝土，形成非常光滑的表面。

rubbed joint Masonry joints simply rubbed smooth with a sack to create a flat mortar joint.
磨平接缝 用粗布打磨石材接缝，形成平整光滑的接缝。

rubber buffer (rubber silencer) A resilient stopper used to prevent doors slamming open on walls and other building furniture.
橡胶门挡 弹性阻挡器，用于防止门打开时撞击到墙壁及其他建筑家具。

rubber cork tile A resilient tile made from a mixture of latex, rubber, and cork.
橡胶软木砖 乳胶、橡胶和软木混合制成的弹性砖。

rubber flooring Resilient latex flooring that provides a smooth non-slip surface.
橡胶地板 弹性乳胶地板，表面光滑防滑。

rubber ring *See* JOINT RING.
橡胶圈 参考“垫圈”词条。

rubbing down Removing rough spots, high spots, and other undulations by sanding the surface ready to receive paint, wax, or varnish.
抛磨 通过打磨掉粗糙凸起的点以及其他不平整的地方，使其适于涂抹涂料、蜡或清漆。

rubbing stone An abrasive block used for grinding concrete, stone, and other ceramic materials.
磨石 用于打磨混凝土、石材和陶瓷材料的磨块。

R

rubbish (builder's rubbish, building rubbish) Debris created from the works on-site including: off cuts, trimmings of timber, brick, rubble from demolition, excess material, waste, and building debris from site-based work.
垃圾（建筑垃圾） 现场施工形成的杂物，包括切削料，木头、砖块的边角料，拆除的碎石，多余的材料，废料以及基于现场工作的建筑碎块。

rubbish pulley Hoist used for lowering building debris to the ground.
垃圾吊车 把建筑垃圾从高处放到地面的吊机。

rubble (rubble ashlar, rubble masonry) A type of stone with an irregular shape utilized in walls and foundations of buildings.
碎石 形状不规则的一种石材，用于建筑物的墙壁和地基处。

rubble wall Wall made from randomly cut stone.
乱毛石墙 用随意切割的石头筑成的墙。

rule off To smooth off plaster or concrete by drawing a smooth float that has two handles for drawing the float over the surface. Excess material is drawn off and the surface is levelled.
抹平 在砂浆或混凝土表面上拉动有两个手柄的光面抹子将表面磨平。多余材料将会被抹掉，表面会变平整。

run 1. A straight or horizontal dimension. 2. A stretch of cables or pipes. 3. A length of bricks laid end to end, also known as a course of bricks. 4. A defect in paintwork where excessive paint has been allowed to drip, forming a thick bead of paint that has flowed down the surface.
1. 水平或横向长度。**2. 一段电缆或管道**。**3. 砌砖层** 一层端接垒砌的砖，也称作一皮砖。**4. 流挂** 油漆的缺陷，因为涂得过多开始向下滴，在表面上形成一层厚厚的珠状油漆。

runby pit *See* LIFT PIT.
电梯底坑 参考“电梯底坑”词条。

run moulding (horsed mould) A fibrous moulding or template that is pulled through the wet plaster to provide the desired shape.
石膏线倒模 将线形模具或模板拉过湿石膏，使其形成预期形状。

runner The rail that guides a drawer, sliding door, or sliding window.
导轨 引导抽屉、推拉门或左右推拉窗开关的轨道。

running bond Walling made entirely of stretcher bricks—bricks laid end to end with the longest face exposed. The joint of the upper course is positioned over the centre of the brick below, ensuring that the joints on different courses are staggered.
顺砖砌合 完全由顺砖砌合的墙壁——端对端铺设砖块，最长边外露。上层砖的接缝位于下层砖的中心处，以确保不同砖层是错缝的。

running mould *See* RUN MOULDING.
石膏线倒模 参考“石膏线倒模”词条。

running tile Patterned or coloured tiles that run a continuous length providing a long border.
瓷砖线 连续铺设一段有图案或不同颜色的瓷砖，形成一道长边。

runway *See* AIRFIELD.
跑道 参考“飞机场”词条。

rupture Something that has broken suddenly.
断裂 某物突然破裂。

rusticated stone Masonry block that has a rough textured (unfinished) face and wide joints to highlight the edge of each block.
砌石 毛面（无饰面）砌石块，接缝很宽以突出每个石块的边缘。

rustication 1. Regularly cut *ashlar stone with blocks of stone that are left angular and protruding while other blocks have smooth faces. The smooth- and rough-faced blocks are randomly distributed to give a rough undulating face. 2. The irregular feature created by stonework that is chiselled, drilled or hammered to provide a rough texture.
1. 粗琢石工 规则切割的方琢石，一些石块切割成有棱角突出的，另一些石块为光面。将光面和毛面石块随意堆砌形成粗糙的有起伏的砌面。**2. 毛石面** 经过凿、钻或锤击的石块形成的不规则的粗糙纹理。

rustic joint A recessed joint set in rough-faced *ashlar stone.
粗接缝 毛面石材中的凹槽灰缝。

rutting The sideway deformation of a road surface to form a sunken track due to the passage of vehicles, exaggerated by heavy vehicles and hot weather.
车辙 道路表面由于车辆通行而造成的变形压痕，有重型车辆和天气炎热的情况下该现象会加剧。

R-value *See* THERMAL RESISTANCE.
R 值 参考“热阻值”词条。

S

Sabin The unit of sound absorption. One Sabin is equivalent to one square metre of 100% absorbing material.
赛宾 吸声量的计算单位。一赛宾等于一平方米的百分百吸收的表面。

sacrificial anode A metal plate, which is electrically connected, used in cathodic protection (a method to control corrosion) of piping or other equipment. The metal plate must be more corrodible than the material to which it is attached.
牺牲阳极 电连接时为管道或其他设备提供阴极保护的金属板。金属板须比与其连接的材料更易被腐蚀。

sacrificial coating (sacrificial layer) 1. A thin coating of metal that oxidizes to protect the main metal from corrosion. A good example of this is galvanized steel (steel coated with zinc). The sacrificial layer oxidizes and protects the main metal from further corrosion. The coating protects the underlying metal so long as a the layer remains in place. 2. A layer of paint, which can easily be covered or removed if defaced with graffiti.
1. 牺牲涂层（牺牲层） 自身氧化以保护主要金属不受腐蚀的一层金属涂层。镀锌钢就是一个很好的例子（镀有锌涂层的钢材）。牺牲层氧化后保护主要金属免受进一步腐蚀。只要有牺牲涂层，底层金属就能得到保护。**2. 易清除涂层** 被涂鸦污损时可被轻易覆盖掉或者除掉的一层涂层。

sacrificial form Permanent formwork that provides a surface for the concrete to be cast on or cast against and remains as an integral part of the structure. Galvanized profiled steel is commonly used as floor formwork and remains in place after the concrete is poured.
一次性消耗模件 为混凝土浇筑提供成型外表的永久性模板，浇筑之后成为结构的一部分。镀锌型钢被广泛用作地板模板，混凝土浇筑后仍留在原位。

saddle back coping A coping stone (or shaped tile) that is cut or formed with a raised central point (apex) and overhangs the sides of the walls which it covers. The coping is shaped so that water runs away from the apex and is cast off away from the face of the wall due to the overhang.
鞍形压顶 切割或加工成的压顶石（或型材瓦），中心点处隆起，悬在其覆盖的墙边上方。压顶经过塑形，如此雨水可从尖顶上流开，且由于其呈悬挑状所以雨水不会流到墙面上。

saddle clip A pipe fitting that is U-shaped with fixing lugs on it, so that it can be positioned over a pipe and clamped to a bracket or fitted to a wall.
管卡 带固定耳的 U 型管件，可固定在管道上并夹紧支架或固定到墙上。

saddle-jib crane A *tower crane with a central mast and a horizontal jib that extends out from the central mast and is used as the lifting arm. A counter jib shorter than the main lifting arm carries a counter-weight to counter-balance objects that the crane lifts. Large tower cranes normally have the cab mounted on top of the mast, giving the operator a central position with the best possible views of the site around which objects are being lifted and moved.
平臂式塔吊 塔吊，带有一塔身和一伸展开作起重用的平臂。平衡臂短于起重臂，并带有一配重以平衡吊起的物件。大型塔吊通常在塔身顶部带有控制室，使驾驶员在中心位置，以对吊起和移动物件周围有最好的视野。

saddle piece A piece of plumber's metalwork that is shaped or fitted over something. The saddle may be clamped to a pipe to provide a seal, or positioned over a ridge to provide weatherproofing.
鞍形件 水管工用于卡在或固定在某物上的金属件。可将鞍形件卡住管子形成密封或安在屋脊上提供防风雨保护。

saddle scaffold Scaffolding that is built over the apex of the roof.
坡屋面脚手架 安装在屋顶处的脚手架。

saddle stone The stone used at the apex of a *gable end of a roof.
山墙压顶石 用于屋面的山墙顶处的石头。

saddle trusses Small roof trusses that sit on top of the main roof trusses to enable the roof to change direction or angle. These are used where a roof turns a corner and sit on the top of existing trusses at right angles to the ones below, ensuring the structure remains stable.
鞍形桁架 安装于屋顶主桁架上方的小屋顶架，使屋顶能够改变方向或角度。用于屋顶拐弯处，与下方桁架成直角安装于现有桁架顶部，以确保结构稳定。

safe 1. Not in danger, secure. 2. A lockable box used to keep valuables secure.
1. 安全 不危险，有保障。**2. 保险柜** 用于保存贵重物品的上锁箱。

safety arch An arch that is used to add strength to the structure and relieve the load placed on smaller arches or lintels below; also known as a relieving arch.
安全拱 用于增加结构强度并将承受的荷载卸到下方小拱或过梁的拱，也称作卸压拱。

safety factor *See* FACTOR OF SAFETY.
安全系数 参考“安全系数”词条。

safety glass (toughened glass) A reinforced laminated glass that is designed to break into small blunt fragments when broken. *See also* TEMPERED GLASS.
安全玻璃（钢化玻璃） 强化型夹层玻璃，打破时会破成不伤人的小块钝角碎片。另请参考“钢化玻璃”词条。

safety helmet *See* HARD HAT.
安全帽 参考“安全帽”词条。

safety ladder (jacket ladder) A steel ladder, permanently fixed to the building with protective hoops that allows a person to pass through and offers additional protection from falling off the ladder.
安全爬梯 永久安装到建筑物上的带防护笼的钢梯，允许一人通行并为其提供额外保护，防止其从梯子上掉落。

safety net A net placed over open areas within partially constructed buildings to prevent workers, materials, and equipment falling through or over the side of buildings.
安全网 在部分建成的建筑物内空旷区域设立的网，用于预防工人、材料和设备从建筑物上方或边上掉落。

safety officer Person appointed by a company to ensure that the statutory provisions of the Health and Safety Act are followed. Under the Health and Safety Act, employers are required to appoint a Health and Safety Officer to meet company legal obligations under Health and Safety Law.
安全员 由一家公司任命的确保严格遵守《健康和安全法》的人员。按照健康和安全法，业主需要雇用一名健康和安全员确保履行《健康和安全法》规定的公司法定义务。

safety railing A *guard rail to prevent people falling from heights.
安全栏杆 防止人员从高空坠落的护栏。

safety signs Signs required on construction sites to warn of dangers and risks. These make employees aware of their legal duties, inform them of protection needed, and ensure correct guidance and management of the site is maintained to ensure the health, safety, and welfare of employees. The legal requirements are available from the *Health and Safety Executive.
安全标志 施工现场需要设立的标志，用于警示危险和风险。这些标志可使工人意识到他们的法定职责与必要防护，并确保现场保持正确的指示和管理，确保雇员的健康、安全和福利。法定要求可从健康和安全部门获取。

safety valve A valve that opens to ensure the pressure (or flow) is not exceeded (or lost) from a system.

S

安全阀 打开后用于确保压力（或流量）不超过系统压力或流量的阀门。

safe working load A load that a structure or foundation can carry safely—normally factored (divided by a *factor of safety) to accommodate a margin of error to ensure failure does not occur.
安全工作荷载 一个结构或基础可安全承载的负载量，通常除以系数（安全系数）调整误差范围以确保不会发生故障。

sag (sagging) Bending downwards. A sagging bending moment is a *bending moment with tensile along the bottom and compression along the top; *See also* HOG.
下垂 向下弯曲。顶部受压底部受拉时的下垂弯矩称作弯曲力矩；另请参考“拱起”词条。

saltation The jumping/bouncing motion of sand and soil particles being carried/ transported by the wind or moving water.
跃移 风、水流携带或运输的沙子、土壤颗粒的跳动/跳跃。

sample A part of a whole, a selection of a few articles where there are many, taken for inspection, reference, or determining, and predicting the nature of the larger sample.
样本 整体的一部分，从许多的量中选一些用于检查、参考或测定，以及预测更大样本的属性。

sample panel (reference panel) A section of brickwork, cladding, roofing, or flooring that is constructed in the required format. It is inspected and improved by the architect or client's representative, and is used as a benchmark of the standard expected throughout the construction.
样本面板（参考样板） 按照指定样式构筑的一段砌砖、挂墙板、屋顶或地板。经建筑师或业主代表的检查和改进后，用作整个施工过程中的预期基准。

sand A *granular material of fine grains (between 0.06 and 2 mm) made up of rocks fragments or mineral particles, usually of quartz origin. It has zero cohesion when dry and an apparent cohesion when wet due to the surface tension of the water within the pore structure.
沙子 由岩石材料或矿物微粒组成的微细颗粒材料（0.06—2 mm），通常来源于石英。干态时没有任何黏着力，湿态下由于细孔结构内水的表面张力而具有表观黏聚力。

sand bedding A 40 – 50mm layer of sand, placed on an excavated area to provide a level surface on which the floor structure can be built. Where the sand is used to create a separation from the substrata below, it may also be called sand blinding.
铺沙层 铺设于开挖区域的一层40—50mm厚的沙子，形成一个平面，可在上方建造楼板结构。沙子用于与下方底基层隔离时，也可称作砂石层。

sand blasting The propelling of sand or other siliceous material from a gun or hose, fired under high pressure and used to clean, remove the surface or contaminant, or reshape a surface. There are many different types of blasting, including hydro-blasting, micro-blasting, bead-blasting, and wet abrasive blasting, which are all variations of the sand-blasting process.
喷砂 高压作用下从枪或软管中喷射沙子或其他硅酸材料，用于清理或移除表面或污染物，或者对表面塑形。有许多不同的喷砂方式，包括高压水喷砂、微喷砂、喷丸、湿喷砂，以上都是喷砂处理的变形。

sand boil *See* LIQUEFACTION.
沙沸 参考“液化”词条。

sand box A box or tube that can be filled with sand to act as a support for a scaffold *prop.
沙箱 装满沙子的箱子或管子，用作脚手架支柱的支撑。

sand catcher A device to collect and monitor the amount of sand in suspension in a flowing river or wind stream.
捕砂器 用于收集并监控水流或气流中悬浮的沙子总量的装置。

sand drain (sandwich drain) A drainage channel filled with *sand.
排水砂井 填满沙子的排水渠道。

sander (sanding machine) Mechanical equipment that is fitted with *sandpaper and rotates on a belt, moves backwards and forwards, or moves in a circular motion, and is used to smooth and clean surfaces. Sanders can be belt, disc, or orbital, and may be hand-held or floor-standing bench tools.
磨砂机（砂光机） 装配有砂纸的机械设备，围绕传动带旋转并向前或向后移动，或者绕圈，用于磨平和清理表面。磨砂机可以是随皮带、圆盘或轨道运转的，可为手持或落地台式工具。

sand filter A filter made from sand. The sand is placed within a container through which water runs, the impurities and salts are caught within the sand, and the water which passes through is removed of a high proportion of impurities.
砂滤器 沙子制成的过滤器。将沙子装在有水流过的容器内，杂质和盐就会留在沙子内，从容器中流出的水已经被过滤掉了大部分杂质。

sanding The action or process of rubbing sandpaper over a surface to remove high spots, paint, irregularities, or to clean the surface. The abrasive material removes a thin layer of the surface material.
打磨 用砂纸研磨表面的过程，用于移除凸出的点、油漆、不规则处或者清理表面。研磨材料会研磨掉薄薄的一层表面材料。

sandpaper Strong reinforced paper or cloth coated with abrasive silicone material or metals that are graded from fine to coarse. The abrasive paper is used for removing a thin layer of surface material from an object—levelling, shaping, or cleaning. Different abrasive materials include sand, emery, aluminium oxide, silicon carbide, aluminium zirconia, chromium oxide, and ceramic aluminium oxide. The different materials are used for different purposes.
砂纸 韧性加强的纸或布，涂有目数由细到粗的有机硅或金属磨料。磨砂纸用于去除物体表面的薄膜层，（以）整平、塑形或清洁。不同的磨料包括砂、金刚砂、氧化铝、碳化硅、铝氧化锆、氧化铬和陶瓷氧化铝。不同的材料有不同的用途。

sand pile A heavy weight is dropped on silty ground to form an indentation, which is filled with sand. The process is then repeated and a column of compacted sand is formed within the ground.
砂桩 将重物砸到淤泥地上，形成凹洞然后填充砂子。重复该过程就会在土中形成密实沙柱。

sandstone A sedimentary rock with sand grains mostly of quartz, but also includes calcite, clay, or other minerals.
砂岩 包含沙粒（主要为石英）的沉积岩，但是还包含方解石、黏土和其他矿物材料。

sand streak A line of exposed aggregate in the surface of concrete, caused by bleeding.
流沙水纹 由于泛水而在混凝土表面形成的骨料外露水纹。

sandwich beam An H- or I-beam, made of different composite materials, improving the strength-to-weight ratio, therefore increasing the performance.
夹层梁 由不同的复合材料制成的 H 或 I 字梁，其强度重量比增加，性能从而增强。

sand wick A type of *sand drain consisting of a small diameter vertical column. It is typically used under embankments constructed on soft compressible clays, to reduce the drainage path the excess pore pressure has to travel.
袋装砂井 一种排水砂井，包括一个小径垂直柱。通常建在柔软可压缩泥土上的路堤下，用于缩短超孔隙压力需通过的排水路径。

sanitary accommodation A room that contains a sanitary *appliance.
卫生间 包含卫生设备的房间。

sanitary appliance (sanitary fitting) An appliance that receives either soil or wastewater. For instance a *WC, a *bath, a *basin, or a *sink.
卫生设备 接收污水或废水的设备。如厕所、浴盆、洗脸盆或水槽。

sanitary connector *See* PAN CONNECTOR.
坐便器接头 参考“坐便器接头”词条。

sanitary cove 1. A curved tile used as the junction between the wall and the floor to prevent water penetration. 2. A metal * cove installed on a stair between the * riser and the * tread.
1. 阴角瓷砖 用在墙壁和地板相交处，防止水渗透的圆弧瓷砖。**2. 楼梯收边条** 安装在楼梯踢板和踏板之间的弧形金属条。

sanitary fitting *See* SANITARY APPLIANCE.
卫生用具 参考“卫生设备”词条。

sanitary pipework (sanitary plumbing) All of the pipework associated with a sanitary appliance.
卫生管道 与卫生设备相连的所有管道。

sarking A flexible * roofing felt or layer of roof boards that are placed under the * roof covering.
衬垫层 铺在屋面覆盖层下方的一层柔性屋面油毡或屋面板。

sash (window sash) The frame that supports the glazing in a window frame.
窗框 为窗户框架内的玻璃提供支撑的框架。

sash balance A coiled spring integrated into the jambs of a * sash window. Used instead of a cord, pulleys, and weights to operate the window.
推拉窗弹簧 安装进推拉窗的窗框侧壁的螺旋弹簧。代替绳子、滑轮和配重来开关窗户。

sash bar (sash astragal) The vertical glazing bar in a * sash window.
窗框条 推拉窗中的垂直玻璃格条。

sash clamp A small clamp used to hold components together when they are being glued.
窗框夹具 胶粘时把组件夹到一起的小夹子。

sash cord (sash line) A cord attached to each * jamb of a * sash window that supports the * sash weights.
推拉窗绳 安在推拉窗的每个窗边梃上的绳子，支承推拉窗配重。

sash fastener (sash lock) A fastener attached to the sash of a window used to keep the window closed.
窗锁 安装在窗框上的紧固件，用于保持窗户关闭。

sash fillister A * plane used to cut rebates for glazing.
窗框槽刨 用于切割和安装玻璃用的槽口的刨子。

sash haunch A joint traditionally used in the framing of a sash window, where a full tenon would weaken the frame. The size and position of the tenon is altered to maintain strength.
窗框榫卯 推拉窗框架中使用的一种传统接合件，但完全凸榫可能会使窗框不牢固。凸榫的大小和位置是可变的，以维持强度。

sash lift A handle on the inside of a * sash window used to open or close it.
推拉窗拉手 推拉窗内侧上的拉手，用于打开或关闭窗户。

sash pulley A pulley set within the jamb of a * sash window. The * sash cord passes over the pulley and supports the * sash weights.
推拉窗滑轮 推拉窗边梃内的滑轮组。推拉窗绳穿过滑轮支撑推拉窗配重。

sash ribbon A thin band of steel used to support the * sash balance.
推拉窗滑条 用于支撑推拉窗弹簧的薄钢带。

sash stop A moulding attached to the inside face of a jamb that keeps the sliding portion of the * sash window in place.
推拉窗卡子 安装在窗边梃内面的限位模制件，使推拉窗的滑动部分保持在恰当位置。

sash weight A weight set within the jamb of a * sash window and hung from the * sash cord. Used to counterbalance the weight of the sliding portion of a * sash window.
推拉窗配重 推拉窗边梃内的配重，吊在推拉窗绳上。用于平衡推拉窗的滑动部分的重量。

sash window (double-hung sash window, hanging sash, vertical sliding window) A window comprising

two sashes (window casements) that opens by sliding one of the sashes vertically. *See also* SLIDING WINDOW.
推拉窗(双悬上下推拉窗、上下推拉窗) 包含两个框格的窗户，通过垂直拉动其中一个框格可打开窗户。另请参考“左右推拉窗”词条。

satin paint A type of paint with a delicate shiny finish, known for its durability and stain resistance.
缎光漆 饰面精细发亮的一种油漆，以其耐用性和抗污染性著称。

saturate Something that is completely full, for instance, the pore spaces in soil with water.
浸透 完全被填充的某物，例如土壤空隙的空间充满了水。

saturated A material, liquid, or container that has absorbed or contains the maximum quantity of another fluid or material that it can hold.
饱和的 材料、液体或容器已经最大限量吸收或装满另一种液体或材料。

saturated solution A liquid that has dissolved or suspended the maximum amount of solid material without separation or segregation.
饱和溶液 液体中已最大限量溶解或悬浮固体物质而不会离析或分层。

saturation coefficient The volume of water that can be absorbed by a brick Different coefficients and absorption tests exist for the UK and brick.
饱和系数 砖能吸收的水量。英国和美国有不同的饱和系数和吸收测试。

saturation chroma The intensity of a hue when describing a colour compared with a neutral grey that is of a similar lightness.
色度 在相近光度下与中性灰相比用于描述颜色的色彩浓度。

saturation point The point at which a given sample of air contains the maximum amount of water vapour that it can contain.
饱和点 给定的空气样本能容纳其最大限量水蒸气时的点。

saucer dome A *domelight.
碟形穹顶 半球形天窗。

saw bench A table or platform with a circular saw mounted within it. The blade of the saw protrudes up into the table allowing wood to be passed over the table and cut by the protruding blade.
锯台 内装有圆锯的桌子或平台。锯的刀片向上从桌面凸出来，在木头经过锯台时突出的刀片可进行切割。

saw doctor A professional skilled at resetting and sharpening the teeth of *saws.
修锯工 精于复位和磨尖锯齿的专业人员。

sawfile A small file, circular in section, measuring approximately 1.5 mm in diameter. The blade is held in place by a jig or hacksaw frame. The file is used for cutting intricate shapes in timber, plastic, tile, clay, metals, and other hard materials.
磨锯锉 小锉刀，截面为圆形，直径约为1.5 mm。刀片由夹具或钢锯架固定。该锉刀用于在木材、塑料、瓷砖、黏土、金属和其他硬质材料中切出复杂形状。

saw horse Bench, trestle, or stool with four legs providing a long thin platform approximately 1 m long and 100 mm wide that can be used alone for working small lengths of timber, or more commonly used in pairs to support pieces of timber that are to be cut or worked.
锯木架 四条腿的工作台、搁架或凳子，提供一个大约1 m长、100 mm宽的长条工作平台，单独一个可用于加工小段木材，通常两个并用来支撑需要被切割或加工的木条。

sawing The cutting of timber using a saw.
锯 用锯子锯木板。

sawn damp course The cutting of a masonry course after the wall is constructed to allow the insertion of a

S

* damp-proof course. The cut is made into the horizontal joint in small sections and the damp-proof course is inserted. Slight settlement of the wall above between 1 – 3 mm is to be expected even when the joint is refilled with mortar or expanding grout. In remedial and refurbishment work it is more common to have the masonry drilled and injected to create a dense impermeable damp-proof course.
后嵌防潮层 墙砌好后切开砖层嵌入防潮层。在水平接缝进行小截面的切割然后嵌入防潮层。尽管接缝中又重新灌有灰浆或膨胀灰泥，但墙体一般还是会产生 1—3 mm 的轻微沉降。修补和翻新工作中更常见的是把砖墙钻开并嵌入形成一层密实不渗水的防潮层。

sawn stone Stone that has been cut to shape for * walling, * cladding, and * ashlar stonework.
石材毛板 锯开可筑成墙壁、外墙挂板和琢石工程的石材。

sawn veneer A veneer cut with a thin saw as opposed to being sliced or rotary-cut.
锯切单板 用薄锯锯开而非刨片或旋切的单板。

saw set (swage) A tool used to give the correct set for saw teeth.
整锯器 用于调整锯齿的工具。

sawtooth roof (northlight roof) A roof comprising a series of triangular sections that are placed in parallel with one another. The result is a roof with a profile similar to that found on the teeth of a saw. The steeper face of each triangular section is usually glazed and orientated to the north.
锯齿形屋顶 包含许多平行布置的三角形面的屋顶。屋顶的截面与锯齿的形状类似。每个三角形截面的陡边通常为玻璃且面向北方。

scabbler A power tool used for breaking up and removing the upper surface of concrete. The rough surface then provides a mechanical key for new concrete to adhere to. The tool scabbles the surface, removing any weak finish that may exist on the surface. This ensures new concrete cast against the surface can form a good mechanical fix.
打毛机 用于打破并移除混凝土上表层物质的电动工具。粗糙的表面提供新浇混凝土的机械黏结。打毛机打磨表面，移除掉表面上所有不够牢固的饰面。这样可确保新浇筑的混凝土能与表面形成良好的机械黏结。

scabbling (hacking) The removing of a weak concrete surface and roughening up of the surface in order to provide a strong rough surface and key for new concrete. When concrete is poured and vibrated, excess water carrying the aggregate fines and cement comes to the surface of the concrete, creating a weak film called laitance. When new concrete is cast against existing concrete, the laitance should be removed. The surface of the concrete can be hacked by a * scabbler, a pointed tool, brush hammer, or comb. The scabbler breaks off the surface creating a rough undulating surface, exposing the concrete that is strong and not affected by the laitance. The rough concrete provides a good mechanical key for the adjoining concrete; this is not as strong as the chemical bond that would be formed between two elements of newly laid concrete, but does provide a strong bond.
打毛（凿毛） 移除不牢固的混凝土面，把表面磨成粗糙状形成粗糙的底层灰泥面以方便黏结新浇混凝土。混凝土浇灌并振动时，带有细骨料和水泥的过剩水分会流到混凝土表面上，形成薄膜，称作浮浆。当新的混凝土浇筑在现有混凝土上时，应先将浮浆去掉。可用打毛机、尖头工具、凿石锤或铁耙对混凝土表面进行打毛。打毛机打碎混凝土面形成粗糙不平整的表面，将牢固的未受浮浆影响的混凝土裸露出来。粗糙的混凝土面为新浇筑混凝土提供了具备良好的机械黏结能力的灰泥底层；尽管可能不如两个新浇筑的混凝土成分之间的化学胶合那样牢固，但也是非常牢固的接合。

scaffold (scaffolding) A temporary structure that is erected to enable work to be carried out at an elevated level. Scaffoldings are used to provide access at heights so that walls can be erected, and ceilings, floors, and roofs can be constructed, decorated, and repaired. The framework provides access for a worker, a place to temporarily store materials, and if providing temporary support, it can act as * falsework. Various scaffolding systems exist, including quick assembly patent systems, tower scaffolds, independent scaffolds, platform scaffolds such as birdcage assemblies, and putlog scaffolds that are partially supported by the building. Scaffolding was traditionally made from timber tied or bolted together. The tubes, poles, and standards of today are more

commonly aluminium and steel, though in some countries, bamboo remains a material that is commonly used.
脚手架 立起来的临时性结构，用于确保高处工作的实施。脚手架用于提供高处通道，以保证墙壁、天花、楼板和屋顶的构筑、建设、装饰和修理的进行。该框架为工人提供临时通道，且提供临时储存材料的地方；且提供临时支撑时可作为膺架使用。脚手架系统有很多种，包括快拆脚手架、塔式脚手架、双排脚手架、平台式脚手架（如满堂脚手架）以及部分由建筑物支撑的单排脚手架。传统的脚手架通常由木材系到或栓接到一起制成。现在的管子、杆和支柱通常为铝和钢，但是在某些国家，最常见的脚手架材料是竹子。

scaffolder (scaffold hand) A competent and skilled worker capable of safely erecting, changing, and dismantling scaffolding. Scaffolders should be properly trained and must be competent in the erection of scaffolding. The Construction Industry Training Board provides training courses in the erection and checking of scaffolding.
架子工 能够安全地架立、改变并拆卸脚手架的技术娴熟的工人。脚手架工人应接受过适当的培训并能绝对胜任脚手架的架立。建筑业培训委员会会提供有关于脚手架的安装检查方面的培训课程。

scaffold fittings The fittings used to tie and link *scaffolding poles together. Scaffolding fittings include: couplings, clips, base-plates, adaptors, and toe boards.
脚手架配件 用于把脚手架的杆件连系到一起的配件。脚手架配件包括扣件、夹子、底座、顶托底盘和挡脚板。

scaffolding boards (scaffold planks) Planks and boards of graded timber that are used to provide the main platforms and working decks of the scaffold. The scaffold planks are a minimum of 40 mm thick, 3.70 m long, and 230 mm wide.
脚手板（踏板） 用于提供脚手架主要平台和工作板的分级的木板和板子。脚手板至少应为 40 mm 厚、3.70 m 长、230 mm 宽。

scaffolding clip A fitting used to link two scaffolding tubes together; this could be a vertical standard, linked to a horizontal ledger or brace.
脚手架夹子 把两个脚手架管子连到一起的配件；可以是连接到水平横杆或斜撑上的竖直立杆。

scaffolding coupler A fitting used to link two tubes such as standards and ledgers, or standards and putlogs together.
脚手架扣件 将两根管子，如立杆和横杆，或立杆和连墙杆连到一起的配件。

scaffold ties *Scaffolding tubes used to link and hold an independent scaffold to a building. The ties often pass through or are braced into *window or *door openings, providing a secure connection between the building and scaffolding. The ties resist lateral wind loads and prevent the scaffolding pulling away from the building.
脚手架穿墙杆 用于连接并将双排脚手架固定到建筑物上的脚手架管子。穿墙杆通常穿过或撑到窗户和门的开口处，在建筑物和脚手架之间形成稳固连接。穿墙杆能够抵抗横向风载并防止脚手架从建筑物上脱离。

scaffold tube Any of the tubes used in a *scaffold, such as a putlog, brace, ledger, standard, transom, etc.
脚手架管子 脚手架中用到的各式管子，如连墙杆、斜撑、横杆、立杆、小横杆等。

scale The ratio between a model or drawing, and the actual building or structure that it represents; for example, 1 : 20 means that the actual building is twenty times greater in size than the model or drawing.
比例 一个模型或图纸与其代表的实际建筑物或结构之间的比例；如 1: 20 指实际建筑是模型或图纸上的尺寸的二十倍。

scallop A carved or casted ornament, fixed to a building or structure, that resembles a scallop shell.
扇贝造型 安装在建筑物或结构上，雕刻或浇铸的类似扇形贝壳的装饰物。

scalloping A decorative pattern made by a row of half circles that join or link together.
扇形装饰 由一排连在一起的半圆形组成的装饰性图案。

S

scantling Wood of non-standard size but between 47 – 100 mm thick and 50 – 125 mm wide.
小块木材 非标准大小但厚度介于47—100 mm，宽度介于50—125 mm的木材。

scarf (scarf joint, scarfed joint) A joint made in laminated timber, formed by bevelling each strip of veneer at an angle of 1 : 6 to 1 : 24 and gluing the angled edges together.
斜面指接 胶合木材中的接缝，通过在每块木板上以1: 6到1: 24的角度切斜边并将斜边胶粘到一起而形成。

schedule 1. A document listing the doors, ironmongery, finishes, reinforcement, windows, or other fittings, finishes, or components within a building. The document lists the component or finish, quantity, special instructions or quality, and the place to be used. 2. A *programme of events, showing the time, duration, and sequence of activities; the chart may also provide notes on the associated resources.
1. 材料清单 列明了门、五金器件、饰面材料、钢筋、窗户或建筑物内的其他配件、饰面或组件的文件。该清单列明了组件或饰面、数量、特殊说明或质量以及使用位置。**2. 时间表** 展示了时间、期限以及事件顺序的表；该表也可就相关资源给出说明。

schedule of defects (schedule of outstanding works, snagging list, punch list) A list of defects identified or works still outstanding that need to be completed before the certificate of practical completion is awarded.
缺陷清单 认定的缺陷或未完工工程的清单，需要在颁发实际竣工证书之前完成。

schedule of prices (schedule of rates) List of prices placed against materials, plant, and labour, used where the nature or quantity of work is unknown. Where the exact quantities of work are unknown, a list of prices, plant, and labour rates can be made available, and the work quantified and costed once complete.
价格清单 对材料、设备、人工的报价清单，工程性质及工程量未知时需要用到。如不知晓具体的工程量，应制定价格清单、设备及人工费率表，完工后应制定工程量及费用清单。

scission The process of cutting is termed scission.
切断 切割的过程。

scissors lift A Mobile Elevated Working Platform (MEWP) with a concertina lifting system. The scissor support structure contracts and extends to raise the working platform. The lifts come in different sizes with large lifts capable of lifting multiple workers and substantial materials. The lifts come with different wheels and tyres making them suitable for navigating in different conditions, e. g. suitable for smooth flat indoor, or external site terrain.
剪刀式升降机 有风琴式升降系统的升降工作平台（MEWP）。剪刀支撑系统通过收缩和伸展来升高工作平台。升降机有不同的大小，大型升降机能够起重多个工人和大量物料。升降机的轮子和轮胎也有不同的类型，可适用于不同条件下的操作，如平整室内或室外的现场地形。

scope of works A description of the extent and nature of work to be undertaken. The specification gives details of the works, limitations, and any work that is excluded.
工程范围 对要承担工程的范围和性质的描述。规范对工作细节、限制以及不包括在内的任何工作进行详细说明。

score (scoring and snap) The cutting of a rebate into sheet material, such as copper, tile, or steel, and manipulating until the material separates. The groove is cut into the material by running a sharp cutting tool along a straight edge and bending it along the groove until it breaks.
刻痕切割 在铜板、瓷砖或钢板等板材中切割出槽口，重复切割直到断开。使用锋利的切削工具沿着直边进行切槽，再沿着槽口将其掰断。

scraper 1. Essentially a large blade, spade, or bucket on wheels, which can be either self-propelled or towed. The front edge of the bucket has a cutting edge, which is lowered (200 mm) into the ground and then dragged through the ground thus loading the bucket. Used in *earthworks. 2. A bladed tool for removing paint from surfaces. The paint may be softened using heat or chemicals, and the blade is run across the surface forcing itself between the layers of paint and the substrate, causing the paint to peel away.

1. 铲运机 主要指下面装有轮子的破土刀、铲子或挖斗，可以自拖动或者被拖动。挖斗的前缘有切削刃，可挖入地下 200mm 然后在地上拖动直到挖斗被装满。用于土方工程中。**2. 刮片** 把油漆从表面移除掉的刀片工具。通过加热或化学物质软化油漆，然后用刀片刮表面，迫使油漆层与底层分开从而使油漆剥落。

scratch coat The coat of plaster with an etched or keyed surface that provides a good bond before the finishing coat is applied. The scratch coat helps level undulations in the substrate and provides a good key for subsequent layers of plaster.
底涂层 拉毛或毛面的灰泥涂层，在涂抹饰面层之前提供良好的黏合。底涂层帮助磨平了底基层的凹凸不平，为随后的灰泥层提供良好的接合。

scratcher comb A plasterer's float with nails projecting from the surface, also called a devil float, used to etch the surface of plaster to help provide a key.
拉毛钉耙 抹灰工用的带有从表面突出的钉子的抹子，也称作带钉抹子，用于在灰泥层表面拉毛以提供接合。

scratching The making of a shallow mark on a surface.
划痕 在表面上留下的浅层痕迹。

scratch tool Plasterer's tool used for etching decorative features and enrichments into the surface of the plaster.
刮刀 抹灰工用的工具，用于在灰泥层表面上刻入装饰性图案和装饰。

screed The top layer poured over insulation or concrete to which floor finishes can be applied.
砂浆找平层 浇灌在保温层或混凝土上的顶层砂浆，在上边可进行地面终饰。

screed board (rule, straight edge) A straight board used for levelling the surface of screed. The board is longer than the screed dabs, guides, or rails so that it can level between them.
刮杆（刮尺） 用于刮平砂浆找平层表面的直板。刮杆长于砂浆板、导板或滑道，方便抹平砂浆。

screeding The laying of concrete screed.
刮平砂浆 刮平混凝土砂浆层。

screed pump Mechanical pump used for distributing concrete screed to the place where it is needed.
砂浆泵 用于把混凝土砂浆配送到需要之处的机械泵。

screed rail Boards, rails, or tubes that are set level and parallel in bays so that screed can be levelled between them using a *screed board.
砂浆滑道 并排平置于凹处的板子、滑道或管子，方便用刮杆抹平砂浆层。

screed to falls A screed which is laid to a gradient falling towards a drain or gulley, allowing water to be discharged from the surface.
找坡 到排水口或集水口逐渐向下倾斜铺设的砂浆层，方便从表面排水。

S

screen 1. A frame with a mesh grid built into it. 2. A framed mesh sieve used for separating fine and coarse aggregate. 3. A door with a mesh panel that allows natural ventilation, and prevents insects and other animals entering the dwelling. 4. A framed mesh ventilation hole used to allow ventilation into roof eaves, under suspended floors, and above windows.
1. 纱网 嵌有编织网的框架。**2. 筛框** 用于将精细骨料和粗骨料分离开来的框筛。**3. 通风纱门** 带网眼板的门，允许自然通风，防止昆虫及其他动物进入住所。**4. 通风网** 带框网状通风孔，可使自然风吹进屋檐内、架空地板下方以及窗户上方。

screens passage A passage at the end of a medieval hall, where screens conceal the main entrance.
屏风通道 中世纪末的大厅通道里，屏风挡住的主要通道。

screw A fixing with a helical threaded shank and pointed end that self-taps into wood as it turns and penetrates the timber. Screws can have a single flat groove in the head for a flat-bladed screw driver to be inserted or can have a crossgroove to receive a Phillips head screw driver.

螺钉 有外螺纹杆身和尖头的固定件，可旋转自动穿入木板。螺丝钉钉头上有一字形槽口供一字形螺丝刀插入或者十字形槽口供十字形螺丝刀插入。

screw cap 1. Screw fixing for lightbulbs. 2. A decorative dome head that can be screwed or fixed on top of the screw once the screw has been inserted into the timber.
1. 螺丝灯头 灯泡的螺丝固定。**2. 螺帽** 螺钉嵌入木板后可拧在或固定在螺丝钉顶部的装饰性圆头。

screw cup A decorative brass or steel cup with a central hole, positioned under a screw so that the underside of the countersunk screw fits into the cup.
螺帽盖 安装于螺钉下的装饰性铜盖或钢盖，中间有孔，可使沉头螺丝钉的底面与螺帽盖充分贴合。

screw-down valve A valve that is operated using a threaded spindle.
螺旋阀 由螺旋杆控制的阀门。

screwdriver A hand-held manual or powered tool used for turning and driving *screws into materials. The screwdriver can have a flat-bladed or Phillips cross-head for inserting into the screws.
螺丝刀 手持式手动或电动工具，用于将螺丝钉旋转钻入材料中。螺丝刀有一字形或十字形刀头，可嵌入螺丝钉内。

screwdriver bit A replacement tip for a screwdriver. Some screwdrivers come with detachable heads, the shafts of such screwdrivers having a hexagon recess into which the hexagonal bits can be inserted. Different sized and shaped bits can be inserted.
螺丝刀头 螺丝刀的替换刀头。有些螺丝刀有可拆卸式刀头，其刀轴上有六角形凹槽，可嵌入六角形刀头。可嵌入不同大小和形状的刀头。

screwed and glued joint A joint which is rebated, glued, and screwed to make a strong reliable joint.
螺钉加胶接合 开口、胶粘然后再用螺钉拧紧而形成的牢固接合。

screw eye A loop fixed to the head of a screw, used as an anchor or fixing for ropes or hooks once the screw is in position.
羊眼螺丝 螺丝头上的固定环，螺丝钉到位后用于锚固或固定绳索或挂钩。

screw gun A power tool used to drive screws, also called an electric screwdriver.
螺丝枪 用于拧紧螺丝的电动工具，也称作电动螺丝刀。

screw plug Two discs with a rubber seal between them used for sealing drains. The discs are engineered so that they can be threaded inside a threaded pipe, creating a sealed access to the drain.
螺旋堵头 中间有橡胶密封的两个圆片，用于密封排水口。圆片是经过特殊设计的，可以旋塞进螺纹管道内，用于密封排水管通道。

S

scribe To cut a light groove into the surface with a scriber. The scriber is used to mark the surface of the material.
画线 用画线器在表面划一道浅槽。画线器用于在材料表面做标记。

scribed joint (cut and fit) Joint made by offering the item to be cut or joined against the surface to which it is to be fitted. For an exact fit the joint position is scribed while in place and then removed for cutting. When returned to the desired position the material should provide a snug fit.
拼缝接合 把需要沿着表面切割或接合的物体接到安装处形成的接合。为了精确匹配，到位后对结合位置画线标记，之后移走并进行切割。重新放到安装位置后能够紧密贴合。

scriber A sharp pointed tool for accurately marking the surface of materials.
画线器 在材料表面上做精确标记的尖头工具。

scrim Woven cloth used to provide a reinforcement and help bridge the joints of plasterboards. The material is used with tapered edge plaster board, plaster, or adhesive, and placed across the joint. The joint is then skimmed with fine plaster. The joint tape comes in rolls and can be used for flush and corner joints. The scrim is normally cloth or hessian, but can be cotton or metal—all being used to provide reinforcement to plaster.

嵌缝带 用于加固并连接石膏板接缝的织布。与坡口边石膏板、石膏或胶粘剂一起使用，跨过接缝铺设，然后刮腻子。成卷的嵌缝带可用于平缝和角缝的接合。嵌缝带通常为布或麻布，也可以是棉花或金属，都用于加固石膏。

scrimming The applying of *scrim over joints.
贴绷带 在接缝上铺设嵌缝带。

scroll A decorative spiral providing a feature to the end of a shelf or under a *handrail in a stairway or hall.
卷状装饰 为架子末端、楼梯扶手或大厅内改良造型的螺旋状装饰物。

scroll saw An electrically powered *fretsaw.
曲线锯 电动线锯。

scrubber An apparatus for removing impurities in gas.
涤气器 清除气体中杂质的设备。

Scruton number An aerodynamic number—the higher the number the less the oscillations will be; used to assess the pedestrian excitation of bridges.
斯柯顿数 空气动力值——数值越高，振幅越小；用于评估桥梁的人行激振。

scullery A room set aside where cooking utensils and dishes were washed.
洗碗房 留出的一间用于洗厨房用具的房间。

scum A fine thin layer of dirt on the surface of a liquid.
浮渣 液体表面上薄薄的一层尘土。

scumble glaze（scumble stain） Transparent paints used for modifying the colour, texture, or appearance of the finish.
丙烯酸清漆 用于修饰饰面的颜色、纹理或外观的透明漆。

scupper An opening in a *parapet wall that enables the rainwater collected in a *box gutter to be discharged to a *downpipe.
排水口 女儿墙上的一个开口，确保雨水能集中到方形排水槽并排入落水管中。

scutch A bricklayer's hammer with a chiselled cross-peen on both ends of the head, used for fair and rough cutting.
扁凿石锤 瓦工用的锤子，锤头的两端都有凿状十字尖头，用于修整或粗糙切削。

seal 1. A material that is used between two components to prevent leakage. 2. The process of filling a gap or space with material to prevent leakage. 3. The water found in a *trap that prevents foul air from the drainage system entering the building.
1. 密封剂 两个组件之间使用的材料，预防泄漏。**2. 密封** 用防漏材料填满缝隙或空间的过程。**3. 液封** 存水弯中的水，防止排水系统中的污浊空气进入建筑物。

sealants A substance (such as an *epoxy resin) used to seal something. A material used in small gaps or openings to prevent the passage of water, air, and noise, i. e. silicone.
密封剂 用于密封的物质（如环氧树脂）。用在小缝隙或开口中防渗水、空气和噪音的材料，如硅树脂。

sealed system A type of central heating system where the expansion of the water occurs within a sealed expansion *vessel.
封闭式水箱系统 水在密闭的膨胀水箱内发生膨胀的一种集中供暖系统。

sealed unit A multiple-glazed unit where the panes of glass have been sealed together in a factory.
中空玻璃 多层玻璃的装置，其中玻璃片在工厂时已被密封在一起。

sealer A *sealer coat that prevents the passage of water and air or gases as well as protects the base onto which it is applied.
密封层 防止水、空气或（其他）气体进入，同时保护其下方底基层的密封层。

S

sealer coat Finish used to seal the surface. Sealer coats can be used on concrete, paint, wood, plaster, and other materials. They are often used to seal porous surfaces, but may be used to protect or enhance a finish.
密封涂层 密封表面用的涂层。密封涂层可应用于混凝土、油漆、木材、石膏和其他材料。通常用于密封多孔材料，也可用于保护或改良饰面。

sealing compound A viscous liquid that is used to fill and seal joints. The material fixes to the sides of the joint and bridges and fills the gap, preventing the ingress of water, debris, dust, and wind. Sealants can be two-part epoxy or elastomeric products or one-part mastic sealants.
填缝剂 用于填充和密封接缝的黏性液体。把填缝剂涂在接缝两边，连起来填满缝隙，预防水、杂物、灰尘和风渗入。密封剂可以是双组分环氧树脂或橡胶制品，或者单组分胶粘密封剂。

sealing ring A elastomeric ring used to provide a water- and air-tight seal around adjoining pipes. The ring is fitted over the sprocket or end of one pipe so that the collar of the coupling pipe or coupling ring can slide over the pipe with the sealing ring trapped and compressed between the pipes.
密封圈 用于在邻接管道周围提供水密和气密密封的弹性环。密封圈固定于链轮或管端，所以连接管的管口或连接圈能够滑过管子，而密封圈被挤压在管子之间。

sealing strip Compressible foam or elastomeric strip applied to movement joints.
密封条 应用于变形缝的可压缩泡沫或弹性长片。

seam (welted seam, welt, lock joint) A joint between two sheets of roof metal where the joining edges of the sheet metal are placed back to back, up standing, and then folded over to create a seal. A single fold is called a single lock and double fold a double lock. Seams that slope down the roof may be left upstanding, whereas horizontal seams are folded down the roof to help improve the seal and allow rain water to flow over the seam.
咬合接缝（咬边、锁边） 两片金属屋顶之间的接缝，其中金属板的缝合边是背对背、向上立起然后折叠形成密封。单折叠称作单锁边，双折叠称作双锁边。沿屋顶向下倾斜的接缝可保持直立，而水平接缝应折向屋顶下方以增强密封，并且使雨水从接缝上流过。

seamer Electrically powered pliers used for rolling and folding sheet-metal roof seams.
锁边机 卷动和折叠金属板屋顶接缝用的电动钳。

seamless tube A pipe made by extruding metal from a solid.
无缝管 从固体中挤压金属制成的管道。

seam roll A *hollow roll where the edges of sheet metal are placed against each other in a standing seam and simply rolled together to create a water resistant roof or wall joint.
咬边卷 金属板的空心卷边互相叠在一起形成直立咬边，用作防水屋面或墙接缝。

S

seasoning The drying out of timber by natural drying or air drying.
风干 木材因自然晾干或风干而失去水分。

seating The surface that an object, equipment, or structure rests on.
座面 可在上方放置物体、设备或结构的表面。

sea wall A structure to divide the sea from the land and designed to prevent flooding and erosion.
海堤 将大海与陆地分隔开来用于防止洪灾和（海水）冲刷的结构。

secant piling A below-ground retaining wall that has been constructed from overlapping cast in-situ board piles. Once the piles have been cast the ground can be excavated from one side.
挡土灌注桩 由重叠的现场灌注桩组成的地下挡土墙。桩浇筑好后，可从一面对地面进行挖掘。

SECED Society of Earthquake and Civil Engineering Dynamics.
地震与土木工程动力学协会

second 1. A unit of time that is equal to a 60th of a minute. 2. A unit of angular measurement equal to 3600th of a degree.
1. 秒 时间单位，等于1/60分。 **2. 秒** 角度单位，等于1/3600度。

secondary beam A beam that spans and is supported by other beams.
次梁 跨越其他梁并由其他梁支撑的横梁。

secondary element A non-essential part of a building's structure; includes fittings and finishes.
次级构件 建筑结构的非主要部分；包括配件和饰面。

secondary fixing A fixing that is housed by another bracket, frame, or bolt. The primary fitting is fixed directly to the building or structure in order to carry the secondary unit. Bolts may be fitted to carry the chains to hold air handling units and lights. Primary frames and cables are fitted to carry the frame of a suspended ceiling.
次级固定 由另一个支架、框架或螺栓承载的固定件。主要配件是直接固定到建筑物或结构上的，用于承载次级设备。螺栓固定后可用于承载空气处理机组和灯具的链条。主要框架和电缆固定后用于承载吊顶框架。

secondary glazing A type of double glazing that is formed by fixing a second sheet of glass or translucent material to the inside face of single *glazing.
内层玻璃 一种双层玻璃，在单层玻璃内面再安装一层玻璃或透明材料。

secondary lining The finishing coat over primary lining in tunnelling.
二次衬砌 挖隧道时底层衬砌上的饰面层。

secondary reinforcement *Reinforcement perpendicular to the main reinforcement, often used to prevent cracking.
分布筋 与主要钢筋垂直的配筋，通常用于防止出现裂纹。

secondary sash glazing (applied sash glazing) A single glazed *sash that has been secondary glazed.
双层玻璃窗 加有第二层玻璃的单层玻璃窗。

secondary treatment The middle stage in a *water treatment, typically involves rapid or slow filters in drinking water treatment plants and *trickling filters or activated sludge processes in *wastewater treatment plants.
二级处理 水处理的中级阶段，通常包括水处理厂中的快滤或慢滤，以及污水处理厂中的滴滤池或活性污泥处理法。

second fixing Fittings that are fitted after plastering and at the same time as the internal finishes are applied. Such fixings include electrical sockets, switches, light fittings, water taps, and other sanitary units.
二次装修配件安装 抹灰后涂抹内饰面同时安装的配件。此类配件包括插座、开关、灯具、水龙头以及其他卫生器具。

second foot *See* CUSEC.
立方英尺每秒 参考“立方英尺/秒”词条。

seconds Bricks with slight defects that can still be used in common brickwork where the face of the brick is hidden.
次品砖 有小瑕疵的砖，但仍能用于砖面隐藏起来的普通砌砖工程中。

second seasoning The loose framing of high-quality joinery items allowing the units to adjust to the humidity levels when in place. Units can also be stacked in the new environment to allow movement before fitting. If any excessive movement or defects occur before fitting the units can easily be replaced.
二次干燥 高品质的精细木工制品的活动框架，能使（可调节）组件内外湿度一致。也可于组装前将组件堆在一个新的环境，供其产生位移。如果组装前出现了任何的错位或缺陷，可轻易进行替换。

secret dovetail A dovetail joint that is formed within and hidden in a mitre joint.
全隐燕尾榫 接合隐藏于斜角接榫内部的燕尾榫。

secret fixing Any method of fixing and jointing that cannot be seen and is hidden from view once complete. Examples would include *secret dovetail and *secret nailing.
暗装 完成后不可见的任何固定和连接方法。暗装的例子包括全隐燕尾榫和暗钉。

secret gutter A gutter that is hidden or largely concealed from view. The gutter is almost hidden by the tiles or roof covering. Due to the tiles covering the gutter, hidden gutters are susceptible to blockages.
排水暗管 隐藏式或大部分隐藏式沟槽。此类沟槽大部分隐藏于屋面瓦或屋顶覆盖面之下。由于屋面瓦覆盖住了沟槽，所以暗装沟槽容易堵塞。

secret nailing (blind nail, edge nail) Nailing that is hidden from view. The nails penetrate into the edge of the floor boarding at an angle, the abutting boards hide the nails. Nails can also be hidden in tongue and groove and rebated timber. The nails are inserted into the groove in tongue and groove timber or into the rebate of lapped timber.
暗钉 看不见的钉子。钉子以一定角度钉入地板边内，邻接的木板可将钉子隐藏。钉子也可隐藏于企口缝以及企口木板内。钉子可被嵌入企口板的沟槽内或搭接木板的槽口内。

secret screwing Screw heads are inserted into keyholes with slot-shaped fixings that are used to hold panelling and trim in place.
螺丝暗装 螺钉头嵌入带槽形固定件的锁眼内，保证镶板和镶边稳固。

secret wedging Wedges fixed into the end of a tenon that are hidden from view. The wedges are inserted and loosely tapped into saw cuts in the end of a stub *tenon before being inserted into the mortise. As the tenon is inserted into the mortise the pressure of the mortise bed (also known as the blind end of the mortise) forces the wedges into the tenon. The tenon then opens out against the sides of the mortise.
暗楔加固 嵌入暗榫头末端的楔子。嵌进榫眼之前先将楔子松散地嵌入短榫末端的锯槽中。当凸榫嵌入榫眼时，榫床（也称作不贯通榫眼）的压力迫使楔子嵌进凸榫内。然后凸榫向着榫眼四周撑开。

section 1. A unit of work, e. g. a bay of concrete or an area of construction. 2. A group of workers.
1. 工段 如一段混凝土或一片建筑。**2. 一队工人**。

sectional insulation *Moulded insulation that is specifically designed and manufactured to fit around services, pipes that service and distribute heat (or are part of a refrigeration system), and other irregular objects. The insulation is delivered in sections that effectively insulate the specific component.
分段保温材料 为了安装于设备、供暖（或属于制冷系统一部分的）管道、以及其他不规则物体周围而专门设计制造的模制保温材料。保温材料分段进行运送，能更有效地对特定组件保温。

sectional tank (bolted tank) A storage tank that has been manufactured from standardized panels.
模压水箱 由标准钢板制成的储水箱。

section manager (section engineer) A manager or engineer responsible for or in charge of an area of work and all of the professionals and operatives that work on or within the *section.
工段长（施工员） 负责或主管一区域内的工作及该工段内所有专业人员和其运营的经理或工程师。

section mould A template used for provide the outline or guide for cutting and shaping material.
型模 为切割和塑形材料提供轮廓或参考的模板。

security Measures taken to reduce the risk of theft, unauthorized entry, and vandalism, as well as controlling and monitoring personnel on the site.
安全措施 为减少偷窃、非法进入、故意毁坏的行为，以及监控现场人员而采取的措施。

security fence Hording or barrier placed around the perimeter of the site to prevent unwanted access, protect the public from site-based activities and risks, and control entry.
安全护栏 沿现场周边安装的临时围栏或栅栏，用于防止外部人员进入，保护公众免受现场活动和风险影响，同时控制出入。

security glazing A type of glass or polycarbonate material that is designed to provide security and withstand attack.
安全玻璃 一种玻璃或聚碳酸酯材料，起到保证安全和抵抗攻击的作用。

security lock (thief-resistant lock) A lock that is designed to prevent unauthorized entry.
安全锁（防盗锁） 用于防止他人非法进入的锁。

sediment 1. Eroded material that has been transported by wind and water and then deposited. 2. Fine material that has settled at the bottom of a liquid.
1. 沉淀 由水或风运输的被侵蚀的物质沉积下来的过程。**2. 沉积物** 液体底部沉积的细粒物质。

sedimentation 1. The process of suspended particles settling in liquid to * sediment. 2. A wastewater treatment stage where the solid components to sewage settle out.
1. 沉淀 液体中的悬浮颗粒沉淀下来形成沉淀物的过程。**2. 沉积** 从污水中沉淀出固体成分的一个污水处理阶段。

S E duct A type of shared flue found in multi-storey buildings, where fresh air enters the flue at the bottom and the flue gases are discharged at the top of the flue.
公共烟道 多层建筑物内的一种共用烟道，新鲜空气从烟道底部进入，废气从烟道顶部排出。

seep For a liquid to slowly pass through or leak out of something.
渗漏 液体从某物中缓慢流出或泄漏的情况。

seepage The passage of a liquid through a substance, for example, water flowing through a soil due to a hydraulic gradient.
渗流 液体渗入流过一种物质，例如，由于水力梯度，水可以流过土壤。

seepage pit A porous walled pit that is connected to a * septic tank to allow the liquid waste from the tank to slowly drain into the surrounding ground.
净化坑 多孔的带围墙的坑，与化粪池相连，使池中的废水可以缓慢流入周边土地中。

segment Part of an object, often a regular division of an item.
段 某物的一部分，通常为某物的整齐分割部分。

segregated sewerage system *See* SEPARATE SYSTEM.
分流制排水系统 参考“分流制”词条。

segregation The separation of fine aggregate in a concrete mix due to excess water in the mix or incorrect compaction and placing; for example, over-vibration or dropping the fresh concrete from a height through the reinforcement.
离析 由混凝土混合物中水分过多或者不当的夯实和浇筑造成的混合物中细骨料分离的现象；例如穿过钢筋从高处倒下新浇混凝土或者过振（都可能造成离析）。

seismic Relating to * earthquakes; tectonic movement of the earth's crust.
地震 与地震相关的；地壳的构造运动。

selected tender (US invited bidder) Different from an * open tender, a contractor is nominated or previously listed and allowed or qualified to * bid. A party may be invited to bid or has met prequalification criteria and is part of a framework agreement.
指定投标人（美国 特邀投标人） 与公开招标不同，被提名或之前已入围有资格进行投标名单的承包商。一方可被邀请进行投标或者已达到了预审资格，构成框架协议的一部分。

selective tendering The process of asking those who have been * selected tender to bid. Only those listed or included in the framework can take part in the bidding process; this is different from an open-tendering process where any party is allowed to provide a * bid.
邀标 邀请选中的投标人进行投标的过程。只有入围的或者包含在框架内的承包商可参与投标；与任一方都可投标的公开招标不同。

self-centring formwork Moulds or shuttering for concrete floors, that sit on beams that are extendable (self-centring), allowing the beam to be adjusted to fit within the floor and support the formwork. The term is a little misleading as neither the beam nor the floor formwork is self-centring. The beams used simply open out to fit different widths and are adjustable.
自定心模架 混凝土楼板用的模具或模板，位于可伸展（自动定心）的横梁上，横梁可调整为适用于楼板内并可支撑模架。该词容易让人产生误解，因为横梁和楼板模架都不是自定心的。横梁只是用来展开并进行调节以适应不同的宽度。

self-cleansing Surfaces and services that are designed not to retain dirt, dust, debris, or other solids, therefore being low-maintenance and not requiring regular cleaning. Self-cleansing pipes, drains, and traps allow solids to be flushed away, facades that are self-cleansing are cleaned by the rain and the wind. The surfaces are designed to be non-stick, avoiding adhesion where possible.
自清洁 不留灰尘、尘土、杂物或其他固体的表面和设备，因此维护简易且无须定期清理。自清洁管道、排水管和存水弯应能将固体物质冲走，自清洁型幕墙可由风雨进行清洁。其表面应为不粘物质，尽可能避免黏附。

self-climbing tower crane A tower crane that is fixed to the shaft of a lift, central core, or building, climbing the building as the construction takes place. The crane achieves the climbing by fixing to the new solid parts of the structure as the building rises, with the lower attachments of the crane being removed once upper sections are attached and secure.
自升式塔吊 固定到电梯井、中轴或建筑物上的塔吊，随着施工的进行在建筑物上不断攀升。通过固定到新建成结构的坚实部分而不断升高，上部连接固定后就将下方的塔吊连接移除掉。

self-curing The process of curing without heat application.
自固化 无须加热就固化的过程。

self-drilling screw A *screw with a tapered thread that is turned into a pre-drilled metal sheet, creating a thread within the material as it is screwed in.
钻尾螺钉 放入预钻孔的金属板中，随着螺钉拧入在材料内形成螺纹的带锥形螺纹的螺钉。

self-embedding screw A screw with a drill bit point that cuts into the metal, forming its own hole, and taping a thread into the material as it is rotated by a power drill.
自嵌入螺钉 将带钻头尖的螺钉钉入金属内形成螺钉钻孔，电钻旋着螺钉钻入时在材料内形成内螺纹。

self-extinguishing Property of a material that renders it capable of catching fire; however, the act of initial combustion causes it to extinguish itself very quickly.
自熄性材料 具备燃烧的属性，但是开始燃烧后会立即自动熄灭的材料。

self-finished (factory finished) Material or component that has a surface prepared for final application needing no further treatment. The component needs to be handled with care during transportation and fitting so as not to scratch or damage the surface.
出厂饰面（工厂饰面） 其表面可以作为最终饰面而无须进一步处理的材料和组件。运输和安装组件的过程中应特别小心，避免刮擦或损坏表面。

S

self-finished felt Bitumen felt with a factory-applied finish, often metal foil, that provides a protective surface.
出厂饰面油毡 带工厂饰面的沥青油毡，通常为金属箔，用于提供保护性表面。

self-illuminating exit sign (self-contained exit sign, emergency exit indicator lighting) An illuminated exit sign that operates in an emergency. It has its own power source, so can operate even if there is a power failure.
自发光出口标志（独立出口标志，紧急出口指示灯） 紧急情况下发光的出口标志。有其自用电源，即使停电也可发光。

self-levelling screed (self-smoothing screed) Highly workable screed mixture that is simply spread over an area in sufficient quantity such that the fluid properties of the screed cause it to find its own level across an area.
自流平砂浆 和易性良好的混合物，简单地将大量砂浆摊到一块区域上，由于其流动性，砂浆可在该区域内自找平。

self-protective material 1. Material that protects itself against oxidation or corrosion. 2. Material, such as felt or plastic, with additives that protect it from the ultraviolet rays of the sun. The fillers or covering material reduce breakdown of the material and bonds.

1. 自保护材料 能自我保护免受氧化或腐蚀的材料。**2. 自保护油毡或塑料** 添加有添加剂保护其免受紫外线照射的材料。填充料或覆盖材料减少了材料和胶粘剂的降解。

self-siphonage The removal of the water seal in a *trap by the discharge from a *sanitary appliance.
自动虹吸 通过卫生洁具排水从而移除存水弯处的水封。

self-supporting Material, component, or element that requires no additional support to maintain its shape and form when in position. *Profiled metal sheeting, *cladding, and some roof structures such as *shell structures need no additional support to function.
自承重的 安装到位后无须额外支撑即可保持其形状样式的材料、组件或元件。型材金属板、外墙挂板以及诸如壳体结构的屋顶结构都无须额外支撑即可工作。

self-tapping screw A screw that cuts a thread into the material to which it is inserted.
自攻螺钉 在其钉入的材料内形成螺纹的螺钉。

semi- Prefix meaning half of something; for exmple, *semi-detached means that, from a front elevation, half the building is detached from another property, with the other part being attached to another *dwelling, *semi-circle means half a circle.
半- 指某物一半的前缀；半独立式住宅指从正面看，一半建筑物与另一处物业分离，而另一半与另一个住所相连；半圆指圆的一半。

semi-bonded screed The structural concrete is cast in place and sets, the screed is laid over the top of the matured concrete. The screed makes a mechanical fix, locking into the pores of the concrete, but does not make a chemical bond with the concrete.
新砼砂浆层 现场浇筑和凝固的结构混凝土，砂浆铺设于已硬化混凝土之上。砂浆层形成机械组合，嵌入混凝土气孔中，但是并不会与表面形成化学胶合。

semi-circle Half a circle.
半圆 圆的一半。

semi-conductor A material, such as silicon, that has electrical conductance between that of a conductor and an insulator.
半导体 一种材料，其电导率介于导体和绝缘体之间，如硅。

semi-detached house (US duplex) A pair of houses that are joined together by a common wall.
半独立式住宅（美国 联排住宅） 共用一堵公用墙的两套住宅。

semi-diurnal Having a half-day cycle, such as the tide.
半日 以半日时间为周期，如潮汐。

semi-enclosed fuse *See* FUSE.
半封闭式熔断器 参考“保险丝”词条。

semi-gantry crane A *gantry crane that has one of its sides supported by a leg that runs on a track at ground level.
半门式起重机 一边由支腿支撑、沿地面轨道运行的门式起重机。

semi-skilled worker An operative who has experience and training sufficient enough to ensure that the activities requiring general construction knowledge, skill, and understanding can be competently undertaken. It does not include operatives who require training to master a skill such that they can be described as a *craftsperson.
半熟练技工 具备的经验和培训知识足够胜任需要具备一般建筑知识、技能及理解（能力）的工作的技工。需要经过培训从而熟练掌握一项技能的、被称作工匠的技工不包含在内。

semi-solid core door A door where part of the core is filled with a material such as particleboard or wooden blocks.
半实心门 门芯部分填充有刨花板或木块等材料的门。

S

sensible heat The heat that is applied to an object to change its temperature but not its state.
显热 施加于物体上改变其温度而不改变其状态的热量。

sensitive ratio The ratio between the shear strength of an undisturbed sample and a disturbed sample of soil.
敏感性比率 原状土试样与扰动土样的抗剪强度之比。

sensor A device that is used to measure some physical quantity—temperature, humidity, movement, or pressure.
传感器 用于测定某些物理量——温度、湿度、位移或压力的设备。

separated flow When the flow of a liquid passes round either side of an object, which can result in eddies forming on the downstream side of the object.
分离流 液体流经某物一边时，就会在物体的下游形成漩涡。

separate screed (bonded screed) A screed which has a mechanical bond to the substrate that it rests on. Generally the concrete floor is *scabbled to ensure the screed can lock into the surface below, creating a strong mechanical bond. A bonding agent may also be used to create a chemical link from one material to another.
中层砂浆 与其下方的底基层机械结合的砂浆层。通常对混凝土楼板进行打毛，确保砂浆层可与下表面紧密贴合，形成牢固的机械结合。也可使用胶粘剂在两种材料之间形成化学结合。

separate system A below-ground drainage system that utilizes one drain and sewer to discharge foul water, and one drain and sewer to discharge surface water. Commonly used but installation costs are higher than a combined or partially separate system. *See also* COMBINED SYSTEM and PARTIALLY SEPARATE SYSTEM.
分流制 使用一条沟渠排放污水，另一条沟渠排放地表水的地下排水系统。使用广泛但安装费用比合流制或部分分流制要高。另请参考“合流制”和“部分分流制”词条。

separating floor (party floor) A floor between two different rooms or buildings.
分隔楼板 两个不同的房间或建筑物之间的楼板。

separating layer A debonding agent or a separating membrane used to ensure that layers of material are separate and free to expand and contract when experiencing thermal and moisture movement.
隔离层 确保材料各层分离所用的脱胶剂或分离膜，遇热和潮气移动时也不会膨胀或收缩。

separating wall (party wall, common wall) An intervening wall that divides two or more dwellings. As the wall sits directly between two or more properties, the owners of each property have rights on alterations to the wall under the *Party Wall Act. The party wall in habitable buildings should restrict the passage of sound, moisture, and heat transfer as described by the *Building Regulations.
隔墙（分户墙，共用墙） 分开成两个或多个住所的介于中间的墙壁。由于墙壁处于两个或多个住所之间，按照《共用墙法案》每个业主都有权对共用墙做出更改。按照建筑规程要求，住宅型建筑中的共用墙应能够隔声，防潮并防止热传递。

S

separation The process of keeping things apart. This can be achieved by the use of *geosynthetics which separate course and fine soil materials and/or man-made aggregates, while allowing the free flow of water across the geosynthetic. An example of their use is in the construction of *pavements, where a geotextile placed between the *subsoil and the *aggregate *sub-base prevents the aggregate being forced down into the soil by compaction during construction and through the weight of vehicles.
分离 把物体分离的过程。通过使用土工合成材料达到分离的目的，在允许水流自由流过土工合成材料的时候将层砌和细土材料以及/或者人工骨料分离开来。其中一个应用例子是硬路面铺设，在下层土和粒料层之间放置土工布，防止骨料在铺设过程压实和车辆碾压被迫沉入土壤。

septage The *septic contents of a *septic tank.
化粪池污泥 化粪池中的腐烂物质。

septic An anaerobic bacterial condition, used to decompose raw sewage, specifically in a *septic tank.
腐烂物 厌氧环境，用于分解生活污水，特别是在化粪池中。

septic tank A large tank used for the collection and treatment of sewage. Found in buildings that do not have

a connection to the main sewer system.
化粪池 用于收集和处理污水的大水箱。常见于未和主排污管系统连接的建筑物中。

sequence arrow An arrow linking tasks in a network diagram showing the logical flow of work for the project. The arrow is used to link tasks together showing the preceding and subsequent tasks linked to each activity. Sequence arrows are used to show the links in *arrow, *precedence, and other network diagrams.
序列箭头 标明项目工作的逻辑流程的网络图中连接任务的箭头。用箭头把任务连到一起，来标明整个进程以及每项活动的后续任务。序列箭头用于标明箭头、次序和其他网络图表相互之间的联系。

sequence of operations The order that activities and tasks take place in a project. The order of tasks can be seen in the logic mapping of activity *networks and *Gantt charts. The sequence arrows show the links and order between activities.
操作顺序 项目中活动和任务发生的先后次序。从活动网络图和甘特图中的逻辑图中可看到任务顺序。序列箭头展示了活动之间的连接和次序。

sequence of trades The order in which skilled workers undertake their work on a traditional development.
工种顺序 按照开发惯例，技工承接不同工作的顺序。

series A progression of one thing after another; for instance, the succession of rock strata being formed from different layers of rock.
连续 一件事物接着另一件；如由不同的岩石层组成的连续岩层。

Serpula lacrymans Fungus that causes *dry rot in the UK.
褐腐菌 英国一种引起木材褐腐的菌类。

service Facilities for public needs—water, electricity, toilets.
设施 公共服务设施——水、电、厕所。

serviceability limit state A condition where a structure is deemed as failed due to excessive deformation, for example.
正常使用极限状态 一种结构由于过度形变被认定为失效的状态。

service cable The cable that supplies electricity to individual buildings. It runs from the electricity supply main in the street to the buildings electricity meter.
供电电缆 为单体建筑供电的电缆。从街道的供电总线跑线至建筑物的电表处。

service clamp A **pipe saddle** fitting that produces a watertight damper around a pipe and provides a pipe outlet from the main pipe producing a tee junction.
机械三通 鞍形管夹配件，在管道周围形成防水密封，并从主管处接出一支管出口，形成一三通接头。

service life The timescale for which something is designed to remain operational.
使用期限 某物设计的可用期限。

service lift (goods lift) A lift used to transport goods from one level to another.
载货电梯（货物升降梯） 用于从一层到另一层运输货物的升降梯。

service pipe 1. The water pipe that runs from the external stop valve (located just outside the boundary of the property) to the inside of the property, where it terminates at the internal stop valve. 2. The gas pipe that runs from the service main in the street to the property, where it terminates at the customer's meter control valve.
1. 给水管 从外部截止阀（紧挨物业边界线的外沿）连接到建筑内部的水管，到内部的截止阀截止。**2. 送气管** 从街道上的供气主管连到物业的送气管，到用户的气表控制阀截止。

service reservoir (distribution reservoir) An underground or covered reservoir that holds drinking water to feed into the *water distribution network.
储水池（配水池） 储蓄饮用水的、地下的或有盖的配水池，向配水网供水。

service road A minor road next to a main road, which can be used for access to business premises to deliver goods.
辅助道 主路旁边的辅路，可用做向商业楼面发送货物的通道。

services The collective name given to the electrical and mechanical services within a building.
设备 建筑物内电力和机械设施的总称。

services core (mechanical core) The grouping of building services into a series of vertical shafts in a multi-storey building. The shafts are usually positioned beside the lifts.
设备管井（机电管井） 在多层建筑物内将设备集中一起的一系列管井。管井通常位于电梯旁边。

services duct A duct used to convey building services.
设备管路 用于运送建筑物设施的管道。

set 1. Components, tools, or furniture that share a common link and providing the full compendium of items for the purpose. 2. The chemical hardening of a material. 3. The angle of a saw-tooth blade to make the *kerf, being just wider than the saw blade, enabling the blade to cut and easily run through the narrow gauge of the timber cut. 4. The frame, rebate, and door for a door opening.
1. 一套 有共同关联和用途并为此提供该物项的全部纲要的组件、工具或家具。**2. 凝固** 一种材料的化学硬化。**3. 夹角** 锯出的锯缝宽度比锯片稍微宽一点的锯齿刀片的角度，能使锯片切入木材并轻易移动。**4. 门组件** 安装于门洞的门框、槽口和门。

set back The distance the building line is positioned from the centre of the road. The line is positioned by the building authorities and is there to prevent developers encroaching on the street line.
后退线距离 建筑红线距路中央的距离。后退线由建筑主管机关划定，用于防止开发商侵占街道边界线。

set coat The *finishing coat of plaster.
抹面砂浆层 砂浆的饰面涂层。

setdown A recess in a floor to take a door mat or carpet.
下沉地毯槽 地板上的一块凹处，用以放置门垫或地毯。

set-off An agreed amount that has been deducted for work that has not been carried out according to specification.
抵消 已经对按照规范未完成的工作进行扣除的议定金额。

set screw A screw with a rectangular head that can be tightened by a spanner to create friction and fix another object into place. Headless screws, also known as *grub screws, are available with recessed hexagon heads allowing an *allen key to be inserted into the screw and tightened, locking other units into place.
止付螺丝/机米螺丝 可用扳手拧紧的方头螺丝，可产生摩擦力并将另一个物体固定到位。无头螺丝也称作平头螺丝，有内六角头，可用艾伦内六角扳手嵌入螺丝拧紧，并把其他组件固定到位。

S

set square A drawing instrument in the shape of a right-angle triangle.
三角板 直角三角形的画图工具。

sett A cube of granite that is used for creating roads, paved areas, and hard-standing. A traditional surface for road construction.
铺路块石 铺设路面，已铺面区域以及硬路面所用的花岗岩块。是路面铺设的传统做法。

setting The initial hardening of plaster, concrete, or mortar after which the material should not be disturbed as it will affect the chemical bonds.
硬化 砂浆、混凝土或灰浆的初凝过程，之后不应施加干扰以免影响其化学结合。

setting block *See* GLAZING BLOCK.
垫块 参考“玻璃垫块”词条。

setting coat The final coat creating the finish, *See* FINISHING COAT.

抹灰罩面 构成饰面的最后涂层，参考“饰面层”词条。

setting out (marking out) The establishment of temporary marks, steel pins, and wooden pegs to show where the building or structure is to be positioned.
放线（放样） 用钢钉和木桩作临时标记，用于标明建筑或结构的位置。

setting shrinkage The shrinkage that takes place in concrete between its placement, and during initial set.
干缩 混凝土浇筑后至初凝期间产生的收缩。

settlement The vertical downwards movement (compression) of soil due to an imposed load (from a structure), or a lowering of the *groundwater level. The compression of the soil can be due to several cuases: elastic deformation of the soil grains (known as immediate settlement), which occurs immediately on loading and is recoverable; decrease in the volume of voids, known as *consolidation and is only recoverable to an extent by a reduction in overburden pressure; lateral flow of soil particles, known as plastic deformation of secondary consolidation; and collapse.
沉降 土壤由于受到（来自建筑物的）外加荷载，或地下水位下降而产生的竖直向下的移动（压缩）。有几种原因可导致土壤压缩：土壤颗粒的弹性形变（称作瞬时沉降），受到荷载后立刻产生沉降，是可恢复的；空隙量减少，称作固结，只有减少荷载压力才可有一定的恢复；土壤颗粒侧流，称作次固结的塑性形变；坍塌。

settling When fine suspended material in water sinks to the bottom and forms *sediment.
沉淀 水中的细小悬浮物质沉到水底形成沉淀物的过程。

set up To erect or put something in position.
装配 安装或将某物放置到位。

sewage The waste liquids and solids that are carried away for treatment by drains and *sewers
污水 由排水管和下水道运走以作处理的废弃液体和固体。

sewer A large underground system of pipes and tunnels used to transport *foul and *surface water. *See also* FOUL WATER SEWER and SURFACE WATER SEWER.
下水道 大型的地下管道和隧道系统，用于运输污水和地表水。另请参考“污水下水道”和“地表水下水道”词条。

sewer chimney (US) *See* BACKDROP.
污水竖管（美国） 参考“垂直落水管”词条。

sextant A navigational instrument used to work out latitude and longitude.
六分仪 用于计算经纬度的导航仪器。

shaft A vertical duct for building services.
竖井 建筑物设备用的竖直管道。

shake A separation of wood fibres that occurs when the tree is felled or the wood is seasoned. A shake is a defect that is not normally acceptable in structural wood. *Gross defects such as shakes are not an *allowable defect when grading. Shakes can be *heart, *radial, *ring, or *star shakes.
裂纹 伐倒树木或树木风干后发生的木质纤维的分离现象。在结构木材中裂纹通常是不可接受的缺陷。诸如裂纹类的严重缺陷在质量定级时是不能被接受的。裂缝可以是芯材、径向、环状或星状裂纹。

shaking test A *dilatancy test to determine if the material is a silt or clay; the water will disappear when the pat is pressed if it is a silt.
手捻试验 用于确定材料是粉土还是黏土的剪胀性试验；如果是粉土，按压后水分会消失。

shallow Not very deep, e.g. shallow foundations, which include strip, raft, and pad types of foundations.
浅的 不是太深的，如浅基础，包括条形、筏板和阶形基础。

shanked drill A drill bit that has a reduced diameter shaft, allowing it to be held by a small *chuck.

带柄钻 有缩径柄的钻头，可使用一个小夹头固定住。

shape code A standard numbering system to define different shapes to which reinforcing bars are bent.
钢筋形状编码 用于定义钢筋条弯成的不同形状的标准编号系统。

shape factor A value that can be incorporated into a *bearing capacity calculation to accommodate three-dimensional shearing at the corners of square or rectangular *foundations.
形状系数 计算承载力时为考虑在方形或矩形基础角位的三维剪力而引入的数值。

shaping The second stage of sharpening the teeth of a saw, where a file is used to ensure the teeth are true and uniform in their profile and size.
成形 削尖锯齿的第二个阶段，用锉刀确保锯齿的形状和大小精确一致。

sharp arris The edge of a corner or angle that has been left with a sharp edge or point. Some arrises are rounded, creating a *rounded arris that removes the hard edge.
尖角 拐角或转角的边沿留有一个锐边或尖角。有些棱角线做磨圆处理，除去硬边做成圆边。

sharp-crested weir A *weir with a thin cross-section and a pointed upper edge; used for measuring flow.
锐缘堰 有薄的横截面以及一个突出上缘的堰；用于测量流量。

sharpening Grinding and filling blades and teeth to restore their cutting edge.
削尖 打磨和填充刀片和锯齿，恢复其切削刃的锋利度。

sharp sand A type of sand that has coarse angular grains, suitable for use in concrete mixes.
多角砂 有粗糙角粒的一种沙子，适用于做混凝土拌合料。

shave hook A T-bladed scraper. The blade is drawn towards the user with the handle fixed to the end of the T-blade.
三角片刮刀 T型刀片的刮刀。刀片朝使用者的方向移动，刀柄固定在T型刀片末端。

shear (shearing, shear action) To displace something relative to something else, for example, the upper portion of soil against that of the lower portion of soil in a *shear box.
剪切 将相关的某物替换成其他物项，例如在剪切盒中上部土壤与下部土壤相对立。

shear centre A point in the cross-section of a structural member where the resultant shear force forms *bending without *torsion.
剪力中心 结构件横截面上的一个点，在该点上合成的剪切力只会形成弯曲而不会产生扭转。

shear cone Failure of an anchor bolt, caused by a cone-shaped piece of concrete or brickwork failing (being pulled out) from around the bolt.
锥形破坏 由于锥形混凝土块或砖块从螺栓周围断裂（从中拉出）而导致的锚栓失效。

S

shear connector A connector between, for example, a beam and a slab to prevent longitudinal shear.
抗剪连接件 例如横梁和混凝土板之间的连接件，用于阻止纵向剪力。

shear lag A slow response to the development of shear stress in a material or structural member, e. g. where a non-uniform stress distribution occurs.
剪力滞后 材料或构件对累积剪力的一种慢速反应，例如不均匀应力发生的位置。

shear load The load perpendicular to the normal load, causing a material or element to *shear apart.
剪切负载 与正常负载垂直的荷载，可导致材料或构件被剪开。

shear modulus *See* MODULUS OF RIGIDITY.
剪切模数 参考“刚性模数”词条。

shear reinforcement *Reinforcement bars designed to resist *shear loads.
抗剪钢筋 用于抵抗剪切负载的钢筋。

shear slide A *landslide.
剪切滑移 一种滑坡。

shear test A test to determine shear strength parameters of a soil, such as a shear box andtriaxial test.
剪力测试 用于测定土壤的剪切强度参数的试验，例如剪切盒和三轴压缩试验。

shear wall A structural wall designed to resist swaying forces.
剪力墙 设计用于抵抗摇摆力的结构墙。

sheathed wiring Electrical wiring that is covered by a protective, usually plastic, coating.
铠装线 包有保护套的电线，保护套通常是塑料涂层。

sheathing 1. A layer of boards applied to the studs of a timber-frame wall or to the rafters of a pitched roof to strengthen the construction. 2. The outer protective covering of electrical wiring.
1. 包板 安装到木框架墙的墙骨或斜屋顶的椽条上的一层板子，用于加固结构。**2. 铠装层** 电线的外壁保护层。

sheathing boards Large boarding sheets used for packing, decking, walling, roofing, and *shuttering.
衬板 大件板材，可用作包装板、盖板、墙板、屋面板和模板。

sheathing felt A type of bitumen roofing felt.
油毡保护层 一种沥青屋面油毡。

sheathing paper Building paper.
防潮纸 建筑用纸。

sheave A grooved pulley wheel, for a rope, cable, or belt to run through.
绳轮 带槽滑轮装置，可供绳索、缆线或皮带穿过。

she bolt A metal tie made out of ribbed reinforcement bar, over which winged or large nuts are used to create a nut and bolt tie. The tie bar is often used on wall shuttering placed through the two faces of the wall shuttering, with the bolts either side, enabling the shuttering to maintain the correct width once it is filled with concrete. The tie bar resists the horizontal hydrostatic loads created by the wet concrete.
对拉螺栓 由带肋钢筋制成的金属系件，在螺杆上用翼型螺帽或大螺帽形成螺母及螺栓连接。通常将联系杆穿过墙壁模板的两边，两边都有螺栓，确保往模板内填充混凝土时模板间能够保持合适的宽度。联系杆用于承受湿混凝土导致的横向水静力荷载。

shed dormer A dormer window with a roof that slopes in the same direction as the roof in which the dormer is located.
单坡屋顶老虎窗 屋面朝同一个方向倾斜的屋顶上安装的采光窗。

shedding *See* LOAD SHEDDING.
减载 参考“减载”词条。

S

sheep's foot roller A self-propelled or towed soil compacting roller, which is characterized by having projections like sheep's feet around its *perimeter.
羊足碾压路机 自推进或拖曳式夯土压路机，其特征为机身周围有像羊蹄一样的突出物。

sheet Large thin flat material formed by rolling, pressing, extruding, or cutting down to size. *Sheet glass and *sheet metal are examples of sheet materials.
片材 通过轧制、冲压、挤压或切割形成的合适尺寸的较大薄片材料。片材的例子有玻璃片和金属薄板。

sheet flow A thin layer of liquid flowing over a surface.
漫流 在表面上流动的一层薄薄的液体。

sheet glass Thin glass commonly used for windows, doors, and wall applications.
玻璃片 窗户、门和墙壁中通常采用的薄玻璃。

sheeting Sheet cladding material.
薄片 薄片覆盖材料。

sheeting rail A *CLADDING RAIL.
干挂用横龙骨 参考“干挂用横龙骨”词条。

sheet metal A thin metal plate; in construction sheet metal is primarily used for roofs in buildings.
金属薄片 薄金属片；施工时金属薄板通常用于建筑物的屋顶。

sheet pile A section of steel, concrete, or timber, typically in a U, W or Z shape, that when sunk into the ground, side-by-side, form an underground barrier, such as a *cofferdam.
板桩 截面通常为U，W或Z形的钢材、混凝土或木材，并排沉入地下后形成地下隔水屏障，如围堰。

shelf Horizontal platform used as a store for items such as books, tools, utensils, or as a display for ornaments and trophies.
搁板 用于放置书籍、工具、餐具或装饰品及可作为奖杯展示台的水平平台。

shelf angle Length of galvanized steel bracket, attached to the structural frame, used to provide support for brickwork cladding.
墙面角钢 安装到构架上的一段镀锌钢支架，用于为砖墙覆面提供支撑。

shelf life The time that something can be stored in a usable state, for example, the time adhesives, varnishes, and paints can be stored and used.
货架期 某物保持可用的期限，例如黏合剂、清漆和油漆的可储存和使用期限。

shell The framework of a building.
框架 建筑物的架构。

shell bedding The usual method of bedding hollow blocks with the mortar pasted along the edge of the blocks and the perpendiculars. Thin strips of mortar are pasted along the edges of the block leaving a gap where the hollow section of the block occurs.
刮浆法 沿着砖边及垂直面刷灰浆来填充空心砖的一种常用方法。沿着砖边涂薄条灰浆，在砖的空心截面处留缝隙。

shell door canopy A shell-shaped canopy located above a door.
壳形雨棚 安装于门上方的贝壳状的雨棚。

shield A protective guard.
防护物 防护罩。

shield bolt An anchor bolt made from a threaded tapered bolt and steel sheath. As the anchor bolt is tightened, the tapered bolt pulls back compressing the sleeve against the sides of the masonry or concrete.
背栓 由锥形螺纹螺栓和钢制背栓套构成的锚栓。拧紧锚栓的过程中，锥形螺栓向后拉，将套筒不断压紧到混凝土或大理石边上。

S

shim (packing shim) Strips of metal, plastic, or timber used to lift, space, and hold components so that they can be fixed and secured in the correct position.
薄垫片（填充垫片） 用于提升、隔开和支撑构件的金属条、塑料块或木条，它们可将构件固定并保持在合适的位置。

shingle 1. Wooden or slate roof or wall titles, placed in overlapping rows. 2. Small rounded pebbles found on a beach. 3. A process in the manufacture of wrought iron, where slag is removed.
1. 木瓦/墙面板 木或石板屋面瓦或墙砖，行与行重叠铺设。**2. 卵石** 沙滩上发现的圆的小鹅卵石。**3. 清渣** 锻铁生产时移除炉渣的过程。

shingling (US) The process of placing overlapping layers of roofing felt.
叠盖（美国） 铺设屋面油毡搭接层的过程。

shiplap (shiplap boards, shiplap boarding, US shiplap siding) Overlapping boards of rectangular cross-section that have a rebate on one side of the board and a tongue on the other.

搭接 矩形截面的搭接板，板一端为槽口，另一端为榫舌。

ship spike (boat spike) A square section bar with a wedge point used for fixing and jointing large timbers.
四方船钉 连接和固定大块木板用的有楔形尖的方形截面钢钉。

shoal 1. Shallow water. 2. An underwater sandbank.
1. 浅水。**2. 浅滩**。

shoddy Slang term meaning poor quality, substandard, or rushed.
冒牌货 质量差、不合规或劣质的俗称。

shoe 1. A short bend or angled section of pipe attached to the bottom of a *downpipe to convey the water away from the building and into a drain. 2. A metal plate used to support the end of a rafter. *See also* PILE SHOE.
1. 雨水管引流器 安装于落水管底部的短弯管或者角形截面管，用于把建筑物内的水输送入下水道。
2. 椽条托架 用于支撑椽条末端的金属板。另请参考“桩靴”词条。

shoe mould (shoe mold, base shoe) A quarter rounded bead or strip of timber, used to provide a finish and hide the junction between the skirting board and floor.
踢脚线压条 四分之一圆线脚或木条，起装饰作用并可隐藏踢脚板和地板之间的接缝。

shoot To remove the rough or sharp edge of a mitre joint, removing the rough cut of a saw by planing.
刨平 去除斜角接头的粗糙面或锐边，通过平刨去除粗糙锯痕。

shop A workshop used as a factory.
车间 用作工厂的工作站。

shop drawing A drawing used in a factory rather than on-site.
施工图 工厂用而非现场用的图纸。

shopfitter Craftsperson or carpenter who is skilled in assembling and fitting out offices, shops, and stores. The fitter provides and fits the fittings and furniture.
工厂装配工 精于组装和装配办公室、商店和商场的工匠或粗木工。组装工装备和组装配件和家具。

shop priming Priming that is applied in the factory, generally ensures better adhesion.
工厂底漆 工厂涂抹的底漆，通常为了取得更佳的黏合效果。

shop work Work undertaken in a joiners' shop or factory, not on the construction site.
工厂工作 在工匠的作坊或工厂内而不是在施工现场进行的工作。

shore A prop or support to a building, such as a *flying shore *prop, or a *dead shore.
支柱 对建筑物的支撑或支持，例如横撑柱或顶撑。

shoring Providing temporary support to or propping up a building, such as underpinning, placing *shores against the building.
支撑 为建筑物提供临时支持或支撑，例如基础托换，放置支柱支撑建筑物。

short circuit A fault in an electrical circuit that occurs when an accidental connection is made between the live and neutral conductors.
短路 火线和零线意外接到一起时发生的电路故障。

short-grain timber Also known as *brash timber, which breaks with little resistance.
短纹木材 也称作脆木料，受到很小阻力时即会断开。

short oil A characteristic of paint or varnish with a low-oil base.
短油 低油度的油漆或清漆的特征。

shot blasting Firing small pellets at a surface to clean it before painting.
喷砂清理 上漆前向表面喷射小丸进行清洁。

shotcrete A type of concrete material commonly known as sprayed concrete, whereby layers of concrete are

S

extruded from a gun or hose to form the required thickness.
喷浆混凝土 一种混凝土材料，通常称作喷射混凝土，从喷枪或软管中喷出形成指定厚度的混凝土层。

shotfired fixing An explosive fixing using a nail gun to fire nails into timber, concrete, or steel. Often used for fixing sheet material to steel and concrete.
螺钉枪固定 用螺钉枪把钉子射入木板、混凝土或钢材内的爆破固定。通常用于将板状材料固定于钢或混凝土中。

shothole A hole left by a wood-boring beetle or grub, usually 2 to 4 mm in size.
蛀洞 蛀木甲虫或幼虫留下的洞，通常为2—4mm。

shovel 1. A hand-held tool, consisting of a long wooden or plastic handle and a metal flat end, which is used to move loose material. 2. A bucket-like device that is attached to an excavator and used to excavate material.
1. 铁铲 手持工具，有长的木柄或塑料手柄和一个金属平头，用于铲动松散材料。**2. 铲斗** 安装在挖掘机上的斗形装置，用于挖掘材料。

shower divertor A mixing *valve for a shower.
淋浴分流器 淋浴用的混合阀。

shower enclosure A cubicle in which to shower.
淋浴间 淋浴用的小隔间。

shower head The perforated water outlet that is used to direct the spray of water in a shower.
淋浴头 淋浴中用于喷射水流的多孔出水口。

shower room A room that contains a shower.
浴室 包含淋浴的房间。

shower rose A large shower head that sprays the water out at low pressure.
花洒头 低压喷射水流的大淋浴头。

shower tray (shower base, shower receiver) The prefabricated base of a shower that collects the water and directs it towards the drain.
淋浴底盆 预制的淋浴底座，可汇集水并将水导向下水道。

shrinkage A reduction in size of a material usually due to the loss of water or change in temperature. Excessive shrinkage can lead to cracking or separation from *substrate material.
缩水 材料尺寸的缩小，通常因为失水或温度改变。过度缩水可导致材料裂纹或基底材料分离。

shrinkage limit *See* ATTERBERG LIMITS.
缩限 参考“阿太堡界限”词条。

shrinkage ratio The rate at which the volume of a fine-grained soil decreases in respect to the decrease in its moisture content.
缩水率 细粒土壤容积的减少相对于水分含量的减少的速率。

shuffle glazing A type of glazing system formed by inserting the glass into angled sections at the head and the sill. The angled head section will be of a deeper profile than the sill section, enabling the glass to be manoeuvred into position.
嵌槽玻璃 将玻璃嵌入上槛和窗台的角形截面槽内的一种玻璃体系。上槛处的角形截面槽要比窗台处的深，确保玻璃可以调整到位。

shutter 1. A protective panel that is pulled over a window or a door to provide security, reduce heat loss at night, or provide privacy. 2. A piece of *shuttering.
1. 门窗遮板 安装在窗户或门上的保护板，用于提供安全性，减少夜间的热量流失或保护隐私。**2. 一块模板**。

shutter bar A metal bar used to secure window or door shutters in place.

门窗遮板锁杆 用于固定门窗遮板的金属杆。

shutter bolt *See* ESPAGNOLETTE BOLT.
遮板插销 参考“天地杆插销”词条。

shutter hinge *See* PARLIAMENT HINGE.
工字形铰链 参考“工字形铰链”词条。

shuttering Concrete formwork used as a temporary mould and support for wet concrete.
模板 用作临时支模和支撑湿混凝土的混凝土板子。

shuttering hand A shuttering joiner's labourer or assistant. Assists the joiner, carrying and loading out materials.
模板工助手 模板工的小工或助手。协助模板工装运和装料。

shutting jamb (close or post) The jamb that a window or door is closed against. *See also* HANGING JAMB.
闭锁边梃 关窗或门时会碰到的边梃。另请参考“带铰链的门/窗边梃”词条。

shutting stile The stile that a window or door is closed against. *See also* HANGING STILE.
闭锁竖梃 窗户或门关闭时碰到的竖梃。另请参考“挂窗竖梃”词条。

sick building (sick building syndrome) A building that causes its occupants to feel irritated, uncomfortable, and even become ill. The symptoms are normally a result of poor ventilation, mould growth in the building and services, lack of natural or adequate light, irritants within the water system, and poor drainage. Common symptoms suffered include sore eyes, sore throat, skin rashes, uncomfortably hot or cold, headaches, and stomach irritation.
病态建筑（病态建筑综合征） 让居住者易怒、不舒服甚至生病的建筑物。此类症状通常是由于通风不佳、建筑和设施上霉菌滋生，缺少自然光或充足的光照、供水系统中的刺激物以及排水不良导致的。常见症状包括眼疼、咽喉痛、皮疹、发热或感冒、头痛以及肠胃不适。

side A face of a board or section of square sawn timber.
面 板面或方形锯材的截面。

sidefill Gravel that is laid at the sides of a pipe in a trench.
管道回填砂 堆在沟槽中管道旁边的砾石。

side form The *formwork that is used to provide the edge *shuttering or edge boards for a concrete beam or box.
侧模 为混凝土梁或箱提供角模或边模的模板。

side gutter A small gutter used at the junction where a vertical surface penetrates the roof slope.
侧沟 垂直面穿透屋顶斜面处用的小沟槽。

side-hung door A door that is hinged along one side.
平开门 沿一侧有铰链连接的门。

side-hung window A window that is hinged along one side.
平开窗 沿一边有铰链连接的窗户。

sidelap The amount by which the edge of one component overlaps the edge of another component. Used in roofing and cladding to provide a weather-tight joint.
侧向重叠 一个组件与另一个组件边缘的重叠量。用于屋顶和挂墙板中提供防水连接。

sidelight (wing light) A fixed window placed at the side of a door.
门侧艺术窗 位于门侧的固定窗。

siding *Cladding.
护墙板 外墙挂板。

sieve analysis The determination of the proportion of different-sized aggregates in a sample of mixed aggregate. The aggregate sample is weighed, placed in the top pan (sieve) on a nest of sieves. Each sieve has a different-sized mesh with the top sieve allowing all but the largest grains of aggregate to pass through, the lower level sieves catch the very small particles, and the pan at the bottom of the sieve captures the dust and silt. The sample of aggregate contained within the sieves is then vibrated causing it to drop, filtered out into different sizes, and be captured by the sieve that it cannot pass through. The contents of each of the different-sized sieves are weighed to determine the proportion of the size of aggregate in the aggregate mix.
筛分分析 鉴定混合骨料样品中不同粒度级配的方法。将骨料样品称重，放在筛机顶盘（筛）上。每面筛都有不同大小的网孔，顶层的筛网让最大颗粒外的骨料通过，下层的筛网留下小的颗粒，筛网底盘接住沙尘和粉土。然后振动不同筛网上的骨料样品使其下落，将不同尺寸的样品过滤分离，留下不能通过筛网的骨料。将不同网孔上的骨料称重从而鉴定混合骨料中的级配情况。

sight A *fore sight, *intermediate sight, or *back sight.
视线读数 前视，中视或后视。

sighting distance The furthest distance an object can still be viewed.
视距 能看见一个物体的最远距离。

sight size (daylight size) The size of glazing that is available to admit daylight.
透光尺寸 玻璃能够透过日光的尺寸。

silencer A device that reduces the noise of machinery or engine.
消音器 减少机器或发动机噪音的设备。

silicone A polymeric material comprised of silicon and oxygen, used for its flexibility and heat resistance.
硅树脂 含有硅和氧的聚合物，因其弹性和耐热性而被应用。

silicone paint A type of paint based on *silicone with excellent heat resistence properties.
硅树脂涂料 一种硅基油漆，有极好的耐热性。

sill 1. The horizontal bottom member of a window or external doorframe. 2. A horizontal layer of *igneous rock that has been forced between layers of sedimentary rock due to a *volcanic eruption.
1. 窗台/门槛/基石 窗户或外门框的底部水平构件。**2. 岩床** 由于火山喷发、沉积岩层之间受挤压而形成的水平火成岩层。

sillboard (Scotland) *See* WINDOWBOARD.
窗台板（苏格兰） 参考“窗台板”词条。

sill height The distance from the top of the sill to the finished floor.
窗台高度 窗台顶面距成品地板表面之间的距离。

S

silo A large cylindrical container on a supporting framework, used to store granular material such as cement, grain, and animal feed.
筒仓 支承架构上的大型圆柱状容器，用于储存水泥、谷物和牲口饲料等颗粒材料。

silt A fine-grained soil that lies between sand and clay, between 0.06 and 0.002 mm. It can be subdivided into coarse silt: 0.06 - 0.02 mm; medium silt: 0.02 - 0.006 mm, and fine silt: 0.006 - 0.002 mm.
粉土 粒径在 0.06 - 0.002 mm 之间，介于砂粒和黏粒之间的细颗粒土壤。可以细分为：粗粉粒 0.06—0.02 mm、中粉粒 0.02—0.006 mm 和细粉粒 0.006—0.002 mm。

silting The settling of *sediment on a river bed.
淤积 沉淀物沉淀在河床上。

silver brazing (silver soldering) A *brazing or *soldering process whereby the principal filler metal is a silver alloy.
银钎焊（银焊） 主要焊料是银合金的钎焊或焊接过程。

silver sand Fine rounded sand used for mortar and filling dry joints in block paving.

细沙 精细圆沙，用于制作灰浆和填充砖砌路面的干缝。

simplex concrete pile A *displacement pile that has been cast in-situ, by firstly driving a steel tube with a sealed base plate in the ground. Reinforcement is then inserted into the tube and concrete is poured around the reinforcement. The steel tube is then withdrawn from the ground using a vibrator, and the base plate is left within the ground.
沉管灌注桩 现场浇筑的挤土桩，首先将一个带密封底板的钢管打入地下。把钢筋嵌入钢管内，并在钢筋附近浇筑混凝土。然后用振动器把钢管从地下抽上来，底板留在地下。

simplex control A control system used to operate a single lift.
单控电梯 控制单一升降梯的控制系统。

simulation The use of a computer to model the behaviour or performance of a material, structure, and so on.
模拟 使用计算机模拟一种材料或结构等的行为或性能。

single-coat plaster Plaster that is applied in one coat only.
单层灰泥 只涂抹一层的灰泥。

single door A door that has one leaf.
单开门 只有一扇门叶的门。

single glazing Glazing that comprises a single sheet of glass.
单层玻璃 只有一层玻璃的装配玻璃。

single-hole basin (one-hole basin, single-taphole basin, single-taphole sink) A *basin that contains a single tap.
单孔面盆 只有一个水龙头的洗脸盆。

single-hung window A sash window where only one of the sashes, usually the bottom one, can move.
单悬窗 只有一扇窗可以移动的推拉窗，通常只有底部窗扇可以移动。

single-lap tile (interlocking tile) A tile that overlaps at the bottom, and sides of the tile, and interlocks at the edges; at the centre of the tile there is only one thickness of tile covering the roof. A double-lap tile will have up to three layers of tiles at the top and bottom edges, and a minimum of two overlapping tiles at the centre of the tile. Single-lap tiles have interlocking edges that prevent the rain penetrating and being blown into the joints.
单搭接瓦（互扣瓦） 底部和各边搭接，边缘互扣的瓦片；瓦的中心部分只有一个瓦片的厚度。双层搭接瓦在顶边和底边有多达三层的瓦片，且瓦片中心处至少有两个搭接瓦。单层瓦有互扣边，可防止雨水渗透并流入接缝。

single-lever mixer A *lever mixer.
单杆调温水龙头 一种提拉杆式水龙头。

single-lock welt A sheet steel joint made by folding one sheet of metal over the other.
单咬口 折叠钢板使其相互咬合的钢板连接。

single-outlet combination tap assembly A tap that has a single outlet for both hot and cold water; also known as a **mixer tap**.
双联式水龙头 只有一个可出热水和冷水的出水口的水龙头；也称作混合水龙头。

single-phase (single-phase supply) The distribution of alternating *current through two *conductors. *See also* THREE-PHASE SUPPLY.
单相（单相供电） 通过两个导体配送交流电。另请参考“三相供电”词条。

single pipe (single pipe system) *See* ONE-PIPE SYSTEM.
单管 参考“单管排污系统”词条。

single-pitch roof A roof with just one pitch to it; also called a mono-pitch, and if the top of the roof abuts an adjacent wall it may also be called a lean-to roof.

单坡屋面 只有一个坡面的屋顶；也称作单坡面，如果屋顶与邻墙相接也可称作披屋顶。

single-ply roof A roof with a single covering of waterproof membrane, the membrane is high-performance made from a synthetic rubber material such as ethylene propylene diene rubber, polyvinyl chloride, polyisobutylene, and other modern synthetic coverings, which are then laid on a plywood sheeting. Single-ply roofs have an expected lifetime of 30 years.
单层屋面 只有一层防水膜的屋面，该防水膜由合成橡胶材料如三元乙丙橡胶、聚氯乙烯、聚异丁烯和其他现代合成覆盖材料制成，防水效果极佳，之后铺在胶合板材上。单层屋面的预期使用寿命为30年。

single-point heater *See* INSTANTANEOUS HOT-WATER HEATER.
单点式热水器 参考“即热式热水器”词条。

single-pole switch A switch that simply breaks or connects one circuit. The switch only has one pole.
单刀开关 仅控制一条线路断电或通电的开关。开关只有一个极。

single stack system An above-ground drainage system that comprises a large-diameter single vertical-discharge stack that conveys both soil and wastewater to the drain, as well as providing ventilation. All of the appliances connected to the stack require deep seal traps. The length of the soil and the waste pipe connections from the stack are restricted. Extensively used nowadays in domestic housing. *See also* ONE-PIPE SYSTEM and TWO-PIPE SYSTEM.
单立管系统 包含一个大口径的垂直排污单管的地面排水系统，可将污水和废水排放到排水沟，还可提供通风。与立管连接的所有设备都需有深存水弯。并且连接污水管和废水管的长度有限制。此系统今天广泛应用于民用住宅内。另请参考“单管排污系统”和“双管排污系统”词条。

sink (kitchen sink) A shallow bowl used for cleaning and the preparation of food.
水槽（厨房水槽） 用于清洗物品和准备食物的浅水槽。

sinker drill A hand-held rock drill, used in shaft sinking.
风钻 手持凿岩机，用于凿井工作。

sink grinder *See* GARBAGE DISOSAL SINK.
食物残渣磨碎机 参考“食物残渣处理器”词条。

sinking A hole, recess, or rebate that allows for a door mat, hinge, or plug to sit neatly and flush with the top surface.
凹陷 可以放置门垫、铰链或插头且放置完后与顶面齐平的洞、凹处或槽口。

sinking-in The effect of the substrate or undercoat absorbing the top layer of paint. With gloss and sheen paint this results in the loss of the desired finish.
渗色 基层或底涂层吸收了面层涂料后的影响。如果是光泽涂料或光泽漆则达不到预期效果。

sink plunger (force cup) A hand-held device used to unblock sinks. Comprises a rubber cup attached to a long pole. The flexible cup is pushed over a plug hole and pressed down then released creating suction which draws out the blockage.
水槽疏通器（搋子） 疏通水槽用的手持工具。长杆一端安有一个橡皮碗。把橡皮碗放于放水孔上方，按压再松开进行抽吸，吸出堵塞物。

sink unit A joinery fitting that is designed to house a sink.
水槽柜 用于安放水槽的木工件。

sintering A process utilized to increase the density (reduce the air content) particularly in ceramic materials; usually conducted at elevated temperatures.
烧结 用于增加密度（减少空气含量）的过程，特别是在陶瓷材料中；通常在高温中进行。

siphonage The suction of fluid up or down a pipe. A siphonic drain uses the downward force of gravity to help draw excess water and control the flow of water from the top of a pipe or cistern. Excessive movement of water in some drain or sewer pipes can lead to siphonage of water from traps and seals, which is undesirable.

虹吸作用 液体在管道内向上或向下抽吸的作用。虹吸排水管在向下的重力作用下从管道或水箱的顶部抽吸多余的水并控制水流。排水管或排污管中水的过量移动可导致存水弯或水封中的水产生有害的虹吸作用。

siphonic closet (siphonic water closet) A *WC that contains a double trap.
虹吸马桶 有双存水弯的冲水马桶。

site accommodation The temporary cabins, toilets, storage, and welfare facilities provided on construction sites.
现场生活设施 施工现场提供的临时住处、厕所、仓库和其他生活设施。

site agent The main contractor's lead representative on-site, responsible for control of site-based operations and activities. The term is often used interchangeably with site manager.
工地总管 总承包商在现场的主要代表，负责控制基于现场的作业和活动。该词条通常可与"现场经理"词条互换。

site boundary The perimeter of the land and building that is under the control of the contractor during the construction and development period. The boundary forms the outer edge of where construction operations take place without permission of neighbouring landowners.
工地界线 施工和开发期间受承包商支配的土地和建筑物的边界线。该界线是无须周围业主许可就能进行施工作业的外沿。

site cabin *See* JACK-LEG CABIN.
工地住所 参考"简易活动房"词条。

site clerk (clerk of works) Acts as one of the client's representatives on-site, checking operations and ensuring the quality of work is carried out to the standard specified. The clerk of works will make reference to standards, specifications, and legal requirements, ensuring the contractor is undertaking the works in the manner expected and specified in the contract documents. Where there are any variations costed on the work undertaken, the clerk will monitor activities, man-hours, and operations required to complete the variation. Due to the way the clerk oversees and checks activities, the site staff and operatives on-site may also be referred to the clerk of works as the checker or timekeeper.
监理员（监工员） 是业主在现场的代表之一，检查并确保各项操作和工程质量能达到指定标准。监工员会参考标准、规范和法定要求，确保承包商按照合同文件中规定和指定的方式施工。如果承接的工程需要变更，监工员将会对完成变更所需进行的活动、人工工时和作业进行监督。由于监工监督和检查活动的方式，现场人员和操作者也可将监工员称作检验员或工作时间记录员。

site constraints Aspects or features in and around the site, such as services, adjoining properties, railway lines, and water courses, which restrict the operations that can be carried out. Buildings and structures near to the site may need supporting, underground culverts and drains may need to be bridged, or there may be restrictions preventing construction operations directly over the top of them. Cranes may need permits and agreement to use the airspace above neighbouring properties, and other aspects such as noise and visual pollution may restrict operations.
工地限制 限制作业实施的工地上或工地附近的某些方面或特征，如设备、邻接物业、铁路线和水道。现场附近的建筑物或结构可能需要支撑，地下水道和排水管可能需要架桥，或者可能有不允许直接在其上方进行施工的限制。在周边物业上空运行塔吊需得到许可和同意，还有一些其他因素，如噪音或视觉污染可能会限制作业。

S

site diary Document kept on-site by the *site agent or *site manager to record the daily activities, the operations and labour; incidents, weather conditions, and other relevant site activities are entered into the book or digital document. The diary is an important document, often referred back to in the event of disputes, activities that surround accidents, or other events where there is a need to check or cross-check site operations.
工地日志 由现场总管或现场经理在现场记录的文件，日常活动、作业和工人、事件、天气状况以及其他现场相关的活动都记入纸质版或电子版文档中。日志是重要文件，发生争端、与意外相关的活动或其他有必要确认或复核现场作业的情况时可以参考当时的日志。

site engineer (US field superintendent) The engineer on-site, employed by the contractor to control and check operations. The site engineer may be required to establish the position of structures, set out roads and buildings, determine the levels of the land and structures, calculate and check the volume of materials removed and required. Operations will vary depending on the nature of the site. The engineer's duties will vary depending on whether the site encompasses mainly *civil works or construction. Most of the engineering design work is carried out off-site by the structural or *civil engineering contractor.

现场工程师（美国 现场监督） 承包商聘用来控制和检验作业的现场工程师。现场工程师可能需要确定结构位置，道路与建筑物放线，确定土地和结构的高程，计算和核查需要和移除的材料量。现场的性质不同，作业内容也会有很大不同。工程师的职责取决于现场包含的主要是土木工程还是建筑施工。大多数工程设计工作是由结构承包商或土木工程承包商在现场外完成的。

site instruction (US field order) Written order issued by the client's representative (client's architect, engineer, or clerk of works), identifying work that should be done; the order normally indicates something that has immediate effect, unless stated. Such instructions can be a written order stating a variation to the work, clarification of works described under contract, order to comply with safe working practices, order to stop work, or carry out operations at specified times or in certain conditions.

工地指示 由业主代表（业主建筑师、工程师或监工员） 发出的书面指令，说明需要完成的工作；除另有说明外，该指令说明的事项需立即执行。此类命令可以是说明工程变更的书面指令，对合同中规定工程的澄清，遵守安全操作准则的命令，停工或在指定时间或特定条件下进行作业的指令。

site investigation (ground investigation, site survey) 1. The collection of information through visual inspection, observations, and tests on the ground, to determine site characteristics and ground conditions, and assess the suitability for development. There are various types of site investigation including desk-top study, site reconnaissance, and soil investigation. 2. Desk-top study, collection of all relevant reports, and historic information surrounding the site. 3. Site reconnaissance, exploration of the site recording visual features, walking and travelling around the site collecting samples, and making observations. 4. Soil investigation, the sampling and testing of soil looking at the strengths, chemical characteristics, and make-up of the ground at different levels. Tests may be carried out on-site and in a soils laboratory.

1. 实地勘测 指通过在现场实地检查、观察和测试而确定现场特征及土地状况，并评估其开发适宜度。实地勘测有很多种，包括书面调查，现场勘测和土壤调查。**2. 书面调查** 收集现场周围的相关报告和历史信息。**3. 现场勘查** 勘查现场，记录其外观特征，沿着现场收集样本并观察。**4. 土壤调查** 对土壤取样并测试，观察其强度、化学特征以及不同土层的成分构成。土壤测试可在现场和在土壤实验室进行。

S

site manager *See* SITE AGENT.

现场经理 参考“工地总管”词条。

site meeting (progress meeting, management and design team meeting) An on-site meeting that gathers the relevant professionals and sub-contractors to ensure that operations can be properly coordinated, scheduled, and managed safely, and problems foreseen, or those that have occurred, can be dealt with. Those invited will depend on the agenda of the meeting and the matters to be discussed. Meetings may be internal for those employed directly by the contractor or can involve all of the main parties connected to the site with representatives for the client, contractor, architect, structural and civil engineers, mechanical, electrical, and plant engineers, and all specialist subcontractors. The meeting will look at the progress, operations of the different parties, and the overall plan of operation with a view to coordinating activities and managing changes.

工地会议（进度会议，管理和设计小组会议） 召集相关专业人员和分包商召开的现场会议，确保可以妥善地协调、规划及安全处理各项作业，并可妥善解决预见的或已发生的问题。与会人员应视会议日程和讨论事项而定。参会人员可为承包商直接聘用的内部人员，或者与现场活动直接相关的主要当事人，包括业主、承包商、建筑师、结构和土木工程师、机械 、电气和设备工程师，以及专业分包商。会议将着眼于进度、不同方的作业以及协调活动和处理改变的整体实施计划。

site practice The operations that take place on-site. Good site practice is the sequence of events, operations, and activities required in order to produce the goods to the right quality, and in a safe and professional manner.
现场施工 现场进行的施工作业。好的现场施工是指以安全专业的方式有序地生产高质量产品所需的事件、作业和活动。

site roads The designated through-passes for vehicles and pedestrians on-site. On some sites these may be made-up roads with hardcore, or constructed to highways standards, or may be simple designated and protected ways through and around the site. The safe movement of vehicles and pedestrians around the site should be planned, and the roads and footpaths fenced or guarded to prevent risk of injury.
现场道路 现场指定通行车辆和行人的道路。某些现场的道路可以是人工硬底路，也可以是按照高速公路标准建铺设的，或者只是环绕现场简单设计和防护的道路。车辆和行人围绕现场的安全通行应该进行规划，另外应在道路和人行道两旁加上防护或围护结构，防止意外伤害。

site security The use of *hoardings, gateways, checkpoints, and security personnel to ensure that movement into, out of, and around the site is observed and controlled, to reduce the risk of theft, and control personnel entering the site. Security of the site contributes to controlling the health and safety of those within the site and general members of the public in the vicinity of the site that may be affected by operations. Controlling the flow of people helps to ensure that materials and equipment coming into and leaving the site are as expected, and theft does not occur.
工地安保 采用临时围墙、闸门、检查点和安保人员确保现场的出入及周边情况受到监控和控制，减少被盗事件并控制人员进入现场。现场安保有助于控制现场人员及现场附近可能会受到施工影响的大众人员的健康和安全。控制人员流量可以确保进入和离开现场时材料和设备量与预期一致，杜绝盗窃现象。

site services Services that are temporarily connected to the site to enable the work to be undertaken include electricity, gas, water, drainage, and telephone.
现场设施 临时连接到现场确保工作能够开展的设施，包括电、气、水、排水和电话设施。

sitework Work undertaken on site, including alterations to prefabricated components and adjustment to works.
现场工作 现场实施的工作，包括预制件更改和工作调整。

SI units The international system of units, with the standard convention being the metric system, which is based on decimalization (the number 10). The SI system is the most common system of measurement in the world.
国际标准单位 国际单位制，标准惯例为基于十进制（数字 10）的公制。国际单位制是世界上最常用的单位制。

S

size A sealant used to seal and reduce porosity of timber, plaster, and concrete, making it suitable to have finishes applied to it. The liquid sealer can be made from a thinned-out paint or varnish or watered-down glue, adhesive, or paint.
胶料 用于密封和减少木材、灰泥和混凝土空隙的密封剂，方便在上面涂抹饰面。该液态密封剂可由稀释油漆或清漆，或冲淡的胶水、黏合剂或油漆制成。

sized slates Slates that are of a consistent size. *See also* RANDOM SLATES.
等尺寸石板 大小一致的石板。另请参考“乱砌石板”词条。

sizing To use size to seal a surface, making it less porous and suitable to receive finishes.
上胶 用胶料对表面进行密封，减少空隙使其适于施加饰面。

skeleton construction A method of construction where all of the dead and live loads are supported by a structural frame of beams and columns. The external walls are non-load-bearing.
骨架构造 使所有的静荷载和活荷载都由一个梁柱结构框架支撑的施工方法。外墙为非承重墙。

skeleton core door A type of hollow-core *door.
骨架填心门 一种空心门。

skeleton stair An * open stair with * treads and no * risers.
框架楼梯 一种开敞楼梯，有踏板没有踢脚板。

skelp A piece of metal used to manufacture pipe or tubing.
管材 用于制造管道的金属。

sketch A rough freehand * drawing, produced to illustrate something quickly.
草图 徒手草画的图纸，用于快速说明。

skew Positioned at an angle to the main component.
斜交 与主要构件成一定角度放置。

skewback The upper surface of a stone, brick, block, or top of a * springer that carries an arch. It is the starting point of the arch stones.
拱座 石头、砖、大块物体的上表面，或者承载拱门的起拱石的顶部。是拱石的起始点。

skewback saw A saw with a slightly curved and concaved back.
背弯锯 背部略弯呈凹形的锯。

skew corbel An overhanging element of brick or stone at the foot of a gable that is used to carry the brickwork and stone of the extended gable roof.
戗檐砖 山墙脚处的悬挂砖材或石材，用于承接延伸的山墙屋顶的砌砖或石材。

skewnail (skew nailing, toe nailing, tusk nailing) Inserting nails at alternating angles to securely fix two or more pieces of timber together.
斜交钉 以交错角度钉入钉子，将两块或多块木板牢牢固定在一起。

skid-mounted equipment Plant and equipment that is mounted on skids to allow the equipment to be pushed or pulled into positon.
橇装设备 安装于滑轨上，可将设备推拉到位的装置设备。

skids Ski-like rails placed under heavy plant and equipment, that enable the objects being moved to slide into position.
滑轨 放于重型设备装置下方的像雪橇一样的轨道，能使物体滑动安装到位。

skilled operative (craft operative) A skilled worker on-site with a training in a craft such as plastering, joinery, electrics, plumbing, ground-works, and bricklaying.
技工 接受过抹灰泥、细木工、电气、给排水、地基工作以及砌砖技术培训的现场技工。

skim coat (skimming coat) The final coat of plaster, maximum 3mm thick, made of a fine plaster, is applied with a trowel filling perforations and undulations in the substrate below. It can be applied as a finish to cover over plasterboard joints, or can be applied over the whole area of plasterboard, or to cover a plaster basecoat.
腻子粉层 最后的抹灰层，最多 3 mm 厚，用抹子涂抹细灰泥，填充下方涂层中的孔隙和凹凸。可用作覆盖石膏板接缝的饰面，或者涂抹整个石膏板区域，或覆盖石膏底涂层。

S

skintle A way of placing bricks such that they are stacked out-of-line.
凸砖 不成直线堆砌砖块的方式。

skirt 1. Material placed around the lower edge of an object, and used to dress or protect the lower face. 2. A downstand on a foundation that penetrates into the ground, and used to prevent the ground being washed from under the foundation.
1. 裙座 放于物体下沿的材料，用于装饰或保护底面。**2. 侧板** 打入地面下基础上的挡板，用于预防基础下土壤被水冲走。

skirting (skirting board) A timber or PVC board that is run along the lower edge of a wall to provide a neat aesthetic finish between the plaster and the floor. The board can be fixed with nails or adhesive, or be plugged and screwed.

踢脚板 沿着墙壁底边的木板或聚氯乙烯板，在抹灰层和地板之间提供外表整洁美观的饰面。可使用钉子、胶粘剂，或者用嵌入并拧紧的方式将踢脚板固定到位。

skirting trunking (US plugmold, wireway) A box section panel positioned along the base of a wall and used to run electrical cables through it.
踢脚线线槽 沿着墙基铺设的方通形面板，供电缆线穿过。

sky component (sky factor) The light received directly from the sky. One of the three components required to calculate the * daylight factor.
天空直射光 从天空直接接收的光。计算采光系数所需的三个分量之一。

skylight *See* ROOFLIGHT.
天窗 参考“天窗”词条。

slab 1. A flat block of concrete, slate, or stone that is used for paving footpaths and roads. 2. A * solid concrete floor.
1. 厚板 平整的水泥板、石板或石头，用于铺设人行道和路面。**2. 实心混凝土板**。

slab-on-grade A concrete floor slab resting directly onto the ground or a sub-base of hardcore.
板式基础 直接置于地基上或硬面垫层上的混凝土板。

slag (blast furnace slag) A by-product of steel and copper production that has glassy characteristics. The material can be used as * pozzolans as a partial cement replacement in some concretes and mortars. The introduction of the slag gives concretes and mortars different characteristics; depending on the mixes, the addition can improve the strength and setting qualities. Slag can also be used to make mineral wool with acoustic and thermal properties. The slag residues from the production of iron and copper have slightly different characteristics and properties.
炉渣 钢铁和铜生产过程中的副产品，有玻璃特征。可像火山灰一样在某些混凝土和砂浆中代替部分水泥。加入炉渣的混凝土和砂浆按混入的比例有不同的特性，可增加其强度和凝固质量。炉渣也可用于制造矿棉，有隔音和保温作用。炼钢和炼铜产生的矿渣在特征和性能上有些微小的差异。

slag wool Mineral wool made from fine filaments of blast furnace slag. The dense properties of the mineral give the wool good sound absorption properties, as the wool material traps and holds still air; it also resists the flow of heat.
矿渣棉 高炉炉渣的细纤维制成的矿棉。矿物致密的特性使其有良好的吸音效果，因为棉质材料可留住静止的空气；同时也可以阻止热流动。

slaked lime [calcium hydroxide, $Ca(OH)_2$] Hydrated lime.
熟石灰（氢氧化钙）水化石灰。

S

slam buffer (US mute, rubber silencer) A nail or screw with a rubber dome that is fixed into the jamb or rebate of a door to reduce the noise of the impact if the door is slammed shut. The dome sits just proud of the surface, providing a cushion to the impact of the door.
橡胶防撞头 把带有橡胶圆头的钉子或螺钉钉入门边梃或门的凹缝中以降低门突然关上时的噪音。橡胶圆头从表面微微突出，在门关闭时提供缓冲。

slamming strip (slamming stile) A vertical strip of wood attached to the shutting * stile of a door that the door closes against.
闭锁门碰条 安装于闭锁竖梃上的竖直木条，门关闭时会被碰到。

slat A thin strip of wood used in fencing and on louvred windows. On louvred windows the slats are held in a frame, placed at an angle to provide shade from the sun, and allow light to penetrate through the gaps.
板条 栅栏及百叶窗中使用的窄木条。百叶窗中板条被固定在框架中，以可遮挡阳光的角度放置，同时光线可以穿过缝隙透过来。

slate A fine-grained material derived from sedimentary rock and commonly used as a roofing material. The sedimentary rock is composed of volcanic ash or clay, which is compressed and has metamorphosed in layers of

the deposits. Due to the cleavage and grain of the slate (lines of separation) it can be split into thin sheets and used for roofing. Slate is very durable, with minimal maintenance, and slate roofs can exceed the lifetime of the occupants.
板岩 沉积岩中开采而来的细纹理材料，常用作屋面材料。沉积岩由火山灰或黏土构成，受压缩后变质成层状沉积。由于存在裂缝和石板纹理（分隔线），可将其分成薄板用于屋顶。板岩十分耐用，维护成本极低，且板岩顶板的使用寿命可超过居住者寿命。

slate boarding *Close boarding installed on a roof.
石板瓦 安装于屋顶上的无缝拼板。

slate hanging (weather slating) Slates hung vertically.
挂石板瓦 垂直悬挂的石板瓦。

slating Roofing with slate tiles.
石板屋顶 石板瓦屋顶。

sledge A flat platform-like vehicle on runners, used to pull equipment behind tunnelling machines.
滑橇 滑道上平台式的运载工具，在隧道开掘机后面拉动设备。

sledge hammer A large heavy *hammer used to break up rock, stone, and hardened concrete.
大锤 用于打碎岩石、石头和硬质混凝土的重型大锤。

sleeper 1. A timber with a large cross-sectional area that is used to support the rails of a railway track. 2. A horizontal strip of wood placed in or over concrete to support the floor covering above.
1. 枕木 支撑轨道铁轨的横截面积很大的木材。**2. 地龙骨** 放于混凝土内或其上方的水平木条，用于支撑上层地板。

sleeper clip *See* FLOOR CLIP.
地板卡扣 参考“地板卡扣”词条。

sleeper plate The wall *plate on a *sleeper wall.
地垄墙垫板 地垄墙上的梁垫。

sleeper wall (basement wall) Wall laid to carry the timber floor joists or concrete beams. For timber floors the wall is usually laid in *honeycomb bond with spaces between the bricks allowing air to circulate under the *floor. *See also* DWARF WALL.
地垄墙 用于承接木地板龙骨或混凝土梁而铺设的墙壁。木地板中地垄墙通常以蜂窝式砌合，砖之间的空隙允许空气在地板下流通。另请参考“矮墙”词条。

sleeve (expansion sleeve) A *pipe sleeve.
套筒（膨胀套筒） 管道套筒。

sleeve coupling (sleeve connector, sleeved joint, socket coupling) A small hollow cylinder used to join the ends of two pipes together.
套筒接头 用于把两个管端连接到一起的小的空心圆筒。

sleeve piece (thimble) 1. A *pipe sleeve. 2. A *ferrule.
1. 管道套筒。**2. 金属箍**。

slender Tall and thin.
苗条的 细长的。

slewing The circular or rotational movements of a crane jib. Most of the crane's movements tend in a circular direction as the jib extends out from a central point. The lifting of a *luffing jib or *trolleying along the jib arm makes the movement outwards and inwards from the crane's central point.
回转 吊臂的环形或旋转运动。大多数吊车都是吊臂从一个中心点延伸作圆形旋转。提升动臂或者滑车沿着吊臂移动使吊车向着或远离吊车中心移动。

sliced veneer Veneer made by cutting a thin slice of timber, using a machine with a knife blade up to 5 m

long that slices the timber away from the main timber log (fitch). The timber is often heated in a vat of water, so that it is durable when cut. The veneers generally range in size from 0.2 to 1 mm thick.
平切单板 使用一个带有长达5 m的刀片的机器从圆木木材（fitch）上切割成的薄木板。通常在一桶水中加热木板，这样切割时会更耐用。薄板的厚度从0.2到1 mm不等。

slick A film or polluting liquid such as oil floating on the sea that has resulted from an oil spill, for example.
浮油膜 漂浮在海洋上的膜或石油等污染液体，例如由于油泄漏造成的浮油膜。

slide bolt (thumb slide) A small *barrel bolt usually found at the top and bottom of double doors.
天地插销 常见于双扇门顶部或底部的小的圆插销。

sliding door A door that is opened by sliding it from side to side.
滑动门 通过从一侧滑动到另一侧来打开的门。

sliding doorlock A lock for a *sliding door.
滑动门锁 滑动门使用的锁。

sliding-folding door *See* FOLDING DOOR.
推拉折叠门 参考“折叠门”词条。

sliding window (sliding sash, slider) A horizontally opening *sash window.
左右推拉窗 横向打开的推拉窗。

slip A thin piece of material.
片 薄片材料。

slip brick (brick slip) A thin strip of brick used as a tile for cladding, giving the impression of a solid brick wall.
面砖 用作覆面瓦的薄长条砖，形成一种实心砖墙的效果。

slip feather (slip tongue) A thin strip of timber that fits into grooves located at the edge of panels or boards to keep the boards aligned.
滑榫 嵌入面板或木板边沿槽口的薄木条，保证木板对齐。

slipform A type of formwork that is designed to slowly move in the direction of the pour, continuously or at short intervals, to prevent the formwork from constantly being struck and erected. Used to form concrete shaft linings, towers, and pavements.
滑模 被设计为向倾斜方向缓慢移动的一种模板，它可连续不断地或在短间隔内缓慢移动，以免不断地拆模和支模。用于构成混凝土井筒垫层、塔和路面。

slip joint A joint made between pipes by sliding the end of one pipe into the end of the other.
伸缩节 管道之间的接头，通过滑动一个管端接入另一个管端。

slip mortise (slip mortice) A mortise, groove, or chase ready to receive a *slip tongue or feather.
滑槽榫眼 可以承插滑榫头或斜面榫条的榫眼、槽口或斜槽。

slipper bend A *channel bend.
明沟弯头 参考“明沟弯头”词条。

slip-resistant (non-slip floors) Floors with perforations, embossed patterns, ribs, studs, and inlays of abrasive material that resist slips.
防滑的（防滑地板） 有防滑作用的，带孔隙、浮雕花纹、肋条、纽钉以及黏性材料镶嵌的地板。

slip sill A *sill that is the same length as the width of the door or window. *See also* LUG SILL.
门槛/窗台 与门或窗的宽边等长的槛或窗台。另请参考“窗台边耳”词条。

slope 1. A side of a hill. 2. A surface that is inclined to the horizontal.
1. 山坡。2. 斜面 倾斜于水平面的面。

slope deflection method An analysis method used to define the stiffness of a structure.

倾角挠度法 用于确定结构硬度的分析方法。

slop hopper A large sanitary fitting used in hospitals to dispose of human waste.
地拖池 医院中用于处理废水的大型卫生设备。

slop sink A large deep sink used to dispose of spilled or splashed liquid.
洗涤槽 用于处理溢出或溅出的液体的深水池。

slot A groove or thin rebate cut into a surface. Some screws have a slot that allows a flat-bladed screwdriver to be located and the screw to be rotated and driven into the wood.
狭槽 切于表面的凹槽或狭槽。有些螺丝钉上有可供一字头螺丝刀嵌入的凹槽，螺丝刀嵌入槽后旋转将螺丝钉拧进木板中。

slot diffuser *See* LINEAR DIFFUSER.
条缝型散流器 参考“线型出风口”词条。

slot drain A drain that is hidden beneath a narrow gap in the surface that is being drained.
缝隙排水 隐藏于表面排水窄缝下方的排水管。

slough 1. An area of low-lying muddy ground. 2. A mud hole.
1. 泥沼 一片低洼泥泞的地面。**2. 排垢孔**。

slow bend (long sweep bend) A long shallow bend. *See also* QUICK BEND.
长半径弯头 半径长的弯头。另请参考“短半径弯头”词条。

slow sand filter A graded sand filter for treating drinking water that requires little or no power. The filter involves a bed of fine sand (0.1 – 0.3 mm diameter; typically 0.6 – 1.2 m deep), which overlies a thick gravel bed. Sometimes a coarse layer of charcoal or low-grade coal is used between these two layers to aid colour adsorption. A constant head of water (1.0 – 1.5 m) is allowed to flow through the bed at rates of 0.2 – 0.4 m/hr. The slow filtration rate allows a thin layer of microorganisms (schmutzdecke) to form within the sand bed, which provides microbiological treatment and helps to remove colloidal material. The schmutzdecke layer remains active, providing the sand bed stays wet. Periodically, the schmutzdecke layer becomes overgrown, allowing colloidal material to build up and eventually block the filter. The top region of the sand layer is replaced or cleaned and the schmutzdecke layer allowed to regrow. *See also* RAPID SAND FILTER.
慢滤池 用于处理饮用水的级配沙滤池，只需很少甚至不需要电力驱动。该沙滤池包括一层细砂（直径0.1—0.3 mm；通常为0.6—1.2 m深），下面再铺一层砾石层。有时候还会在这两层中铺一层粗粒炭或者低级煤用于颜色吸收。用恒定的水头（1.0—1.5 m）以每小时0.2—0.4 m每小时的速度流过滤床。缓慢的过滤可在砂床上形成一层薄的微生物（去污菌），通过微生物处理去除胶质物质。去污菌层保持活性，使砂床保持湿润。最终去污菌层过度生长使胶质物质堆积并堵塞滤池。替换或清理顶层的砂层让去污菌层再生。另请参考“快滤池”词条。

sludge The solid waste that has settled out during sewage treatment.
污泥 污水处理中沉淀下来的固体废物。

sludger A bailer.
挖泥筒 一种抽泥筒。

sluice An artificial water channel or floodgate, used to control the flow of water.
水闸 人造水渠或水门，用于控制水流。

slump The collapse or downwards movement of a body under its own weight, e.g. the vertical downwards movement of concrete in a *slump test.
坍落 物体在其自重下倒塌或向下运动，例如坍落度试验中水泥的垂直向下运动。

slumping The movement of a substance such as a sealant or concrete downwards; the material flows downwards under the force of gravity. Most modern sealants applied by gun are designed to be non-slumping. Concrete slumps when piled in its wet state; once the process of hydration has taken place the concrete has set and should not slump.

滑塌 物质如密封剂或混凝土等的向下运动；在重力作用下向下流动。现代用密封枪敷涂的大多数密封剂都是非坍塌型的。混凝土在湿态下堆叠会向下滑塌；发生水合作用后混凝土就会凝固而不再滑塌。

slump test Concrete test used to determine the * workability of concrete. The test uses concrete compacted in three layers within a cone that is upturned and the difference between the cone and slumped concrete is measured. The steel cone is 300 mm tall with a 100 mm closed bottom and 200 mm diameter open top. The concrete is loaded into the cone in 100mm layers, tamped, and compacted at each layer, and once the upper layer of the cone is compacted and levelled, the cone is upturned and the concrete discharged to form an inverted cone. The * slump recorded is the measure of the distance from the top of the steel cone to the top of the slumped concrete.
坍落度试验 测定混凝土和易性的混凝土测试方法。该测试将混凝土在锥桶内分三层压实，接着将锥桶倒转过来后测量锥桶和塌落后混凝土顶平面的高度差。钢锥桶为 300 mm 高，封闭底部直径 100 mm，开口顶部直径 200 mm。混凝土按每层 100 mm 高灌入，每层捣实，当最顶层捣实并抹平后，将锥桶倒转过来使混凝土倒出，形成一个倒锥体。钢锥桶顶部和塌落混凝土顶部的高度差即为坍落度。

slurry A liquid mix of cement and water, or cement, fine aggregate, and water.
泥浆 液态的水泥和水，或者水泥、细骨料和水的混合物。

small-bore system A type of central heating system that uses small diameter pipework, usually around 15 mm, to feed the heat emitters.
小管采暖系统 一种中央供暖系统，使用小口径管道（直径通常为 15 mm 左右）对散热器供暖。

small-bore unit A boiler that can be used to feed a * small-bore system.
小管采暖锅炉 可用于供应小管采暖系统的锅炉。

SMM *See* STANDARD METHOD OF MEASUREMEN.
SMM 参考“标准计量方法”词条。

smoke alarm An alarm that detects smoke and emits an audible alarm when activated.
烟雾报警器 启动后可监测烟雾并发出音响警报的报警器。

smoke chamber The area in a chimney above the fireplace where smoke gathers prior to entering the flue.
烟腔 壁炉上方的烟囱中的一片空间，烟在进入烟道之前在此聚集。

smoke control The process of restricting and controlling the flow of smoke within a building.
排烟控制 限制和控制建筑物内烟雾流动的过程。

smoke detector An electrical device that is designed to detect the presence of smoke and activate a fire alarm.
烟雾探测器 用于探测烟雾并激发火警警报的电气装置。

smoke door (smoke control door, smoke stop door) A door that is designed to prevent the passage of smoke.
防烟门 防止烟雾通过的门。

smoke explosion *See* BACKDRAUGHT.
烟气爆炸 参考“逆通风”词条。

smoke extract fan A fan that is designed to remove smoke from a building.
排烟风机 用于从建筑物中排烟的风扇。

smoke-logging The filling of a space with smoke.
烟雾弥漫 让一个空间充满烟雾。

smoke outlet An opening that allows smoke to escape from a space.
排烟口 从一个空间排出烟雾的开口。

smoke pipe A * flue.

烟管 烟道。

smoke rocket (rocket tester) A rocket-shaped canister filled with smoke used to undertake a *smoke test.
烟雾筒 装满烟雾的火箭状的筒，用于进行通烟测试。

smoke shelf A horizontal ledge located above the fireplace and below the smoke chamber. It prevents down draughts and collects any rain entering the chimney.
烟挡 壁炉上方烟腔下方的水平壁架。可防止倒灌风并收集进入烟囱的雨水。

smoke test A test that can be undertaken on drainage systems to identify leaks. Would be undertaken after an air or water test.
通烟测试 可在排水系统内进行的测试，用于确认是否存在泄漏。在气密或水密测试后进行。

smoke venting *See* FIRE VENTING.
出烟口 参考"排烟"词条。

smoothing (resource smoothing) The balancing out of human and mechanical resources on a project by the relocating of tasks and activities, ensuring that resources are used to maximum effect with minimum breaks in activity.
均衡（资源均衡） 在项目中通过重新部署任务和活动来平衡人力和机械资源，确保最大限度利用资源及最低程度打断活动。

smoothing compound A liquid grout or self-levelling compound.
流平化合物 液浆或自流平化合物。

snagging Ensuring all of the jobs, minor defects, and finishes are checked and rectified before handing the building over.
交付前检修 指交付建筑物之前确保对所有的工作、小瑕疵和饰面都进行了检查和修整。

snagging list A document or schedule containing a list of all the jobs that need finishing, defects that need correcting, and adjustments made to ensure the building and structure meet with the specified standard.
检修清单 包含了为确保建筑物和结构达到规定标准而记录需要完成的工作、需要调整的瑕疵、需要进行的调整的文件或清单。

snake (electric eel, plumbing snake, toilet jack) A flexible *auger that is used to unblock drains.
弹簧通渠器 用于疏通管道的弹性螺旋钻。

snap header (half bat) A brick cut to approximately half its length or 100 mm long that is positioned in a wall with its uncut header exposed, giving the impression that the header brick penetrates fully into the wall.
半砖 切成大概一半砖长或 100 mm 长的砖，砌墙时让未割丁面朝外，留下整块砖完全砌入墙壁的外表面。

snapping line (snap line) A chalk line used for marking a straight line over a surface. A string line is pulled through chalk dust, the chalked string is then pulled taut over the surface, from point to point, where the straight line is required. Gripped by the worker's fingers the string is pulled slightly off the surface and then allowed to snap back into position and onto the surface causing the chalk to leave the string and mark the wall, floor, or board.
弹线 在表面上做直线标记的粉笔线。将绳子粘上粉笔灰，然后将绳子在需要连直线的两点之间拉紧。工人用手指将线微微拉开然后弹回到表面上，把粉笔灰从绳子上弹下并在墙壁、地板或板上留下标记。

snap-ring joint *See* O-RING JOINT.
扣环接头 参考"O 形环接头"词条。

snap tie A tie for formwork, made of notched flat steel, enabling the exposed end of the tie to be broken off once the concrete has been cast and the formwork is removed.
穿墙螺栓 用于模板的螺栓，由锯齿状扁钢制成，当混凝土完成浇筑，模板已经移开，螺栓的裸露端就可以断开。

sneck In stone rubble walls, a small stone less than 75 mm high, distinctly smaller than normally cut stone.
（砌石墙时填补空隙用的）方形小垫石 在毛石墙中，高度小于 75 mm 的明显比正常切割小的小石块。

snecked rubble A rubble wall built of irregular-sized stones or small cut stones.
杂乱毛石 由不规则尺寸的石块或小琢石砌成的毛石墙。

snib (check lock) 1. A hand-operated catch used to alter the position of the latch within a lock. 2. (Scotland) A bolt, catch, or lock for a door or window.
1. 反锁 手动的门闩，用于改变锁内门闩的位置。**2. 插销（保险锁）（苏格兰）** 门或窗户的插销、拉手或锁。

snib latch A *latch that is operated using a snib.
插销锁 使用插销的门闩。

snots Excess mortar on the face of masonry that has been allowed to drip or drop onto the horizontal and vertical surfaces of the masonry.
流挂 砖石表面过剩的灰浆，会滴落到砖石的水平和垂直表面。

snow board [**snow guard, snow slats,** (Scotland) **snow cradling**] A device installed on a pitched roof above the eaves designed to prevent any damage that may be caused by large accumulations of snow sliding off the roof. The board retains the snow, allowing it to either drop off in small amounts or melt completely. Various types of snow board are available including horizontal boards, horizontal slats, wire meshes, horizontal pipes, or a series of individual pads.
挡雪板［拦雪板、雪栏、（苏格兰）挡雪板］ 一个安装在斜屋顶屋檐以上，用于防止大量堆积的雪从房顶滑落可能造成损害的设备。挡雪板只允许少量雪滑落或使其完全融化。有各种类型的挡雪板，如水平木板、水平板条、金属网、水平管或一连串单个垫板。

soakaway A hole in the ground, either open or filled with stone, designed to temporarily hold excess surface water until it can drain away into the surrounding ground.
渗水坑 地面上的一个孔洞，敞开或用石块填充，用于暂时拦住过剩的地表水，直至它排入周围地表。

soaker A small section of flexible sheet metal, usually lead, used to make a watertight joint between a pitched roof and an abutment. The soakers are overlapped and are held in place by turning the top edge of the soaker over the tile on the pitched roof. A stepped *flashing is then used to hold the upstand in place and provide weather tightness.
柔性泛水 一小段软性金属薄片，通常是铅，用于制作斜屋顶和突出结构之间的防水接头。柔性泛水重叠，通过将顶部边缘翻过来覆在斜屋顶的瓦片上，将其固定于适当位置。然后作阶形防水层固定住立铺卷材，提供耐候防护。

soap 1. Release mould oil painted onto formwork to reduce adhesion and improve striking 2. A brick of standard length but with its width equal to its height. The standard dimensions of a soap brick in the UK are 215 mm × 46 mm × 46 mm.
1. 涂脱模油 将脱模油涂在模板上，以减少黏附力和帮助脱模。**2. 长条砖** 标准长度的宽度和高度相等的砖块。在英国，长条砖的标准尺寸为 215 mm × 46 mm × 46 mm。

socket 1. The enlarged end of a section of pipe into which a *spigot is inserted. 2. A hollow cylindrical pipe fitting that contains a thread on the inside. 3. A *socket outlet. 4. An electrical outlet into which a lightbulb is inserted (**light socket**).
1. 管道承口 管子段的扩大端，可插入插口端。**2. 空心圆柱管件** 内里有螺纹。**3. 插座**。 **4. 灯头** 插入灯泡的电源插座（电灯插座）。

socket coupling *See* SLEEVE COUPLING.
套管 参考"套筒联轴器"词条。

S

socket former A hand-held tool used to join copper pipe to form a socket that can be jointed using a * capillary joint.
插口成型机 一个用于连接铜管以形成可用细管接头连接的插口的手提式工具。

socket inlet The electric plug that is inserted into a * SOCKET OUTLET.
进线插头 插入插座的电源插头。

socket iron pipe A cast-iron pipe with a spigot and * SOCKET JOINT.
承插铁管 带承插式接口的铸铁管。

socket joint *See* SPIGOT and SOCKET JOINT.
承插式接口 参考"承插式连接"词条。

socketless pipe *See* UNSOCKETED PIPE.
无承口管 参考"无承口管"词条。

socket outlet (US receptacle, power outlet, socket) An electrical outlet into which the electric plug from an appliance is inserted.
插座（美国 插座、电源插座、插口） 可供电器插头插入的电源插座。

sodium A metallic element, symbol Na, sometimes used as an alloying element in steel and other commercial metals. Sodium is a soft white metal, has a white silverly colour, oxidizes in air, and reacts violently with water.
钠 一种金属元素，符号为 Na，有时是指钢铁中的合金元素和其他工业金属。钠是一种白色软金属，银白色，在空气中发生氧化，遇水剧烈反应。

Sod's law *See* MURPHY'S LAW.
墨菲法则 参考"墨菲定律"词条。

soffit (soffite) The underpart of an overhang or overarching element of a building or structure.
底部（下表面） 建筑物或结构悬垂或拱形构件的下部。

soffit board (eaves lining, US plancier piece) A flat board or lining placed under the overhanging part of the eaves.
底板（挑檐底板，挑檐底面） 在屋檐悬垂底下的平板或衬板。

soffit form (soffit shutter) Formwork used for the underside of a floor or ceiling deck, beam, or overhang.
底板模板 用于地板或顶棚衬板、横梁或飞檐底下的模板。

softboard A soft, porous particleboard used for insulation.
软质纤维板 用于隔热的软质多孔刨花板。

S

soft landscaping (organic landscaping) Live plants, trees, bushes, and shrubs within the external works.
软质景观（有机景观） 室外工程中的活体植物、树木、灌木和灌木丛。

soft sand Sand that has rounded, rather than sharp aggregate, and is used for render and mortar as it is easier to work. It is rarely used in concrete.
细沙 是圆形的沙子而不是有棱角的骨料，用于粉刷和抹灰，更方便作业。很少用于混凝土。

software A computer program or operating system.
软件 电脑程序或操作系统。

soft water Water that has a low concentration of calcium carbonate, a hardness value of less than 60 mg/L, and as such, it is easy to form a lather with soap.
软水 碳酸钙浓度低且硬度值小于 60mg/L 的水，（相比于硬水）与肥皂共用时更易产生泡沫。

softwood The description of timbers that belong to the gymnosperms group, being different from hardwoods. Softwoods are generally faster-growing and cheaper than hardwoods, they are considered less decorative, and generally are less prone to * moisture movement.

软木材 属于裸子植物科的木材，与硬木相对。软木通常比硬木成长更快也更便宜，软木装饰性较弱，通常更不易发生水分渗透。

soil branch A branch *pipe that is connected to the *soil stack.
排污支管 连接至污水立管的支管。

soil cement The binding clay used in *cob construction.
水泥土 在夯土中使用的结合黏土。

soil classification *See* AIRFIDLD SOIL CLASSIFICATION.
土壤分类 参考“飞机场土壤分类”词条。

soil drain A drain that conveys the waste from the *soil pipe to the *sewer.
排污渠 将污水从排污管排至下水道的排水渠。

soil fabric friction The amount of shear resistance developed at the interface between a soil and a fabric, such as a *geotextile.
土壤织物摩擦 土壤和织物（例如土工布）接触面产生的抗剪力。

soil fitment (soil appliance) A *sanitary appliance that is connected to the *soil pipe.
便溺污水设备（排污设备） 连接至排污管道的卫生设备。

soil mechanics The investigation, classification, and determination of a soil's properties and parameters, together with its behaviour.
土力学 对土壤性质、参数和性能的调查、分类和核定。

soil mixing The mixing of a soil with another material, e. g. lime, to improve its behaviour.
土壤混合物 土壤和其他材料，例如石灰混合，以提高其性能的混合物。

soil nail A steel rod that has been inserted into the ground, typically on a *slope, to increase stability.
土钉 插入地面的钢杆，通常用在斜坡上以增强稳固性。

soil pipe A pipe that conveys the liquid and solid waste from a *sanitary appliance.
排污管 排放卫生器具中液体和固体废物的管道。

soil profile The vertical layers of soil at a particular location.
土壤剖面 在某一特定位置土壤的垂直切面。

soil-reinforced wall A mass of soil that has been retained internally by the use of embedded *geotextile layers (or steel strips). Cladding is used to provide an aesthetic finish and prevent soil loss rather than to provide structural support.
加筋土挡墙 用嵌入的土工布层（或钢带）从内部挡住大量土壤的墙。墙面板可作为美观的饰面，也可防止土壤流失，而不是提供结构支承。

S

soil sampler A tube-like device that is inserted into the ground to collect an undisturbed soil sample, e. g. a **split barrel sampler** is used to obtain soil samples from a borehole.
取土器 一个用于插入土地中收集原状土样本的管状设备，例如劈管式采样器，用于从钻探孔中取得土壤样本。

soil stabilization The use of another material, as in *soil mixing, to improve certain characteristics of the in-situ soil, e. g. shrinkage, optimum dry density and shear strength.
土壤加固 在土壤混合中使用其他材料，以提高现场土壤的某些特性，例如收缩性、最大干密度和抗剪强度。

soil stack A vertical *soil pipe that conveys the waste to the *soil drain. It is usually vented at its highest point.
排污立管 将污水排至排污渠的垂直排污管道，通常在最高点通气。

soil-structure interaction The study of how soil and structure will perform together, for example, the amount

and effects of differential settlement.
土与结构的相互作用 对土壤和结构如何一起作用的研究。例如，不均匀沉降的量和影响。

soil vent A * vent pipe used to ventilate a * soil stack.
排污通风管 用于给排污立管通风的通风管。

soil water Water that is discharged from a * soil fitment.
污水 从便溺污水设备中排出的水。

solar collector (solar panel) A device designed to collect solar radiation.
太阳能集热器（太阳能电池板） 设计用以收集太阳辐射的设备。

solar control film A thin film of material that is applied to a window to convert it into * solar control glazing.
玻璃隔热膜 贴在窗户上使窗户变成遮阳玻璃的薄膜材料。

solar control glazing (anti-sun glass) Glazing that filters out certain wavelengths of infrared radiation emitted from the sun, but still allows the transmission of visible light. Used to reduce solar heat gain and control glare.
遮阳玻璃（防晒玻璃） 过滤掉太阳发射出的特定波长的红外线照射但允许可见光透射的玻璃。一般用于减少太阳热量的吸收和控制眩光。

solar dial A 24-hour time switch used to switch lighting on and off automatically throughout the year.
光控定时开关 一个 24 小时的开关，可用于全年自动打开和关闭照明。

solar energy Energy derived from the sun.
太阳能 太阳产生的能量。

solar gain (solar heat gain, passive solar gain) The amount of additional heat gained within a space by the transmission of solar radiation.
太阳辐射（太阳辐射热、被动式太阳能） 在一个空间内通过传输太阳辐射所获得的余热。

solar heating The process of utilizing solar radiation from the sun to heat a building. *See also* PASSIVE SOLAR HEATING.
太阳能供暖 利用太阳辐射为建筑物供暖。也称为被动式太阳能供暖。

solar hot-water system A type of hot-water heating system where solar radiation from the sun is used to preheat a glycol mixture as it passes through a * solar collector. The pre-heated mixture is then passed onto a hot-water storage tank where it may be heated further, usually using a conventional heat source, until it reaches the required temperature.
太阳能热水系统 一种利用太阳能辐射预加热太阳能集热器中的乙二醇混合物的热水供暖系统。预热混合物经过热水箱，可进一步加热传统的热源，直至达到所需的温度。

S

solar load The amount of cooling that is required to counteract solar gain.
太阳能负荷 中和太阳能需要吸收的冷却量。

solar protection A device that is used to protect a building against excessive amounts of solar gain, such as louvres or an overhang. External devices are much more effective than internal devices.
遮阳隔热 使建筑物不吸收过量太阳能的设备，例如百叶或檐挑。室外设备比室内设备效果更好。

solder A lead (Pb) and tin (Sn) alloy with a low melting point that is used to join conductors together.
焊锡 用低熔点的铅（Pb）和锡（Sn）的合金连接导体。

solder balls Spherical parts of * solder.
锡球 焊锡的球形部分。

solder bridging A * solder paste used to make a bridge or path.
锡焊桥 用于桥接的一种锡膏。

soldering A technique used to join conducting objects or metals together using a lead-tin alloy. The solder has a low melting point of 190℃, the joint and solder are heated, and a sound electrical and mechanical joint can be formed.
焊接 使用铅锡合金将导体或金属连接的一种技术。加热接口和焊料，焊料有190℃的低熔点，然后就可以形成一个牢固的电气和机械接头。

soldering iron A piece of metal that can be heated then used for *soldering and used to melt and apply *solder.
烙铁 可加热的金属片，用于焊接，也用于融化焊料。

solder mask A mask used to cover parts not intended for soldering.
焊接掩膜 用于遮盖不打算焊接部分的掩膜。

solder paste A mixture of minute spherical solder particles, activators, solvent, and a gelling or suspension agent.
焊膏 微小的球状焊料颗粒、催化剂、溶剂和胶凝剂或悬浮剂的混合物。

soldier A short upstanding member of timber or masonry. Soldier timbers are positioned upright and level, and used for grounds fixing *skirting boards and other board finishes. Bricks on end can also be described as soldiers; *See also* SOLDIER COURSE.
立件 木材或砌石的直立短构件。立件木材垂直水平放置，用于地板安装、踢脚板和其他面板装饰。竖立着的砖块可称为立砌砖；参考“立砌砖层”词条。

soldier arch A flat arch created from a row of bricks on end.
立砌砖平拱 从一排竖立的砖块开始砌的平拱。

soldier course A row or course of bricks laid on end.
立砌砖层 直立铺排的一排或一层砖块。

solenoid A coil of wire that surrounds an iron core to form an electromagnet.
螺线管 在铁芯外缠绕以形成电磁铁的一盘线圈。

solenoid valve An electrically operated valve that can be either opened or closed automatically.
电磁阀 一个电动操作的阀门，可自动打开或关闭。

sole plate (footplate, ground sill, mudsill, sole piece, US abutment piece) Large pieces of supporting timber or railway *sleepers laid on the ground to support *scaffolding standards, *struts, *props, *shores, or *raking shores.
底板（垫板、卧木、底梁、下槛，美国垫底横木） 铺设在地面以支承脚手架、撑杆、支柱、撑柱或斜撑柱的大块支承木或铁轨枕木。

S

solid bedding A layer of mortar or adhesive that covers the whole area under the brick or block leaving no gaps or voids.
灰浆层 覆盖在砖块或砌块下整个区域，不留空隙或空洞的灰泥层或胶粘层。

solid block A concrete block with no perforations or holes that have been deliberately cut into it.
实心砌块 已经专门切割好并且无孔洞的混凝土砌块。

solid brick A brick without any perforations or holes deliberately cut into it.
实心砖 无专门切出孔眼或孔洞的砖块。

solid bridging (US block bridging) Solid sections of timber used to add bracing between floor joists. *See also* SOLID STRUTTING.
搁栅横撑 用于增强楼板搁栅间支撑力的实心木型材。另请参考“搁栅横撑”词条。

solid core door (solid door) A door where a solid material, such as insulating foam, has been used to fill the space between the internal and external face of the door. *See also* HOLLOW-CORE DOOR.
实心门 用实心材料（例如绝缘泡沫）填充门内部和外部表面之间的门。另请参考“空心门”词条。

solid floor (solid slab) Concrete floor that rests directly onto compacted hardcore that is in contact with the ground. The term also refers to concrete that rests on insulation, but is still in contact with the ground. A solid floor is different from a suspended floor which has a space between the floor structure and the ground.
实心地板（实心板） 直接铺在与地面接触的压实碎石块垫层上的混凝土地板。也指铺在隔热层上的混凝土，但是与地面接触的。实心地板与架空地板不同，架空地板的楼板结构和地面之间留有空间。

solid frame A window or doorframe manufactured from a single piece of wood.
实木框架 由单块木材制成的窗户框架或门框。

solid glass door A door constructed completely from glass.
全玻璃门 完全由玻璃制成的门。

solidification The process whereby a liquid changes to a solid matter on cooling.
固化 液体经过冷却变为固体的过程。

solidification range The temperature range between the liquid and solid state for a material on cooling from the liquid state. Many solids (especially metals) do not have a specific or single melting point; indeed, melting occurs over a range, e. g. 20℃, this is the solidification range. When a metal or material is heated and enters this range, it undergoes incipient melting.
固化范围 材料从液体状态经过冷却时，液态状态和固体状态之间的温度范围。许多固体（尤其是金属）没有特定的或单个熔化点；实际上熔化是超过一个范围发生的。例如，20℃是一个固化范围。当金属或材料加热达到这个范围时，它就会初熔。

solidification shrinkage crack A crack that forms during * SOLIDIFICATION.
凝固收缩裂缝 在固化过程中形成的裂缝。

solid moulding (stuck moulding) A moulding that is cut or carved into the timber rather than being fixed or planted onto the surface of the timber.
雕刻线条 切入或雕入木材中而不是安装在木材表面的线条。

solid newel stairs A stone spiral stair, where the inner section of each step is carved so that it forms a round end that is laid directly over the rounded end of the step below, and directly under the rounded end of the step above. As the steps rotate round, a central newel is created from the overlapping stone steps.
实心中柱螺旋楼梯 石材旋转楼梯，每级阶梯内部都有切刻，这样就可以形成直接铺设在下级阶梯的圆端上的圆端，或形成直接铺设在上级阶梯的圆端下的圆端。因为楼梯旋转，交叉的石阶形成了中柱。

solid partition A partition wall without a cavity or gap for insulation.
实心隔断 隔断墙，没有空洞或空隙，用于隔热。

S

solid plastering In-situ plastering, not * DRY LINING. Plaster mixed on-site, placed, and floated against the wall or ceiling where it is desired. This is different from plastered surfaces that use plasterboard.
现场抹灰 在现场进行抹灰工作，而不是使用石膏板。现场搅拌、放置灰泥，按要求涂上墙壁或顶篷。这与使用石膏板的抹灰表面不同。

solid slab A solid concrete floor.
实心板 实心的混凝土地板。

solid stop A rebate with a stop that is cut into the doorframe rather than fixed to it.
实木门挡线 带有门挡功能的槽口，切入门框而不是安装在门框上。

solid strutting The insertion of sections of timber between floor or ceiling joists that are the same depth as the joists being braced. * Herringbone strutting uses less timber.
搁栅横撑 在地板或顶棚搁栅之间插入的木型材，深度与其支承的搁栅相同。人字形支撑使用的木材更少些。

solid timber A unit of timber that is made from one piece being carved, worked, and with rebates cut out to create the desired shape, rather than additional sections of timber planted or fixed to create the unit.

实木 用一块原木进行切割加工，切成所要求的形状的槽口，而不是用其他木材的部分去加工或安装而成。

solid wall A wall without a cavity or a gap filled with insulation.
实心墙 用隔热材料填充的没有空洞或空隙的墙壁。

solid wood floor A floor made out of strips or panels cut from a single element of timber rather than being built up as a * **laminated**, blockboard, * glulam, or * plywood floor.
实木地板 用从单块木材中切出的木条或板块制成的地板，而不是拼接而成的复合地板、木芯地板、胶合地板或多层板。

sound 1. An audible noise. 2. Goods or materials that are not damaged. 3. To measure the depth of a water body using * sonar. 4. An ocean inlet. 5. A channel between two larger water bodies or between an island and the mainland.
1. 可闻噪音。 **2. 无损的** 货物或材料完好无损。**3. 测量水深** 采用呐纳测量水体深度。**4. 海峡**。 **5. 两大水体或岛屿和大陆间的渠道**。

sound absorption (noise absorption) The process where some of the sound emitted from a source is absorbed as the sound strikes a material. All materials absorb sound to some degree. The absorption coefficient is a measure of the level of sound absorption exhibited by a material.
吸声（消噪） 某一声源发出的声音被吸收的过程。在某种程度上，所有材料都吸声。吸声系数是一个材料的吸声等级的衡量指标。

sound boarding *See* * PUGGING BOARDS.
隔音板 参考“隔声板”词条。

sounder A device used in alarms to make a sound, for instance, a bell, a horn, or a siren.
发声器 在警报中用于发声的设备，例如警铃、喇叭或警报器。

sound insulation (noise insulation) The main method used to control the transmission of airborne and impact sound from one space to another within a building. Four main factors will have an influence on the level of sound insulation achieved. These are completeness, flexibility, heaviness, and isolation. In addition, the quality of workmanship and detailing is also very important. A construction will only achieve its expected sound insulation if it is constructed without defects.
隔音（消噪） 用于抑制气载噪声在建筑物中从一个地方传播到另一个地方的主要方法。有四个主要因素会影响隔音等级，构件的完整度、弹性、重量和隔离度。另外，工艺和细节设计也非常重要。只有当构件无缺陷时，才能达到预期的隔音效果。

sound-level meter A device used to measure noise.
声级计 用于测量噪音的设备。

soundproofing The process of insulating a space against noise.
隔音 隔绝一个地方噪音的过程。

sound-reduction factor A measure of the reduction in the intensity of a sound.
隔音量 降低声音强度的程度。

space heating The process of heating a space with a heater.
小范围取暖 在一个空间内用加热器取暖的过程。

spacer Preformed unit of material that is used to maintain a gap, a distance between two objects, or hold one element of construction in a specific position. Spacers and packing * shims are used to temporarily hold the position of window frames, * SOLE PLATES, glazing bars, etc. while being fitted. Plastic, concrete, and steel spacers are used to maintain the * COVER and position of reinforcement bar in concrete. The spacers are tied to the steel reinforcement and positioned so that a cover of concrete is maintained between the ground and formwork. Spacers are also used to maintain the position and distance between reinforcement within the concrete; and they are used to allow the correct position of parallel sheets of wall formwork.

间隔件/垫片 预成型的材料，用于维持两个物体之间的间隙、距离或在特定位置托住结构组件。例如，垫片和分隔片用于在安装时暂时顶住窗户框架、底板和玻璃隔条的位置。塑料、混凝土和钢间隔件用于维持混凝土中钢筋的位置与保护层。间隔件系于钢筋并且固定，这样混凝土保护层就保持在地面和模板之间。间隔件也用于保持混凝土内钢筋的位置和间距；用于保持墙体模板平行板的正确位置。

spacer-lug tile A tile with spacing lugs attached to maintain the correct distance between adjacent tiles.
间隔凸缘瓷砖 附有定位凸缘的瓷砖，以保持相近瓷砖间的正确间距。

spacing The distance between two objects, typically in reference to their centre lines or * CENTRES.
间距 两个物体间的距离，尤其是指中心线或中心之间的距离。

spade 1. A hand tool used for digging, composed of a handle, shaft, and rectangular flat blade. The blade or spade can be easily and firmly pushed into stiff soil or clay enabling chunks to be extracted. The smaller the spade cutting end, the easier it is to insert and cut through and extract the soil. 2. A bulldozer with a large flat blade capable of pushing material over the ground in order to create a flat plane.
1. 铁锹 手工工具，用于挖掘，由手柄、杆和矩形平叶片组成。叶片或铲子可以轻易铲入坚硬的土壤或黏土中并铲出大块。叶片的刀刃越小，越容易铲入、刺穿和挖取土壤。 **2. 推土铲** 有大平叶片的推土机，可以推开地面上的材料，以便形成一个平坦的平面。

spall 1. To break rough edges of stone, masonry, and concrete with a * chisel or * spalling hammer. 2. For concrete or stone to break away from the main structure due to frost attack, aggregate, or material expansion and contraction caused by moisture or thermal movement.
1. 把（矿石等）破成碎片 用凿刀或碎石锤破碎石块、砌块和混凝土的边缘。**2. 剥落** 混凝土或石块由于潮湿或热运动造成的霜冻、骨料或材料伸缩而引起从主结构体脱离的现象。

spalling The breaking away or removing by force the edge or face of concrete or masonry; *See* * SPALL.
开裂 被迫从混凝土或砌体边缘或表面所产生的裂纹；参考“剥落”。

spalling hammer (spall hammer) A heavy hammer with a chiselled head used to break and shape concrete by hitting the surface and causing it to * spall.
碎石锤 带凿刻锤头的重锤，通过锤击混凝土表面致其裂开而进行破坏和塑形。

span The horizontal distance between two supports.
跨度 两个支撑件间的水平距离。

spandrel 1. The walled area between the arches in an arcade, which is an inverted triangular shape. 2. The triangular boxing out under a stairway to make a cupboard, store, or simply box out the area. 3. The area under a window, * spandrel panel.
1. 拱肩 在拱廊内拱门之间呈倒三角形的有墙的区域。**2. 楼梯下三角空间** 在楼梯下封起来的三角区域，形成橱柜、储物间或只是单纯将该区域封板。 **3. 窗槛下方** 窗户下的区域，窗槛墙。

spandrel panel The solid or opaque panel under a window on one floor which extends down to the window on the floor below.
窗槛墙 在一层楼中，窗户下的厚镶板或不透明板，延伸至下一层楼的窗户。

spandrel step Triangular stone stairs, which when interlocked together have a smooth * soffit.
三角形踏板 三角形石阶梯，互锁以形成平滑的底面。

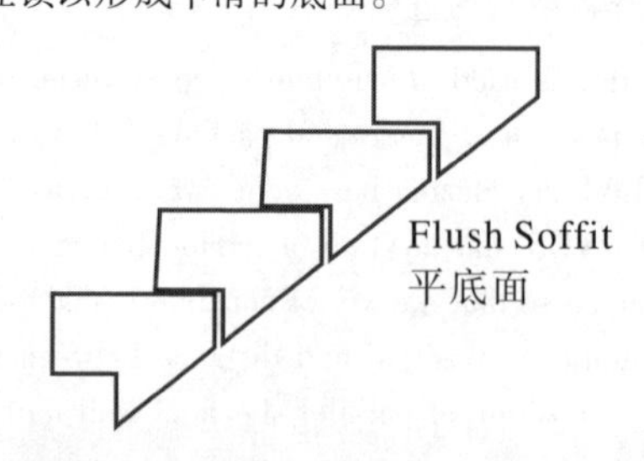

Spandrel step Interlocks creating a smooth soffit.
三角形踏板 互锁形成平滑的底面。

Spandrel The triangular area under a stair and above an arch.
拱肩 楼梯下或拱门上的三角形区域。

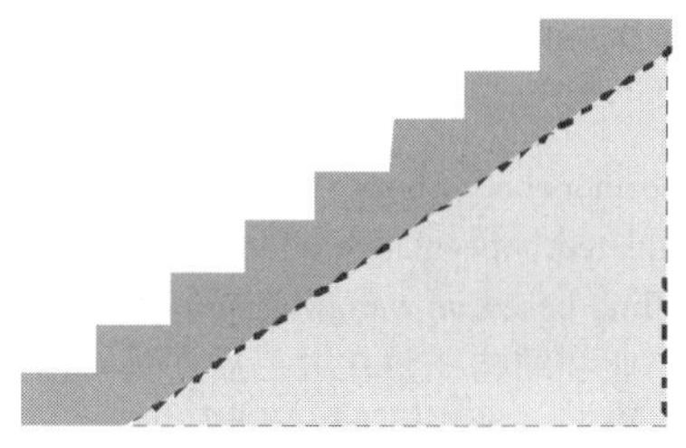

Spandrel and spandrel step
拱肩和三角形踏板

Spanish tile Half-cylinder-shaped clay roof tile that is wider at the bottom than it is at the top.
西班牙瓦 半圆柱形黏土屋面瓦，底部比顶部宽。

span roof A standard pitched roof.
等斜双坡屋顶 标准的斜屋顶。

spar angular Limestone and white rock chippings used as ballast to hold down roofing felt and protect it from degradation caused by the sun.
屋面碎石层 石灰岩和白色碎石，用做压载物，以压紧屋面油毡，并使其不因太阳照射而退化。

spar dash A render coated in or exposing limestone or white angular pebbles. Dry dash or render pebble dashed with limestone.
石碴砂浆面 涂在或外露于石灰岩或白色有角卵石上的涂层。干黏石或掺有石灰岩的卵石。

sparge pipe A horizontal perforated pipe that produces a spray of water for flushing urinals.
尿厕喷水管 用于冲洗便池的喷水水平多孔管。

spatterdash (spatter dash) A rich mix 1 : 1. 5 to 1 : 2 cement and sand, which is flicked onto a surface in small globules by a machine (roughcast machine). It is applied to surfaces that have minimal suction in order to provide a key.
撒砂仔（甩浆） 混合比例达 1 : 1. 5 至 1 : 2 的水泥和砂浆，用机器（喷浆机）将其以小水珠形式喷在表面。涂于表面，具有最小的吸水性以提供黏结。

S

special (special brick) A brick of non standard dimensions. Special bricks are normally manually cut and expensive as they are not mass produced.
特殊砖（异型砖） 非标准尺寸的砖。特定砖通常手动裁切，因为不是批量生产，所以价格较贵。

special attendance Charge for use of contractors' equipment by a subcontractor for a short period, usually charged at an hourly rate.
总包设备配合费 分包商在短期内使用承包商设备需付的费用，通常按小时计费。

specialists 1. Expert or professional who has very specific skills and knowledge, and has a reputation for carrying them out reliably. 2. Contractor, subcontractor, or consultant selected for their skill and knowledge in a specific area, e. g. structural, mechanical, electrical, and environmental engineers.
1. 专家 具备某些技术和知识，且声誉良好并可贯彻执行任务的专家或专业人员。**2. 专业人员** 具备某些领域技术和知识的承包商、分包商、顾问或技术人员，如结构、机械、电气和环境工程师。

special risks *See* *FORCE MAJEURE.
特殊风险 参考“不可抗力”词条。

specification A description of the works and/or workmanship required. The written description forms part of the *contract documents and normally includes reference to the workmanship, standards, or qualification necessary, tests that may be required to ensure performance, quality, or work, key indicators, and other measures that ensure the work is done properly.
规范 所要求的工程或工艺的说明。书面格式的说明，作为合同文件的组成部分，通常包括为确保性能、质量或工程、关键指标和确保工程正确完成的其他措施所需的工艺、标准或资格、测试。

spectrum A range, particularly the distribution of colour light.
光谱 一个范围，尤其指彩色光的分布。

speculative builder A builder or developer who buys land and builds properties and structures in the anticipation that they will sell on once completed. Speculative building is different from a builder who sells properties "off-plan", which is selling the building based on a drawing prior to it being completed.
投机性建筑商 购买土地及建造物业和结构期望在竣工后立马售出的建筑商或开发商。投机建筑商与售卖期房的建筑商不同，后者在竣工之前是基于图纸进行售卖。

spigot The end of a pipe that is inserted into the *socket end of another pipe to form a *spigot and socket joint.
插口 插入另一管道承口以形成承插式接头的管道端。

spigot and socket joint (US bell-and-spigot joint) A type of pipe joint made by inserting the *spigot end of one pipe into the *socket end of another pipe.
喇叭口（美国 承插式）接头 一种管道接头，由一个管道的插口端插入另一管道的承口端组成。

spike A sharp pointed end coming from a shaft.
尖端 机轴的尖端。

spile A wooden peg or timber supporting post that is driven into the ground.
木桩 插入地面的木钉或木支柱。

spillway An overflow structure, in the form of a *BELLMOUTH overflow or a side channel, that excess water is safely discharged through, in order to prevent water from flowing over the top of a *DAM.
溢洪道 一个溢流结构，呈钟形溢流口式或侧槽状，多余的水可以安全排出，以防止水漫过堤坝的顶部。

S

spindle 1. A piece of wood that has been turned to create a *baluster. 2. A square section rod that passes through a door from one handle to another, going through the latch mechanism. The rotation of the bar operates the latch. 3. The axel at the centre of a tap that rotates to open and close the valve.
1. 栏板 安装成栏杆柱的木块。**2. 门锁方杆** 方形截面杆，从门把手穿过锁扣装置连接另一边门把手。杆的转动可操控门插销。**3. 阀轴** 位于阀门中心的轴，旋转以控制阀门的开关。

spine wall A wall inside the building, normally in a central position, that can also offer lateral support.
纵向承重墙 建筑物内部的墙壁，通常在中心位置，可以提供纵向承重。

spiral A continuous curve that increases around a central point.
螺旋 在一个中心点向外扩散的连续曲线。

spiral duct (spiro duct) A tubular duct formed by winding a sheet of metal around in a spiral.
螺纹导管 金属薄片以螺旋形式缠绕而成的导管。

spiral stair A *stair that is helical, not truly spiral, constructed with winders radiating from a central newel.
旋转楼梯 （像）螺旋形的楼梯，并不是真正的螺旋形，而是由从中柱向外辐射的斜踏板形成的。

spire A tall, elongated, and pointed roof structure.
尖塔 高高细长的尖屋顶结构。

spirit level A straight and flat length of metal with a single bubble captured in a tube of liquid. The position of the bubble reveals whether a surface is level.
水平尺 在液体管中有单个气泡笔直扁长的金属尺。气泡的位置反映了表面是否水平。

spirit stain A dye based in methylated spirits, used for colouring the surface of wood.
醇溶染剂酒精着色剂 以甲基化乙醇为基的染料，用于木材表面染色。

splashback (backsplash) A vertical area of tiles, or other easily cleaned material, located behind a sink or cooker to protect the wall from splashes.
防溅挡板（后挡板） 砖块或其他易清洁的材料的垂直面，位于水槽或炊具后面，以避免墙壁受到喷溅。

splashboard 1. A scaffold board laid on edge on the insider face of the scaffolding platform to protect the building wall from mortar, concrete, and other materials. 2. An angled moulding placed on the foot of an external door to discharge the rain away from the face of the doorway, helping to ensure water does not enter the building. Also called weather moulding.
1. 挡泥板 在脚手架平台内表面边缘的脚手板，用以保护建筑墙壁免受水泥、混凝土和其他材料喷溅。**2. 挡水板** 位于外门门底的成角装饰线条，用于排掉门口的雨水，确保雨水不会进入屋内。也称为泄水线条。

splash lap The lap of sheet metal joint that extends onto the flat surface of the sheet metal below.
屋面防溅搭接 延伸至下面的表面平整的金属薄片搭接。

splat A strip that is used to cover over the joints of wall boards.
墙面压条 用于盖住墙板接头的装饰条。

splay A sloping cut at an angle to the main surface or edge.
斜面 在主表面或边缘按一定角度进行的斜切。

splay brick (cant brick) A special brick that has been cut with either a slant, chamfered, corner edge, or face that is shaped to make a return or form a finished stop at the end of a wall.
斜面砖 一种用斜面、斜削角、切角边或成形表面进行切割以在墙壁末端形成转向延续的特殊砖块。

splayed coping A stone that is used to provide a cover to a wall that is cut at an angle or chamfered.
压顶斜面 一块用于墙壁压顶的石头，成一定角度进行切割或倒棱。

splayed grounds Timber grounds that are shaped and angled to provide a key for plaster or render.
斜面冲筋条 成一定角度为石膏或灰泥提供黏结的木条。

splayed heading joint A joint between timber boards where the joint is cut at 45° so that the top of the joint overlaps the bottom.
八字接头 木板之间的接头，45°角切割，这样接头的顶部就可与底部重合。

splayed skirting A bevelled skirting board.
八字形踢脚板 边缘斜切的踢脚板。

splay knot A timber knot that has been cut along its length, the length of the knot is exposed on the face of the timber.
条状节 沿着长度切割的木结，木结的长边外露于木材表面。

splice An end-to-end butt joint, where steel plates, splice bars, and crimped couplers are used to strengthen the joint so that it is capable of resisting lateral loads, tension, and compression. The aim is to make the column or beam act as a single material.
拼接 首尾相连的对接接头，钢板、鱼尾板型钢和蜷曲联结器可用于强化接头，这样接头就可以抵抗横向荷载、张力和压缩。目的是使柱子或横梁成为单个材料。

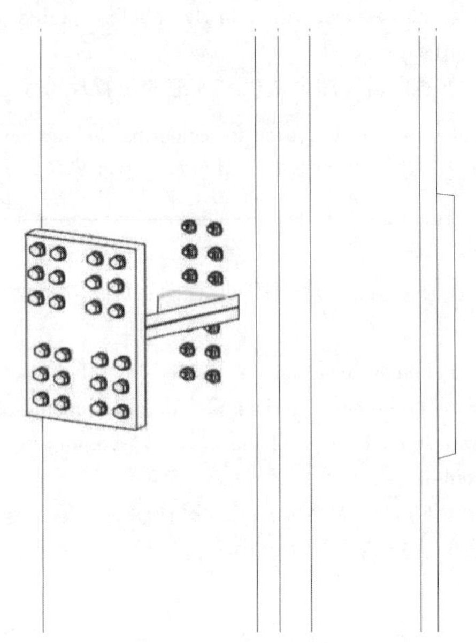

Column to column spliced joint
柱与柱的拼合接头

splice bar A bar used to form a connection between two butt ends, half of the length of the bar penetrates into one half of the joint, with the other half penetrating into the other element to be joined. The bar may be tightly fixed, glued, or mechanically fixed into each adjoining part.
鱼尾板 连接两个粗端之间的板，一半长的板穿入接头的一半，另外一半穿入将被连接的另一构件中。板可以紧紧固定、粘贴或机械固定入各邻接的构件中。

split course A course of split bricks.
劈裂砖层 一层劈裂砖。

split-face block A block that is made to be split in half, giving one rough face and one fair face. The block is normally made to twice the normal width of a standard block.
劈裂砖 劈成两半的砖块，一面是粗糙表面，另一面是平滑表面。通常砖块宽度是标准砖块宽度的两倍。

split gasket Two strips of rubber (gaskets) used at either side of a pane of glass in *gasket glazing.
玻璃封边胶条 在玻璃镶边中玻璃块任一面所使用的两条橡胶（垫圈）。

split head A prop with a U-shaped head that is used to carry timber or steel beams. The supporting head is formed to allow the beams to simply slot into place and be carried by the prop.
U 型卡支撑 带有 U 形头的支柱，用于支承木梁或钢梁。可将横梁轻易放入狭槽，由支柱支承。

split pin A pin with two legs that is used to secure fixings. The pin is made from a single length of metal that has been doubled over to form a single pin that has a looped head and two legs. The legs sit together so that they can pass through a single hole. The wider part of the looped head prevents the pin passing straight through the hole. Once inserted the pin legs can be splayed out to hold the pin securely in position.
开口销 有两条支腿的插销，用于固定配件。插销由单条金属制成，对折形成单个插销，有一圆圈头和两条支腿。支腿组在一起，这样就可以穿过单个孔。圆圈头较宽的一边防止插销直接穿过孔洞。一旦插入插销，支腿可以张开，以牢牢固定住插销。

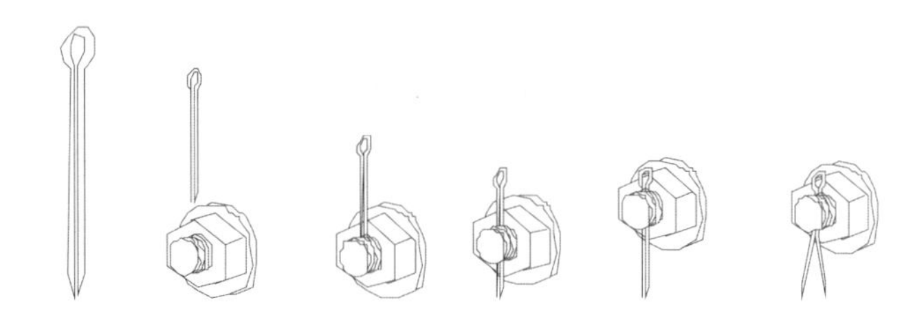

1. The nut is securely fixed in position.
螺母牢牢固定到位。
2. The split pin is inserted through the predrilled hole in the bolt.
开口销插入螺栓中预钻的孔。
3. The pin is pushed into the hole.
将插销按入孔中。
4. The pin is inserted until the head reaches the bolt.
直至头部触及螺栓，插入插销。
5. The legs of the pin are splayed apart, ensuring the pin cannot fall out.
插销支腿张开，确保插销不会掉落。

Split pin used to secure a nut on a bolt, preventing the bolt from coming undone through vibration
开口销用于固定螺丝上的螺母，防止螺栓因为震动解开。

split pipe A pipe that has been cut along its length.
对开管 沿着长度剖开的管道。

split system A type of air-conditioning system that comprises an indoor *fan coil unit and an outdoor *refrigeration unit.
分流系统 一种由一个室内风机盘管装置和一个室外制冷装置组成的空调系统。

splittable activity An activity within a programme that is capable of being split into two or more parts and does not have to be completed before the next activity starts.
A non-splittable activity has to be delivered in one unit of work and cannot be broken up by other activities.
可拆分活动 在进度计划中，可拆分成两部分或以上的活动，并且该活动不必在下一活动开始之前完成。不可拆开的活动必须以整个活动为单位交接，不得被其他活动打散。

splitter damper A damper used to divide the air between two ducts.
分流调节器 用于两个管道之间空气分流的风门。

splitting test (splitting tensile test) A compression test on a concrete cylinder to determine the tensile strength. The cylinder can either be cast or cut, and is placed on its side and loaded through its diameter. The tensile splitting strength is then determined from:

$$T = 2P/\pi dl$$

where P = load, d = diameter and l = length.
劈裂试验（劈裂抗拉强度试验） 对混凝土柱进行的抗压试验，以确定其抗拉强度。混凝土柱可以是浇铸或切割而成，倾向一边，通过直径承载。劈裂抗拉强度由以下公式计算得出：

$$T = 2P/\pi dl$$

其中，P = 荷载，d = 直径，l = 长度。

spoil (muck) Material and earth from excavation activities.
弃土（淤泥） 挖掘活动所产生的材料和泥土。

spokeshave A plane that has handles extending from each side of the cutting blade (like bike handles), enabling the blade to be drawn over curved surfaces to shape them.
弯刨 一种刨，刀片两边有伸出的把手（像自行车把手），使得刀片能够从弯曲表面上刨削。

sponge-backed rubber flooring (foam-backed rubber flooring) Flooring that is backed with resilient ma-

terial to reduce impact sound and provide a cushioned effect.
海绵衬里橡胶地板（泡沫衬里橡胶地板） 以弹性材料为衬里以降低撞击声，提供缓冲效果的地板。

sponging Using a sponge to give a textured finish to paint. The sponge can be used to apply the paint or texture wet paint that is already applied.
海绵上色法 用海绵上漆形成有纹理的表面。海绵可以用于上漆或为已涂好的未干油漆表面形成纹理。

spontaneous combustion Combustion that occurs when a material self-ignites.
自燃 材料自行点着燃烧。

spot board (mortar board) Square surface with a handle fitted to the underside enabling bricklayers and plasterers to carry plaster or mortar and easily apply it to the surface.
抹泥板（托灰板） 方形表面，下方有一把手，使瓦工和抹灰工可以托住灰泥或砂浆并轻松将其涂抹在墙或天花的表面。

spot gluing Fixing *plasterboard with small *dabs of plaster or adhesive. Small amounts of adhesives are placed at regular intervals around the perimeter and in the centre of the plasterboard panel. The panel is offered to the adhesive and pressed into position so that it is flush and level with surrounding boards.
点胶 用少量灰泥或胶粘剂胶点安装石膏板。每隔一定间距，将少量胶粘剂涂在石膏板的四边和中心。石膏板涂上胶粘剂，然后按入适当位置，使其与周围板块齐平。

spot item Small unit of defective work that cannot be linked to just one trade, may require a labourer to break out a hole, a bricklayer to lay blocks, and a plasterer to replaster the area.
瑕疵项 与不止一个工种相关的少量瑕疵工程，可能需要一个工人开一个洞、一个瓦工铺砖或一个抹灰工重新抹灰。

spotlight (spot) A *luminaire which produces a concentrated beam of light that is used to illuminate a narrow area.
聚光灯 发出聚集光束的灯具，用于照明一小块区域。

spotting An area of ***paint** that has a slightly different colour or texture to the rest of the paintwork.
斑点 一小块地方的涂漆颜色或纹理与其他地方稍微有些不同。

spotting in (spot finishing) The repairing and blending in of small defects; may involve filling in, smoothing off, sanding down, or rubbing up and then painting or finishing, to ensure the repair matches with its surroundings.
小面积修补 小瑕疵的修补和调色；包括填充、磨滑、用砂纸磨光表面或者擦亮，然后涂漆或涂饰，以确保修补的地方与周围相配。

S

spot weld A joint formed by *welding at various points.
点焊 在不同点进行焊接而形成的接缝。

spout An outlet on a pipe through which a fluid is poured.
喷嘴 管道上喷出液体的出口。

spray A fine jet of water particles moving through the air.
喷雾 喷射的细微水粒在空气中浮动。

sprayed mineral insulation (firespray) Insulation material that is applied by spraying. Used for thermal insulation purposes or for fire protection.
喷涂的矿物绝缘物（消防喷雾） 喷涂形式的绝缘材料，用于隔热或消防。

sprayer (spray nozzle) A shower head or the fixing at the end of a hose used to provide the desired direction or dispersion of the liquid.
喷雾器（喷头） 淋浴喷头或装在软管末端的装置，用于按要求的方向或散布方式喷洒液体。

spray gun Machine with a pumping unit that is used to disperse and direct paint or other liquid.

喷枪 带泵机组的机器，可以用于喷洒和定向喷漆或其他液体。

spraying The process of projecting and dispersing liquid, such as paint or stiffer liquids, such as plaster and render, through a * spray gun or hose.
喷出 通过喷枪或软管喷洒液体的过程，如喷洒油漆或更黏稠的液体（例如室内外砂浆）。

spray painting The application of paint using a spray gun rather than brush or roller.
喷涂 使用喷枪而不是刷子或滚辊进行喷漆。

spray plastering The application of plaster using a plaster pump and gun, enables the plaster to be easily applied to the wall. The plaster still needs to be smoothed and levelled by hand using a float.
喷涂抹灰 使用砂浆泵和喷枪喷涂砂浆的方式进行的抹灰，这种方式能轻易涂砂浆上墙。砂浆仍需要使用抹泥板手动抹平滑。

spray tap A type of very low flow rate tap that discharges the water as a fine spray.
喷雾式水龙头 一种低流量可以喷出细雾般的水的水龙头。

spread To disperse or level out a material over an area.
涂布 在某块区域散开或抹开某材料。

spread and level (US wasting) To take excess material or soil, or to remove material from a high part of the site, using it to fill in undulations and lower areas of the site. The material is moved around creating a flatter and more level site.
调配 运走多余的材料或土壤，或清理现场较高部分的材料，用于填平现场较低或高低不平的区域。调配材料使得现场更加平整。

spreader bar 1. A strut used to keep elements of a structure apart, for example, a strut placed at the bottom of a pressed metal doorframe, used to keep the jambs of the door at the correct distance apart. The bar may be positioned below the floor covering or concrete screen, and hidden from view. 2. A beam used to distribute and spread the loads coming from concentrated or point loads above.
1. 撑杆 用于保持结构构件分离的撑杆，如压型金属门框底部的撑条，用于使门边梃保持正确的间距。撑杆可以位于地板覆盖层或混凝土砂浆层下面，肉眼不可见。**2. 撑梁** 用于均布和分散来自集中荷载或点荷载的横梁。

spreading The distribution and levelling out of a material over an area.
铺展 在一块区域内分布和平铺材料。

spreading rate The area that can be covered by a unit volume of material or paint, applied in one coat or at a specified thickness.
涂布率 区域内可覆盖的材料或油漆的单位容积，按一层或者按规定的厚度涂。

spread of flame *See* FLAMESPREDA.
火焰蔓延 参考“火焰传播指数”词条。

sprighead roof nail A galvanized steel nail used for fixing down corrugated roofing; the nail has a steel cup under the head to suit the profile of the sheeting.
伞钉/瓦楞钉 镀锌钢钉，用于固定波纹屋面；在钉头下有一钢帽以契合板材的形状。

spring 1. A natural water course that arrives at the surface of the ground. 2. A metal helix with elastic properties that when compressed, or stretched, and released will return to its original shape. The coiled metal is used to cushion the impact of vibration and create suspension systems.
1. 泉水 在地表露出的自然水层。**2. 弹簧** 金属螺旋形，有弹性，可压缩、可拉伸，当释放时可恢复原样。盘卷的金属可用于缓冲震动冲击，形成悬浮系统。

springer A stone or brick on which an arch rests or the arch starts.
拱脚石 拱结构放置或拱开始的第一块石头或砖块。

spring hanger A type of pipe support that uses coiled springs to support the pipe from above, while allowing

vertical movement.
弹簧吊架 一种管道支承，使用螺旋弹簧从上方支撑管道，可垂直移动。

spring hinge A hinge with a spring to assist or enable the door to self-close.
弹簧铰链 带弹簧以协助或使门自动关闭的铰链。

springing The intersection of the lower side of the arch and the pier or wall which it abuts.
起拱点 拱结构的较低边和与其对接的桥墩或墙壁的交接处。

springing line A horizontal line that joins the two intersections between the arch and the pier or wall carrying the arch.
起拱线 一条水平线，连接拱结构和承载拱的桥墩或墙壁之间的两个交接点。

springing point The point at which stairs are set out, the start of the incline.
起步点 楼梯段开始的点，斜边的起点。

sprinkler system A system that is designed to extinguish fires. Comprises a series of water pipes and spray nozzles that automatically operate when a fire is detected.
自动喷淋灭火系统 用于灭火的系统。由一系列水管和喷淋头组成，当探测到火灾时，可自动启动。

sprocket A length of timber, often cut diagonally on section and used to lift the edge of the last row of eaves tiles.
飞椽 一段木材，通常截面为斜边三角形，用于支持最后一行檐口瓦的边缘。

sprocket eaves (sprocketed eaves) The eaves tiles that are tilted upwards by a *SPROCKET.
飞椽瓦 通过飞椽而向上倾斜的檐口瓦。

spun pipe Pre-cast concrete pipes manufactured using a spinning mould.
离心铸管 使用离心铸模制作的预制混凝土管道。

spur An electrical socket that is connected to the ring main by a single cable. Often used to add additional sockets.
支线 通过单支线连接到环形电路的电插座。通常用于增加插头。

square A triangular or T-shaped engineering tool, set at 90°. The tool allows 90° angles to be set and lines to be scribed at 90°to a surface.
直角尺 三角形或T形工程工具，为90°角。该工具可用来设90°角和划与表面成90°角的直线。

squared log A baulk timber log square sawn.
方形木料 锯成方形的木材。

S

squared rubble A rubble wall made from stones that have been squared but are different sizes. The courses are lined up every 3 or 4 courses to give the wall a uniform appearance.
毛方石墙 用方形但尺寸不一的石头砌成的毛石墙。每3或4层对齐，使墙面外观整齐统一。

square-edged timber (square-sawn timber) Sawn timber that has a cross-section that has been cut at right angles.
方材（方形锯材） 横截面按直角锯成的木材。

square hook An L-shaped hook with a threaded screw.
方形挂钩 L型挂钩，配一个螺钉。

square joint A joint that is butted up at right angles, 90°.
直角拼接 90°角平接的接缝。

square-turned baluster (square-turned newel) A newel or baluster that has not been turned but has moulding cut into the faces.
方形栏杆柱（方形大柱） 不是车削的而是模制切入表面的大柱或栏杆柱。

squat To make use of a building or land without permission or ownership.

非法占据　在没有经过允许或没有所有权的情况下使用建筑物或土地。

squatting closet (Asiatic closet, squat pan, squat toilet)　A WC that is set into the floor and requires the user to squat.
蹲厕　安装在地板上的便器，要求使用者蹲下使用。

squint (squint brick)　A special brick made with a chamfered face or cut corners. The cut brick allows it to be used to create a wall with changed direction, without parts of the brick projecting from the surface.
斜边砖　一种有倒棱面或切角的异型砖。切边砖可用于砌成变相的墙壁，并且砖块不会从表面凸出。

squint corner (squint quoin)　A corner of a building that is not a right angle, where the bricks or stone project out from the main face.
斜隅　建筑物中不是直角的角落，砖块或石头会从主表面凸出。

SS (Special Structural)　Grade timber that has been visually graded for situations that demand high-quality timber, capable of sustaining high loads. MSS or Machine grade Special Structural is the equivalent.
SS级（特殊结构）　按照高质量木材的要求状态目测分等的分级木材，能够承受高荷载。MSS级或机器级别特殊结构的含义与其同等。

stability　The ability of an element or unit to resist breakdown or collapse. Stability is a term used in fire resistance as an element's capability to resist collapse during a fire.
稳定性　构件或装置抗破裂或倒塌的能力。在耐火性能中，稳定性一词是指在大火中构件抗倒塌的能力。

stabilizer　1. A foot that projects out from a digger, crane, or concrete pump that ensures the vehicle remains safe, does not move, or overturn, also called an * OUTRIGGE. 2. Additive used to prevent a liquid or chemical from breaking down into its separate parts; adds a specific strength, or quality to a material.
1. 稳定器　从挖掘机、起重机或混凝土泵下伸出来的支脚，确保机器安全，使得机器不会移动或翻转，也称为支腿。**2. 稳定剂**　用于防止液体或化学物分解的添加剂；为材料增加一定的强度或质量。

stable　A building or shed that has been used or is used as a shelter for horses.
马厩　用作养马居所的建筑物或棚子。

stable door (US Dutch door)　An external door that is divided into two horizontal leafs, with each leaf being able to be opened independently.
两截门（美国 荷兰式门）　室外的门，分为两个水平门扇，能够单独打开。

stack　1. A vertical drainage pipe that conveys liquid and solid waste to a drain. 2. A * **chimney stack**. 3. Bonding bricks or blocks simply stacked directly on top of each other, where no attempt is made to stagger the joints. Without ties and brick reinforcement the wall is inherently weak as it is only the mortar that binds the wall together.
1. 立管　用于排放液体和固体废物到下水道的排水立管。**2. 烟囱立管**。　**3. 堆垛砖**　简单互相堆叠的砖或砌块，不刻意错缝。因为只有砂浆将墙壁黏合在一起，所以没有拉结和砖块加固，墙壁本身就会很脆弱。

stack effect　The movement of air into, out of, and within a building or space due to natural buoyancy. In the case of a building, air inside the building is heated by solar gains, heating appliances, people, and equipment. As the internal air increases in temperature, it becomes less dense and rises due to convection. This results in higher pressure air at the top of the building, relative to the external air pressure, leading to exfiltration. As the warm air exits the building, cooler denser air from outside infiltrates the building through cracks and gaps in the construction.
烟囱效应　由于天然浮力，空气在建筑物或空间内流入或流出。如在建筑物内，建筑物内的空气由于太阳辐射、加热器、人和设备而升温。当室内空气温度升高时，空气密度变小产生对流上升。这样与室外空气压强相比，建筑物顶部的空气压强更大，造成空气渗出。当暖空气流出建筑物时，室外较冷密度较大的空气通过结构中的裂缝和空隙流入建筑物中。

S

stack vent *See* VENT STACK.
下水道排气管 参考“排气管”词条。

stadia lines (stadia hairs) Two horizontal lines, one above and the other below the cross-hairs of a * **theodolite** or * level, which are used in * TACHEOMETRY.
视距丝 两条平行线，一条在经纬仪或水准仪的十字丝的上面，另一条在下面，用于视距测量。

Staffordshire blue brick A dense clay * engineering brick that is blue in colour, either pressed or wire cut with very high strength and water resisting properties. The brick has one of the highest strength characteristics in the UK. The density of the brick helps it to resist water penetration, enabling it to be used for * damp-proof courses.
斯塔福德郡皇室砖 密实的黏土工程砖，颜色为蓝色，经过加压或线切割，具有很高的强度和防水性能。在英国，这种砖是具备最高强度性能的砖块之一。砖块的密度使其能够防止水渗透，可用于防潮层。

staggered joints (break joints) Brick and blocks laid so that the centre of one brick or block sits over the centre of the joint below.
交错搭接（错缝接合） 砌砖时，砖或砌块的中心对准下面接缝的中心。

staggered-stud partition A double-skin timber stud wall with a gap between the skins to improve acoustic insulation. Each wall is lined with acoustic insulation and acoustic board. The panels are staggered to reduce the chance of penetration through joints.
错列龙骨隔断 双层木龙骨墙，在两层墙之间有空隙以提高隔音性能。两面都有隔音材料和吸声板。墙板交错以减小通过裂缝渗透的概率。

staging Low level platform * scaffolding or an add-on to a scaffolding platform that takes the surface to a slightly higher level, creating a step up to access higher levels.
脚手架楼梯 低层工作平台脚手架或脚手架平台的附件，可使表面稍微抬高，形成进入更高层的阶梯。

stain A solution used for changing the colour of another material. The liquid is commonly used on timber to change the surface colour.
着色剂 改变一种材料颜色的溶剂。通常用于木材上，改变其表面颜色。

stained-glass window A window that is constructed from pieces of coloured glass (stained glass). The pieces of glass are usually held together with lead.
彩色玻璃窗 由彩色玻璃组成的窗户。玻璃块通常用铅条联结。

S

stainless steel A type of * steel with excellent resistance to corrosion and degradation due to the high chromium content, which is typically 15%. One of the most popular metals; however, due to the relatively high cost of chromium, galvanized steel is a cheaper alternative in the construction industry and is thus more widely used.
不锈钢 钢的一种，因为铬含量高（通常为15%），具备良好的抗腐蚀性和抗退化性，是最普遍的金属之一；但是，由于铬的成本相对较高，在建筑行业中，镀锌钢的成本较低，因此也被广泛使用。

stainless-steel plumbing Pipes and fittings made from stainless steel. Used due to their high resistance to corrosion.
不锈钢给排水 不锈钢制成的管道和管件。由于其良好的抗腐蚀性而被使用。

stair Steps that are linked together enabling safe vertical movement from one level to another. To enable safe movement, and considering that the steps will be used by people of differing physical ability, the steepness of the stair flight is restricted and landings are used to provide a break point or change direction.
楼梯 连在一起的台阶，可以安全地从一层往另一层垂直移动。为了人能安全走动，并且考虑到台阶是供各种不同体能的人使用，梯段的陡度有所限制，平台用于提供歇脚点或改变方向。

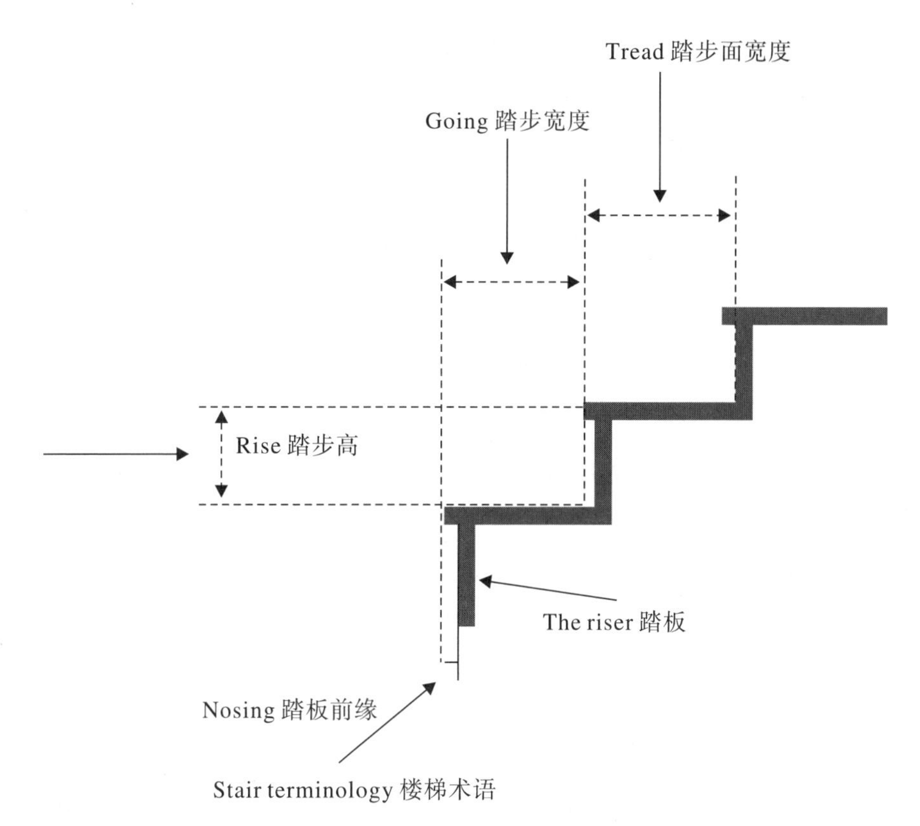

Stair terminology 楼梯术语

staircase A unit that contains a series of steps used to gain access from one level to another. For a timber staircase, it is series of * steps that are encased by * strings that carry and contain the steps. The staircase is often prefabricated and brought to site preassembled. The strings are * rebated and the * steps and * risers are inserted, glued, and wedged into position. The staircase is then delivered to site where it is fixed to the upper and lower floors and the newel posts and handrails are fitted.
楼梯通道 由一串台阶形成的从一层进入另一层的通道。例如，木楼梯是由斜梁所包围的一串台阶。楼梯通常是预制，然后运至现场预装配的。斜梁是矩形切口的，可供台阶和踢板插入、黏结，然后楔入适当的位置。楼梯运至现场，连接上一层和下一层楼层，安装楼梯栏杆支柱和扶手。

stairlift (inclinator) A power-operated chair that is fixed to the side of a stair, enabling people with disabilities or those who have difficulty using stairs, to be mechanically lifted, by the chair, from one level to another.
座椅电梯（椅式升降梯） 安装在楼梯一边的电动座椅，供残疾人或不便使用楼梯的人使用，座椅自动升降，从一层到另一层。

stairway A stair that is used as a main circulation route or part of a fire escape.
阶梯 作为室内主要出入通道或消防通道的一部分的楼梯。

stairwell The space or gap between flights of stairs that run parallel to each other.
楼梯井 相互平行的单梯段之间的空间或间距。

stake A timber post with a pointed end for driving into the ground.
桩 有尖端的木柱，尖端可钻入地面。

stall board In a traditional shop front, the sill that supports the shop window.
橱窗下槛 在传统的店面，撑起店面窗户的下槛。

stall board light A * PAVEMENT light installed below a * STALL BOARD.
橱窗下采光窗 安装在铺面下槛下的路面采光窗。

stall riser In a traditional shop front, the material installed between the window sill and the ground.

橱窗下竖壁 在传统店面中，安装在窗台和地面间的材料。

stanchion A vertical column, pole, or strut, typically made of steel.
立柱/排队柱 竖直的柱子、杆或撑柱，通常由钢制成。

stanchion base A *BASEPLATE.
立柱底座 一柱脚底板。

stanchion casing Concrete *ENCASEMENT to a *STANCHION.
支柱壳 立柱的混凝土包裹。

standard details A drawing, for example of a *COLUMN or *BEAM, that has sufficient and replicable detail it can be used on many different projects with minimal or no alteration.
标准节点图 一种图纸，例如柱子或横梁的图纸，有充分可复制的细节，可以用于很多不同项目中，只需做少许修改或无须修改。

standard form (standard form of contract) A contract accepted and used by the industry, with the main terms pre-written, allowing for drawings and information that deals specifically with the project or development to be inserted. The standard terms have the advantage that they are known and understood by the industry, have been tested through cases in court, and the implications of the terms, and a breach are largely understood. Typical standard forms of contract include those produced by the Joint Contracting Tribunal: JCT Minor Works Contract, JCT Intermediate Form, JCT Management Contract, JCT Measured Term Contract, JCT Standard Form with Contractors' Design, JCT Prime Cost Contract, JCT Standard Form of Contract. FIDIC (Federation Internationale des Ingenieurs-Conseils) produces engineering and building contracts for major international projects. The Institute of Civil Engineers publishes standard forms including ICE Conditions of Contract, ICE Minor Works Contract, ICE Design and Construct Contract, and the New Engineering Contract.
标准格式（格式合同） 行业接受并采用的合同，有预先制定的主要条款，涉及单独附上的项目或开发的图纸与信息。标准条款的优势已被行业熟知并理解，已经通过法庭案例的检验，并且条款的含义及违约已被广泛了解。典型的格式合同包括合同联合审定会（JCT）编写的合同：JCT 小型工程建设合同、JCT 中型工程建设合同、JCT 管理合同、JCT 测量期合同、JCT 承包商设计的标准合同、JCT 主要成本合同、JCT 标准格式合同。FIDIC（国际咨询工程师联合会）为主要国际项目编写工程和建筑合同。土木工程师学会（ICE）出版的标准格式，包括 ICE 合同条款、ICE 小型工程合同、ICE 设计和建设合同以及新工程合同。

standardization (standardisation) To unitize and make components that have regular sizes, fixings, and dimensions, enabling them to fix together. Modules often have regular dimensions to enable them to fit together.
标准化 统一化，使组件具有规范的大小、附件和尺寸，使其能够组装在一起。模块通常具有规范的尺寸以便能够组装在一起。

S

Standard Method of Measurement (SMM) A common method of measuring and describing works. It is accepted as the construction standard and enables multiple parties to describe and price work with a good degree of commonality and ensures consistency.
标准计量方法（SMM） 测量和描述工程中常用的方法。其被认可为建筑标准，使得各方能够用高度共同性描述和定价工程，确保其一致性。

standard overcast sky A model of a completely cloudy sky. Used when calculating daylight factors.
标准全云天 完全阴天的模型。计算采光系数时需用到。

standard penetration test (SPT) An in-situ test to determine the shear strength of *granular soil. A metal rod is hammered into the ground by dropping a 63 kg mass through 760 mm onto it. The number of blows "N" is counted for the rod to penetrate the ground by 300 mm. The angle of shear resistance "Φ" can then be read off a graph defining the relationship between Φ and SPT N value.
标准贯入试验（SPT） 一个确定颗粒土壤的剪切强度的现场试验。63 kg 重量的物体从 760 mm 的高度自由落下，将金属杆锤击入地面。冲击数"N"计算为金属杆击入土中 300 mm 所需的锤击数。然后抗剪切角“Φ”可以从定义 Φ 和 SPT N 值之间关系的图表中读出。

standards A set of documents that define a standardized set of methods, procedures, etc. *See* BRITISH STANDARD, INTERNATIONAL STANDARDS ORGANIZATION, and CEN.
标准 定义标准化方法、程序等的一套文件。参考“英国标准”“国际标准化组织”和“欧洲标准化委员会（CEN）”词条。

standard special brick A brick made to a different specification to the standard brick. The brick could be angled, chamfered, skew cut, short, or long. *See* SPECIAL BRICK.
标准异型砖 按与标准砖不同规范定制的砖块。砖块可为有角的、倒棱的、斜切的、短的或长的。参考“异型砖”词条。

standard wire gauge (swg, imperial swg) An old method of specifying the thickness of metal wire, tube, nails, and sheet material. An swg of 1 =7.62 mm, 2 = 7.01 mm, and 3 =6.401 mm.
标准线规（swg 英国标准线规） 规格化金属线、管子、钉子和板材厚度的老方法。单位 swg 中 1 = 7.62 mm, 2 = 7.01 mm 和 3 = 6.401 mm。

standing leaf A folding door leaf that is fixed shut. Often found on double doors, where one leaf is a standing leaf and the other is an *OPENING LEAF.
固定门扇 固定的折叠式门扇。通常可在双开门上见到，一扇门是固定门扇，另一扇是活动门扇。

standing seam A vertical joint in a sheet metal roof that brings together two pieces of metal to form a water-resisting joint. The sheet edges are placed against each other and turned upwards providing a 40mm upstand. The edges that are now back-to-back are then folded together in the same direction through 180 °, making a vertical upstand with the edges both bent in the same direction so that they overlap. The material that has been bent to overlap is turned again through 180 °. The joint between the two sheets runs from the ridge to the eaves providing a weather resisting connection.
直立缝 金属板屋面的垂直接缝，使两片金属接合在一起，形成防水接缝。金属板边缘相互对着，向上翻转，形成40mm 的竖立高度。边缘紧挨着，向同一方向一起折叠 180°，形成直立构件，边缘向同一方向弯曲，这样就可重叠。将弯曲折叠的材料再翻转 180°。两块金属板间的接缝从屋脊延伸至屋檐，形成耐候连接。

standpipe (US) *See* FIRE RISER.
立管（美国） 参考“消防立管”词条。

stank To stop the flow of water—to prevent something from leaking.
水坝 截止水流以防止某物泄漏。

staple 1. A metal U-shaped nail. 2. A loop for a *hasp and staple gate or door lock.
1. U 型钉 一种 U 形金属钉。**2. 扣环** 搭扣锁门或门锁上的环。

staple fibres Short-length fibres used in the manufacture of nonwoven needle-punched *geotextiles.
短纤维 在生产无纺针刺土工布中使用的短纤维。

staple gun (stapling machine) A hand-held machine, manually, electrically, or pneumatically powered for firing *staples into materials to fix them together.
U 型钉枪（压钉机） 手持式机器，有手动、电动或气动式，用于将 U 型钉打入材料中，以组装在一起。

starling *See* CUTWATER.
分水桩 参考“分水角”词条。

starter (starter bar) A steel reinforcement bar that is used to overlap with and connect to adjoining reinforcement. The bars protrude out from one concrete pour so that they can be fixed to the reinforcement cage in the next pour or bay. The length of the bar protruding is dependent on the length of cover that is required.
锚筋 用于搭接和连接至邻近钢筋的钢筋。钢筋从混凝土浇筑块中伸出，这样就可以固定到下一浇筑块的钢筋笼上。伸出的钢筋长度取决于所需覆盖的长度。

S

start time The commencement of an activity or programme. The commencement of the first activity in a * network diagram or * Gantt chart.
启动时间 活动或计划开始。网络图或甘特图中第一次活动的开工。

stat An abbreviated version of * thermostat.
stat 恒温器的缩写。

statically determinate structure A structure that can be analyzed by * statics alone, i. e. the resolving forces, taking moments, etc.
静定结构 可以单独依靠数据分析的结构，例如，受力分析、作用力偶等。

statically indeterminate structure A structure that cannot be analyzed by * statics alone, because there are too many unknowns. Analysis using, for example, * virtual work is required.
超静定结构 因为有太多未知因素，不能仅靠数据分析的结构。采用的分析如：所要求的虚功。

static electricity Electrical charges generated from the rubbing together of synthetics, the force fields created from electrical equipment, for example, visual display units. The effect can be reduced and prevented by having floors and equipment * earthed.
静电 合成物摩擦产生的电荷，电气设备产生的力场，例如可视显示器。可通过地板和设备接地减弱以及消除。

static head (static pressure) The level or pressure of water above a certain point in a stationary water body.
静水头（静压） 在静态水体中某一点以上的水位或水压。

static penetration test A penetration test in soil which is pushed into the soil (*See* CONE PENETROMETER) rather than hammered (as in the * standard penetration test).
静力触探试验 土壤中的触探试验，推入土壤中（参考“圆锥贯入仪”词条）而不是锤击入土壤(如标准贯入试验)。

static positioning The determination of a position on the earth using the * global positioning system.
静态定位 使用全球定位系统在地面确定位置。

station A position where specific tasks are performed, e. g. a **railway station** is where trains stop and people embark and disembark; a **surveying station** is where a peg, * theodolite, or * level is placed to enable readings to be obtained.
站点 执行特定任务的位置，例如，火车站是火车停车及旅客上下车的地方；测量站是底脚、经纬仪或水平仪放置以获取读数的地方。

stationary Not moving, is static in one place.
静态的 不移动，在一个地方保持静止。

S

station roof (umbrella roof) An umbrella-shaped roof that is supported on a central column or a row of columns. Often found in railway stations.
伞形屋面 由中心柱或一排柱子支撑的伞形屋面。通常可在火车站见到。

statistics To analyze large amounts of numerical data to define a population revealing significant trends or patterns of behaviour.
统计 分析大量的数值数据，以划定反映群体的显著趋势或行为模式。

stay 1. A horizontal bar used in a window to strengthen a mullion. 2. A * casement stay.
1. 横梃 窗户中用于加固竖梃的横杆。**2. 平开窗风钩**。

stayed-cable bridge A bridge whose deck is supported by cables that are directly hung from masts.
斜拉索桥 桥面板由直接从桥塔上悬挂的绳索支撑的桥。

steam Water that has been boiled and vaporized.
蒸汽 沸腾和蒸发的水。

steam stripper (wallpaper stripper) Piece of equipment used to remove wallpaper and paper finishes from

walls, using steam to soak the paper and soften the adhesive.
壁纸蒸汽机 用于从墙壁上剥落壁纸和纸装饰的工具，用蒸汽浸润壁纸，软化黏合剂。

steel A ferrous alloy with a carbon content between 0.1 and 1.0 weight %. Such steels are often referred to as plain carbon steels. Carbon has a very marked effect on the properties of steel. The effect of an addition of a fraction of 1.0 weight % is considerable. Increasing the carbon content of steel increases the tensile strength, the yield strength, and the hardness. However, increasing the carbon content decreases the ductility of the steel. Most constructional steels contain between 0.15% and 0.4% of carbon as this combines *ductility and *tensile strength. This type of metal is used exclusively in structures for civil engineering applications.
钢 铁合金，碳含量为0.1%～1.0%。这种钢通常也称为普通碳钢。碳对钢的性能有显著效果。增加钢1.0%的碳含量效果是非常显著的。增加钢的碳含量能够增强钢的抗拉强度、屈服强度和硬度。但是，增加碳含量会减弱钢的延展性。大部分结构钢的碳含量为0.15%～0.4%，因此可保持其延展性和抗拉强度。这种金属专用于土木工程结构部分。

steel casement 1. A steel casement window. 2. The opening part of a steel window.
1. 钢框窗户。 2. 钢窗开口部分。

steel conduit Conduit made from steel.
钢管 用钢制成的管子。

steel erector A skilled worker who erects the steel frames to structures.
钢架安装工 安装钢架结构的熟练工人。

steel-fabric-reinforced concrete Concrete that has the added benefit of steel reinforcement to provide the concrete with tensile as well as compressive strength. The steel is placed where tensile forces are exerted, and the concrete takes the compressive loads.
钢筋网混凝土 添加钢筋以增加混凝土的抗张强度和抗压强度的混凝土。钢放置在发挥张力的地方，混凝土承受压缩负荷。

steel fixer A skilled worker who cuts and bends reinforcing bars to the correct size and shape, and places them in the correct position.
钢筋工 将钢筋切割和弯曲至正确尺寸和形状，并将其放入适当位置的熟练工人。

steel fixing (US bar setting) To tie together and assemble steel reinforcement mats and cages for reinforced concrete.
钢筋绑扎 为钢筋混凝土绑扎和组装钢筋网和钢筋笼。

steel frame The support skeleton of certain structures.
钢架 某些结构的支承架。

steel section A general term used to refer to *universal sections.
型钢 通用型材的泛称。

steel square (framing square, roofing square) A square which is graduated to enable it to be used for calculating the cutting angles and the length of roof rafters.
钢角尺（木工角尺） 刻有刻度的角尺，能够用于计算切角和屋面椽条的长度。

steel window A window made from steel.
钢窗 用钢制成的窗户。

steelwork A general term used to describe a *steel frame or a number of *steel sections.
钢制品 钢框架或型钢的泛称。

steeple A tall tower or *spire.
尖塔 高塔或尖塔。

steeplejack A tradesperson skilled at accessing and maintaining tall buildings, structures, and towers.
高空作业工人 检修和维护高层建筑物、结构与塔的技术工人。

stem 1. A threaded rod that is used to open and close the gate of a valve. 2. A central shaft or column from which other objects or structures extend.
1. 阀杆 用于开启和关闭阀门的螺杆。**2. 主杆** 从其他物体或结构延伸的中心轴或柱子。

step A platform used by pedestrians to gain access to a higher level. Intermediate platforms may be used to break up different levels to improve the ability of people to move from one level to another with greater ease and safely. A series of steps make up a *stair and is contained within a *staircase.
台阶/梯级 供行人步向另一层的平台。中间平台可用于分解不同层，以使人们更方便安全地从一层移向另一层。一组台阶可组成楼梯，并被楼梯通道包围。

step flashing A *flashing that is built into raked out joints, providing a weather-resistant junction where a roof abuts a wall. The lead flashing is inserted into the wall and folded so that it laps over and under roof tiles. Individual flashings are cut into the joint or a sheet of lead is cut in steps so that it can be cut into each course of brickwork.
阶梯式泛水板 嵌入刮缝内的泛水板，在屋顶交接墙壁处形成耐候连接。铅皮泛水板嵌入墙壁中然后折叠，这样就可重叠搭接在屋面瓦下。单个泛水板切入接缝中或一张铅皮切成阶梯形，这样就可以切入各砖层。

step ladder Foldout ladder with shelf-like platforms to position feet.
人字梯 可折叠式梯子，有阶梯式台阶可登上。

stepped Something that has a sudden change.
阶形的/礓礤 突然变化的某物。

stepped flashing *See* STEP FLASHING.
阶梯式泛水板 参考"阶梯式泛水板"词条。

stepped footing (stepped foundations) Strip foundation that is stepped to follow the contour of the land. Stepping the foundation reduces the amount of concrete needed to create a stable platform onto which the building loads can be transferred to the ground.
阶梯式基脚（阶梯式基础） 循着等高线的阶梯形的条形基础。基础阶梯化，可减少建筑荷载传到地基平台所需的混凝土用量。

stepped scarf joint (splayed scarf joint) A scarf joint between the ends of veneer or board, formed by lapping the joints diagonally, with a step in the middle enabling the joint to cope with compression and tension.
阶梯式斜接（八字形斜接） 单板或板材端点间的斜接，通过斜面搭接而成，中间有一梯级以使接口能够应对压缩和张力。

step-tapered pile (Raymond pile) A cast in-situ pile formed by driving a steel casing, which is wider at the top, into the ground. Concrete is then cast with or without reinforcement—the casing remains in the ground.
锥形桩（雷蒙桩） 现浇桩，通过将顶部较宽的钢护筒钉入地面形成。然后浇筑带有钢筋或不带钢筋的混凝土，护筒留在地下。

stereographic The representation of three-dimensional objects on a two-dimensional plane.
立体平面法的 在二维平面上展示三维物体。

stiff Rigid; something that is not easily deformed.
坚硬的 刚硬的；不易变形的。

stiffness The resistance to deformation under load. It is calculated by considering EA/l, where E = *modules of elasticity, A = area, and l = length.
刚度 在荷载下的抗变形性。计算表达式为EA/l，其中E = 弹性模数，A = 面积，l = 长度。

stiffness matrix (K) An array of mathematical elements that generalizes the stiffness of *Hooke's law, used in finite element analysis.
刚度矩阵（K） 数学元素排列，推导了胡克定律的刚度，用于有限元分析。

stiffness methods A structural analysis method where forces are expressed as displacements.

刚度法 位移时力的结构分析法。

Stillson Adjustable grips and wrench in one, used by hand to clamp objects such as pipes and rods, while nuts or other objects can be released.
管钳 可调式把手和扳钳，为一体式，可手动松开螺母或其他物体，以夹紧物体，如管道和杆。

stillwater Water with no flow, as would be found in a pond.
静水 不流动的水，在池塘中可见到。

stimulation The implementation of a process, technique, or product to increase the yield or output from something else.
刺激 执行一个过程、技术或产品以增加产出或输出量。

stipple Rough-textured finish formed by using a rough-textured brush, *roller or pad that draws the *paint or *Artex out from the surface creating a undulating pointed finish.
拉毛漆面 粗糙饰面，通过使用粗糙刷子、辊子或漆板涂油漆或 Artex（涂料品牌）修饰表面，形成起伏的饰面。

stippler (stippling brush) Tool used to create a *stipple finish.
拉毛刷 用于做拉毛饰面的工具。

stippling Creating a stipple finish, using a brush, roller, or pad.
油漆拉毛 使用刷子、辊子或漆板做成拉毛饰面。

stirrup 1. A rope or hanger to support something e.g. a beam. 2. A piece of *reinforcing bar to resist a shear force.
1. 镫形吊架 用于支撑如横梁等的绳索或吊架。**2. 箍筋** 抗剪切力的钢筋。

stitching The replacement of damaged bricks with new bricks in existing brickwork.
换砖 在原有的砌砖中，用新的砖块更换受损砖块。

stochastic Random; involving probability or guesswork.
随机的 任意的；包含可能性或猜测。

stock Those items that are available and in store.
存货 商店里有备货的能购买得到的商品。

stock brick Standard *bricks that are available ready for dispatch.
砖堆 准备好可砌的标准砖。

stockpile To store large quantities of items or equipment, ensuring that sufficient quantity is readily available.
储备 储存大量物品或设备，确保有足够的量可供使用。

Stokes' law An expression that defines the settling velocity of spherical particles in liquid. It is used for sedimentation analysis for soil particles, such as silts and clays that have a particle diameter between 0.2 to 0.0002 mm.
斯托克斯定律 定义液体中球形颗粒的沉降速度的表达式。用于土壤颗粒的沉积分析，例如颗粒直径为 0.2 – 0.0002 mm 的粉土和黏土。

stone Rock and other hard naturally occurring, non-metallic material, quarried or collected from the ground. A few examples of rock or stone include: limestone, sandstone, mudstone, slate, shale, granite, and quartz.
石头 岩石和其他自然存在的坚硬的非金属物质，从地面开采或采集。岩石或石头包括石灰岩、砂岩、泥石、板岩、页岩、花岗岩和石英。

stone facing Wall or building that is covered with a veneer of stone or is clad in stone.
石料饰面 覆上一层石板或石材覆面的墙壁或建筑物。

stonemason Craftsperson who works with stone. The mason will cut, shape, and lay stones.

S

石匠 做石头的工匠。石匠会切割、塑造和铺设石头。

stonework Walls that are made or clad in stone.
石墙 石头制成或覆盖的墙壁。

stop 1. An ornamental end to a decorative moulding. 2. A *special brick that forms the *stop end to a top course or decorative course of bricks.
1. 饰头 装饰线条的装饰端。**2. 装饰砖** 形成砖墙压顶层或装饰层的异形砖。

stop bead (plaster stop) A metal or plastic straight edge that is fixed level with the wall to provide a neat break for a plaster finish, or a break where the finishes change.
墙压条 金属或塑料直棱，与墙壁齐平安装，使砂浆饰面断开处或饰面改变处整齐划一。

stopcock (stop cock) A valve that is used to regulate or stop the flow of a gas or fluid in a pipe.
旋塞阀 用于调节或关闭管道里气体或液体流动的阀门。

stop end A moulded or splayed special brick that forms a neat end to a top course of bricks or a decorative course of bricks.
压顶收尾砖 形成砖块压顶层或装饰层整齐端口的模制或八字形异形砖。

stopped chamfer (stop chamfer) A spayed cut on an edge or a corner, which results in a triangular shape as it spays towards or away from the arris.
局部倒角 在边缘或角落的切角，因其向着或远离棱线斜切而成三角形。

stopped mortise A blinded or sub mortise, where the mortise does not penetrate all the way through the timber. One side of the timber has a rebate and the other side of the timber has no sign of the joint.
暗榫眼 不贯通或闷榫眼，榫眼不能完全贯穿木材。木材的一边有一个凹槽，另一边没有榫接痕迹。

stopper A plug designed to block the flow through a pipe.
堵头 用以阻塞管内水流的塞子。

stopping 1. Filling holes in timber, concrete, masonry, or other building materials. 2. The material used to fill holes.
1. 嵌填 填塞木材、混凝土、砌体或其他建筑材料的孔洞。**2. 填充料** 用于填充孔洞的材料。

stopping knife (putty knife) A blunt knife with a rounded edge used by *glazers for smoothing the putty used to fit windows.
油灰刀（腻子刀） 有圆边的钝刮泥刀，玻璃镶嵌工人用来抹平油灰以装配窗户。

stopvalve (stop valve) *See* STOPCOCK.
截止阀 参考“旋塞阀”词条。

S

storage cistern A *CISTERN.
储液池 一种水箱。

storage tank A *TANK.
储液箱 一种箱。

storage water heater An appliance that heats water and stores it in a tank until it is required.
贮水式热水器 加热水然后储存在水箱中供需要时使用的电器。

storey A full level of a building, the height of which is measured between one floor and the next. An eleven-storey building would mean that there were eleven full floors all enclosed within the building.
楼层 建筑物完整的一层，其高度为上一个楼层与下一个楼层的距离。11 层的建筑是指建筑物内有完整的 11 个楼层。

storey rod (gauge rod) A length of timber or material that is cut to the full height of a *storey and used to measure off the storey height to ensure all measurements are consistent.
楼高标杆（皮数杆） 一段按一个楼层完整高度切割的木材或材料，用于测量楼层的高度，以确保测量的一致性。

storm cellar (cyclone cellar) A room under the house or dwelling, normally contained below ground, that is built to offer protection during periods of severe weather conditions and storms.
防风地窖(地下避风室) 房子或住所下面的房间，通常在地下，能在恶劣的天气条件和风暴期间提供防护所。

storm clip A clip used to retain glazing in position.
玻璃夹 用于固定玻璃的夹子。

storm door A door installed in front of the main external door into a building that is designed to provide additional protection from the elements.
防风门 安装在进入建筑物的主要外门的前面，为抵御恶劣天气提供额外保护的门。

storm-proof window A window that is designed to resist the wind-driven rain, hail, sleet, and snow experienced in a storm. It usually incorporates double rebates and additional seals.
防暴风雨窗 设计用以抵抗暴风雨中的风夹雨、冰雹、雨夹雪和雪的窗户。通常包含双槽口和额外的密封条。

storm window A window installed in front of the main window of a building that is designed to provide additional protection from the elements.
防风雨窗 安装在建筑物主窗户前面的窗户，为抵御恶劣天气提供额外保护。

straddle scaffold A *saddle scaffold that is built over the ridge of a roof, often used to access chimney stacks or abutting walls.
跨式脚手架 安装在屋脊上的坡屋面脚手架，通常用于检修烟囱或邻接墙壁。

straight arch *See* FLAT ARCH.
平拱 参考“平拱”词条。

straight edge A length of metal or timber that has a straight edge, used as a profile for producing straight and flat surfaces or scribing lines on materials.
平尺 有直边的一段金属或木材，作为生产笔直平坦表面的板形或在材料上画线条的型线。

straight flight A flight of stairs that goes directly from one level to another without a change in direction.
直行单跑楼梯 可以直接从一层走向另一层而不需转向的一跑楼梯。

straight joint 1. Two pieces of material where the ends are simply butted together; a *butt joint. 2. A defect in brickwork where the joints or one course sit directly above the joints of the course below, rather than being staggered, sitting partway across the top of the brick below, for example, the joints in stretcher bond sit midway above the brick in the course below.
1. 对接接头 两片材料端点直接简单对接；即对接接缝。**2. 竖向通缝** 砌砖中的错误做法，接缝或一砖层直接与下一砖层的接缝相连，而不是错缝，与下一砖层错开，例如，顺砖式砌合中的接缝位于下面砖层的砖块中间。

straight-joint tiles *Single lap tiles designed so that the joints between successive courses are not staggered but straight, running in a straight line from the top of the *ridge to the *eaves.
屋面通缝瓦 单搭接瓦，连续瓦片之间的接缝不是错缝而是通缝，使其从屋脊顶部至屋檐一直保持直线。

straight-peen hammer A hammer with a double head where one head has a chisel or wedge-shaped profile.
直头尖顶锤 双头锤子，一个锤头有凿形或楔形形状。

straight tee (bullhead connector) A T-shaped duct or pipe fitting that has three openings.
直三通 T形管子或管件，有三个开口。

straight tongue A straight projection from the edge of a board, such as *tongue-and-groove boarding.
榫齿 边缘直线条凸出的板材，例如企口接合的板材。

straight tread A regular rectangular tread on a stair (*flier), where the edges of the tread are all parallel to each other, unlike a *winder, *tapered tread, or *kite winder.
直梯踏步 楼梯（水平支撑）上的统一规格矩形踏步板，所有踏步的边缘相互平行，不像螺旋形楼梯、扇形踏步或三角踏步楼梯。

strain A measure of how much a material has extended during deformation when subjected to a tensile force. Strain (ε) has no units and is usually expressed as a percentage; the value is calculated as follows:

$$\varepsilon = \frac{\Delta L}{L}$$

where ε is strain in measured direction,
ΔL is the original length of the material,
L is the current length of the material.

张拉 当测量材料承受拉力时，材料变形的程度。张拉无单位，通常用百分比表示；按以下公式计算：

$$\varepsilon = \frac{\Delta L}{L}$$

其中，ε 是指受测量方向的拉力
ΔL 是指材料的原本长度
L 是指材料的现长

strain energy The energy stored in an elastic element under *strain.
应变能 在张拉作用下，弹性元件储存的势能。

strainer A coarse mesh filter installed on the intake of a pump that is designed to prevent solid particles passing through and damaging the pump.
粗过滤器 安装在泵吸入口的粗效过滤网，用于阻止固体颗粒通过，防止对泵造成损害。

strain gauge A device such as a *Wheatstone bridge that measures *strain.
应变仪 一个设备，例如测量张拉的惠斯通电桥。

strain hardening *See* COLD WORKING.
应变硬化 参考"冷成型"词条。

straining beam A horizontal strut.
横梁/系梁 水平支柱。

strain rate The rate at which strain changes in respect to time.
应变速率 相对时间内应变的速率。

S

S-trap An S-shaped *trap commonly used on baths, sinks, wash hand basins, and WCs.
S 型存水弯 通常用于浴室、水槽、洗手盆和厕所的 S 形存水弯。

strap bolt A bolt with a flat metal bar with pre-drilled holes welded directly to the head of the bolt. The threaded bar is used to firmly secure the panels or unit that the strap is fixed to.
扁头栓 一种螺栓，带预钻孔的扁平金属棒，直接和螺杆焊接。用螺纹杆将扁头牢固地安装于面板或装置上。

strawboard A compressed straw slab or panel. The compressed panel is formed through hot-pressing straw. The material is lightweight with good thermal resistance, and can be lined with plasterboard or paper producing a more workable product with a finish. The board can be made with ducts and rebates within it for cables and other services.
草纸板 草制纸板。草纸板由热压制稻草制成。材料轻，具有良好的耐热性，可以作为石膏板或纸品衬里，制成更切实可用的带饰面产品。纸板可以制成内藏的管子和槽口，用于电缆和其他设施。

straw thatching *Thatching made from the yellow straw of red wheat, the most common type of thatching material.

茅草屋顶 用红小麦的黄色稻草制成的茅草屋顶，是茅屋材料最常使用的类型。

stream 1. A small river. 2. A constant flow of gas or liquid.
1. 小河流。 **2. 流** 不间断的气流或液流。

stream gauge A device used to determine the depth of water in respect to a *datum, typically used along a riverbank or at a *fjord.
水标尺 一个用于测定相对于基准点水深的设备，通常沿着河堤或峡湾使用。

street furniture *Lights, *lamps, benches, road boards, and signs that are used in *external works.
街区公共设施 室外工程中使用的灯具、路灯、长凳、路牌广告和指示牌等。

strength 1. The capacity to withstand a force, pressure, or stress. 2. The physical power available.
1. 强度 承受施加力、压力或应力的能力。**2. 可用体力**。

Streptococcus A genus of bacteria that are characterized by being spherical, and used to indicate faecal contamination in drinking water.
链球菌 一类细菌，其特征是球形，用于测试饮用水中的粪便污染。

stress A measure of the average amount of force exerted per unit area. It depicts the intensity of the total internal forces acting within a body across imaginary internal surfaces, as a reaction to external applied forces and body forces. In general, stress, σ, is expressed as:

$$\sigma = \frac{F}{A}$$

where σ is the average stress, also called engineering or nominal stress, and F is the force acting over the cross sectional area A.
应力 测量每一单位面积所作用的力的平均值。描述了穿过假想内部平面作用于物体内的内力的总强度，作为对外部施加的力和体积力的反应。总之，应力 σ按以下公式表达：

$$\sigma = \frac{F}{A}$$

其中，σ是平均应力，也称为工程应力或名义应力，F 是指作用于横截面 A 上的力。

The SI unit for stress is the pascal (Pa), which is a shorthand name for one newton (force) per square metre (unit Area). The unit for stress is the same as that of pressure, which is also a measure of force per unit area. Engineering quantities are usually measured in megapascals (MPa), newtons per squared millimetres (N/mm^2), or gigapascals (GPa). In Imperial units, stress is expressed in pounds-force per square inch (psi) or kilopounds-force per square inch (ksi). As with force, stress cannot be measured directly but is usually inferred from measurements of strain and knowledge of elastic properties of the material. Usually, stress is calculated from the results of a tensile test.
应力的国际单位是帕斯卡（Pa），是每一平方米（单位面积）一牛顿（力）的简写。应力的单位与压力单位相同，也是对每一单位面积力的测量。工程计量通常以兆帕斯卡（MPa）、牛顿/平方毫米（N/mm^2）或吉帕斯卡（GPa）为单位。在英制单位中，应力通常以磅力/平方英寸（psi）或千磅力/平方英寸（ksi）为单位进行测量。正如力一样，应力也不能直接测量，但是通常可根据张拉的测量值和对材料弹性性能的认识来推断。通常，应力是根据拉伸试验的结果进行计算。

stress concentration A localized area where the stress in an element increases; this may be due to a crack or defect or a change in the cross-sectional area.
应力集中（区） 在一构件上应力增加的局部区域；这可能是由于横截面裂缝或瑕疵或变化造成的。

stress contour plot A multicoloured diagram to represent how the stress changes over the area of something (e. g. the cross-section of a beam); different colours are used to represent different stress values.
应力等值线图 彩色图表，反映某面积内（例如，横梁的横截面）应力的变化情况；使用不同颜色表示不同的应力值。

stress curve (S-N curve) A graph that indicates failure due to *FATIGUE, i. e. stress versus the number of fatigue cycles.

S

应力曲线（S-N 曲线） 反映疲劳破坏的图表，例如，应力对疲劳循环次数。

stressed skin panel A structural timber panel, with a timber frame and a skin of plywood on one or both sides of the panel. The plywood takes the stress when used as a floor, roof, or wall unit, and transfers them across the timber frame. The unit acts as a whole with the frame and skin, transferring the stress and sustaining the loads.
受力蒙皮板 结构木板，有木框，木板一边或两边都有一层胶合板表皮。当作为地板、屋顶或墙体时，胶合板承力，然后将其传给木框。木框和表皮及其木板作为一个整体，转移应力，承受负荷。

stress graded Timber that has been classified for its strength capabilities, either through visual stress grading or machine stress grading.
应力分级 木材根据其强度性能进行分级，既不是通过外观应力分级也不是机器应力分级。

stress graded timber Timber that has been * stress graded.
按应力分级的木材 已按应力分级的木材。

stress grading The process of grading the strength of timber.
应力分级 根据木材强度分级的过程。

stress grading machine Machine that grades the strength of timber.
应力分级机器 对木材强度进行分级的机器。

stress intensity factor (*K*) A scale factor used to define and gauge the magnitude of the crack-tip for a stress field, being dependent on the configuration of the cracked component and the way the loads are applied. Such factors are important in the mechanics of metal fracture.
应力强度因子（*K*） 用于确定和测量应力场裂纹尖端量级的比例因子，取决于裂化组分的布局和施加负荷的方式。这些因子在金属断口机理中很重要。

stress relaxation The reduction in tensile stress with time when subjected to constant strain. It occurs in polymeric materials, such as a * geosynthetic used as basal reinforcement under an embankment. It is not to be confused with * creep, which is a time-dependent increase in strain under constant load, as would occur for a geosynthetic in a reinforced wall.
应力松弛 当承受恒定应变时，应力随着时间的增长而减小。这种现象通常发生在高分子材料上，例如作为路堤基部加固的土工合成材料。不可与蠕变混淆，蠕变是张力在恒定荷载下按时间增加，通常发生在钢筋墙的土工合成材料上。

stress strain curve A graph that represents the relationship between stress and strain of a material. It is similar to a load displacement plot, however, with load and displacement, no indication of the size of the sample is given, thus stress (load/area) and strain (change in length/original length) is often used.
应力应变曲线 反映材料应力和张力之间关系的图表。与荷载位移曲线相似，但标识荷载和位移，没有标出样本的尺寸，因此通常使用应力（荷载/面积）和张力（长度变化/原本长度）。

S

stretcher A brick or stone laid lengthways, if laid with other bricks or stones in the same direction, a * stretcher course will be created. Multiple courses of stretchers create a * stretcher bond wall. * Headers and * closers will be used to finish the wall and create other courses.
顺砖 纵向铺设的砖块或石块，如果与其他砖块或石块同一方向铺设，则需形成一个顺砖层。多层顺砖则砌成全顺砖墙。需用丁砖和过渡砖完成墙壁和其他砖层铺设。

stretcher bond (common bond, half bond, running bond, stretching bond) Multiple courses of * stretchers laid on top of each other. The centre of the joint on one course sits directly over the centre of the brick or stone on the next course.
顺砖式砌法 多层顺砖互相堆砌。一层接缝的中心直接落在下一层砖块或石块的中间。

stretcher course Stretchers laid end to end throughout the course.
顺砖层 整层顺砖首尾相连铺设。

stretcher face The long face of a * stretcher brick which is exposed.

顺砖面 顺砖外露的较长的一面。

strike To remove the formwork from recently cast concrete. The *formwork should be removed cleanly without pulling concrete away from the structure. A clean strike is achieved when the formwork has been cleaned, properly prepared with *release oil, is free from *defects, and the concrete has been properly *vibrated and left to *set.
拆模 从刚浇筑的混凝土上拆除的模板。模板应该完整移除但不能把混凝土拉离结构。要干净拆模，需提前清洁好模板，正确涂上拆模油，让混凝土无瑕疵且已适当振实并让其凝固。

striking The dismantling of formwork and other temporary supports.
拆除 模板和其他临时支承的拆卸。

striking off The removal or skimming of any excess, mortar, or concrete from a mould or formwork.
刮平 移除或撇去模具或模板上多余的砂浆或混凝土。

striking plate (keeper, strike, strike plate) A plate fixed to the jamb of a door that has a rectangular hole to receive the door latch.
锁舌片（导向片） 固定到门边梃的一块板，有一矩形孔，以插入门闩。

striking time The time that must be allowed for the concrete to mature and maintain its strength before *striking—when the formwork is removed. The concrete must be sufficiently *set so that it can transfer its own load without fracturing the bonds within the newly set concrete.
拆模时间 模板被拆除前，等混凝土成熟并且能维持其强度所需要的时间。混凝土必须充分凝固，这样在刚凝固的混凝土内在不破坏黏结的情况下就可以转移其荷载。

striking wedges Wedges used to prise and lever the formwork from the face of the concrete, the triangular wedges are carefully tapped between the concrete and formwork. The wedges should be used only where necessary, since levers and wedges can leave marks on the face of the concrete.
拆模楔块 用于从混凝土表面撬松模板的楔块，将三角形楔块在混凝土和模板之间小心敲入。只有在必要的时候才能使用楔块，因为撬棍和楔块会在混凝土表面留下痕迹。

string (stringer) The diagonal boards that carry the *treads and *risers in a staircase. Strings can be closed and rebated out to carry each tread, or can be rough and cut to the profile of the stair with the tread resting on the top of the string. The string is a diagonal beam that supports the staircase, transferring the loads to the upper and lower floors.
斜梁 楼梯中承载踏板和踢脚板的斜板。斜梁可以是暗步的和开凹槽以承载各踏板，或者可以粗略裁切至楼梯剖面形状，踏板位于斜梁顶部。斜梁支承楼梯，将负荷转移到上下层楼板。

string course (band course, belt course) A course in brick that either projects or is indented from the surrounding courses. Courses run in different-coloured brick may also be referred to as a band or belt course.
挑出层（腰线） 砖块中突出来或者从周围砖层缩进的一层。砖块颜色不同的砖层也可称为腰线。

string line A line pulled between two points that is used for setting out works. String lines are used between the ends of new walls to provide a guide for *courses of brick or blockwork; lines can be pulled between setting out pegs to provide a guide for excavating *trenches and *footings; they are used between steel pins to create a guide when *setting out and laying concrete *kerbs and *edgings in road works, and have many uses when marking lines in *formwork *joinery.
放样线 用于放样工作中，在两点之间拉开的线条。放样线用于新墙壁端点之间，以便为砖层或砌块提供标记；放样线可以在放样钉之间拉开，以便为挖掘沟渠和基脚提供标记；也可以用于钢针之间，以便在道路工程中放样和铺设混凝土路缘时提供标记，在模板细木工做标志线时也有很多用途。

string wall A wall that is used to support the treads of a staircase.
梯墙 用于支承楼梯踏板的墙壁。

strip board *See* BLOCKBOARD.
木芯板 参考“木芯板”词条。

strip diffuser (slot diffuser) *See* LINEAR DIFFUSER.
条形散流器（条缝型散流器） 参考“线型出风口”词条。

strip flooring Long narrow strips of solid tongue-and-grooved wood flooring.
条形木地板 长窄条企口实木地板。

strip foundation A foundation that is excavated and cast in long lengths, used to carry longitudinal loads such as external walls and walls to houses. The foundations can be wide strip, reinforced to distribute the loads over a greater area, reducing the load per surface area; narrow or deep strip footings are slightly wider than the loads being carried and transfer the loads down to good load-bearing strata.
条形基础 挖掘和浇筑的长条基础，用于承受纵向负载，例如外墙和连接房子的墙壁。基础可以是扩展形，加钢筋以在更大面积内分散负荷，减少每单位表面积上承受的负荷；窄条或深条基脚比需承重的面积稍宽些，将负荷转移至承重地层。

The foundation runs under continuous loads such as brick walls. Where the ground has good load bearing strata, this is the most economical way to found walls.
基础在连续荷载（例如砖墙）下铺设。地基有良好的承重地层，对于建墙来说是最经济的方式。

Strip foundations are used for continuous loads such as walls.
条形基础用于连续荷载，例如墙壁。

strip lamp A long narrow fluorescent luminaire.
灯带 狭长的荧光照明装置。

stripper 1. A labourer or formwork joiner who strips formwork. 2. Chemical paint remover, used to break down and soften *paint so that it can be scraped away from the surface it has been applied to.
1. 拆模工 拆除模板的小工或模板工匠。**2. 脱漆剂** 化工脱漆剂，用于分解和软化油漆，这样就可从其表面刮除油漆。

strip sealant Preformed strip of compressible material used around door and window frames to exclude draughts; solid rigid material is also used in glazing to position, space, and hold glass.
密封条 预成型的条形压缩材料，用于门框和窗框周围以密封；在玻璃镶嵌中也会用坚硬的刚性材料去定位、分隔和固定玻璃。

strip tie A *twist tie that is stood vertically.
垂直止水拉结片 直立止水的拉结片。

strong back A proprietary metal support used for resisting high loads on concrete formwork. The beam-type unit can be strapped vertically, or fixed horizontally to the secondary supports that need to transfer the loads to a strong unit.
大模板 用于承受施加于混凝土模板上高荷载的专用金属支承。梁式装置可以垂直捆扎或水平固定至需将荷载转移至支撑装置的二次支承上。

struck capacity The amount of material that can be held in an excavator's bucket which has been levelled off. It is equivalent to the volume of water a bucket would hold if it was watertight.
平装斗容量 挖掘机铲斗中可装的刮平材料的容量。相当于铲斗可装的水容量（如果铲斗水密性良好的话）。

struck joint *See* WEATHER-STRUCK JOINT.
刮斜缝/平缝 参考“防雨斜缝”词条。

structural Relating to a *structure; something that is built such as buildings, bridges, frameworks that carry loads and resist forces.
结构的 与结构相关的；建造的可承受荷载和抵抗施力的建筑物、桥梁和模板等。

structural analysis The mathematical examination of structures and their behaviour to loads, displacement, stresses, strains, and moments.
结构分析 对于结构及结构的承受荷载、位移、压力、张力和力矩能力的数学分析。

structural clay tile *See* HOLLOW CLAY BLOCK.
空心黏土砖 参考“空心黏土砖”词条。

structural connection The joining of two members that carry loads. The joint transfers the loads and forces.
结构连接 两个承受荷载的构件的连接。接缝可转移荷载和受力。

structural design The design of structures, particularly in relation to carrying loads and minimizing displacement.
结构设计 结构的设计，尤其是与承受荷载和将位移最小化相关的设计。

structural drawings The *structural designs and *drawings produced by the *structural engineer or design engineer that show all the *structural elements and the structural frame in detail.
结构图纸 由结构工程师或设计工程师编纂的结构设计和图纸，详细标示了所有的结构构件和结构框架。

structural element *See* STRUCTURAL MEMBER.
结构元件 参考“结构构件”词条。

structural engineer A professional engineer who specializes in structural design and is a chartered member of the Institution of Structural Engineers (*IStructE).
结构工程师 专长结构设计的专业工程师，是结构工程师学会（IStructE）的特许会员。

structural gasket A *GASKET used in curtain-walling to hold glass against the mullion.
玻璃垫 幕墙中用于将玻璃固定于竖梃的垫片。

structural glass Glass, usually coloured, applied to masonry or plastered walls as a decorative finish.
墙面玻璃 玻璃，通常是有色玻璃，用于砌体或抹灰墙上用作装饰饰面。

structural glazing Glazing that has some load-bearing capacity. It can either be bonded together using structural sealants such a silicone (unsupported system), or it can be supported on a concealed framework (supported system), which is fixed back to the structure. It is used to create a smooth all-glass facade.
结构性玻璃 有一定承重能力的玻璃。可以使用结构密封胶，例如用硅胶进行黏合（非支承系统），或由安装在结构背部的密封框架支承（支承系统）。可用于做成平滑的全玻璃幕墙。

structural integrity The ability of a structure to remain fit for purpose, in respect to carrying loads and stress, and minimizing displacements.
结构完整性 使结构保持适用于其用途的性能，与承受荷载与压力，以及最小化位移相关。

structural member A beam, column, etc., that forms an important part of the structure, such that it carries loads, and provides support or stiffness.
结构构件 横梁、柱子等形成结构的重要组成部分，可承受荷载以及提供支承或刚度。

structural steel Steel that has a carbon content of less than 0.25% by weight.
结构钢 碳含量小于0.25%的钢。

structural steelwork *Structural members manufactured from *structural steel that have been connected to each other to form a *steel frame.
钢结构工程 由相互连接形成框架的结构钢制成的结构构件。

structural timber Beams and trusses that are manufactured from wood and used in a structure to provide support and carry loads.
结构木材 由木材制成的横梁和桁架，在结构中使用，以便提供支承和承受荷载。

structural trades All of the trades that contribute to the structural components and members of the building; includes the ground workers who excavate and cast the footings, bricklayers who erect structural masonry, structural frame erectors who erect either steel, concrete, or timber, and any other trades that contribute to components or elements that are structural.
结构类工种 与建筑物的结构组件和构件相关的所有工种；包括挖掘和浇筑基础的地基工人、建造结构砌体的瓦工、建造与结构性组件或元件相关的钢、混凝土、木材以及其他工种相关的结构框架安装工。

structure The load-bearing frame of a building, bridge, or other structure; something that is constructed.
结构 建筑物、桥梁和其他结构的承重框架。

structure-borne sound Noise that is transmitted through the fabric of a structure, through a wall or floor, as opposed to an air-borne sound.
结构声 通过结构的组织，以及通过墙壁或地板传播的噪音，与空气传声相对。

strut A structural member inserted to act in compression and is used to hold or brace something; it can be vertical or horizontal.
支柱 嵌入的结构构件在抗压中发生作用，用于支承或支撑；可以是垂直的，也可以是水平的。

strutting System of struts used to hold or brace an element; herringbone or solid strutting used to brace floor joists, or struts used to prop up floor formwork.
支撑 用于支承或支撑构件的支承系统；用于支承楼板托梁的人字形或实心支柱，或用于支撑楼板模板的支柱。

stub A short element or shorter version of something, for example a *stub tenon.
短粗的东西 短的构件或某些东西的短小版，例如短榫。

stub tenon A tenon that is used for a *blind mortise; as the *tenon does not need to penetrate all of the way through the mortise, it is a shorter tenon and described as a *stub.
短榫 用于不贯通榫的榫头；因为榫头不需要完全穿透榫眼，所以榫头可以稍短些，可描述为短粗。

stucco A material used for render, but has its origins in more classical forms of render. Traditionally, it was smooth with mouldings to represent columns and ornamental structures. Pre-nineteenth century it was made from lime-cement mortar and since that time cement mortars have become much more common. The sustainability movement argues for the reintroduction of lime-cement mortars due to their ability to be reused.
灰泥粉饰 用于抹灰的材料，但词源来自更古老形式的灰泥。传统上，是通过模具抹平来展现柱和装饰结构。19 世纪以前由石灰水泥砂浆制成，之后水泥砂浆变得更加常见。现在有可持续发展的声音认为应重新使用石灰水泥砂浆，因为其可重复使用。

stud Vertical timber (or more recently, pressed steel) supports, used in *stud partition walls. The vertical timbers and horizontal connecting pieces, including *head and *sole plate, and *strutting, act as a structural frame onto which plasterboard can be fixed and supported.
立柱 垂直的木（或现代的压制钢）支承，用于立筋式隔断墙。垂直木材和水平连接件，包括上槛和下槛，以及支撑，作为固定和支撑石膏板的结构框架。

S

stud partition (stud wall) A *partition wall where the *structure is made of *steel or *timber *studs, creating a *frame onto which *plasterboard finish is applied. Originally the partition wall was just considered a divider of space and either *load-bearing or *non-load bearing. Today partition walls have various roles to perform in addition to dividing space; they are often required to perform acoustic, thermal, and fire-resisting functions. While in most cases partition walls are not thermally insulated, it is common practice to insert acoustic insulation, and if a fire-rating is required, the walls will have fire protection in the form of fire board, lapped joints, and insulation to delay the passage of fire.
立筋式隔断墙 由钢或木龙骨结构制成的隔断墙，形成石膏板饰面覆盖的框架。原先，隔断墙只是作为空间的分隔，分为承重式和非承重式。现今，隔断墙除了分隔空间之外，还有各种性能；通常，隔断墙需具备隔声、隔热和防火性能。在大部分情况下，隔断墙并不隔热，常见的是嵌入隔音设备，如果有防火等级要求，隔断墙应具备防火板、搭接接缝以及隔绝材料等形式的防火性能，以延缓火势。

styrene/butadiene copolymer A thermoplastic copolymer comprising styrene and butadiene. One of the outstanding characteristics of styrene/butadiene thermoplastic copolymers is their combination of high transparency and impact resistance. The good miscibility allows adjustment to the desired toughness, while at the same time reducing material costs.

苯乙烯/丁二烯聚合物 一种热塑性塑料聚合物，由苯乙烯和丁二烯组成。苯乙烯/丁二烯热塑性塑料聚合物最突出的性能是其高透明度和抗冲击性及良好的混溶性，可调至需要的韧性，同时降低材料成本。

sub-base The lowest layer in the formation of a *pavement, normally constructed of crushed stone or gravel. Its function is to form a working platform and provide strength.
底基层 路面铺装的最底层，通常由碎石或砾石建成。其功能是形成工作平面，提供强度。

subbie (subby) A slang word used to define a *subcontractor.
分包商 用于称呼承包商的俚语。

sub-board *See* DISTRIBUTION BOARD.
配电盘 参考“配电箱”词条。

subcontract A contract that is placed by the main contractor to a third party to carry out a proportion of the works. In order to introduce a subcontractor into a contract, agreement should be sought from the client; it is usual for the main contract to state that work cannot be subcontracted without prior agreement.
分包合同 总承包商签发给第三方，由第三方履行一部分工程的合同。将分包商加入合同中，应该获得业主的同意；通常，总包合同会明确指出，在没有获得业主事先同意的情况下不得分包工程。

subcontractor Company or individual who is employed by the main contractor to undertake work. The work is part of the contract that has been awarded by the client to the main contractor under a *subcontract.
分包商 受总承包商雇佣来执行工程的公司或个人。在分包合同中，工程内容为业主授予总承包商的合同的一部分。

sub-floor The smooth floor base onto which the finished material is placed.
底层地板 光滑的地板底板，在上面铺设装饰材料。

sub-frame Any frame or surround used to carry or support finishes or materials to be attached to the main structural component.
副架 用于承受或支撑将连接至主要结构组件的装饰物或材料的框架或环绕物。

subgrade The natural ground below the formation level of a *pavement—*See* FLEXIBLE PAVEMENT.
路基 路面平整面标高下的天然地基—参考“柔性路面”词条。

subletting The *subcontracting of work to another party.
分包 将工程分包给另一方。

sub-level The level below the main level. When discussing the ground it is the strata below that at the top of the ground.
基层 主层下面的层。在地基里面是指在顶层下面的岩层。

S

submersible pump A *pump that operates under water, at the bottom of a *borehole to *DEWATER a site, for instance.
潜水泵 在水下运行的泵，例如，位于钻井的底部以现场排水。

subset A subdivision of a set, e. g. a smaller set of numbers that also belong to a larger set of numbers.
子集 一个集合的分部，如一个较小数目的集合也属于一个较大数目的集合。

subsidence (settlement) The movement of ground as it compresses, slips, or deforms. Settlement is a problem when buildings are placed on the ground and the structure moves downwards as the ground moves. If the whole building moves downwards at the same rate only the incoming services will have to cope with the movement. If the movement is differential, acting at different rates of fall across the building, then cracks will appear and in severe cases the building may collapse.
地表下陷（沉降） 地表压缩、滑坡或变形运动。当建筑位于地基上并且其结构随地基移动而向下移动时，沉降是一个问题。当整栋建筑以相同的速率向下运动，则只有输入设备需应付沉降问题。如果移动不均匀，即建筑的沉降速率不同，则会出现裂缝，严重情况下建筑会倒塌。

subsill A *sill inserted below a *threshold or window sill, designed to ensure that any water that drips down from the threshold or window sill is thrown clear of the wall. A range of materials can be used to form a subsill, including brick, concrete, plain tiles, stone, and timber.
底槛 嵌入门槛或窗台下的基石，用于确保从门口或窗台滴落的水不会渗入墙壁。底槛可以用不同材料制成，包括砖块、混凝土、平瓦、石头和木材。

subsoil The compact soil that lies directly beneath the *topsoil.
底土 直接位于表层土以下的紧实土壤。

substance Material matter; from which people and things are made.
物质 构成人和物的材料。

substation An electrical facility that contains transformers and other types of electrical equipment that decrease, increase, and regulate the voltage of the electricity that is sent through the power lines.
变电站 包含变压器和其他类电器设备的电力设施，可减小、增大和调节通过电线输送的电压。

substrate 1. The surface onto which finishes are applied, such as plaster stuck on a block substrate. 2. The ground on which the building rests.
1. 衬底 涂上饰面的表面物质，例如涂在砌块衬底上的灰泥。2. **基底** 建筑物已建的地基。

substructure The parts of a building that are constructed below ground level or below that which will be the finished ground level once the building is completed. The substructure elements include the foundations and basement. The substructure elements are critical activities, and as they have to be completed before other works can start, they will almost certainly appear on the *critical path of a *Gannt chart.
下部结构 建筑物在地表水平下或完工后地面水平下的部分。下部结构构件包括基础和地下室。下部结构构件是关键构件，因需要在其他工作完工后才能开始，所以在甘特图上几乎一定以关键路径标示。

subtense bar A bar of standard length typically 2 m, manufactured from *invar, and used in *tacheometry.
视距尺 标准长度通常为 2 m 的杆，由殷钢制成，用于视距测量。

suburb Town development that is built against the edge of an existing city or town; it is an *urban area but is not at the heart of the main town.
郊区 位于已有城镇边缘的城市周边；属于城市范围但不在市中心。

subway An underground passage for pedestrians to walk from one side of the road to the other. In the US the term is used to refer to an underground railway.
地下通道/地铁 从道路一边到另一边的地下行人通道。在美国是指地下铁路。

S

successful tender (success bidder) The bid selected to enter into the contract for the works. Traditionally, the bid would be selected on price alone; today the qualification process and selection has to ensure the capabilities of the contractors, their ability to perform the works, financial stability, level of expertise, and quality of work expected.
中标（中标者） 被选为签订工程合同的投标者。传统上，投标只取决于价格；如今，资格审查过程和选择必须确保承包商的实力、执行工程的能力、财务稳定性、专业水准和工程所期望达到的质量。

suction The process of moving a fluid from one place to another due to differences in pressure. For instance, a fluid will flow from an area of high pressure to an area of low pressure.
抽吸 液体由于压力不同从一个地方流向另一个地方的过程。例如，液体将从高压区域流向低压区域。

sugar soap A paint remover made from alkaline solution.
糖皂 由碱性溶液制成的油漆清洁剂。

suite A set of matching or complementing parts, for example a set of keys matching doors in a hotel, a matching set of furniture, a set of interconnecting rooms making one designated area.
套件 一套配套的或补充的部件，例如一套配对酒店所有门的钥匙，一配套的家具，一互相连通的房间构成特定的区域。

sullage 1. Wastewater from domestic activities. 2. Sediment or silt deposited by flowing water.
1. 污水 日常生活产生的废水。**2. 淤泥** 水流动时沉积的沉淀物或粉土。

sulphate attack A reaction between the tricalcium aluminate in concrete and the salts contained in some clays, soils, or flue condensates. The reaction leads to the production of a chemical called hydrated calcium sulphoaluminate (ettringite) which has a greater volume than the cement material which it is replacing. The expanse of material within the concrete or mortar causes it to crack and degrade, losing structural strength, and affecting the finish of the concrete.
硫酸盐侵蚀 混凝土中铝酸三钙和一些黏土、土壤或烟道冷凝水中盐分之间的反应。反应产生一种叫作水化硫铝酸钙（钙矾石）的化合物，其体积要比原来的水泥物质要大。引起水泥或砂浆中物质体积的膨胀造成裂缝和退化，失去其结构强度并影响水泥的表面。

sulphate resisting Portland cement A Portland cement with a chemical composition that is resistant to *sulphate attack.
抗硫酸盐水泥 含有能够抵抗硫酸盐腐蚀的化学成分的波特兰水泥。

sump A small pit or well used to drain and collect water.
集水坑 用于排水和集水的小坑或井。

sump pump A small electrically driven pump used to remove water from a *sump.
集水坑水泵 用于排掉集水坑里的水的小电动泵。

sunken fence *See* HA-HA.
哈哈墙 参考“哈哈墙”词条。

sunken gutter A *box gutter.
暗天沟 一种方形排水槽。

sun shade A device designed to reduce *glare and *solar gain.
遮阳物 用于减少眩光和太阳辐射的设备。

superelasticity (pseudoelasticity) A phenomenon whereby a material exhibits a large elastic (non-permanent) response to relatively high stress and usually returns to its original shape or dimension on removal of the load.
超弹性 一种材料对于相对高的压力显示出较大弹性（非持久），并在荷载撤销后回弹到原来形状或尺寸时的现象。

superheated steam (anhydrous steam, steam gas, surcharged steam) Steam heated to a temperature that is beyond its saturation point at a given pressure. Used in power generation.
过热蒸汽 在特定压力下蒸汽被加热到超过其饱和点的高温。用于发电。

superheater A device used to generate superheated steam.
过热器 用于产生过热蒸汽的装置。

superimposed load *See* LIVE LOAD.
附加荷载 参考“活荷载”词条。

superplasticizers A component added to concrete to increase its workability, usually used as an additive or alternative to water.
高效减水剂 加入到混凝土中以增加其和易性的成分，通常作为添加剂或水分的替代。

superstructure In general it refers to all the parts of the building that are above ground level, but in specific terms it addresses the structural components that sit on top of the *substructure. The superstructure works include the frame, floors, and walls above the ground, and the roof, finishes, and fittings to all above ground works.
上层结构 通常是指位于地平面以上的建筑物，但是具体来说，它强调位于下部结构之上的结构组件。上层建筑工程包括位于地面上的框架、地板和墙壁、屋顶、饰面和所有位于地基以上的配件。

supervision The overseeing and hands-on management of workers or a package of work, offering guidance, checking work done, and coordinating activities.
监督 监视和在现场对工人或工作进行管理，提供指引，审核工作和协调活动。

supplier Company that supplies materials, labour, plant, or services for a specific area. The term is more usually associated with companies that provide physical products and materials.
供应商 为特定区域供应材料、劳动力、设备或服务的公司。该词条更常指供应实体产品和材料的公司。

supply air The air that is supplied to a room or space from the air-conditioning, ventilation, or HVAC system.
送风 由空调、通风设备或暖风空调（HVAC）系统供给一个房间或空间的空气。

supply and fix (furnish and install) A contract for the supply and installation of a product and component. The * **contractor** or * **supplier** delivering the unit also takes responsibility for ensuring the installation is correct and the product operates correctly.
供应安装合同（提供和安装） 关于产品及组件供应和安装的合同。交付装置的承包商或供应商也应确保产品安装正确并且正常运行。

supply only Designates a contract that is for the supply of a product or material only, and does not include the installation of the unit.
供应合同 特指仅供应产品或材料，安装装置并不包括在内的合同。

supply pipe The pipe from the building to the site boundary, stop valve, or highway; it is under the consumer's ownership. Beyond the boundary and the stop valve, it is the property of the utility provider. The ownership of the service is important as it designates responsibility for maintenance and repair.
供给管道 从建筑物到现场边界、截止阀或公路的管道，属于业主所有。超出现场范围和截止阀，则属于设备提供者所有。设备的所有权很重要，因为这关系到维护和修理的责任归属。

support To bear the weight or hold up something.
支撑 承受重量或支承某物。

supported sheet-metal roofing (flexible metal roofing) Sheets of metal roofing that are supported by * **roof deck**.
柔性金属屋面 由屋顶板支撑的金属屋面。

suppressed weir A * weir that is the full width of the channel; *See also* CONTRACTED WEIR.
无侧收缩堰 堰宽与渠道宽度一样的堰。另请参考“有侧收缩堰”词条。

S

surcharge Additional loading of the ground, for example, behind a retaining wall or as a form of ground improvement, a temporary loading of the ground to reduce the amount of * consolidation once the structure or embankment has been built.
加载 地面的附加荷载，例如，在挡土墙后面或作为地基加固的一种形式，当建成结构体或路堤时，地面的临时荷载可减小固结量。

surface condensation Condensation that occurs on surfaces that are at or below the * dew point temperature of the air immediately adjacent to them. It tends to occur on the internal surface of external elements of a building, and on cold pipes and cisterns within a building.
表面冷凝 表面在接触温度达到或低于露点的空气时的凝结。表面冷凝倾向于在建筑外围构件的内表面，以及建筑内冷管和水箱上产生。

surface damp-proof membrane (surface damp-proof course) A liquid-based damp-proof membrane or * damp-proof course that is applied to the surface of a material.
表面防潮膜（表面防潮层） 液态防潮膜或防潮层，覆于材料表面。

surface dry A stage in paint drying, where the surface has a thin film of dried paint and feels dry to touch,

although the underlying surface is not dry.
表面干燥（表干、指干） 油漆干燥的一个阶段，即表面有一层干燥的油漆薄膜，并且摸上去感觉是干燥的，但是里面并未干燥的阶段。

surface filler A fine-grained filling material suitable for creating a smooth finish.
表面填料 适用于形成光滑表面的细粒的填充材料。

surface-fixed hinge A hinge that is mounted directly onto the surface rather than being recessed or sunk into it.
表面安装式铰链 直接安装在表面而不是内嵌或埋入式安装的铰链。

surfacing The treatment or finishing to a surface by smoothing, texturing, or painting.
表面修整 光滑化、纹理化或涂漆等表面处理或装饰。

surfacing material Finishing material, such as paints, bitumens, lining, coverings, chipping, etc. Used to protect against water, provide skid resistance, and present an aesthetical finish.
装饰材料 例如油漆、沥青、衬里、包覆层、削片等。用于防水、防滑和呈现装饰性饰面。

surface spread of flame *See* FLAMESPREAD.
火焰表面蔓延 参考“火焰传播指数”词条。

surface tension A measure of the ability of the molecules contained within a liquid to stick together. *See also* CAPILLARTITY.
表面张力 测量液体内凝聚到一起的分子能力。另请参考“毛细现象”词条。

surface water Rainwater; also water from streams, rivers, and lakes, as opposed to *groundwater.
地表水 雨水；来自溪流、河流、湖泊的水也称为地表水，与地下水相对。

surface water drain A *drain that takes the water collected from roofs, the faces of buildings, driveways, and paved areas, and conveys it to the *surface water sewer.
地面排水沟 排干从屋顶、建筑物表面、车道和人行道聚集的水，将其排到地表水下水道的排水沟。

surface water sewer A *sewer that conveys *surface water.
地表水下水道 输送地表水的下水道。

surround A frame, moulding, or trim that is placed around a component or an opening, such as a door.
边饰 在组件或开口周围的框架、线条或饰边，例如门。

survey 1. An inspection of a building to determine its structural soundness. 2. To measure an area of land to enable a scale map or plan to be produced.
1. 检测 检查建筑物以确定其结构稳固性。**2. 测量** 测量一块土地区域，以编撰比例图或平面图。

S

surveyor A professional who performs an evaluation or assessment of property or land: a land surveyor describes the nature, level, and lie of the land; a property surveyor determines and evaluates the physical characteristics of a building; or a quantity surveyor assesses the value of land, buildings, work packages, and materials for commercial purposes.
测量员 对物业或土地进行评价或评估的专业人员：土地测量员负责测绘土地的自然状况、水准和地形地貌；物业测量员负责确认和评价建筑物的物理特性；而工料测量员则负责对土地、建筑、工作包和商用材料估价。

suspended ceiling A ceiling finish that is supported or hung from the structural floor above.
悬吊式吊顶/吊顶天花 受支承或从结构楼板上悬吊的天花板装饰。

suspended floor A floor that is raised and supported above the ground; *See also* SUSPENDED TIMBER GROUND FLOOR.
架空地板 地面以上架空支承的地板；另请参考“架空木地板”词条。

suspended scaffolding A *flying scaffold, *projecting scaffold, or *swinging scaffold that is not supported from scaffolding standards or props, but is supported or hung from a building or surrounding structures.

悬挂式脚手架 悬空脚手架、悬挑脚手架或吊篮脚手架，无支柱或支撑物支承，而是由建筑物或周围结构体支承或悬挑。

suspended slab Reinforced concrete slab that is raised above the ground, and supported by beams and walls.
悬板 钢筋混凝土板，悬空于地面上，由梁和墙壁支承。

suspended timber ground floor A floor constructed from timber joists at ground level, where the joists span across external and intermediate *honeycomb walls, raising the floor above the ground. The gap below the floor should be ventilated to prevent moisture forming on the timber and causing mould growth.
架空木地板 由木龙骨构成的地板，位于地面，当龙骨跨越蜂窝状墙表面和墙中间时，使地板架空于地面上。地板下面的间隙应该保持通风，以防止潮气入侵木材，而导致发霉。

suspended timber upper floor A floor above ground level that is made from timber joists spanning between walls or beams.
上层架空木楼板 位于地面层以上的楼板，由木龙骨制成，横跨墙或梁之间。

suspension 1. To temporarily stop something. 2. The dispersion of fine particles in a liquid.
1. 暂停 暂时停止某事。**2. 悬浮** 液体中微粒的离散。

suspension bridge A *bridge, such as the *Forth Road Bridge, that is suspended by cables, which are carried by support towers and anchored at either end of the bridge.
悬索桥/吊桥 如福斯桥的桥梁，由缆索悬吊、索塔支承，锚固于桥梁的任意一端。

suspension cable The main cable, made of steel wire, on a *suspension bridge.
悬索 主缆，由钢丝制成，用于悬索桥上。

suspension of works Instruction given to halt the flow of works.
工程中止 发出指令中止工作流程。

Sussex garden wall bond *See* FLEMISH GARDEN WALL BOND.
苏塞克斯花园墙砌法 参考“荷兰花园墙式砌法”词条。

sustainability The ability to be able to maintain without destroying a natural balance.
可持续性 在不破坏生态平衡情况下的持续能力。

sutro weir A contracted weir that has one of the following cross-sections:
苏特罗式堰 有以下横截面之一的有侧收缩堰。

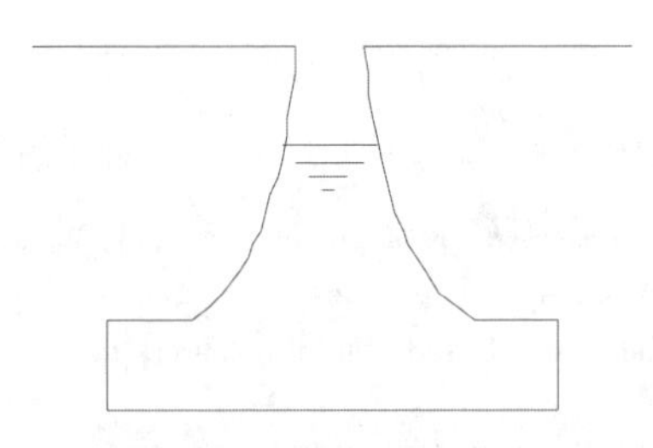
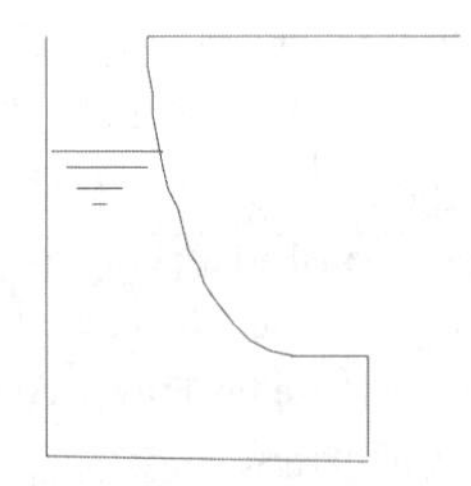

swale A swale is a broad, shallow channel that water runs along when it has been raining.
集水渠 下雨时排水的宽浅沟渠。

swallow hole A vertical hole or shaft in the ground that connects with an underground passage or tunnel.
落水洞 地下的垂直洞穴或竖井，连接地下通道或隧道。

swan neck An S-shaped pipe or drain, used to create an air trap in drainage preventing the escape of foul smells from sewers.
鹅颈管 S形管道或排水沟，用于在排水中形成水封，防止下水道臭味外泄。

S wave (secondary wave, shear wave) A type of seismic wave that travels in a perpendicular direction to the direction the wave propagates in.

S 波（次级波、剪力波） 地震波的一种，震动方向与震波传递方向垂直。

sweep tee (sweeptee, swept tee) A duct or pipe fitting that has three openings, where a branch joins a slight curve. *See also* STRAIGHE TEE.
弯三通 管子或管件有三个开口，支管稍微弯曲。另请参考“直三通”词条。

swelling An increase in volume, particularly in relation to clays as they absorb water.
膨胀 体积增大，尤其是指黏土吸水后。

swept-sine testing A dynamic loads test.
正弦振动试验 一动荷载试验。

swept valley *See* LACED VALLEY.
搭瓦斜沟 参考“搭瓦斜沟”词条。

swg The traditional method of describing the thickness of wire, tubes and sheet material, known as the *STANDARD WIRE GAUGE.
标准线规 描述电线、管子和板材厚度的传统方法，全称为“standard wire gauge”。

swing To move backwards and forwards.
摇摆 前后移动。

swing door (double-action door) A door that can open in either direction.
双向开启门 可往内外两边开启的门。

swinging jamb or post A *hanging jamb or post.
带铰链的门/窗边梃 带铰链的门/窗边梃。

swinging scaffold A cradle scaffolding platform that is hung or slung from outriggers that reach out over the edge of the building. The cradle, which can carry the operatives, equipment, and materials, is hung by cables or ropes. Machine operated winches allow the platform to be raised and lowered so that the workers can access the desired area of the building.
吊篮脚手架 吊篮式脚手架平台，悬吊或悬挂于建筑边缘伸出的悬臂支架上。吊篮可以装载作业人员、设备和材料，用缆索或绳索悬吊。机动式绞盘可以使平台升起和降下，这样工人就可以进入其所需进入的建筑物区域内。

swing-up door (up-and-over door, overhead door) A door that opens by lifting up the bottom of the door until it is in a horizontal position. Commonly found on garage doors.
上开门（上翻门、卷帘门） 通过升起门的底部直至其处于水平位置的门。通常为车库门。

switch An electrical device used to open or close an electrical circuit.
开关 用于开启和关闭电路的电气装置。

S

switch and fuse *See* SWITCH-FUSE.
开关熔断器 参考“带保险丝开关”词条。

switchboard 1. A panel comprising a number of switches that are used to control power and lighting. 2. A device that is used to connect telephone callers with one another.
1. 开关柜 一个具有多个开关的面板，用于控制电力和照明。**2. 交换机** 用于连接各电话来电的设备。

switchbox A box, usually mounted on a wall, that contains an electrical switch.
配电箱 一个箱子，通常安装在墙壁上，有一个电子开关。

switch-fuse (switch and fuse) A switch in which a fuse is placed in series with the contact.
带保险丝开关（开关熔断器） 内有保险丝的开关，与触点串联。

switchgear The electrical switches, circuit breakers, and other items of electrical equipment required to control and protect electrical circuits.

开关设备 用于控制和保护电路的电气开关、断路器和其他电气设备。

switchroom A room that contains *switchgear or a telephone *switchboard.
配电室/电话交换机房 内有开关设备或电话交换机的房间。

symbol (symbols) Drawing shapes that represent real manufactured components, materials, or functions. The drawings and diagrams provide a common language that is used to easily interpret the components on a drawing.
图例 代表实际生产的组件、材料或功能的图案符号。图案和图标作为一种共同的语言可用于解读图纸上的组件。

syncline A downward *fold in *sedimentary rock, caused by *tectonic movement.
向斜 由于地壳运动引起的沉积岩向下折叠。

synthetic Not naturally occurring; something that is man-made, such as a *geosynthetic product made from a polymer, *polyester, *polyamide, *polypropylene, and *polyethylene, rather than a vegetable fibre such as *coir or *jute.
合成的 非自然产生的；人造的某物，例如土工合成材料，由聚合物、聚酯、聚酰胺、聚丙烯和聚乙烯制成，而不是由植物纤维如椰壳棕丝或黄麻制成。

synthetic fibre (man-made fibre) A strand of material that is not naturally occurring in its final form.
合成纤维（人造纤维） 并非自然生成其最终形态的一束材料。

system building The use of prefabricated, off-site, or factory-built components that are made to fit together as a complete building. The components are made off-site and the assembly designed so that the building becomes a kit of parts that can be quickly bolted and fitted together. The need for wet trades, such as bricklaying, concreting, and plastering are eliminated or drastically reduced, removing drying out time that is required in traditional buildings. Some buildings are volumetric, built off-site as large boxes that contain fully fitted-out bedrooms, hospital rooms, or plant rooms that can be simply bolted together on-site, whereas other buildings are designed to be delivered flat (flat-pack construction) and built up on-site.
模块化建筑 使用预制的、现场外的或工厂制好的组件组装成完整的建筑物。组件是在现场外制成，装配流程也已设计好，这样建筑物就由一套能够快速固定和组装的零件组成。无须或大量减少湿施工，例如砌砖、浇混凝土、抹灰，省去传统建筑中所需的干燥时间。有一些建筑物是大体积的，在现场外制成大箱子，包含完全配套的卧室、病房或机房，可以在现场简单固定，而其他建筑需要平板式包装运输，在现场组合。

T

table form (table formwork) Horizontal mould used for supporting *in-situ reinforced concrete when casting structural floors. The formwork is delivered partly preassembled in a table shape and is then quickly adjusted to the correct height and length so that it can be craned quickly into position. Large table formwork systems are sometimes called flying forms as they can be easily craned and moved into position.
台模 卧式模具，当浇筑结构地板时，用于支撑现浇的钢筋混凝土。模板分部分预装成台式运送，可快速调整至正确的高度和长度，这样就可以快速起吊到适当位置。大的台模系统有时也称为飞模，因为可以将其轻松起吊和移动至所需位置。

tacheometry A surveying method that is used to rapidly measure a distance without the need of a chain or tape. The *stadia lines of a theodolite are sighted onto a graduated staff making a fixed *parallactic angle.

When the staff is held perpendicular to the collimation line, the collimation distance (from the instrument to the staff) can be computed.
视距测量 无须测量链或卷尺就能快速测量距离的方法。经纬仪上的视距丝瞄准视距尺形成一固定的视差角。当视距尺与十字丝垂直时，就可计算出视距距离（仪器到视距尺）。

tack A short and very sharp flat-headed nail.
大头钉 短而锋利的平头钉子。

tack-free The point at which a surface that is in the process of drying is no longer tacky. *See also* TACKY.
无黏性 在干燥过程中当表面不再具有黏性的点。参考“发黏的”词条。

tack rag Cotton rag or cheese cloth with a small quantity of varnish on the cloth, which is used for dusting or smoothing a surface after it has been rubbed down; it is applied before the main coat. To avoid the cloth hardening, tack rags should be placed in an airtight container.
除尘黏性抹布 布料上有少量凡立水的棉擦布或纱布，在表面打磨后使用，可除尘或使表面变平滑；在主涂层之前使用。为避免布料变硬，除尘黏性抹布应该放置于密闭容器中。

tacky A term used to describe the stickiness of a surface as it is drying.
发黏的 用于描述当表面变干燥时的表面的黏性。

tail 1. A short section of electrical cable that connects the electricity meter to the consumer unit. 2. The leading edge of a roofing tile or slate.
1. 尾线 一小段电缆，将电表连接至配电箱。 **2.（屋瓦）外露部分** 屋面瓦片或板岩的前缘。

tailings The waste (fines) remaining from the mining industry after the ore and minerals have been extracted from rock. A tailing dam is an embankment that has been constructed using the fines from such material.
尾矿 采矿时，从矿料中提取矿石和矿物后剩下的废料（细料）。尾矿坝就是用这类材料的细料筑成的堤坝。

tailpiece (US) *See* TRIMMED JOIST.
短撑（美国） 参考“短撑”词条。

take-off The compiling of descriptions and recording of quantities of work from drawings and schedules; this is normally the first stage in preparing a *bill of quantities. *Standard Method of Measurement rules should be followed so that the quantities of the work measured are consistent and pricing of the work is comparable.
编制清单 编写从图纸和材料清单中获取的工程量的描述和记录；通常是准备工程量清单的第一步。应该遵循标准计量方法的规则，这样计量出来的工程量才具统一性，工程报价才有可比性。

taking-off The process of compiling the *take-off.
清单编制 编写清单的过程。

tall boy (tallboy) 1. A long, narrow chimney pot that is designed to prevent down draughts. 2. A tall cupboard.
1. 高烟囱帽 长又窄的烟囱帽，可防止倒灌风。 **2. 高柜**

tally 1. To check a record or score. 2. To count something.
1. 核数 检查记录或记数。 **2. 计数**

tally slates Regular-sized slates that are purchased by the number required, rather than by their weight.
论块出售的岩板 按照所需数量购买的统一规格的岩板，而不是按重量计算。

tambour Anything that is cylindrical or drum-shaped; French for drum, it applies to the wall of the structure whether it is situated on the ground or within a ceiling. For example, a cylindrical ceiling with a dome on top.
鼓形物 圆柱状或鼓状的东西；法语是指鼓状，用于墙体结构，无论是位于地板上还是天花内。例如，圆柱形天花顶部为穹顶。

tamped finish Undulating surface finish formed in concrete by putting a straight edge (*tamper) across the surface of the concrete and taping the surface. By lifting the straight edge off the surface of the concrete, the tension forces the concrete to raise as it attempts to adhere to the tamper. The operation is repeated causing the surface of the concrete to undulate.

拉毛路面 通过在混凝土表面放一把直棱（手夯锤）然后轻拍表面，在混凝土中形成波状的表面。将直棱从混凝土表面拿起，因为混凝土粘着手夯锤产生张力被提起。重复操作，使混凝土表面呈波浪形。

tamper Piece of concreting equipment used to create a rough undulating finish on the surface of wet concrete. The straight edge, which sometimes has handles at each end, is raised and lowered onto the surface of the concrete by two * operatives.

手夯锤 一个混凝土设备，用于在未干的混凝土表面上形成粗糙的波状表面。直棱，有时在各端会有把手，可以由两个操作员操作，在混凝土表面抬起和放下。

tamping The process of creating a * tamped finish.

捣实 形成拉毛路面的过程。

tang The tapered section of a metal hand-tool that is fixed into the tool's handle.

柄脚 固定到工具把手的金属手持工具的锥形部分。

tangential cut (plain-sawn, slab-cut) Cuts made in wood that are at a tangent to the grain.

弦切（弦锯） 木材切割时顺着纹理切。

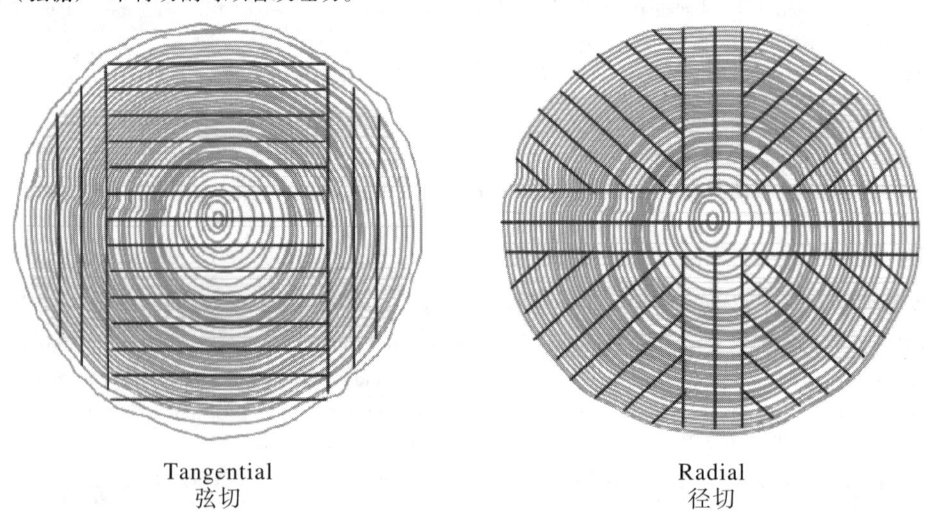

Wood cut in the tangential and radial direction.
木材按弦切和径切方向切割。

tangent modulus The slope of a line on a non-linear stress-strain curve. Until the * proportional limit is reached, the tangent modulus is equivalent to * Young's modulus.

切线模量 非线性应力－应变曲线上的斜率。在到达比例极限前，切线模量与杨氏模量相等。

tank A large, usually rectangular container that is used to store liquids or gases.

箱 通常为矩形的大容器，用于储存液体或气体。

tanking A waterproof membrane used to prevent groundwater penetrating into a basement. Can be applied to the inside or the outside of the walls and floor.

防水层 防水膜，用于防止地下水渗入地下室。可用于墙和地板的内外面。

tap 1. A valve connected to the end of a pipe that is used to control the flow of a liquid or gas. 2. The process where a connection is made to a water, gas, or electrical main in order to draw from it. 3. A tool used to produce an internal screw thread on an existing hole.

1. 龙头 连接至管道端的阀门，用于控制液体或气体的流动。**2. 分接** 连接至水、气或电气干线以汲取的过程。**3. 螺丝攻** 用于在原有洞口形成内螺纹的工具。

tape 1. A *joint tape. 2. A *measuring tape.
1. 接缝胶带。 **2. 卷尺**。

taper 1. To narrow gradually. 2. A duct or pipe that gradually reduces in size.
1. 逐渐收窄。 **2. 锥形管** 尺寸逐渐缩小的管子或管道。

taperbend A duct or pipe bend that gradually reduces in size along the bend.
锥形弯管 弯曲的管子或管道，沿着弯曲的地方尺寸逐渐缩小。

tapered tread A *stair tread that reduces in size along its length.
扇形踏步 沿着长度方向，楼梯踏板尺寸逐渐缩小。

taper thread A screw thread that gradually reduces in size along its length.
锥形螺纹 沿着长度方向，尺寸逐渐缩小的螺纹。

tap holder 1. A bracket used to support a tap. 2. *See* TAP WRENCH.
1. 攻丝夹头 用于夹持螺丝攻的支架。**2.** 参考“丝攻扳手”词条。

tap-off unit An electrical device that is connected to a busway run to provide power to a branch connection.
分线设备 连接至母线槽，提供电力给支线的电气设备。

tap wrench Tool used to cut (tap) a thread. The tool often has a T-shaped handle to ensure even pressure is applied while cutting the thread.
丝攻扳手 用于切（攻）螺纹的工具。该工具通常有T形把手，以确保当切割螺纹时，均匀施压。

tare weight The weight of something unloaded, for example the unladen weight of a lorry.
空载重量/皮重 空载时的重量，例如，货车的空载重量。

target-price contract A contract where the costs of the project is estimated and a price is fixed which the contractor aims for. Any savings made on the 'target price' normally entitle the contractor to a percentage bonus. This contract is different from *fixed-price and reimbursement contracts.
目标价格合同 工程价格已被估算，并且承包商的目标价格已固定的合同。如果在目标价格上有所节约，那么通常会授予承包商一定百分比的奖金。这种合同不同于固定价格合同和补偿合同。

tarmac (tarmacadam) A composite material that is used for surfacing roads. It consists of *macadam (uniform broken stone) and either *tar or tar plus *bitumen as the binder.
柏油碎石 用于铺路的复合材料。由碎石和柏油或柏油加沥青作粘合剂组成。

task An item of work, activity, or an event that has been planned, and normally forms part of a network of events (tasks) that are linked together.
任务 某项已经计划的工作、活动或事件，通常作为事件（任务）网络的一部分连接在一起。

taut Stretched or pulled tight.
绷紧的 拉紧的。

T

T-beam A structural steel section that has been manufactured with a cross-section in the shape of a "T".
T型梁 结构钢型材，生产出来的横截面是T形的。

TBM A *temporary benchmark.
临时水准点

technical assistant An assistant to a quantity surveyor who does *taking off, or an assistant to an architect who does the technical detailing.
技术助理 工料测量师的助手，负责清单编制，或者是作为建筑师的助手负责技术细节。

technology The application of tools, methods, and technical knowledge to solve practical problems.
技术 为解决实际问题所使用的工具、方法和技术知识。

tee A T-shaped duct or pipe fitting. *See also* STRAIGHT TEE and SWEEP TEE.
三通 T形管子或管件。另请参考“直三通”和“弯三通”词条。

tee bar 1. A T-shaped metal bar. 2. A type of suspended ceiling that uses T-shaped bars to support ceiling panels.
1. T形材 T型金属条。**2. T形龙骨** 一种吊顶天花，使用T型龙骨支撑天花板。

tee-hinge A T-shaped hinge.
T型铰链 T字形铰链。

teeming 1. Very heavy rain. 2. To pour molten metal.
1. 暴雨。 **2. 浇注** 浇铸熔融金属。

tegula A flat rectangular roof tile that has two upturned edges running along its longest sides.
平瓦 扁平的矩形屋面瓦，最长的两边边缘向上翘。

telescopic boom A hydraulically operated crane *jib that extents outwards and folds back into itself, and as such, is typically used on mobile cranes.
伸缩式吊臂 液压操作的起重机吊臂，可向外伸展，然后向后折叠恢复原位，因此通常安装于移动式起重机上。

telescopic centring Adjustable *formwork.
伸缩式模板 可调节的模板。

telescopic prop An adjustable supporting strut. The adjustable strut can be used to provide support for concrete formwork during casting and curing. Where the vertical support for walls, roofs, and floors has to be removed and replaced, and where it is necessary to provide temporary support, adjustable props can be used. When undertaking remediation, remedial, and other temporary works where it is necessary to use temporary supports, adjustable supports are quick to install to support *pins and beams.
伸缩式支柱 可调节支柱，可用于在浇筑和养护期间为混凝土模板提供支承。当必须移除和更换墙壁、屋顶和地板的垂直支承，以及当有必要提供临时支承时，可使用可调节支柱。当需要使用临时支承进行修复、修补和进行其他临时工程时，可使用可调节支承，可调节支承能够快速安装以支承销和梁。

Telford pavement A road pavement that has no *binder and is formed from a layer of large stones laid on the *subgrade, with smaller stones place on top of this, which are hard-rolled to form a smooth surface.
泰尔弗路面/碎石路面 无黏合剂的路面，由一层铺设在路基上的大石子组成，顶部有小石子，再滚压形成光滑的表面。

telltale A building survey device used to check for or measure the movement of cracks in existing buildings. The monitoring devices, which are often made out of glass, can be fixed over a crack to check for movement. The ends of the glass strip are securely fixed, using epoxy resin, to the wall; if the crack increases, the glass will break. The movement can be measured by measuring the break in the glass. Modern telltales may be made of two strips of Perspex with an incremental scale marked on each strip. One strip is fixed to one side of the crack and the other strip is fixed to the other side with the scales overlaying each other. If the crack moves, the degree of movement can be measured on the scales.
裂缝监测标尺 用于检查或测量原有建筑物中裂缝位移的建筑检查设备。通常是由玻璃制成的监测设备，可以安装在裂缝上以检查位移。使用环氧树脂将玻璃条的端点牢牢固定在墙上；如果裂缝增大，玻璃将破裂。可以通过测量玻璃的破裂程度来测量位移。现代指示器可以由两条有机玻璃组成，每条玻璃上都有增量刻度尺。一条安装在裂缝的一边，另一条安装在另一边，刻度重叠。如果裂缝移动，移动的距离可以通过刻度进行测量。

tellurometer A surveying device to undertake *electronic distance measurement.
微波测距仪 执行电子测距的测量仪器。

temperature A measure of the how hot or cold a material or element is in comparison to another.
温度 测量一材料或元素与其他事物的冷热程度比较。

temperature gradient The difference in temperature across a material. Materials with the highest thermal re-

sistances will have the steepest temperature gradients. Materials with the lowest thermal resistances will have the shallowest temperature gradients. Used to calculate the risk of interstitial condensation.
温度梯度 材料内部的温度差。材料热阻越高，其温度梯度越大。材料热阻越低，其温度梯度越小。用于计算分子内缩合的风险。

temperature movement *See* THERMAL MOVEMENT.
热运动 参考“热运动”词条。

temperature rise The rise in the temperature of a material or component.
升温 材料或组件温度的上升。

tempered glass A type of glass modified to exhibit better toughness. Also known as toughened glass, this type of glass is at least twice as strong as annealed glass. This type of glass is safe as it disintegrates into small fragments upon failure and is commonly utilized for many applications in buildings, i. e. doors, facades, and in bathrooms.
钢化玻璃 经改良的一种玻璃，具有更好的硬度。也称为强化玻璃，这种玻璃强度至少比退火玻璃高两倍。这种玻璃也很安全，因为当它掉落时会分解成小碎片，通常用于建筑上的很多地方，例如，门、幕墙和浴室中。

tempering Reducing the brittleness of steel by heating. When steel has been hardened by *annealing (heating and rapid cooling) it is often too hard and brittle for practical use. The steel is then heated to a specific heat, but below its hardening temperature, then it is held at that temperature and slowly allowed to cool. Cooling in the air reduces the stresses in the steel.
回火 通过加热减弱钢的脆性。当钢通过退火（加热然后快速冷却）硬化时，对于实际使用来说通常会太硬和易脆。将钢加热到一个特定的温度，但是要小于其淬火温度，然后保持该温度，慢慢冷却下来。在空气中冷却可以减小钢的应力。

template Something that serves as a full-size pattern, for example to allow other things to be cut out to the same size and shape.
模板 作为全尺寸样板的东西，例如可使其他东西裁切成与之相同的尺寸和形状。

temporary Something that is not permanent; *See also* TEMPORARY BENCHMARK, TEMPORARY PROTECTION, and TEMPORARY WORKS.
临时的 并非永久不变的某物。参考临时水准点，临时防护和临时工程。

temporary benchmark (TBM) A horizontal platform that is established to provide a known reference point with a set level. Once the level is established, the temporary reference point can be used to check and fix other levels around the site. Using *automatic levels, the temporary benchmark can check ground levels, excavation levels, and heights of structures etc. The height of the temporary benchmark is established by calculating its level in relation to an *ordnance benchmark, which is a permanent establishment with a known level.
临时水准点 水平平台，建立以提供已知的有设定水准的参考点。一旦建立了水准，临时参考点可用于检查和确定现场的其他水准。使用自动水准仪，临时水准可以检查地面高程、开挖水准和结构高度等。临时水准点的高度通过计算其与基准点相关的水平高度来确定，基准点是永久性确立，有已知水准。

T

temporary protection Resilient materials, such as wood, cardboard, and bubble wrap, that are used to protect the finishes of fittings that have been installed while building work is still ongoing. Staircases are often installed early during the construction programme to allow safe movement between levels; however, the treads must be protected by protective wooden slats to ensure that the impact of safety boots and other traffic does not damage the staircase.
临时防护 弹性材料，例如木材、硬纸板和泡沫包装，建筑工程施行时，用以保护已安装的配件表面。在施工进度中，楼梯通常较早安装以方便在各层之间安全移动；但是，踏板必须有保护性木板条保护，以确保安全靴和其他通行不会对楼梯造成损坏。

temporary works Additional structural work that is needed to secure, stabilize, make safe, or provide ac-

cess to the works during the construction or refurbishment stage. Examples include scaffolding, facade retention, earth works support, shoring and propping, hoarding, concrete crane bases, and platforms. Temporary works are priced under the preliminary items in the *bills of quantities.
临时工程 在施工或整修阶段，需要固定、稳固、提供安全或检修的附加结构工程。例如，脚手架、幕墙挡板、土方工程支撑、支承和支柱、临时围墙、混凝土起重机基脚和平台。工程量清单中，临时工程估价计入开办费中。

tenacity The tensile failure strength of a fibre, yarn, textile, or *geosynthetic product. In the textile industry, the cross-section of the material is normally not uniform or solid, for instance a fibre may be hollow and taper toward the ends. Thus, its cross-sectional area is hard to define and its stress value (load/area) is impossible to calculate, so tenacity is often used. Tenacity has a unit of N/tex. Tenacity refers to the value at the breaking load condition and specific stress (N/tex) is used for values up to the breaking load.
韧度 纤维、纱线、纺织品或土工合成材料的拉伸断裂强度。在纺织品工业中，材料的横截面通常不一致或不完整，例如纤维可能是中空的且沿着端点变细长。因此，很难确定其横截面面积，也无法计算其应力值（荷载/面积），所以常采用韧度来表述。韧度的单位为 N/tex。韧度是指材料断裂时荷载的数值，而此应力（N/tex）是由断裂荷载决定的值。

tender (bid) An offer submitted by a contractor to undertake the works for a set price or agreed remuneration. The offer is normally competitive where the contractor competes against other contractors for the work. While the tender establishes cost, it is not normally the only factor that is considered when the client evaluates the bid. Selection may be done on company reputation, financial security, quality, and timeframe for completing the works.
标书 承包商提交的履行工程所需的固定价格或协议报酬的报价。承包商间互相竞争，因此报价通常具有竞争性。虽然报价中确定了费用，但是这并不是业主估算标书时会考虑的唯一因素。业主会根据公司信誉、财务安全、完成工程的质量和时间表来做出选择。

tendering (bidding) The process or act of sending out requests for tenders, supported with drawings, bills of quantities, and performance criteria that allow the contractor to estimate and price the works. All of the *tenders are returned before a set date and evaluated in terms of price, time, and quality of work. Company reputation is important for establishing the potential quality of the work that it could deliver.
招标 向投标者发出邀请，并且发出图纸、工程量清单和性能标准以便承包商估价的过程或行为。所有投标者在规定的日期前交标并且估算价格、时间和工程质量。公司信誉对于交付工程的潜在质量很重要。

tendon A high tensile steel bar, cable, or rope that is stressed. Used to introduce a pre-stress to concrete, in order to carry and transfer loads.
钢筋束 一种预张拉的高应力钢筋、电缆或绳索。用于将预应力引入混凝土，以承载和转移荷载。

tenement A dwelling comprising several *apartments or flats.
经济公寓 包含几套公寓的住宅。

tenon A rectangular projecting tongue at the end of a component that is shaped to fit into a mortise to form a joint.
凸榫 位于组件端点的矩形凸出的榫舌，形状适于嵌入榫眼以形成连接。

tenon saw A small hand saw with a stiffening rib that runs along the top edge of the saw, the strengthening rib makes the saw easier to control enabling more accurate and precise cutting.
开榫锯/夹背锯 一种小手锯，锯子顶部边缘有加劲肋，使得锯子更易操控，能够进行更精确的切割。

tensegrity A structure that contains a collection of continuous tensile members that hold discontinuous compression members together.
张拉整体结构 包含一批连续的，将不连续受压构件组在一起拉伸构件的结构。

tensile Relating to tension; being stretched or pulled.
拉力的 与张力相关；受拉伸。

tensile strength The maximum stress a material can endure under a resultant tensile force/stress. Also usually known as *ultimate tensile strength (UTS). If a material is put into tension, this is the stress that must be exceeded to cause tensile fracture. Tensile strength is calculated as:

$$\text{Tensile Strength (UTS)}\ (\text{N/mm}^2) = \frac{\text{Tensile Load to cause Fracture (N)}}{\text{Cross-sectional area}\ (\text{mm}^2)}$$

Units are either N/mm^2 or MPa.

抗拉强度 一个材料在张力/应力合力下能够承受的最大应力。也被称为极限抗拉强度（UTS）。如果一个材料在受拉状态下，超过该应力必然造成拉伸断裂。抗拉强度计算公式如下：

$$\text{抗拉强度（UTS）}\ (\text{N/mm}^2) = \frac{\text{造成断裂时的张拉荷载（N）}}{\text{横截面积}\ (\text{mm}^2)}$$

单位为 N/mm^{2}或 MPa。

tension A force that pulls or stretches something.
张拉力 使某物拉伸的力。

terminal 1. A device used to connect electrical cables to an electrical appliance. 2. An ***air terminal unit**. 3. The end of a duct, pipe, flue, or lightning conductor.
1. 接线端子 用于连接电缆和电器的设备。**2. 空气终端装置**。**3. 末端** 管子、管道、烟道或避雷针的末端。

termination The end of a contract of employment.
终止 雇佣合同结束。

termite (white ant) A wood-eating insect found in tropical and warm temperate regions of the world. Properties where termites exist need to be protected with *termite shields.
白蚁 一种在热带和温带地区发现的食木昆虫。有白蚁的物业需要用防蚁板保护。

termite shield (ant cap) Metal sheet that is laid along foundations or within the foundation wall to prevent termites building mud tunnels up to any of the building's timbers. The cap consists of a long metal sheet with edges turned down at 45° preventing *termites climbing the shield. The sheet is laid within the mortar bed of a wall or on top of foundations.
防蚁板 沿基础或在基础墙内安装的金属片，可防止白蚁在建筑物木材内建通道。防蚁板有一长的金属边并向下45°角弯曲，防止白蚁爬上防蚁板。防蚁板安装于墙的砂浆层内或基础的顶部。

terrace 1. Raised area where the ground has been levelled to create a platform. 2. Flat roof designed so that it can act as a floor accommodating people and light traffic.
1. 平台/露台 地面上抬高的区域，平整形成一个平台。**2. 平屋顶** 平屋面，这样屋面也可以作为一个楼层，少量通行。

terrace blind *See* AWNING.
遮阳篷 参考“雨篷”词条。

terrace house Row of houses positioned side-by-side sharing the intermediate *party walls that separate each dwelling.
连栋房屋 并排布置，通过中间分户墙分隔每套住宅的一排房屋。

terrain A stretch of land viewed in terms of its surface or physical features.
地形 按照其表面或自然特征观察的一片土地。

tertiary treatment The final stage in a water-treatment plant, occurring after *secondary treatment, that normally includes a *disinfection stage, such as *chlorination, *ultraviolet light, or *ozone oxidation.
三级处理 废水处理厂的最后阶段，发生在二级处理之后，通常包括消毒阶段，例如氯化处理、紫外线或臭氧氧化。

tessellated An area filled with shapes or tiles, with no gaps between the tiles. The tiles need not be square but are normally the same shape, fitting together with no gaps, and filling the area. A mosaic wall or floor may also be considered tessellated.

网格状 填满图形或贴满瓷砖的区域，瓷砖之间无间隙。瓷砖无需为方形，但是通常为同一形状，贴合在一起，无缝隙，填满整个区域。马赛克墙或地板也是这种类型。

tessera (plural tesserae) Cubes used in mosaic work.

色块 马赛克工作中使用的方块。

test To evaluate something (or someone) by examining its (their) performance under certain conditions; for instance, to determine the tensile strength of a piece of steel, or to assess the knowledge of a person by their ability to answer correctly a series of questions on a particular subject.

试验/测试 在某一特定条件下通过测验评估某物（某人）的性能/表现，例如，确定一条钢铁的抗拉强度，或通过某特定科目下一系列问题的回答情况评估一个人的知识水平。

testing machine Any machine that is capable of introducing a load to a test sample to determine its physical response, e. g. compression or tensile strength.

拉力试验机 通过施加荷载到试样上以确定其物理性能（如抗压或抗拉强度）的任何器械。

tex The mass in grammes for one km of the fibre/yarn. Also known as linear density. *See also* TENACITY.

特克斯 1 km 长纤维/纱线重量的克数。又称为线密度。另请参考“韧度”词条。

textile A fabric structure produced by a woven, nonwoven, or knitted process. When used in intimate association with the earth (*geo), it is referred to as a *geotextile.

纺织品 纺织的、无纺的或针织加工生产的织物结构。当与土地（土）紧密联系时是指土工织物。

theatre A building made to accommodate musical, dramatic, and theatrical productions.

剧院 用于音乐、喜剧和剧场作品表演的建筑。

theodolite An optical surveying instrument used for measuring vertical and horizontal angles.

经纬仪 用于测量垂直和水平角度的光学测量仪器。

theory The use of scientific principle(s) to explain something.

理论 运用科学原理去解释某物。

thermae A Roman spa or public bath.

浴场 罗马浴场或公共浴池。

thermal Relating to or involving heat.

热的 含热的或与热有关的。

thermal break A material with a low thermal conductivity that is placed within a component to minimize conductive heat loss. For instance, the insertion of polyurethane as a spacer bar between the panes of multiple glazed units.

热障 放于组件内以将导热损失降至最小的低热导率的材料。例如，嵌入聚氨酯作为多层玻璃窗板间的垫片。

T

thermal bridge (cold bridge, heat bridge) An area of the building fabric that has a higher thermal transmission than the surrounding parts of the fabric, resulting in a reduction in the overall thermal insulation of the structure. It occurs when materials that have a much higher thermal conductivity than the surrounding material (i. e. they are poorer thermal insulators) penetrate the thermal envelope or where there are discontinuities in the thermal envelope. Heat then flows through the path created—the path of least resistance—from the warm space (inside) to the cold space (outside). The higher thermal transmission of this part of the fabric results in a reduction in the thermal performance (an increase in *U*-value) as heat flows through the fabric, and the surfaces of the interior side of the bridge become cooler. The use of the term “thermal bridge” is somewhat misleading as it implies that the thermal envelope must be “bridged” in some way for a thermal bridge to occur; this is, in fact, not the case. Thermal bridges can occur in unbridged construction where discontinuities exist in the thermal envelope.

热桥（冷桥） 一个建筑的区域结构，比周围的其他结构有更高的热传导，使该建筑的整体保温性能下降。其发生原因为该材料比周围材料的热导率要高（如它们是不良的热绝缘体）并穿透保温围护，或者在保温围护上有中断。然后热量通过通道传递——通过最小阻力的路径——从较暖的空间（内部）到较冷的空间（外部）。该部分结构的热传导越高，热量传递会使其保温性能下降（*U* 值上升），桥内表面也会变冷。“热桥”的名称有某程度的误导，因为暗示了保温围护产生热桥需要某种形式的“桥”存在，但事实并非如此。保温围护中断处没有桥的结构也能产生热桥。

thermal capacity (heat capacity, thermal inertia, thermal mass) A measure of the ability of a material or structure to store heat. Measured in J/K. All buildings have a thermal capacity that is determined by the quantity and types of materials that have been used to construct the building. The thermal capacity of a building is important as it not only determines how stable the internal temperatures will be, it also determines the time taken for the building to heat up and/or cool down. Buildings with a high thermal capacity will have a high thermal stability, while those with a low thermal capacity will have a low thermal stability. Buildings with a low thermal capacity will respond rapidly to control and will warm up quickly, they will respond more quickly to surrounding temperature changes, and they may overheat in the summer. Buildings with a high thermal capacity will require long pre-heating times, will respond slowly to surrounding temperature changes, and will cool down slowly.

热容量（热惯量、热质量） 计量材料或结构储藏热量的能力。单位为 J/K。所有的建筑物都有其热容量，由其体量和建筑用材决定。建筑物的热容量非常重要，因为其不但决定了室内温度的稳定性，还决定了加热/降温建筑物所需的时间。有高热容量的建筑物即有高的热稳定性，低热容量的其热稳定性越低。低热容量的建筑物对于温控反应迅速，能较快升温，对于周围环境的温度变化反应快速，因此在夏天会过热。高热容量的建筑物需要较长的预热时间，对于周围环境的温度变化反应较慢，降温也较慢。

thermal comfort A subjective state of well-being that exists when the occupants of a building are unaware of their surroundings. Thermal comfort is influenced by a number of individual and environmental factors. Individual factors include activity, age, and clothing while environmental factors include air temperature, mean radiant temperature, humidity, and air movement.

热舒适 当建筑内的住户感受不到周围环境时的一种主观舒适感觉。热舒适受个体和环境因素影响。个体因素包括行为、年龄和衣着，环境因素包括空气温度、平均辐射温度、湿度和空气流动。

thermal conductivity (heat-transmission value, *K*-value) The amount of heat flow in watts through a material, which has a thickness of 1 m and a surface area of 1 m^2, for each one degree of temperature difference between the inside and outside surfaces. Measured in W/m^2K, the thermal conductivity of a material is a property of the material, and is independent of the thickness of the material. Materials with a high thermal conductivity, such as metals, are good conductors of heat and poor insulators, while materials with low thermal conductivities, such as insulation materials, are poor conductors of heat and good insulators.

热导率（导热系数、*K* 值） 热量通过材料传递的数值，在 1 m 厚表面积为 1 m^2 的条件下，内外表面相差一度时的瓦特数。单位为 W/m^2K。材料的热导率是该材料的一种特性，与材料的厚度无关。高热导率的材料，如金属，是良好的导体和不良的绝缘体，而低热导率的材料，如保温材料，是不良的导体和良好的绝缘体。

T

thermal degradation of timber A degradation mechanism for timber material. When timber undergoes prolonged exposure to elevated temperatures, there is a reduction in both strength and toughness. There is some uncertainty about what minimum temperature will degrade timber, in some cases temperatures as low as 60℃ can induce degradation over many years of exposure. The rate of degradation will rise markedly with an increase in temperature and time of exposure; hardwoods appear to be more susceptible to thermal degradation than softwoods. Thermally degraded timber develops a brown colour and breaks easily with a brittle failure.

木材热降解 木材的降解机制。当木材长时间暴露于高温下时，强度和硬度都会有所降低。对于降解木材的最低温度还不确定，在一些情况下，多年暴露的木材 60℃ 即可引发降解。温度和暴露时间增加时，降解率将明显上升；硬木似乎比软木更易发生热降解。热降解的木材会变成棕色，更易发生脆性破坏。

thermal diode A device where heat can flow more easily in one direction than the other.
热二极管 一种更易将热量单方向传递的装置。

thermal expansion The process where a material or component changes its size (area, length, or volume) due to an increase in temperature.
热膨胀 材料或组件因温度的上升而造成尺寸（面积、长度或体积）发生变化的过程。

thermal fusion joint (heat fusion joint) A joint between two components that is created by applying heat.
热熔连接 由加热形成的两个组件间的接合。

thermal insulation 1. A material that has a low thermal conductivity, thus reducing the amount of heat that will flow through it. 2. The process of applying insulation material to a building to reduce its heat loss.
1. 热绝缘体 低热导率的材料，能减少热量传递。**2. 隔热** 在建筑物中采用隔热材料以减少热量损失。

thermally broken A term used to describe a component that incorporates thermal breaks.
隔热 用于描述包含热障的组件的术语。

thermal movement (temperature movement) The expansion or contraction of a material due to a change in temperature.
热运动 材料由于温度改变而膨胀或收缩。

thermal resistance (*R*-value) A measure of a material's resistance to heat transfer. It is calculated by dividing the thickness of the material by its * thermal conductivity. The thicker the material, the greater the thermal resistance. The lower the thermal conductivity of a material, the greater the thermal resistance. All materials present a
resistance to the flow of heat through the material. Measured in m^2K/W.
热阻（*R* 值） 材料对热传递阻挡的计量。由材料厚度除以导热率计算所得。材料越厚，热阻越大。材料导热率越低，热阻越大。所有材料都会对热量传递有阻挡。单位为 m^2K/W。

thermal resistivity The inverse of thermal conductivity.
热阻系数 热导率的倒数。

thermal shock Failure of a material due to a sudden increase in temperature. This is highly prevalent in glasses and ceramics, which have a very poor resistance to thermal shock; conversely, porous materials and metals have good resistance. In general materials with low toughness are susceptible to this type of failure.
热冲击 材料由于温度忽然升高而发生的破坏。这在玻璃和陶瓷中很常见，因为这两种材料抗热震性能很差；相反，多孔材料和金属有良好的抗热震性能。通常，韧性低的材料更易发生这类损坏。

thermal stress breakage The breakage of glass due to tensile stresses. The stresses result from temperature gradients across the glass, which occur when the glass is non-uniformly heated. If the tensile stresses exceed the edge strength of the glass, thermal breakage occurs.
热应力开裂 由热应力造成的玻璃破裂。当玻璃不均匀受热时，温度梯度造成的应力作用于玻璃，导致玻璃破裂。如果张应力超出玻璃的边缘强度，则会发生热应力断裂现象。

T

thermal transmittance (*U*-value) The rate of heat flow in watts, through 1 m^2 of an element, when there is a temperature difference across the element of 1℃ (or K). Measured in W/m^2K.
传热系数（*U* 值） 当组件存在 1℃（或 K）的温差时，热量通过 1 m^2 面积组件传递的瓦特数。单位为 W/m^2K。

thermal wheel (heat-recovery wheel) A type of * heat exchanger that comprises a large rotating wheel positioned in between the supply and extract ducts in an air-conditioning or ventilation system.
转轮式热交换器（轮式全热换热器） 一种热交换器，由空调或通风系统中位于送风管和排风管之间的大转轮组成。

thermic boring A method that uses an * oxy-acetylene flame at the end of a lance to cut through steel or concrete.

气焊开孔　一种使用枪端氧乙炔火焰切割钢或混凝土的方法。

thermistor　A device for measuring temperature—when the temperature rises the electrical resistance of the device falls.
热敏电阻　一种温度测量装置——当温度升高时装置的电阻下降。

thermit welding　A chemical heat reaction welding process that uses aluminium and iron oxide; used to weld large sections such as railway lines.
铝热焊　一种化学热反应焊接过程，使用铝和氧化铁；用于焊接大型材，例如铁轨。

thermocouple　A device for measuring temperature. Two wires of different metals are joined to give a small potential difference, which is proportional to the temperature of something they touch.
热电偶　一种测量温度的装置。两种不同的金属丝连接在一起造成微小的电势差，与其接触的物件温度成比例。

thermodynamics　The branch of physics that deals with energy conversion involving heat, pressure, volume, mechanical action, and work.
热力学　与能量转换相关，包括热、压力、体积、机械作用和做功的物理学分支。

thermofusion welding　The fusing of plastics by heating edges and holding the melted soft edges together.
热熔焊接　通过加热边缘和将熔化的软边接在一起熔合塑料。

thermohygrograph　A chart recorder that measures and records air temperature and relative humidity simultaneously.
温湿计　能同时测量和记录空气温度和相对湿度的记录仪。

thermometer　A device that is used to measure temperature.
温度计　用于测量温度的装置。

thermo-osmosis　The passage of water in a porous medium due to a temperature differential.
热渗透性　由于温度差异造成的多孔介质中的水分通道。

thermoplastics　A category or group of polymers that soften when heated and stiffen when cooled without enduring a physical or chemical change in property (a reversible process). As a result they are very ductile and workable, and they can be easily recycled. Structurally, thermoplastics consist of long carbon chains tangled together and intertwined, without any cross-linking. This accounts for their flexibility and resilience as a group. Examples include polyethylene, PVC, nylon (polyamide), polystyrene, and polycarbonate. Uses in construction include: polyethylene sheet for dpc and vapour barrier systems, UPVC window frames and doors, etc. They can deform elastically, yield, and can deform plastically to a considerable extent (nylon will give over 300% plastic strain). Before yielding they also show non-linear visco-elasticity, a form of behaviour that is partly elastic and partly viscous flow. This accounts for the high levels of *ductility as a group. Thermoplastics can easily be made into complex shapes to make products such as window frame sections, waste pipe systems, water cisterns, and simpler things such as polyethylene sheeting, etc. Thermoplastics are often supplied in the form of small pellets of material, which can be fed into either a moulding or an extrusion process. All polymers can be classified as being either a thermoplastic, a *thermoset, or an *elastomer, according to their properties.
热塑性塑料　一类或一组聚合物，当加热时会变软，冷却时会变硬，其物理或化学属性没有发生变化（可逆过程）。因此这种材料很具可塑性和易用性，易于循环利用。结构上，热塑性塑料由长碳链缠绕在一起组成，无交联。因此其聚合物具有柔韧性和弹性。例如聚乙烯、聚氯乙烯、尼龙（聚酰胺）、聚苯乙烯和聚碳酸酯。建筑中用于防潮层和隔气层用的聚乙烯薄膜、UPVC 窗框和门等。可以弹性变形、屈服，以及可塑性变形至一定程度（尼龙具有超过 300% 的塑性应变）。在屈服前具有非线性黏弹性，即一种兼具部分弹性和部分黏性的特性。因此其聚合物具有高延性。热塑性塑料容易被制成形状复杂的产品，如窗框型材、废水管道系统、水箱和更简单的物件如聚乙烯薄膜。热塑性塑料通常以颗粒状做产品，可供料至模具或挤压加工。所有的聚合物都可以根据其特性分类为热塑性塑料、热固性塑料和弹性体。

thermosets A group or category of polymers that do not soften when heated to an elevated temperature; instead, they disintegrate with a substantial decrease in the mechanical properties and dimensional stability. The molecular structure of thermosetting polymers is heavily cross-linked. This cross-linking gives them great rigidity, *hardness, and wear resistance. In general, nearly all thermosets are hard, wear-resistant, stiff (high elastic [Young's] modulus, E), and have no ductility. They tend to be strong and *brittle, and thus have to be shaped and polymerized in one operation. Once *polymerization has taken place no further shaping or alteration is possible. Thermosets lend themselves to the production of accurately moulded components such as switch and socket plates, boxes, etc. for electrical products. They are also used for kitchen surfaces. In practice, they are compression moulded, i. e. moulded and polymerized in one. All polymers can be classified as being either a *thermoplastic, a thermoset, or an *elastomer, according to their properties.
热固性塑料 一组或一类聚合物，加热至高温时，不会变软；相反，材料会分解，其机械性能和尺寸稳定性会发生大幅降低。热固性聚合物的分子结构为交联链接，使其具有很大的刚度、硬度和耐磨度。通常，几乎所有热固性塑料都很坚固、耐磨和刚硬的（高弹性［杨氏］模量，E），无延展性。这种材料通常都比较坚固，有脆性，因此要在一次性操作中塑形和聚合。一旦聚合完成，就无法再进一步塑形或更改。热固性材料用于生产出精确铸模的组件，例如开关和插座板、配电箱等电气产品。也可以用于厨房表面。实际上，它们是压模式，例如铸模和聚合一体。所有聚合物都可根据其特性分类为热塑性塑料、热固性塑料或弹性体。

thermosetting *See* THERMOSETS.
热固的 参考“热固性塑料”词条。

thermosetting compound A compound that cures by forming a chemical reaction or as a result of the release of a solvent.
热固性复合材料 通过化学反应或通过溶剂释放固化而成的复合物。

thermosiphon *See* GRAVITY CIRCULATION.
热虹吸管 参考“重力循环”词条。

thermostat An electronic or mechanical device that is used to control the temperature of an appliance or a system.
恒温器 用于控制电器或系统温度的电子或机械设备。

thermostatic mixer A thermostatically controlled mixing *valve.
恒温水龙头 恒温控制的混合阀。

thicknessing machine (panel planer, thicknesser) Machine for reducing the thickness of wood to a set size, producing square, level, and flat surfaces.
木工刨床（刨板机、压刨床） 用于降低木材厚度至指定尺寸，形成方形、平整和平坦表面的机器。

thief-resistant lock A security *lock.
防盗锁 一种安全锁。

T

Thiessen polygon A graph that represents the rainfall intensity over a particular area from data collected from rainfall gauges over that area.
泰森多边形 通过搜集一特定区域雨量计数据，表示该区域降雨强度的图形。

thimble *See* SLEEVE PIECE.
套筒 参考“管道套筒”词条。

T-hinge (tee-hinge) A hinge that is T-shaped when opened.
T 型铰链 打开时是 T 字形的铰链。

third angle Drawing an object where the view of the top of the object is drawn from above, and the view from the left of the object is drawn to the left, the view from the right is drawn to the right, and the view from the rear is drawn to the extreme of the right perspective.
第三角视图 当绘画一物体时，物体俯视图放上方，物体左视图放左边，物体右视图放右边，物体后视图放最右边。

third fixing (second fixing) Any work carried out after plastering.
三次整修（二次装修配件安装） 在抹灰之后进行的任何工作。

third rail An additional rail, either running alongside or between the existing track, which carries an electric current that powers the train that runs on the track.
第三轨/供电轨 附加的轨道，位于原有轨道边上或之间，带电，给运行于轨道上的列车供电。

thread The spiral-shaped ridge found on the outer face of a screw or bolt or the inner face of a nut.
螺纹 螺丝或螺栓外表面或螺母内表面的螺旋形凸纹。

threaded rod A long cylindrical bar that is threaded at both ends, and may even be threaded along its entire length. Has a variety of uses, such as a fastener, a hanger, or a tie rod.
螺杆 长的圆柱形杆，两端都有螺纹，甚至全长都有螺纹。有各种用途，例如用于紧固件、挂件或系杆。

three-coat work The application of three separate coats of plaster. The first coat is termed the *scratch coat, the second coat the browning coat, and the third the *finishing coat.
三道抹灰工作 涂上三层抹灰。第一层称为底涂层，第二层称为二道抹灰，第三层称为面层。

three-hole basin A basin that incorporates three holes; one for the hot water valve, one for the cold water valve, and a one for the spout.
三孔盥洗盆 有三个孔洞的盥洗盆；一个用于装热水阀，一个用于装冷水阀，还有一个用于装龙头。

three-moment equation An equation that incorporates the internal bending moments in adjacent spans of a continuous elastic beam.
三弯矩方程 包含多跨连续弹性梁上内弯矩的方程式。

three-phase supply An AC electricity supply that comprises three alternating voltages that are spaced 120° apart. *See also* SINGLE-PHASE.
三相电源 交流电源，由三个交流电压以120°隔开。另请参考“单相”词条。

three-pipe system A type of pipework system that has two separate flow supply pipes and a common return.
三管系统 一种管道系统，有两个单独的供水管道和一个同程回水管。

three-prong plug A plug that has three contacts; one for the live, one for the neutral, and one for the earth.
三线插头 有三个触点的插头；一个接火线，一个接零线，还有一个接地线。

three-quarter bat A brick that has been reduced in length by a quarter.
长度减少四分之一的砖块。

three-quarter header A brick that has been reduced in width by a quarter.
宽度减少四分之一的砖块。

three-quarter-round channel An open drainage channel that is three-quarters round.
四分之三排水沟 四分之三圆的排水明沟。

three-way valve (twin-port valve) A valve that is capable of directing the fluid or gas to one of two outlets.
三通阀（二通阀） 能够将液体或气体导至两个出口的阀门。

threshold The *sill of an external door.
门槛 外门的基石。

threshold strip A strip of material, usually metal, positioned under the closed leaf of a door that hides the joint between the floor finishes in different rooms.
门槛条 通常为金属条的材料，位于闭合的门扇底下，用于遮盖不同房间地面装饰间的接缝。

throat 1. The narrow portion of a chimney situation between the *gathering and the *flue. 2. A groove cut along the length of a sill to provide a *drip.

1. 吸烟口 烟囱内狭窄的部分，位于集烟罩和烟道之间。**2. 滴水槽** 沿着窗台的槽型切口的滴水构件。

throated sill A sill that incorporates a * throat.
有滴水槽的窗台。

through lintel A lintel that spans the full thickness of an opening.
等厚过梁 横跨整个开口厚度的过梁。

through stone A stone that penetrates the entire thickness of a wall.
穿墙石 贯穿整个墙壁厚度的石块。

through tenon A * tenon that penetrates through the entire thickness of the timber that contains the mortise.
贯穿榫 贯穿整个木材（含有榫眼）厚度的凸榫。

thumb latch A door latch that is operated by pressing the lever with your thumb.
指按门闩 用拇指按下锁销杆来操作的门闩。

thumb slide *See* SLIDE BOLT.
天地插销 参考“天地插销”词条。

tie 1. Fasten, attach. 2. A beam, rod, rope, or strap that fastens a material or components together. Designed to operate in tension.
1. 系紧 紧固，系上。**2. 连系材** 紧固材料或连接组件的横梁、杆、绳索或带子。使用时需拉紧。

tie beam A horizontal structural member used at the foot of rafters or between two walls to prevent them spreading under load.
系梁 用于椽条基脚处或两堵墙壁之间，以防止其在荷载下散开的水平结构构件。

tie wire *See* BINDING WIRE.
扎线 参考“绑扎钢丝”词条。

tight knot A * **knot** that is firmly fixed in a piece of timber.
紧密节 牢牢位于木材上的节。

tight sheathing (close sheathing, closed sheathing) Horizontal or vertical boards laid tight against one another. Used as a retaining wall in excavations.
镶拼挡板（密拼底板） 挨个紧靠的水平或垂直板材。在挖掘中作为挡土墙应用。

tight size (full size, rebate size) The size of the * REBATE.
槽口尺寸

tight tolerances Allowable deviation is less than usual, meaning that setting out, manufacture, and casting must be precise.
精密公差 比通常的小的允许偏差，意味着放线、生产和浇铸必须精确。

T

tile A thin regularly shaped piece of material used to cover or finish floors, walls, and roofs.
砖/瓦块 形状规则的薄片材料，用于覆盖或装饰地板、墙壁和屋面。

tile-and-a-half tile A roof tile that is the same length but one-and-a-half times the width of the other tiles used on the roof.
阔瓦 长度跟其他屋面瓦一样但宽度是其他屋面瓦 1.5 倍的屋面瓦。

tile batten *See* TILING BATTEN.
挂瓦条 参考“挂瓦条”词条。

tile clip An aluminium, plastic, or stainless steel clip used to secure roof tiles to the roof structure. Designed to prevent wind uplift and tile chatter. Three main types are available: eaves clips, tile-to-tile clips, and verge clips.
砖夹 铝、塑料或不锈钢夹，用于将屋面瓦固定至屋顶结构，以防止风吸力和瓦片震颤。主要有三种类型：檐瓦夹、瓦对瓦夹和檐口夹。

tile creasing *See* CREASING.
压顶瓦 参考“压顶砖”词条。

tiled valley A *valley created using *valley tiles.
贴瓦斜沟 用斜沟瓦铺成的屋面斜沟。

tile fillet (tile listing) A tile whose upper edge is fully bedded in mortar at an abutment. Used instead of a cover ***flashing**.
屋面瓦封口 在对接处，上边缘完全埋入砂浆的砖瓦。可用于替代泛水板。

tile hanging (vertical tiling, weather tiling) The process of fixing tiles to a wall.
挂瓦法 将瓦片铺贴于墙壁的过程。

tile listing *See* TILE FILLET.
屋面瓦封口 参考“屋面瓦封口”词条。

tile peg (tile pin) A *peg used to secure a *peg tile.
瓦栓钉 用于固定带孔平瓦的栓钉。

tiler (tile layer) A tradesperson who lays *tiles.
铺瓦工 铺砖瓦的工人。

tiling The process of applying *tiles.
贴砖 铺贴砖瓦的过程。

tiling batten (roofing batten, tile batten) A *timber batten used to secure roof tiles to the roof. Usually 25 mm × 38 mm or 25 mm × 50 mm in size.
挂瓦条 用于将屋面瓦固定于屋面的木板条。通常尺寸为 25 mm × 38 mm 或 25 mm × 50 mm。

tilt-and-turn window A window that can tilt inwards at the top to provide ventilation, and open inwards or outwards on a hinge.
内开内倒窗户 顶部可以通过铰链向内倾斜以提供通风，也可向内或向外打开的窗户。

tilting fillet (doubling piece, skew fillet) A piece of timber inserted at the eaves of a roof to raise the bottom row of tiles.
披水条（檐口板） 嵌入屋檐，以提高底排瓦块的木条。

tilting level An optical surveying instrument used for *levelling. It is characterized by having a tilting telescope on a pivot; *See also* DUMPY LEVEL. The initial horizontal plane is obtained by using the levelling screw; the horizontal line of collimation is then sighted by finely adjusting the tilting screw on the telescope. *See also* AUTOMATIC LEVEL.
微倾式水准仪 用于水准测量的光学测量仪器。特点是枢轴上有一微倾望远镜；另请参考“定镜水准仪”词条。用水平螺旋调得粗平；然后细微调节望远镜上的微倾螺旋以观察水平瞄准直线。另请参考“自动水准仪”词条。

tiltmeter An instrument for measuring very small changes from the horizontal level of the ground—it can also be used in structures. Used primarily for monitoring volcanoes, but can be used to monitor changes due to excavation, tunnelling, dewatering, and so on.
倾角测量仪 测量地面水平处非常微小的变化的仪器，也可用于结构中。主要用于监测火山，但是也可用于监测由挖掘、挖隧道、排水等所引发的变化。

timber A natural and extensively utilized material extracted from trees. Timber has been used for centuries and it is still an important construction material today, because of its versatile properties, its diversity, and aesthetic qualities. At least one-third of all timber harvested goes into construction, the rest goes into paperproduction, is used as fuel, or is wasted during the logging process. The earth's forests, besides being a source of timber, also act as a carbon sink. That is, they absorb carbon dioxide from the atmosphere, and very importantly,

breathe out the oxygen. Timber is also a renewable resource, and so can make an important contribution towards the achievement of sustainable building. Because of the way that timber grows, the material has different properties in different directions, thus timber is anisotropic. Some species of timber (often the hardwoods) display good resistance to degradation, while others have little resistance (many softwoods). The agencies of degradation are biological, chemical, photochemical, thermal, fire, and mechanical; however, timber, especially in the UK, is most susceptible to insect and fungal attack.
木材 由树木取得的天然且广泛应用的材料。人类利用木材已有数个世纪，其由于多功能、多样化及美观特性，至今仍然是重要的建筑材料。至少三分之一的木材被用于建材，其余的用于制纸、燃料或在砍伐过程中作废料。地球上的森林，除了作为木材来源外，也是一种碳汇。即从大气中吸收二氧化碳，更重要的是呼出氧气。木材也是可再生能源，所以能为可持续建筑做出重大贡献。由于木材的生长方式，在不同的方向上材料质地不同，因此木材是非均质的。某些品种的木材（如硬木）有很强的抗降解特性，另一些（很多的软木）就弱很多。降解的媒介包括生物、化学、光化学、热力、火、机械，但是木材最容易受虫害和霉菌影响，特别在英国。

timber-frame construction The use of timber studs, rails, and sheathing to produce a load-bearing structural frame. Can be site-built or, more commonly, prefabricated in a factory. There are two main types of timber-frame construction: platform frame and balloon frame.
木材框架 使用木龙骨、栏杆和衬板生产的承重结构框架。可以现场建造或更通常的在工厂预制。主要用于建筑的木材框架有两种：平台型底架和轻型木构架。

timber-framed building A building that has the main structural support system constructed of timber, rather than the inter-leaf of a cavity wall in conventional UK brick-built houses. In the UK, brickwork is most often used as the external cladding to the timber structure.
木框建筑 某建筑的主要结构支撑系统为木材，而不是英国传统砖砌建筑中夹心墙的内叶。在英国，砌砖多用于木结构的外层覆面。

timbering 1. The use of timber. 2. Timber formwork.
1. 使用木材。 2. 木模板。

Timber Research and Development Association (TRADA) Established in 1934 to protect the use of timber within the construction industry and progress research and development.
木材研究和发展协会（TRADA） 成立于1934年，致力于在建筑业内保护木材并推动其研究和发展的组织。

time-and-a-half An overtime rate that pays 50% more for the equivalent period of normal work.
一倍半加班费 比正常工作时间多费用50%的加班费。

time distribution (time system, master clock system) Network or system that ensures that every clock in the building or organization works and is constantly synchronized with the master clock.
时钟分配（时间系统、主时钟系统） 确保建筑物或组织工程中的每一个时钟都与主时钟随时同步的网络或系统。

time for completion (construction time, contract time) The period allowed for the construction of the building and site operations. The project usually has a start date and finish date, which are stated in the contract.
竣工时间（施工时间、合同时间） 建造建筑和现场施工所允许的时间。项目通常有开工日期和竣工日期，在合同中注明。

timekeeper Sometimes used to describe the clerk of works, especially if this person is monitoring working hours and breaks.
工程监工员 有时候用于描述工程监督员，尤其是如果这个人负责监督工时和休息时间。

timer An electrical device used to switch something on or off after a period of time, e. g. a switch on lights to delay the lights turning off in stairways and corridors, or a unit on a cooker that sets the start and finish times of cooking.

定时器 经过一段时间后打开或关闭开关的电子设备，如楼梯和走廊的灯延迟开关，或灶头上设定开始和完成烹饪时间的部件。

time scale The x axis on a programme where the days, weeks, months, or years are represented in a linear scale. Vertical lines corresponding to the dates allow progress of activities to be easily checked and monitored.
时间标尺 进度表的 x 轴标注为日、周、月或年的刻度。竖线为活动进度对应的数据，方便核查和监控。

tin A non-ferrous metal with symbol Sn. Tin is alloyed with copper to make bronze, which is commonly used in the construction industry; furthermore, it is used in *solders. It has excellent resistance to corrosion, thus can be used to coat other metals to provide corrosion resistance.
锡 一种有色金属，符号为 Sn。锡和铜的合金为青铜，广泛用于建筑业；另外也用于焊锡。锡有极强的抗腐蚀性，因此可以镀在其他金属上以抗腐蚀。

tine A prong or pointed projections such as teeth on a excavator's bucket.
尖齿 叉状或尖锐的突出物，例如挖掘机铲斗上的斗齿。

tingle A strip of metal used to secure edges of sheet-metal at drips, rolls, and seams.
夹片 用于固定滴水处、卷曲处和接缝处的金属薄片边缘的金属条。

tin snips Scissors-like tool used for cutting thin metal.
铁皮剪 剪刀似的工具，用于剪切薄金属

tinted glass (window tint) A thin film applied to the interior part of windows to reduce the amount of ultraviolet (UV), infrared, and visible light passing through.
有色玻璃 贴于窗户内部的薄膜，以减少紫外线、红外线和可见光通过量。

tinters *See* STAINERS.
染色剂 参考“染色剂”词条。

tinting *See* TINTED GLASS.
着色 参考“有色玻璃”词条。

tip 1. The sharpened end of an object, such as a nail. 2. The process of dumping something, such as rubbish. 3. To tilt at an angle.
1. 尖端 物体的尖端，例如钉子。**2. 倾倒** 倾倒某物的过程，如倾倒垃圾。**3. 倾斜**。

titan crane A large crane mounted on a *portal frame, having a *jib that can rotate around its vertical axis.
巨型起重机 装在门式刚架上有一起重臂可绕垂直轴旋转的大型起重机。

titanium A non-ferrous metal, symbol Ti. Titanium is a strong, lightweight metal with excellent resistance to corrosion. It has similar tensile strength to steel yet is 45% lighter. Due to the cost, its application is limited to the aerospace industry, otherwise it could replace steel in nearly all applications in civil engineering.
钛 一种有色金属，符号为 Ti。钛是一种坚硬、轻质金属，有极强的耐腐蚀性。它的抗拉强度接近钢，但重量轻45%。由于成本原因钛仅限用于航空领域，否则可以在土木工程的所有应用上替代钢铁。

titanium dioxide Also known as titania or titanium (Ⅳ) oxide, chemical formula TiO_2. Titanium dioxide is usually used as a pigment called titanium white. This pigment is particularly used to both colour polymers white, and more importantly, protect polymers from degradation due to UV sunlight.
二氧化钛 又被称为金红石或氧化钛（Ⅳ），化学式为 TiO_2。二氧化钛通常用作称为钛白的颜料。该颜料常用于染白聚合物，更重要的是不会被阳光紫外线降解。

title block The bordered space within a drawing to add information and text about the drawing. The information contained within the title block usually states the drawing name, the drawing number, date created or updated, scale, and the organization responsible for creating the drawing.
图纸标题栏 图纸上用来添加相关信息和文字的带框表格。标题栏内的信息通常注明图纸名称、图纸号、创建或更新日期、比例和负责创建图纸的组织。

T

toe 1. The portion of a lintel that projects beyond the head of a window or door. 2. The lowest part of a shutting *stile. *See also* HEEL. 3. The lower end of something, e. g. the toe of an embankment where the slide slopes meet the ground.
1. 过梁凸边 过梁的一部分，超出窗楣或门楣凸出的部分。**2. 闭锁竖梃脚** 闭锁竖梃的最低部分。另请参考“柱脚”词条。**3. 脚** 某物的下端，例如堤脚，就是边坡与地面相接的地方。

toe board (guard board) A board placed around the outer edge of scaffolding or a roof to prevent equipment, tools, and people from falling over the edge.
挡脚板（安全挡板） 位于脚手架或屋顶外部边缘的板块，用于防止设备、工具或人从边缘跌落。

toe nailing *See* SKEWNAIL.
斜交钉 参考“斜交钉”词条。

toe recess The recess at the base of a kitchen unit.
踢脚板 厨房家具底部的凹处。

toggle bolt A type of fastener used to fasten heavy objects to plasterboard walls. Comprises a spring-loaded toggle that is attached to a threaded bolt.
系墙螺栓 一种紧固件，用于将重物紧固于石膏板墙上。包括连接螺纹栓上装弹簧的两翼。

toggle mechanism A linked hinged section of a device that can be used to apply a large pressure onto something from a small amount of force; used in jaw crushers.
肘节机构 用铰链连接的装置部分，用少量的力即可在某物上施加较大的压力；用于颚式破碎机上。

toggle switch *See* TUMBLER SWITCH.
肘节开关 参考“倒扳开关”词条。

toilet A *water closet (WC).
厕所 抽水马桶（WC）。

tolerance (permissible deviation) The discrepancy allowed between an exact location or fit, and one that deviates slightly, but is still acceptable and functions. When setting out, cutting, manufacturing, and fitting, it is normal to attempt to obtain total accuracy but, in practice, the process often results in slight variation. As long as the variation is within the acceptable tolerance, then functionality will still be achieved.
公差（允差） 准确位置或装配与细微偏离之间的允许差距，但仍在可接受和正常工作范围内。当放线、切割、生产和安装时，虽然会尽量要求精确，但实际操作中仍会有些微偏差。只要该偏差在公差范围内，仍可以正常运作。

toll 1. The fee for driving across a particular road or bridge. 2. The cost or damage sustained by something, as a consequence of a disaster, for instance, the toll being the number of people killed, or the financial cost to repair property and infrastructure damaged.
1. 过路费（过桥费） 开车经过某条路或某座桥的通行费用。**2. 重大损失（或伤亡）** 由于灾难所蒙受的损失，例如伤亡人数或者修复受损的财产和基础设施所花费的财务成本。

tomb A burial chamber or grave, which may include a monument or headstone.
坟墓 墓室或墓穴，通常包括纪念碑或墓石。

tong tester An instrument for measuring electrical current without the need to disconnect the circuit. The jaws of the instrument are placed around one wire, the instrument then measures the electromagnetic force or field that is created.
钳形电流表 无须断开电路即可测量电流的仪器。仪器的钳口放在电线周围，然后仪器就可测量出其形成的电磁力或场。

tongue A long, narrow piece of timber that projects from the side of a board.
榫舌 从木板的一边突出的狭长的一部分，插入榫眼。

tongue-and-groove (tongued and grooved joint) A type of joint formed by inserting the *tongue of a

board into the corresponding * groove on another board.
企口接合（舌槽式接合） 由将板块的榫舌插入另一板块相应的凹槽所形成的接合。

tool (hand tool) Any small tool that can be used for forming, shaping, making holes, texturing, and setting out that is not machine-operated; e.g. a hammer, saw, or screwdriver.
工具（手动工具） 任何可以成型、塑性、开孔、织构和放线的非机械操作的小工具，例如锤、锯或螺丝刀。

tooled joint A mortar joint that has been finished or shaped by a pointing tool or trowel. The joint is finished and shaped as the work proceeds. Typical joints include: flush, tooled, recessed, keyed, and weathered. All joints, regardless of finish that have been shaped by a pointing tool are tooled joints, but the joint that is specifically a 'tooled' joint has a semi-circular finish that can be formed by the rounded handle of a pointing tool, also called a bucket-handle joint.
工具勾缝 用勾缝工具或抹子塑形的灰缝。灰缝随工程进展进行勾缝。典型的灰缝包括齐平的、用工具的、凹陷的、企口的和斜缝的。所有的灰缝，不管是哪种完成面，只要是用工具进行勾缝的都是工具勾缝，但是由圆形柄勾缝工具勾出来的半圆形完成面，也称为半圆凹缝。

tooling The process of forming and finishing mortar joints in brick and blockwork.
擦缝 形成和装饰砌砖工程中灰缝的过程。

tool pad (pad, tool holder) A combination tool capable of holding various fittings and tools, such as saws, awls, screwdrivers, and sockets.
工具包 装入各种配件和工具，例如锯子、尖锥、螺丝刀和套接口的组合工具箱。

toother A stretcher brick that projects at the end of a wall providing a projection that can be used for bonding subsequent walls.
马牙搓 顺砌砖，在墙壁末端凸出，提供用于黏结下一堵墙的凸出端。

toothing (indenting) Alternate courses are left projecting from the end of a wall so that brick or blockwork can be bonded directly to the wall. Indenting is normally the reverse of toothing, where bricks (half-batts or headers) are left out or cut out to allow new brickwork to be inset and bonded to the wall.
槎接（缩进） 从墙壁末端突出的交变层，这样砖块或砌块就可以直接砌合于墙壁上。缩进通常与槎接相反，切去砖块（半砖或丁砖）以嵌入新的砖块并连接墙壁。

toothplate connector (bulldog plate connector) A steel nail ring, used for bonding or securing two pieces of timber together. The steel ring has sharp teeth that project from both sides of the circular plate. When sandwiched and clamped between two pieces of timber a mechanical bond is formed.
齿环 钢钉环，用于连接或紧固两片木材。钢环有从圆板两边凸出的尖齿。当夹入两片木材间时，形成机械结合。

top-course tiles *See* UNDER-RIDGE TILES.
顶层砖 参考"平脊瓦"词条。

top-down construction (downwards construction) A method of construction that enables the basement and the superstructure to be constructed simultaneously.
逆作法施工 地下室和上层建筑同时施工的方法。

top-hung window A window that comprises a horizontally hinged openable * sash at the top.
上悬窗 顶部装铰链窗框可水平打开的窗。

top lighting The process of lighting an object from above.
顶部照明 从物体上方照明的过程。

topping 1. The final layer applied to the top of a screed or concrete floor. 2. The process of applying the top finish to a screed or concrete floor.
1. 顶层 砂浆找平层或混凝土地板顶部的最后一层。**2. 加顶** 在砂浆找平层或混凝土地板涂上顶层的过程。

topping out The completion of the building marked by the ceremonial laying of a stone or brick, the planting of a tree, or the unveiling of a plaque. Dignitaries are often invited to help mark the topping out ceremony.
封顶仪式 地铺上石头或砖块、植一棵树或揭开牌匾，象征着建筑竣工的仪式。通常会邀请达官贵人出席封顶仪式。

topple To fall forwards or overbalance.
倾倒 向前倒或失去平衡。

top rail The horizontal highest rail of a door or a *sash.
上冒头 门或窗框上最高的水平构件。

topsoil The fertile upper (surface) layer of soil that supports vegetation.
表层土 支撑植被的肥沃的上层（表面）土壤。

torbar Deformed and twisted bar.
变形和扭转的钢筋

torch A gas burner used for heating and working materials, and bringing them to the point of melting. Often used on flat roofs for softening asphalts and bituminous-based materials so that they can be bonded together.
喷灯 用于加热和加工材料的煤气喷灯，可使材料达到熔点。通常用于平屋顶以软化沥青和沥青基材料，这样它们就可以黏结在一起。

torque A rotating force that causes rotation.
转矩/扭矩 造成旋转的旋转力。

torsion The action of twisting an object by applying an equal and opposite torque at either end.
扭转 通过在两端施加同样大小但是方向相反的转矩而扭曲物体的动作。

total float The allowable slippage or delay of an activity that can occur without affecting subsequent tasks.
总时差 在不影响接下来的任务的情况下，一个活动可延迟或延误的时间。

total going The sum of the horizontal length of a step; the length of the going is measured from the nose of one step to the nose of the next step.
踏步总宽度 踏步的水平总宽度；踏步的宽度是指从一踏板前沿至下一踏板前沿的长度。

total rise The vertical rise from the foot of the stair to the top of the stair, also the sum of all the stairs.
踏步总高 从楼梯底部到楼梯顶部的垂直高度，也是所有梯级之和。

touch dry The point at which a painted surface is sufficiently dry such that it is no longer *tacky to the touch, and the application of light pressure to the surface does not leave any marks. The paint will not be completely dry.
表干 指涂漆表面充分干燥，摸起来不再黏黏的，轻轻按压表面也不会留下痕迹的特征。油漆还没完全干燥。

touchless controls Electronic controls that operate automatically using presence detectors. Used to open and close doors, switch lights on and off, switch taps on and off, and in flush urinals.
非接触式控制 使用存在型检测器自动操作的电子控制器。用于开门和关门、开灯和关灯、打开和关闭水龙头和小便冲洗阀。

touch up The process of repainting small missed or damaged areas of paintwork.
补漆 为小部分漏刷或受损的油漆表面重新上漆的过程。

toughened *See* TEMPERED GLASS.
钢化的 参考“钢化玻璃”词条。

toughness A material property indicating the energy required to fail a material; in other words how much energy a material can absorb up to fracture. In general materials tend to be the toughest. The area under a stress-strain plot for a material is a way of measuring toughness.
韧度 一种材料性能，即破坏材料所需的能量；换句话说，即是材料可吸收多少能量才会断裂。通常，材料应选最坚韧的。材料应力应变图的面积是测量韧度的一种方法。

tourelle A tower that is circular on plan, can have bricks or stone that project out from the main wall to form a corbelled turret.
回转炮塔 平面是圆形的塔楼，可从主墙上伸出砖或石块形成一悬挑炮塔。

tower Tall structure, usually slender and higher than surrounding buildings.
塔楼 高层建筑结构，通常比周围建筑更细长更高。

tower bolt A large *barrel bolt.
圆管形插销 一种加大的圆插销。

tower crane A tall steel-truss framed structure with a lifting jib mounted on top of the tower. Particularly tall cranes are often tied to the building structure. The lift shaft of a building is often constructed first and provides a rigid stable structure against which the tower crane can be tied. The rotating jib of the crane may be fixed horizontally, or in the case of luffing jib cranes, can rotate up and down assisting lifting. Luffing jib cranes can also be used where the boundary of the site is tight and where permission has not been granted to over-sail surrounding property. Where over-sailing rights have not been granted, the luffing jib is raised to avoid over-sailing the property. On most tower cranes counterweights are used at the back of the jib.
塔式起重机/塔吊 在顶部安有一吊臂的一种高大钢桁架结构。超高起重机通常系在建筑结构上。通常先建好建筑物里面的电梯筒井以为塔式起重机提供稳定的支撑。起重机的旋臂可以水平安装，如果是动臂起重机，则可以上下移动以帮助提升。动臂起重机亦可用于场地窄小和周围物业的上空通过权未被批准的地方，动臂起重机向上提升以避开经过附近物业。大部分的塔式起重机都在后方有配重。

tower scaffold Prefabricated aluminium and steel frame scaffold system. The system is easy to assemble with quick-fit *braces, *ledges, and *outriggers.
塔式脚手架 预制铝和钢框脚手架系统。用快速安装的支柱、平台和悬臂支架可轻易组装。

town A collection of dwellings larger than a village and smaller than a city.
城镇 比村庄大比城市小的住宅群。

town gas (coal gas) Gas made by burning coal. Superceded by natural gas.
城镇燃气（煤气） 燃烧煤制成的气体。已被天然气取代。

town house A terraced house close to the centre of a city or large town. Traditionally, townhouses were expensive, large, and tall terraced houses. As land in towns was expensive, the properties were often built with three or more storeys. The house was called the town house as the rich often had a house in the town and one in the country.
市区房 接近市中心或大型城镇的连栋房屋。传统上，市区房又贵又大且是连栋高楼。因为城镇土地昂贵，房屋通常建为三层或更高。称为市区房是因为有钱人通常在市区拥有一套房子，在郊区也拥有一套。

town planning Legislation exercised under the Town and Country Planning Act to control and regulate development.
城镇规划 在城镇规划法案下立法控制和规范城镇开发。

trace heating (cable heating) The use of an electric element to provide sufficient heat such that the required temperature can be maintained in pipes and vessels.
管道加热保温 使用电气元件提供充足热量，这样管道和容器中就可以维持所需的温度。

tracer (warning tape) *See* MARKER TAPE.
警示物（警示胶带） 参考“警示带”词条。

tracery (warning tape, marking sand, marker tape) Material laid above buried services used to alert people working in the area that services are positioned under the marker.
警示带 铺设在埋地设施上的材料，用于警示在此区域工作的人，让他们知道设施位于警示带下面。

track 1. A mark left by a person, animal, or vehicle travelling over soft ground, such as a footprint. 2. A path or road. 3. A set of parallel rails upon which a train or tram travels along. 4. A particular course of action during an investigation.
1. 足迹 人、动物或车辆在软地面经过留下的痕迹，例如脚印。**2. 小道** **3. 轨道** 火车或有轨电车沿着行使的平行路轨。**4. 追踪** 侦查时的一套特定的行动方案。

traction 1. The amount of friction that allows movement to occur. 2. The action of pulling something along a surface.
1. 牵引力 移动发生所需的摩擦力。**2. 牵引** 沿着表面拉动某物的动作。

tractive force The force available from a vehicle to pull something.
驱动力 车辆能产生拖曳某物的力。

tractor A vehicle used for pulling heavy loads.
拖拉机 用于拖曳重负荷的车辆。

TRADA *See* TIMBER RESEARCH AND DEVELOPMENT ASSOCIATION.
TRADA 参考“木材研究和发展协会”词条。

trade Work undertaken by craftspeople or professionals rather the general public. A tradesperson describes a skilled person connected to the construction industry.
工种/专业 由工匠或专业人士而不是普通大众进行的工作。技术工人是指在建筑行业受过相关专业训练的人员。

trade collection A summary of all of the trade work contained within the bills of quantities. The quantity of work contained for each different trade can be summarized.
工种总和 工程量清单中包含的所有工种工作的总结。不同工种的工程量也可总计。

trade foreman * Charge hand or group leader who supervises the work of trades such as bricklayers, joiners, and ground workers.
工种工长 负责监督各工种工作（如瓦工、细木工和地基工）的负责人或组长。

tradeless activity A job that does not require any particular skill and could be undertaken by any tradesperson or worker.
非技术工作 可由任何技术工人或普工执行，无须特殊技能的工作。

trade-off The relaxation of one regulatory requirement or contractual condition subject to it being met by other agreed means.
折中 规章要求或合同条款因为要满足另行议定的条件而做出放宽。

trade rubbish Waste materials collected for disposal from developments, construction sites, manufacturing, commercial properties, industry, and other non-domestic activities.
城市商业垃圾 开发、建筑工地、生产、商业地产、工业和其他非生活产生的废弃垃圾。

T

tradesperson (craftsperson) Trained and skilled workers that can undertake work within their field with minimum supervision. Recognized trades within construction include: electricians, joiners, bricklayers, ground workers, plumbers, decorators, and plasterers.
技术工人（工匠） 无须过多监督，能够在其领域内完成工作并经过培训的熟练工人。建筑行业的工匠包括电工、细木工、瓦工、地基工、水管工、装修工和抹灰工。

trades union An organization of people who work in a particular field or industry, that form to represent the common good of the people within the organization. Such organizations gain their power through the number of members signed up and the potential or actual action they take.
工会 在特定领域或行业工作人员组成的组织，代表组织成员的利益。该类组织通过其会员数及其采取或可能采取的行动获得权力。

traffic The movement of vehicles, goods, and information along a particular route—cars along a road, the illegal trade of drugs from one country to another, the transfer of data along a computer network, and so on.

交通 车辆、货物和信息通过特定路径的移动，如汽车通过道路、一个国家到另一个国家的非法毒品交易、电脑网络内的数据传输等。

trafficable roof A roof that is designed to be walked on. *See also* NON-TRAFFICABLE ROOF.
可上人屋面 可在上面行走的屋面。另请参考“不上人屋面”词条。

trailing dredge A *dredge that contains a suction pump or cutter to suck up water and sediment as the dredge moves along.
耙吸式挖泥船 具有一个抽吸泵或切刀的挖泥船，可边航行边吸水吸泥。

training wall A wall to divert or contain the flow of a river.
导流墙 分流或容纳河流的墙壁。

trammel 1. An instruction used to draw ellipses. 2. A device to check curvature.
1. 椭圆规 用于画椭圆形的仪器。**2. 量规** 量度弯曲度的设备。

tramway A light railway system that uses trams.
电车轨道 使用有轨电车的轻型轨道系统。

transducer A device that converts a physical quantity such as displacement, load, temperature, etc into another form, such as an electrical signal, and in turn can be recoded by a data logger.
传感器 将物理量（例如位移、荷载、温度等）转换成另一种形式（例如电信号）的设备，可由数据记录器记录。

transferred responsibility Duty changed from one person to another by agreement.
可转移的责任 通过协议，将责任从一个人转给另一个人。

transformation piece A connection or adaptor that serves to take one pipe or duct from one size, shape, or fitting to another, for example, from a rectangular duct to a square duct.
转接头 将一条管道或槽管从一种尺寸、形状或配件转换成另一种的连接头或转换头，例如从矩形槽转到方形槽。

transformer A device that changes the voltage, current, phase, or impedance of an alternating current.
变压器 改变交流电的电压、电流、相位或电阻的设备。

transition 1. The period or activity of change between one operation and another. 2. Fittings or materials used to change one material or component to another. May also describe the fitting used to change the size or shape of a pipe or duct to another, *See also* TRANSFORMATION PIECE.
1. 过渡 从一操作到另一操作的改变时期或活动。**2. 转换** 用于将一种材料或组件变为另一种的配件或材料。亦可用于形容改变管道或槽管尺寸或形状的配件，参考“转接头”词条。

transition temperature (metals) A temperature for metals where a change in property takes place, e. g. ductile to brittle transition temperature.
转变温度（金属） 金属属性发生变化时的温度，例如韧性 - 脆性转变温度。

transit method A method to distribute the closing error on a closed traverse.
中天法 从闭合导线求出闭合误差的方法。

translucent An object that allows light to pass through it; however, without optimal clarity, not fully transparent.
半透明的 容许光透过的物体，但如果不是最佳透明度，则不是完全透明。

transmission The movement of heat, light, or sound from one place to another.
传递/传输/传播 热量、光线或声音从一个地方传到另一个地方。

transmissivity 1. A measure of a material's ability to transmit flow, for example the flow of water through an aquifer. 2. The volumetric flow of water per unit width of a *geosynthetic per unit gradient in a direction parallel to the plane of the product, as defined in BS 6906 – 7 1990.
1. 透过率 物体能被透过能力的测量，如含水层水流的透过率。**2. 导水系数** 每单位宽度土工材料

T

中与产品平面方向平行的单位梯度下水流的通过量，在 BS 6906 - 7 1990 中有说明。

transmittance 1. The amount of radiation (visible, UV, total solar, etc.) that passes through a material or construction. In the case of glazing, it is stated as the percentage of radiation. 2. Thermal transmittance.
1. 透射比 辐射（可见光、紫外线、总日光等）通过材料或建筑的量。在玻璃中，经常表述为辐射的透过率。**2. 传热系数**。

transmitter A device to generate and transmit radio-frequency waves, for example, broadcasting equipment.
发射机 能产生和传播无线电波的设备，如广播系统。

transom The horizontal member that subdivides a * window frame.
中冒头 细分窗户框架的水平构件。

transom window (transom light) A small rectangular * window located above a door. May be fixed or horizontally hinged.
亮子/气窗 位于门顶上的矩形窗户。可为固定式或水平铰链式。

transparent See-through, clear; having the property that light passes through it almost undisturbed, such that one can see through it clearly.
透明的 能被看透的、清澈的；能让光几乎无障碍地透过的，有这种特性就能清楚看穿。

transport Carrying someone or something from one place to another.
运输 运载某人或某物从一个地方到另一个地方。

transverse Crosswise across something or at right angles to something.
横向的 交叉穿过某物或与某物成直角。

trap A section of pipe located below a drain that is designed to provide a water seal, thus preventing the flow of foul air back up through the drain. Can be J-, P-, S-, or U-shaped.
存水弯 位于排水管下面的一小节管道，用以提供水封，防止浊气通过排水管回流。可以为 J 形、P 形、S 形或 U 形。

trapezoid tear test A method to determine the tear strength of a * geosynethtic.
梯形抗撕裂强度测试 确定土工合成材料抗撕裂强度的方法。

trapped waste (waste trap) A trap connected to the waste pipe of an appliance.
污水存水弯 连接至卫生器具排水管的存水弯。

trap vent A * branch vent.
通风支管。

travel The allowed movement of an object, such as the distance that a sliding door moves.
行程 物体能够进行的移动，例如推拉门移动的距离。

T

traveller 1. In surveying and setting out, a traveller is a T-shaped implement that can be sighted and levelled between two profiles of known height. The profiles are set with horizontal timbers at specific levels to provide a horizontal or sloping sight line. The top of the T-shaped traveller is then sighted so that it is in line with the top of the other profiles. The T-shaped traveller is set at a length so that when it is lined-in, the bottom of the T marks the depth of dig or position of the top of a pipe. 2. Piece of equipment that moves backwards and forwards along a groove or rod.
1. 定深板 在测量和放线中，定深板是一种 T 形的工具，可在两个已知高度的控制桩中间观测和抄平。控制桩由水平的木头定在固定的高度上，作为水平或斜坡的视准线。然后 T 形定深板的顶部对准其他控制桩的顶部。T 形定深板的长度固定，当其对准时，底部即为开挖的深度或者是管道顶部的位置。**2. 滑块** 在凹槽或杆上来回移动的设备。

traverse A type of survey formed by a series of straight lines and horizontal angles between them. A closed traverse forms a closed loop whereas an open traverse does not.
导线 由一连串直线和水平角形成的一种测量类型。闭合导线可形成闭环，而非闭合导线则不行。

tray (shower tray) A shallow open container used to collect the water from a shower.
淋浴底盆 一个浅的敞口容器，用于汇集淋浴水。

tray aerator A method of aerating water by causing it to fall through a series of perforated trays.
盘式曝气器 通过多个有孔盘跌水而曝气的方法。

tread The horizontal part of a stair where each foot is placed. The tread is measured from the front of the step, the *nose, to the back of the step, and is the full length of the horizontal surface upon which a foot can be placed.
踏板/踏步 楼梯供脚踏上的水平部分。踏板长度是从楼梯的前边、前缘至背面测量，是脚可放置的水平表面的全长。

tread plate Embossed metal sheet with a raised profile to create a slip-resistant surface, often placed on stairs and steps.
花纹板 花纹金属板，有凸起以形成防滑表面的花纹，通常位于楼梯和梯级中。

treatment A method to alter the physical, chemical or biological character or composition of something, e. g. water treatment removes physical, chemical, and biological impurities from water such that it is fit for human consumption.
处理 改变某物物理、化学或生物属性或组分的方法，例如，水处理是除去水里的物理、化学和生物杂质，使其适合人类使用。

tree A perennial woody plant that provides the material *timber.
树木 多年生木质植物，可作为木材材料。

treenail (trenail) A large dowel-shaped peg that is driven into a hole that has been bored through two pieces of timber to form a joint. This jointing system has considerable historic use.
木钉 大的销形木钉，钉入两片木材中已钻好的孔洞中，形成连接。该连接方式已有很长的历史。

tremie (tremie tube) A funnel-shaped *hopper containing a large pipe at the bottom. It is used to pour concrete, particularly under water.
混凝土导管 漏斗形的料斗，底部有大的管道。用于浇筑混凝土，尤其是在水下。

trench 1. A long narrow groove cut into a material. 2. A strip of earth excavated to form a long narrow hole in the ground. The trench is often excavated in order to lay foundations or drainage.
1. 沟槽 在材料上切成的狭长凹槽。**2. 沟渠** 在地上挖掘的狭长槽。挖掘的沟渠用于铺设基础或排水系统。

trench duct Tube or pipework laid in the ground to provide a void along which cables and pipes can be threaded through. Once the ducts are in position and the trench is *backfilled, any cables housed in the ducts are hidden neatly underground.
电缆槽 铺设在地下的管子或管道，用于提供一个空间供电缆和管道穿过。一旦电缆槽铺设到位并且挖沟已经回填，封装在管子中的电缆将被整齐地埋藏在地下。

trench-fill foundation A long strip of earth is excavated down to a good load-bearing strata and the remaining trench is simply filled with concrete to provide a foundation. The foundation is suited to carrying longitudinal loads such as walls.
带形回填基础 挖掘至承重好的地层的狭长地带，剩下的沟槽仅用混凝土简单填充以作为基础。该基础可承受纵向荷载，例如墙体。

trenchless technology A method of installing service, pipes, etc. underground without the need of ***open cut**.
非开挖技术 一种无须明挖就可安装（如地下设施、管道等）的方法。

trench support A supporting structure positioned against the walls of a trench to prevent them collapsing while people are working within the trench.
沟槽支护 位于槽壁的支承构件，当人在沟槽中施工时，防止其塌陷。

trepan 1. A tool for boring holes/shafts in rock. 2. A machine, containing cutting wheels, used for cutting and loading coal from along a coal seam. 3. To cut a circular groove into a surface.
1. 钻孔器 用于在岩石上钻孔/井的工具。**2. 钻煤机** 一种有切割轮的器械，用于在煤层上切下并运送煤。**3. 环锯** 在表面上开圆槽。

Tresca yield envelope A criterion to denote the state of stress at yield in a material, with the maximum shear stress being the decisive factor for yield; *See also* VON MISES YIELD.
特雷斯卡屈服包络面 材料在屈服状态下应力的准则，最大剪应力是屈服的决定性因素；另请参考"冯·米塞斯屈服"词条。

trestle A supporting tower-like structural framework for bridges, which consists of a horizontal beam held up with raking legs.
架柱 桥梁的塔式支承结构框架，由斜支脚支承的水平横梁组成。

trial hole (trial pit, test pit) A small hole excavated either by hand or by an excavator to collect information on the soil, strata, and/or services below the ground. Used during a * site investigation to reveal details of the strata and groundwater condition at shallow depth. For soil investigations samples can be taken from the various levels of strata that will be visible on the sides of the excavation. When attempting to locate the position ofservices, the exploratory hole is carefully excavated by hand.
试验孔（探坑） 用手或挖掘机挖掘的小孔，用于收集土壤、地层和/或地下设施的信息。试验孔在现场勘查期间使用，以反映浅层地层和地下水条件等具体信息。土壤调查样本可以从挖掘各边可见的各层地层获取。当尝试定位设施位置时，要小心地手动挖掘探孔。

triangulation A method used in surveying that divides an area of land into triangles. The angles of the triangles are measured as well as one side (termed the baseline). The length of the other sides can then be calculated by trigonometric relationships.
三角测量 在测量中，将土地范围划分为三角形的方法。测量三角形的角和一边（称为基线），然后根据三角关系可算出其他边的长度。

trichloroethane An organic chemical that is monitored in drinking water supplies because it is hazardous to health.
三氯乙烷 饮用水供应中需监测的有机化学物，因为其对健康有害。

trickling filter A method of treating wastewater, by passing it slowly through a granular filter, where it is subject to aeration and microbial degradation.
滴滤池 处理废水的一种方法，污水缓慢流过颗粒层过滤器，进行曝气和微生物降解处理。

trihalomethane (THMs) A chemical compound with three halogen atoms, derived from methane; it is formed as a by-product during the chlorination of drinking water and is considered carcinogenic.
三卤甲烷（THMs） 由甲烷衍生而来，有三个卤素原子的化学物；是在饮用水氯化消毒时的副产品，有致癌性。

trilateration A surveying technique using a network of triangles, where the sides of the triangles are measured using a * tellurometer, as well as the internal angles; *See also* TRIANGULATION.
三边测量 三角测量网中使用的测量技术，使用微波测距仪测量三角形各边和内角；另请参考"三角测量"词条。

trim 1. The collective name given to decorative finishes such as architraves and skirting boards. 2. To cut to the correct size.
1. 饰条 装饰性饰面，如门头线和踢脚板等的统称。**2. 修剪** 裁切成正确的尺寸。

trimmed door A door leaf that has been reduced in size once hung.
门扇修整 当门挂上铰链时，将门扇尺寸缩减。

trimming 1. The collective name given to the trimming joist, the trimmer joist, and the trimmed joists used to form an opening. 2. The process of cutting an object to the correct size.

1. 洞口龙骨 用于形成开口的横筋龙骨、竖筋龙骨和短撑的统称。**2. 裁剪** 将物体裁切成正确尺寸的过程。

trimming joist (trimming, Scotland bridling) A floor joist that runs parallel to the *bridging joist to support the end of a trimmer joist at an opening in the floor. The trimmer joist runs perpendicular to the bridging joist to support the trimmed joists. The trimmed joists are short bridging joists that are used to create an opening in the floor, usually for stairs.
横筋龙骨（洞口龙骨） 与横搁栅平行的地板龙骨，以支撑楼板开口处的竖筋龙骨。竖筋龙骨与横搁栅垂直以支承短撑。短撑是短的搁栅，用于形成楼板开口，通常用于楼梯。

trimming machine (mitring machine) Guillotine used for trimming mouldings and cutting angles.
修边机（斜切机） 用于修剪线条和切割角的裁切机。

tripod A frame or stand with three legs, usually collapsible, used to support a surveying instrument, for example a *theodolite.
三脚架 有三条支腿的框架或支架，通常可折叠，用于支撑测量仪器，如经纬仪。

triumphal arch Arch or monument built for the return of a victorious army or group.
凯旋门 用于迎接获胜军队或团队凯旋回师的拱形或纪念建筑。

trolley (crab) Wheeled frame that runs along the jib of a crane; used to carry and position the lifting hooks.
绞车（卷扬机） 沿着起重机吊臂滑行的轮框；用于搬运和定位吊钩。

trolleying The movement inwards and outwards of the *trolley on the jib of a crane. As the trolley comes inwards, the radius of the hook is reduced and as the trolley is positioned further out on the jib the radius is increased.
滑动 起重机吊臂上的绞车向内和向外移动。当绞车向内移动时，吊钩的半径缩减；当绞车向吊臂外延伸时，半径增大。

Trombe wall A thermally massive, usually masonry, south-facing wall that is clad in glazing, with a small air gap between the glazing and the wall. It is designed to maximize the collection of solar radiation during the day and slowly release it at night to the inside of a building. A mass Trombe wall is one in which the thermal mass used in the wall is masonry, while a water Trombe wall is one where the thermal mass used is water.
特隆布墙 隔热墙，通常为南向砌体墙，包覆着玻璃，在玻璃和墙壁之间有小的气缝。该墙设计可在白天最大化吸收太阳辐射，然后在晚上缓慢释放到建筑物内。蓄热特隆布墙是指墙体中蓄热体使用的是砌体，而水墙则是用水。

trough A channel, trench, or drain.
槽 沟渠或排水沟。

troughed sheeting *See* PROFILED SHEETING.
压型钢板 参考“压型钢板”词条。

trough gutter *See* BOX GUTTER.
槽形沟 参考“方形排水槽”词条。

trowelling The final smoothing of plaster, screed, or concrete with the edge of a trowel. The final stage of smoothing takes place after the initial set has taken place. For concrete floors, trowelling takes place after ***power-floating**.
用抹子抹平 用抹子边缘最后抹平灰泥、砂浆或混凝土。抹平的最后阶段发生在初凝之后。对于混凝土地板，用抹子抹平发生在电动抹平之后。

truck A general term used for a vehicle (such as a *dump truck) for hauling loose material.
卡车 用于搬运松散材料的车辆（如自卸货车）的统称。

true bearing The horizontal angle made between *true north and any line; *See also* MAGNETIC BEARING.
真方位角 真北方向和任何线之间的水平角；另请参考“磁方位角”词条。

true north The direction north according to the earth's axis.
真北 根据地球轴方向确定的北向。

true strain The value of the actual *strain of a material subjected to a tensile force at any specific point of load. The value of strain between any two specific points.
真实应变 在任意特定点的荷载下，承受张力的材料的实际应变值。任何两个特定点之间的应变值。

true stress The value of the actual ***stress** of a material subjected to a tensile force at any specific point of load. The value of stress between any two specific points.
真实应力 在任意特定点的荷载下，承受张力的材料的实际应力值。任何两个特定点之间的应力值。

truncated solid A solid figure, such as a cuboid, with its top cut off making another shape, for example, a frustum
截断的固体 一立体型（如立方体截去顶端）形成别的形状体，如平截头体。

truncated truss A trussed *rafter that has had the top section removed.
截顶桁架 顶部被截去的桁架椽条。

trunk The main stem of something e. g. a tree. A trunk road is a large main road. A trunk sewer is a large main sewer.
主干 某物的主干部分，例如树木。干道是指主干道。污水干线是指排污主干线。

trunking A protective duct for cables or pipes that has a removable cover.
线槽 电缆或管道的保护性槽管，有可拆开的盖子。

trunk lift *See* FREIGHT ELEVATOR.
货梯 参考“货梯”词条。

trunnion A cylindrical projection used as a pivoting point on optical instruments, such as theodolites, where they are mounted on either side of the telescope to allow it to rotate in the vertical plane.
耳轴 圆柱形凸出物，用作光学仪器（如经纬仪）的枢轴点，可安装在望远镜的任一边，以使其在垂直平面内旋转。

truss A structural framework of beams, posts, and struts that is designed to support a structure, such as a roof, or to span an opening.
桁架 横梁、柱子和支柱的结构框架，用以支承一个结构体，例如屋顶或横跨开口。

truss clip A stainless steel clip designed to secure a roof truss to the wall plate.
桁架夹具 不锈钢夹具，用于将屋面桁架固定至梁垫上。

trussed beam A *beam made in the form of a *truss.
桁架梁 桁架形式的横梁。

T

trussed purlin A *purlin made in the form of a *truss.
桁架式檩条 桁架形式的檩条。

trussed rafter A prefabricated structural framework, triangular in shape, that is used to support a roof. Commonly used in domestic construction. A wide range of trussed rafters are available including attic, cantilever, fan, fink, Howe, king post, mono, queen post, and scissor.
桁架式椽条 预制的结构框架，为三角形，用于支撑屋面。通常用于住宅建筑工程。有各种不同的桁架式椽条，包括阁楼式、悬臂式、扇形、芬克式、豪式、桁架中柱、单坡桁架、桁架副柱和剪刀式。

truss element A two-node member that defines a line, in a finite element model, which can only resist or transmit an axial force.
桁架单元 双节构件，在有限元模型中确定一条线，仅可抵抗或传输轴向力。

truth table A method to record values, in rows and columns, for all the possibilities in a logic relation.
真值表 记录数值的方法，（将数值）排成行和列，列出逻辑关系的全部可能性。

try plane A long plane used for truing or final smoothing of wood. Try planes are usually over half a meter in length.
半精刨 用于修整或最终刨平木材的长刨。半精刨长度通常超过半米。

try square A square, used to check that the wood is true; having angles of 90°.
曲尺 用于检查木材是否精确的直角尺；成90°角。

T-square A drawing instrument in the shape of a "T" used with a drawing board, to draw horizontal straight lines.
丁字尺 T形的绘图工具，和绘图板一起使用，画出水平直线。

tube A pipe, cylinder, hose, or something that has a lateral dimension (typically diameter) far less than its longitudinal dimension, and is typically used to pass liquid through.
管子 管道、圆筒、软管或侧面尺寸（通常是直径）远小于纵向尺寸的物体，通常用于输送液体。

tubing bay A *manhole or backdrop inspection chamber.
检修间 检修孔或垂直落水管的检查井。

tubular In the shape of a tube.
管状的 管子形状的。

tubular mortise lock *See* BORE LOCK.
管状插锁 参考"球形锁"词条。

tubular scaffolding *Scaffold made using tubular steel tubes.
钢管脚手架 用钢管制成的脚手架。

tubular trap A *trap made using tubular pipe.
管式存水弯 用管材制成的存水弯。

tuck A small recess made in the mortar joint prior to *tuck pointing.
凸缝槽 在凸嵌缝之前的灰缝处做成的小凹处。

tuck pointing 1. A type of decorative *pointing that projects slightly from the face of the masonry courses. 2. The process of removing and replacing old and damaged mortar between masonry courses; *See* REPOINTING.
1. 凸嵌缝 稍微从砌砖层表面凸出的一种装饰性勾缝。**2. 重勾** 移除和更换砌砖层之间老旧受损的砂浆的过程。参考"重新勾缝"词条。

Tudor style Architecture from the Tudor period 1485—1603, typified by buildings constructed between the Gothic and Renaissance periods. The Tudor arch was a feature of this period. In domestic construction, wattle-and-daub was often used. External features associated with this period include exposed wood structure with in-fill white render. Tall decorative chimneys were often constructed on houses of this period.
都铎风格 都铎王朝 1485－1603 年间的建筑，以哥特式和文艺复兴时期之间的建筑为代表。都铎拱是该时期的一个特色。在住宅建筑上，常使用抹灰篱笆墙。与该时期相关的外观特色包括外露木架构填入白色的墙壁。房子上通常会建有高大的装饰性烟囱。

tumbler The portion of a *lock that enables it to be operated only when the correct key is inserted.
锁簧 锁的一部分，只有插入正确的钥匙时，才能打开。

tumbler switch (toggle switch) A spring-loaded electrical switch with a small level.
倒扳开关（肘节开关） 弹簧电气开关，配有一个小钮子。

tumbling in (tumbling courses) A sloping course of brickwork laid across the top of a gable wall. The bricks are laid at 90° to the gable wall course.
斜嵌砖层 铺设在山墙顶部的倾斜砖层。砖块与山墙形成90°角。

tundish An open container used to collect condensate.
集水盘 用于汇集冷凝水的开口容器。

tuned mass damper (TMD) A mechanical system, consisting of an *oscillating *spring and a ***dashpot**; used to reduce the effects of *motion.
调谐质块阻尼器 一种机械系统，包含一个振荡弹簧和一个缓冲器；用于减少摆动。

tungsten A non-ferrous metal with the symbol W. Tungsten has a very high melting point (3422℃) and is used mainly in electrical applications, such as lightbulb filaments.
钨 有色金属，符号为W。钨有非常高的熔点（3422℃），主要应用于电气用途，例如灯泡灯丝。

tungsten-filament lamp An incandescent lamp.
钨丝灯 一种白炽灯。

tungsten-halogen lamp *See* HALOGEN LAMP.
卤钨灯 参考“卤素灯”词条。

tungsten-tipped drill (hard metal tungsten drill, masonry drill) Drill bit with a tungsten carbide tip that provides a harder cutting edge; used to drill into masonry.
碳化钨钻（石工钻） 钻头为碳化钨刃，刀口更坚硬；用于钻入砖石。

tunnel An underground channel or passage.
隧道 地下通道或过道。

tunnel-boring machine (TBM) A tunnelling machine, which consists of a rotary cutting head that occupies the full face of the tunnel and a system of conveyors (or pumps) to remove the excavated material.
隧道掘进机 一种挖隧道机械，包含一个占据整个隧道断面的旋转刀头和一个移除挖掘材料的传输系统（或泵）。

tunnel form Pre-assembled floor and wall formwork allowing the casting of concrete walls and floors in a single pour. The forms are suited to buildings where the room sizes are consistently similar, allowing multiple uses and considerable efficiency in using the forms. Where rooms are the same size, there is minimal setup and adjustment required between pours. Hotels, apartment blocks, and lodges are particularly suited to this type of formwork. The forms are partially constructed with two walls joined by a folding deck. The units can be rolled on wheels, and quickly erected to form the shuttering for the walls and floors. Once the concrete is cast the units are removed by collapsing the foldable deck and rolling the form out on wheels.
隧道模 预装配的地板和墙壁模板，可使混凝土墙壁和地板一次性浇筑。隧道模适合于尺寸类似的房间连续排列的建筑，可使模板更有效率地重复利用。当房间尺寸相同时，浇筑之间只需少量的设置和调整。酒店、公寓和旅馆尤其适用这类模板。模板部分由折叠板连接的两堵墙壁建成。装置可由轮子滚动，快速安装成墙壁和地板的模板。混凝土浇筑后可折起折叠模板并将模板通过轮子移出。

tunnelling A process of forming passageways under the ground, through the sides of mountains, below rivers or channels, so that vehicles and trains can pass through.
挖隧道 在地底、山的两边、河流或渠道下面形成过道的过程，隧道可使车辆和火车通过。

T

tupper Local term in some areas in the north of England to describe a bricklayer's labourer.
砌砖工 英格兰北部某些地区称呼瓦工的本地用语。

turbid A reference to the cloudiness of water caused by suspended particles. It is measured using a nephelometer, which determines how much light is scattered though a column of water to give a turbidity value, quoted in Nephelometric Turbidity Units (NTU). The WHO drinking guidelines allow up to 5 NTU to be present in drinking water supplies.
浑浊的 与悬浮颗粒引起水体模糊相关的。可用浊度计来度量，可用光线通过水柱被散射的程度得出浊度值，以比浊法浊度单位（NTU）来计量。世界卫生组织规定饮用水供应中NTU最高为5。

turbidities A geological deposit formed under water, e.g. from an landslide.
浊度 水底下形成的土体沉积，如由于山崩滑坡造成的。

turbine A machine that converts fluid energy into mechanical energy. The fluid energy is typically from wind or water, and imparts rotation to the turbine's vane or blades, which in turn drives a generator to create, for ex-

ample, electricity.
涡轮机 将流体能量转化成机械能的机器。流体能量通常来自风或水，旋转传给涡轮叶片，转而使发电机产生机械能，例如电力。

turnbuckle (stretching screw, bottle screw, tension sleeve) A sleeve-like connector used for adjusting tension ropes, cables, etc. The sleeve is threaded and has two threaded nuts or eyelets, which are threaded in the opposite direction to each other, at either end connected in line with the rope or cable to be tensioned. The tension is adjusted by rotating the turnbuckle.
松紧螺丝扣 类似套筒的连接器，用于调整拉绳、缆绳等。套筒有螺纹，有两个螺母或孔眼，在各自相反方向都有螺纹，两端与受拉伸的拉绳或缆绳连接。通过旋转松紧螺丝扣调整张力。

turn cock A *stopcock.
旋塞阀

turn-down *See* HOOK INTAKE.
电缆套管 参考"电缆套管"词条。

turned bolt A bolt that has been machined to produce a shank dimension of close tolerance.
精制螺栓 加工使螺杆尺寸达到精密公差的螺栓。

turning Working or forming an object by rotating and cutting it on a turning lathe. Wood, steel, and other materials can be cut to form more circular sections by rotating it on a lathe and shaping the edges with chisels.
车削 通过在车床上旋转或切割以加工材料。木材、钢材和其他材料都可以通过在车床上旋转切割并用凿子加工其边缘以做成更圆的型材。

turning piece (centering) Temporary support that provides support to a brick, stone, or other form of segmental arch. The support is often made from plywood with the sheets being cut to the same shape as the arch. The centre or temporary support is then held in position using *telescopic props or temporary adjustable props until the arch is constructed and set.
拱架 为砖块、石头或其他形式的平圆拱提供支撑的临时支承。该支承件通常由胶合板和切成与拱门相同形状的板材制成。用伸缩支柱或临时可调支柱固定拱架或临时支承，直至拱门建成并固定。

turning point 1. A crossroads. 2. A change point in surveying. 3. The maximum and minimum points of a curve.
1. 交叉路口。 **2. 转点** 测量中的转换点。 **3. 拐点** 曲线的最高点和最低点。

turnkey project A contract where the whole design, build, and furnish is undertaken by a contractor. The client signs the contract and expects to be "turning the key" and opening the door to a fully functional building.
交钥匙工程 整个设计、建造和安装都由一个承包商负责的合同。业主签订合同，然后期望"转钥匙"开门就得到功能全部完善的建筑物。

T

turnpike A toll gate; a place where a *toll is collected on a section of road.
收费公路 一个收费站；在道路某处收费的地方。

turn-up The adjustment of material where a horizontal surface meets a vertical surface so that the material can be laid flat along the horizontal and angled to travel up the vertical surface. *Flashings, *asphalt to up-stand walls, and *skirtings are used to form turn-ups.
阴角线 水平表面与垂直表面交接处的材料修整，使材料能与水平面和垂直面都能平滑过渡。常用泛水、竖墙沥青和踢脚线来做阴角线。

turret A small circular tower, normally quite notably smaller than the main structure.
角楼 小圆塔，通常比主建筑体更小。

turret roof Pitched roof to tower or round structure.
尖塔屋顶 塔楼或圆形结构的坡屋面。

turret step Triangular stone steps, traditionally used within *turret towers. The thin end of the triangular stone steps are often rounded, which creates a rounded *newel when the steps are placed on top of each other.
螺旋梯 三角形石阶，传统上用于塔楼中。三角形石阶的细端通常是圆边的，当台阶一级一级铺设时，形成圆形的螺旋梯中柱。

tusk nailing (skew nailing, angle nailing) Nails that are installed at an angle into timber to provide extra strength to the joint.
斜钉 按一定角度将钉子钉入木材中以增强接合强度的方式。

tusk tenon A kind of mortise and tenon joint that uses a wedge-shaped key to hold the joint together.
牙销紧固闭口贯通榫 一种用牙状木楔紧固的榫卯接合。

TV distribution (US master antenna system) The collective name given to all of the components required to provide TV reception in a building, such as the aerial, an amplifier, the wiring, the connectors, and the socket outlets.
共用天线系统 在建筑物内需要提供电视接收功能的所有组件的统称，例如天线、放大器、线路、连接器和电缆插座。

twin and earth Sheathed mains electrical cable that contains two conductors and an earth. Used for lighting and power circuits.
双绞股屏蔽电缆 铠装电源电缆，包含双导线和接地电缆。用于照明和电源电路。

twin cable Sheathed mains electrical cable that contains two conductors.
双芯电缆 包含双导线的铠装电源电缆。

twin tenon (divided tenon) Two *tenons arranged in a row on the same piece of timber. Often confused with a double tenon.
双榫 在同一块木材上排成一排的两个榫头。通常与双夹榫相混淆。

twin-thread screw A *screw that contains two separate, deep interlinked threads.
双头螺杆 有两个不同的互连螺纹的螺杆。

twin-tube fluorescent A luminaire that incorporates two fluorescent tubes.
双管荧光灯 有两个荧光灯管的灯具。

twist The warping of timber. If timber is seasoned or dries out too quickly it has a tendency to warp in a helical screw-like manner.
扭曲 木材翘曲。如果木材烘干或干燥过快，就容易变翘曲，出现类似螺旋上升的形状。

twist drill Hardened steel drill bit with steep helical cutting edge.
麻花钻 硬化钢钻头，有陡峭的螺旋刃片。

T

twisted fibres Overlapping and interlocking grain found in timber.
交错纹理 木材中重叠和交叉的纹理。

twitcher A small trowel cut or shaped to an angle for pointing.
角抹子 勾缝用的形状成一定角度的小抹子。

two-bolt lock A lock that incorporates two bolts.
双锁头锁 有两个锁栓的锁。

two-brick wall A solid wall two bricks thick.
两砖墙 两砖厚的实心墙。

two-coat work (render and set) Plaster with an undercoat or basecoat, and topcoat or finish. The basecoat helps to level out the surface of the substrate and the finish coat ensures a smooth true surface.
两层抹灰 包括底涂层和面层的抹灰。底涂层使得基底表面变平整，面层确保其表面光滑。

two-peg test *See* COLLIMATION TEST.

准直测试 参考“准直测试”词条。

two-piece cleat A two-piece *PIPE RING.
两件式管夹 两件式的喉箍。

two-pipe system (dual system, dual-pipe system) An above-ground drainage system that comprises two discharge pipes; one that discharges the soil, and another that discharges the waste. *See also* ONE-PIPE SYSTEM and SINGLE-STACK SYSTEM.
双管排污系统 地上排水系统，由双排水管组成；一条排放污水，另一条排放废水。另请参考“单管排污系统”和“单立管系统”词条。

two-stage alarm system A type of fire alarm system that activates an alarm in the area in which a fire is detected, and at the same time activates an alert in the remainder of the building.
两级报警系统 火灾报警系统，当在一个区域内检测到火灾时，触发该区域的警报，在建筑物其他区域也同时发出警报。

two-stage tendering First tender is submitted with approximate quantities; once accepted negotiations commence on the final price and quantities.
两阶段招标 第一阶段招标是提交大概的工程量；如中标后即开始最终的价格和工程量谈判。

two-tailed test A statistical test in which the distribution of the *null hypothesis is considered by looking for any change in the parameter.
双尾检验 在寻找参数中的任意变量时所参考的原假设分布的一种统计方法。

two-way slab A slab that spans, and hence has reinforcement, in two directions.
双向板 向两个方向跨的板，因此也在两个方向有支撑。

two-way switch A switch that allows lights to be switched on from two separate locations. For instance, at either end of a corridor or stairway.
双路开关 可从两个不同位置开灯的开关。例如，走廊或楼梯的任意一端。

tying wire *See* BINDING WIRE.
系结钢丝 参考“绑扎钢丝”词条。

typical floor A repetitive floor plan in a multi-storey building.
标准楼层 在多层建筑中重复的楼层平面。

U-bend A U-shaped *trap. *See also* P-TRAP and S-TRAP.
U 型弯管 一种 U 形存水弯。另请参考“P 型存水弯”和“S 型存水弯”词条。

U-bolt U-shaped bar with threads on each end of the bar. The threads enable the U-bolt to be bolted to or fitted with a cross piece, making it suitable for securing pipes, fittings, and appliances.
U 型螺栓/U 型管卡 两端都有螺纹的 U 形杆。螺纹使 U 型螺栓能够拴牢或安装夹片，使其能够固定管道、配件和装置。

UCATT (Union of Construction, Allied Trades and Technicians) A trade union which pursues the rights of workers in construction and allied industries.
UCATT（建筑、协作专业和技工联盟） 为建筑及相关行业工人争取权益的专业联盟。

U-duct A ducted flue with intake and outlet at roof level.
U 型管 屋面上有进出口的管形烟道。

UF *See* UREA FORMALDEHYDE.
脲醛 参考“脲醛”词条。

U-gauge Clear glass or plastic U-tube filled with water and used for measuring the effective seal of drainage pipes. Plugs are placed at each end of a drain run. One of the plugs has two ports; one to which a U-gauge is connected, and one to which a pump is attached. Air pressure is applied to the pipe run using the pump until the water in the U-gauge rises to a set level. If the water holds at this level then the drainage run is effectively sealed.
U 型压力计 透明玻璃或塑料 U 形管子，管内有水，用于测量排水管的有效水封。排水口各端有塞子。其中一个塞子有两个端口；一个连接 U 型压力计，另一个连接泵。使用泵向管道内输入气压，直至压力计内的水上升至指定的水平。如果水保持在该水平，则排水管是被有效密封的。

U-liner A high-density polyethylene pipe, which has been deformed during manufacture into a U-shape so that it can be inserted into existing pipes that are damaged. It can then be remoulded into a circle cross-section within the existing pipe. The result is a structurally sound pipe within an old damaged pipe, and the need for excavation is eliminated.
U 型衬管 高密度聚乙烯管道，在生产过程中加工成 U 型，这样就可以插入现有的受损管道中。然后可以在原有管道中变成圆形截面管。这样就可以在旧的受损管道内形成结构结实的管道，而无须开挖。

ullage 1. The volume of empty space above a liquid in a container. 2. The volume of lost liquid from a container due to evaporation or leakage.
1. 空距 容器内液体上方未装满的空间。**2. 缺量** 由于蒸发或泄露造成的容器中流失的液体容量。

ultimate tensile strength (UTS) *See* TENSILE STRENGTH.
极限抗拉强度（UTS） 参考“抗拉强度”词条。

ultra-filtration A membrane filtration technique in which hydrostatic pressure is used to force a liquid through a very fine filter.
超滤 利用液体静压力迫使液体通过非常细密的滤器膜过滤技术。

ultra-lightweight aggregate *See* AGGREGATE.
超轻骨料 参考“骨料”词条。

ultrasonic Very high sound frequencies above the human audio range, which are about 20 kilohertz.
超声波 超过人类听力范围的（频率）非常高的声波，大约为 20 千赫。

ultra very high-strength cement A special type of cement material that imparts very high strength, for example, in high-strength concrete.
超高强度水泥 一种特殊的水泥材料，例如在高强混凝土（使用的水泥）。

ultraviolet (UV) An electromagnetic wave of light that is emitted by the sun and is invisible to the human eye. Though the atmosphere blocks most of the UV radiation, some gets through and it can be potentially harmful for many materials, for example, polymer materials degrade under exposure from ultraviolet radiation thus requiring protection.
紫外线（UV） 由太阳发出且人肉眼不可见的一种光线电磁波。虽然大气能阻挡大部分的紫外线辐射，但仍有部分能穿透并能对很多材料造成损害，例如高分子聚合物材料暴露在紫外线辐射下会降解，因此需要进行保护。

ultraviolet radiation (UV radiation) Electromagnetic radiation, with wavelengths from about 5 to 400 nm, that can cause degradation of certain materials.
紫外线辐射 电磁波辐射，波长在 5 到 400nm 之间，能导致某些材料降解。

umber A ferric oxide material usually utilized as a *pigment.

棕土 一种氧化铁材料，通常作为颜料使用。

unbonded Something that has failed to stick or has become unstuck.
无黏结 无法黏结或变得不黏的东西。

unbonded screed Improper application of * screed resulting in debonding of the applied layer.
剥离的砂浆层 由于不正确涂抹砂浆层导致已涂抹的砂浆层脱粘。

unburnt brick * Brick material that has not been exposed to very high temperatures for a prolonged period during the manufacturing process.
欠火砖 在生产过程中没有经过长时间高温焙烧的砖块。

uncased fan A * fan where the rotating blades are not enclosed within a case.
无壳风机 旋转叶片没有装在壳内的风机。

uncased steelwork Structural steelwork that is left exposed and not encased.
外露钢结构 外露没有封装的结构钢。

unconfined Not restricted, for example an unconfined triaxial compression test, where zero lateral pressure is applied.
不受限的 不受限制的，例如不受限三轴压缩试验，其施加侧压力为零。

uncoursed Stone walling that is not laid to a set course.
乱砌的 不按层铺设的石墙。

uncoursed rubble (broken range ashlar) Stone of random size that is not laid to set courses.
乱砌毛石 不按层铺设的任意尺寸的石块。

uncovering The instruction and action to expose work that is suspected of being defective. Where an engineer or * clerk of works responsible for checking the standard of construction suspects that there is defective work, they may give an instruction to remove finishes, coverings, or other materials so that the works can be properly inspected. Where no defects are found, then the costs and delays associated with exposing the structure can normally be reclaimed.
重新打开检查 对疑似有瑕疵的工程打开检查的指示和行动。当负责检查建筑标准的工程师或监理员怀疑工程有瑕疵时，可以下指示移除饰面、覆盖层或其他材料以便检查工程。如果没有发现瑕疵，那么与此相关的费用和延误时间可要求相关部门赔偿。

undercloak 1. Tiles or compressed cement board laid at the * verge of a * roof and positioned along the top of the wall so that the ridge tiles slope slightly inwards, thus preventing water dripping over the edge of the roof. 2. The lower sheet metal of a drip, roll, or seam.
1. 檐口卷边层 铺设在屋顶山墙檐口和墙壁顶部的砖瓦或压缩水泥板，这样屋脊瓦就可稍微向内倾斜，防止水滴落在屋檐内。**2. 垫层挡水板** 滴水槽、绲边或咬合接缝较低处的金属片。

undercoat The penultimate coating applied in the painting process. The undercoat is applied prior to the final top coat. The function of the undercoat is to provide a smooth surface, good covering power, and good adhesion for the finishing coat. Undercoats usually contain a large amount of pigment to give good covering (hiding) power, and most are based on * alkyd resins or * acrylic emulsions. Undercoats and * primers will not survive exposure well without the top coat.
底涂层 在涂漆过程中涂抹的倒数第二层涂层。底涂层涂在最终的表面涂层之前。底涂层的作用是为面层提供一个光滑的表面、良好的遮盖力和良好的黏结力。底涂层通常含有大量的颜料以提供良好的遮盖力，大部分颜料是醇酸树脂或丙烯酸乳液。如果没有面层，底涂层和底漆则不能外露。

under cooling A procedure whereby a metal is cooled from an elevated temperature (below its melting point) to achieve the desired material chemistry. This procedure is commonly used in the manufacture of many commercial metals, such as steels.
过冷 金属经高温（低于其熔点）冷却以获得所需材料的力学性能。该过程通常用于生产金属，例如钢等。

undercroft A cellar or underground storage; it can also be used to describe an open area at the base of a building (street level) that is covered by the first floor of the building.
穹窿顶地下室 地窖或地下储存室；亦可以用于描述建筑底层（临街）处并被第一层覆盖的开阔场地。

undercuring A material that has not achieved full * curing.
欠熟 没有进行充分养护处理的材料。

undercutting The process of trimming the bottom of a door leaf while it is still hung.
底切 当门悬挂时，修整门扇底部的过程。

under-eaves course The lower course of tiles at the eaves of the roof.
檐口瓦层 屋顶檐口的较低层瓦层。

underfelt Underlay placed under carpets.
地毯垫毡 地毯下的衬垫。

underfired brick A type of brick that has not been subjected to the full * firing process during manufacture. This results in lower strength of the material.
欠火砖 一种在生产过程中没有经过完全烧结的砖块。这使得材料的强度降低。

underfloor heating (floor heating, heated floor) A type of central heating system comprising a series of electric cables (dry systems), or water pipes (wet systems) that are embedded in or fixed below the floor. In wet systems, water is circulated through the pipes at between 30 – 60℃, resulting in a floor surface temperature of around 25 – 28℃. High surface temperatures should be avoided (≤29℃) to avoid discomfort to the occupants. Surface temperatures must also be compatible with the floor covering. Most types of floor covering can be used with underfloor heating, including ceramic tiles, carpets, vinyl, laminate flooring, and wood flooring.
地暖设施（地板采暖） 一种中央供暖系统，由埋设或安装在地板下的一系列电缆（干式系统）或水管（湿式系统）组成。在湿式系统中，水在管道中循环流通，水的温度为30—60℃，使得地板表面温度保持在25—28℃。应避免过高的表面温度（≤29℃），以防止住户感觉不舒适。表面温度必须与地板覆盖层相容。大部分地板覆盖层可与地暖一起使用，包括瓷砖、地毯、乙烯基地板、复合地板和木地板。

underfloor space The void that is created between the * suspended floor and the structure that is in contact with the ground.
地板下的空间 架空地板和与地面接触的结构体之间的空间。

underground Below the surface of the ground.
地下的 地面以下的。

underground services Those services that are buried underground.
地下设施 埋设在地下的设施。

U

underground stop valve A * stop valve buried underground that is accessed via a surface box.
地下截止阀 埋设在地下的截止阀，通过地面操纵箱操作。

underlay Any sheet material that is placed under another material. Resilient sheet material can be placed underneath carpets to offer extra comfort or sheet-felt, such as * sarking felt, can be placed under tiles, to prevent the wind and any moisture penetrating into the roof space. Both could be classed as underlay materials.
衬底 铺设在其他材料底下的材料。弹性板材可以铺设在地毯下以提供舒适感或毛毡感，例如衬垫油毡可以铺放在瓦块下，以防止风和潮气渗入屋面。这些都可以归类为衬底材料。

underpinning The addition of extra foundations underneath existing foundations to provide extra support. The underpinned foundations are provided where the building loads are to be increased or where the existing foundations have started to settle and are no longer able to provide the support.
基础托换 在现有基础的下部增加新的基础以提供额外支承。当建筑荷载增加或现有基础开始沉降，不能再提供支承时，应进行托换基础。

under-reaming A process where the bottom of a *bored pile is made wider than the shaft diameter to increase end-bearing resistance and/or prevent uplift.
扩底 将钻孔桩的底部扩宽至比桩身直径宽，以增加端承阻力和/或防止隆起的过程。

under-ridge tile (top course tile) The course of tiles located directly under the *ridge.
平脊瓦（顶层瓦） 直接位于屋脊下的瓦层。

under-tile An ***eaves tile** or an *under-ridge tile.
底瓦 檐口瓦或平脊瓦。

undertone The presence of another or different colour on a white background that can also be viewed under the presence of transmitted light.
底彩 在白色背景下透射光下可见的其他颜色或不同颜色的显现。

under-vibration Failure to vibrate concrete sufficiently to ensure the concrete compacts, fills the formwork, excludes air bubbles, and achieves its full design strength.
欠振 未能充分振捣混凝土以确保混凝土振实、填充模板、排除气泡和达到充分的设计强度。

undisturbed sample A sample of soil that represents in-situ conditions, for example, it has not been allowed to be removed or dried out. Such samples are collected during a site investigation and tested using laboratory equipment to determine stress strength parameters, amount of consolidation, and density.
原状土试样 可展现现场条件的土壤样本，例如不能移除或干燥的土壤。在现场勘查阶段收集这些样本，并且使用实验室设备进行测试，以确定应力－强度参数、固结量和密度。

undrained test (immediate test, quick test) A shear strength test where drainage of water is prevented, such that no dissipation of pore pressure is possible. Undrained parameters c_u and ϕ_u are obtained, which are applicable to immediate bearing-capacity issues regarding foundations placed on saturated clays.
不排水试验 阻止排水的剪切强度试验，这样就不会消耗空隙水压力。获得不排水参数 c_u 和 ϕ_u，适用于位于饱和黏土上的基础的即时承载力。

undressed timber (unwrought timber) Unplaned timber that is left in its rough cut or sawn state.
未刨光木材（未加工木材） 未刨光的木材，仍是粗削或粗锯状态。

undulation A continuous to and fro movement.
波动 持续的往返运动。

unequal angle An *angle section that has unequal legs.
不等边角钢 有不等边的角钢。

uneven grain (uneven texture) Timber grain that shows considerable contrast and difference between the summer and spring growth rings.
不均匀纹理 在夏天和春天年轮间有鲜明反差和不同的木材纹理。

unfixed material Materials and goods ordered and on-site ready for construction that have yet to be used or installed. Partial payment for the unfixed materials is made available within *interim certificates.
到场材料 已下单并且已在现场等待使用或安装的材料或货物。可在中期付款通知书时期内支付部分到场材料款项。

unframed door *See* MATCHBOARD DOOR.
无框门 参考“企口板门”词条。

ungauged lime plaster Sand and lime plaster. Traditional plaster without gypsum tends to have a slightly lower strength than gypsum plaster, but has the benefit that it can be removed and reused at a later date. Such plasters are increasingly being used in sustainable building projects.
无石膏灰泥 沙子和石灰泥。传统的灰泥没有石膏，强度略低于石膏灰泥，但是有一个优点，就是便于日后移除并且重新使用。在可持续性建筑项目中越来越多地使用这种灰泥。

unhanging Lifting a door off its hinges or removing the door from the doorframe.
解扣 从铰链上取下门或从门框中拆除门。

uniaxial Relating to one axis, for example, the reflection of light in a single direction or a force along a tie in a truss.
单轴的 跟一轴相关的，例如光向单一方向反射或桁架杆上的力。

unidirectional Relating to one direction.
单向的 与一个方向相关的。

uniform Of the same consistency/size throughout.
均匀的 整体相同的一致性/尺寸。

Uniform Building Code (UBC) Australian equivalent to the * Building Regulations.
统一建筑规范（UBC） 澳大利亚等同于建筑条例的规范。

uniform corrosion Chemical degradation affecting the whole surface of a material generally. Uniform corrosion is highly prevalent in metals, especially ferrous alloys, and is exemplified by the rusting of steel. Other examples are the formation of a green patina on copper and the tarnishing of silver. When designing a component for a particular environment it is usual and relatively straightforward to adopt measures to prevent uniform corrosion. These can include protective coatings (galvanization), inhibitors, cathodic protection, and appropriate choice of materials.
均匀腐蚀 影响材料整个表面的化学降解。均匀腐蚀在金属中非常常见，尤其是铁合金，常见例子是铁的生锈。其他例子还有铜锈绿的形成和银的锈蚀。当为某一特定环境设计组件时，通常相对明确的是要采取防止均匀腐蚀的措施，包括保护性涂层（镀锌）、缓蚀剂、阴极保护以及选择适当的材料。

uniformly distributed load (UDL) A load that is distributed over a portion or the full length of a beam and expressed as kN/m.
均布荷载（UDL） 分布在梁部分或全长的荷载，单位为 kN/m。

uninterruptible power supply (UPS) A battery-operated backup power source that is designed to provide power to specific devices for a short period of time during a power failure.
不间断电源 电池供电的备用电源，用于在断电期间短时间内为指定设备提供电源。

union A type of threaded pipe joint comprising a male and female end that are joined together and then the joint is sealed using a nut. Enables pipes to be disconnected easily.
活接头 一种螺纹管接头，包括连接在一起的公头和母头，然后用螺母封住接头。使管道能够轻易断开。

union bend A pipe bend that incorporates a * union on one end.
活接弯头 在一端有个活接头的管道弯头。

unit 1. A discrete part into which something can be divided. 2. A standard measurement denoting a quantity, e.g. metre, kilogram, degree, volt, hour. 3. A single element, component, or module. 4. A specified proportion of work.
1. 零件 可拆卸的离散零件。**2. 单位** 表示标准量的名称，例如米、千克、度、伏、时。**3. 单元** 单一元件、组件或模块。**4. 单位** 工程的某一部分。

unit air conditioner (packaged unit) An air-conditioning unit where all of the components are housed within a self-contained unit.
单元式空调机组 所有组件都封装在一个独立装置内的空调机组。

unit costing A price given per metre, square metre, hour, or item. Listed against standard descriptions in the bill of quantities, prices may be obtained from previous site outputs or from a price index.
单位成本 按每米、每平方米、每小时或每项所定的价格。在工程量清单中列为标准的一部分，价格可从原先的现场获得或指导价得出。

unit heater (air heater) A self-contained space heating system that circulates warm air using a fan or fans.

Used to heat non-domestic buildings such as factories and warehouses.
暖风机（空气加热器） 独立的空间供暖系统，使用风机或风机组循环暖空气。用于给非住宅建筑物（例如工厂和仓库）供暖。

unit of bond The pattern of bricks within a course that repeats itself.
砌砖图案 同一皮砌层内重复使用的砖块样式。

unit price (rate) *See* UNIT COSTING.
单价 参考“单位成本”词条。

universal section A steel section manufactured to a standard size. A universal beam is an *I-sectionand a universal column is an *H-section.
通用型材 按标准尺寸生产的型钢。通用钢梁是工字形型材，通用立柱是 H 形型材。

universal set A ***set** that contains all objects, elements, processes, events, or categories.
通用集合 包含所有物体、元素、过程、事件和种类的集合。

Universal Transverse Mercator (UTM) A 60-zones grid-based coordinate system, measured in metres, that is used to define locations on the flattened surface of the earth.
通用横轴墨卡托（UTM） 分成 60 条带的网格坐标系统，以米来量度，用于在地球平面表面上定位。

unloading The removal of load.
卸载 卸除荷载。

unplasticized polyvinyl chloride [PVC-U (formerly uPVC), rigid PVC] *See* POLYVINL CHLORIDE (PVC).
未增塑聚氯乙烯 参考“聚氯乙烯”词条。

unreinforced masonry (URM) Buildings or houses whose walls have not been reinforced with steel bars/rods. These walls can be load-or non-load-bearing.
无筋砌体（URM） 墙壁没有经过钢筋/钢条加固的建筑物或房子。这些墙壁可以是承重型，也可以是非承重型。

Unsafe theorem *See* UPPER BOUND THEOREM.
上限定理 参考“上限定理”词条。

unsaturated Not completely full of water, for example, where the pores in the ground have both air and water, as in *vadose water; *See also* SATURATED.
不饱和的 没有完全充满水，例如，地面孔隙有水也有空气，如同渗流水；另请参考“饱和的”词条。

unsocketed pipe (socketless pipe, US hubless joint pipe) A plain-ended pipe that is joined together using a ***sleeve coupling**.
无承口管 使用套筒接头连接在一起的平口管。

unsound Structurally unstable or not satisfactory.
不牢固的 结构上不稳定或不符合要求。

unsound knot (rotten knot) A knot that is softer than the surrounding wood.
松软节（腐朽节） 比周围木料软的木节。

unvented system (mains pressure system) *See* SEALED SYSTEM.
闭式系统（承压式系统） 参考“封闭式水箱系统”词条。

unwrought timber Undressed lumber.
未刨光木材 毛材。

up-and-over door (overhead door, swing-up door) A large door that swings outwards and upwards to

open. Commonly used for garages.
上翻门（卷帘门，上开门） 向外和向上翻转以打开的大门。通常在车库中使用。

updating (rescheduling) The monitoring of progress against the project programme and making necessary changes to ensure that the delivery comes in on time and within budget.
进度更新（重定进度） 根据项目进度表监管进度，并且可做必要更改以确保准时交付并且控制在预算范围内。

uplift restraint strap Galvanized steel strap attached to the *wall and the *wall plate, or roof *truss to prevent the wind lifting the roof.
抗风拉件 系于墙壁、梁垫或屋架以防止风吹翻屋面的镀锌钢带。

uplighter A ***luminaire** that is designed to project its light upwards.
上射灯 灯光向上投射的灯具。

Upper Bound theorem (Unsafe theorem, plastic collapse) The collapse mechanism cannot be less than the collapse load for a given structure and loading condition.
上限定理（塑性破坏） 破坏机构不能少于给定结构和荷载条件下的破坏荷载。

upper chord The uppermost member in a *truss.
上弦杆 桁架最上面的构件。

upper floor Floors above ground level.
上层楼面 地面以上的楼面。

upset The forming of metals with a swage.
镦锻 用冲模制金属。

upside-down roof *See* INVERTED ROOF.
倒置式屋面 参考"倒置式屋面"词条。

upstand (upturn) Upward projection of a roof that is usually designed to prevent rainwater flowing over the edge of the roof. The upward projection may also be used at the edge of floors, roofs, and stairs as a safety barrier, or to offer protection.
向上翻（上翘） 屋面向上凸出，通常设计用以防止雨水流入屋面边缘。向上翻可用于地板、屋面和楼梯的边缘，作为安全屏障或提供防护。

upstand beam A floor beam that projects at its ends above floor level.
上翻梁 位于楼板上面，末端从楼板凸出的楼板梁。

upstand flashing The under part of lead, copper, or aluminium flashing that is turned up against a wall or an abutment. To fully waterproof the abutment, additional pieces of flashing are embedded in the wall and turned down so that it covers the upturn.
上翻泛水 向上翻至墙壁或连接部位的铅、铜或铝泛水的下部。为使连接部位完全防水，在墙壁内埋设附加泛水板，泛水板向下翻折，这样就可使遮盖上翘。

upsurge A rapid rise in something, for instance, an increase in water flow.
剧增 某物急剧上升，例如，水流的上升。

uptake 1. An upward air current. 2. A chimney vent. 3. The absorption of something into a living organism.
1. 上升气流。 2. 烟囱排气道。 3. 摄入 活体吸收某物。

uptime Time during which plant is available for use, different from downtime where the plant is standing idle because it cannot be used.
（机器）正常运行时间 设备可使用的时间，与设备停止运行时间不同，停止运行时间是指因为设备无法使用而闲置的时间。

uPVC *See* POLYVINYL CHLORIDE (PVC).

未加增塑剂的聚氯乙烯 参考“聚氯乙烯”词条。

upward flow filter A filter in which the direction of water flow is upwards so as to reduce the amount of backwashing.
上向流过滤 水流方向向上的过滤器，这样可减少反冲洗。

upwelling A process of rising up from a lower depth, for example, where colder water rises from the ocean floor to the surface.
上升流 从较深处向上流的过程，例如，较冷的水从海底上涌至表面。

upwind Against the wind, facing the direction in which it is blowing. *See also* WINDWARD.
逆风 顶风，面向风吹来的方向。另请参考“顺风的”词条。

urban The built-up areas such as towns and cities.
城市 已建筑的区域，如城镇和城市。

urbanization 1. The changing of an area, developing it into a town or city. 2. The accommodation of attitude, culture, and other changes that enable city living.
1. 城市化 区域改变，开发成城镇或城市。**2. 都市化** 迎合都市生活的观念、文化和其他的改变。

urea formaldehyde (UF) A *thermosetting plastic (polymer), made from urea and formaldehyde. Urea formaldehyde is a widely used polymer especially in construction for electrical switch plates and electrical sockets (white plastic plates), and is highly prevalent in dwellings and municipal buildings. UF is characterized by desirable mechanical and physical properties, —*tensile strength, low water absorption, and ***hardness**. Urea formaldehyde was previously used in cavity walls, however, its use was discontinued in the 1980s due to the emission of toxic formaldehyde. Currently its main use is for electrical sockets, however, UF resin is used in adhesives, finishes, and moulded objects.
脲醛（UF） 一种热固性塑料（聚合物），由尿素和甲醛构成。脲醛是一种用途广泛的聚合物，特别在建筑行业，用于开关面板和电源插座（白色塑料盖板），在住宅和市政建设中也非常常见。脲醛有良好的机械和物理特性——高抗拉强度、低吸水性和硬度。脲醛以前用于夹心墙中，但因为会释放有毒甲醛而在20世纪80年代被停用。现在其主要用于电源插座，但脲醛树脂也用于胶粘、饰面和模制器件上。

urea formaldehyde foam Low-density *urea formaldehyde polymer in foam form typified by the presence of porosity or holes in its internal structure. Urea formaldehyde foam was previously used in cavity wall insulation but was discontinued due to health and safety concerns. *See also* UREA FORMALDEHYDE.
脲醛泡沫 低密度的呈泡沫状的脲醛聚合物，特点是其多孔或有孔洞的内部结构。脲醛泡沫原先用于空心墙隔热，但是出于健康和安全考虑已被停止使用。另请参考“脲醛”词条。

urethane Carbamic acid, *polymerized to form *polyurethane.
尿烷 氨基甲酸，聚合形成聚氨酯。

urinal Bowl or gully used in non-domestic toilets to capture urine. Most commonly used in male toilets although similar female appliances are being developed.
小便池 非住宅厕所内用于接住尿液的池或沟。常用于男厕内，不过现在正在开发类似的女用器具。

usable life Length of time indicative of the durability of a material, object, or component.
使用寿命 指材料、物体或组件耐用性的使用时间。

use factor The number or diversity of uses an item can facilitate.
利用率 物件能使用的数目和多样性。

U-tie A heavy wall tie.
拉结杆 重型墙拉结。

utilities 1. Services and appliances 2. Service providers that were traditionally within the public sector, such as water, gas, electricity, telecoms, and sewers.
1. 设施和设备。 2. 公共服务 传统上在公共部门，例如水、燃气、电、电信和排污等范围内的服务供应商。

utilization factor (*U*) The amount of light given off by a light that reaches the working plane it is designed to illuminate. The surface to be lighted is called the reference plane. The light on the surface is measured in lux.
利用系数（*U*） 灯具发出的到达其欲照明的工作平面的灯光量。照亮的表面称为参考平面。表面的灯光以勒克斯计量。

UV (ultraviolet) *See* ULTRAVIOLET.
紫外线 参考“紫外线”词条。

***U* value (air-to-air heat transmission coefficient)** *See* THERMAL TRANSMITTANCE.
***U* 值（空气对空气热传导系数）** 参考“传热系数”词条。

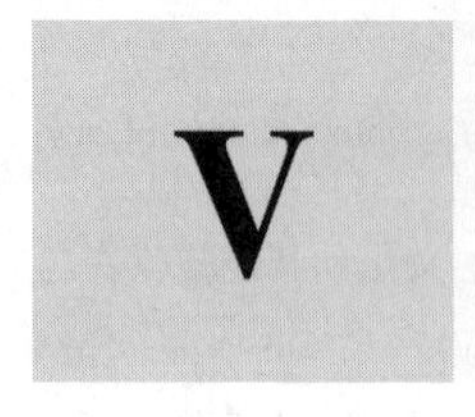

vacuum An enclosed space that is empty of all matter.
真空 没有任何物质的密闭空间。

vacuum cleaning plant (central vacuum cleaner, central vacuum system) A vacuum cleaning system that is built into a building and comprises a lightweight hose that can be connected into a series of outlets distributed throughout the building. Pipework concealed within the building is then used to transport any picked up material from the hose to a central externally vented vacuum.
真空吸尘装置（中央真空吸尘器，中央真空吸尘系统） 建于建筑物内的真空吸尘系统，包括可连接至分布在整个建筑物内的多个插口的轻量软管。埋设在建筑物内的管道用于将软管吸入的物件运至向外排放的中央真空系统。

vacuum concrete A type of concrete containing a vacuum mat to remove excess water in order to achieve higher compressive strength.
真空混凝土 一种混凝土，用真空模板以去除多余水分以获得更大的抗压强度。

vacuum mat A metal screen used to suck out excess water and air from concrete.
真空模板 用于吸出混凝土中多余水分和空气的金属网。

vacuum-pressure impregnation Saturation of timber by extracting the air and flooding with preservative.
真空压力浸渍 用防腐剂排除空气和挥发物而使木材饱和。

vacuum pump A device that removes air or vapour molecules from a sealed volume of space in order to maintain a pressure below atmospheric, to cause a partial vacuum.
真空泵 排除密封容器内的空气或蒸汽分子以保持低于大气压的压强，形成局部真空的设备。

valley The internal angle created when two separate pitched roofs intersect.
屋面斜沟 两个倾斜屋面相交所形成的内角。

valley board A board used to support a *valley gutter.
天沟底板 用于支撑排水斜沟的底板。

valley gutter A *gutter used to drain water from a *valley.
排水斜沟 用于排掉屋面斜沟内的水的排水沟。

valley jack (US) A *jack rafter.
屋面斜沟短椽 一种小椽。

valley rafter A rafter used to form a * valley.
屋面斜沟椽条 用于形成屋面斜沟的椽条。

valley tile A specially shaped roof tile used to construct a tiled valley.
斜沟瓦 用于构建贴瓦斜沟的特殊形状的屋面瓦。

value 1. The degree to which an item, product, service, or event meets the desired performance criteria required by the client. 2. The degree of lightness of a colour.
1. 价值 物体、产品、服务或事件达到客户所要求的性能标准的程度。**2. 明度** 色彩的明亮程度。

Value Added Tax (VAT) The government levy added to the cost of goods and services to support governance of the country. VAT rates for new build, demolition, and refurbishment often vary.
增值税（VAT） 国家为支撑政府运作对商品和服务所征收的税赋。新建、拆除和翻新建筑的增值税率经常会改变。

value cost contract A cost reimbursement contract where the contractor is entitled to a higher fee if the final costs are less than expected.
价值成本合同 一种成本补偿合同，如果最终成本小于预计的成本，将给予承包商更高的费用。

value management The process of determining the function and performance criteria required, and identifying products and services at the lowest cost that meets the criteria.
价值管理 确定所要求的功能和性能标准，并且确认产品和服务以最低价格达到标准的过程。

valve A fitting that is capable of stopping or reducing the flow of gases or liquids through a pipe.
阀门 能够停止或减小管道中气体或液体流动的配件。

valve boss The raised area around a valve.
阀座 阀门上凸出的区域。

vane test A simple in-situ test to determine the * shear strength (cohesion) of saturated clay at the bottom of a trial pit or * borehole. The vane is driven or pushed into the soil and a measured torque applied to it until it rotates. The failure surface is the curved surface plus the flat ends of the cylinder of soil whose diameter and height are that of the vane.
十字板剪切试验 一个测定探坑或钻孔底部饱和黏土剪切强度（黏聚力）的简单现场试验。十字板插入或压入土壤中，施加实测扭矩直至十字板旋转。破裂面是弯曲表面加土壤圆柱体的平端，圆柱体的直径和高度等于十字板的直径和高度。

vanity basin A bowl-shaped basin recessed into a * vanity top for washing the face and hands.
台式洗脸盆 嵌入台面板，用于洗脸和洗手的碗状洗脸槽。

vanity cabinet A mirror-fronted cupboard installed in a bathroom.
浴柜 安装在浴室内的前面装有镜子的柜子。

vanity top A worktop installed in a bathroom.
卫生间台面 安装在浴室内的台面。

vanity unit A floor-mounted bathroom cupboard that has a * vanity top and * vanity basin.
组合式盥洗台 落地式浴柜，带有一个卫生间台面和台式洗脸盆。

vaporizing liquids Liquids that are stored under pressure which vapourize when the pressure is released.
汽化式液体 当释放压力时，在压力下汽化储存的液体。

vapour barrier (vapour control layer) A material that resists water vapour transmission, and is usually made of polythene sheeting. Should always be installed on the warm side of an insulation layer to avoid the risk of * interstitial condensation.
隔汽层 可防止蒸汽渗透的材料，通常由聚乙烯板制成。通常安装在隔热层较暖的一边以防止缝隙冷凝。

vapour blasting A jet spraying water onto surfaces, normally used to clean or remove debris.

蒸汽喷射器 向表面喷射水的喷射器，通常用于清理或移除残渣。

vapour check A material that is similar to a * vapour barrier, but has a higher * vapour resistance.
隔气带 与隔气层相似的材料，但水蒸气阻隔性能更佳。

vapour compression cycle A type of refrigeration cycle where a refrigerant continuously changes state from a liquid to a gas, and back again. In the evaporator, the refrigerant evaporates at low temperature and pressure, absorbing heat from its surroundings resulting in cooling. The refrigerant is then compressed, increasing its pressure and temperature. The refrigerant then passes through a condenser where it condenses, releasing heat and reduces its pressure. It then passes back to the evaporator where the process is repeated.
蒸气压缩式制冷循环 制冷剂持续从液态变为气态，再从气态变成液态的制冷循环过程。在蒸发器中，制冷剂在低温低压下蒸发，从周围环境中吸热使其冷却。然后制冷剂经压缩，压力和温度升高。制冷剂经过冷凝器冷凝，释放热量，降低压强。然后又回到蒸发器，循环往复。

vapour control layer *See* VAPOUR BARRIER.
隔气层 参考“隔气层”词条。

vapour resistance A measure of the ease with which a given thickness of a material will resist vapour transmission. Calculated by multiplying the * vapour resistivity of a material by the thickness of that material in metres. Measured in meganewton seconds per gram (MNs/g).
透湿性 对已知材料厚度阻隔水蒸气渗透的容易性的测量。通过材料的隔汽率乘以材料厚度（以米为单位）计算得出。以兆牛顿秒每克为单位（MNs/g）。

vapour resistivity A measure of the ease with which a material will resist vapour transmission. It is a characteristic of the material. Measured in meganewton seconds per gram per metre (MNs/gm).
透湿率 对材料阻隔水蒸气渗透的测量。它是材料的一种特性。以兆牛顿秒每克每米为单位（MNs/gm）。

variable air volume box (VAV box) A type of * air terminal unit that uses dampers and thermostats to control the volume of conditioned air entering a space. Reheat coils can be incorporated into the VAV box to provide additional heating if necessary.
变风量风箱（变风箱） 使用风门和恒温器以控制进入空间的经调节的空气量的空气终端设备。如有需要，变风量风箱中可装入再热线圈以提供额外加热。

variable air volume system (VAV system) A type of centralized all-air air-conditioning system that supplies conditioned air to the space through a variable air volume (VAV) box. These are capable of providing zone control of the temperature from the central air-conditioning plant. However, humidity is still controlled centrally, so humidity conditions may vary widely between different zones. In a VAV system, air is supplied at a constant temperature and relative humidity to all parts of the building. Different cooling requirements are achieved by varying the volume of air supplied to each zone from the central plant. Control of the air volume is performed by a thermostatically controlled VAV box, which is fitted in the air supply ducts. VAV systems are common in new open-plan buildings.
变风量系统（VAV 系统） 集中式全空气空调系统，通过变风量风箱向空间内提供经调节的空气。该系统能够提供来自中央空调装置的空气并能进行温度分区控制。但是，湿度依然是中央控制，这样不同区域的湿度可能大不相同。在 VAV 系统中，空气是以恒定的温度和相对湿度提供给建筑物内的所有区域。不同的冷却需求可通过改变中央设备提供给各区域的风量达成。通过安装在送风管内的恒温控制的变风量风箱控制风量。VAV 系统通常用于开敞式的建筑物。

variation (change) A modification or alteration to that agreed under the terms of the contract, or that described in the bill of quantities.
变更（更改） 在合同条款下或在工程量清单描述下协定的修改或变更。

variation notice Request from the contractor to the client's representative that a written * variation order is required.
变更通知 承包商要求业主代表给出书面变更令。

variation order A written order from the client's representative authorizing a variation under the terms of the contract.
变更令 业主代表给出的书面变更令，批准合同条款下的变更。

varnish A transparent, glossy, and hard coating applied to timber and other materials, mainly for gloss and protection.
清漆 透明的、有光泽的硬涂层，涂于木材或其他材料上，主要是为有光泽度和防护性。

vault 1. An arched ceiling or roof constructed from masonry. 2. A room, usually underground, that is used to store valuables. 3. An underground chamber used to bury the dead.
1. 拱顶 砌体砌成的拱形顶棚或屋顶。**2. 地下室** 一种房间，通常位于地下，用于储存贵重物品。**3. 墓穴** 用于埋葬死者的地下洞穴。

vault light *See* PAVEMENT LIGHT.
地下室采光窗 参考“路面采光窗”词条。

VB consistometer test A laboratory test designed to determine the workability of concrete. The test measures the time in seconds for wet concrete to fill a standard cone.
维勃稠度试验 用于确定混凝土和易性的实验室试验。试验测量湿混凝土填充标准的坍落度筒所需的时间，以秒为单位。

V-belt (vee-belt) A flexible belt with a cross-section V-shape. The belt sits in V-shaped pulleys helping to ensure good contact between the belt and the pulley.
三角带（V型皮带） 横截面为V形的弹性带子。三角带用于V形滑轮上，以确保带子和滑轮间良好接触。

V-braced frame A structural frame with diagonal braces that provide resistance to lateral forces.
V形支撑框架 有斜撑的结构框架，抗侧向力。

vee joint A joint made by cutting a groove into the end of the timber resembling the letter "V", the male joint is also cut to the same profile to fit into the V. The nature of the joint masks the effects of shrinkage on the joint.
V形接头 通过在木材末端切出形似字母“V”的凹槽而形成的接头，公头也切成相同剖面以接入V形凹槽。接头的性质可掩盖接头的收缩效果。

velocity The rate at which a body moves: distance divided by time.
速率 物体移动的速度：距离除以时间。

velocity head A static head of pressure from a fluid which would produce a certain flow rate. Also defined as:

$$v^2/2g$$

where v = velocity and g = acceleration due to gravity.
速度水头 产生一定流速的流水压力的静水头。定义为：

$$v^2/2g$$

其中v为速度，g 为重力加速度。

veneer Thin slices of wood glued onto panels.
薄木贴面/木皮 胶粘至面板的薄木板。

veneered wall A wall with a facing that is not chemically bonded to it, but is mechanically bonded by wall ties; for example, a brick facing on a timber frame.
干挂墙 不是化学黏合，而是通过墙拉结机械接合的饰面的墙壁；例如，木材框架上的砌砖面层。

veneer finish plaster A coating applied to provide a hard, protective, abrasion-free surface.
薄面抹灰 涂抹的涂层，用以提供坚硬的防腐蚀保护性表面。

veneering The process of affixing two layers of a substance (usually wood), together.
薄片胶合 将两层物质（通常是木材）粘贴在一起的过程。

veneer tie A wall tie for holding a skin of masonry onto a timber frame.
干挂件 将砌体饰面固定在木材龙骨的墙拉结。

Venetian door A central door, often arched, and flanked by flat, square windows on either side.
威尼斯联窗门 中开门，通常有顶拱，两侧有方形平面窗。

Venetian window A central window, often arched, flanked by flat, square windows on either side.
威尼斯式窗 中开窗户，通常有顶拱，两侧有方形平面窗

vent 1. A pipe or opening that allows gases to enter a space or be discharged to the outside. 2. The process of discharging gases to the outside.
1. 通风口 将气体排入一个空间或排放到外面的管道或开口。**2. 排气** 将气体排放到外面的过程。

ventilated lobby A naturally ventilated *fire lobby.
通风前室 天然通风的消防前室。

ventilating brick *See* AIR BRICK.
通风砖 参考“空心砖”词条。

ventilation The process of supplying or removing air, either by natural or mechanical means, to or from an enclosed space.
通风 通过自然或机械手段，向密闭空间内送风或排风的过程。

ventilation branch *See* BRANCH VENT.
通气支管 参考“通风支管”词条。

ventilation duct A *duct that carries air for *ventilation.
通风管 输送空气以通风的管道。

ventilation pipe *See* VENT PIPE.
通风管道 参考“排气管”词条。

ventilation rate The rate at which air within an enclosed space is removed or replaced with air from outside the space. Measured in air changes per hour (ach).
通风率 密闭空间内排出空气或用空间外空气置换的速率。以换气次数/小时（ach）为单位计量。

ventilator A device that allows air to enter into or exit an enclosed space. It may take the form of a mechanical device, such as an *extract fan, or may be a passive device, such as an *air brick.
通风设备 使空气能够进入或排出密闭空间的设备。可以是机械设备，如排气扇，也可以是被动装置，如空心砖。

vent light (night vent, ventilator) A small casement window or *fanlight, which sits above the *transom, is top-hung, and opens outwards.
通风窗（气窗、通风设备） 位于横楣上的小平开窗或腰头窗，上悬，向外打开。

vent pipe (ventilating pipe, US vent stack) 1. The top of a stack drain system that allows the foul drain smells to be discharged at a level higher than any openings (windows and doors) in the house. 2. An open discharge pipe that sits over the top of a cistern. The pipe is used to discharge steam and hot water, preventing pressure from building up within the system.
1. 排气管（通风管道，美国 通气立管） 将臭味排放到比房子内任何开口（窗户和门）高的地方的排水立管系统顶部。**2. 排水管** 水箱顶部的排放管，用于排放蒸汽和热水，以防止增强系统内的压力。

V

vent soaker *See* PIPE FLASHING.
出屋面防水套管 参考“出屋面防水套管”词条。

venture A reduction, at a particular section, in a pipe diameter, which is designed to cause a drop in pressure when a fluid or gas flows though it. Used for measuring the flow in closed pipes.
缩径 特定截面管道直径缩小，这样当液体或气体流经管道时，会造成压力下降。用于测量封闭管道

中的流量。

veranda An open-sided roofed structure providing shade to a part of a building.
边廊 侧边敞开，为建筑物局部提供遮阳的屋顶结构。

verdigris The green corrosion—oxidation—that forms on the surface of copper.
铜锈/铜绿 铜表面形成的绿色腐蚀物—氧化锈。

verge The sloping section of a pitched roof at the intersection with the *gable wall. The verge may be finished flush or may overhang the gable wall.
山墙檐口 与山墙相交处的坡屋顶的倾斜部分。檐口可与山墙齐平或挑出。

verge abutment The portion of a *gable wall that rises above the roof slope at the *verge.
山墙压顶 从屋顶坡度檐口处凸出的山墙部分。

verge board (vergeboard, verge rafter) *See* BARGE BOARD.
封檐板 参考“博风板”词条。

verge clip A metal clip attached to the *tiling batten to hold the *verge tile in place.
檐口夹 连接至挂瓦条以固定山墙檐瓦的金属夹。

verge fillet A triangular-shaped piece of timber that is fixed to the end of the *tiling battens at a *gable wall.
山墙檐口角撑 固定至山墙挂瓦条末端的三角形木材。

verge flashing (verge trim) 1. A flashing used to waterproof the verge of a built-up roof. 2. The section of a roof that projects beyond the *gable wall.
1. 檐口泛水 用于组合屋顶檐口防水的泛水板。**2. 悬山** 从山墙凸出的屋面部分。

verge tile A roof tile that is used to form a *verge. The verge tile should project over the gable wall by at least 38 mm to prevent watermarks occurring down the face of the gable wall.
山墙檐瓦 用于形成山墙檐口的檐瓦。山墙檐瓦应该至少凸出山墙 38 毫米，以防止在山墙表面出现流下的水渍。

vermiculite A hydrated silicate mineral, which expands on heating. It is used in insulation and as a medium for planting in horticulture.
蛭石 一种硅酸盐矿物，受热时会膨胀。用于保温以及在园艺种植中作基质。

vermiculite-gypsum plaster A plaster with vermiculite used as an aggregate. The vermiculite is a lightweight material that has been expanded at high temperatures, forming a light material, which when used as an aggregate in plaster can improve the fire resistance and thermal performance of the structure.
蛭石灰浆 以蛭石作骨料的灰浆。蛭石是一种轻质材料，会在高温下膨胀，形成轻型材料，当在灰浆中作为骨料时，可以提高结构的防火性能和隔热性能。

vertex The highest point of a pyramid, where the intersection of the faces meet.
顶点 金字塔的最高点，各面相交点。

vertical angle The angle formed from the intersection of a pair of lines in a vertical plane, for example, the angle measured on the vertical circle in a *theodolite.
垂直角 垂直平面上一对线条相交形成的角度，例如在经纬仪上地平经圈上测量的角度。

vertical axis wind turbine A turbine with blades that rotate about a vertical axis, perpendicular to both the direction of the wind and the surface of the ground. The prevailing feature which distinguishes the vertical from horizontal axis turbine, along with its appearance, is that the gearing and generator are normally located at the base of the turbine, which is at ground level.
垂直轴风力涡轮机 叶片绕着垂直轴旋转的涡轮机，与风的方向和地面垂直。除了外观之外，垂直轴涡轮机与水平轴涡轮机之间显著的区别还在于垂直轴涡轮机的齿轮箱和发电机通常位于涡轮机的底部，即与地面持平。

vertical control A process of using different *benchmarks to check the vertical *closing error of a levelling survey.
高程控制 使用不同基准检查水准测量中垂直闭合误差的过程。

vertical curve Typically a parabolic curve, in the vertical plane, which is used to relieve gradient changes in roads and railways.
竖曲面 通常是垂直平面内的抛物曲线，用于减小道路和铁路的坡度变化。

vertical irregularity Any form of discontinuity (with regards to stiffness, strength, geometric, and mass) in one storey of a building in relation to the other storeys. Vertical irregularities are defined FEMA 368 Section 5. 2. 3. 3 and Table 5. 2. 3. 3.
竖向不规则结构 建筑物内一层与其他层之间的任何不连续形式（关于刚度、强度、几何特征和质量）。竖向不规则在 FEMA 368 第5. 2. 3. 3 节和表5. 2. 3. 3 中有定义。

vertical lift bridge (lift bridge) A bridge where the deck moves up and down, while the rest remains horizontal.
垂直升降桥（升降吊桥） 桥面板可以上下移动，其他部分保持水平的桥。

vertical lift gate A gate that moves in a vertical direction to allow retained water through.
直升式闸门 可沿垂直方向移动以挡水的闸门。

vertical sash A vertically sliding *sash window.
竖拉窗 竖向推拉的推拉窗。

vertical shingling (hanging shingling, weather shingling) Tiles hung on a wall or tile cladding.
竖挂板瓦 悬挂在墙壁或砖瓦覆面的瓦片。

vertical shore The props or studs used to support a *needle that provides temporary support to a wall or beam.
竖向支撑 起支撑作用，为墙壁或横梁提供临时支承的临时托梁的支柱或立柱。

vertical sliding window (vertical slider) A vertical *sash window.
垂直推拉窗 垂直上下拉动的推拉窗。

vertical tiling The use of vertically hung tiles as an external cladding.
竖贴砖 作为室外覆面垂直悬挂的砖块。

vertical transport Lift, escalator, and other mechanical equipment used to provide movement from one level to another.
垂直运输 用于从一层移至另一层的电梯、升降机和其他机械设备。

vertical twist tie (strip tie) A cavity wall tie, formed by twisting galvanized steel metal. The twist provides the *drip, preventing water passing across the cavity. Steel wall ties with sharp protruding legs are considered a health and safety hazard. As the wall ties are fixed into the internal wall, the legs are left protruding as the external wall is built up; the protruding legs have been responsible for a number of injuries to workers' eyes and faces.
垂直止水拉结片 夹心墙墙拉结，由扭曲的镀锌钢金属制成。扭绞形成滴水构件，防止水流过夹心墙。有尖锐凸出支腿的钢系墙铁应考虑其健康和安全隐患。因为系墙铁固定至内墙，当外墙建成时，支腿凸出；凸出的支腿是一些工人眼睛和脸部受伤的原因。

vertical work Bill of quantities description used to describe the application of hot asphalt to vertical surfaces. It is harder and more time-consuming to trowel asphalt onto vertical surfaces than it is to horizontal surfaces.
竖向工程 工程量清单中用于描述在垂直表面涂上热沥青的描述栏。将沥青抹于垂直平面比抹于水平平面更加困难和费时。

vestibule 1. An open area or court in front of a building. 2. A small area within a building, providing a room from which doors open out into other rooms. A larger room is often called a lobby.
1. 门廊 建筑物前面的露天区域或庭院。**2. 门厅/玄关** 建筑物内的一小片区域，能让门打开进入其他房间的一个房间。大一点的房间叫大厅。

vestry An abutting or adjacent room to the church's main area in which clothes for the clergy and choristers are kept. These are called the vestments.
（教堂的）祭衣室 与教堂主区域相邻或连接的房间，用于保存神职人员或圣诗班的衣服，统称为祭衣。

vetting of tenders The reviewing of tenders to ensure all the works have been properly priced, all of those tendering qualify and are capable of performing the works, and finally, the selection of the tender that is most suitable for the work. The work may be awarded on best value or most suitable contractor.
评标 审查投标者以确保所有工程都已合理报价，投标资格和投标者具备实施工程的能力，最后，选择最适合实施工程的投标者。工程将被授予最优价或最适合的承包商。

viaduct An arched bridge carrying a road or railway. The bridge is constructed from a series of adjacent arches, and particularly tall viaducts may have arches built on top of each other.
高架桥 承载道路或铁路的拱桥。桥梁由一系列桥拱建成，尤其是高墩高架桥由桥拱互相堆叠建成。

vibrated concrete Concrete that has been subjected to some form of vibration in order to compact it.
振捣混凝土 经过某种形式振捣以碾实的混凝土。

vibrating float A vibrating device, typically long and flat, used to finish in-situ concrete.
振捣尺 一种用于现浇混凝土的振捣设备，通常是长的和扁平的。

vibrating pile driver (vibrating pile hammer) A device that is mounted on top of a sheet pile, which vibrates to help drive the pile into the ground or extract it from the ground. Typically used for driving piles into saturated granular soils.
振动打桩机 安装在板桩顶部的设备，振动以帮助桩打入地面或将其从地面抽出。通常用于将桩打入饱和粒状土中。

vibrating plate A hand-operated device, which has a flat, square, or rectangular vibrating surface, and is used to compact asphalt, concrete, and earth surfaces over small areas.
振捣板 手动操作的设备，有扁平、方形或矩形振动表面，用于压实小范围内的沥青、混凝土和土地表面。

vibrating screed board A long horizontal beam that is used to vibrate and level concrete used for roads and floor slabs.
振捣梁 长的水平梁，用于振动和整平道路和楼板用的混凝土。

vibrating table A device in the form of a table that vibrates, and is typically used to compact small pre-cast concrete products.
振捣台 振动的台式设备，通常用于压实小型的预浇混凝土产品。

vibration 1. The unwanted shaking of the building and fabric caused by plant and equipment. Flexible and resilient anti-vibration devices may be used to stop the vibration and noise. 2. Method used to compact concrete and remove air bubbles trapped within the concrete.
1. 震动 由不必要的设备和装置造成的建筑物和结构晃动。可使用柔性和弹性防震设备防止震动和噪音。**2. 振捣** 用于压实混凝土和清除混凝土内部气泡的方法。

vibrator (vibrating poker) A cylindrical probe that is inserted into in-situ concrete, which vibrates to expel entrapped air. Air is trapped into concrete during mixing and placing. If it is not expelled it reduces the cast strength of the concrete. The vibrating part of the poker is typically 450 mm long; with diameters ranging from 20 - 180 mm. Pokers are either powered by electricity, diesel engine, or compressed air and have cycle speeds between 90 and 250 vibrations per minute.
振捣器（振捣棒） 插入现浇混凝土的圆柱棒，振动以排出截留的空气。在搅拌和浇筑混凝土过程中，空气截留在混凝土内。如果不把空气排尽，则会降低混凝土的浇筑强度。振捣棒的振动部分通常长为450毫米；直径为20—180毫米。振捣棒可为电动、柴油机发动或压缩空气式，循环速度为每分钟振动 90—250 次。

vibrocompaction *See* VIBROFLOTATION.
振动密实 参考“振冲”词条。

vibroflotation An in-situ ground improvement technique used to compact loose granular soils to a depth of 70 m. A crawler crane is used to lower a Vibroflot, a cylindrical probe (450 mm diameter, 3 - 5 m long), into the ground under a vibration/ jetting action (using water and/or compressed air). This action causes the surrounding ground to form a denser configuration. The Vibroflot is normally inserted in the ground at regular intervals, on regular grid spacing, over the site. The treated ground will settle less and have improved bearing capacity.
振动水冲法/振冲 用于将松散的粒状土压实至深度 70 米的现场地基加固技术。用履带起重机在振动/冲作用下将振冲器（圆柱棒，直径为 450 毫米，长度为 3—5 米）放入地面。这会使得周围土地更密实。振冲器通常在现场按固定间隔或固定网格间距插入地面。经处理过的地面将更少发生沉降，其承载力也有所提升。

vice (US vise) A clamp fitted to a workbench used to hold firmly materials that are to be worked. The clamp has a screw fitting, which when tightened secures the material firmly in place.
虎钳/台钳 安装到工作台以牢牢固定将加工材料的夹具。夹具有一个螺丝，拧紧时可将材料牢牢固定住。

Vickers hardness test Sometimes known as the Diamond Pyramid Hardness test, it is a popular method of determining the * hardness value of materials, especially metals.
维氏硬度试验 这是测定材料，尤其是金属的硬度值的常用方法，有时也称为正四棱锥体金刚石硬度试验。

Victorian architecture Buildings and structures constructed during the period from 1837 to 1901. Buildings were constructed on a scale and level that had not been experienced in any previous era. During this period, Paxton's Crystal Palace was constructed in 1851 and back-to-back housing emerged.
维多利亚式建筑 在 1837 年至 1901 年期间建设的建筑和结构。该时期建筑的规模和水平都超过之前任何时期。其中，帕克斯顿的水晶宫于 1851 年建成，背靠背房屋开始出现。

Vierendeel truss A structural frame that does not have any diagonal (triangulated) members. Openings between members are therefore rectangular. Fixed joints are required between members to resist vertical shear/bending, rather than pin joints, which are associated with conventional triangulated trusses.
空腹桁架/佛伦第尔桁架 没有斜构件（三角构件）的结构框架。构件之间的开口因此是矩形的。构件之间需要有固定连接，而不是传统三角桁架所采用的枢接，以抵抗垂直剪切/弯矩。

villa 1. A Roman landowner's residence. 2. Dwelling on the outskirts of the town, a country house, or a holiday home.
1. 庄园 罗马地主的住所。**2. 别墅** 坐落于郊区的住宅，乡间别墅或度假寓所。

village A collection of dwellings in a rural area with a church. A collection of small homes in the country without a church is known as a hamlet.
村庄 在乡郊地区带有教堂的住宅群。没有教堂的乡郊住宅群称为小村。

vinyl Shiny, tough, and flexible plastic, used especially for floor coverings.
乙烯基 发亮且柔韧的塑料，通常用于地板覆面。

virtual work The work done on a structure that has resulted from either a virtual force acting through a real displacement, or a real force acting through a virtual displacement.
虚功 虚拟的力作用于实质上的位移，或者是实质上的力作用于虚拟的位移，所产生的对结构所做的功。

viscoelastic A material that exhibits both viscous and elastic properties. For example, asphalt could deform under the influence of stress or high temperature to such an extent that it would flow (become a liquid). How-

ever once the heat/stress has been removed, it could recover elastically to its original "solid" shape. If a material has viscoelasticity, it will exhibit both viscous and elastic characteristics when under deformation.
黏弹性的 同时具有黏性和弹性的材料。例如，沥青可以在应力或达到可流动（变液体）高温的影响下变形。但是一旦热/应力消除，沥青可以恢复成其原来"固体"的形状。如果材料具有黏弹性，那么它在变形时同时展现出黏性和弹性。

viscometer An instrument for determining the *viscosity of a liquid.
黏度计 用于确定液体黏度的仪器。

viscoplasticity A material that exhibits both viscous and plastic properties. Within the plasticity phase, the material would be subjected to a time-dependent stress/strain response, while in the viscous phase the material could flow like a liquid.
黏塑性 同时具有黏性和塑性的材料。在塑性相时，材料可承受随时间变化的应力/应变响应，而当黏性相时，材料可以像液体一样流动。

viscosity Coefficient of resistance to flow; a measure of how easily a liquid flows.
黏度 流动阻力系数；对液体流动容易性的测量。

viscous damping The reduction of energy when the velocity of a particle is resisted by an opposing proportional force.
黏滞阻尼 当粒子速度受到成比例的反作用力阻碍时，能量减小。

viscous impingement filter *See* IMPINGEMENT FILTER.
黏滞撞击滤尘器 参考"撞击滤尘器"词条。

visible defect A defect in timber, such as a knot or shake that is identified during *visual grading. The defect may make the timber unsuitable for work that is to have aesthetic qualities. The defects may also affect the structural properties, but the structural properties of timber are now determined by mechanical grading.
可见缺陷 在目测分等过程中可辨的木材缺陷，例如木节或环裂。这些缺陷可能会使木材不适于具有审美质量要求的工程使用。这些缺陷也可能影响结构性能，但是木材的结构性能现在是由机械分等决定的。

vision-proof glass Glass that is obscured preventing one from seeing through it.
防透视玻璃 防止别人看透的玻璃。

visual display unit (VDU) A computer screen.
直觉显示器（VDU） 一种电脑屏幕。

visual grading The grading of timber by visual examination of the gross features, the defects within the timber.
目测分等 通过目测检查木材的宏观特征和缺陷给木材分等。

vitiated air 1. Air in which the oxygen content has been reduced. 2. Air that is not fresh.
1. 缺氧空气 氧气含量较少的空气。**2. 污浊空气** 不新鲜的空气。

vitreous china A ceramic material manufactured at very high temperatures, thus imparting excellent impermeability properties, making it ideal for bathroom applications since it prevents the ingress of bacteria through the surface.
玻化瓷 在高温下生产的陶瓷材料，具有良好的抗渗能力，因为它可以防止细菌通过表面渗入，所以是浴室用具的理想选择。

vitreous enamel A coating material produced by fusing glass that has been ground to a powder. The glass is heated until it melts at temperatures of 750 – 800℃; in its liquid state it is applied to surfaces and used to provide a protective cover to materials such as metal, providing a smooth vitreous glass finish.
搪瓷 通过热熔磨成粉末状的玻璃生产的涂层材料。加热玻璃至750—800℃的高温，直至玻璃熔化，当其为液体状态时，涂于表面，为材料（例如金属）提供保护性涂层，以及提供光滑的玻璃饰面。

vitreous glass A type of glass that is popular for mosaic tile work. This type of glass is durable, resistant to

chemical attack, and comes in an array of colours.
马赛克玻璃 一种普遍用于马赛克工程的玻璃。这种玻璃耐用、耐化学腐蚀，并且有各种颜色。

vitrification During firing of a ceramic body, the formation of a liquid phase that upon cooling becomes a glass-bonding matrix. Vitrification is one of the key reactions that occurs during the fabrication of glass.
玻璃化 在陶瓷坯体烧结时形成的液体相，在冷却时变成玻璃相黏结的基质。玻璃化是制作玻璃时发生的关键反应。

vitrify To change into glass, usually by heat fusion. Vitrified clayware is a ceramic material that has been subjected to a temperature around 1100℃ to form a material that has very low water absorption properties; it is used for floor tiles and drain pipes.
使玻璃化 转变成玻璃的过程，通常用热熔手段。玻化黏土制品是一种陶质材料，加热到1100℃左右形成一种吸水非常少的材料；用于地板砖和排水管。

Vitruvius Roman architect whose work had a significant influence on Renaissance structures and buildings. Vitruvius defined the Vitruvian Man, as drawn by Leonardo da Vinci, as ideal proportions of the human body defined within geometry.
维特鲁威 罗马建筑师，他的作品对文艺复兴建筑和结构有重要的影响。维特鲁威认为维特鲁威人是几何学中最佳的人体比例，并由列奥纳多·达·芬奇绘画出来。

V-joint A joint formed in a "V" shape. *Vee joint.
V 形接头 V 字形的接头。

V-notched trowel A trowel with a "V" cut into it for applying even amounts of adhesive to a surface.
锯齿抹子 有一个 V 形切口，以将黏合剂均匀涂抹在表面上的抹子。

void A space or gap; voids occur in concrete due to air bubbles that occur during mixing and hydration and are not vibrated out, or between coarse aggregate where there are no fine particles, such as sand and cement, to fill the gaps.
孔隙 空间或间隙；混凝土中的孔隙是由于在搅拌和水化过程中产生气泡并且没有振捣排尽，或者在粗骨料之间没有细颗粒，例如沙子和水泥填充孔隙。

void ratio (*e*) The ratio between the volume of voids and the volume of solids in a material, particular soil.
孔隙比（*e*） 材料（尤其是土壤）中孔隙体积和固体颗粒体积之比。

volt (V) The *SI unit of electrical potential difference or electromotive force. One volt is the electromotive force that is required to move one ampere of current through one ohm of electrical resistance.
伏特（V） 电位差或电动势的国际单位。一伏特是指一安培的电流穿过一欧姆电阻所需的电动势。

voltage A measure of the electrical potential difference or electromotive force. Measured in *volts.
电压 电位差或电动势的测量。以伏特为单位。

voltage drop The drop or difference in voltage that occurs between two points in an electrical circuit.
电压降 电路中两点间的电压降落或电压差的伏特数。

voltage overload (excess voltage, overvoltage) A voltage load on a component that is in excess of what it was designed to carry. Electrical devices are usually protected from voltage overload by *protection gear.
电压过载（过电压） 组件的电压负荷超过其设计可承载的电压。电气设备通常有保护装置以防电压过载。

volume The space that a material or area occupies, measured in cubic meters m^3.
体积 材料或区域占用的空间，以立方米（m^3）为单位计量。

volumetric building (volumetric systems) A preassembled building unit or system. The module can be a component of a building, such as a toilet pod, bathroom pod, chimney stack, or other similar unit, or could be a whole building built in a factory and delivered to site.
模块化建筑 预装配的建筑单元或系统。模块可以是建筑物组件，例如厕所间、浴室间、烟囱筒身或其他类似单元，或者可以是在工厂预制然后运送到现场的整座建筑。

volumetric efficiency The actual volume of liquid that is pumped by an engine or piston compared to its theoretical maximum.
容积效率 与理论最大值相比，引擎或活塞能泵送的液体的实际容积。

volumetric strain The ratio between the change in volume to its original volume, which occurs when a material deforms.
体积应变 当材料变形时，材料变化的体积与原来体积之比。

volume yield 1. The volume of concrete that can be determined from the weight of cement, aggregates, and water. 2. Volume of lime putty obtained from the weight of * quicklime.
1. （混凝土）体积产率 可根据水泥、骨料和水的重量确定的混凝土体积。**2. （石灰）体积产率** 根据生石灰重量获得的石灰膏体积。

volute A spiral that forms part of the uppermost part of a column (column capital).
螺旋形 形成柱子最上面部分（柱头）的螺旋。

von Mises yield A criterion to denote that failure will occur when the strain energy reaches the same energy for yield to occur. It is frequently used to calculate the approximate yield of ductile materials and is also known as the maximum distortion energy criterion, octahedral shear stress theory or Maxwell-Huber-Hencky-von Mises theory.
米塞斯屈服准则 表示当应变能达到与屈服产生时一样便会发生破坏的准则。通常用于计算塑性材料的屈服应力，也被称为最大剪应变能准则、八面体剪应力理论或麦克斯韦-胡贝尔-汉基-冯·米塞斯理论。

vortex 1. A circular motion or forced vortex, approximating to the flow pattern created by a mechanical rotor or by a body of fluid being whirled around in a container. 2. A circular motion or free vortex, which occurs naturally; for example, a flow down a drain hole or around a river bend. 3. A forced vortex or compound vortex, surrounded by a free vortex; for example, the core of the vortex rotates as a solid body. 4. A combination of radial flow and a free vortex, or free spiral vortex, occurs in the volute of a * centrifugal pump or between the guide vanes of a * Francis turbine.
1. 强制涡流 圆周运动或强制漩涡，近似于机械转子或容器内旋转的流体产生的流态。**2. 自由涡流** 自然发生的圆周运动或自由旋涡；例如，从排水孔流下的水流或河湾附近的水流。**3. 复合涡流** 自由涡流环绕的强制涡流或复合涡流；例如，涡核像固体般旋转。**4. 自由螺旋形涡流** 在离心泵蜗壳内或混流式水轮机的导流叶片间产生的辐流和自由涡流的组合，或称为自由螺旋形漩涡。

voussoir V-or wedge-shaped bricks used to form an arch.
楔形拱石 用于建拱的 V 形或楔形砖石。

voussoir arch An arch in a bridge that contains wedge-shaped blocks, which are placed in such a way as to distribute the load efficiently to maximize the compressive strength of the stone.
楔形桥拱 桥上的拱，包括按照有效荷载分布放置石块以使其压强度最大的楔形砌块。

vulcanization A chemical process to treat rubber. Sulphur is added to rubber at high temperatures to form crosslinks between individual polymer chains, resulting in a product which is harder, stronger, and more elastic, for example, hoses.
硫化 一个橡胶加工的化学处理过程。橡胶在高温下加入硫，在单个高分子链之间形成交联，使产品更加坚硬、强韧和更有弹性，例如软管。

WADGPS (wide-angle differential GPS) A network of ground-based stations that track satellite signals to improve GPS accuracy.
广域差分全球定位系统（广域差分 GPS） 一个能追踪卫星信号，以增强 GPS 的准确性的地面站点网络。

waffle floor A suspended floor slab employing a square grid construction of ribs with deep coffers (recesses). Large spans are possible due to reduced dead load.
密肋楼盖 采用有凹槽的方形网格构造密肋的悬挂式楼板。由于降低恒荷载，大跨距也是可能的。

wagon drill A pneumatic or percussive-type rock drill, mounted vertically on three or four wheels, used to sink shafts from the surface of the ground.
钻机车 垂直安装在三个或四个轮子上的气动或冲击式凿岩机，用于在地面凿井。

wailing (US wale) A horizontal timber beam, used to provide support to the side of an excavation or formwork.
横档 水平木梁，用于为挖掘或模板侧边提供支撑。

wainscot (wainscoting) Timber wall panelling or lining used on an internal wall that extends from the *** skirting** board up to * dado rail height. Previously used to cover the lower half of internal walls that were prone to damp penetration, nowadays used for decorative purposes.
护壁板 内墙使用的木壁板或墙衬板，从踢脚板延伸至护墙板线条高。原先是用于遮盖内墙可能有潮气渗入的较低部分，现在则用于装饰。

waiver Voluntarily giving up a right or privilege.
弃权 自愿放弃权利或特权。

walking line The central line on a stairway where it is expected that people will use the stairs. It is the point at which the going is measured for * kite winders or stairs that are diagonally shaped.
楼梯行走线 楼梯上预计行人使用最多位置的中线。是测量转向三角踏步楼梯或斜踏步梯段的点。

wall A vertical element used to enclose or subdivide a space, it is one of the primary elements of a building.
墙壁 用于封闭空间或将空间分区的垂直构件，是建筑物的主要构件之一。

wall chaser A power tool for cutting a * chase in brickwork, blockwork, and concrete.
砖墙开槽机 在砖墙、砌体和混凝土上开槽的电动工具。

wall friction The shear resistance that has been generated at the interface between the back of a retaining wall and the soil that helps to stabilize the wall.
井壁摩擦 在挡土墙背面，帮助稳固墙壁土壤之间的接触面产生的抗剪力。

wall-hung Attached to a wall.
壁挂式 附于墙壁上。

wall panel (panel wall) A prefabricated section of wall.
墙板 预制的墙体部分。

wall plate A structural element (connection) that distributes a load onto a wall, and acts like a mini * lintel.
梁垫 分散荷载到墙壁的结构构件（连接），作用相当于小过梁。

wall string *See* STRING.
楼梯斜梁 参考“斜梁”词条。

wall tie A fixing element that is used in cavity wall construction. It spans the cavity and is embedded within the mortar joints on both sides to hold the two leafs together. Usually made from stainless steel, galvanized steel, or plastic, and will be designed to prevent water passing from one leaf to the other.
墙拉结 夹心墙结构中使用的固定构件。墙拉结横跨墙体空腔，埋设在两边的灰缝内，以固定内外两叶墙。通常由不锈钢、镀锌钢或塑料制成，以防止水从一叶流向另一叶。

ward The area or grounds of a castle that are surrounded by the castle walls.
(城堡内的) 广场 由城堡围墙围起来的区域或场地。

warm roof A *pitched roof where the insulation is placed in line with the roof slope.
保温屋面 配合屋面斜度铺设保温层的坡屋面。

warning pipe An *overflow connected to a sanitary appliance that discharges to an obvious position outdoors.
露天溢水管 与卫生器具相连，排放液体至室外显眼位置的溢流管道。

warp The distortion or twisting of something in the longitudinal direction, e. g. yarns running in the machine direction of a loom, which become crimped due to the insertion of the *weft (cross-machine direction) yarns.
挠曲 某物的纵向变形或扭曲，例如沿着织布机纵向运行的纱线，由于插入纬纱（机器横向）而变卷曲。

Warren girder (Warren truss) A triangulated truss that has only sloping members of approximately equal sides between the top and bottom *chords (it has no vertical members).
华伦式大梁（华伦式桁架） 只在上弦杆和下弦杆之间接近等边的斜构件（无垂直构件）的三角桁架。

WASC (water and sewage companies) UK water utilities companies that both supply drinking water and treat sewage.
水务公司 英国水务公司，既供应饮用水，也处理污水。

wash boring A drilling technique that employs a jet of water to displace the soil in front of a pile or casing being installed.
水冲式钻探 在打桩或安装套管前端采用水的喷射置换土壤的钻探技术。

washdown closet (washdown pan) A WC (*water closet) where the contents are removed by a flush of water running down the pan.
冲落式坐便器 借冲洗水的冲力直接排出污物的冲水马桶。

washer A thin, usually flat disk that is used to prevent leakage, distribute pressure, reduce friction, or as a spacer.
垫片 扁平的薄垫圈，用于防止泄漏、分散压力、减小摩擦力或者作为间隔片。

washout closet An early type of *water closet where the human waste was deposited in a shallow bowl filled with water. When the toilet was flushed, the contents of the bowl were washed out. Replaced by the *washdown closet.
旧式冲水马桶 早期的马桶，人类排泄物排在装满水的浅槽中。当马桶满了时，槽内的排泄物会被冲洗掉。已被冲落式坐便器取代。

washout valve A valve used to drain (empty) something completely.
冲洗阀 用于完全排尽（清空）某物的阀门。

waste 1. Rubbish, something that has been disgarded. 2. To use something carelessly, without effect or purpose.
1. 垃圾 毫无价值被丢弃的物件。**2. 浪费** 随便使用，毫无效果或目的。

waste disposal unit A device that flushes away food waste. It is typically fitted beneath a kitchen sink, on

the underside of the plughole. Its purpose is to shred food waste into small pieces such that they can be passed through the drainage pipework into the sewage system.
食物垃圾处理装置 冲掉食物垃圾的设备。通常安装在厨房洗涤盆下面，排水孔的下部。用于将食物垃圾切成碎片，这样就可经排水管排入排污系统。

waste pipe A pipe connected to a basin, bath, or sink to carry discharge water to the * soil stack.
废水管/排水管 连接至水槽、浴室或水池的管道，将水排放至排污立管。

waster A mason's chisel characterized by a claw-type cutting head or with a cutting head 19 mm wide.
齿凿 石匠的凿刀，有爪形刀头或19毫米宽的刀头。

wastewater Any water that has been used, either domestically or industrially, and now requires water treatment before reuse.
废水 已经使用过的水，无论是生活用水还是工业用水，在再次使用前须进行废水处理。

wasteway (waste weir) *See* SPILLWAY.
溢流渠 参考“溢洪道”词条。

water bar 1. A strip of material, usually metal, plastic, or rubber, that is designed to prevent water ingress. Can be located in a groove between two components, on the threshold of a door, or can be cast into the construction joints of basement walls to prevent the passage of moisture through the joint. 2. A water barrier (waterbar) in the form of a strip or channel. It can be used across a construction joint to prevent the ingress of water or across steep-sloping earth roads to prevent erosion from fast-flowing surface water.
1. 止水条 条状材料，通常为金属、塑料或橡胶条，用于防止水进入。可以安装于两个组件间的凹槽、门槛或者也可以埋入地下室墙壁的施工缝以防止潮气透过。**2. 拦水栅** 条状或槽形拦水栅。可以穿过施工缝以防止水进入或穿过陡峭的土路，以防止表面湍流的侵蚀。

water catchment *See* CATCHMENT AREA.
集水流域 参考“集水区”词条。

water/cement ratio The ratio of water and cement/pozzolan (by weight) present in both concrete and mortar. This ratio is very important as it influences both the compressive strength and durability. In most concretes and mortar mixes, an optimum ratio corresponds to the maximum compressive strength.
水灰比 在混凝土和砂浆中水和水泥/火山灰的（重量）比例。因为该比例会影响抗压强度和耐用性，所以很重要。在大多数混凝土和砂浆混合物中，最佳配比对应最大的抗压强度。

water closet (WC) A sanitary appliance that collects human waste (excreta and urine), and then uses water to flush the waste to another location. Usually constructed from * vitreous china. *See also* WASHDOWN CLOSET, SIPHONIC CLOSET, and WASHOUT CLOSET.
抽水马桶 收集人体废物（排泄物和尿），然后将其用水冲到另一个地方的卫生洁具。通常用玻化瓷制造。另请参考“冲落式坐便器”“虹吸马桶”和“旧式冲水马桶”词条。

water content *See* MOISTURE CONTENT.
含水量 参考“水分含量”词条。

water gauge 1. An instrument that measures the flow of water in a stream or river. 2. An instrument for measuring the water pressure in a tank or boiler.
1. 水位标尺 测量溪流或河流中水流的仪器。**2. 水压表** 测量水箱或锅炉中水压的仪器。

water hammer The noise caused by a sudden change in the flow of a fluid in a pipe.
水锤 管道中液体流动骤然变化而造成的噪音。

water jet A pressurized stream of fluid forced out of a small nozzle; *See* PELTON WHEEL.
喷水 从小喷嘴中喷出的加压水流；参考“培尔顿水轮机”词条。

water level The surface level of a body of water.
水位 水体的表面高度。

water main The main underground pipe in a water distribution system.
总水管 配水系统中的地下总管道。

water meter An instrument that measures the amount of water flowing through it. Used to record for billing purposes how much water has been consumed by a household.
水表 测量流经的水量的仪器。记录下来用于计费，计量一个家庭的用水量的费用。

water of capillarity * Groundwater that is held above the * water table due to * capillarity.
毛细水 由于毛细作用而位于地下水位以上的地下水。

waterproof Objects or materials that resist water penetration.
防水的 可防止水渗透的物体或材料。

water purification *See* WATER TREATMENT.
水净化 参考“水处理”词条。

water quality The physical, chemical, and biological characteristics of water, which are determined by the end use; *See* DRINKING WATER.
水质 水的物理、化学和生物属性，由其最终用途决定；参考“饮用水”词条。

water reducer *See* PLASTICIZER.
减水剂 参考“塑化剂”词条。

water-related diseases Illnesses that stem directly from consuming or indirectly from being in contact with contaminated water. The contaminant can be either biological, or chemical, or a mixture of both. The majority of diseases resulting from microbiological pollution are essentially contracted from water contaminated with human faecal matter and include cholera, typhoid, and other diarrhoeal or non-diarrhoeal infections such as schistosomiasis, skin infections, and yellow fever. Contamination of chemical origin include arsenicosis, which is caused by prolonged low-level exposure to arsenic, and can lead to skin keratosis. Such illnesses are prevalent in developing countries where water treatment is lacking, and are so widespread that they cause more deaths in the world than malaria and HIV/AIDS.
水污染疾病 来源于直接使用或间接接触受污染水而产生的疾病。污染源可以为生物的或化学的，或者是两者混合。由微生物污染引起的疾病主要因为水受人体排泄物污染，包括霍乱、伤寒和其他腹泻类疾病，或者其他非腹泻类感染如血吸虫、皮肤感染和黄热病。化学来源的污染包括地方性砷中毒，由长期摄入低浓度砷引起，可导致皮肤角化。此类疾病常见于缺乏水处理的发展中国家，在世界范围内造成的死亡数比疟疾和艾滋病更多。

water repellent (waterproofing paints) Special types of paint that are water repellent and can be applied to porous surfaces including brick, concrete, stone, and renderings to prevent damp penetration. Such treatment will not prevent rising damp, but it will allow the continued evaporation of moisture within the masonry.
憎水性涂料（防水涂料） 憎水的特殊涂料，可以涂在多孔表面，包括砖块、混凝土、石块和抹灰上，以防止潮气渗入。这种防水处理不能防止上升的潮气，但是可以使砌体内的水分持续蒸发。

water-repellent cement * Ordinary Portland cement can contain a water-repellent agent that has been added during the manufacturing process. Such a cement is used to produce impermeable concrete.
憎水水泥 在生产过程中加入憎水剂的普通的波特兰水泥，这种水泥可用于生产抗渗混凝土。

water retention The ability to retain water.
保水性 保留水的能力。

watershed The boundary between two * catchment areas.
分水岭 两个集水区之间的分界线。

water softener A chemical or device that is used to reduce the hardness of water by removing calcium and magnesium salts.
软水剂/机 用于去除钙盐和镁盐以降低水硬度的化学物或设备。

water-soluble The ability to dissolve in water.

水溶性 在水中溶解的能力。

water supply The distribution of *drinking water to consumers by a *water utility.
供水 通过自来水公司供应饮用水给住户。

water table (phreatic surface) The level of water in the ground below which the pore structure is totally saturated; *See* GROUNDWATER.
地下水位（潜水面） 地下孔隙结构完全饱和的水高度；参考“地下水”词条。

water test A hydraulic test used to check the seal in drains. The drain is filled with water to a predetermined head of water and allowed to stand. The level of the water is marked and checked after a period of time to see if the water level has fallen. If the hydrostatic pressure of the water has caused the water level to fall, there is likely to be a leak in the drain.
水压试验 用于检查排水密封的水力测试。排水管道中充满水至一个预定的水头并保持。记录下水位后隔一段时间检查看水位是否下降。如果因水的流体静压力导致水位下降，则排水管道可能有渗漏。

water tower An elevated tank containing water to be feed into a water distribution system.
水塔 装有注入配水系统的水的高位水箱。

water treatment The removal of impurities from raw water in order to make it fit for human consumption. There are a variety of techniques available to treat raw water, which can typically be represented in three main stages: *primary treatment, *secondary treatment, and *tertiary treatment.
水处理 除去生水中的杂质，使其能够供人使用。有各种处理生水的技术，通常主要有三个阶段：一级处理、二级处理和三级处理。

water turbine A *turbine that is driven by a head of water; *See also* FRANCIS TURBINE and KAPLAN TURBINE.
水轮机 水头驱动的涡轮机；参考“弗朗西斯式水轮机”和“卡普兰水轮机”词条。

water utility A company that supplies drinking water to consumers and/or treats wastewater. *See also* WASC.
自来水公司 为用户供应饮用水和/或处理废水的公司。另请参考“水务公司”词条。

water vapour Water in a gaseous state.
水蒸气 气态的水。

waterway 1. A river or canal used by boats and ships. 2. A channel to drain water away.
1. 航道 小船和轮船使用的河道或运河。**2. 排水沟** 排水的沟渠。

waterworks The location where raw water is treated, stored, and distributed into the supply network.
自来水厂 将生水处理、储存和配送至供水网络的地方

watt (W) The *SI unit of electrical power in terms of one *joule per second.
瓦特（W） 电功率的国际单位，即1焦耳每秒。

wattle Sticks or stakes that have been interwoven with branches and twigs to form fences, walls, and roofs.
编条 枝条与树枝和细枝相互编织形成栅栏、墙壁和屋面。

wattle and daub A type of wall construction comprising interwoven twigs, sticks, and branches (wattle) that are covered with clay, cow dung, or mud (daub). Traditionally used as an infill in post-and-beam timber-frame buildings.
抹灰篱笆墙 由编织的细枝、枝条和树枝（编条）构成的墙壁，涂有黏土、牛粪或泥土（泥浆）。传统上用于填充桩梁木框架建筑。

wave energy Harnessing energy from sea waves.
波浪能 由海波浪形成的能源。

waybeam A beam that runs beneath the rail to provide support and a place to secure base plates when a bridge section of the rail track is being repaired.

轨道纵梁 当在修理铁轨的桥梁部分时，在轨道下面穿过的横梁，用于支撑和固定基板。

waypoint A signpost on a route noting a change of direction.
转折点 道路上用于提示方向改变的路标。

WC *See* WATER CLOSET.
WC 参考“抽水马桶”词条。

wear Damage caused by continuous rubbing or friction.
磨损 持续摩擦造成的损坏。

wear failure A type of degradation mechanism caused by continuous rubbing and friction, especially common in metals. Usually there are three types of wear failures: adhesive, abrasive, and erosive. Most wear failure is characterized by loss of material (from the surface).
磨损失效 一种由于持续摩擦造成的退化机制，在金属中尤其常见。通常有三种类型的磨损失效：粘着、磨粒和腐蚀。大部分磨损失效都具有材料（表面）损耗的特征。

weather To allow materials to be exposed to the elements to give them a used and tired look.
风化 让材料暴露在自然环境中让其外观变老旧。

weather bar Steel or aluminium strip of metal bar that is rebated in the sill of window and doorframes to prevent water being blown into the property over the sill. The bar provides a physical barrier that prevents water entering the building by either capillary action or driving rain.
挡水条 钢或铝条，嵌入窗台或门框中，以防止水被吹进窗台屋内。挡水条可作为物理屏障阻挡水通过毛细作用或风雨进入建筑物内。

weatherboard 1. A piece of *weatherboarding. 2. A horizontal board fixed to the bottom of an external door to prevent rain penetration.
1. 外墙护墙板。 **2. 挡雨板** 安装在外门底部的水平板，以防止雨水进入。

weatherboarding (clapboard) External timber cladding comprising a series of long thin feather-edged horizontal timber boards.
外墙护墙板（木隔板） 外部木材贴面，由一组细长薄边水平木板组成。

weathercock (weather vane, wind vane) A device, usually located on the top of a building, that is used to indicate the direction of the wind.
风向标 一个用于指示风的来向的设备，通常位于建筑物顶部。

weathered joint *See* WEATHER-STRUCK JOINT.
斜勾缝 参考“防雨斜缝”词条。

weathering The deterioration due to the exposure to climatic conditions. It is particularly used in reference to the breakdown of rocks, caused by water, ice, chemicals, and changing temperature.
风化作用 因暴露于自然条件下产生的退化。特别是指岩石因为水、冰、化学和温度变化而破碎。

weatherstrip A thin strip of material used around windows and doors to prevent air infiltration through cracks and gaps.
挡风雨条 安装在窗户和门周围以防止空气通过裂缝和缝隙渗入的薄条材料。

weather-struck joint (struck joint, weathered joint, weathered pointing) A horizontal masonry joint that slopes outwards from the top of the joint to shed rainwater away from the wall.
防雨斜缝 从接缝顶部向外倾斜以泄水的水平砖缝。

web The vertical part of a beam or rail, which adjoins the top and bottom horizontal flanges. If the web experiences high compressive stresses and moves out of line web buckling can occur.
腹板 工字梁或钢轨的垂直部分，连接顶部和底部水平的翼缘。如果腹板承受了高压应力并且不成直线，则会发生腹板屈曲。

weber The SI unit of magnetic flux. It measures the amount of magnetic field passing through a conducting

surface and is equal to one joule per ampere.
韦伯　磁通量的国际单位。测量通过传导面的磁场量，1 韦伯等于 1 伏特秒。

wedge　A tapered solid block of wood or steel that is used to separate two objects or to split timber to form two separate pieces.
楔子　用于将两个物体分开或将木材拆分成两块的锥形实心木块或钢块。

wedge anchor　A * rock anchor that uses a mechanical wedging effect for fixing.
楔形锚固件　利用机械楔入作用以固定的岩石锚杆。

wedge cut　A drill hole pattern used in "drill and blast tunnelling" that forms a "V" or wedge shape.
楔形掏槽　在"隧道开挖和爆破"中采用的形成 V 形或楔形的钻孔模式。

wedge theory　A method to analyze the force behind a retaining wall, based on the theory that a wall needs to support the weight of a wedge of soil that would move if the wall fails.
楔体理论　分析挡土墙后面的力的方法，基于如果墙壁倒塌，墙壁需要支撑楔体土壤的理论。

weep hole　A small water drainage hole. In masonry wall construction, it is used above a * cavity tray and * damp-proof course.
泄水孔　一个小的排水孔。在砖墙结构中，一般位于空腔泛水和防潮层以上。

weeping　Something that is leaking fluid.
滴水的　正泄漏液体的某物。

weft　The crosswise yarns that pass through the * warp yarns on a loom to produce a woven structure (for example, a * geotextile).
纬纱　织布机中穿过经纱形成编织构造的横向纱线（例如土工织物）。

weigh batcher　A concrete batching plant where the quality of each material in the mix (excluding water) is measured by weight.
混凝土配料机　按重量测定混合物（水除外）中各材料质量的混凝土配料器。

weighting　To add or put a bias on something so that it prevents or counteracts something occurring, e. g. the use of additional loads to make an object heavier so that it does not float.
加重　增加或放一偏压在某物上以防止或抵消某事发生，例如增加荷载使某物变重使其不上浮。

weir　A small dam or wall built across the full width of a river to regulate flow. It has a horizontal crest or notch to allow water to continue to flow over it but causes the water level to rise upstream. The weir head is the depth of water between the weir crest or bottom of the notch, and the upstream water level.
堰　横跨整个河流宽度以调节水流的小水坝或墙壁。有一个水平堰顶或堰槽能使水流持续经过，但是会使上游水位增高。堰上水头是堰顶或堰槽底部和上游水位之间水的深度。

Weisbach triangle　A surveying setup used for shafts.
魏思贝奇三角形法　一种用于井筒的测量设置法。

weld　*See* WELDING.
焊接　参考"焊接"词条。

weldability　The ability of a metal or other material to be welded under specified conditions.
可焊性　金属或其他材料在特定条件下的焊接能力。

weld decay　Corrosion that occurs in some welded stainless steels at regions adjacent to the weld.
焊缝腐蚀　在一些焊接的不锈钢中邻近焊接的区域发生的腐蚀。

weld defects　Imperfections occurring during or after the welding process.
焊接缺陷　在焊接过程中或焊接之后产生的缺陷。

welder　A tradesperson who specializes in welding materials together.
焊工　专门焊接材料的技术工人。

welding A technique for joining metals in which actual melting of the pieces to be joined occurs in the vicinity of the bond. A filler metal may be used to facilitate the process. In this process both similar and dissimilar metals can be used; several welding techniques exist including arc welding, gas welding, brazing, and soldering.
焊接 将连接处实际熔化的金属连接在一起的技术。可以使用填充金属加速进程。在这个过程中，可以使用同种金属，也可以使用异种金属；有几种焊接技术，包括弧焊、气焊、钎焊和锡焊。

welding work angle In arc welding, the angle between the electrode and one of the joints.
焊接角度 在弧焊中，电焊条和其中一条焊缝之间形成的角度。

well A hole or shaft that is dug or drilled into the ground to extract water, oil, gas, and so on.
井 挖或钻入地面以提取水、油、气等的洞或井筒。

well-conditioned triangle A triangle used in surveying that is, more or less equilateral, such that any error in the measurement of the angle has minimal effect when computing length.
小三角 在测量中用到的三角形，几乎等边，这样计算长度时，角度测量误差的影响就可忽略不计。

Welsh arch A small arch spanning less than 12 in. or 300 mm.
威尔士拱 跨距小于 12 英寸或 300 毫米的小拱。

welt (welted seam) A raised seam found in sheet-metal roofing.
绲边 金属板屋面上的凸起接缝。

welted drip A drip used at the eaves of built-up felted roofs.
摺边滴水线 在油毡屋面的屋檐处使用的滴水线。

westing The position of movement westwards in reference to * longitude coordinates.
西行 按经度坐标向西方向移动。

wet analysis A particle-size sieve analysis that uses water to wash the particles through the various sized sieves. It is normally used for very fine particles that would otherwise agglomerate when dry and would not pass through the aperture.
水洗筛分 使用水冲洗颗粒通过各种尺寸筛网的粒度筛分分析法。通常用于微细颗粒，不然当其变干时会凝成块并且无法通过筛孔。

wet and dry bulb thermometer A * hygrometer containing two * thermometers. The bulb on one of the thermometers is wrapped in a muslin cloth soaked in distilled water (wet bulb), while the other thermometer is left unwrapped (dry bulb). The drier the air the quicker the water will evaporate from the wet cloth and in doing so will cool the bulb. Atmospheric humidity can be measured from the difference between the two thermometer readings.
干湿球温度计 包括有两支温度计的湿度计。其中一支的球部用蒸馏水浸润的棉布包住（湿球），另一支则没有包裹（干球）。空气越干燥则棉布的水分会越快蒸发并令球降温。通过对两支温度计读数的差别计算出大气的湿度。

wet cube strength The strength of a concrete cube after it has been completely saturated with water. The wet strength is lower than the dry strength.
湿立方体抗压强度 立方体混凝土在浸饱水之后的抗压强度。湿抗压强度低于干抗压强度。

wet dock A * dock in which water levels are kept at high tide by dock gates.
湿船坞 水平面被坞闸保持在高潮水位的船坞。

wet drilling The use of water on the drill bit when drilling to reduce dust.
湿式钻眼 在钻孔时用水在钻头上降低粉尘。

wet edge How easily a painted wet edge can be integrated in areas of overlap.
湿边 已上漆的湿边融入重叠区域的难易程度。

wet galvanizing A galvanizing technique whereby the metal is introduced into a bath of zinc via molten flux.
湿式熔剂镀锌法 通过熔剂将金属引入镀槽的镀锌技术。

wet mix Concrete or mortar containing too much water, immediately evidenced by a runny consistency.
过湿 混凝土或砂浆含有太多的水，可由过稀的稠度马上看出。

wet rot A decay-affecting timber, caused by alternate wetting and drying.
湿腐 由于干湿交替而导致的腐烂木材。

wetted perimeter The sum of the depth of water at each side of an open channel plus the width of the base of the channel.
湿周 明渠每边的水深加上渠道底宽之和。

wharf A landing area for ships and boats that is sometimes sheltered from the elements. *See also* BERTH or JETTY.
码头 轮船和小船停靠的地方，有时可躲避风雨。另请参考"泊位"或"突堤式码头"词条。

Wheatstone bridge A device used to measure the change in electrical resistance of a flat coil of very fine wire that is glued to the surface of an object.
惠斯通电桥 用于测量粘到物体表面的细丝扁平线圈的电阻变化的仪器。

wheelabrating A form of * SHOT blasting that utilizes steel grit.
喷丸处理 使用钢砂来进行喷砂清理。

wheeling step (Scotland wheel step) A * winder for a stair.
扇形踏步 一种螺旋形踏步的楼梯。

Whipple-Murphy truss Similar to the * PRATT TRUSS but has diagonals that extend across the base of at least two panels.
惠普尔-墨菲桁架 类似于普拉特桁架，但是有穿过至少两个面板底部的斜杆。

Whirley crane A large crane that can rotate 360 °.
回转式起重机 可以360°旋转的大型起重机。

white and coloured Portland cement A type of * PORTLAND cement known for the high degree of whiteness in appearance, although other colours are also possible to obtain.
白色和彩色硅酸盐水泥 外观上很白的硅酸盐水泥，尽管也有其他颜色的。

Whitney stress diagram A plot showing the stress distribution in a concrete reinforced beam based on the theory of ultimate load.
惠特尼应力图 基于极限荷载理论，钢筋混凝土梁的应力分布图。

whole brick wall (one brick wall) A solid wall with one brick thick.
整砖墙（一砖墙） 一块砖厚度的实心墙。

whole-circle bearing A horizontal angle measured in a clockwise direction from a fixed point, which is usually true north.
全圆方向角 从固定点按顺时针方向量度的水平角度，通常为正北向。

Wichert truss A multi-span truss in which vertical members above intermediate supports have been omitted to make the truss statically determinate.
维歇特桁架 多跨桁架，省去中间以上支承的垂直构件，使桁架静定。

wicket door A small access door located in the leaf of a larger door.
便门 位于大门门扉上的小门。

wicking The * capillary action that occurs in fabrics and yarns.
芯吸效应 在纤维和纺纱中出现的毛细作用。

winch A device for hoisting loads using a cable or rope wound round a cylinder, which is turned by hand or by a motor.
绞车 使用缠绕在滚筒上的缆绳或绳索提升荷载的设备，手动或发动机操作。

wind brace A support to a structure that is designed to resist wind loads.
抗风支撑 用于抵抗风荷载的对结构的支撑。

wind effect Air moving around and over a building induces pressure differences across the building and generally results in a small negative pressure internally. Outdoor air will enter the building through cracks and gaps in the construction (*infiltration) on the windward side of the building that is under positive pressure, and indoor air will exit the building (*exfiltration) on the leeward side of the building that is under negative pressure.
风效应 建筑物周围和上方的空气流动产生了压力差，通常会导致内部轻微的负压。在处于正压下的建筑物迎风面，室外空气将通过结构中的裂缝和缝隙进入建筑物（渗入），而在处于负压下的建筑物背风面，室内空气将流出建筑物（渗出）。

wind energy (wind power) Harnessing the energy from the wind through the use of wind turbines.
风能 通过风力涡轮机利用风能。

winder (winding stair) A tapered or triangular-shaped stair. The triangular or kite-shaped tread is used to enable a flight of stairs to change direction while still continuing to rise.
螺旋形阶梯 锥形或三角形楼梯。使用三角形踏步，使梯段连续上升时能够改变方向。

wind generator (wind turbine) A device for capturing *wind energy.
风力发电机（风力涡轮机） 产生风能的设备。

windlass A lifting device similar to a *winch used to raise and lower an anchor.
绞盘 类似于绞车的起重装置，用于提升和放落锚固件。

wind load The force exerted on a structure by the *wind.
风荷载 风施加在结构上的力。

window An opening formed in a wall or roof to admit daylight through a transparent or translucent material. The most common transparent or translucent material used is glass, although other materials such as perspex are used for rooflights. The glass is fixed in a *window frame that may be subdivided by *mullions and *transoms, and may contain a number of *casements. Many types of windows are available, including *casement windows, *pivot windows, *sash windows, and composite action windows.
窗户 墙壁或屋面上的开口，让日光可透过的透明或半透明材料。虽然天窗也可使用其他材料（例如有机玻璃），但是窗户最常使用的透明或半透明材料是玻璃。玻璃固定在窗框内，可被竖梃和中冒头分隔，还可包含多个窗扉。有各种类型的窗户，包括平开窗、旋转窗、推拉窗和复合窗。

window back *See* WINDOW BOARD.
窗台板 参考“窗台板”词条。

window bar *See* GLAZING BAR.
窗格条 参考“玻璃窗格条”词条。

window board (elbow board, elbow lining, Scotland sillboard, window back, US window stool) The panelling on the internal face of an external wall under a window sill.
窗台板 在窗台下面的外墙内表面上的镶板。

window frame The members that form the perimeter of a window. Consists of a *sill, a *head, and two *jambs. The frame may be subdivided by *mullions and *transoms.
窗框 形成窗户周边的构件。包括窗台、窗楣和两个边梃。窗框可被竖梃和中冒头分隔。

window sill The horizontal shelf at the bottom of a window frame.
窗台 窗框底部的水平搁板。

window stool (US) *See* WINDOW BOARD.
窗台板（美国） 参考“窗台板”词条。

wind shear A variation in wind velocity or direction over a short distance in the atmosphere.

风切变 大气中风速或风向在短距离内的改变。

windshield A screen or wall used to provide protection from the wind.
风挡 挡风的屏障或墙壁。

wind speed The rate at which air moves in the atmosphere.
风速 大气中风移动的速率。

wind tunnel A tunnel-shaped testing chamber in which air is blown through at different speeds. Objects or scale models are placed in the tunnel to assess how they disrupt the air flow.
风洞 以不同的速度吹风的管道状的测试室。将物件或等比例模型放置在风洞中以测试其对气流的扰动。

windward Facing the wind—towards the direction the wind is blowing.
迎风的 面向风的，正对风吹来的方向。

wiped joint A joint prevalent in lead pipes facilitated by the soldering process.
焊接点 焊接过程中铅管常见的接缝。

wire A strand of metal used to carry electric current.
电线 用于带电流的金属绞线。

wobble-wheel roller A roller that has a system of pneumatic tyres suspended on springs.
轮胎式压路机 用弹簧悬挂的充气轮胎辗压的压路机。

wood A naturally occurring material derived from trees, also known as timber.
木材 由树上来的自然材料，也被称为木料。

wood adhesive *See* ADHESIVE.
木材胶粘剂 参考“黏接剂”词条。

wood block floor Flooring constructed from individual rectangular wooden blocks. Used in areas where heavy traffic is expected.
镶木地板 由单独矩形木块组成的地板。在频繁通行的区域使用。

wood bonding The joining of pieces of wood.
木材胶合 木材的连接。

wood-boring insects Species that bore through timber and wood resulting in degradation of the material. In the UK the predominant insects are beetles.
木材钻孔虫 在木材上钻孔导致木材降解的物种。在英国，主要的钻孔虫是甲壳虫。

woodworm A beetle larva that bores into wood.
木蛀虫 钻入木材的甲壳虫幼体。

workability The ease of placing and compacting concrete, measured by a *slump test or *compaction factor test.
和易性 浇筑和压实混凝土的容易性，通过坍落度试验或压实系数测试。

work hardening A procedure used in metallurgy to increase the ultimate tensile strength of a metal by modifying the crystal structure. The process involves the repeated plastic deformation of the material. During work hardening, the dislocation density in metals increases through straining with an applied stress. This technique is also commonly known as strain hardening or cold work.
加工硬化 在冶金中通过改变晶体结构以增强金属极限抗拉强度的过程。这个过程包括材料的重复塑性变形。在加工硬化过程中，经过施加的应力应变，金属的位错密度增强。这种技术通常也称为应变硬化或冷加工。

working drawings The drawings that are the most current and those which are being used to construct the building. Many drawings may be superseded and would not form the working drawings. The working drawings

can often be different from the contract drawings, but would be important in deciding contract variations.
施工图 最新的图纸以及将用于施工建筑的图纸。很多图纸都是作废的，不能成为施工图。通常施工图与合同图纸不同，但是在决定合同变更方面同样很重要。

working load (working stress) The maximum safe load or stress that a structure can withstand.
工作荷载（工作应力） 结构可承受的最大安全荷载或应力。

workmanship 1. The skill of a manual worker. 2. The quality of the work produced.
1. 手艺 体力劳动者的技术。**2. 做工** 所生产的产品的质量。

workmate Person working alongside or within a team; a colleague.
工友 一起工作或在同一组内的人；同事。

world geodetic system A geographical reference frame for the earth, used in *global positioning systems.
世界大地坐标系 用于全球定位系统中的地球的地理参考框架。

wreath A curved section of stair handrail that also inclines or declines.
鹅颈 倾斜或下降的楼梯扶手的弯曲部分。

wrought alloy A metal alloy that is suitable for mechanical forming below melting point temperatures.
锻制合金 适合在低于熔点的温度下机械成型的金属合金。

wrought iron A ferrous metal that is essentially pure iron, chemical formula Fe. Wrought iron used to be made and utilized in very large quantities for construction and general engineering purposes in Victorian times; applications included bridges and fences. However, currently it is not produced any more, except in small amounts made for demonstration purposes in industrial museums. A distinction is made between cast and wrought metals. Metals that are so brittle that forming or shaping by appreciable deformation is not ordinarily possible, are cast. On the other hand, those that are amenable to mechanical deformation are termed wrought alloys.
锻铁 本质上为纯铁的黑色金属，化学式为 Fe。在维多利亚时期，锻铁在建筑和一般工程中大量被使用；用途包括桥梁和栅栏。但是，目前除了工业博物馆中少量用于展览目的的锻铁外，已不再生产锻铁。铸造金属和锻制金属之间有一个区别。如果金属易脆，无法通过明显变形来形成或成型的则是铸造金属。另外，能够承受机械形变的是可锻合金。

w-shape A wide-flange *I-BEAM.
宽翼缘工字钢 宽翼缘的工字钢。

wye Something that has a Y-shape, such as a drainage fitting.
Y 型 Y 字形的某物，例如排水管件。

X-bracing *See* CROSS-BRACING.
X 型支撑 参考“剪刀撑”词条。

Xestobium rufovillosum (death watch beetle) Wood-boring insect that bores through wood making a ticking sound.
红毛窃蠹（报死虫） 钻孔时发出嗒嗒声的木材钻孔虫。

X-rays High-energy electromagnetic radiation, with very short wavelengths (between 0. 01 and 10 nanometres) which can penetrate substances such as concrete and metal. The X-ray technique is a very useful method

of ascertaining the physical properties of materials, especially physical metallurgy.
X 射线 高能电磁辐射，波长非常短（0.01 到 10 纳米），能穿透物质如混凝土和金属。X 射线技术对探明材料物理特性非常有用，特别是物理冶金学。

ray scattering techniques Non-destructive analytical tests and techniques that reveal information on the chemical composition, crystallographic structure, and some of the physical properties of the materials. The techniques observe the scattered intensity of an X-ray beam hitting a material sample.
射线散射技术 非破坏性分析检验和技术，可探明材料的化学成分、晶体结构和某些物理特性。该技术原理是观察 X 射线光束冲击材料样品产生的散射强度而得。

x-section *See* CROSS-SECTION.
剖截面 参考“剖面图”词条。

xylonite A *thermoplastic polymer registered in 1870 as Celluloid. It can be easily moulded and shaped and has been used for photo-elastic models. It is also highly flammable and decomposes easily, and is therefore no longer widely used.
硝酸纤维素塑料 热塑性聚合物，在 1870 年登记商标名为赛璐珞。可以轻易模制和塑形并被用于光测弹性。因其高度易燃及易分解，所以不再广泛使用。

yard (yd) Unit of length, defined within the *imperial units system equal to 36 inches or 3 feet, which is equivalent to 0.9144 metres in the *metric system.
码 长度单位，在英制单位系统中等于 36 英寸或 3 英尺，在国际单位系统中等于 0.9144 米。

yard trap *See* GULLY.
下水道进口截污设备 参考“雨水口”词条。

Y-branch (Y-bend, Y-fitting) *See* WYE.
支管（Y 型弯头、Y 型管件） 参考“Y 型”词条。

yield To give way or provide no further resistance.
屈服 放弃或不再提供阻力。

yielding The start of *PLASTIC DEFORMATION when a material is deformed.
屈服 当材料变形时，塑性变形的开始。

yield point (yield stress) The point (stress level) at which a material changes from elastic to plastic behaviour. If a slight increase in loading is applied beyond the elastic limit (where stress is proportional to strain), the material will now start to deform without any increase in loading.
屈服点（屈服应力） 材料从弹性变为塑性的点（应力水平）。如果稍微增大施加的荷载，超过其弹性极限（应力与应变成正比），材料不用增加任何荷载即开始变形。

yield strength The stress required to permanently deform a material, the point where *plastic deformation commences. On a stress-strain graph the yield strength corresponds to the maximum value of stress in the linear region of the graph. Yield units of strength are usually measured in MPa or N/mm^2. In some cases the yield point may not be easily discernible from the stress-strain, thus, as a consequence, a convention has been established where a straight line is constructed parallel to the elastic portion of the stress-strain curve at some specified strain offset, usually 0. 002. The stress corresponding to the intersection of this line and the stress-

strain curve as it bends over in the plastic region will determine the yield strength.
屈服强度 使材料发生永久变形所需的应力，塑性变形开始的点。在应力－应变图中屈服强度对应图的线性区中的应力最大值。屈服强度单位通常为 MPa 或 N/mm^2。某些情况下在应力－应变线上并不容易辨识屈服点，因此习惯上在与特定应变相距 0.002 处作和弹性阶段平行的直线。该直线与应力－应变曲线弯向塑性阶段的相交点对应的应力即为屈服强度。

yoke 1. A wooden beam to harness two draught animals. 2. A clamp to hold/control the movement of parts on a machine. 3. A strengthening framework which is temporarily fixed around the *perimeter of the *formwork to a square/rectangular column during casting.
1. 轭 给两头役畜上挽具的木梁。**2. 叉臂** 用于控制机器上零件移动的夹具。**3. 柱箍** 在浇筑过程中，将模板周边暂时固定至方形/矩形柱的加固框架。

Yorkshire bond (Monk bond, flying Flemish bond) The pattern of brickwork where two *stretchers are placed between each *header and successive courses are consistent with equal *lap, to bring alternative headers into regular arrangement.
跳丁砖砌法（四分之三梅花丁砌法） 一种砌砖模式，两块顺砖与一块丁砖相间，下一层砖与其错开距离一致，使交替的丁砖按规则排列。

Yorkshire light A horizontally sliding *sash window where one half moves horizontally and the other half is fixed.
单推拉窗 一种水平推拉窗，一扇窗扉水平推动，另一扇则固定。

Young's modulus *See* MODULUS OF ELASTICITY.
杨氏模数 参考“弹性模数”词条。

Ytong A commercial trademark for *aircrete (autoclaved aerated concrete).
伊通 充气混凝土（蒸压加气混凝土）的商业品牌。

yttrium (Y) A silvery metallic element, occurring in nearly all rare-earth minerals, but not a rare-earth element, and used in various metallurgical applications, notably to increase the strength of other metal alloys and materials.
钇（Y） 银色金属元素，存在于几乎所有的稀土矿中，可应用于各种冶金，特别是增加其他合金和材料的强度。

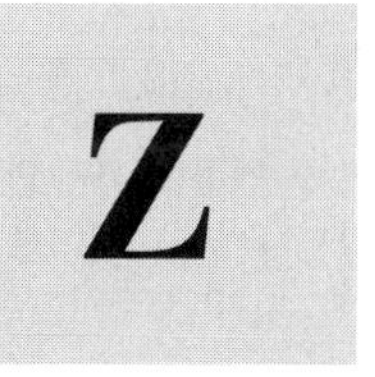

zalutite A type of steel that is coated for corrosion protection by a mixture of zinc and aluminium, present in an approximate 50－50 ratio.
镀铝锌钢板 一种涂有锌和铝（大约 50：50 的比例）防腐蚀涂层的钢。

zax (slater's axe) Roofer's axe with a blade at one end and point at the other. Used for cutting roofing slates and making holes in the slates for nails and fixings.
瓦刀 修整屋顶的工匠用的斧头，一端有刀片，另一端是尖的。用于切割屋面石板瓦和在石板瓦上打洞以打钉子和固定。

Z-beam (zed beam) A beam with a cross-section in the shape of a letter Z, often used as a *Z-purlin.
Z 型钢（乙字钢） 横截面为 Z 字形的梁，通常用作 Z 型檩条。

zenith Highest point within the celestial sphere that is directly above the observer.

天顶 观察者头顶上方天球的最高点。

zeolite One of a large group of silicate minerals, occurring in weathered igneous rocks that have reacted with alkaline groundwater. Zeolite is often used to soften water due to its ion exchange properties.
沸石 硅酸盐矿物中其中一大类，由风化火山岩和碱性地下水反应生成。沸石因其离子交换特性通常用于软化水。

zinc (Zn) A non-ferrous alloy with a relatively low melting point. Zinc is used to a small extent for roofing and cladding, but its main importance in construction is as corrosion protection for steel as galvanizing. A thin coating of zinc is applied to steel items to protect them against wet corrosion. This process is cheap and it significantly extends the life of steel items, and improves the durability of steel.
锌 (Zn) 一种有色金属，合金熔点相对较低。锌可小范围内使用于屋面和外墙覆面，但其建筑方面最重要的用途是以镀锌形式保护钢铁防腐蚀。在钢件上涂上一层薄的锌即可对抗潮湿腐蚀。该处理过程既廉价又可大幅延长钢件的寿命，提高其耐用性。

zinc chromate primer A *primer of zinc chromate used as a protective coating in metal components.
锌铬黄底漆 在金属组件中用作保护性涂层的铬酸锌底漆。

zinc oxide (Chinese white, zinc white, ZnO) An amorphous white or yellowish powder, used as a pigment in compounding rubber.
氧化锌（中国白，锌白，ZnO） 一种无定形白色或淡黄色粉末，在复合橡胶中用作颜料。

zinc phosphate primer A type of primer used for its excellent resistance to the formation of rust.
磷酸锌底漆 一种具有良好防锈性能的底漆。

zinc-rich paint A type of paint with a high concentration of *zinc.
富锌涂料 一种锌含量高的涂料。

zinc roofing Weather protection on a roof that is formed out of zinc or zinc alloy sheets.
锌板屋面 由锌板或锌合金板制成的屋面防候层。

zinc silicate primer A *primer based on zinc silicate used in anti-corrosive paint systems to promote long life of the surface.
硅酸锌底漆 硅酸锌基底漆，用于防腐蚀涂料系统以延长表面的使用寿命。

zirconium (Zr) A non-ferrous metal relatively abundant in the earth's crust. Zirconium is primarily used as an alloying element due to its excellent resistance to corrosion.
锆 (Zr) 一种有色金属元素，广泛存在于地壳中。锆因其出色的耐腐蚀性能主要用作合金元素。

zoning 1. A system of land-use planning based on boundaries, inside which land can only be used for specific purposes, such as agriculture, dwellings, green belt, industry, or recreation. 2. The subdivision of the space within a building into separate areas to enable individual control of factors such as air conditioning, building services, fire detection systems, heating, lighting, noise, ventilation, and so on.
1. 分区规划 以边界为基础划分的土地使用规划系统，在分区内，土地只能用于特定目的，例如用于农业、住宅、绿化带、工业或娱乐。**2. 功能分区** 将建筑物内的空间划分为不同区域，使各分区各自控制不同因素，例如空调、屋宇设备、火灾监测系统、供暖、照明、噪音、通风等。

Z-purlin (US Z bar) A cold-formed *steel *purlin in the shape of the letter Z. Used as part of a roofing support system or as side rails for vertical cladding.
Z 型檩条（美国 Z 型钢） Z 字形的冷弯型钢檩条。用作屋面支撑系统的一部分或垂直覆面层的纵梁。